New Frontiers in SCIENCES, ENGINEERING and the ARTS

Vol. III-A

The Chemistry of Initiation of Ringed, Ringed-Forming and Polymeric Monomers/Compounds

Sunny N.E. Omorodion

Professor of Chemical Engineering
University of Benin.

authorHOUSE®

AuthorHouse™
1663 Liberty Drive
Bloomington, IN 47403
www.authorhouse.com
Phone: 1 (800) 839-8640

Published by AuthorHouse 04/12/2018

ISBN: 978-1-5462-2897-4 (sc)
ISBN: 978-1-5462-2896-7 (e)

Contents

Preface

Volume (III) titled "The chemistry of initiation of Ringed, Ring-forming and Polymeric Monomers/ Compounds" completes the initiation of compounds for chemical and homopolymeric reactions (Section D). The Volume which is a section contains six chapters, and is indeed a continuation of Volume (II).

Based on the unique characters of rings, for more numerous than non-ringed compounds, new concepts were introduced in this Volume. The characters of rings, the conditions favoring their openings, expansions and reduction in size and formations, the manners by which rings are opened or closed, etc. have been clearly explained. Having covered virtually every type of rings, all the different types of resonance stabilization and molecular rearrangement phenomena and other pheno-mena present in these systems would have been identified. For the first time for example, how meta, ortho-and para-substitutions take place in benzene rings, 2-, 5- positions or 3-, 4- positions substitu-tions take place in pyrrole, furan and thiophene rings etc. have been clearly explained based on the types of resonance stabilization phenomena favored by them. Interestingly enough, none of the substitution reactions is ionic in character as has been thought to be the case over the years. So also are most other reactions undergone by these members of fully unsaturated rings. For the first time, it will also be interesting to note that Silicon cannot carry ionic bonds or ionic charges and can therefore not undergo charged reactions. All the reactions favored by silicon are radical in character.

Chapter 10 which is the first chapter in this Volume deals with cycloalkanes and only cyclo-propenes. These are rings without functional centers. For the first time, the chemistry of the compounds is being clearly explained from a different point of view. Chapter 11 completes all types of only carbon-containing rings. Why some favor being used as monomers and others do not, have been clearly explained. Nucleophilic, electrophilic, radical-pushing and radical-pulling capacities of the compounds and their substituted groups (new) have been provided. Chapters 10 and 11 form Part (I) of this Volume.

Chapters 12 and 13 which form Part (II), deal with cyclic ethers, acetals, esters, anhydrides, amides, oxaazocyclo-propane, pyrrole, carbon anhydrides and imino compounds. All the chemistry provided for these compounds are all new. These are compounds or rings with functional centers. Chapter 14 concludes with cyclosiloxanes, cyclic sulfides, cyclic disulfides, rhombic sulphur, and inorganic or semi-inorganic cyclic monomers etc. These are unique rings, some of which have functional centers which cannot be used. The routes favored by all the rings which favor opening of the rings, have been provided. Why and how some rings do not favor any existence or favor isolated existence, have been explained. All the concepts and chemistry provided here are complete departures from the past. Chapters 14 and 15 form Part (III) of this Volume.

Chapter 15 deals with ring forming and polymeric monomers. All the cases above have largely been Inter-addition types of monomers, while the cases here include Intra-addition with Inter-addition. Different and extended new classifications for the groups of these monomers were introduced here, based on the mechanisms of ring formation. Some undergo Inter-Intra-additions, Di-inter-additions, or only Intra-additions. The routes favored for cyclization were also identified. This Volume has never been intended for copolymerization systems. However, in view of the need to provide convincing and unquestionable evidences for the new developments, some unique copoly-merization reactions were considered beginning from Volume (II). As usual, at the end of every chapter, rules are proposal in order to provide convincing and clear picture of the entire concepts. Most of the new concepts, which were developed right from the beginning and in Volume (I), have been put into rules beginning from Volumes (II) into (III) and in particular in chapter 15 which is the last chapter in this volume.

In Volume (II), more than five hundred rules were proposed. In Volume (III), more than one thousand rules have been proposed bringing the total to one thousand seven hundred and twenty rules. Without the rules, no foundations can be established. With the rules already herein proposed, the chemical properties of any compound can readily be obtained without the further need for chemical analysis, once the structure of the compound can be ascertained. This completely eliminates the use of empirical rules, rule of thumb, trial and error methods, etc. methods on which the developments of most disciplines have largely depended on since the beginning of our time.

Before understanding this Volume, Volumes (I) and (II) must be read. This Volume will find most useful applications to the medical scientists, biochemists, chemical and other related disciplines, where little or nothing is indeed known about ringed compounds.

While all the works are original to the author, one will not continue to forget to thank many publishers whose works have helped to lay these new foundations for scientist, Engineers and the Arts. This Volume like the others still remains dedicated to humanity.

University of Benin

Sunny N. E. Omorodion.
(ETG)

SECTION D

CHAPTER 10

TRANSFER OF TRANSFER SPECIES IN CYCLOALKANES AND CYCLOPROPENES

10.0 Introduction

Ring-opening monomers are to be considered in a manner different from the past. From the analysis to be provided, ring-opening monomers are Addition monomers where the initiation steps are uniquely different from what has been known so far. Rings are like springs in Physics with strains in them. To open them, stresses in different forms must be applied. The stresses can be physical or chemical or both in character depending on size of the ring and what the ring is carrying.

Since the developments of strain theories have been largely limited to few monomers e.g., cyclic ethers, sulphides, ethylamine, the choice of cycloalkanes, cycloalkenes etc. as the starting point for consideration of ring-opening monomers is apparent. With these monomers, the followings have been fully established-

(i)   The order of presence of strain energy with the size of the ring[1-4]
(ii)   Substituents other than hydrogen on the ring structure generally increase its stability, although the reason for this is not well understood[5].
(iii) That the ease of formation of cyclic compounds, that is, tendency for intramolecular reaction (cyclization) depends on the proximity of the atoms being joined in the reaction[4].
(iv)   That if the atoms were forced to remain in an extended chain, the chances for intramolecular reaction or intermolecular reaction would decrease or increase respectively as more and more carbon atoms are made to separate the reacting groups (in the absence of an initiator)[4].
(v)   That the introduction of a double bond into ring system increases the strain, the effect being greatest for cyclopropane and decreasing with increasing ring size.

From the order of presence of strain energy with the size of the ring, it has been observed, that for cycloalkanes, the strain energy released is greatest (27 Kcal/mole)[4] for the three-membered ring (cyclopropane); for which the followings can be used as reference points.

$$
\begin{array}{ccc}
\underset{\substack{|\\ CH_2}}{\underset{|}{H-\overset{\overset{H}{|}}{C}-\overset{\overset{H}{|}}{C}-H}} &
H-\overset{\overset{H}{|}}{\underset{\underset{H}{|}}{C}}-\overset{\overset{H}{|}}{\underset{\underset{H}{|}}{C}}-\overset{\overset{H}{|}}{\underset{\underset{H}{|}}{C}}-H &
\overset{\overset{H}{|}}{C}\overset{\pi}{\underset{\delta}{=}}\overset{\overset{H}{|}}{C} \\
& & \underset{H}{|}\qquad\underset{H}{|}
\end{array}
\qquad\qquad 10.1
$$

cyclopropane	propane	ethene
(C-C bond = 56kcal)	(C-C bond = 83Kcal)	(C=C π-bond = 56Kcal)

According to the equation above, the energy required to activate the π-bond of ethene is the same as that required to open the three-membered cyclopropene ring. Then, the question that arises is why is it that cyclopropane has never been very popularly commercially used as a monomer for linear polymerization or Inter-molecular reactions? It is however known that where the strain energy is zero and the heat of polymerization is zero, the monomer cannot be polymerized. This has been the case with six and fourteen-membered rings of cycloalkanes,[5] and six-membered ring of cyclic ethers and acetals[6-8]. The strains in cycloalkane structure have been observed to be very high for the 3- and 4- membered rings, decrease sharply for 5-, 6- and 7- membered rings, increases for 8- to 11- membered rings and then decreases again for larger rings.[5]

Since the ease of polymerization of a given cyclic monomer depends on the following factors;

(a) The types of functional center in the ring e.g. $\overset{\frown}{O}$o, $\overset{\frown}{O}$c = o, $\overset{\frown}{O}$N, etc. when present.

(b) The reactivity of the functional center.

(c) The type of initiators involved.

(d) The size of the ring and amount of strain energy in the ring.

(e) The presence or absence of double bonds in the ring.

(f) The equilibrium state of recyclization (intramolecular) versus linear polymerization (intermolecular) reactions, which exists for some of these monomers.

The strain in cyclic monomers will differ for their rings from family to family. While cycloalkanes and the others here can be said to have no functional center, the other monomers have at least one functional center.

While the strain energy required to be released for 3-membered cycloalkane ring is 27.4Kcals, it can be shown that those for 4-, 5- and 6- membered rings are 25.2, 6.0, and approximately 0Kcals respectively. Therefore, the C-C bond energies are 56.0 57.0, 77.0 and 83.0Kcals respectively for 3-, 4-, 5-, and 6- membered cycloalkane rings. These are energies which are equal to or greater than the C=C π-bond in ethene (ethylene) or equal to or less than the C-C σ- bond in alkanes. The larger the strain energy, the less difficult it is to open a ring, noting that there are minimum strain energies that a ring must possess for the ring to be opened instantaneously when initiators with electrostatic forces are around its vicinity. If the ring does not possess this minimum strain energy, it cannot be opened. The ring will remain closed. Cycloalkanes, Cyclodienes and Cyclotrienes do not have any point of attachment or functional center by which the minimum strain energy can be attained, if not present in the original monomer. Hence ring-opening with these families of carbon-carbon monomers is not as common as with hetero-chain monomers. When one double bond is present in the rings, the strain energy in the monomer is increased, decreasing the strength of the C-C or C-O or C-N single bond in the ring, making it easier to open the ring, except when other factors prevail.

In most of the proposed mechanisms in the literature and textbooks, there have been quite a lot of impossible reactions, for which there is no need to make any reference. The reasons why this is so, is clearly becoming obvious is view of the current developments and methods of analysis, using natural laws.

10.1 Cycloalkanes

Unlike the epoxides, cycloalkanes do not have any functional center, through which additional strain can be added to the monomer to attain the minimum required strain energy. It is said that although unprotonated ethers do not react with nucleophiles (that is, do not favor "anionic" attacks), the 1,2 – epoxides are said to be very susceptible to nucleophilic attack because of the release of strain in the three-membered ring when it is opened.[9] In order words, anionically when 1,2-epoxide is attacked, the following are obtained.

$$:R:^{\ominus} \quad + \quad H-\overset{\overset{\displaystyle H}{|}}{\underset{}{C}}-\overset{\overset{\displaystyle H}{|}}{\underset{O}{C}}-H \quad \xrightarrow{\text{(Strain Release)}} \quad R:^{\ominus} \quad + \quad \overset{\oplus}{\underset{\underset{\displaystyle H}{|}}{\overset{\overset{\displaystyle H}{|}}{C}}}-\overset{\overset{\displaystyle H}{|}}{\underset{\underset{\displaystyle H}{|}}{C}}-O^{\ominus}$$

Non-free   (Strained)

anion   27.28Kcal/mole

$$\xrightarrow{\hspace{2cm}} \quad R-\overset{\overset{\displaystyle H}{|}}{\underset{\underset{\displaystyle H}{|}}{C}}-\overset{\overset{\displaystyle H}{|}}{\underset{\underset{\displaystyle H}{|}}{C}}-O^{\ominus}$$

10.2

Due to the paired radicals (negative charges) carried by the initiator, electrostatic forces are released to open the ring instantaneously. The ring is already strained in view of the fact that it has strain energy (SE) high but less than the minimum required strain energy (MRSE). That is the energy added to the SE to unzip the ring. The non-free negative charge cannot attach itself to the oxygen center due to electro-dynamic forces of repulsion. A free negative charge cannot react with the activated monomer. When opened, it is the oxygen center that carries the non-free negative charge. The oxygen center can never carry real cationic or positive charge.

Since cyclopropane is strained, one should expect nucleophilic attack to release strain as it did for 1, 2-epoxide above. Hence the followings should be expected.

$$R:^{\ominus} \quad + \quad H-\overset{\overset{\displaystyle H}{|}}{\underset{}{C}}-\overset{\overset{\displaystyle H}{|}}{\underset{CH_2}{C}}-H \quad \xrightarrow{\text{Strain release}} \quad R:^{\ominus} \quad + \quad \oplus\overset{\overset{\displaystyle H}{|}}{\underset{\underset{\displaystyle H}{|}}{C}}-\overset{\overset{\displaystyle H}{|}}{\underset{\underset{\displaystyle H}{|}}{C}}-\overset{\overset{\displaystyle H}{|}}{\underset{\underset{\displaystyle H}{|}}{C}}\ominus \quad \xrightarrow{\hspace{1cm}}$$

<u>Free</u>   (I) (Strained) -27.43Kcal/mole

<u>Negative Charge</u>

(Note: The negative charge above is not an anion and cannot therefore be isolated)

$$R-\overset{\overset{\displaystyle H}{|}}{\underset{\underset{\displaystyle H}{|}}{C}}-\overset{\overset{\displaystyle H}{|}}{\underset{\underset{\displaystyle H}{|}}{C}}-\overset{\overset{\displaystyle H}{|}}{\underset{\underset{\displaystyle H}{|}}{C}}^{\ominus}$$

10.3

In view of the absence of molecular rearrangement of **a third kind** chargedly that which can never take place radically or chargedly, there is no transfer species to prevent initiation noting that only negatively charged paired initiators can be used. The reactions are favored, since cyclopropane and many of its derivations give open-chain compounds with most of the reagents that react with olefins.[4, 10, 11] However, notice that while the monomers considered so far are Nucleophiles or FEMALES in character, the initiators used above are not Natural to them.

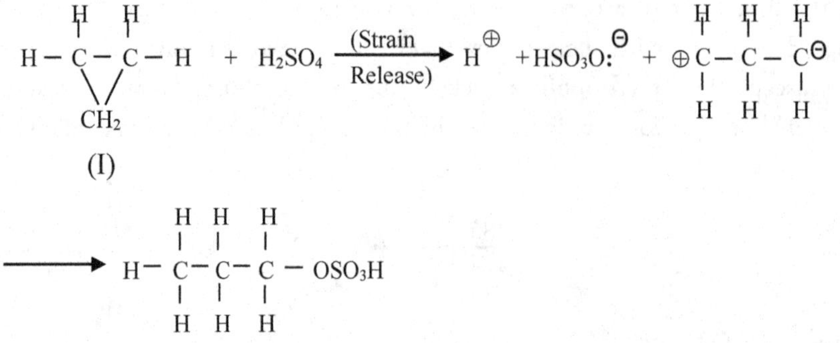

$$H-\underset{\underset{CH_2}{\diagdown\diagup}}{\overset{H\ \ \ H}{\overset{|\ \ \ \ |}{C}}}-C-H\ +\ H_2SO_4\ \xrightarrow[\text{Release}]{\text{(Strain}}\ H^{\oplus}\ +HSO_3O:^{\ominus}\ +\ {}^{\oplus}C-C-C^{\ominus}$$

(I)

$$\xrightarrow{\hspace{2cm}}\ H-\underset{\underset{H\ \ H\ \ H}{|\ \ |\ \ |}}{\overset{H\ \ H\ \ H}{\overset{|\ \ |\ \ |}{C}}}-C-C-OSO_3H$$

10.4

Radically, this reaction is possible particularly when $TiCl_3$ is used. Fuming H_2SO_4 cannot be used. Used above is concentrated sulfuric acid. Like dilute H_2SO_4, stable products are obtained.

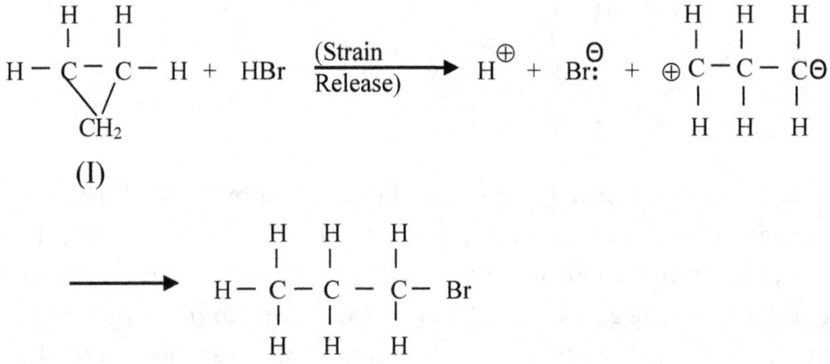

$$H-\underset{\underset{CH_2}{\diagdown\diagup}}{\overset{H\ \ H}{\overset{|\ \ |}{C}}}-C-H\ +\ HBr\ \xrightarrow[\text{Release}]{\text{(Strain}}\ H^{\oplus}\ +\ Br:^{\ominus}\ +\ {}^{\oplus}C-C-C^{\ominus}$$

(I)

$$\xrightarrow{\hspace{2cm}}\ H-\underset{\underset{H\ \ H}{|\ \ |}}{\overset{H\ \ H\ \ H}{\overset{|\ \ |\ \ |}{C}}}-C-C-Br$$

10.5

$$H-\underset{\underset{CH_2}{\diagdown\diagup}}{\overset{H\ \ H}{\overset{|\ \ |}{C}}}-C-H\ +\ Br_2\ \xrightarrow[\text{Release}]{\text{(Strain}}\ Br:^{\bullet en}\ +\ n\bullet C-C-C\bullet e\ +\ :\overset{\bullet\bullet}{\underset{\bullet\bullet}{Br}}\bullet nn$$

$$\xrightarrow{\hspace{2cm}}\ Br-\underset{\underset{H\ \ H\ \ H}{|\ \ |\ \ |}}{\overset{H\ \ H\ \ H}{\overset{|\ \ |\ \ |}{C}}}-C-C-Br$$

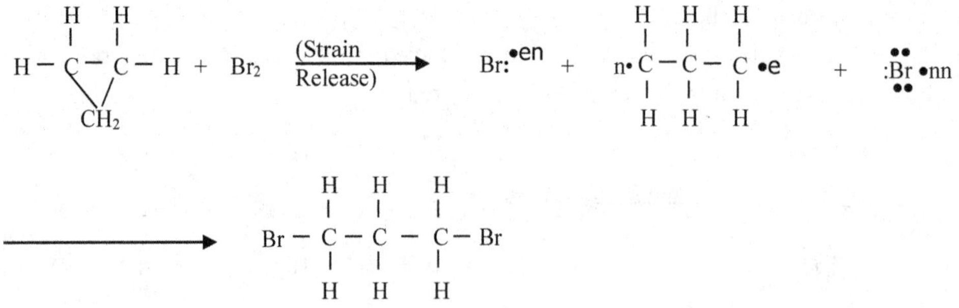

10.6

Since (I) is symmetric, any of the centers can carry any radical or charge. (I) cannot favor molecular rearrangement of **a third kind** as shown below.

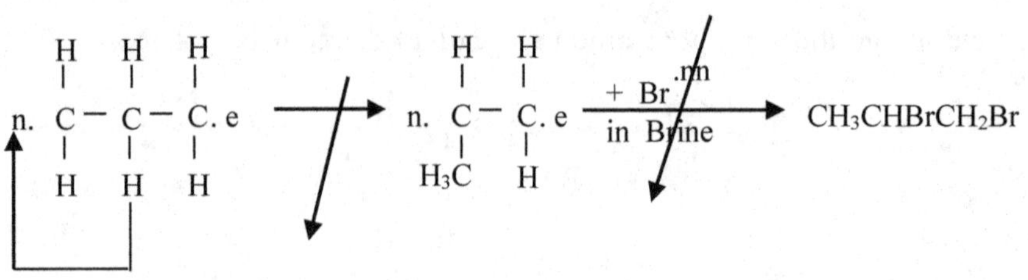

$$n.\ \underset{\underset{H\ \ H\ \ H}{|\ \ |\ \ |}}{\overset{H\ \ H\ \ H}{\overset{|\ \ |\ \ |}{C}}}-C-C.e\ \ \ \ \ \ \ \ n.\ \underset{\underset{H_3C\ \ H}{|\ \ |}}{\overset{H\ \ H}{\overset{|\ \ |}{C}}}-C.e\ \xrightarrow[\text{in Brine}]{+\ Br^{\bullet nn}}\ CH_3CHBrCH_2Br$$

10.7

All the last three reactions above are however instantaneous (very rapid), noting that the nucleo-non-free- radical cannot attack these female monomers in the presence of an electro-free or non-free -radical. Hence, all the reactions above take place via Equilibrium mechanism.

Cyclopropane can also be reduced catalytically at 80^0C as follows to n-propane. The mechanism is a single stage Equilibrium mechanism in which the electro-free-radical first attacks or adds, followed by closing with the nucleo-free-radical.

$$\underset{\underset{CH_2}{\displaystyle\vee}}{H-\overset{\overset{H}{|}}{C}-\overset{\overset{H}{|}}{C}-H} \; + \; H_2 \;\; \xrightarrow[\substack{80^oC\\ (Heat\ scission)}]{Pt} \;\; e.\overset{\overset{H}{|}}{\underset{\underset{H}{|}}{C}}-\overset{\overset{H}{|}}{\underset{\underset{H}{|}}{C}}-\overset{\overset{H}{|}}{\underset{\underset{H}{|}}{C}}.n \; + \; H^e \; + \; H^n \;\; \longrightarrow$$

$$H-\overset{\overset{H}{|}}{\underset{\underset{H}{|}}{C}}-\overset{\overset{H}{|}}{\underset{\underset{H}{|}}{C}}-\overset{\overset{H}{|}}{\underset{\underset{H}{|}}{C}}-H \hspace{8cm} 10.8$$

It is as a result of presence of minimum required strain energy provided for the ring by heat that the ring was opened instantaneously. In the reaction above, the rings cannot be opened by $H^{.e}$ since $H^{.e}$ does not have paired unbonded radicals in the last shell. Pt.en, the passive catalyst could assist in opening of the ring with presence of paired unbonded radicals in the last shell.

Cyclobutane ring is much less reactive than cyclopropane ring, since more energy is required to make it possess the minimum required strain energy and there is no way of introducing it into the monomer with no functional centers. Cyclobutane can therefore not be opened easily with sulfuric acids, hydrobromic acid or bromine without adding additional forces. However, it can be reduced catalytically at 120°c to n-butane, a temperature higher than that for cyclopropane.

$$\begin{matrix} H_2C \!-\!\!-\!\!-\! CH_2 \\ | \qquad\quad | \\ H_2C \!-\!\!-\!\!-\! CH_2 \end{matrix} \; + \; H_2 \;\; \xrightarrow[\substack{120^oC\\(Higher\ heat\\scission)}]{Pt} \;\; n.\overset{\overset{H}{|}}{\underset{\underset{H}{|}}{C}}-\overset{\overset{H}{|}}{\underset{\underset{H}{|}}{C}}-\overset{\overset{H}{|}}{\underset{\underset{H}{|}}{C}}-\overset{\overset{H}{|}}{\underset{\underset{H}{|}}{C}}e. \;\; H^{.e} \; + \; H^{.n} \;\; \longrightarrow$$

$$CH_3CH_2CH_2CH_3 \hspace{8cm} 10.9$$

It is important to note that the reduction reactions are free-radical reactions, not ionic or charged in the presence of non-ionic metallic catalysts, Pt, called hydrogen catalyst.

Higher cyclanes, which cannot be opened instantaneously in view of the decreasing influence of the strain energy in their larger rings, can however be reduced catalytically, but at increasing temperatures up to and above 200°C. Thus, only cyclopropane can be made to favor the charged-paired route, since it seems to be the only member with the least minimum required strain energy. From the analysis of the reactions, the influence or the effect of the strain energy can be observed to decrease from three-membered ring progressively downwards with increasing size of ring. Only cyclopropane, cyclobutane and their derivatives can therefore be considered for further homopoly-merization studies. In order words the minimum required strain energy must be greater than particular values (SE)-of the order of 20 – 30 Kcal/mol depending on the family of monomer, before the ring can be opened. This is in the range of the minimum activation energies of typical commercial free-radical initiators (60°C), used in free-radical polymerization systems.

10.1.1 Charged Characters of Cyclopropanes

Of all cyloalkanes, only cyclopropanes and some substituted cyclobutane and cyclopentane can be used as monomers for homopolymerization and copolymerizations. Beginning with the first member cyclopropane, the followings are obtained "cationically".

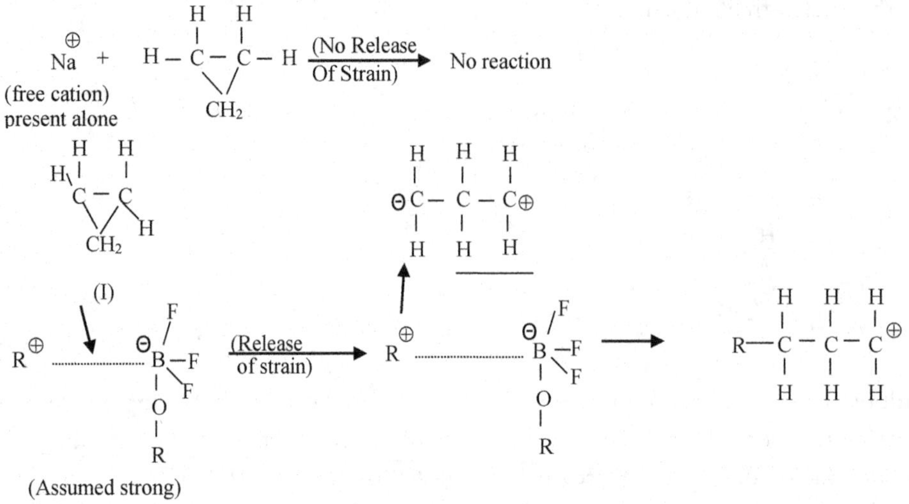

10.10

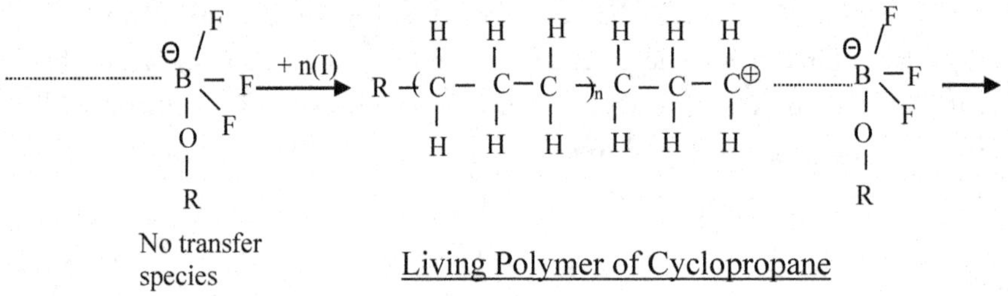

No transfer species

Living Polymer of Cyclopropane

10.11

The first reaction is not favored because apart from the absence of paired unbonded radicals required for the opening of the ring instantaneously on the cation, the equation will not be chargedly balanced. Non-free positively charged paired- media initiators do not exist and cannot also be used. As shown by the second reaction, the electrostatic forces can be provided with the presence of coordination centers of covalent or electrostatic types, if the center is strong enough. The positively charged-paired initiator above is "assumed" to be strong enough to open the ring instantaneously after providing the minimum required strain energy. It is important to note the name of the polymeric product obtained from the cyclopropane. It is polycyclopropane and not polyethylene. It is important to note that the activated state cannot be deactivated chargedly or radically as shown;

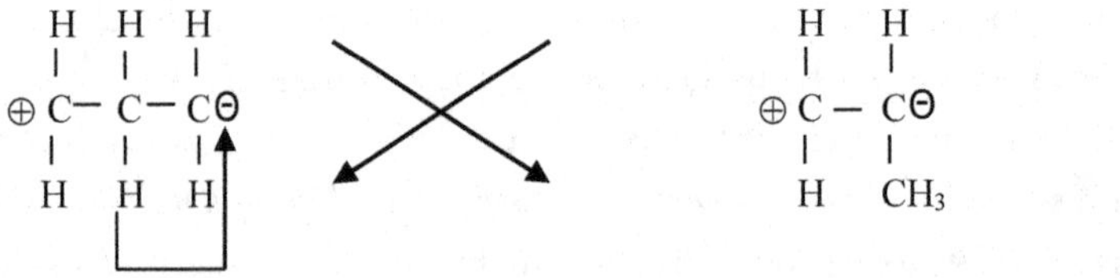

(Impossible activated state)

10.12

The positively charged route being natural to the monomer, transfer species of the second kind cannot be released from the growing polymer chain of Equation 10.11. Polymethylene from diazomethylene

(Present-day name-Diazomethane), polyethene (Present-day name-Polyehylene) from ethene (Present-day name-ethylene) and polycyclopropane may look the same structurally and probably in properties, but not for substituted members in view of the manners by which transfer species are involved for the different monomers.

Now consider presence of one alkyl group on the monomer.[10, 11] With the use of paired negative charges as initiators, the following will be expected.

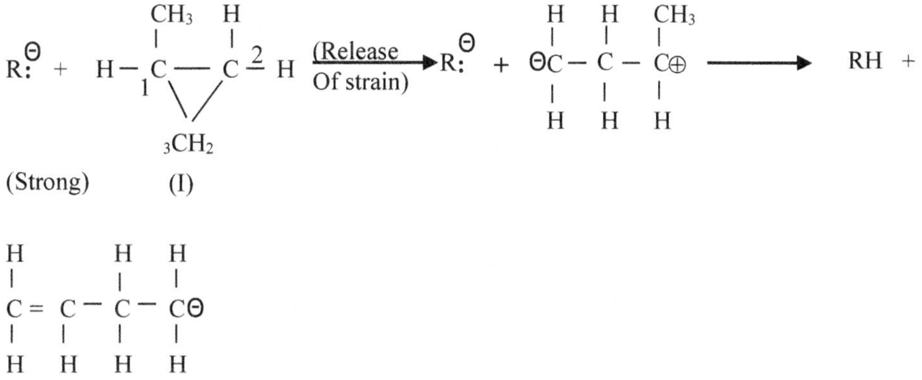

$$
\text{R:}^{\ominus} + H-\underset{1}{\overset{\overset{CH_3}{|}}{C}}\overset{\overset{H}{|}}{\underset{^3CH_2}{-}}\overset{2}{C}H \xrightarrow[\text{Of strain}]{\text{(Release}} \text{R:}^{\ominus} + {}^{\ominus}C-\underset{H}{\overset{H}{-}}C-C^{\oplus} \longrightarrow RH +
$$

(Strong)   (I)

$$
\overset{H}{\underset{H}{|}}C=\overset{H}{\underset{H}{|}}C-\overset{H}{\underset{H}{|}}C-C^{\ominus}
$$

10.13

As will become clear in subsequent developments, scission takes place on the 1,3 or 1,2 single bond center and not 2,3 bond. For this reason, the negatively charged route is not favored. If two hydrogen atoms are replaced with alkylane groups, minimum required strain energy will be very easy to attain, and the negatively charged route will still not be favored as shown below.

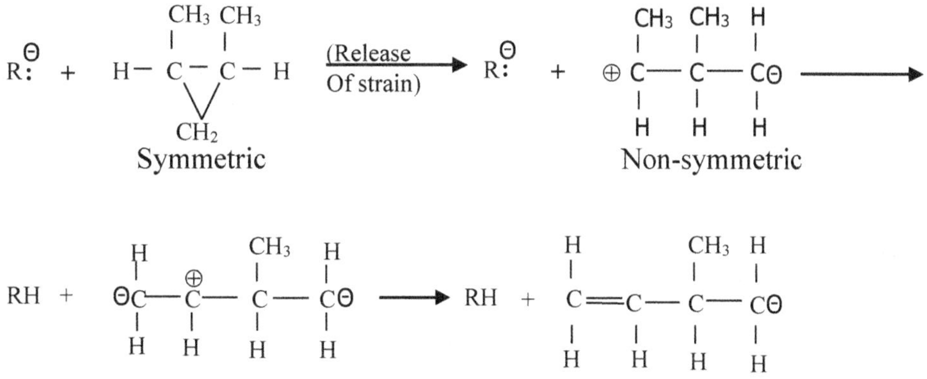

$$
\text{R:}^{\ominus} + H-\overset{\overset{CH_3}{|}}{C}\overset{\overset{CH_3}{|}}{\underset{CH_2}{-}}C-H \xrightarrow[\text{Of strain}]{\text{(Release}} \text{R:}^{\ominus} + {}^{\oplus}C-C-C^{\ominus} \longrightarrow
$$

Symmetric   Non-symmetric

$$
RH + {}^{\ominus}C-\overset{\oplus}{C}-C-C^{\ominus} \longrightarrow RH + C=C-C-C^{\ominus}
$$

10.14

Using positively charged-paired or Zeigler-Natta (Z/N) coordination initiators, all the monomers will favor polymerization with dead terminal double bond polymer produced for the two cases above as shown below for the last case.

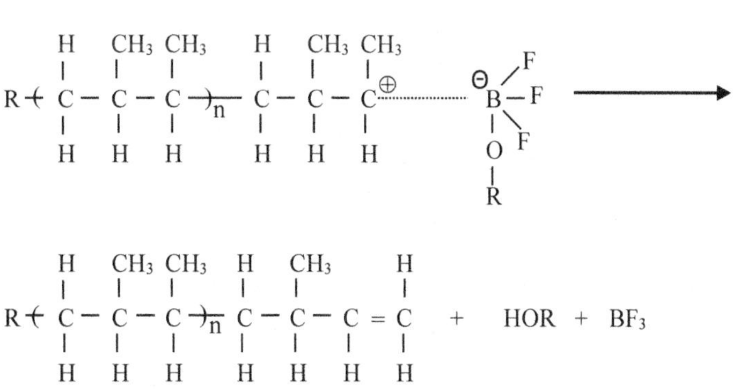

$$
R \left[C-C-C \right]_n C-C-C^{\oplus} \cdots\cdots {}^{\ominus}B\overset{F}{\underset{O}{-}}F \longrightarrow
$$

$$
R \left[C-C-C \right]_n C-C-C=C + HOR + BF_3
$$

10.15

The transfer species involved above is of the first kind of the first type. From the reactions above, as will be further shown, the more the substituent groups of radical-pushing character carried, the greater the nucleophilic character of the monomer and the more strained is the ring even to the point where the ring cannot exist as a ring anymore. In general, only one replacement of H for CH_3 for example is required to change the character of the first members, that is,

$$
\begin{array}{ccc}
\underset{\underset{CH_2}{\diagdown\diagup}}{\overset{\overset{H\ \ \ \ H}{|\ \ \ \ |}}{H-C---C-H}}
&>&
\underset{\overset{|\ \ \ \ |}{H\ \ \ \ H}}{\overset{\overset{H\ \ \ \ CH_3}{|\ \ \ \ |}}{C=C}}
\ ;\
\underset{\underset{CH_2}{\diagdown\diagup}}{\overset{\overset{CH_3\ \ H}{|\ \ \ \ |}}{H-C-C-H}}
&>&
\underset{\overset{|\ \ \ \ |}{H\ \ \ \ CH_3}}{\overset{\overset{H\ \ \ \ CH_5}{|\ \ \ \ |}}{C=C}}
\end{array}
$$

Order of Nucleophilicity   10.16

The same will obviously apply for other radical- pushing groups such as OR and NH_2, if the minimum required strain energy can be provided and the initiators are strong.

$$
R\!:^{\ominus} \ + \ \underset{\underset{CH_2}{\diagdown\diagup}}{\overset{\overset{R\ \ \ \ R}{\overset{|\ \ \ \ |}{\overset{O\ \ \ \ O}{|\ \ \ \ |}}}}{H-C-C-H}} \xrightarrow[\text{Of strain)}]{\text{(Release}} R\!:^{\ominus} \ + \ \underset{\underset{R\ \ \ R}{\overset{|\ \ \ |}{O\ \ \ O}}}{\overset{\overset{H\ \ H\ \ H}{|\ \ |\ \ |}}{{}^{\oplus}C-C-C^{\ominus}}} \longrightarrow R_2
$$

Free anion   (Assumed)

$$
\underset{\underset{R}{\overset{|}{O}}}{O=}\underset{\overset{|}{O}}{\overset{\overset{H\ \ H\ \ H}{|\ \ |\ \ |}}{C-C-C^{\ominus}}}
$$

                                                                10.17

With non-free-negative charge, there will also be no initiation. Instead, the ring will be closed, leaving the negative charge hanging on the O center.

$$\longrightarrow$$
                                                                10.18

For radical-pulling groups, consider replacing just one H atom in the first member with CX_3, where X is a halogen). It is "assumed" that with their presence, the minimum required strain energies will be more than when CH_3 was there on the monomer, since its radical-pushing capacity is less than that of H or chargedly it is charge-pulling group.

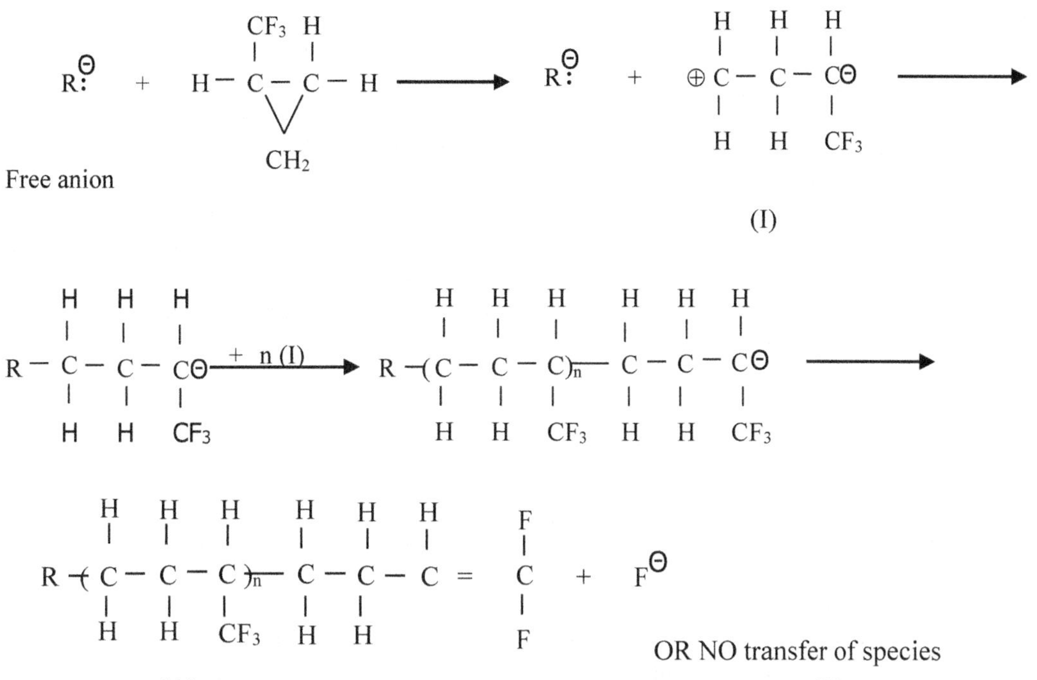

Free anion

(I)

(A)   (B)   10.19

Transfer species of the first kind of the first type cannot be released, the monomer being a nucleo-phile. It is (B) that is favored. With positively charged paired initiator, initiation is not favored as shown below.

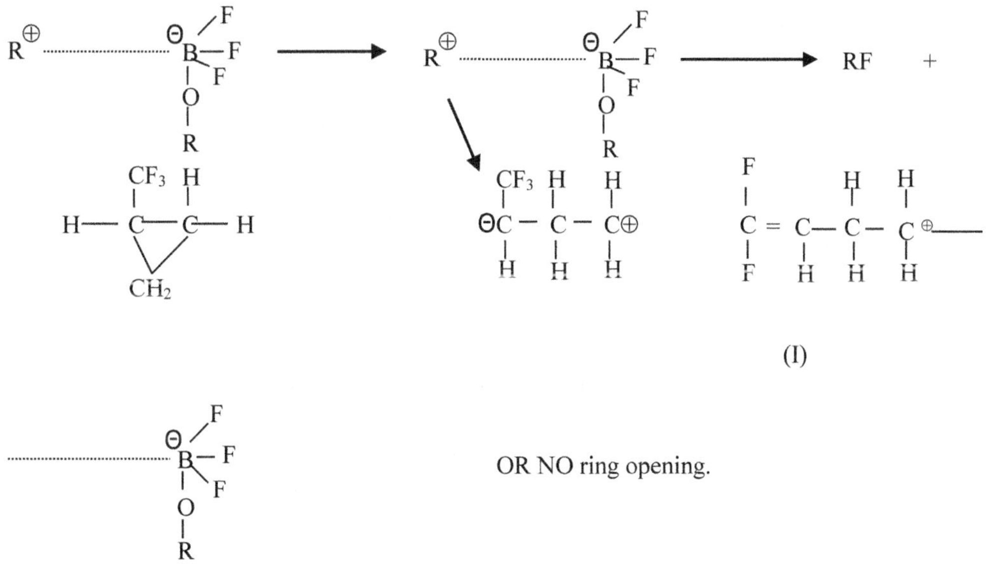

(I)

OR NO ring opening.   10.20

Though the route is natural to the monomer, transfer species of first kind can be observed to be abstracted. It is important to note that like alkenes, cyclopropane have nucleophiles and electrophiles. The real electrophiles are yet to be identified. With two hydrogen atoms on two carbon centers replaced with CF_3 type of groups, the followings are obtained.

$$R^{\oplus} \cdots\cdots\cdots \overset{\overset{F}{\underset{|}{\underset{}{}}}\overset{F}{\diagup}}{\underset{\underset{|}{O}}{\underset{|}{F}}{\overset{\ominus}{B}}}\diagdown F \longrightarrow R^{\oplus}\cdots\cdots\cdots \overset{\overset{F}{\underset{|}{}}\overset{F}{\diagup}}{\underset{\underset{|}{O}}{\underset{|}{F}}{\overset{\ominus}{B}}}\diagdown F \longrightarrow RF \quad +$$

$$\underset{CH_2}{\overset{\overset{CF_3}{|} \overset{CF_3}{|}}{H-\underset{}{C}-\underset{}{C}-H}}\qquad\qquad \overset{\overset{CF_3}{|}\ \overset{CF_3}{|}\ \overset{H}{|}}{\underset{\underset{H}{|}\ \underset{H}{|}\ \underset{H}{|}}{\ominus C-C-C\oplus}}$$

$$\overset{\overset{F}{|}\quad \overset{CF_3}{|}\ \overset{H}{|}}{\underset{\underset{F}{|}\ \underset{H}{|}\ \underset{H}{|}\ \underset{H}{|}}{C=C-C-C\oplus}}\cdots\cdots\cdots \underset{\underset{R}{|}}{\overset{\overset{F}{|}\diagup F}{\underset{O}{\overset{\ominus}{B}}\diagdown F}} \longrightarrow \text{No polymerization}$$

$$\text{10.21}$$

$$R\!\!\left(\!\!\overset{\overset{H}{|}\ \overset{CF_3}{|}\ \overset{H}{|}}{\underset{\underset{H}{|}\ \underset{H}{|}\ \underset{CF_3}{|}}{C-C-C}}\!\!\right)_{\!\!n}\!\!\overset{\overset{H}{|}\ \overset{CF_3}{|}\ \overset{H}{|}}{\underset{\underset{H}{|}\ \underset{H}{|}\ \underset{CF_3}{|}}{C-C-C\ominus}} \longrightarrow R\!\!\left(\!\!\overset{\overset{H}{|}\ \overset{CF_3}{|}\ \overset{H}{|}}{\underset{\underset{H}{|}\ \underset{H}{|}\ \underset{CF_3}{|}}{C-C-C}}\!\!\right)_{\!\!n}\!\!\overset{\overset{H}{|}\ \overset{CF_3}{|}\ \overset{H}{|}}{\underset{\underset{H}{|}\ \underset{H}{|}\ \underset{CF_3}{|}}{C-C-C\ \ominus}}$$

$$\text{10.22}$$

Based on the types of group carried, the rings here are more difficult to open. However, with higher operating conditions, they can be opened using initiators which in addition are carrying paired unbonded radicals. The monomer here can be observed to be more electrophilic than the case before this, which has only one CF_3 group. In the last reaction above no transfer species could be released, because the route is not natural to it.

The same method of analysis above will apply to COOR, $CONH_2$ etc. types of groups more so than CF_3 types, used above; noting however that with former groups, the monomers become Electrophiles.

(i) The minimum required strain energy (MRSE) cannot very readily be provided for their rings.

(ii) Strong initiators are also involved to open the ring, since there is no functional center on the monomer.

The activated monomers of equations 10.19 and 10.21 cannot undergo molecular rearrange-ment of the third kind chargedly as follows.

$$\overset{\overset{CF_3}{|}\ \overset{H}{|}\ \overset{H}{|}}{\underset{\underset{H}{|}\ \underset{H}{|}\ \underset{H}{|}}{\ominus C-C-C\oplus}} \qquad\qquad \times \qquad\qquad \underset{\underset{CF_3}{|}}{\overset{\overset{H}{|}\quad\overset{H}{|}}{\oplus C-C\ominus}}$$

(I) Nucleophile $\qquad\qquad$ (A) More stable and stronger Nucleophile
  $\qquad\qquad\qquad\qquad\qquad\qquad$ Wrong activated state.

$$\text{10.23}$$

$$\overset{\overset{CF_3}{|}\ \overset{CF_3}{|}\ \overset{H}{|}}{\underset{\underset{H}{|}\ \underset{H}{|}\ \underset{H}{|}}{\ominus C-C-C\oplus}} \qquad\qquad \times \qquad\qquad \underset{\underset{CF_3}{|}}{\overset{\overset{H}{|}\quad\overset{CF_3}{|}}{\oplus C-C\ominus}}$$

(II) Nucleophile $\qquad\qquad$ (B) More stable and stronger Nucleophile

$$\text{10.24}$$

The last case above may be favored chargedly, but not radically.

Indeed, chargedly $H^{\mathring{A}}$ unlike H.e cannot be moved chargedly, but only ionically. It is important to note that when this kind of molecular rearrangement takes place, it should only be radically. When it does, the character of the monomer obtained should be more stable, favoring route not natural to it. However, though the transfer species involved is that which is not natural to the monomer, the hydrogen atom cannot be moved as an anion. For an electrophile, for molecular rearrangement of first kind, a radical-pulling group should be involved. In molecular rearrangement of the first kind, the transfer species involved must be less than or equal to those of the same character carried by the receiving center. The same applies in the molecular rearrangement of the **second and third kinds**. The transfer species involved must be less than or equal to those of the same character carried by the receiving center.

With COR type of groups, the situation is slightly different from all the cases above. With only one COR, both routes are favored. With two COR groups on two different carbon centers, both routes are also favored as shown below, with or without molecular rearrangement of a particular kind, if and only if the ring can be opened instantaneously. With radical-pulling groups, the rings are more difficult to open instantaneously.

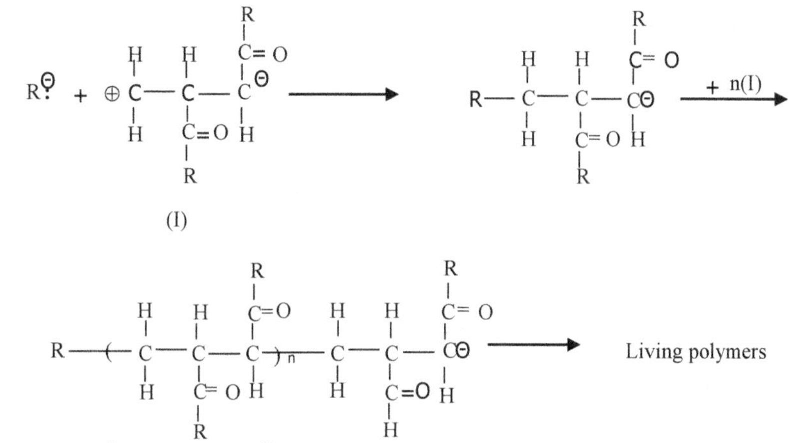

(I)

$$\text{(10.25)}$$

Living polymers

$$HOR + BF_3 +$$

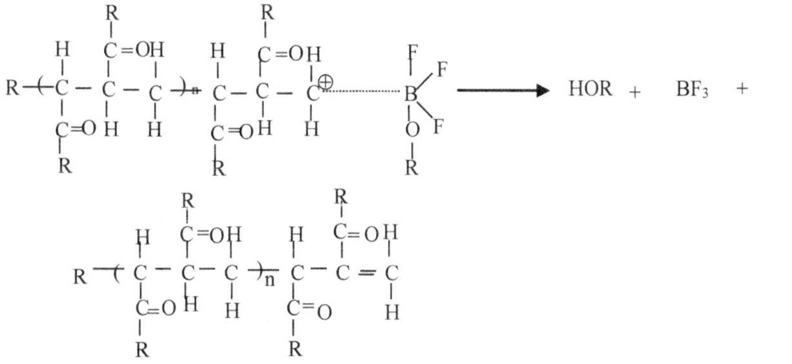

$$\text{10.26}$$

In the positively charged route, dead terminal double bond polymers are produced, since the monomer is an electrophile. Similar pattern of characters of the substituted groups can be observed, with cyclopropanes being however of higher or lower nucleophilicity or electrophilicity in character than the alkanes, depending on what the ring is carrying. As it may seem, their C = O centers can be attacked as the X center, being more nucleophilic than the invisible π-bond in the ring, the Y center. When attacked electro-free-radically or positively (Paired-media), the ring can never be opened. What are being considered so far are indications of how Nature operates. The rings considered so far have points of scission. If no point of scission exists in a ring, nothing can be done to open the ring without weakening it, by for example hydrogenation as we shall see downstream. Use groups such as CN as an exercise, noting the strong nucleophilic character of the $C \equiv N$ center.

As will be shown, it is believed that the presence of radical pushing groups in cyclopropanes decreases the stability of the ring in terms of providing it with more strain energy than the presence of radical-pulling groups.[12] In view of the use of strong initiators and most importantly the fact that the transfer species involved is not of the same kind before and after propagation, molecular rearran-gement of the **first and third kinds** will never take place with these monomers as shown below for some cases when the ring is forced to be opened.

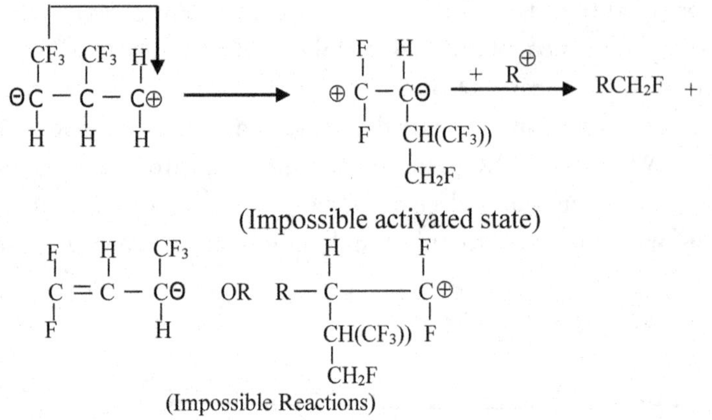

$$\text{(Impossible activated state)}$$

$$\text{(Impossible Reactions)} \tag{10.27}$$

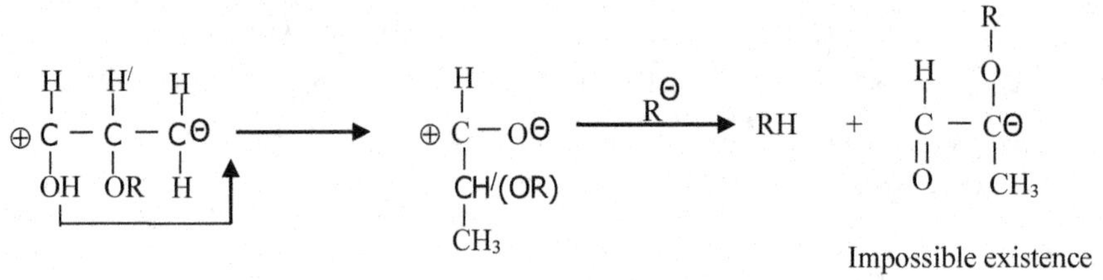

$$\text{Impossible existence} \tag{10.28}$$

Nevertheless, like activated cyclopropane of equation 10.12, substituted nucleophilic or electrophilic cyclopropane cannot rearrange as already shown and repeated below for one of them.

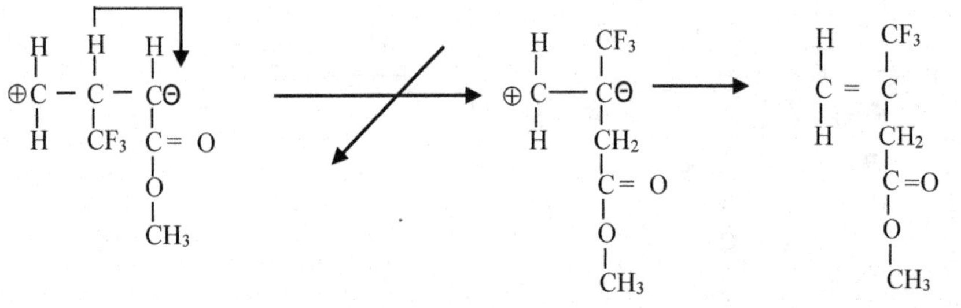

$$\tag{10.29}$$

Notice that the opened ring above looks like an Electrophile, where the transfer species should be OCH_3 group to give a Ketene. Based on the Laws of Conservation of transfer of transfer species, these rearrangements are not possible though the ring is an Electrophile. While existence of perfluoro-cyclopropane is favored, activation chargedly is impossible, due to electrostatic forces of repulsion.

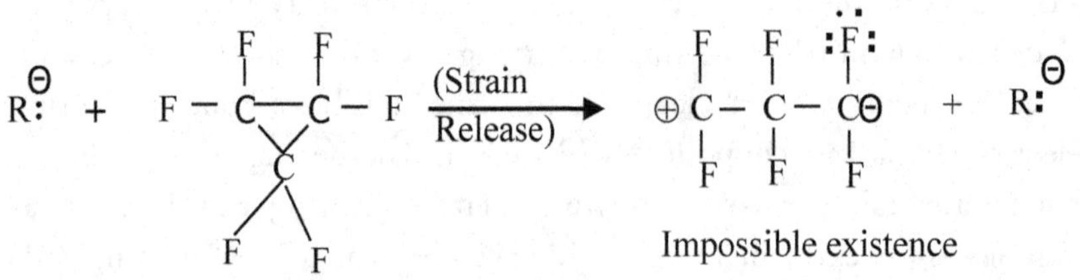

$$\text{Impossible existence} \tag{10.30}$$

12

No matter what means are provided to shield the initiating center, activation chargedly cannot be favored as shown below, if minimum required strain energy can be provided.

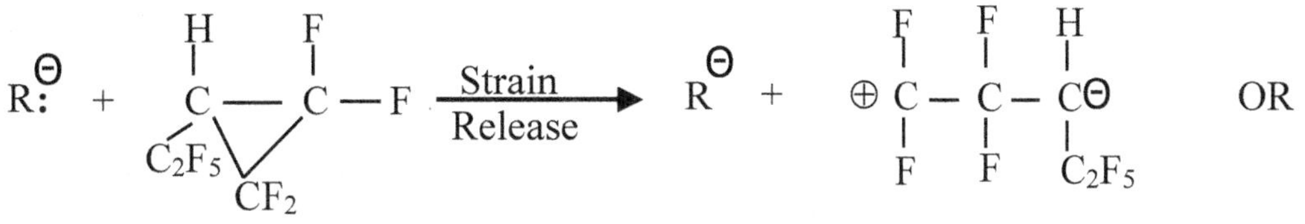

$$R : \overset{\ominus}{} \;+\; \underset{\underset{CF_2}{\overset{|}{C_2F_5}}}{\overset{\overset{H \quad F}{\overset{|}{}\quad\overset{|}{}}}{C - C - F}} \;\xrightarrow{\overset{\text{Strain}}{\text{Release}}}\; R^{\ominus} \;+\; \underset{\overset{|}{F}\;\;\overset{|}{F}\;\;\overset{|}{C_2F_5}}{\overset{\overset{F\quad F\quad H}{\overset{|}{}\;\overset{|}{}\;\overset{|}{}}}{\oplus C - C - C\ominus}} \qquad \text{OR}$$

(I) Impossible existence

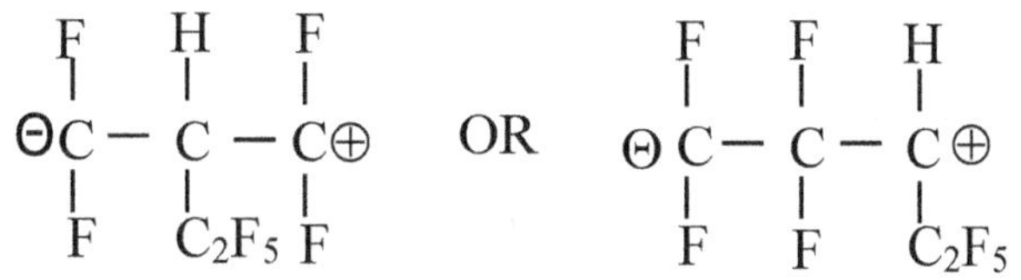

$$\underset{\overset{|}{F}\;\;\overset{|}{C_2F_5}\;\overset{|}{F}}{\overset{\overset{F\quad H\quad F}{\overset{|}{}\;\overset{|}{}\;\overset{|}{}}}{\ominus C - C - C\oplus}} \qquad \text{OR} \qquad \underset{\overset{|}{F}\;\;\overset{|}{F}\;\;\overset{|}{C_2F_5}}{\overset{\overset{F\quad F\quad H}{\overset{|}{}\;\overset{|}{}\;\overset{|}{}}}{\ominus C - C - C\oplus}}$$

(II) Impossible existence   (III) Impossible existence   10.31

Only (III) above is the real activated state which cannot exist due to electrostatic forces of repulsion. Nevertheless, though the instantaneous opening of these rings carrying radical-pulling groups will be impossible chargedly, they have been used to illustrate some basic principles. If ever they are opened, this can only be done radically, for which only the electro-free-radical route will be favored.

10. 1.2. Radical Character of Cycloalkanes

Free-radicals cannot provide the electrostatic forces needed to bring the SE to the MRSE for instantaneous opening of the rings, because the free ones do not have paired unbonded radicals in the last shell. Hence, they cannot be involved alone for initiation of ringed monomers unless when high temperatures are involved. Non-free radicals which have electrostatic forces will activate cyclo-propane, but cannot add to it, since the equation cannot be chemically balanced. It is all logic.

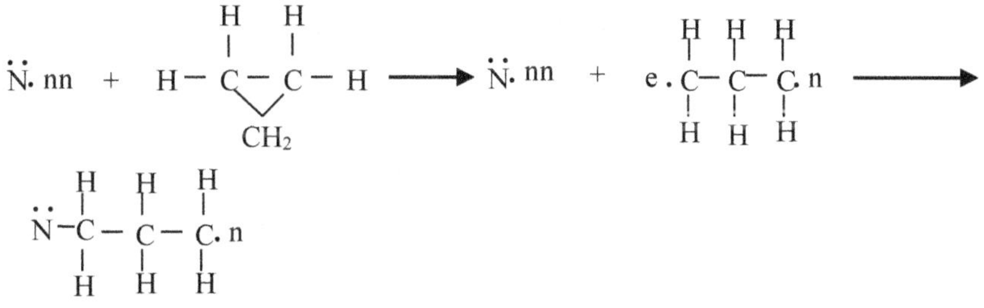

$$\overset{..}{N} . \, nn \;+\; H - \underset{CH_2}{\overset{\overset{H\quad H}{\overset{|}{}\;\overset{|}{}}}{C - C}} - H \;\longrightarrow\; \overset{..}{N} . \, nn \;+\; e . \underset{\overset{|}{H}\;\overset{|}{H}\;\overset{|}{H}}{\overset{\overset{H\;H\;H}{\overset{|}{}\,\overset{|}{}\,\overset{|}{}}}{C - C - C} . } n \;\longrightarrow$$

$$\overset{..}{N} - \underset{\overset{|}{H}\;\overset{|}{H}\;\overset{|}{H}}{\overset{\overset{H\;\;H\;\;H}{\overset{|}{}\;\overset{|}{}\;\overset{|}{}}}{C - C - C} . } n$$

<u>Not balanced</u>   10.32

When the nucleo-non-free-radical is used in the presence of an electro-free-radical (e.g. H $^{.e}$ Cl.nn), polymerization is not favored free-radically, because Cl.nn will close the first addition. A look at cyclopropane shows that it is isomeric with propene. When decomposed, it cannot molecu-larly rearrange to propene. But when forced to be in Equilibrium state of existence, it decomposes to propene as shown below. The phenomenon here is called Electro-radicalization.

Stage 1:

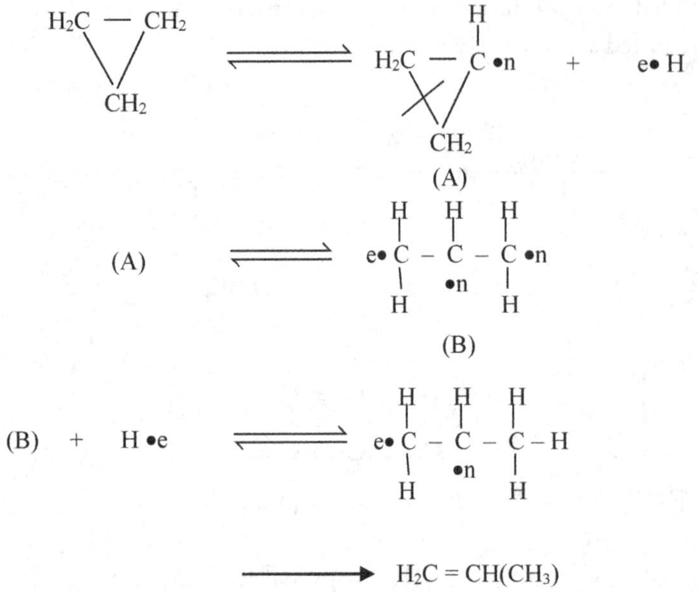

$$H_2C \text{—} CH_2 \text{ (cyclopropane)} \rightleftharpoons H_2C \text{—} \underset{\overset{|}{CH_2}}{\overset{\overset{H}{|}}{C}} \bullet n \quad + \quad e\bullet H$$

(A)

$$(A) \rightleftharpoons e\bullet \underset{\overset{|}{H}}{\overset{\overset{H}{|}}{C}} - \underset{\overset{|}{\bullet n}}{\overset{\overset{H}{|}}{C}} - \underset{\overset{|}{H}}{\overset{\overset{H}{|}}{C}} \bullet n$$

(B)

$$(B) \quad + \quad H\bullet e \rightleftharpoons e\bullet \underset{\overset{|}{H}}{\overset{\overset{H}{|}}{C}} - \underset{\overset{|}{\bullet n}}{\overset{\overset{H}{|}}{C}} - \underset{\overset{|}{H}}{\overset{\overset{H}{|}}{C}} - H$$

$$\longrightarrow \quad H_2C = CH(CH_3) \qquad \qquad 10.33a$$

Overall Equation: Cyclopropane $\longrightarrow$ Propene $\qquad\qquad$ 10.33b

Cyclopropane has been used with aluminum tribromide ($AlBr_3$) as catalyst to produce low molecular weight ill-defined oils. It is thought that the most likely carbonium ion generated from the cyclopropane is an unstable primary cation which probably rearranges immediately to a variety of low-molecular-weight branched products. In contrast, the dimethyl derivative does not rearrange and is said to give uniform polymers as shown below.[13-15]

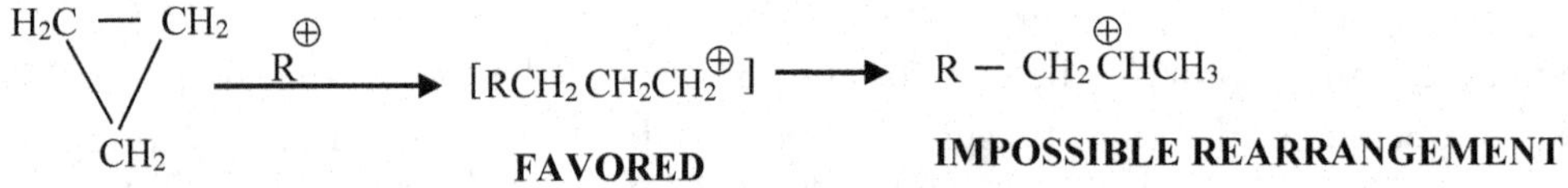

$$H_2C \text{—} CH_2 \text{ (cyclopropane)} \xrightarrow{\overset{\oplus}{R}} [RCH_2CH_2CH_2^{\oplus}] \longrightarrow R - CH_2 \overset{\oplus}{C}HCH_3$$

$$\textbf{FAVORED} \qquad\qquad \textbf{IMPOSSIBLE REARRANGEMENT}$$

10.33c

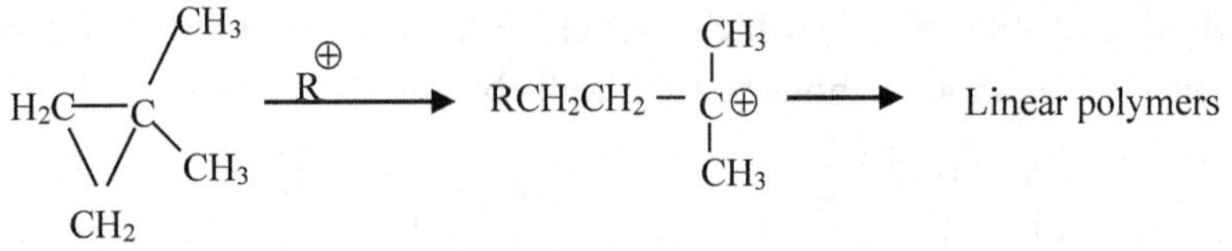

$$H_2C \text{—} \underset{\overset{|}{CH_2}}{\overset{\overset{CH_3}{|}}{C}} \text{—} CH_3 \xrightarrow{\overset{\oplus}{R}} RCH_2CH_2 - \underset{\overset{|}{CH_3}}{\overset{\overset{CH_3}{|}}{C}}\oplus \longrightarrow \text{Linear polymers}$$

10.34

With the use of $AlBr_3$, either the presence of positively charged-paired initiators or radical initiators or both are required.

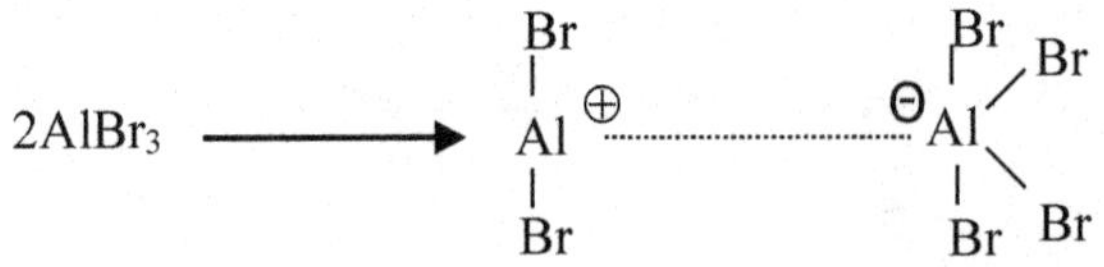

$$2AlBr_3 \longrightarrow \underset{\overset{|}{Br}}{\overset{\overset{Br}{|}}{Al}}{}^{\oplus}\text{--------}{}^{\ominus}\underset{\overset{|}{Br}\ Br}{\overset{\overset{Br}{|}\diagup Br}{Al}}$$

positvely-charged-paired

10.35

$$AlBr_3 \rightleftharpoons BrAl\,.\,e \quad + \quad :\overset{..}{Br}\,.\,nn$$

10.36

By the nature of the reaction in equation 10.33 above, radical initiators may be said to be involved as shown below.

14

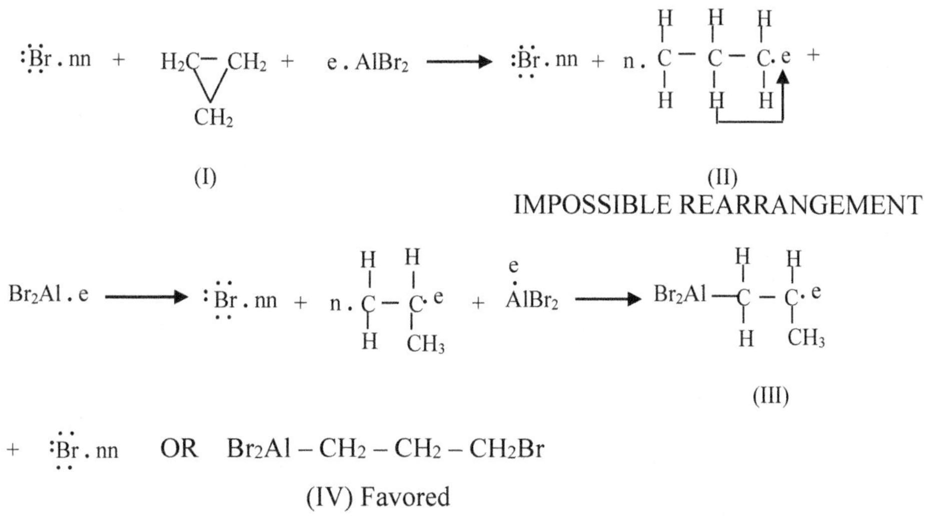

$$\ddot{Br}\cdot nn \;+\; H_2C\!-\!CH_2 \;+\; e\cdot AlBr_2 \;\longrightarrow\; \ddot{Br}\cdot nn \;+\; n\cdot \overset{\displaystyle H}{\underset{\displaystyle H}{C}} - \overset{\displaystyle H}{\underset{\displaystyle H}{C}} - \overset{\displaystyle H}{\underset{\displaystyle H}{C}}\cdot e \;+$$

with cyclopropane ring (CH₂) below

$$\text{(I)} \qquad\qquad\qquad\qquad \text{(II)}$$

IMPOSSIBLE REARRANGEMENT

$$Br_2Al\cdot e \;\longrightarrow\; \ddot{Br}\cdot nn \;+\; n\cdot \overset{H}{\underset{H}{C}} - \overset{H}{\underset{CH_3}{C}}\cdot e \;+\; \overset{e}{AlBr_2} \;\longrightarrow\; Br_2Al - \overset{H}{\underset{H}{C}} - \overset{H}{\underset{CH_3}{C}}\cdot e$$

$$\text{(III)}$$

$$+ \;\; \ddot{Br}\cdot nn \qquad OR \qquad Br_2Al - CH_2 - CH_2 - CH_2Br \qquad\qquad 10.37a$$

$$\text{(IV) Favored}$$

The ring is instantaneously opened by Br . nn which cannot react with it alone. This is then followed by molecular rearrangement of (II) which was thought to be favored free-radically and not chargedly, but found to be impossible which ever center the H is moved to, noting however that H can never be moved as a nucleo-free-radical. It is (IV) that now exists in Equilibrium state of Existence, with e·Al(Br) CH₂CH₂CH₂Br being the initiator that adds only some few monomer units to be closed by Br·nn. This is how the telomers are obtained via Equilibrium mechanism. (IV) was produced in the first stage. It is now used in the second stage as follows.

Stage2: (IV) $\rightleftharpoons$ Br•nn + e•Al(Br)CH₂CH₂CH₂Br

$$\text{(V)} \qquad\qquad\qquad \text{(VI)}$$

(VI) + (I) $\underset{\textit{of ring}}{\overset{\textit{Activation}}{\rightleftharpoons}}$ BrCH₂CH₂CH₂(Br)Al CH₂CH₂CH₂ •e

$$\text{(VII)}$$

(VII) + (V) $\longrightarrow$ BrCH₂CH₂CH₂(Br)Al(CH₂CH₂CH₂)Br

$$\text{(VIII)} \qquad\qquad 10.37b$$

The telomer (VIII- n = 2) was produced just in two stages via Equilibrium mechanism instead of the usual Combination mechanism. It could not add more than one monomer unit in one stage, unless if the monomer was self avtivated to form rings. Addition continues in more stages until there is no Br carried by the Al center in (VIII). If the initiator of Equation 10.35 had been formed, then high molecular weight polymers will be obtained.

The type of molecular rearrangement in (II) of Equation 10.37a above which was not favored, can however be favored by a cycloalkane such as the dimethyl derivative as shown below radically.

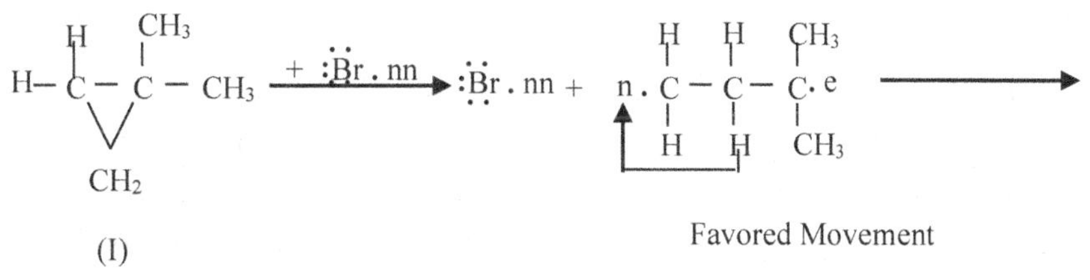

$$H-\overset{H}{\underset{\;}{C}} - \overset{CH_3}{\underset{\;}{C}} - CH_3 \;\xrightarrow{+\; \ddot{Br}\cdot nn}\; \ddot{Br}\cdot nn \;+\; n\cdot \overset{H}{\underset{H}{C}} - \overset{H}{\underset{H}{C}} - \overset{CH_3}{\underset{CH_3}{C}}\cdot e \;\longrightarrow$$

with cyclopropane ring (CH₂) below (I)

$$\text{(I)} \qquad\qquad\qquad\qquad \text{Favored Movement}$$

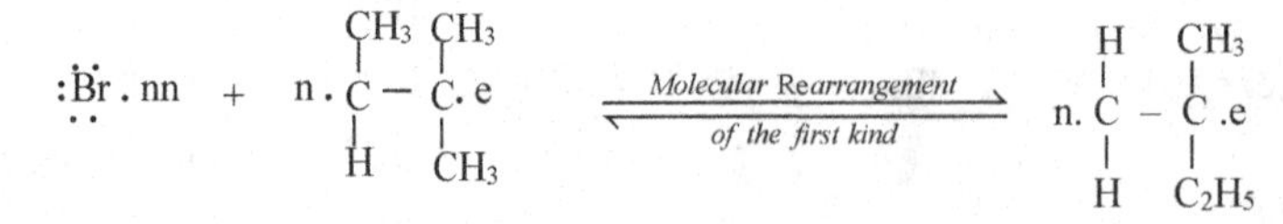

$$:\overset{..}{\underset{..}{Br}} . nn \quad + \quad n . \overset{\overset{CH_3}{|}}{\underset{\underset{H}{|}}{C}} - \overset{\overset{CH_3}{|}}{\underset{\underset{CH_3}{|}}{C}} . e \quad \xrightleftharpoons{\underset{\textit{of the first kind}}{\textit{Molecular Rearrangement}}} \quad n . \overset{\overset{H}{|}}{\underset{\underset{H}{|}}{C}} - \overset{\overset{CH_3}{|}}{\underset{\underset{C_2H_5}{|}}{C}} . e$$

$$\underset{\text{(III) \underline{Favored}}}{} \qquad\qquad\qquad\qquad\qquad\qquad \underset{\text{(IV)}}{} \qquad\qquad 10.38$$

(III) above can be observed to be a strong Nucleophile, which can also undergo further molecular rearrangement of the first kind to give (IV). When positively charged-paired initiator from $AlBr_3$ is used for the monomer above, the followings are obtained in the absence of molecular rearrangements.

$$2AlBr_3 \quad + \quad n . \overset{\overset{H}{|}}{\underset{\underset{H}{|}}{C}} - \overset{\overset{H}{|}}{\underset{\underset{H}{|}}{C}} - \overset{\overset{CH_3}{|}}{\underset{\underset{CH_3}{|}}{C}} . e \quad \longrightarrow \quad Br_2Al - \overset{\overset{H}{|}}{\underset{\underset{H}{|}}{C}} - \overset{\overset{H}{|}}{\underset{\underset{H}{|}}{C}} - \overset{\overset{CH_3}{|}}{\underset{\underset{CH_3}{|}}{C}}^{\oplus} .... \,^{\ominus}AlBr_4$$

$$\underset{\text{(I)}}{}$$

$$\xrightarrow{+ n \, (I)} \quad Br_2Al - \left(\overset{\overset{H}{|}}{\underset{\underset{H}{|}}{C}} - \overset{\overset{H}{|}}{\underset{\underset{H}{|}}{C}} - \overset{\overset{CH_3}{|}}{\underset{\underset{CH_3}{|}}{C}}\right)_n \overset{\overset{H}{|}}{\underset{\underset{H}{|}}{C}} - \overset{\overset{H}{|}}{\underset{\underset{H}{|}}{C}} - \overset{\overset{CH_3}{|}}{\underset{\underset{CH_3}{|}}{C}}^{\oplus} .... \,^{\ominus}AlBr_4$$

$$\longrightarrow \quad Br_2Al - \left(\overset{\overset{H}{|}}{\underset{\underset{H}{|}}{C}} - \overset{\overset{H}{|}}{\underset{\underset{H}{|}}{C}} - \overset{\overset{CH_3}{|}}{\underset{\underset{CH_3}{|}}{C}}\right)_n \overset{\overset{H}{|}}{\underset{\underset{H}{|}}{C}} - \overset{\overset{H}{|}}{\underset{\underset{H}{|}}{C}} - \overset{\overset{CH_3}{|}}{\underset{\underset{H}{|}}{C}} = \overset{\overset{H}{|}}{\underset{}{C}} \; + \; HBr \; + \; AlBr_3$$

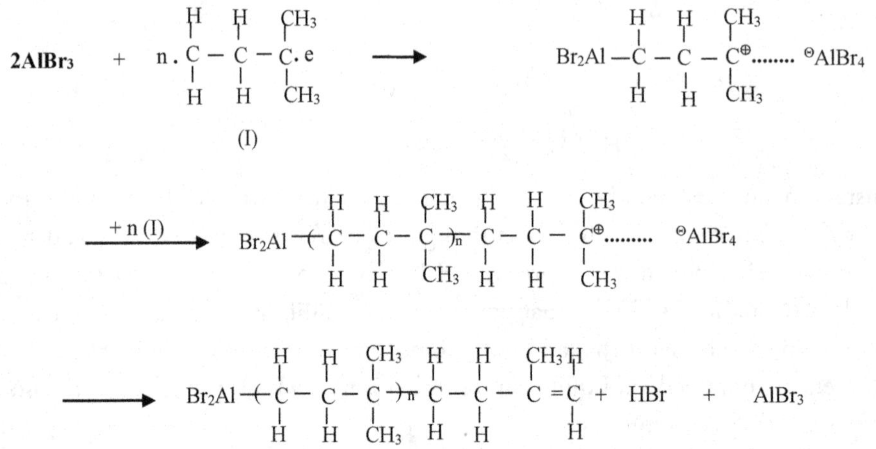

Transfer species of 1st kind of 1st type $\qquad\qquad$ 10.39

It is important to note that when positively charged-paired route or electro-free-radical route is involved, $AlBr_2$ group is always a part of the polymeric product. With the dead terminal double bond polymer above, branching sites can be formed when an electro-free-radical growing chain attacks it.

Shown below is another substituted cycloalkane where molecular rearrangements of the first and third kinds take place in the presence of a weak initiator to give (III).

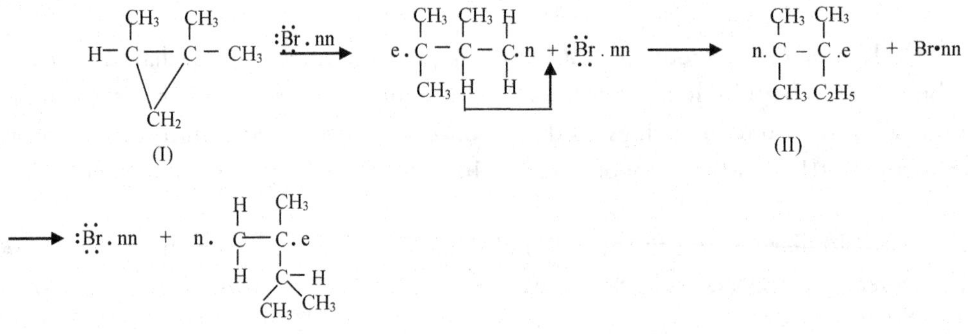

$$H - \overset{\overset{CH_3}{|}}{\underset{\underset{}{}}{C}} - \overset{\overset{CH_3}{|}}{\underset{\underset{CH_2}{\diagdown}}{C}} - CH_3 \quad \xrightarrow{:\overset{..}{Br} . nn} \quad e . \overset{\overset{CH_3}{|}}{\underset{\underset{CH_3}{|}}{C}} - \overset{\overset{CH_3}{|}}{\underset{\underset{H}{|}}{C}} - \overset{\overset{H}{|}}{\underset{\underset{H}{|}}{C}} . n + :\overset{..}{\underset{..}{Br}} . nn \quad \longrightarrow \quad n . \overset{\overset{CH_3}{|}}{\underset{\underset{CH_3}{|}}{C}} - \overset{\overset{CH_3}{|}}{\underset{\underset{C_2H_5}{|}}{C}} . e \quad + \; Br \bullet nn$$

$$\underset{\text{(I)}}{} \qquad\qquad\qquad\qquad\qquad\qquad\qquad\qquad \underset{\text{(II)}}{}$$

$$\longrightarrow \quad :\overset{..}{\underset{..}{Br}} . nn \quad + \quad n . \overset{\overset{H}{|}}{\underset{\underset{H}{|}}{C}} - \overset{\overset{CH_3}{|}}{\underset{\underset{\underset{CH_3}{}\overset{|}{C}\overset{}{}\,CH_3}{|}}{C}} . e$$

$$\underset{\text{(III)}}{} \qquad\qquad\qquad\qquad\qquad\qquad\qquad 10.40$$

Thus, one can observe why (I) above is more nucleophilic than (I) of Equation 10.38, where two types of molecular rearrangement is favored for both of them. The second molecular rearrange-ment is well known- the first kind of the first type. The first molecular rearrangement is very unique, unlike the case in Carbenes which is of the second kind. ***The molecular re-arrangement seen so far is molecular rearrangement of third kind of the first type.*** It is the electro-free-radical that does the first attack and not the nucleo-non-free-radical halogen center (such as Br.nn above). When paired, polymers are produced; but when not paired, telomers or ordinary compounds are produced.

Where nucleo-free-radicals are present in the presence of nucleo-non-free-radicals or radical-donating molecules to open the ring instantaneously and minimum required strain energy (MRSE) can be provided, initiation is not favored due to the nucleophilic characters of the radical-pushing cyclo-propanes, and presence of transfer species of first kind.

It will be very instructive at this point in time to digress by considering olefinic monomers with ringed groups, as substituted groups. The "cationic" polymerization with catalyst such as $SnCl_4$ or $AlBr_3$, of vinylcyclopropane is said to involve some measure of 1,5-polymerization as indicated below.[16-18]

(I) Major product

(II) Minor product – 1,5 - enchainment.

$$10.41$$

The structure of the 1, 5-minor product obtained is said to be identical with that obtained by ring cleavage polymerization of cyclopentane in the presence of certain catalysts of the Ziegler-Natta type. This last statement is true as will be shown in the next chapter. With $AlBr_3$, the following are obtained. Though it is the paired one of Equation 10.35 that can be used, *one has been using the free one for simplicity.*

$$10.42$$

Transfer species of the 1st kind of the ninth type.

The transfer species involved here is of the first kind, but of the ninth type, since it is located on a ring. This is the same transfer species that will prevent the monomer from undergoing nucleo-free-radical or negatively-charged routes. The ring cannot provide resonance stabilization for the electro-free-radical center and the CH=CH- group cannot also provide resonance stabilization for the ring as against what was thought to be the case in Equation 10.41 in the past and as will be shown downstream. Between the two rings shown below, (I) is more nucleophilic than (II) via instantaneous opening of the ring, since the -CH=CH group is of equal or higher capacity than H.

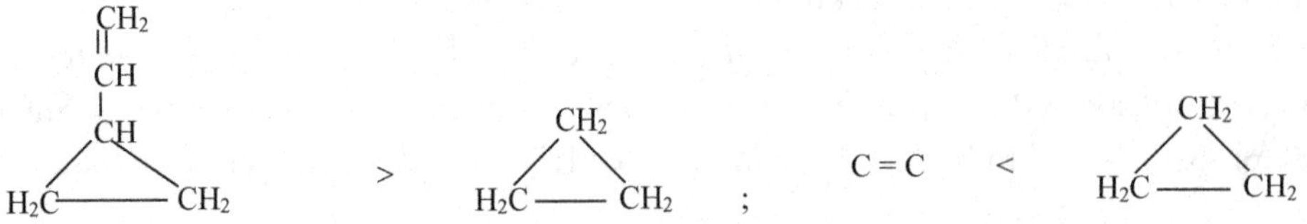

$$
\begin{array}{ccccc}
\underset{\text{(I)}}{
\begin{array}{c}
CH_2 \\ \parallel \\ CH \\ \mid \\ CH \\ \diagup \diagdown \\ H_2C \!-\!\!-\!\!- CH_2
\end{array}}
& > &
\underset{\text{(II)}}{
\begin{array}{c}
CH_2 \\ \diagup \diagdown \\ H_2C \!-\!\!- CH_2
\end{array}}
& ; \qquad C = C \quad < \quad
\begin{array}{c}
CH_2 \\ \diagup \diagdown \\ H_2C \!-\!\!- CH_2
\end{array}
& \qquad 10.43
\end{array}
$$

Order of Nucleophilicity

(I) is slightly more strained than (II) for the same reason ***and most importantly, the invisible π-bond is more nucleophilic than the visible π-bond (C = C) when present as above in a compound (i.e.. with*** ***$H_2C=CH_2$).*** Hence, a stronger initiator or more energy will be required to open (II) than (I). When a stronger initiator is involved, the following is obtained.

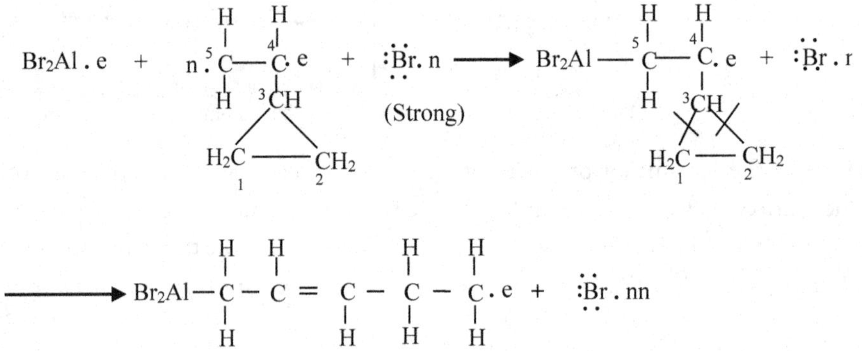

$$
Br_2Al \cdot e \; + \; n \overset{5}{\cdot} C \!-\! \overset{4}{C} \cdot e \; + \; : \overset{..}{Br} \cdot n \longrightarrow Br_2Al \!-\! \overset{5}{C} \!-\! \overset{4}{C} \cdot e \; + \; : \overset{..}{Br} \cdot r
$$

(Strong)

$$
\longrightarrow Br_2Al \!-\! C \!-\! C = C \!-\! C \!-\! C \cdot e \; + \; : \overset{..}{Br} \cdot nn \qquad 10.44
$$

The "Br$^\ominus$" or Br.nn seen so far is indeed $^\ominus$AlBr4 or nn.Br and not nn•AlBr4, for the use of the paired charged or radical initiator respectively.

$$
\begin{array}{ccc}
\begin{array}{c}
Br \qquad\qquad Br \;\; Br \\ \mid \qquad\qquad\; \mid \diagup \\
Al\bullet^e \text{------}^{nn}\bullet\, Al \!-\! Br \\ \mid \qquad\qquad\; \mid \\ Br \qquad\qquad Br
\end{array}
& ; &
\begin{array}{c}
Br \qquad\qquad Br \;\; Br \\ \mid \qquad\qquad\; \mid \diagup \\
Al^\oplus \text{-------}\, ^\ominus Al \!-\! Br \\ \mid \qquad\qquad\; \mid \\ Br \qquad\qquad Br
\end{array} \\
\text{"Paired Radical"?} & & \text{Paired Charged} \\
\text{[NOT FAVORED]} & & \qquad\qquad 10.45
\end{array}
$$

The existence of "paired radical" initiator electrostatically is questionable as will be shown downstream, noting however that before pairing can take place, it is not true that the centers must only be charged. Covalently and ionically, pairing can take place radically when the conditions exist.

Note the change in the numbering of the activated monomer in Equation 10.44. Scission can only take place at 1,3 or 2,3 carbon-carbon bond in the ring, points having higher and same radical-pushing potential difference. It is important to note that the externally located activation center conjugatedly placed is the first to be activated before the ring is opened since the activation center is less nucleophilic than that inside the ring. As can be observed there is no radical rearrangement taking place as was thought to be the case-in the past. However, it is only resonance stabilized when the ring is opened instantaneously *that which is impossible without first activating the vinyl center.*

For the two monomers shown below, the situations are different for them.

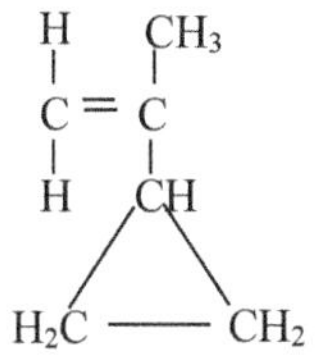

(I) C_6H_{10} 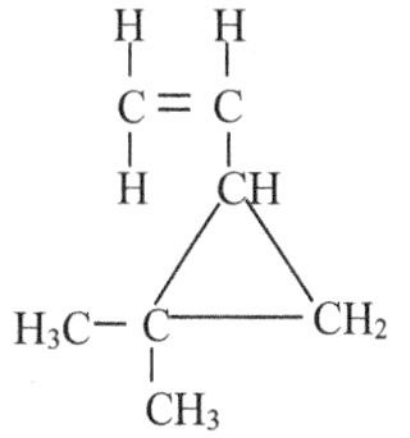   (II) C_7H_{11}

Derivatives of vinylcyclopropane   10.46

Since $H_2C = C(CH_3)$- group is less radical-pushing than $H_2C = CH$-group, the ring in (I) will require more energy to open than that in (II). Shown below is the opening of (II)-using $SnCl_4$ as the source of initiator. The paired unbonded radicals on the Sn not shown have been blocked.

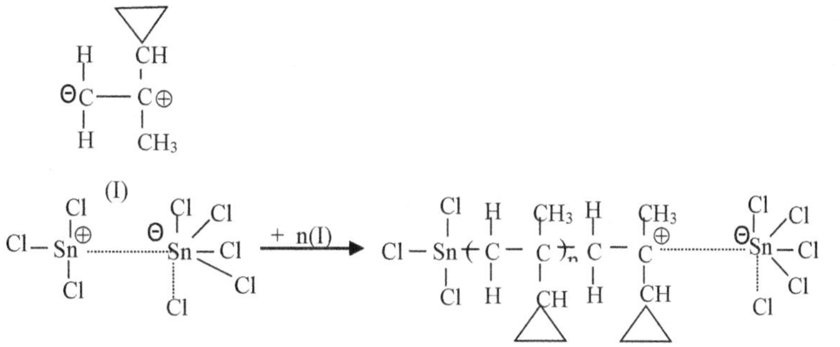

10.47

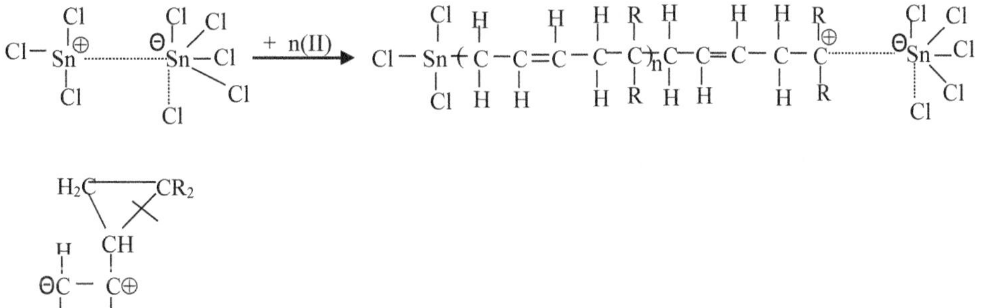

10.48

The point of scission can be observed to be that with the largest potential radical-pushing difference. In view of the presence of CR_2, a carbon center with two radical-pushing groups in the ring, the ring is more strained than cyclopropane. Nevertheless, it is important to note that, it is the external activation center conjugatedly placed to the ring that is first attacked. Finally, one can observe why the product via the $C = C$ center without opening of the ring is far more than (Major) that obtained when the ring is opened (Minor) in Equation 10.41. The same will apply to the first case above, but not to the second case. In the second case above (the last equation), the ring will more readily be opened being highly strained after attack on the vinyl center.

Vinylepisulfide which is structurally closely related to vinylcyclopropane, is said to undergo rings-opening rearrangement polymerization under the influence of Friedel-Crafts halides, but not under anionic conditions, for example with $Et_2Zn.H_2O$.[19] The particular monomer is hereby used to illustrate the importance of one of the driving forces determining points of scission, being that with the largest potential radical-pushing difference. It is also used to illustrate the fact that since S and C atoms truly have the same electronegativity (2.5), the sulfur atom cannot be made to carry a positive ionic charge whether positively paired charged initiators are involved or not in the presence of a carbon center.

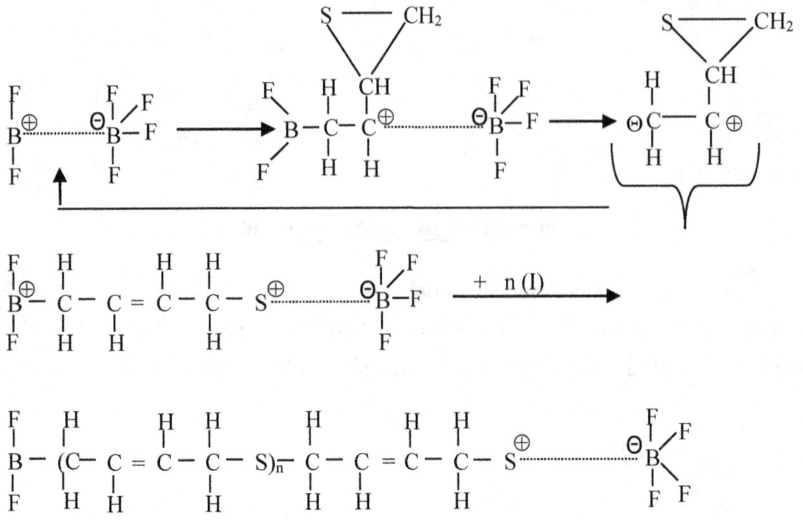

$$10.49$$

(Not Favored)

When an electro-free-radical initiator is used, the followings seem to be expected.

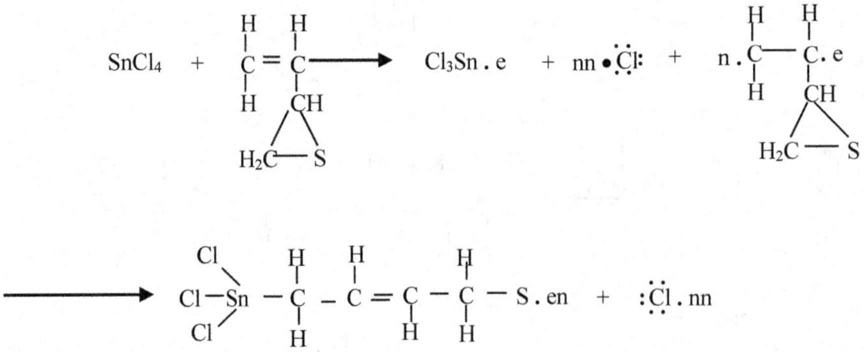

$$10.50$$

(Not Favored)

Radically and chargedly, the equations above are not favored. There is no doubt that the radicals or charges carried by the centers are the wrong ones. But however, the exploration continues. From the reactions above it seems that vinyl episulfide is more strained than vinylcyclo-propane and that they don't belong to the same family. It is the order shown below that is valid for the parent monomers, as will become clear downstream in Chapter 14 when the foundations must have fully been laid.

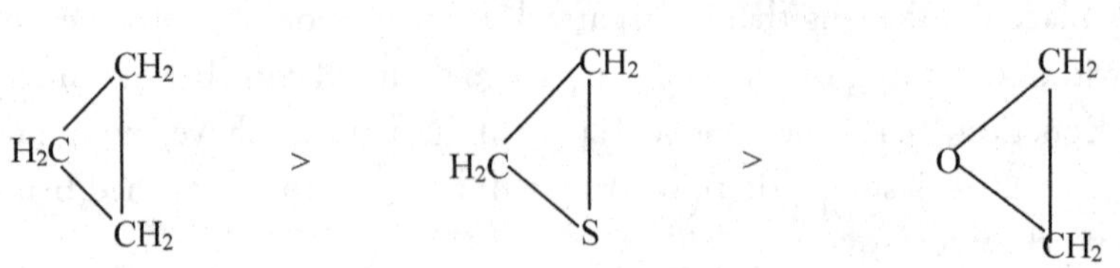

Order of Strain Energy   10.51

The order of Nucleophilic character is different, indeed the reverse.

While the reaction above seem not to be favored for vinylepisulfide, the same will also apply to vinyl epoxide, noting that the oxygen center is more electronegative than S which in turn is equi-electropositive with C.

Vinylepisulfide   Vinylepoxidede   10.52

The functional center should be the first to be attacked if C=C activation center is more nucleophile than the O functional center in the ring. Cyclic ether should be more nucleophilic than cyclic sul-fide (S). However as will be shown downstream, the following is valid.

C = C   <   S   <   O

(Olefins)   (Cyclic sulfides)   (Cyclic ethers)

Order of nucleophilicity

10.53

Hence the reactions of Equations 10.49 and 10.50 are not favored as will become obvious down-stream. Meanwhile from the above, the two families are different from vinylcyclopropane (Vinyl-cycloalkanes), because the Vinylcyclic sulfides and Vinylcyclic oxides just considered for specific reasons are not Nucleophiles as one will show downstream, but Electrophiles.

At high temperature of up to $70 - 80^{\circ}C$, full polymerization of substituted cyclo-propane will be favored in the presence of electro-free-radicals such as $TiCl_3$.

10.54

Nucleo-free-radicals cannot be used, since the monomers are nucleophiles. Unlike cyclo-propane, there may be enough time for the activated monomer above to undergo ***molecular rearrangements of both kinds*** if the initiator is weak.

With cyclobutane, its activation will require more energy than with cyclopropane, in view of the size of ring wherein the same SE as in cyclopropane is being shared in a four-membered ring (Cyclobutane). However, with radical-pushing substituted cyclobutane, activation is very readily favored since they have different and higher strain energies (SEs).

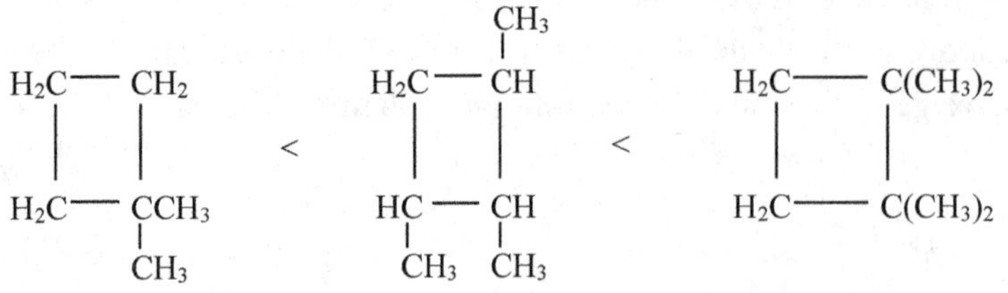

$$
\begin{array}{ccc}
\underset{\text{H}_2\text{C}\underset{|}{-}\text{CH}_2}{\overset{\text{H}_2\text{C}-\text{CH}_2}{\big|\quad\big|}}\underset{\underset{\text{CH}_3}{|}}{\text{CCH}_3} & < & \underset{\text{HC}\underset{|}{-}\underset{|}{\text{CH}}}{\overset{\text{H}_2\text{C}-\overset{|}{\text{CH}}}{\big|\quad\big|}} \\
\end{array}
$$

H2C—CH2   H2C—CH (CH3)   H2C——C(CH3)2
| | < | | < | |
H2C—CCH3   HC——CH   H2C——C(CH3)2
      | | |
      CH3   CH3 CH3
(I)   (II)   (III)

Substituted cyclobutanes – Order of Strain Energy   10.55

Notice here that, the order of their SEs is the same as the order of their Nucleophilicities

In the presence of $AlBr_3$, the followings are obtained.

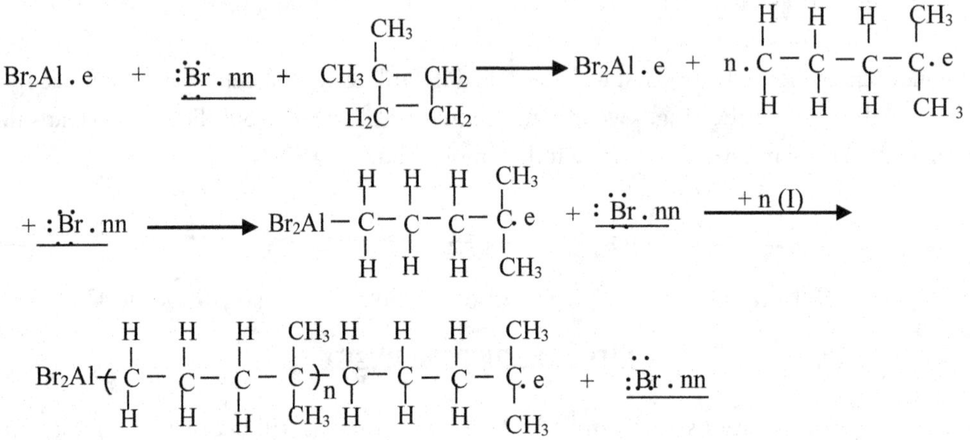

10.56

In the reactions above, a single molecule or telomers are largely obtained. But when the paired initiators are used, polymers are obtained as shown below.

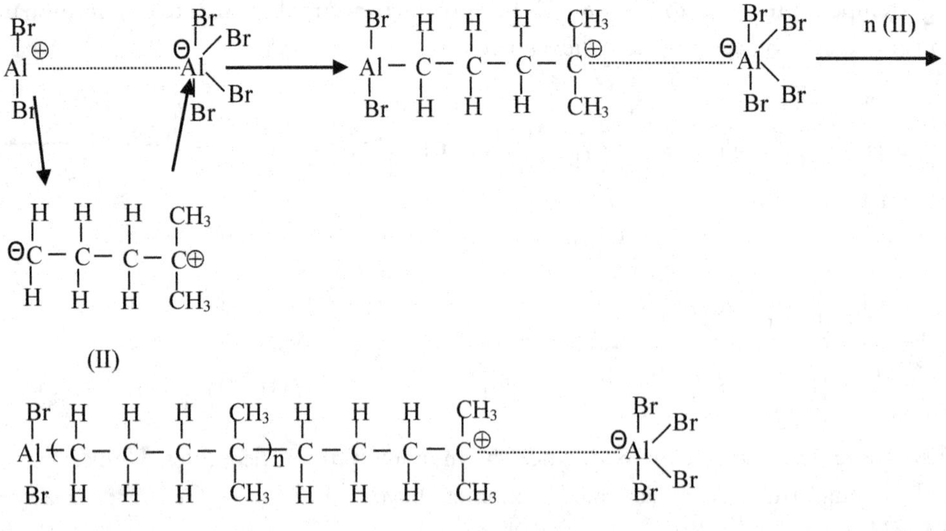

(II)

10.57

Thus, the minimum required strain energy (MRSE) can very readily be provided for such monomers to favor the instantaneous opening of their rings. If the MRSE can be provided radically for the first member, then the followings shown below can never be obtained, since as has been maintained so far, H can never be moved as a nucleo-free-radical.

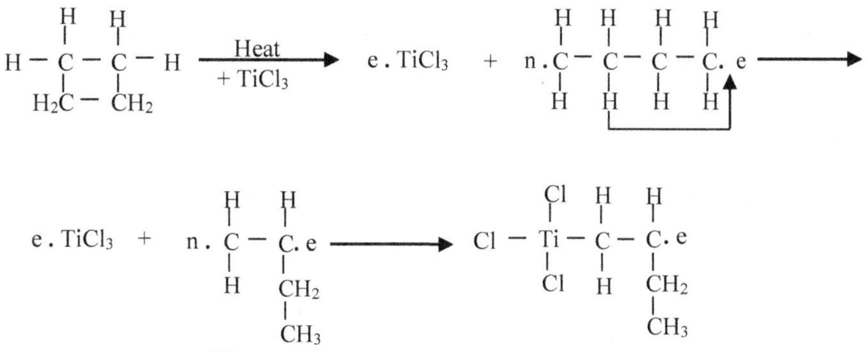

$$H-\underset{\underset{CH_2}{|}}{\overset{\overset{H}{|}}{C}}-\underset{\underset{CH_2}{|}}{\overset{\overset{H}{|}}{C}}-H \quad\xrightarrow[\text{+ TiCl}_3]{\text{Heat}}\quad e\cdot TiCl_3 \;+\; n\cdot\overset{\overset{H}{|}}{C}-\overset{\overset{H}{|}}{C}-\overset{\overset{H}{|}}{C}-\overset{\overset{H}{|}}{C}\cdot e \longrightarrow$$

$$e\cdot TiCl_3 \;+\; n\cdot\underset{\underset{\underset{CH_3}{|}}{CH_2}}{\overset{\overset{H}{|}}{C}}-\underset{H}{\overset{\overset{H}{|}}{C}}\cdot e \longrightarrow Cl-\underset{\underset{Cl}{|}}{\overset{\overset{Cl}{|}}{Ti}}-\underset{H}{\overset{\overset{H}{|}}{C}}-\underset{\underset{\underset{CH_3}{|}}{CH_2}}{\overset{\overset{H}{|}}{C}}\cdot e$$

(II)

IMPOSSIBLE REACTIONS 10.58a

As can be observed so far, the first members cannot undergo molecular rearrangement of the third kind. What is actually obtained is as shown below, a growing chain which cannot be killed from within.

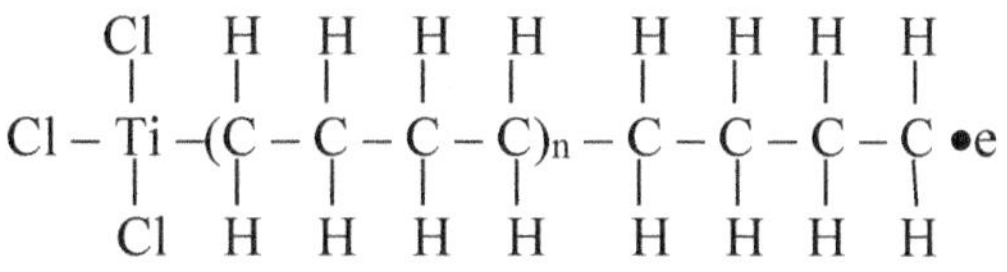

$$Cl-\underset{\underset{Cl}{|}}{\overset{\overset{Cl}{|}}{Ti}}-(\underset{H}{\overset{\overset{H}{|}}{C}}-\underset{H}{\overset{\overset{H}{|}}{C}}-\underset{H}{\overset{\overset{H}{|}}{C}}-\underset{H}{\overset{\overset{H}{|}}{C}})_n-\underset{H}{\overset{\overset{H}{|}}{C}}-\underset{H}{\overset{\overset{H}{|}}{C}}-\underset{H}{\overset{\overset{H}{|}}{C}}-\underset{H}{\overset{\overset{H}{|}}{C}}\bullet e$$

10.58b

Obviously, one should expect vinyl cyclobutane to be less strained than cyclobutane.

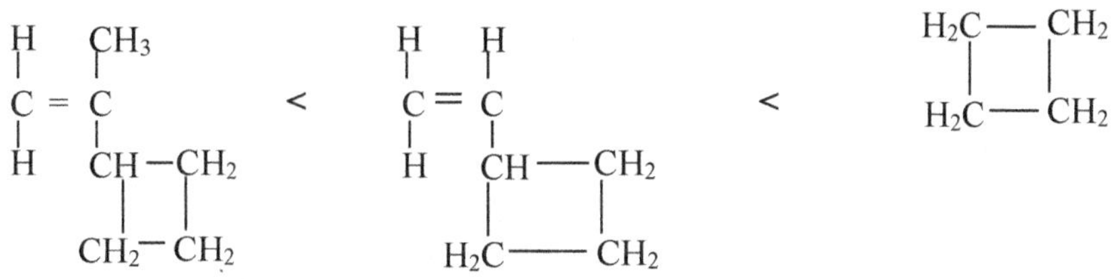

$$\underset{H}{\overset{\overset{H}{|}}{C}}=\underset{\underset{\underset{CH_2-CH_2}{|\quad|}}{CH-CH_2}}{\overset{\overset{CH_3}{|}}{C}} \quad<\quad \underset{H}{\overset{\overset{H}{|}}{C}}=\underset{\underset{\underset{H_2C——CH_2}{|\qquad|}}{CH——CH_2}}{\overset{\overset{H}{|}}{C}} \quad<\quad \underset{\underset{H_2C——CH_2}{|\qquad|}}{\overset{H_2C——CH_2}{}}$$

Order of Strain Energy 10.59

Notice that, the order of SEs here is the reverse of the order of Nucleophilicity, because a Di-ene is more nucleophilic than a Mono-ene. These members of monomers are like Di-enes, but not Di-enes.

But, when radical-pushing substituted groups are introduced into the ring, depending on where located, the ring will be opened instantaneously after the C=C bond has been activated and added to the initiator.

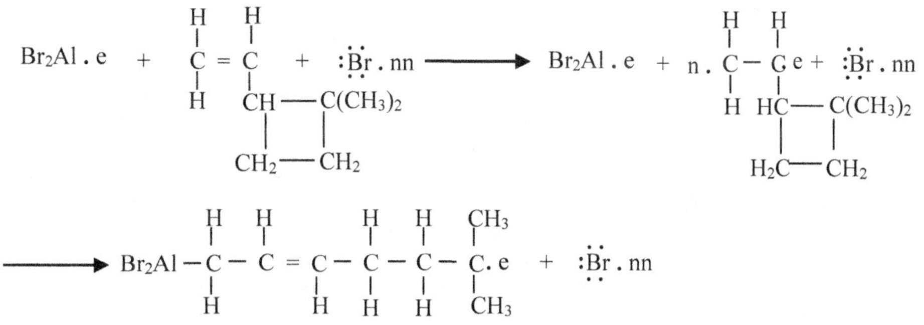

$$Br_2Al\cdot e \;+\; \underset{H}{\overset{\overset{H}{|}}{C}}=\underset{\underset{\underset{CH_2——CH_2}{|\qquad|}}{CH——C(CH_3)_2}}{\overset{\overset{H}{|}}{C}} \;+\; :\ddot{\overset{\cdot}{B}r}\cdot nn \longrightarrow Br_2Al\cdot e \;+\; n\cdot\underset{H}{\overset{\overset{H}{|}}{C}}-\underset{\underset{\underset{H_2C—CH_2}{|\qquad|}}{HC——C(CH_3)_2}}{\overset{\overset{H}{|}}{\ddot{G}}}e \;+\; :\ddot{\overset{\cdot}{B}r}\cdot nn$$

$$\longrightarrow Br_2Al-\underset{H}{\overset{\overset{H}{|}}{C}}-\overset{\overset{H}{|}}{C}=\underset{H}{\overset{\overset{H}{|}}{C}}-\underset{H}{\overset{\overset{H}{|}}{C}}-\underset{\underset{CH_3}{|}}{\overset{\overset{CH_3}{|}}{C}}\cdot e \;+\; :\ddot{\overset{\cdot}{B}r}\cdot nn$$

10.60

$$Br_2Al \cdot e \; + \; \underset{\substack{H \\ | \\ C \\ | \\ H \; HC——CH_2 \\ \quad | \qquad | \\ \quad H_2C——C(CH_3)_2}}{\overset{H}{\underset{}{C}}} = \underset{}{\overset{H}{\underset{|}{C}}} \; + \; :\ddot{B}r \cdot nn \longrightarrow Br_2Al \cdot e \; + \; n \cdot \underset{\substack{H \\ | \\ C \\ | \\ H \; HC——CH_2 \\ \quad | \qquad | \\ \quad H_2C——C(CH_3)_2}}{\overset{H}{\underset{|}{C}}} - \overset{H}{\underset{|}{C}}e \; + \; :\ddot{B}r \cdot nn$$

$$+ \; :\ddot{B}r \cdot nn \longrightarrow Br_2Al \cdot e \; + \; n \cdot C - Ce + :\ddot{B}r \cdot nn \longrightarrow \quad \text{Ring cannot be opened} \qquad 10.61$$

$$Br_2Al \cdot e \; + \; C = C \; + \; :\ddot{B}r \cdot nn \longrightarrow Br_2Al \cdot e \; + \; n \cdot C - Ce + :\ddot{B}r \cdot nn$$

$$\longrightarrow Br_2Al - \overset{H}{\underset{|}{\underset{H}{C}}} - \overset{H}{\underset{|}{C}} = \overset{H}{\underset{|}{C}} - \overset{H}{\underset{|}{\underset{H}{C}}} - \overset{H}{\underset{|}{\underset{H}{C}}} - \overset{CH_3}{\underset{|}{\underset{CH_3}{C}}} \cdot e \; + \; :\ddot{B}r \cdot nn \qquad X \equiv CH_3 \qquad 10.62$$

In the second reaction above, the ring cannot be opened, because there is no point of scission. The two possible points of scission do not have the highest radical-pushing potential difference. With vinylcyclopropane, this limitation is not posed. This will increase with increasing size of ring. Nevertheless as will be fully shown, the following is valid, noting that the linear isomer of each one of them, whether the ring rearranges to it or not when opened instantaneously, is the basis of comparison.

Order of Nucleophilicity 10.63

Notice here that, this is the order of the MRSE, i.e. the yield point required to open their rings. This is the same as the order of their linear isomers.

The points of scission of their rings when they carry radical-pushing groups can be obtained from the following analysis shown in Fig. 10.1.

Figure 10.1 Points **of scission when strain energy is released instantaneously**

24

The number of points of scission have been indicated above in parenthesis. This is different from when an activation center such as $H_2C = CH-$ is externally conjugatedly located to the ring in place of one hydrogen atom as shown clearly in Figure 10.2.

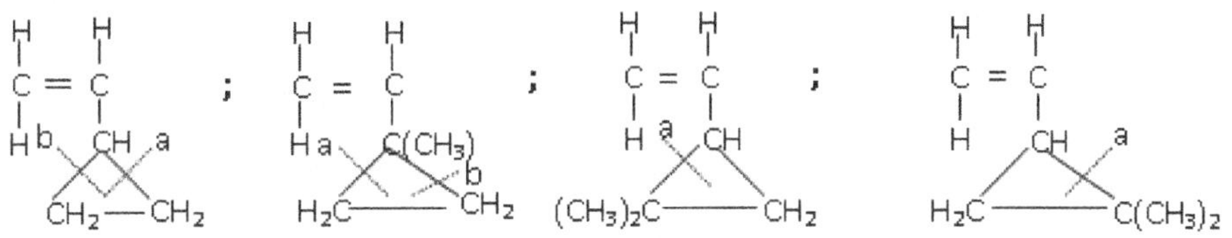

(I) (2)   (II) (2)   (III) (1)   (IV) (1)

Three-membered rings

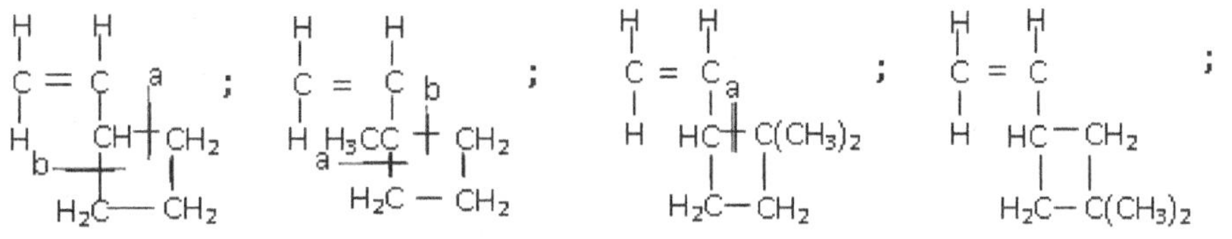

(V) (2)   (VI) (2)   (VII) (1)   (VIII) (None)

Four-membered rings

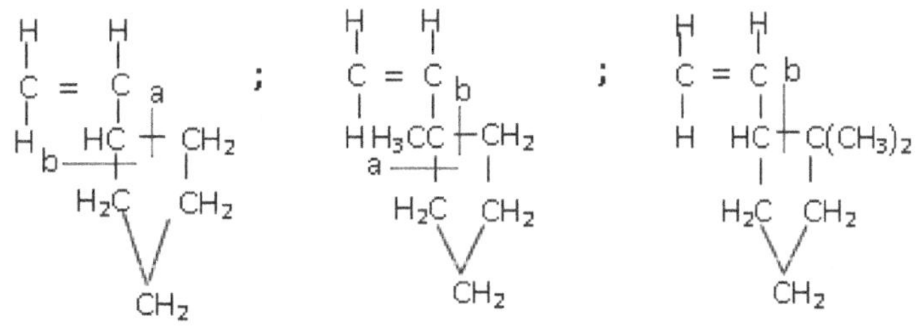

(IX) (1)   (X) (2)   (XI) (2)   (XII) (1)

(XIII) (None) ; (XIV) (1) ; (XV) (None)

<u>Five-membered rings</u>

Fig 10.2. <u>Points of scission for 3 and 4- membered rings (- 5- membered ring assumed).</u>
Only the three-membered [(I) – (II)] and four-membered rings [((II), (III), (IV))] can be opened as indicated above. The five and higher membered rings cannot be opened instantaneously using initiators, since to increase the strain energy from say 10 Kcals//mole to 20 – 25 Kcals//mole using radical-pushing groups will be impossible. The SE of 20 – 25 Kcals/mole in the three membered ring has been made to look like 10 Kcals/mole in the five-membered ring (Less strained), because of the increased size of the ring. ***When the SE is Zero, no rings can be formed, because there is no force to hold it. When the SE is MaxRSE, a ring cannot exist. But when the SE is the MinRSE or MRSE (The Yield Point), the ring which originally existed is instantaneously opened (Zero SE), if the ring has a point of scission. If it has no point of scission, the ring can never be opened.***

Thus, it can be observed here that the order of the magnitude of strain energy in these mono-mers is the reverse of their order of nucleophilicity, but not for the members of a sub-family as shown in Equation 10.55. Now, consider introducing three radical-pushing groups separately on three carbon centers of cyclopropane and cyclobutane.

$$:\overset{..}{\underset{..}{Br}}.nn \; + \; \text{(cyclopropane deriv.)} \; + \; e.AlBr_2 \longrightarrow n.\overset{H}{\underset{CH_3}{C}} - \overset{CH_3}{\underset{H}{C}} - \overset{H}{\underset{CH_3}{C}}.e \longrightarrow$$

$$Br_2Al - \overset{CH3}{\underset{H}{C}} - \overset{H}{\underset{CH_3}{C}} - \overset{CH_3}{\underset{H}{C}}.e \; + \; :\overset{..}{\underset{..}{Br}}.nn \; ; \qquad\qquad 10.64$$

$$R_3N: \; + \; N.n \; + \; n.\overset{H}{\underset{CH_3}{C}} - \overset{CH_3}{\underset{H}{C}} - \overset{H}{\underset{CH_3}{C}}.e \longrightarrow NH \; + \; \overset{H}{\underset{H}{C}} = \overset{H}{\underset{H}{C}} - \overset{H}{\underset{CH_3}{C}} - \overset{CH_3}{\underset{H}{C}}.n \; + \; NR_3$$

Used to open ring
instantaneously

$$\qquad\qquad 10.65$$

Free radically, only electro-free-radical polymerization is favored. The same applies chargedly positively. If one CH_3 group is replaced with C_2F_5 groups, no free- or paired-media charged route will be favored.

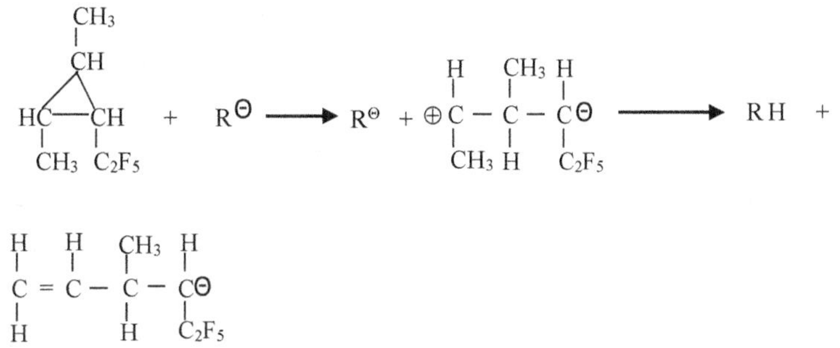

$$10.66$$

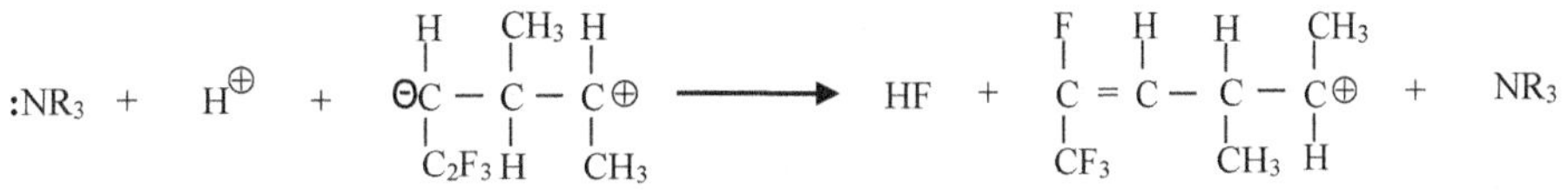

$$10.67$$

NR$_3$ has been used here to open the ring instantaneously. Only negatively charged-paired initiator will polymerize the monomer above, depending on how the groups are placed, if the ring can ever be opened instantaneously using initiators and heat.

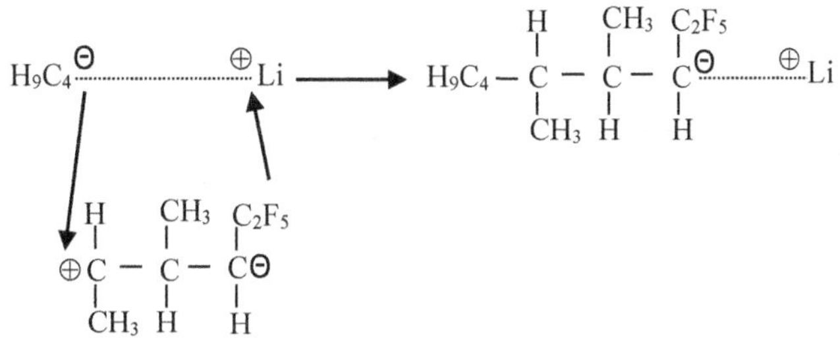

$$10.68$$

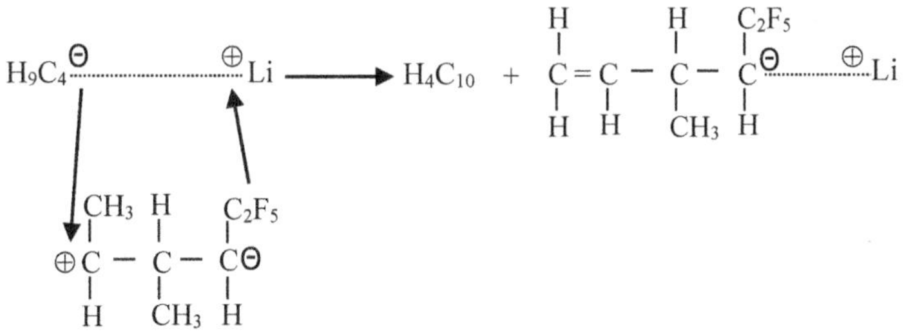

$$10.69$$

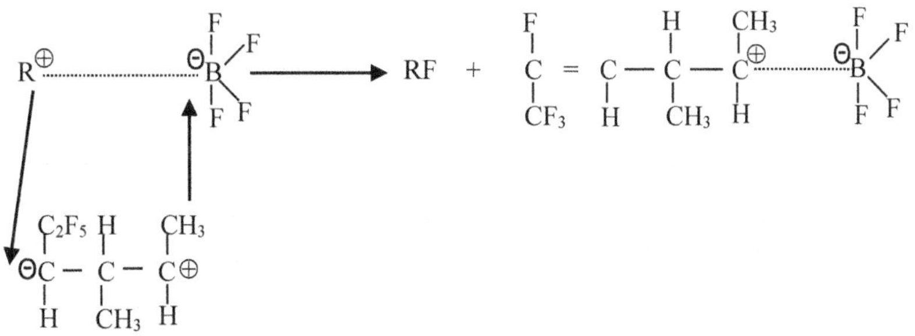

$$10.70$$

The initiator used in the first two equations above (LiC$_4$H$_9$) has been "assumed" to be negatively charged-paired in character as is wrongly believed to be the case today (so-called Anionic ion-paired initiator). As will be shown downstream, it is dual in character only radically, for which the first equation may not be favored, since the monomer is a Nucleophile.

Though the ringed monomer may not be easy to open instantaneously, it has been used to illustrate some basic principles. The orientations of the activated monomers in the vicinity of the coordination center are important to note.

$$
\ddot{\overset{..}{Br}} . nn \;+\; \begin{matrix} CH_3 \; CH_3 \\ | \quad\; | \\ HC - CH \\ | \qquad | \\ HC + CH_2 \\ | \\ CH_3 \end{matrix} \;+\; AlBr_2 \overset{e}{\longrightarrow} Br_2Al . e \;+\; n . \begin{matrix} H \; H \; CH_3 H \\ | \; | \quad | \; | \\ C - C - C - C . e \\ | \; | \quad | \; | \\ H \; CH_3H \; CH_3 \end{matrix} \;+
$$

$$
\ddot{\overset{..}{Br}} . nn \longrightarrow Br_2Al - \begin{matrix} H \; H \; CH_3 H \\ | \; | \quad | \; | \\ C - C - C - C . e \\ | \; | \quad | \; | \\ H \; CH_3 \; H \; CH_3 \end{matrix} \;+\; \ddot{\overset{..}{Br}} . nn \;;
$$

10.71

Like the corresponding cylcopropane of Equation 10.64, the electro-free-radical route is favored here. As the size of the ring becomes bigger, the order of nucleophilicity increases, with the ring becoming more difficult to open.

Replacing one of the CH_3 groups above with C_2F_5, the followings are obtained.

$$
R^{\ominus} \;+\; \begin{matrix} CH_3 \; C_2F_5 \\ | \quad\; | \\ HC + CH \\ | \qquad | \\ H_2C - CH \\ | \\ CH_3 \end{matrix} \longrightarrow R^{\ominus} \;+\; \begin{matrix} CH_3H \; H \; C_2F_5 \\ | \; | \quad | \; | \\ \oplus C - C - C - C \ominus \\ | \; | \quad | \; | \\ H \; H \; CH_3 \; H \end{matrix} \longrightarrow
$$

(I)

$$
RH \;+\; \begin{matrix} H \qquad\; H \; H \; C_2F_5 \\ | \qquad\; | \; | \quad | \\ C = C - C - C - C \ominus \\ | \qquad | \; | \quad | \; | \\ H \qquad H \; H \; CH_3 \; H \end{matrix}
$$

10.72

(I) above will not favor any charged route. Even if the external substituent groups are trans-placed, the negatively charged-paired route may not also be favored, the route being unnatural to it.

$$
R \overset{\ominus}{:} \;+\; \begin{bmatrix} CH_3 \; CH_3 \\ | \quad\; | \\ HC - CH \\ | \qquad | \\ H_2C - CH \\ | \\ C_2F_5 \end{bmatrix} \longrightarrow \begin{matrix} H \qquad\; H \; H \; C_2F_5 \\ | \qquad\; | \; | \quad | \\ \oplus C - C - C - C \ominus \\ | \qquad | \; | \; | \\ CH_3 \quad CH_3 \; H \; H \end{matrix} \longrightarrow
$$

(II)

$$
RH \;+\; \begin{matrix} H \qquad\; H \; H \; C_2F_5 \\ | \qquad\; | \; | \quad | \\ C = C - C - C - C \ominus \\ | \qquad | \; | \quad | \; | \\ H \qquad H \; CH_3 \; H \; H \end{matrix}
$$

10.73

(II) where the external substituent groups are trans-placed will not favor the use of Electrostatically negatively-charged paired initiators. The routes favored will depend on the capacity of the radical-pulling group relative to the radical-pushing groups and the placement of the groups. These will depend on the types of groups involved.

Free-radically, the monomer of Equation 10.66 will favor no route, while for the correspond-ing case of cyclobutanes, the followings are obtained.

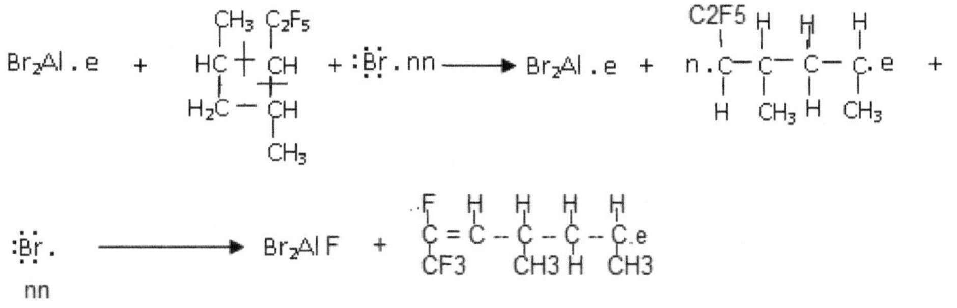

$$Br_2Al.e \;+\; \text{[monomer]} \;+\; :\overset{..}{Br}.nn \longrightarrow Br_2Al.e \;+\; n.C-C-C-C.e \;+$$

$$:\overset{..}{Br}. \longrightarrow Br_2AlF \;+\; \text{[monomer]}$$

10.74

$$Br_2Al.e \;+\; \text{[monomer]} \;+\; :\overset{..}{Br}.nn \longrightarrow Br_2Al.e \;+\; n.C-C-C-C.e \;+$$

$$:\overset{..}{Br}.nn \longrightarrow \text{[monomer]} \;+\; :\overset{..}{Br}.nn \;+\; AlBr_2F$$

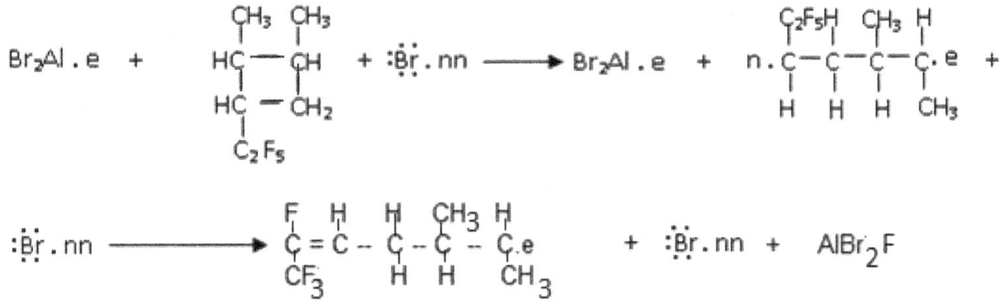

10.75

Even the use of paired initiators will not favor the polymerizations of these monomers whether the groups are trans-placed or not. If CF_3 is replaced with radical pulling groups, the route natural to the monomers will be favored when the groups are trans-placed. This no doubt is due to the fact that the groups controlling the orientation of monomers in the vicinity of the coordination center are the radical-pulling group-MOTHER NATURE. From all indications, it can be observed why first members of cycloalkanes are more nucleophilic than their isomeric olefins in general. While the monomers with charged-pulling groups such as CF_3, C_2F_5 cannot be opened chargedly, they can be opened radically, since the groups are free-radical-pushing groups of lower capacity than H. The same will apply more so to groups such as CH_2CN, CH_2COOCH_3, CH_2F etc., all allyl groups and radical-pushing in character and indeed of greater capacity than H.

Finally, consider introducing CH_3 and C_2H_5 groups on two carbon centers of ethene and cyclopropane in place of its H atoms.

$$N\overset{\cdot}{}^{n} + \underset{\underset{CH_3}{|}}{\overset{\overset{H}{|}}{C}} = \underset{\underset{H}{|}}{\overset{\overset{C_2H_5}{|}}{C}} \longrightarrow N\overset{\cdot}{}^{n} + e\cdot\underset{\underset{\underset{CH_3}{|}}{\underset{CH_2}{|}}}{\overset{\overset{H}{|}}{C}} - \underset{\underset{H}{|}}{\overset{\overset{CH_3}{|}}{C}}\cdot n \longrightarrow N\overset{\cdot}{}^{n} +$$

$$n\cdot\underset{\underset{CH_3}{|}}{\overset{\overset{H}{|}}{C}} - \underset{\underset{H}{|}}{\overset{\overset{C_2H_5}{|}}{C}}\cdot e \longrightarrow NH + \underset{\underset{CH_3}{|}}{\overset{\overset{H}{|}}{C}} = \underset{\underset{H}{|}}{\overset{\overset{H}{|}}{C}} - \underset{\underset{H}{|}}{\overset{\overset{CH_3}{|}}{C}} n$$

10.76

$$E\cdot e + \underset{\underset{CH_3}{|}}{\overset{\overset{H}{|}}{C}} = \underset{\underset{H}{|}}{\overset{\overset{C_2H_5}{|}}{C}} \longrightarrow + E - \underset{\underset{CH_3}{|}}{\overset{\overset{H}{|}}{C}} = \underset{\underset{H}{|}}{\overset{\overset{C_2H_5}{|}}{C}}\cdot e$$

10.77

$$N\overset{\cdot}{}^{n} + :NR_3 + H\underset{\underset{CH_2}{|}}{\overset{\overset{CH_3}{|}}{C}} - \underset{}{\overset{\overset{C_2H_5}{|}}{C}}H \longrightarrow N\overset{\cdot}{}^{n} + :NR_3 + n\cdot\underset{\underset{H}{|}}{\overset{\overset{H}{|}}{C}} - \underset{\underset{CH_3}{|}}{\overset{\overset{H}{|}}{C}} - \underset{\underset{H}{|}}{\overset{\overset{C_2H_5}{|}}{C}}\cdot e$$

$$\longrightarrow NH + \underset{\underset{CH_3}{|}}{\overset{\overset{H}{|}}{C}} = \underset{}{\overset{\overset{H}{|}}{C}} - \underset{\underset{CH_3}{|}}{\overset{\overset{H}{|}}{C}} - \underset{\underset{H}{|}}{\overset{\overset{H}{|}}{C}}\cdot n$$

10.78

$$E\cdot e + :NR_3 + n\cdot\underset{\underset{H}{|}}{\overset{\overset{H}{|}}{C}} - \underset{\underset{CH_3}{|}}{\overset{\overset{H}{|}}{C}} - \underset{\underset{H}{|}}{\overset{\overset{C_2H_5}{|}}{C}}\cdot e \longrightarrow :NR_3 + E - \underset{\underset{H}{|}}{\overset{\overset{H}{|}}{C}} - \underset{\underset{CH_3}{|}}{\overset{\overset{H}{|}}{C}} - \underset{\underset{H}{|}}{\overset{\overset{C_2H_5}{|}}{C}}\cdot e$$

(I)

10.79

Thus, it can partly be observed why cycloalkanes are more nucleophilic than olefins. The olefins above along with the corresponding substituted cyclopropane favor the electro-free-radical route with some differences. The cyclic ones unlike the olefins do not readily molecularly rearrange. This will however depend on the choice and types of groups.

With these monomers of cycloalkanes which do not have functional centers located on the ring, one can see the influence of minimum required strain energy (MRSE) in the opening of their rings, and the factors determining points of scission on the rings.

10.2 Cyclopropenes

Cycloalkenes are cases where one double bond is present in the ring and cyclo-propene is the first member of cycloalkenes. Since cyclopropane is strained, that is, has strain energy close to the minimum required strain energy, one should expect cyclo-propene to be more strained in view of the presence of a double bond in the ring. However, the ring well known to exist, cannot be opened in the absence of molecular rearrangement *or other forces*, because there is no point of scission in the ring as shown below radically or chargedly, when suppressed.

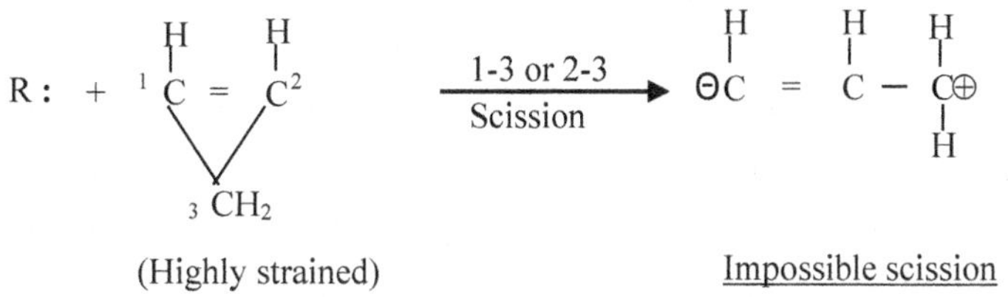

$$R: \quad + \quad {}^1C = {}^2C \underset{3\ CH_2}{\diagdown\diagup} \qquad \xrightarrow[\text{Scission}]{\text{1-3 or 2-3}} \qquad \ominus C = C - C\oplus$$

(Highly strained)   Impossible scission   10.80a

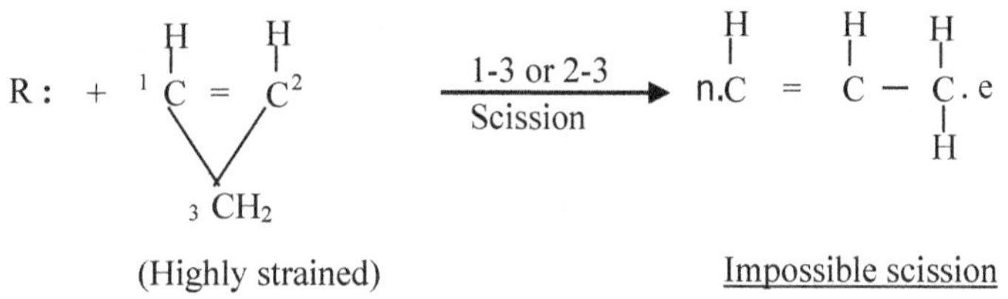

$$R: \quad + \quad {}^1C = {}^2C \underset{3\ CH_2}{\diagdown\diagup} \qquad \xrightarrow[\text{Scission}]{\text{1-3 or 2-3}} \qquad n.C = C - C.e$$

(Highly strained)   Impossible scission   10.80b

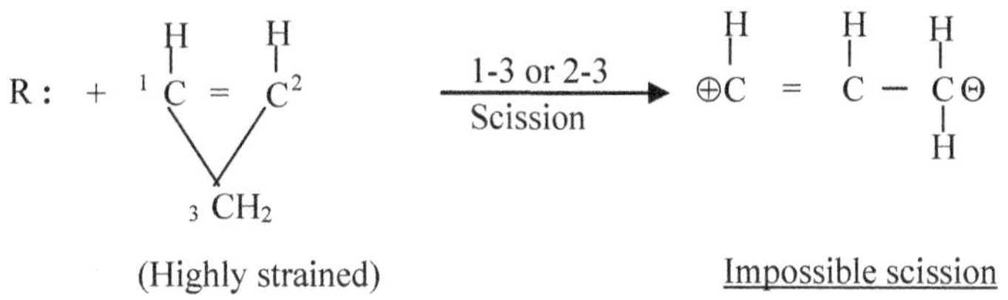

$$R: \quad + \quad {}^1C = {}^2C \underset{3\ CH_2}{\diagdown\diagup} \qquad \xrightarrow[\text{Scission}]{\text{1-3 or 2-3}} \qquad \oplus C = C - C\ominus$$

(Highly strained)   Impossible scission   10.80c

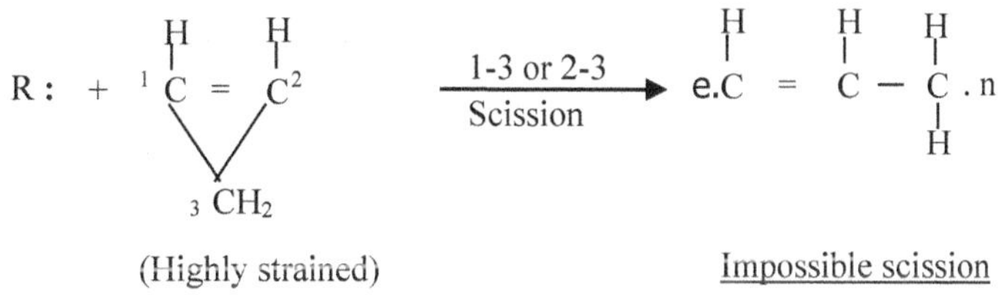

$$R: \quad + \quad {}^1C = {}^2C \underset{3\ CH_2}{\diagdown\diagup} \qquad \xrightarrow[\text{Scission}]{\text{1-3 or 2-3}} \qquad e.C = C - C.n$$

(Highly strained)   Impossible scission   10.80d

First and foremost, is the fact that a ring cannot be scissioned next door to a double bond, otherwise higher rings such as Benzene, Naphthalene, Anthracene and their condensed forms can never exist. These are the origins of building blocks of Living systems- Lungs, Heart, muscles, arteries, veins and much more as will be seen downstream. The double bonds in the cases above must be desaturated, before the rings can be opened, just as is done with benzene via in particular hydrogenation. The scissions above are not favored, because chargedly electrostatic forces of repulsion from the π-bond will not allow the negative charge to be placed on the C center in the first equation and radically, a cumulene would suddenly be obtained. ***Most important to note however is that rings which carry more than the MRSE but less than the MaxRSE, usually have no points of scission in them, otherwise they will not exist***. That does not mean that such rings cannot be opened. Some of them can while some cannot. For example, when the cyclopropene is not in a suppressed state, i.e., stable state, under hearse operating conditions, i.e. temperature of the order of 425^0C^{20}, the followings take place via Equilibrium mechanism.

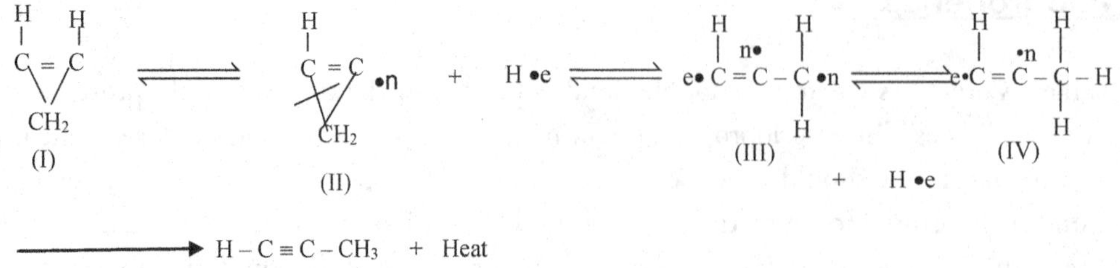

$$H_2C=C(CH_2) \rightleftharpoons \overset{H}{C}=\overset{H}{C}(CH_2)\cdot n + H\bullet e \rightleftharpoons e\bullet C=C-C\bullet n \rightleftharpoons \bullet e C=C-\overset{\bullet n}{C}-H$$

(I)   (II)   (III)   (IV)

                                             +   H $\bullet$e

$$\longrightarrow H-C\equiv C-CH_3 + Heat$$

Propyne or Methyl acetylene

<u>ELECTRORADICALIZATION PHENOMENON</u>   10.81

When in Equilibrium state of existence due to the presence of higher operating conditions, (II) is scissioned as shown above to form a triple bond (IV). ***The invisible electro-free-radical did not grab the visible nucleo-free-radical to form a bond. One can see that the invisible π-bond is imaginary.*** It was after the H held in Equilibrium state of existence had added to the new center, that deactivation followed to release heat and give propyne. This is not the type of Molecular rearrangements we have been encountering so far, but another phenomenon also called ELECTRO- RADICALIZATION as we already know and will be fully high-lighted down-stream. It is another form of rearrangement, but under Equilibrium state of existence. It was the nucleo-free-radical carried by (II) that necessitated the scission of the bond at that point. This cannot take place with benzene ring for example to form butadiene acetylene ($HC\equiv C - CH = CH - CH = CH_2$), because no point of scission still exists as there is no CH_2 adjacently located to the double bond. If any species is to be abstracted or transferred from this monomer, it is the hydrogen atom connected to 3-carbon center, and this is only possible free-radically with no change taking place.

    Consider using the double bond in the ring as follows-

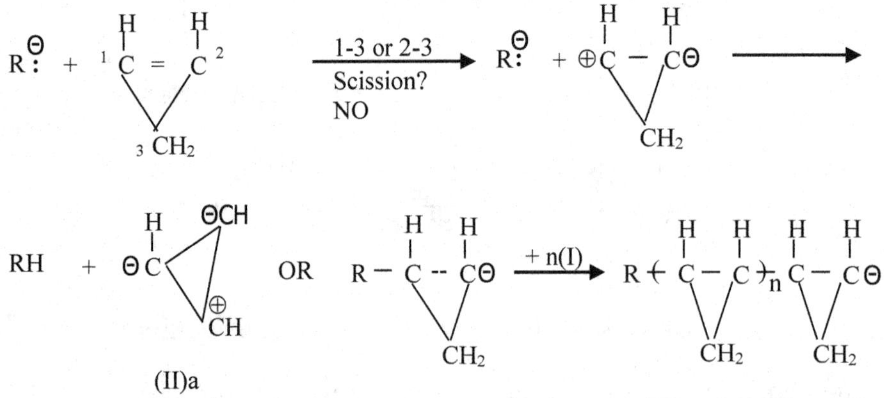

$$R:^{\ominus} + \overset{1}{C}=\overset{2}{C}(_3CH_2) \xrightarrow[\text{NO}]{\text{1-3 or 2-3 Scission?}} R:^{\ominus} + \oplus C - C\ominus(CH_2) \longrightarrow$$

$$RH + \ominus C \overset{\ominus CH}{\underset{\oplus CH}{\triangle}} \quad \text{OR} \quad R-C--C\ominus(CH_2) \xrightarrow{+n(I)} R(C-C)_n C-C\ominus(CH_2)(CH_2)$$

(II)a   (III)a   (NOT favored)

<u>Impossible existence</u>   10.82

$$R^{\oplus} + \ominus C --- C\oplus(CH_2) \longrightarrow R-C-C\oplus(CH_2) \xrightarrow{+n(I)} R(C-C)_n C-C\oplus(CH_2)(CH_2)$$

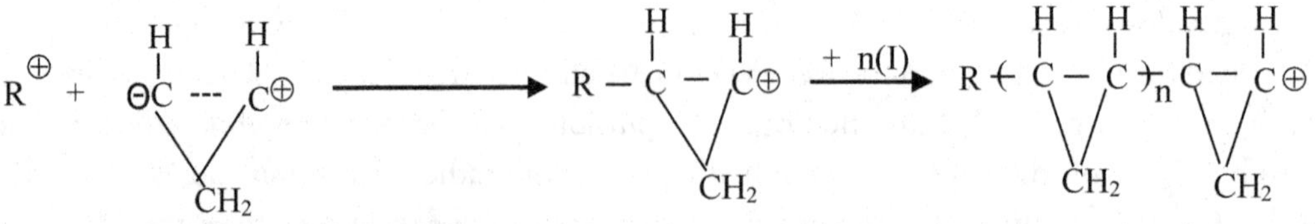

(I)   (III)b

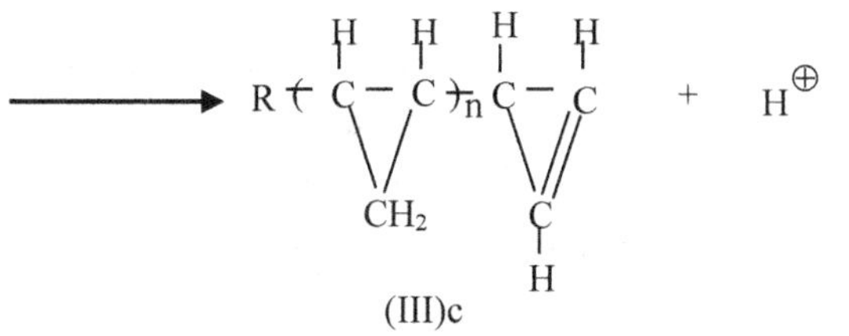

$$\longrightarrow \quad R \left(C - C \right)_n C - C \quad + \quad H^{\oplus}$$

(III)c

10.83

Transfer species of first kind of the tenth type

The transfer species involved in (II)a, and (III)b above is of ***the first kind of a different type. It is of the tenth type*** because it is located inside an unsaturated ring and as such shared by two active centers as opposed to the other types of transfer species of the first kind. It can be observed to obey the law of conservation of transfer species. However presence of (II)a is not possible due to electro-dynamic/static forces of repulsion. **Hence radically,** only the electro-free-radical route is favored to produce dead terminal double bond cycloalkene-polymers-(III) c. This can only take place at very low temperatures when the monomer is stable. Invariably, it cannot be activated chargedly.

With a CH_3 group replacing one hydrogen atom, the followings are obtained.

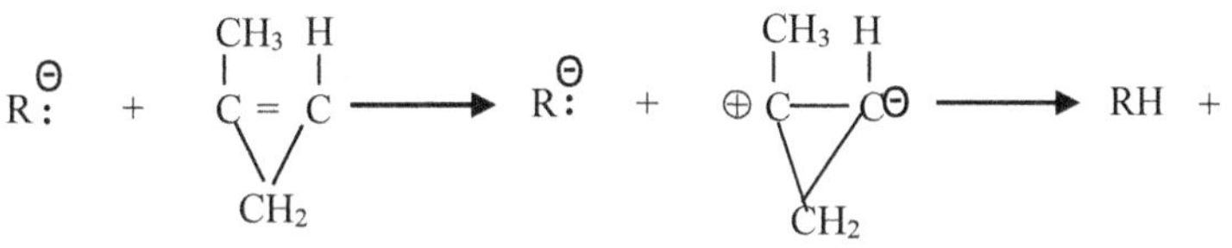

$$R:^{\ominus} \quad + \quad \underset{CH_2}{C} = C \quad \longrightarrow \quad R:^{\ominus} \quad + \quad ^{\oplus}C \underset{CH_2}{-} C^{\ominus} \quad \longrightarrow \quad RH \quad +$$

(I)

$$^{\ominus}\underset{H}{C} - C = CH \qquad OR \qquad ^{\ominus}C \cdots C^{\ominus}$$

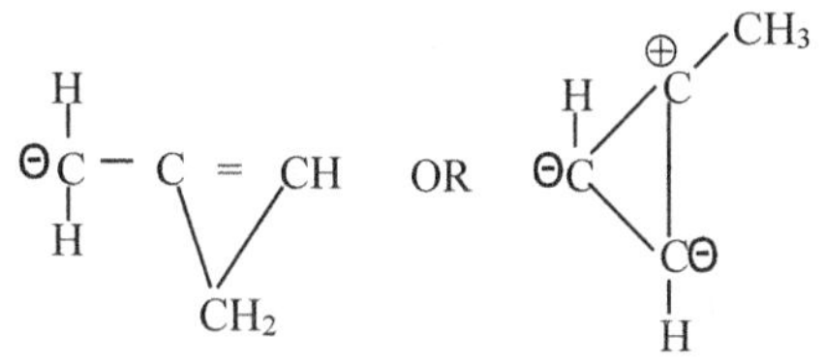

(II)   (III) Impossible existence

10.84

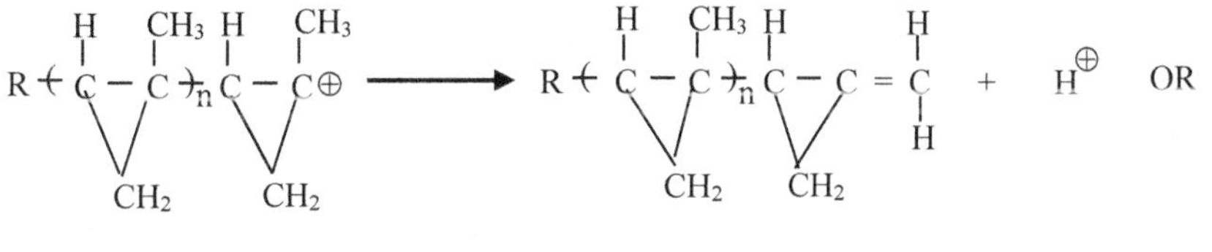

$$R \left(\underset{CH_2}{C} - \underset{CH_2}{C} \right)_n C - C^{\oplus} \quad \longrightarrow \quad R \left(\underset{CH_2}{C} - \underset{CH_2}{C} \right)_n C - C = C \quad + \quad H^{\oplus} \quad OR$$

(IV)

Transfer species of the first kind of the first type (Not Favored)

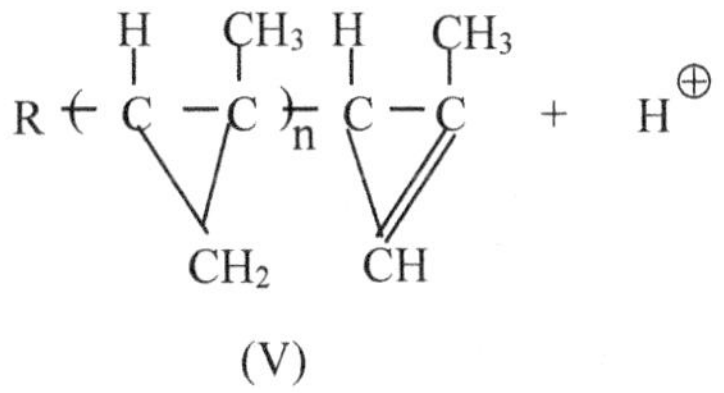

$$R \left(\underset{CH_2}{C} - \underset{}{C} \right)_n C - C \quad + \quad H^{\oplus}$$

(V)

10.85

Transfer species of the first kind of the tenth type (Favored radically)

33

Since the ring is more radical-pushing than the CH$_3$ group, it is (V) that should be favored wherein the same character of monomer is obtained. From the reactions above, one can observe the difference between the two types of transfer species. The dead terminal double bond of the second reaction (V) does however identify with the original monomer because of the absence of CH$_2$ group in the ring, wherein transfer species of the first kind cannot be released due to electro-dynamic/static forces of repulsion making (III) non-existent. Free-radically, this will not be the case and the polymerization is only favored at very low temperatures. ***Hence, it is believed that cyclopropene unlike other larger members of the cycloalkene family but like the acetylenes cannot be activated via the visible activation center chargedly because there is electrostatic force of repulsion between the negative charge which though is real and the invisible π-bond inherently located on it which is also real.*** Hence, it cannot be polymerized chargedly as shown in Equations 10.82 to 10.85, where the Laws of Conservation of transfer of transfer species cannot hold. Nevertheless, the monomers with all radical-pushing substituent groups are strongly nucleophilic, since only electro-free-radical route is favored by three-membered ring, while electro-free-radical and positively charged route is favored by four and larger membered rings, where the strength of the invisible π-bond keeps decreasing.

With radical-pushing groups of lower capacity than H, consider CF$_3$ types which chargedly is charged-pulling.

Electrostatic forces of repulsion 10.86

10.87

10.88

LiC$_4$H$_9$ will not favor the last reaction above, because the monomer is a Nucleophile. The initiator above as will become clear downstream is dual in character only radically. It has been used as above because of what it is still called today-"Anionic-ion-paired initiator". Downstream as we move along, the name will fully be revealed. It is the Li center that will do the attack only radically, the monomer being a Nucleophile and when it does, the route will not be favored with this initiator. The monomer can be observed to be neither nucleophilic nor electrophilic in character. The same will not apply when groups such as COOR, $CONH_2$ are used in place of CF_3. For COR type of groups, electro-free-radical polymerization will be favored via C=O center. These are Electrophiles.

However, the monomer cyclopropene being almost similar to methyl acetylene (Its isomer), may require one or the two hydrogen atoms as will be shortly become apparent be replaced before there can be any full polymerization. Molecular rearrangement with transfer species of the ninth type can readily be favored by some of these monomers, if weak initiators are involved.

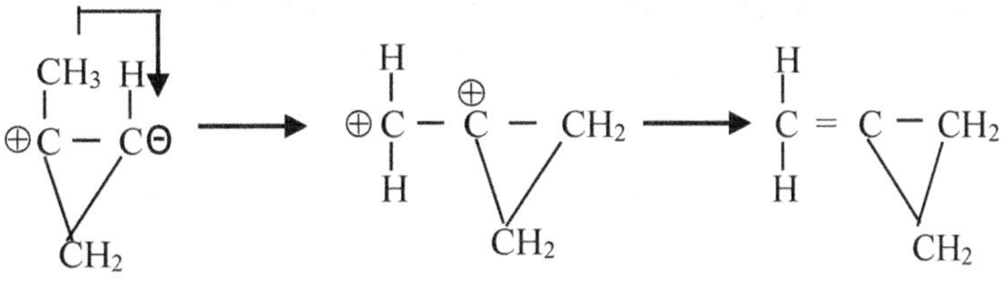

(I)   (II)   (III)

Non-favored Rearrangement

10.89a

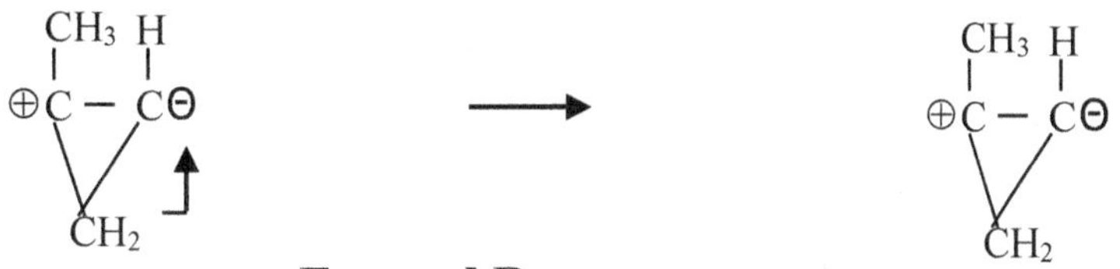

Favored Rearrangement

10.89b

Since the monomer looks like propene which rearranges to itself, it is the second reaction above that is favored. On the other hand, notice that (III) above looks like a cumulene which in many cases rearranges to acetylenes. Hence the presence of (III) is not favored. (III) which is methylene cyclo-propane has in recent years been well known to also exist. (III) cannot be obtained from Methyl cy-cyclopropene by molecular rearrangement. Via Vibrational overtone activation of methyl cyclopro-pene, it has been shown that methyl cyclopropene does not produce methylene cyclopropane.[21] If the Equilibrium state of Existence of methyl cyclopropene was as shown below, then, methylenecyclo-propane would have been obtained.

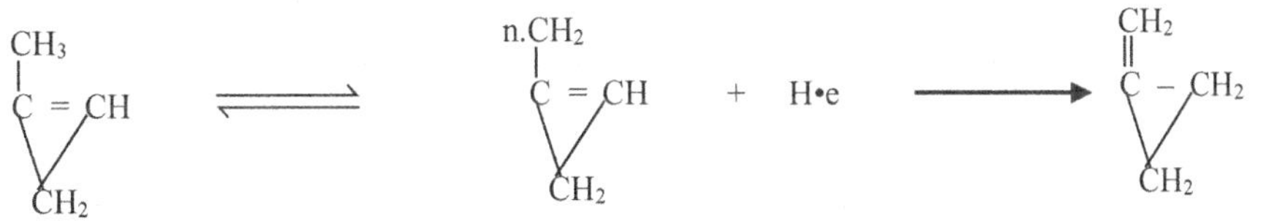

Methylene cyclopropene

**ELECTRORADICALIZATION PHENOMENON
(NOT FAVORED)**

10.90a

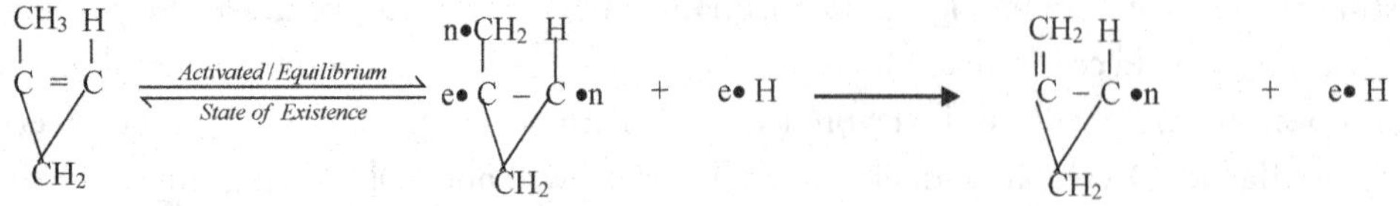

$$\underset{\text{Methyl cyclopropene}}{\overset{\displaystyle CH_3\ \ H}{\underset{CH_2}{\overset{|\ \ \ \ |}{C=C}}}} \quad \xrightleftharpoons[\text{State of Existence}]{\text{Activated / Equilibrium}} \quad \underset{CH_2}{\overset{n\bullet CH_2\ \ H}{\overset{|}{e\bullet C - C}\bullet n}} \quad + \quad e\bullet H \quad \longrightarrow \quad \underset{CH_2}{\overset{CH_2\ \ H}{\overset{||\ \ \ |}{C - C}\bullet n}} \quad + \quad e\bullet H$$

$$\longrightarrow \quad \underset{CH_2}{\overset{CH_2}{\overset{||}{C - CH_2}}}$$

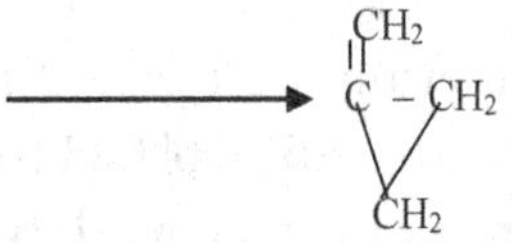

Methylene cyclopropane

ACTIVATED/EQUILIBRIUM STATE OF EXISTENCE
REARRANGEMENT PHENOMENON
(NOT FAVORED)

10.90b

All these including Molecular rearrangements of many kinds are just some parts of Tautamerism. Vinyl alcohol which is unstable at STP, molecularly rearranges to acetaldehyde. Under another operating condition, the same acetaldehyde rearranges back to vinyl alcohol via Activated/ Equilibrium state of existence phenomenon. Diazomethane ($H_3CN=NCH_3$) rearranges to give $H_2C=N - NH(CH_3)$ via Electro-radicalization phenomenon under particular operating conditions. What controls everything in life are **OPERATING CONDITIONS**. The reactions above are said not to favored, because the real Equilibrium state of existence of methyl cyclopropene is as shown below.

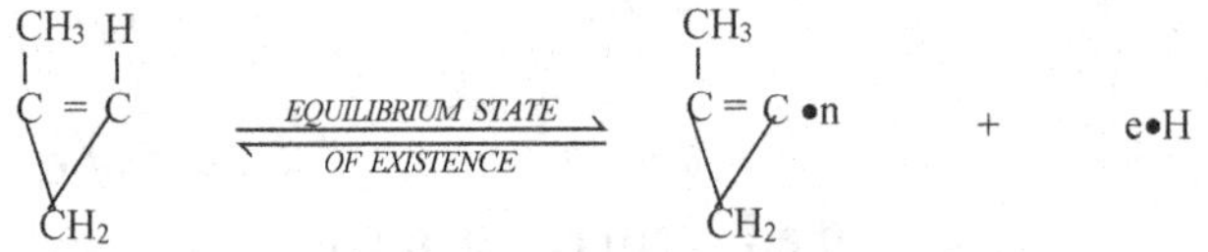

$$\underset{CH_2}{\overset{\displaystyle CH_3\ \ H}{\overset{|\ \ \ \ |}{C=C}}} \quad \xrightleftharpoons[\text{OF EXISTENCE}]{\text{EQUILIBRIUM STATE}} \quad \underset{CH_2}{\overset{\displaystyle CH_3}{\overset{|}{C=C}\bullet n}} \quad + \quad e\bullet H$$

EQUILIBRIUM STATE OF EXISTENCE OF
METHYL CYCLOPROPENE

10.91a

With this Equilibrium state of existence, it is impossible for the phenomena above to take place to give methylene cyclopropane. From what has been seen so far, methyl cyclopropene cannot undergo any type of molecular rearrangement phenomena to methylene cyclopropane. If it rearranges when activated, it rearranges back to itself, because what is in the ring is far more radical-pushing than CH_3 as will shortly be shown. *Also, if in Equilibrium state of existence and heated at above 400°C like the case shown in Equation 10.81, it will decompose to give $H_3CC \equiv CCH_3$ as shown below via Equilibrium mechanism.*

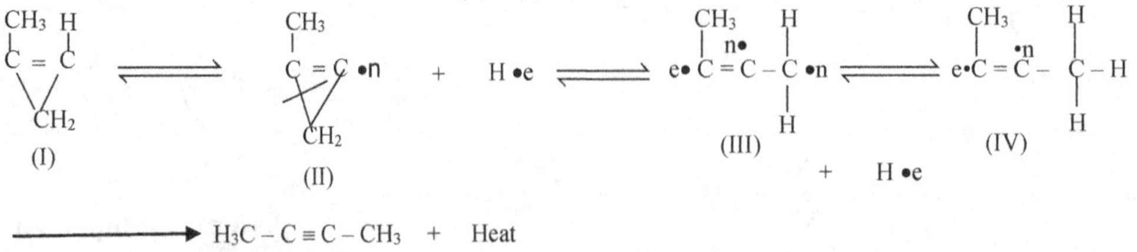

$$\underset{(I)}{\underset{CH_2}{\overset{CH_3\ \ H}{\overset{|\ \ \ |}{C=C}}}} \rightleftharpoons \underset{(II)}{\underset{CH_2}{\overset{CH_3}{\overset{|}{C=C}\bullet n}}} + H\bullet e \rightleftharpoons \underset{(III)}{\overset{CH_3\ \ \ H}{e\bullet C = C - C\bullet n}} \rightleftharpoons \underset{(IV)}{\overset{CH_3\ \ \ H}{e\bullet C = C - C - H}} $$
$$+ \quad H\bullet e$$

$$\longrightarrow \quad H_3C - C \equiv C - CH_3 \quad + \quad \text{Heat}$$

Butyne or Di-methyl acetylene

ELECTRORADICALIZATION PHENOMENON

10.91b

Note that this is valid for all sizes of alkylane group, with the group remaining intact during rearrangement. The SE in methyl cyclopropene is 53.5 k cal/mole[22], while that in methylene cyclo-propane is 32.7 k cal/

mole[22]. The SE in cyclopropene is 52.2 kcal/mole[23], while that of cyclopropane is 27.5 kcal/mole[24]. The difference between methylene cyclopropane and cyclopropane (5.2 kcal/ mole) clearly shows the strength of the radical pushing capacity of $H_2C=$ group. The difference between methyl cyclopropene and propene (1.3 kcal/mole) clearly shows that CH_3 group is far weaker in radical-pushing capacity than $H_2C=$ group which are all opened. Indeed as already found and established, $H_2C=$ group is greater than one R group of any size in radical-pushing capacity.

Because the Equilibrium state of existence of methylene cyclopropane is very strong as shown below, it cannot open by itself. Yet, it seems to have two points of scission. But, when heated at 400^0C^{25}, a condition where its equilibrium state of existence is strong, it decomposes to give acetylene and ethene as shown below.

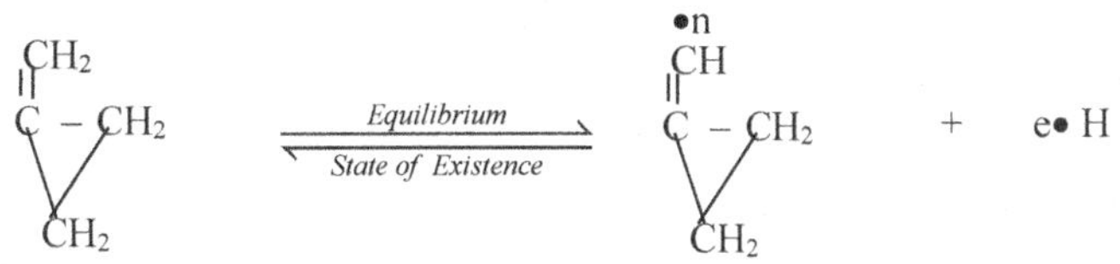

EQUILIBRIUM STATE OF EXISTENCE OF
METHYLENE CYCLOPROPANE

10.92a

With this Equilibrium State of Existence, Electro-radicalization can still take place, but with a difference. ***When H atom on a double bonded carbon is held in Equilibrium state of existence, the double bond can never at the same time be activated.*** When the methylene cyclopropane is activated, it cannot undergo molecular rearrangement to methyl cyclopropene as cumulene does to acetylene, because apart from the fact that the SE in the methyl cyclopropene is higher than that in methylene cyclopropane and the Conservation laws will be broken, ***the radical pushing capacity of any of the H atoms in the three-membered ring is greater than the radical pushing capacity of the H atom outside the ring.***

H in Ring   >   H outside the same ring

<u>Radical-pushing capacity of H inside and outside a three-membered ring only</u>

10.92b

Methyl cyclopropene is like methyl propene or isobutylene $[(CH_3)_2C=CH_2]$ and dimethyl cumulene $[(CH_3)_2C= C=CH_2]$. While the first two can molecularly rearrange back to itself (Induction) when the initiator or other force is weak, the second to the last cannot.

<u>Stage 1:</u>

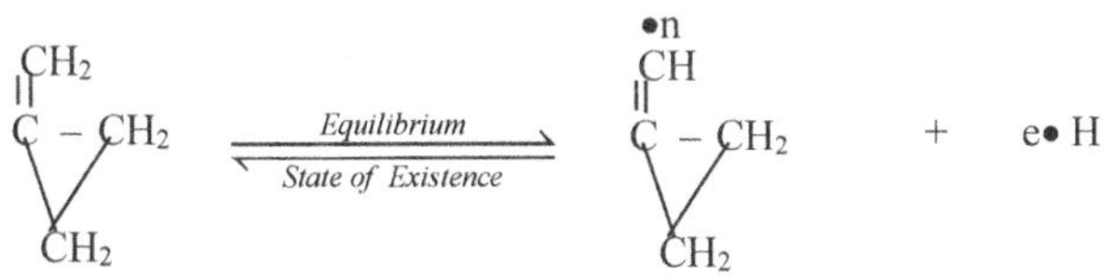

Methylene cyclopropane   (A)

            (A)

37

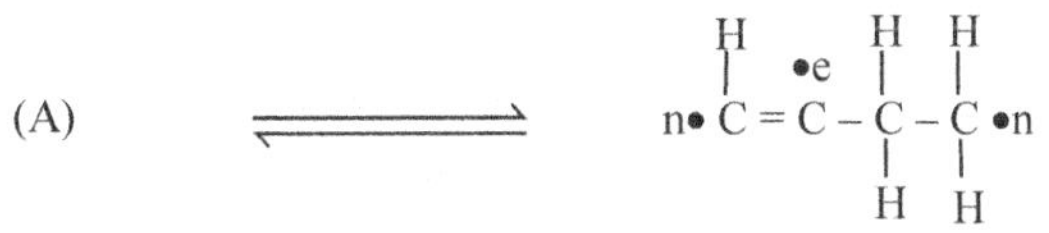

$$\rightleftharpoons \quad \underset{(B)}{H-C \equiv C - \overset{\overset{\displaystyle H}{|}}{\underset{\underset{\displaystyle H}{|}}{C}} - \overset{\overset{\displaystyle H}{|}}{\underset{\underset{\displaystyle H}{|}}{C}} \bullet n}$$

$$\rightleftharpoons \quad \underset{(C)}{H-C \equiv C \bullet n} \quad + \quad \underset{(D)}{e \bullet \overset{\overset{\displaystyle H}{|}}{\underset{\underset{\displaystyle H}{|}}{C}} - \overset{\overset{\displaystyle H}{|}}{\underset{\underset{\displaystyle H}{|}}{C}} \bullet n}$$

$$(C) \quad + \quad e \bullet H \quad \rightleftharpoons \quad H - C \equiv C - H$$

$$(D) \quad \xrightarrow{\text{Deactivation}} \quad H_2C = CH_2 \quad + \quad \text{Heat} \qquad \text{10.93a} \qquad 10.93a$$

Overall Equation: Methylenecyclopropane $\longrightarrow$ HC≡CH + $H_2C{=}CH_2$ 10.93b   10.93b

Note that this is valid for all $H_2C{=}$, $HCH_3C{=}$, and $HC_2H_5C{=}$ types of groups for as long as there is one H atom. The case above looks similar to the case of Equation 10.81 also a one stage Equilibrium mechanism system, except that decomposition as opposed to rearrangement took place above. It is important to note how the ring was scissioned in both cases. *It also looks as if just as scission cannot take place close to a double bond in a ring when internally located, so also it cannot take place when the double is adjacently externally located to the ring.* However, the last case is not the case. Methylene cyclopropane can readily be obtained from methylene and cumulene as follows.

Stage 1:

$$e \bullet \overset{\overset{\displaystyle H}{|}}{\underset{\underset{\displaystyle H}{|}}{C}} \bullet n \quad + \quad H_2C = C = CH_2 \quad \xrightleftharpoons{\text{Activation}} \quad n \bullet \overset{\overset{\displaystyle H}{|}}{\underset{\underset{\displaystyle H}{|}}{C}} - \overset{\overset{\displaystyle H}{|}}{\underset{\underset{\displaystyle H}{|}}{C}} - \overset{\overset{\displaystyle}{}}{\underset{\underset{\displaystyle CH_2}{||}}{C}} \bullet e$$

$$\longrightarrow \quad H_2C = \overset{\displaystyle C - CH_2}{\underset{\displaystyle CH_2}{\diagdown \diagup}}$$
            10.94a

Overall equation: H_2C + $H_2C{=}C{=}CH_2$ $\longrightarrow$ Methylenecyclopropene   10.94b

One can see the close relations of this compound. It is a one stage Equilibrium mechanism system.

Based on the strong Equilibrium states of existence of methyl cyclopropene (MCP) [Mistakenly universally called 1- or 2- methyl cyclopropene] (B pt. = -12⁰C) and methylene cyclopropene, it is therefore no surprise why these and similar compounds, *have been found to be synthetic plant growth regulators[26]. They help to slow down the ripening of climacteric fruits and to help to maintain the freshness of cut flowers and shedding of leaves, prevent premature wilting, leaf yellowing, premature opening and premature death[27-29].* They help to block the effect of ethylene a natural plant hormone. This, they probably do as follows.

Stage1:

$$\underset{\textbf{Methyl cyclopropene}}{\overset{\overset{\displaystyle CH_3}{|} \quad \overset{\displaystyle H}{|}}{\underset{\underset{\displaystyle CH_2}{\diagdown \diagup}}{C = C}}} \quad \xrightleftharpoons[\text{State of Existence}]{\text{Equilibrium}} \quad \underset{(A)}{\overset{\overset{\displaystyle CH_3}{|}}{\underset{\underset{\displaystyle CH_2}{\diagdown \diagup}}{C = C}} \bullet n} \quad + \quad e \bullet H$$

$$H \bullet e \quad + \quad H_2C = CH_2 \quad \xrightleftharpoons{\text{Activation}} \quad \underset{(B)}{H - \overset{\overset{\displaystyle H}{|}}{\underset{\underset{\displaystyle H}{|}}{C}} - \overset{\overset{\displaystyle H}{|}}{\underset{\underset{\displaystyle H}{|}}{C}} \bullet e}$$

$$(A) \quad + \quad (B) \quad \longrightarrow \quad \begin{array}{c} CH_3 \quad C_2H_5 \\ | \qquad | \\ C = C \\ \diagdown \diagup \\ CH_2 \end{array}$$

$$(C)$$

10.95a

Overall Equation: Methyl cyclopropene + Ethylene $\longrightarrow$ (C)   10.95b

The ethylene (either of the Carbene family, after methylene or ethene used above is very readily consumed by so-called 1-MCP to form (C) which is still unstable ready to consume another ethene as will be shown downstream. In the absence of steric limitations, (C) above is not still a monomer. But when suppressed, only the electro-free-radical route will be favored by it.

Now consider replacing the second H atom on methyl cyclopropene with CH_3 group like (C) above.

$$\begin{array}{ccc}
\begin{array}{c} CH_3 \quad CH_3 \\ | \qquad | \\ e\bullet C - C \bullet n \\ \diagdown \diagup \\ CH_2 \end{array}
&
\begin{array}{c} CH_3 \quad H \\ | \qquad | \\ e\bullet C - C \bullet n \\ \diagdown \diagup \\ C(CH_3)H \end{array}
&
\begin{array}{c} CH_3 \quad H \\ | \qquad | \\ C = C \\ \diagdown \diagup \\ C(CH_3)H \end{array}
\end{array}$$

More Stable

FAVORED REARRANGEMENT   10.96a

The cases above are Nucleophiles, wherein their Natural routes are the use of electro-free-radical initiators. Even the first member of the family here- cyclopropene cannot favor all routes like the case of ethene as has partly been shown. One can observe so far, that no two families can be the same, clear indication of how Nature operates. Thus, the CH_2 connected to two C centers in a ring is more radical pushing than CH_3 group externally located to a C center in the same ring.

$$H_2C\diagdown \qquad > \qquad H_2C = \qquad >>> \qquad - CH_3 \text{ [Any R]}$$

(Closed)   (Opened)   (Opened)

Order of Radical-pushing capacity   10.96b

$$\begin{array}{ccc}
\begin{array}{c} CH_3 \\ | \\ O \\ | \\ C=O H \\ | \qquad | \\ n C - C \bullet e \\ \diagdown \diagup \\ CH_2 \end{array}
&
\begin{array}{c} O \qquad\quad H \\ \| \quad \bullet n \quad | \\ e\bullet C - C - C(OCH_3) \\ \diagdown \diagup \\ CH_2 \end{array}
&
\begin{array}{c} O \qquad\quad H \\ \| \qquad\quad | \\ C = C - C(OCH_3) \\ \diagdown \diagup \\ CH_2 \end{array}
\end{array}$$

(I) (less stable)   (II)   (III) (more stable)

ELECTROPHILE   10.97a

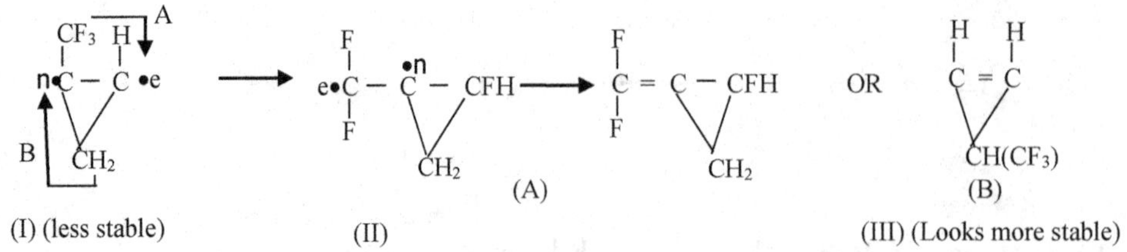

(I) (less stable)　　　　　(II)　　　　　　(A)　　　　(III) (Looks more stable)　(B)

NUCLEOPHILE　　　　　　　　　　　　　　　　　10.97b

The first case above is favored. For the second case which is a Nucleophile, it is (B) that should have been favored. It is not favored, because H is greater than CF_3 in radical-pushing capacity. The factors determining the stability here are not fully the same as exist with olefins. For all the cases above and other members, their rings cannot be readily opened radically. The (II) of Equation 10.97a above is sterically hindered for continuous addition or for other reasons such as electro-dynamic forces of repulsion. Hence, telomers are largely expected to be produced for this family of monomers. Thus, it seems that when these three-membered rings are used as monomers, different polymeric products are obtained.

For the first members, the followings were thought to be obtained radically.

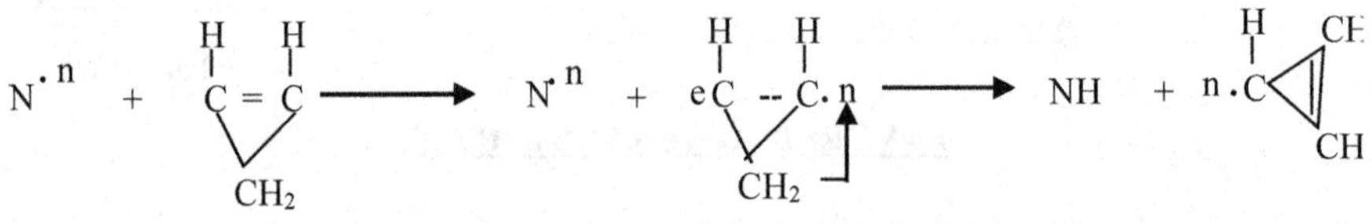

Transfer species of 1st kind of the tenth type

10.98a

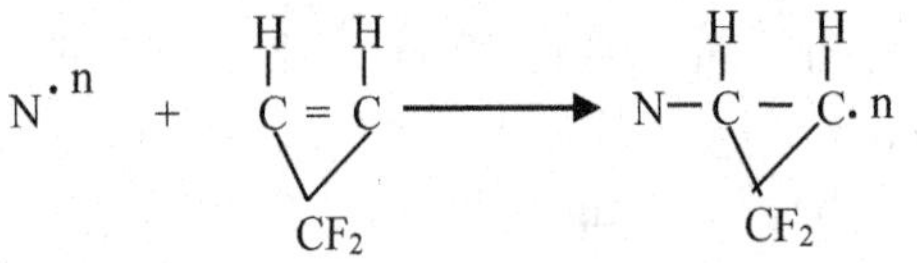

10.98b

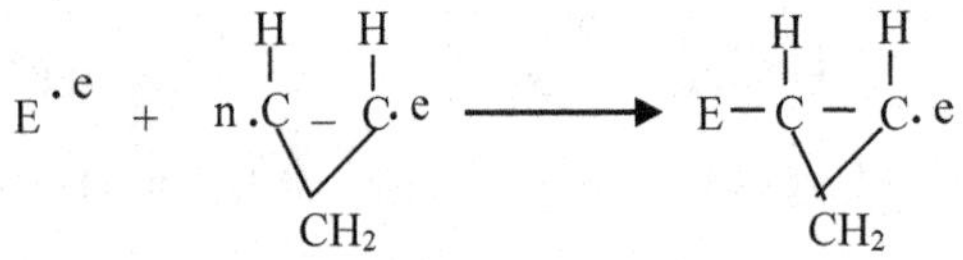

10.99a

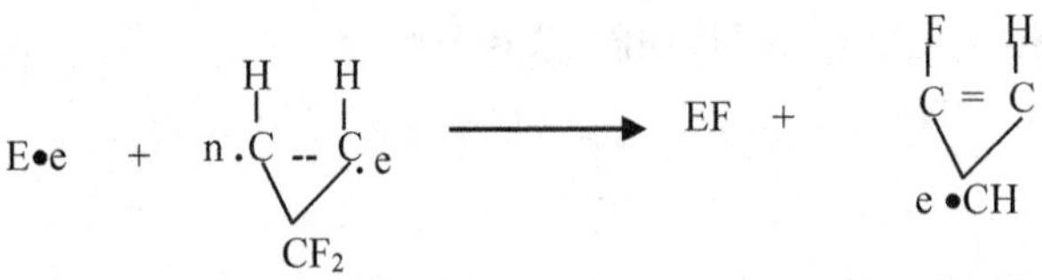

Transfer species of the first kind of the tenth type　　10.99b

This is unlike olefinic first members as has already been said. Now, replacing one of the hydrogen atoms with radical-pushing groups of greater capacity, the followings are to be expected.

40

$$N^{\cdot n} + e\cdot\underset{\underset{CF_2}{|}}{\overset{\overset{CH_3}{|}}{C}}=\underset{}{\overset{\overset{H}{|}}{C}}\cdot n \longrightarrow NH + n\cdot\underset{\underset{CF_2}{|}}{\overset{\overset{H}{|}}{C}}-C=CH$$

10.100a

$$E^{\cdot e} + n\underset{\underset{CF_2}{|}}{\overset{\overset{H}{|}}{C}}-\overset{\overset{CH_3}{|}}{C}\cdot e \longrightarrow E-\underset{\underset{CF_2}{|}}{\overset{\overset{H}{|}}{C}}-\overset{\overset{CH_3}{|}}{C}\cdot e \quad OR \quad EF + \underset{e\bullet C-F}{\overset{\overset{H\quad CH_3}{C=C}}{}}$$

(Favored) 10.100b

Only the use of paired electrostatic initiators would have favored the polymerization of this monomer in the route not natural to it, if charged activation of this three-membered ring had been possible. As a Nucleophile, the natural route is electro-free-radical route (and in addition positively charged route for four- and larger-membered rings).

Now, replacing the CH_3 group above with radical-pulling substituent group, the following are obtained.

$$n\cdot\underset{\underset{CF_2}{|}}{\overset{\overset{\overset{\overset{CH_3}{|}}{O}}{|}}{\underset{C=O\ H}{}}}\underset{}{\overset{}{C}}-\overset{}{C}\cdot e + N^{\cdot n} \longrightarrow N-\overset{}{C}-\overset{}{C}\cdot n$$

10.101a

$$E^{\cdot e} + n\cdot\overset{}{C}-\overset{}{C}\cdot e \longrightarrow E-\overset{}{C}-\overset{}{C}\cdot e \quad OR \quad EOCH_3 + e\bullet\overset{O}{\overset{||}{C}}-C=C$$

(FAVORED) 10.101b

With the remaining H atom replaced with radical-pushing substituent group, the followings are obtained.

$$n\cdot\overset{}{C}-\overset{}{C}\cdot e + N^{\cdot n} \longrightarrow NH + n\cdot\overset{}{C}-C=C$$

10.102a

$$n\cdot\overset{}{C}-\overset{}{C}\cdot e + E^{\cdot e} \longrightarrow E-\overset{}{C}-\overset{}{C}\cdot e \quad OR \quad EOCH_3 + e\bullet\overset{O}{\overset{||}{C}}-C=C$$

(FAVORED) 10.102b

41

From the reactions of Equations 10.101a and 10.101b above, the strong Electrophilic character of the monomer of the type used is obvious. When F in CF_2 is replaced with CF_3, no free-radical route will be favored like the case of Equations 10.102a and 10.102b. It is important to note the types of network system that abound in nature and chemical systems. Current observations in chemical system can be extended to other concepts and other areas of discipline.

10.2.1 Molecular rearrangement of the third Kind

Considering the Cumulenic cyclopropane, the followings are obtained.

$$\begin{array}{ccc}
\underset{\substack{|| \\ \text{C}-\text{CH}_2 \\ \diagdown \diagup \\ \text{CH}_2}}{\text{CH}_2} & ; & \underset{\substack{|| \quad | \\ \text{C}-\text{CH} \\ \diagdown \diagup \\ \text{CH}_2}}{\text{CH}_2 \ \text{CH}_3} \qquad ; \text{Versus} \quad \underset{\substack{| \quad\ | \quad\ | \\ \text{H} \quad \text{H} \quad \text{H}}}{\text{H} \quad \text{H} \quad \text{H}} \ \text{C}=\text{C}=\text{C} \ \text{and} \ \underset{\substack{| \quad\ | \quad\ | \\ \text{H} \quad \text{H} \quad \text{H}}}{\text{H} \quad \text{H} \quad \text{CH}_3} \ \text{C}=\text{C}=\text{C} \\
\textbf{CYCLIC CUMULENES} & & \textbf{OLEFINIC CUMULENES} \qquad 10.103
\end{array}$$

It has already been established-

i)   That methylene cyclopropane above is very unstable, i.e., it is always in Equilibrium state of existence. So is methyl cyclopropene its isomer.

ii)   That when $H_2C=$ group is cumulatively placed to a ring or when a double bond is inside the ring, the ring cannot be scissioned at the bond next to it under Stable state of existence conditions via Decomposition mechanism, but only under Equilibrium state of existence conditions via Equilibrium mechanism.

iii)   That the three-membered ring is distinctly different from other larger sized-membered rings for all families of rings, because of the very strong capacity of H_2C closed group in three, than in four than in five and others.

Now, let us see if the products obtained in the decomposition of methylene cyclopane shown in Equation 10.93a via Equilibrium mechanism can be obtained via Decomposition mechanism *in the presence of a passive catalyst* to suppress the Equilibrium state of existence of methylene cyclo-propane.

The $CH_2=$ group on the rings is no functional center, but radical-pushing substituent group of great capacity. Since the capacity is great, the rings above should unzip as follows. Indeed, this can only be done radically. It is being shown chargedly and radically for exploratory purposes. It is also being shown, in order to see the types of isomers than can be obtained from them.

Case (a)

$$\underset{\substack{|| \\ \text{C}-\text{CH}_2 \\ \diagdown \diagup \\ \text{CH}_2}}{\text{CH}_2} \longrightarrow \underset{\substack{| \quad | \quad || \\ \text{H} \quad \text{H} \quad \text{CH}_2}}{\overset{\text{H} \quad \text{H}}{\ominus\text{C}-\text{C}-\text{C}\oplus}} \longrightarrow \underset{\substack{| \\ \text{CH}_2 \\ | \\ \text{CH}_3}}{\overset{\text{H}}{\ominus\text{C}=\text{C}\oplus}}$$

(Impossible existence- chargedly)   10.104

Radically the product above would be obtained via Decomposition mechanism under certain operating conditions. Though the product is different from that obtained via Equilibrium mechanism in Equation 10.93a, the same product would have still been obtained if the temperature had been less than 400°C.

See the third step of the stage of Equation 10.93a. However, via decomposition mechanism, the products obtained via Equilibrium mechanism can never be obtained.

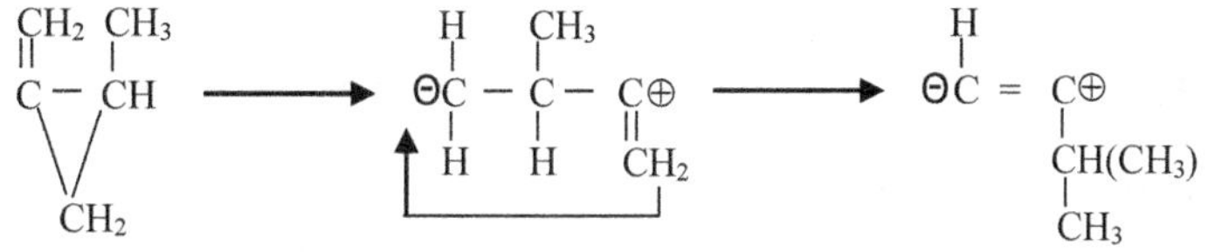

(Impossible existence chargedly)   10.105

Due to electrostatic forces of repulsion, the last components above cannot exist, clear indication that the transfer of transfer species above cannot take place chargedly. On the other hand, the movement of the H above is in line with the Laws of Conservation of transfer of transfer species. The transfer species involved in the molecular rearrangement shown below in Case (b) is the same abstracted or rejected.

Case (b)

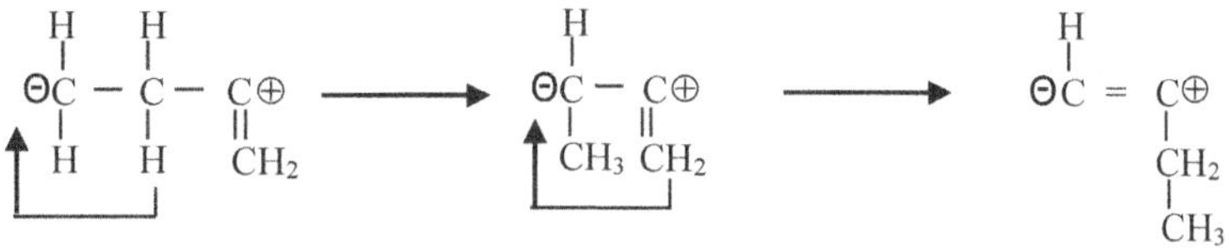

(Impossible existence chargedly)   10.106

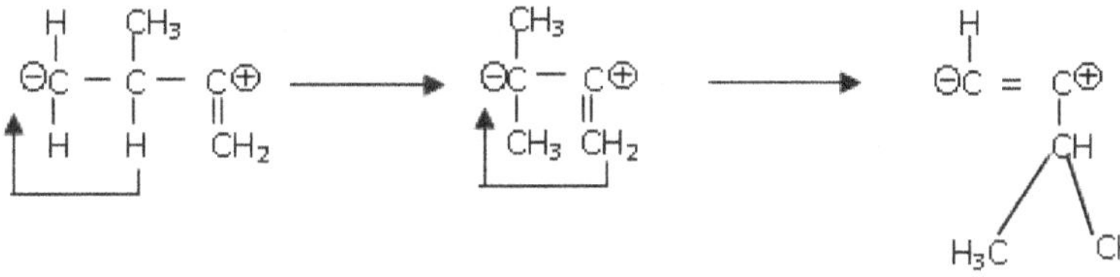

(Impossible existence –chargedly)   10.107

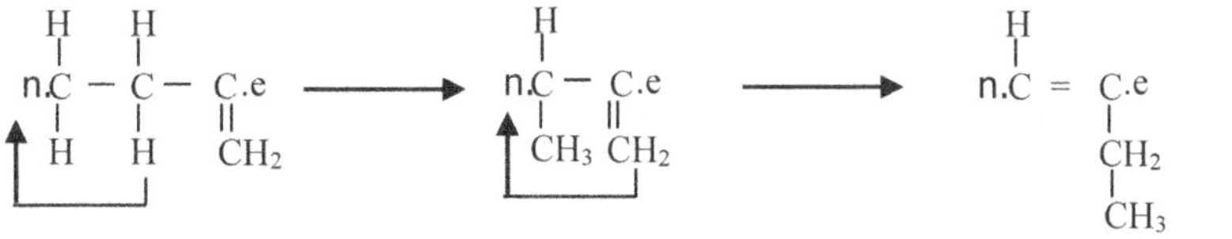

.(A) More Stable   (B) Most Stable

(FAVORED)   10.108a

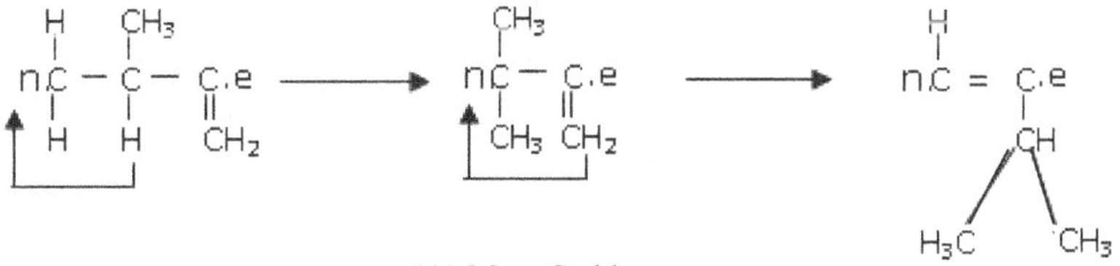

(A) More Stable

(B) Most Stable

(FAVORED)   10.108b

It is said to be favored, because the ring has been scissioned at the right bond. Just as cumulene rearranges to methyl acetylene and methyl cumulene rearranges to di-methyl acetylene so also the cyclic cumulene rearranges to isopropyl acetylene when instantaneously opened. When not opened instantaneously, it will not rearrange to methyl cyclopropene and vice visa (See Equations 10.89a and 10.89b). In addition to several reasons already given, it will not rearrange to methyl cyclopro-pene, because the following is valid. This is an extension of Equation 10.43. The ring will first be

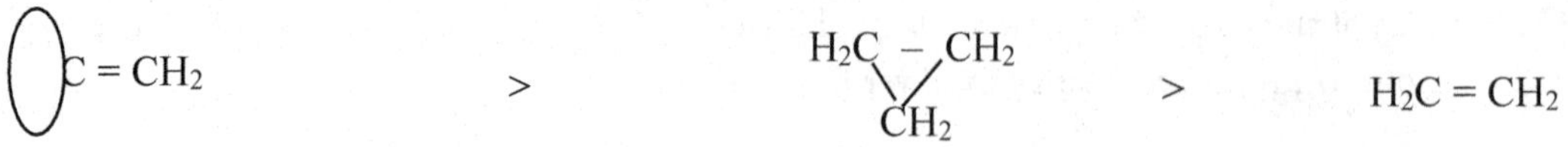

$$\overset{\bigcirc}{C} = CH_2 \qquad > \qquad \underset{CH_2}{\overset{H_2C - CH_2}{\diagdown \diagup}} \qquad > \qquad H_2C = CH_2$$

ORDER OF NUCLEOPHIICITY FOR THREE-MEMBERED RINGS 10.109

activated before the conjugatedly placed double bond gets activated.

The molecular rearrangement here in Equations 10.108a and 10.108b is carried out in two steps in Case (b). This molecular rearrangement can be observed to be of a different type, since two types of transfer species are involved and this can only take place radically. As a matter of fact, it was astonishing to note that these cumulenes exist, because it was thought that the three-membered ring on which $H_2C=$ group is cumulatively placed, is too strained to exist. Hence, methyl cyclopropene of Equation 10.89a was said not to molecularly rearrange as indicated therein. This will not be the case for larger sized rings-from four and above.

Now, consider the following wherein one is going to gradually reduce the capacity of H_2C group in the three-membered ring by steadily replacing the H with F as one has started doing.

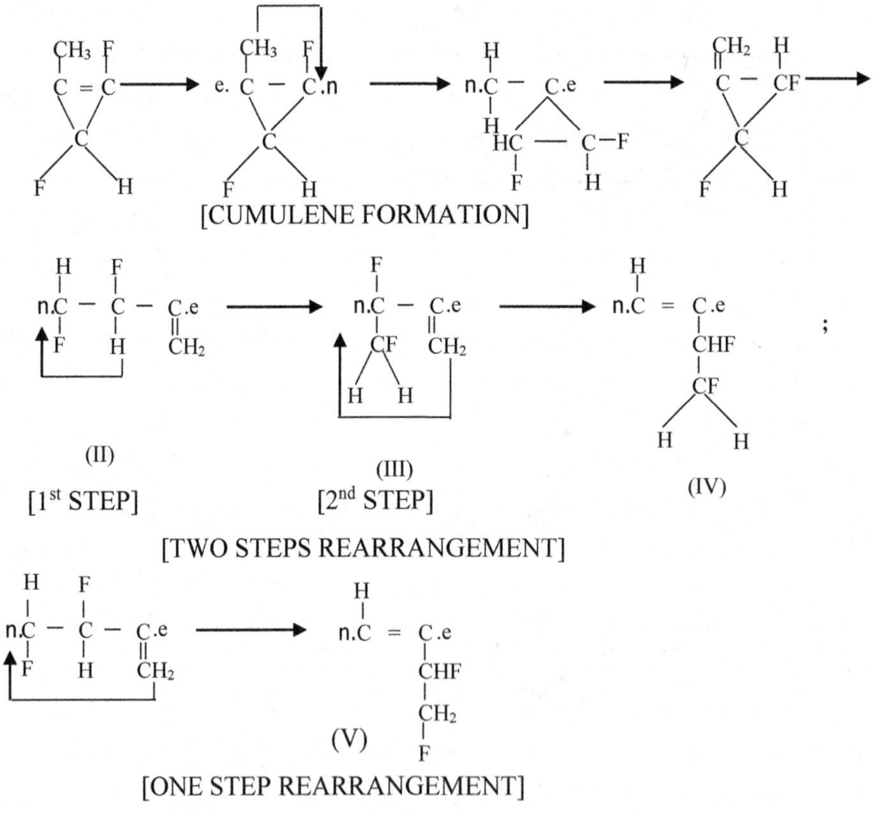

[CUMULENE FORMATION]

[TWO STEPS REARRANGEMENT]

[ONE STEP REARRANGEMENT] 10.110a

Though (IV) and (V) look the same, this type of molecular rearrangement can indeed be carried out only in two steps and not in one step since it is only with the two steps movement that the Laws of conservation of Transfer of transfer species can be seen to be obeyed. However, the case above is still questionable, because it is believed that the following is valid.

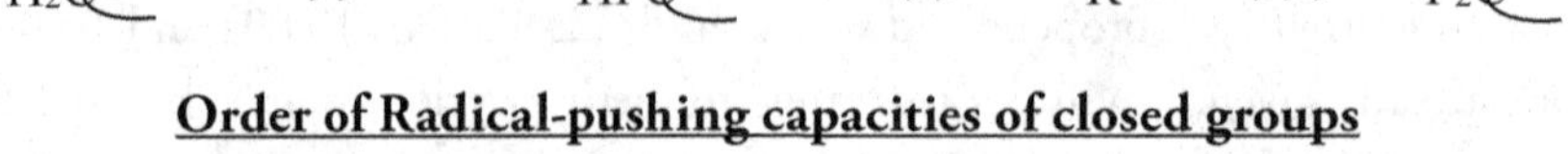

$$H_2C\overset{\frown}{} \qquad >> \qquad HFC\overset{\frown}{} \qquad >> \qquad R \qquad >>> \qquad F_2C\overset{\frown}{} \qquad \qquad 10.110b$$

Order of Radical-pushing capacities of closed groups

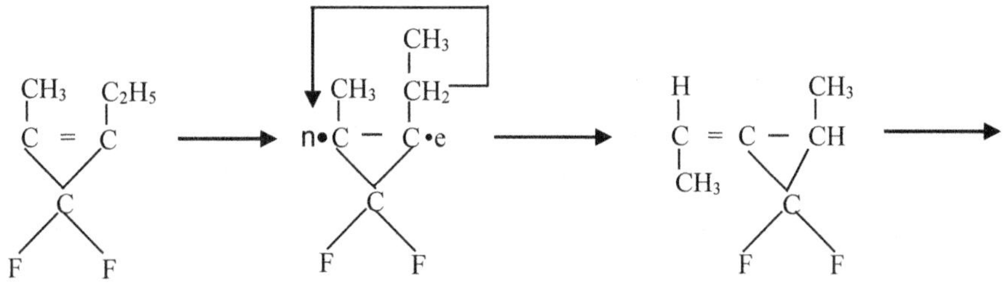

[CUMULENE FORMATION]

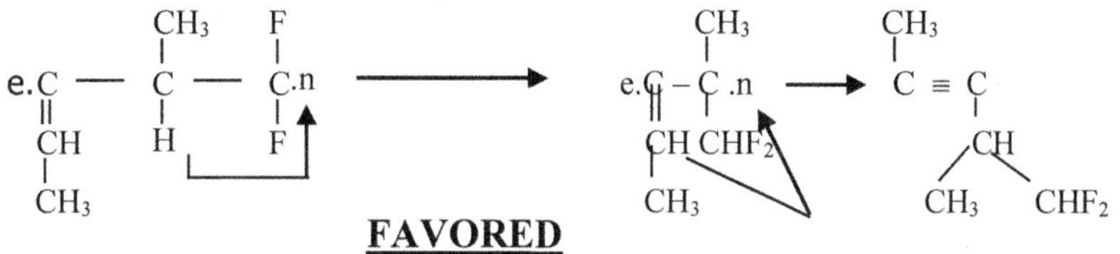

FAVORED

10.111

Of the three monomers shown below[4], though none will in reality undergo molecular rearrangement, artificially only one of them will not favor molecular rearrangements. Let us identify their linear isomers.

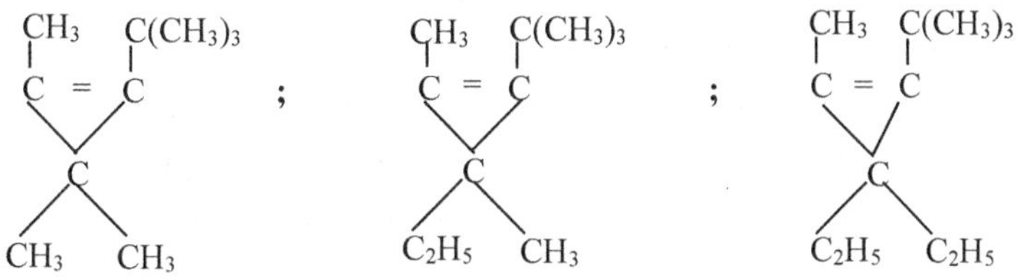

(I)   (II)   (III)

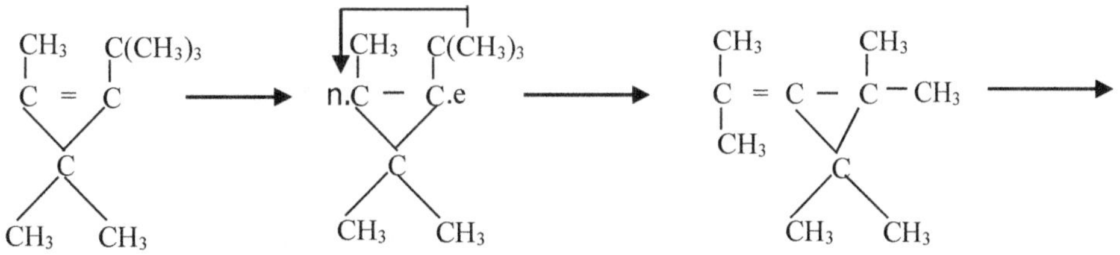

(I)

10.112

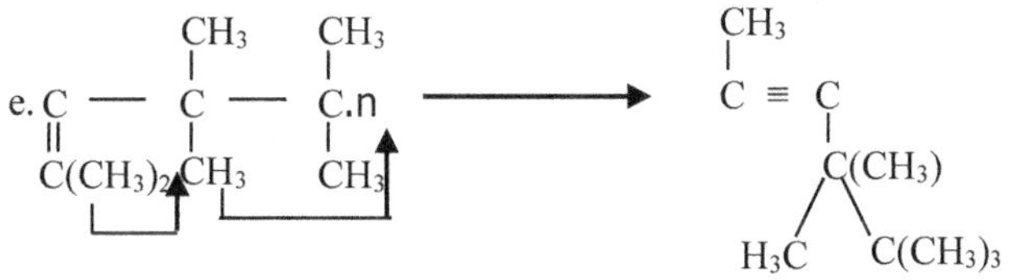

NOT FAVORED

10.113

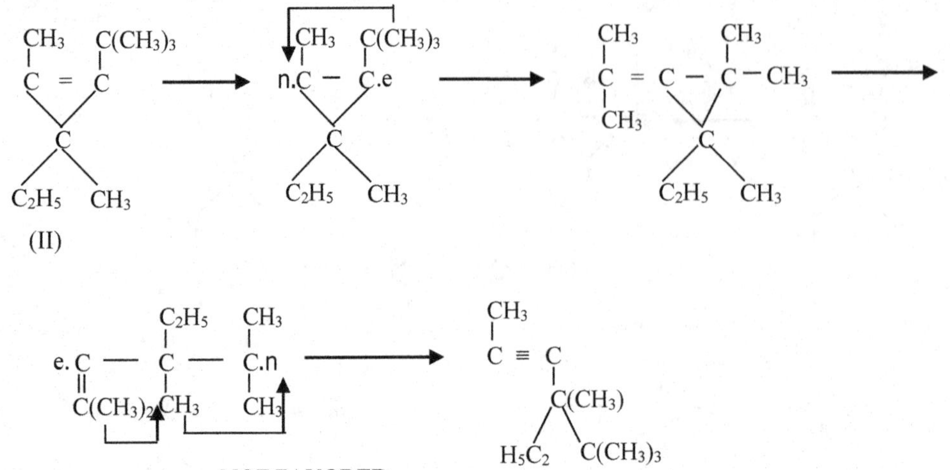

$$
\begin{array}{ccc}
\underset{C}{\overset{CH_3}{|}}=\underset{|}{\overset{C(CH_3)_3}{|}}C & \longrightarrow & n.\underset{|}{C}-\overset{CH_3\ \ C(CH_3)_3}{C}.e \\
\underset{C_2H_5\quad CH_3}{\overset{C}{\diagdown}} & & \underset{C_2H_5\quad CH_3}{\overset{C}{\diagdown}}
\end{array}
$$

(II)

$$
\begin{array}{cc}
\underset{|}{\overset{CH_3\qquad CH_3}{C}}=C-\underset{|}{\overset{|}{C}}-CH_3 & \longrightarrow \\
\underset{C_2H_5\quad CH_3}{\overset{C}{\diagdown}} &
\end{array}
$$

$$
e.\underset{\|}{\overset{C_2H_5\quad CH_3}{C}}-\underset{|}{\overset{|}{C}}-\overset{|}{C}.n \quad \longrightarrow \quad \underset{\underset{H_5C_2\quad C(CH_3)_3}{C(CH_3)}}{\overset{CH_3}{\overset{|}{C}\equiv C}}
$$

C(CH_3)_2 CH_3 CH_3

NOT FAVORED

10.114

With (III), the molecular rearrangement of the first kind will not be favored, since C_2H_5 is more radical-pushing than the CH_3 groups carried by the receiving center. ***All the above are for explora-tory purposes deliberately done in order to reveal some basic fundamental principles and one of their linear isomers. The transfer species moved for (III) not shown above is giving more than what should be given (Overdose), that which should not be. The real transfer species is that located inside the ring.***

In continuation of the journey, shown below are three monomers whose existences like the last cases above are favored[4], despite the fact that they are more strained than methyl cyclopropene. They are very stable. Unlike methyl cyclopropene, there is no H atom directly connected to the ring. Therefore, they cannot readily exist in Equilibrium state of existence.

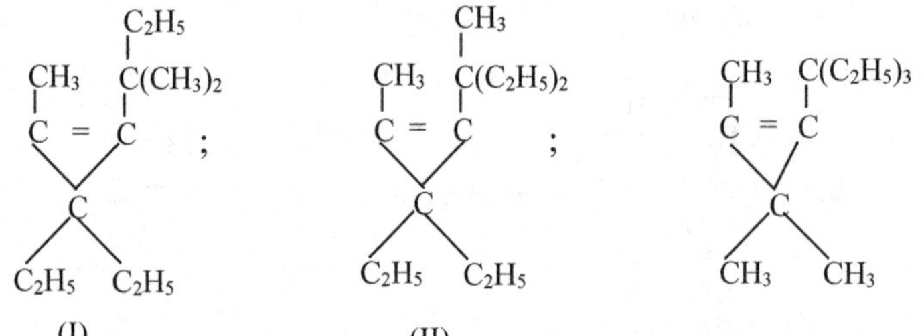

$$
\begin{array}{ccc}
\underset{|}{\overset{CH_3\quad C(CH_3)_2}{C}}=C & ; & \underset{|}{\overset{CH_3\quad C(C_2H_5)_2}{C}}=C & ; & \underset{|}{\overset{CH_3\quad C(C_2H_5)_3}{C}}=C
\end{array}
$$

(I)   (II)   (III)

$$
(I) \longrightarrow n.\underset{|}{C}-\overset{C_2H_5}{\overset{CH_3\quad C(CH_3)_2}{C}}.e \longrightarrow \underset{|}{\overset{C_2H_5\qquad CH_3}{C}}=C-\overset{|}{C}-CH_3 \longrightarrow
$$

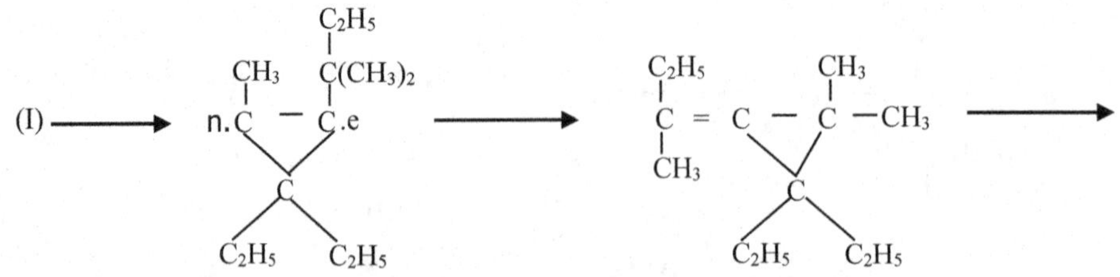

$$\text{e. } \underset{\substack{\| \\ C(CH_3) \\ | \\ C_2H_5}}{C} - \underset{\substack{| \\ C_2H_5}}{\overset{C_2H_5}{C}} - \underset{\substack{| \\ CH_3}}{\overset{CH_3}{C.n}} \longrightarrow \text{ No Transfer species}$$

10.115

(II) $\longrightarrow$ e.C $-$ C.n $\longrightarrow$ C $-$ C(CH$_3$)$_2$ $\longrightarrow$

10.116

$$\text{e. } \underset{\substack{\| \\ C(C_2H_5)_2 C_2H_5}}{C} - \underset{\substack{| \\ }}{\overset{C_2H_5}{C}} - \underset{\substack{| \\ CH_3}}{\overset{CH_3}{C.n}} \longrightarrow \text{ No Transfer species}$$

10.116

All these are taking from the Poor and giving to the Rich, that which is naturally impossible. How-ever (III) cannot artificially undergo molecular rearrangement of the first kind. For the three cases above, the rings cannot be opened. They can however readily be used as monomers (Nucleophiles), via electro-free-radical polymerization. ***In reality, while (III) will rearrange to give same monomer, (I) and (II) will not undergo any rearrangement, when R_2C group is placed in the ring.***

Most primary and secondary alkylane groups carried by active carbon centers cannot readily molecularly rearrange to acetylenes either via Equilibrium or Decomposition mechanisms. Tertiary alkanes were used above to illustrate some basic principles which are natural in character. For primary and secondary alkanes groups, consider the two cases below.

SECOND MOVEMENT FAVORED

10.117a

In reality, neither will the transfer above take place nor will the ring ever be opened. Transfer species can only come from inside the rings. But, notice what would have happened if inside the ring was F_2C in place of $(CH_3)_2C$.

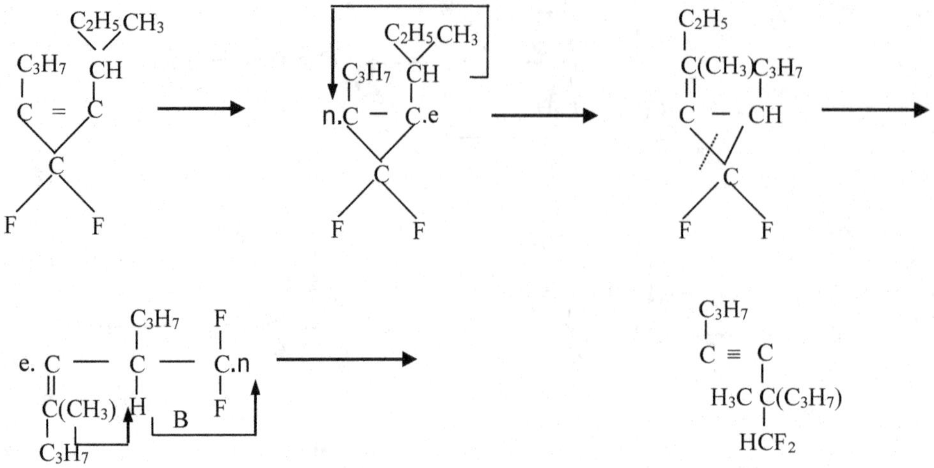

.(A) NOT FAVORED

10.117b

In reality, like all the cases above, no transfer, no ring opening, and in many cases, no rearrangement can take place.

One can begin to observe the conditions favoring the existence of cyclopropenes of the nucleophilic types. Based on the exploratory cases above, for these monomers to favor molecular rearrangement to produce acetylenes, one of the groups in the internal carbon center in the ring must have a capacity equal to or less than those carried by the carbon center on the double bond, if conditions for favoring molecular rearrangement of the first kind exist. Nevertheless, the conditions under which the monomers above can be made to undergo molecular rearrangement to give acetylenes do not exist.

For (I) of Equation 10.97a, after molecular rearrangement, the ring cannot be opened since O = C = group is a strong radical-pulling group. For (I) of Equation 10.97b, if any attempt is made to open the ring, in view of the weak radical-pushing capacity of F_2C = group as will shortly be shown, the following is obtained.

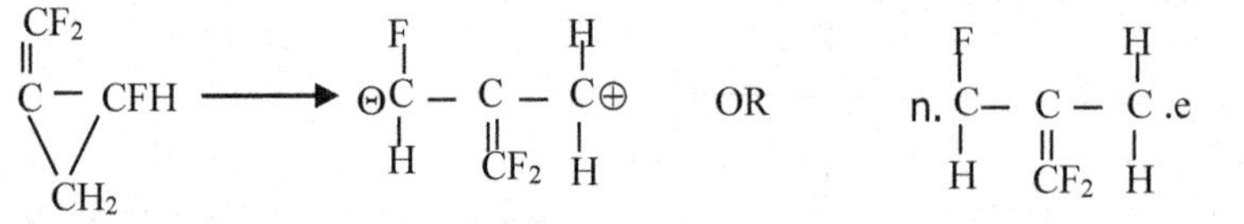

Impossible existence   Impossible existence

10.118

Though its presence was not favored in Equation 10.97b, it has been used to show the importance of F_2C= group as a radical-pushing group of lesser capacity than H, for which the point of scission above is wrong. Comparing $H_2C=C=CH_2$ with $F_2C=C=CH_2$ and $F_2C=C=CF_2$, the followings are obtained.

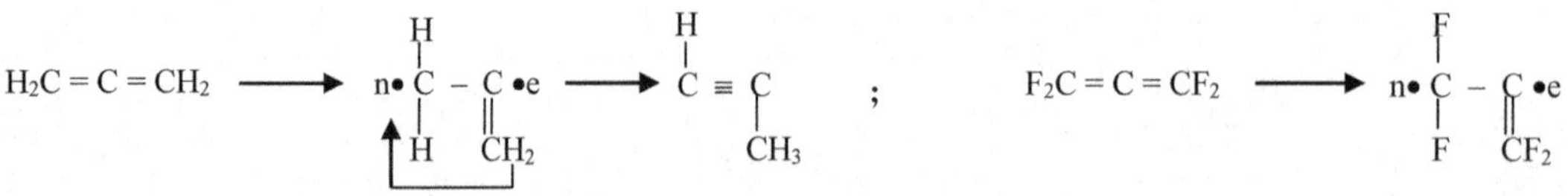

NUCLEOPHILE   "BOTH"

10.119a

$$F_2C = C = CH_2 \longrightarrow n\bullet \overset{\displaystyle F}{\underset{\displaystyle F}{C}} - \overset{\displaystyle \bullet e}{\underset{\displaystyle CH_2}{C}} \longrightarrow \overset{\displaystyle H}{\underset{\displaystyle CF_2H}{C \equiv C}} \quad ; \quad F_2C = C = CH_2 \longrightarrow e\bullet \overset{\displaystyle H}{\underset{\displaystyle H}{C}} - \overset{\displaystyle \bullet n}{\underset{\displaystyle CF_2}{C}}$$

(III)   (D)   (E)   (III)   (F)

               X-center   Y-center

AN ELECTROPHILE
10.119b

For the first time, one is observing a different type of Electrophile (III), cumulenic in character. It has both X and Y centers cumulatively placed as against being adjacently placed. This is where for the first time we are beginning to see how the halogen F in particular shows its strong Electrophilic character. Unlike, (I) and (II), the X-center in (III) favors only the electro-free-radical route, while the Y-center favors only the nucleo-free-radical route. Only the X-center can be made to molecularly rearrange to give a more nucleophilic acetylene (E), *if (III) is a Nucleophile.* As an Electrophile, the Y-center cannot rearrange, because $F_2C=$ group is a radical pushing group, with transfer species which cannot be used for rearrangement. Like Electrophiles where the two centers are adjacently placed, wherein the X center is more nucleophilic than the Y center and Y center only rearranges, the same applies with Electrophiles where the two centers are cumulatively placed whether Y center is more nucleophilic than the X center or not. Hence, the rearrangement in the last equation from (D) to (E) is not favored. (I) is a Nucleophile in which the two activation centers are alike just like in (II). As has been shown before, it can be made to readily molecularly rearrange to give a more nucleophilic acetylene. Since, *cumulene after rearrangement is more nucleophilic than acetylene, this also means that cumulene is more unsaturated than acetylene.* This is very visible from the number of double bonds. (II) unlike (I), favors the route not natural to it, the nucleo-free-radical route. It is a Nucleophile, but behaving like an Electrophile, since there can be no Y without an X. Hence, the two similar centers in (II) are both X and Y. In (III), the X center is the first to be activated being less nucleophilic. The Y center cannot rearrange; for which if the X center is allowed to rearrange, then (III) becomes a Nucleophile. Based on the considerations above, the followings are valid.

$$R_2C= \; > \; RHC= \; > \; H_2C= \; > \; RR_FC= \; > \; HR_FC= \; > \; (R_F)_2C= \; > \; R_FC= \; > \; FHC= \; >> \; R \; >>$$

$$R \;\; >> \;\; H \;\; > \;\; R_FFC= \;\; > \;\; (F)_2C= \;\; > \;\; R_F$$

$$\text{Where } R \equiv C_nH_{2n+1} \quad ; \quad R_F \equiv C_nF_{2n+1}$$

ORDER OF RADICAL-PUSHING CAPACITIES OF RADICAL-PUSHING GROUPS
10.119c

Notice that all the groups above are opened. However in general, notice that though -COOR is a radical-pulling group, $(COOR)_3C-$ is a radical-pushing group of lower capacity than H, because there is no H amongst the groups carried by C. $(COOR)_2HC-$ is a radical-pushing group of greater capacity than H but of lower capacity than F_2HC-, because COOR is more radical-pulling than F. The same applies to the cases above. One can observe the enormous power of the C atom and its elements.

Opening of the ring in Equation 10.118 can only take place radically but not as shown. In fact, it is even believed that the first member favors only free-radical existence, like acetylene, in view of the size of the ring, and high unsaturation (carbon to hydrogen ratio equals 1:4/3) favored by the presence of the ring, as indicated below.

49

$$\underset{\substack{H \\ | \\ H}}{C} \equiv \underset{\substack{| \\ H}}{C} \quad < \quad \underset{\substack{H \quad H \\ | \quad | \\ \diagdown \diagup \\ CH_2}}{C = C} \quad < \quad \underset{\substack{H \qquad H \\ | \qquad | \\ C = C = C \\ | \qquad | \\ H \qquad H}}{}$$

H : C ratio: <u>1:1</u>   <u>1.3 : 1</u>   <u>1.3 : 1</u>

<u>ORDER OF UNSATURATION</u>   10.120

Hence, cyclopropene can be written as follows. Thus, while allene readily molecularly rearranges to methyl acetylene, cyclopropene does to remain the same like propene. However, it readily exists in Equilibrium state of existence, its intensity higher than that of acetylene (HC≡C •n + H•e).

$$\underset{\substack{H \quad H \\ | \quad | \\ \diagdown \diagup \\ CH_2}}{C = C} \quad \rightleftharpoons \quad \underset{\substack{H \\ | \\ \diagdown \diagup \\ CH_2}}{C = C \cdot n} \quad + \quad H^{\bullet e}$$

       10.121

Presence of a suppressing agent or an activated monomer may be required before the double bond in the ring can be activated. Hence, if polymerization of the monomer is to be fully favored, the two H atoms on the two carbon centers carrying double bond must first be replaced with substituent groups. Hence also, the reactions indicated by Equations 10.98a and 10.99a are not favored in the absence of suppression, and should therefore be as follows-

$$N^{\cdot n} + \underset{\substack{H \\ | \\ \diagdown \diagup \\ CH_2}}{C = C \cdot n} H^{\cdot e} \longrightarrow NH + \underset{\substack{H \\ | \\ \diagdown \diagup \\ CH_2}}{C = C \cdot n}$$

       10.122

$$E^{\cdot e} + \underset{\substack{H \\ | \\ \diagdown \diagup \\ CH_2}}{C = C \cdot n} H^{\cdot e} \longrightarrow H^{\cdot e} + \underset{\substack{E \quad H \\ | \quad | \\ \diagdown \diagup \\ CH_2}}{C = C}$$

       10.123

This does not apply to the cases of Equations 10.98b and 10.99b, because of the presence of CF_2 in place of CH_2 in the ring. This case is Stable. Even after the first replacement of one H atom in the two cases above, the second H atom is still held in Equilibrium state of existence. This is one of the reasons why cyclopropene is not known to readily undergo polymerization. Nevertheless, it can be observed that cyclopropenes are more nucleophilic than cyclopropanes or cycloalkanes. ***Based on current observations, cyclopropane should be called cyclopropene, while cyclopropene be called cyclopropyne. In order words, there is no cyclopropane or cycloalkanes, since what is inside the ring is an invisible π-bond which is nucleophilic. It can be a Y or an X depending on what the ring is carrying particularly with respect to cyclopropane (now called cyclopropene). Nevertheless, one will continue the use of the names as used today.***

Coming back to free-radical aspect of the analysis following Equation 10.111, the followings are obtained when H is replaced with CH_3 and CH_2 replaced with CF_2.

$$N \cdot n \; + \; e.\overset{\overset{\displaystyle CH_3 \; H}{|\quad|}}{\underset{\underset{\displaystyle CF_2}{\diagdown\diagup}}{C = C}} \cdot n \; \longrightarrow \; N \cdot n \; + \; n \cdot \overset{\overset{\displaystyle H}{|}}{\underset{\underset{\displaystyle H}{|}}{C}} - \overset{\overset{\displaystyle e}{}}{\underset{\underset{\displaystyle CF_2}{\diagdown\diagup}}{C}} - CH_2 \; \longrightarrow$$

$$N \cdot n \; + \; \overset{\overset{\displaystyle H}{|}}{\underset{\underset{\displaystyle H}{|}}{C}} = \overset{\overset{\displaystyle H}{|}}{\underset{\underset{\displaystyle CF_2}{\diagdown\diagup}}{C}} - C - H \quad OR \quad NH \; + \; n \cdot \overset{\overset{\displaystyle H}{|}}{\underset{\underset{\displaystyle H}{|}}{C}} - \overset{}{\underset{\underset{\displaystyle CF_2}{\diagdown\diagup}}{C}} = CH$$

(I)   (II)-Favored   10.124

$$E \cdot e \; + \; e.\overset{\overset{\displaystyle CH_3 \; H}{|\quad|}}{\underset{\underset{\displaystyle CF_2}{\diagdown\diagup}}{C = C}} \cdot n \; \longrightarrow \; E \cdot e + \overset{\overset{\displaystyle CH_2}{\|}}{\underset{\underset{\displaystyle CF_2}{\diagdown\diagup}}{C}} - CH_2 \quad OR \quad E - \overset{\overset{\displaystyle H}{|}}{\underset{\underset{\displaystyle H}{|}}{C}} - \overset{\overset{\displaystyle e.}{}}{\underset{\underset{\displaystyle CF_2}{\diagdown\diagup}}{C}} - \overset{\overset{\displaystyle H}{|}}{C} - H$$

Weak   (I)   (II)   10.125

$$Br_2Al.e \; + \; :\ddot{B}r.nn \; + \; \overset{\overset{\displaystyle CH_2}{\|}}{\underset{\underset{\displaystyle CF_2}{\diagdown\diagup}}{C}} - CH_2 \; \longrightarrow \; Br_2Al.e \; + \; :\ddot{B}r.nn \; + \; e.\overset{\overset{\displaystyle H \quad F}{|\quad|}}{\underset{\underset{\displaystyle CH_2 \; H \; F}{}}{C} - C - C} \cdot n$$

$$\longrightarrow \; Br_2Al.e \; + \; :\ddot{B}r.nn \; + \; n.\overset{\overset{\displaystyle H}{|}}{\underset{\underset{\underset{\displaystyle CF_2H}{|}}{CH_2}}{C}} = C.e \; \longrightarrow \; Br_2Al - \overset{\overset{\displaystyle H}{|}}{\underset{\underset{\underset{\displaystyle CF_2H}{|}}{CH_2}}{C}} = C.e \; + \; Br.nn$$

10.126

To open the ring, electrostatic forces provided by the Br atoms have been used, noting that the initiator cannot be paired radically. It is better to use that from AlR_3/H_2O (Paired) combination as will be shown to exist downstream. After opening by decomposition followed by molecular rearrangement of third kind, polymerization is favored. All these depend on if the first rearrangement of the first kind takes place from "cyclo-acetylene to "cumulene", that which will obviously take place if the initiator is weak as has already been shown. Indeed, electro-free-radical route is favored after rearrangement.

$$E \cdot e \; + \; e.\overset{\overset{\displaystyle CH_3 \; CH_3}{|\quad\;|}}{\underset{\underset{\displaystyle CF_2}{\diagdown\diagup}}{C = C}} \cdot n \; + \; :NR_3 \; \longrightarrow \; n.\overset{\overset{\displaystyle H}{|}}{\underset{\underset{\displaystyle H}{|}}{C}} - \overset{\overset{\displaystyle e}{}}{\underset{\underset{\displaystyle CF_2}{\diagdown\diagup}}{C}} - CH \; + \; E \cdot e \; + \; :NR_3$$

(I)

$$\longrightarrow \; e.\overset{\overset{\displaystyle CH_3 \; F}{|\quad|}}{\underset{\underset{\displaystyle CH_2 \; H \; F}{}}{C} - C - C} \cdot n \; + \; E^{\cdot e} \; \longrightarrow \; :NR_3 \; \longrightarrow \; n.\overset{\overset{\displaystyle H}{|}}{\underset{\underset{\underset{\displaystyle CH_3 \; CF_2H}{\diagup\diagdown}}{CH}}{C}} = C.e \; + \; E^{\cdot e} \; + \; R_3N:$$

$$\longrightarrow \; E - \overset{\overset{\displaystyle H}{|}}{\underset{\underset{\underset{\displaystyle CH_3 \; CF_2H}{\diagup\diagdown}}{CH}}{C}} = C \cdot e \; + \; :NR_3 \; \xrightarrow{n(I)} \; [\text{Low mol. wt. Polymers}]$$

10.127

51

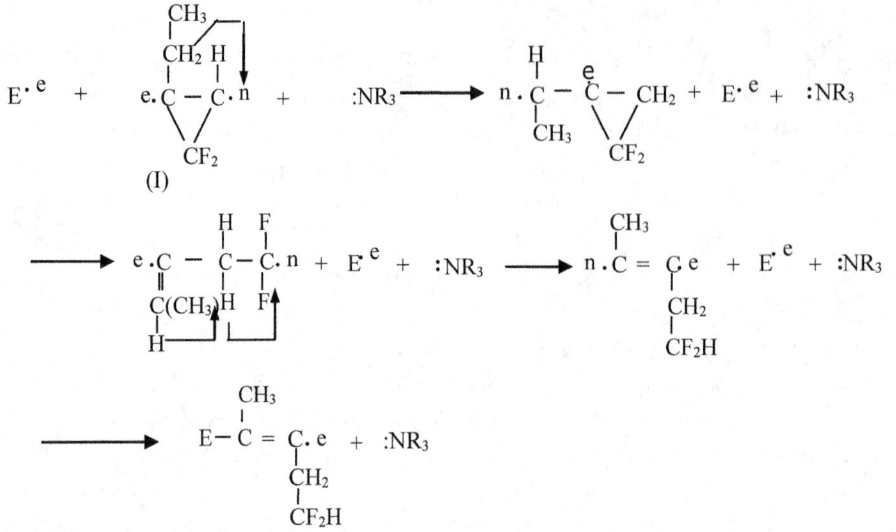

$$E\cdot e \;+\; \underset{(I)}{\overset{\overset{\displaystyle CH_3}{\overset{|}{\overset{\displaystyle CH_2}{|}}\;H}}{e\cdot C - C\cdot n}}\;+\; :NR_3 \longrightarrow n\cdot \overset{H}{\underset{CH_3}{C}} - \overset{e}{C} - CH_2 \;+\; E\cdot e \;+\; :NR_3$$

$$\longrightarrow e\cdot C - \overset{H\;\;F}{C} - C\cdot n \;+\; E\overset{e}{} \;+\; :NR_3 \longrightarrow n\cdot C = \overset{CH_3}{\underset{\underset{CF_2H}{CH_2}}{C}}e \;+\; E\overset{\cdot e}{} \;+\; :NR_3$$

$$\longrightarrow E - \overset{CH_3}{\underset{\underset{CF_2H}{CH_2}}{C}} = C\cdot e \;+\; :NR_3 \qquad\qquad 10.128$$

Being Nucleophilic, only electro-free-radical polymerization is possible if the ring can be opened by Decomposition mechanism and the monomer is stable. Chargedly, polymerization is not possible due to electrostatic forces of repulsion, resulting from molecular rearrangement of the third kind as shown below apart from the fact that the ring cannot be opened chargedly.

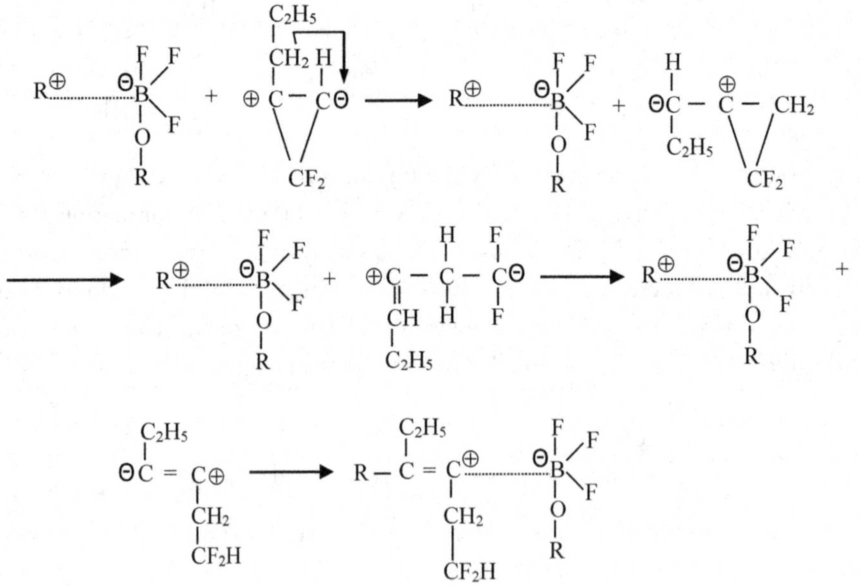

$$\text{Impossible existence} \qquad\qquad 10.129$$

Apart from the impossible existence of the activated acetylene above, the last equation is not possible because of electrostatic forces of repulsion. Radically, the rearrangement is not favored, because H_5C_2 is more radical-pushing than $C_2H_3F_2$. Therefore, the R group carried cannot exceed CH_3.

Thus, for these members of the family shown below, the following are worthy of note

$$\underset{CF_2}{\overset{R\;\;\;\;H}{C = C}} \qquad\qquad \underset{H_2C - CH_2}{\overset{R\;\;\;\;H}{C = C}}$$

52

(i)   The presence of CF_2 in place of CH_2 in the ring in the three-membered ring.

(ii)   The maximum size of R group for free-radical polymerization is CH_3.

(iii)   If $R \geq C_2H_5$, acetylenes cannot be formed (where the Rs are alkylane group).

(iv)   That Chemistry can be seen as a complex network system so ordered and too much to comprehend for which the use of the EYE OF THE NEEDLE is inevitable (Not the physical eyes).

These can simply be seen by the simple example shown below.

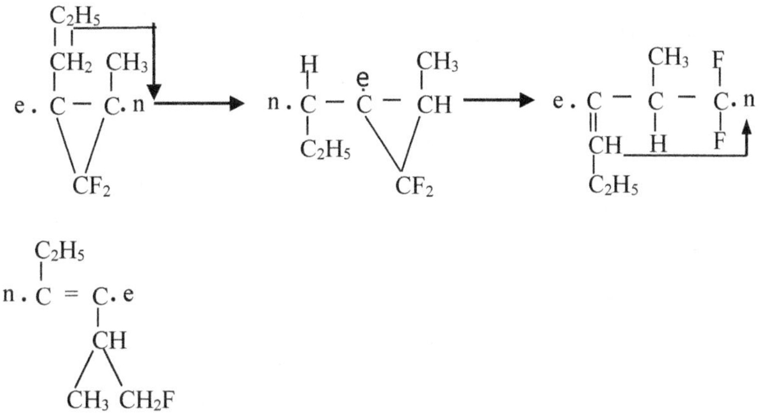

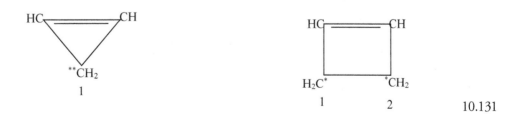

10.130

While the observations above will readily take place for three-membered rings, it will not take place when CH_2 is put in place of CF_2. For the four-membered ring shown above, since acetylene is more nucleophilic than cyclobutenes, instead of CH_3 for three-membered ring above, with extended limitations R cannot exceed C_2H_5. Secondly, the need to change the CH_2 to CF_2 for four-membered and larger membered rings does not arise. Based on Equations 10.127 and 10.130, when the H atoms in the ring were changed to F atoms, the situation changed drastically, since a balance must be maintained. This is one of the unique characters of NATURE. The use of alkylanes substituent groups can be observed to be important to the understanding of the characters of these monomers.

When one H atom in cyclopropanes is replaced with OH group, the followings are obtained, noting that the radical-pushing capacity of OH group is "assumed" to be less than that in the ring, wherein a C center is shared between two carbon centers, unlike the case in larger membered rings. Hence, two- and three-membered rings whether they exist or not are in general very important.

53

$$\text{(I)(Nucleophile)} \xrightarrow{\text{Activation}} \text{nn} \cdot O - \overset{e}{\underset{\underset{\displaystyle CF_2}{\diagdown\diagup}}{C}} - CH_2 \longrightarrow \overset{\displaystyle O}{\underset{\underset{\displaystyle CF_2}{\diagdown\diagup}}{\overset{\|}{C}}} - CH_2 \xrightarrow[\text{opening)}]{\text{(Ring}}$$

(I)(Nucleophile)   (Strained) (I)a

$$e \cdot \underset{\underset{\displaystyle O}{\|}}{C} - \underset{\underset{\displaystyle H}{|}}{\overset{\overset{\displaystyle H}{|}}{C}} - \underset{\underset{\displaystyle F}{|}}{\overset{\overset{\displaystyle F}{|}}{C}} \cdot n \longrightarrow \mathbf{e} \cdot \underset{\underset{\displaystyle O}{\|}}{C} -- \underset{\underset{\displaystyle CF_2H}{|}}{\mathbf{C}} \cdot \mathbf{n} \quad \text{(Electrophile)}$$

(II)   (III) (Favored)

10.132

Chargedly, the reaction above is not possible. It is favored radically, because the transfer species involved during rearrangement is the same abstracted nucleo-free-radically. This is like the case of acetylene [HO-C≡CH] when allowed to take place. Molecular rearrangement of the first kind only takes place to produce (I)a strained ketonic ring (Cyclopropanone) which like those encountered in Chapter 7 (E.g., Cyclo-heptanone) cannot readily be opened though strongly strained, 0 = group being a strong radical-pushing group. This will not take place for any ring size even if the CF_2 is changed to CH_2. ***Hence cycloketone can be opened to form ketenes since the Laws of conservation of transfer of transfer species will not be broken.*** On the other hand, such rings can be easily expanded in size without opening of the ring. The compound (I) above will favor being used as a monomer if it exists and not made to exist in Stable state of existence all the time. For larger sized ring, they will rearrange to give cyclo-ketones,

With OR groups, the followings are obtained-

$$e \cdot \underset{\underset{\displaystyle CF_2}{\diagdown\diagup}}{\overset{\overset{\displaystyle R}{|}\,\,\,\,CH_3}{C}} - \underset{}{C} \cdot n \longrightarrow \text{nn} \cdot O - \overset{e.}{\underset{\underset{\displaystyle CF_2}{\diagdown\diagup}}{C}} - \overset{\overset{\displaystyle R}{|}}{C}(CH_3) \longrightarrow \overset{\displaystyle O}{\underset{\underset{\displaystyle CF_2}{\diagdown\diagup}}{\overset{\|}{C}}} - \overset{\overset{\displaystyle R}{|}}{C}(CH_3) \longrightarrow$$

(I)

$$e \cdot \underset{\underset{\displaystyle O}{\|}}{C} - \underset{\underset{\displaystyle CH_3}{|}}{\overset{\overset{\displaystyle R}{|}}{C}} - \underset{\underset{\displaystyle F}{|}}{\overset{\overset{\displaystyle F}{|}}{C}} \cdot n \quad (\text{R} \le CH_3)$$

(II) (Not Favored)

10.133

It is not favored because it will break the Laws of Conservation of transfer of transfer species, unless when R is CH_3 or less.

With NH_2 group, the followings are obtained.

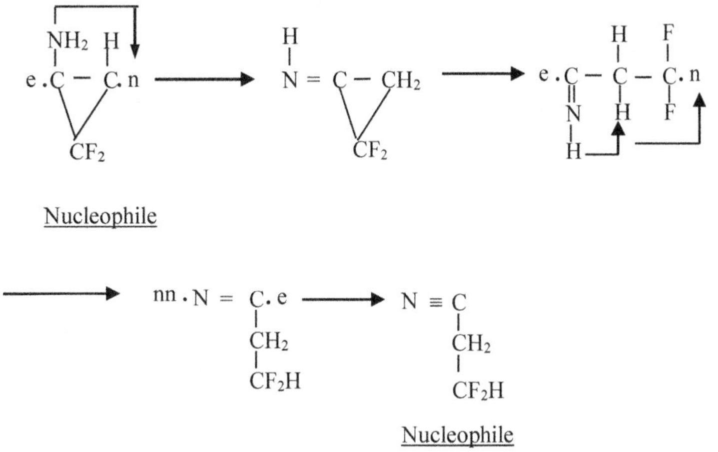

$$
\begin{array}{c}
\overset{\displaystyle NH_2\ \ H\downarrow}{e\cdot C - C\cdot n} \\
\diagdown\ \diagup \\
CF_2
\end{array}
\longrightarrow
\begin{array}{c}
\overset{\displaystyle H}{N = C - CH_2} \\
\diagdown\ \diagup \\
CF_2
\end{array}
\longrightarrow
\begin{array}{c}
\overset{\displaystyle H\ \ \ F}{e\cdot C - C - C\cdot n} \\
\ \ \parallel\ \ \ \ |\ \ \ \ | \\
\ \ N\ \ \ H\ \ \ F \\
\ \ | \\
\ \ H
\end{array}
$$

<u>Nucleophile</u>

$$
\longrightarrow
\begin{array}{c}
nn\cdot N = C\cdot e \\
|\\
CH_2 \\
|\\
CF_2H
\end{array}
\longrightarrow
\begin{array}{c}
N \equiv C \\
|\\
CH_2 \\
|\\
CF_2H
\end{array}
$$

<u>Nucleophile</u>   10.134

Unlike the OH group, molecular rearrangement of the third kind is favored here to produce nitriles. This will not be the case with NHR and NR$_2$ groups as shown below.

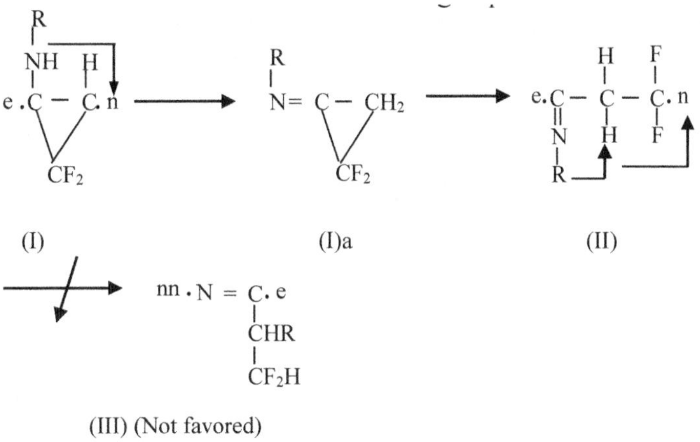

$$
\begin{array}{c}
\overset{\displaystyle R}{\underset{}{\overset{|}{NH\ \ H\downarrow}}} \\
e\cdot C - C\cdot n \\
\diagdown\ \diagup \\
CF_2
\end{array}
\longrightarrow
\begin{array}{c}
\overset{\displaystyle R}{N = C - CH_2} \\
\diagdown\ \diagup \\
CF_2
\end{array}
\longrightarrow
\begin{array}{c}
\overset{\displaystyle H\ \ \ F}{e\cdot C - C - C\cdot n} \\
\ \ \parallel\ \ \ \ |\ \ \ \ | \\
\ \ N\ \ \ H\ \ \ F \\
\ \ | \\
\ \ R
\end{array}
$$

(I)   (I)a   (II)

$$
\longrightarrow\!\!\!\!/\ \
\begin{array}{c}
nn\cdot N = C\cdot e \\
|\\
CHR \\
|\\
CF_2H
\end{array}
$$

(III) (Not favored)   10.135

Since the transfer species in (III) is different from R, molecular rearrangement of the third kind is not favored. If R is of lower capacity than H, such as CF$_3$, then it will be favored. When only one movement takes place just as with OH group, then an Electrophile is obtained [RN=C=CH(CHF$_2$]-(See Equations 10.38 and 10.40). One can thus observe very clearly why NH$_2$, NHR, NR$_2$ groups are more radical-pushing than OH or OR groups and the differences in character between the groups here. While NH$_2$ will favor molecular rearrangement of the third kind, NHR and NR$_2$ will not, if Electrophiles are to be obtained.

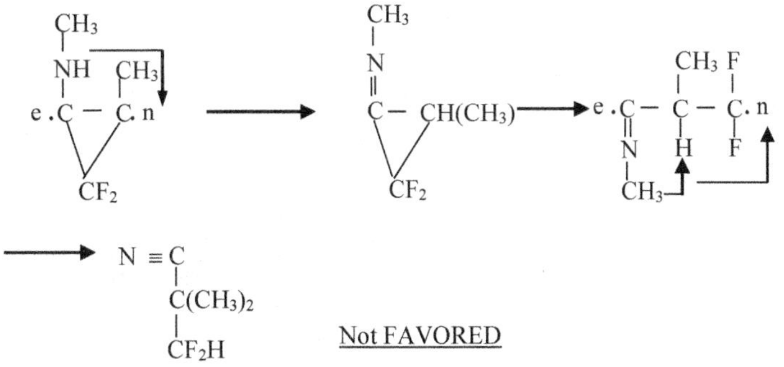

$$
\begin{array}{c}
\overset{\displaystyle CH_3}{\underset{}{\overset{|}{NH\ \ CH_3\rceil}}} \\
e\cdot C - C\cdot n \\
\diagdown\ \diagup \\
CF_2
\end{array}
\longrightarrow
\begin{array}{c}
\overset{\displaystyle CH_3}{\underset{}{\overset{|}{N}}} \\
\parallel \\
C - CH(CH_3) \\
\diagdown\ \diagup \\
CF_2
\end{array}
\longrightarrow
\begin{array}{c}
\overset{\displaystyle CH_3\ \ F}{e\cdot C - C - C\cdot n} \\
\ \ \parallel\ \ \ \ |\ \ \ \ | \\
\ \ N\ \ \ H\ \ \ F \\
\ \ | \\
\ \ CH_3
\end{array}
$$

$$
\longrightarrow
\begin{array}{c}
N \equiv C \\
|\\
C(CH_3)_2 \\
|\\
CF_2H
\end{array}
$$

         <u>Not FAVORED</u>   10.136

The case above is not favored, because the transfer species abstracted nucleo-free-radically will then become CHF$_2$.

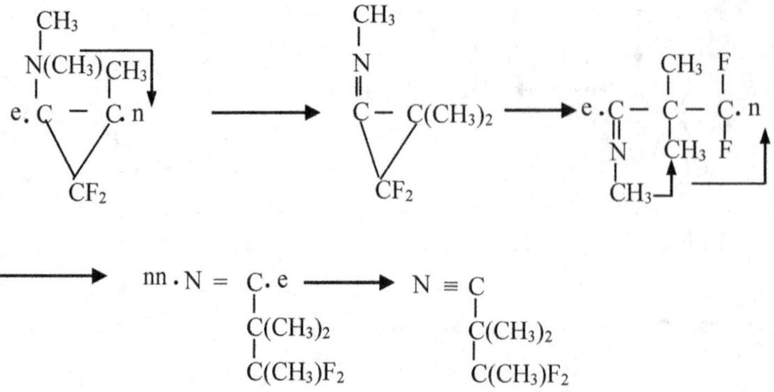

$$
\begin{array}{ccc}
\underset{\text{CF}_2}{\overset{\overset{\displaystyle\text{CH}_3}{|}}{\underset{|}{\overset{\text{N(CH}_3)\overline{\text{CH}_3}|}{\underset{e.\,C\;-\;C.\,n}{\triangle}}}}} & \longrightarrow & \underset{\text{CF}_2}{\overset{\overset{\displaystyle\text{CH}_3}{|}}{\underset{C\;-\;C(\text{CH}_3)_2}{\triangle}}} \;\;\longrightarrow\;\; e.\,C\;-\;\overset{\overset{\displaystyle\text{CH}_3}{|}}{\underset{\underset{\text{CH}_3}{|}}{C}}\;-\;\overset{\overset{\displaystyle\text{F}}{|}}{\underset{\text{F}}{C.\,n}}
\end{array}
$$

$$
\longrightarrow \quad nn\cdot N = \underset{\underset{\text{C(CH}_3)\text{F}_2}{|}}{\overset{\underset{\text{C(CH}_3)_2}{|}}{C.\,e}} \quad\longrightarrow\quad N \equiv \underset{\underset{\text{C(CH}_3)\text{F}_2}{|}}{\overset{\underset{\text{C(CH}_3)_2}{|}}{C}}
$$

<u>FAVORED</u>   10.137

It is said to be favored, because $-CF_2(CH_3) > -CH_3$ in radical-pushing capacity. From the analysis, it can be observed that the two transfer species involved in molecular rearrangement of the third kind, do not have to be the same. ***If the law of conservation of transfer of transfer species must be obeyed, it is the group carried in the last step that matters.*** Based on the different types of nitriles obtained, the following which has in the past been ascertained is therefore obvious.

$$
NR_2 \qquad > \qquad NHR \qquad > \qquad NH_2
$$

Order of radical-pushing capacity   10.138

Based on the activated states obtained when OH and OR are involved, that is, (I), of Equations 10.132 and 10.133, the following as has already been ascertained is also obvious.

$$
OH \qquad > \qquad OR
$$

Order of radical-pushing capacity   10.139

Now, when radical-pulling groups are involved radically, what is to be expected for for example COOR group, has already been shown in Equations 10.97a, 10.101a to 10.102b. Equation 10.97a is recalled below radically.

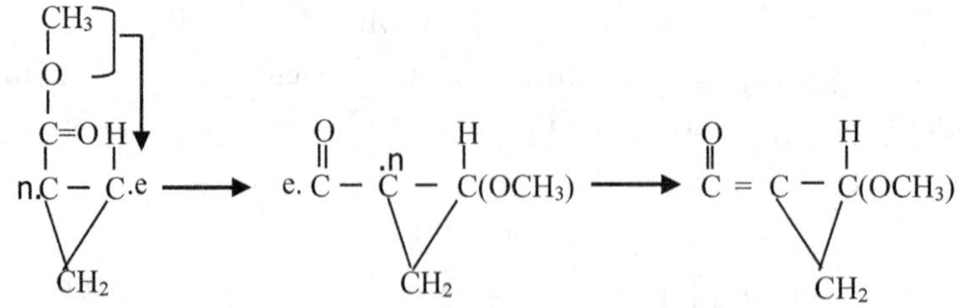

$$
\underset{\underset{\text{(I) (Less stable)}}{\text{CH}_2}}{\overset{\overset{\overset{\overset{\displaystyle\text{CH}_3}{|}}{\underset{\text{O}}{|}}}{\underset{\underset{n.C\;-\;C.e}{\triangle}}{C=O\;H}}}{\triangle}} \;\longrightarrow\; \underset{\underset{\text{(II)}}{\text{CH}_2}}{\overset{\overset{\displaystyle\text{O}}{\|}}{\underset{e.\,C\;-\;\overset{.n}{C}\;-\;C(OCH_3)}{\triangle}}} \;\longrightarrow\; \underset{\underset{\text{(III) (More stable)}}{\text{CH}_2}}{\overset{\overset{\displaystyle\text{O}}{\|}}{\underset{C\;=\;C\;-\;C(OCH_3)}{\triangle}}}
$$

<u>ELECTROPHILE</u>   10.140

Molecular rearrangement of the first kind is favored with the monomer still retaining its Electrophilic character. For the case of the monomer shown below from Equation 10.97b, the following are to be expected radically.

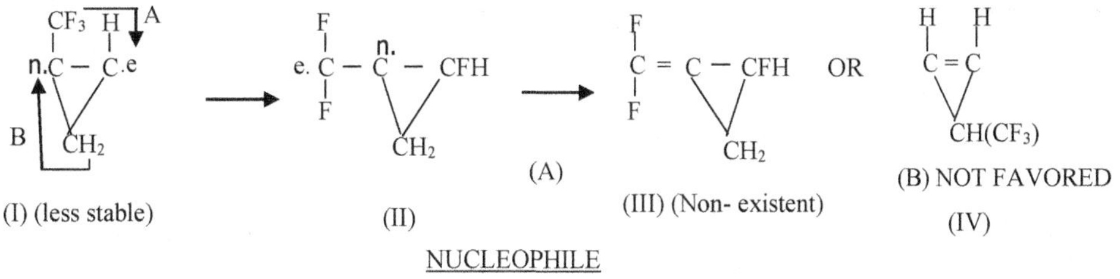

$$
\begin{array}{ccccc}
\underset{\text{(I) (less stable)}}{
\begin{array}{c}
\overset{CF_3\; H}{|\quad|}A\downarrow \\
n.C - C.e \\
\diagdown\diagup \\
B\;\;CH_2
\end{array}}
&\longrightarrow&
\underset{\text{(II)}}{
\begin{array}{c}
\overset{F}{|} \qquad \overset{n.}{} \\
e.\,C - C - CFH \\
\overset{|}{F}\;\diagdown\diagup \\
CH_2
\end{array}}
&\longrightarrow&
\underset{\text{(III) (Non- existent)}}{
\begin{array}{c}
\overset{F}{|} \\
C = C - CFH \\
\overset{|}{F}\;\diagdown\diagup \\
CH_2
\end{array}}
\;\;OR\;\;
\underset{\substack{\text{(B) NOT FAVORED}\\ \text{(IV)}}}{
\begin{array}{c}
\overset{H\;\; H}{|\;\;|} \\
C = C \\
\diagdown\diagup \\
CH(CF_3)
\end{array}}
\end{array}
$$

(A)

NUCLEOPHILE

$$10.141$$

It is (B) above that should be favored only radically, being a Nucleophile. But it is not favored because H is greater in radical-pushing capacity than CF_3. In the absence of molecular rearrangement, transfer species exist nucleo-free- and electro-free-radically. Even when paired initiators are involved in the absence of rearrangement, nothing can be done to polymerize this type of monomer. **Not all monomers with ACTIVATION CENTER(S) can be polymerized no matter what the operating conditions are. However, they can still be used advantageously or non-advantageously for specific applications.**

When CF_3 is replaced with C_2F_5, the following is expected. H has been replaced with CF_3.

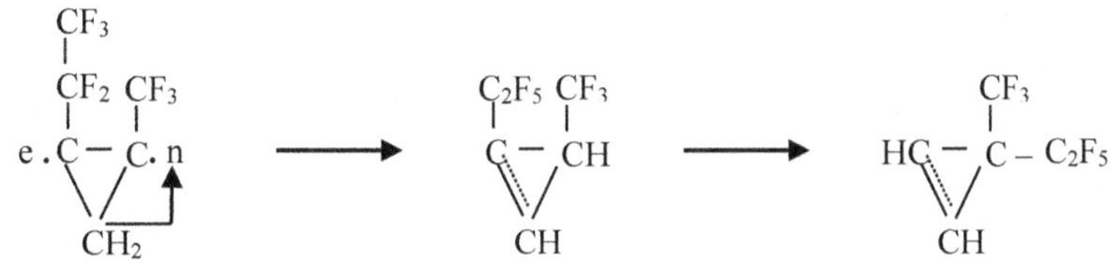

$$
\begin{array}{ccccc}
\begin{array}{c}
\overset{CF_3}{|} \\
\overset{CF_2\; CF_3}{|\quad|} \\
e\,.\,C - C\,.\,n \\
\diagdown\diagup \\
CH_2
\end{array}
&\longrightarrow&
\begin{array}{c}
\overset{C_2F_5\; CF_3}{|\quad\quad|} \\
C - CH \\
\diagdown\diagdown\diagup \\
CH
\end{array}
&\longrightarrow&
\begin{array}{c}
\overset{CF_3}{|} \\
HC - C - C_2F_5 \\
\diagdown\diagdown\diagup \\
CH
\end{array}
\end{array}
$$

$$10.142$$

No rearrangement of the first kind can take place here since H is more radical-pushing than CF_3. Shown below, is an example of a case where transfer is possible.

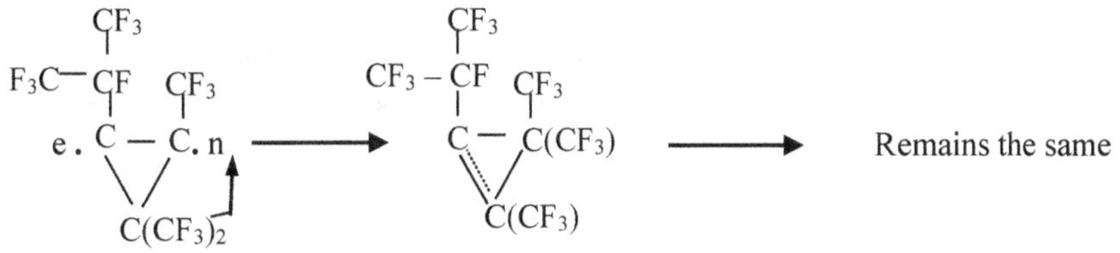

$$
\begin{array}{ccccc}
\begin{array}{c}
\overset{CF_3}{|} \\
F_3C - \overset{CF}{}\;\;CF_3 \\
\qquad\quad|\quad\;\; | \\
e\,.\,C - C\,.\,n \\
\diagdown\diagup \\
C(CF_3)_2
\end{array}
&\longrightarrow&
\begin{array}{c}
\overset{CF_3}{|} \\
CF_3 - CF\;\;CF_3 \\
\qquad\quad|\quad\;\; | \\
C - C(CF_3) \\
\diagdown\diagdown\diagup \\
C(CF_3)
\end{array}
&\longrightarrow&
\text{Remains the same}
\end{array}
$$

$$10.143$$

Rearrangement takes place to produce the same monomer. With no hydrogen atom in and outside the ring, the movement of CF_3 is still favored. In molecular rearrangement of the third kind, the two transfer species involved are of the first kind of the third type (Real) and transfer species of the first kind of the first type (Auxiliary). Chargedly, the monomer above cannot be polymerized as already indicated using Equation 10.97b. However, radically, it can be polymerized. When the $-CF(CF_3)_2$ group is replaced with $-C(CF_3)_2$, the situation remains the same. For the case below, the situation remains the same.

$$
\begin{array}{ccc}
\begin{array}{c}
\overset{CF_3}{|} \\
\overset{C(CF_3)_2\; H}{||\qquad\;|} \\
n\,.\,C - C\,.e \\
\diagdown\diagup \\
CH_2
\end{array}
&\longrightarrow&
\begin{array}{c}
\overset{CF_3}{|} \\
\overset{CCF_3\quad H}{||\qquad\;|} \\
C - C - CF_3 \\
\diagdown\diagup \\
CH_2
\end{array}
\end{array}
$$

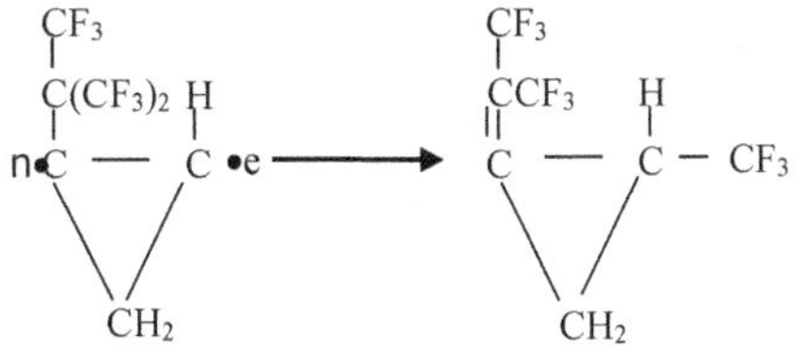

Impossible transfer

$$10.144$$

The transfer above cannot take place. In general, one can observe how strongly nucleophilic cyclopropenes are and the advantages offered in considering this unique family of monomers.

10.2.2 Resonance Stabilization in cyclopropenes

10.2.2.1 Vinyl cyclopropenes:

There is need to consider the effect of resonance stabilization in cyclopropenes. For this purpose, one will begin with vinyl cyclopropene.

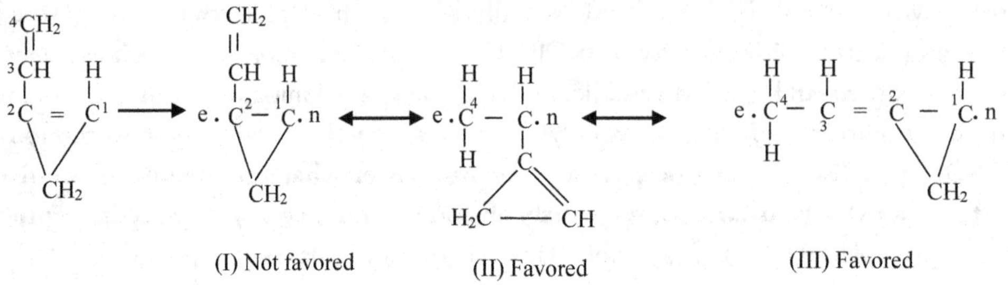

$$10.145$$

(I) Not favored   (II) Favored   (III) Favored

Since it is non-symmetric, the radicals shown on the active centers, which are readily obvious from (II) and (III), are fixed. It should be noted that the $>CH_2$ group is partly or fully resonance stabilized. Whether it is partly or fully resonance stabilized will shortly be shown. The mono-forms [(II) and (III)] above are resonance stabilized, with them showing nucleophilic tendency particularly after replacing the remaining H atom in position #3 with a radical-pushing or pulling substituent group, the group to be shielded. The resonance stabilization group provider is the ring which is less-radical-pushing than the ethene group, because the CH_2 group in the ring is resonance stabilized. The ring is more nucleophilic than the alkene. Since $>CH_2$ group is partly or fully shielded, hence the monomer above can be said not to have a transfer species. However, the monomer is unstable, because it will most of the time be in Equilibrium state of existence.

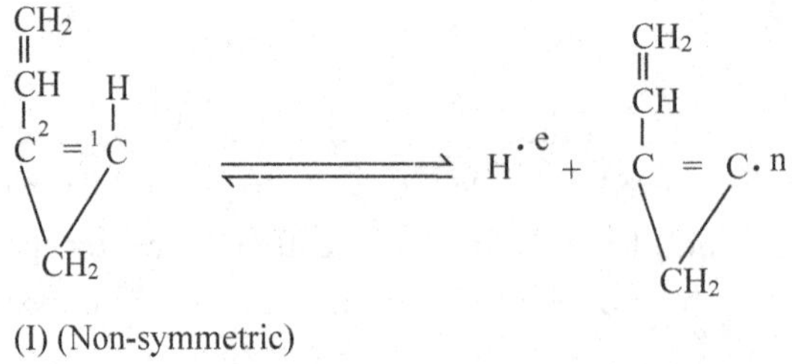

(I) (Non-symmetric)

$$10.146$$

This is like the case of alkene group resonance stabilized by an alkyne group, wherein only one H atom is still loosely bonded when not passively suppressed. Like the alkyne group which is less radical pushing than H unlike the ethene group, the same applies to cyclopropene group here. Since it is still unstable, as shown below the followings take place when the vinyl cyclopropene is decomposed.

Stage 1:

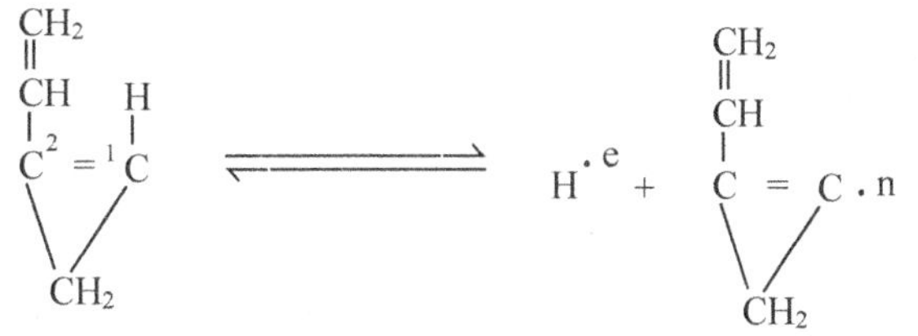

$$e \bullet H \ + \ (B) \longrightarrow H_2C = CH - C \equiv C - \overset{\overset{\displaystyle H}{|}}{\underset{\underset{\displaystyle H}{|}}{C}} - H \ + \ Heat$$

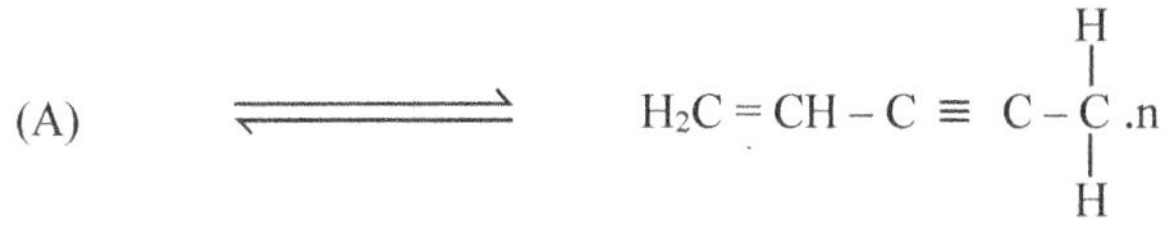

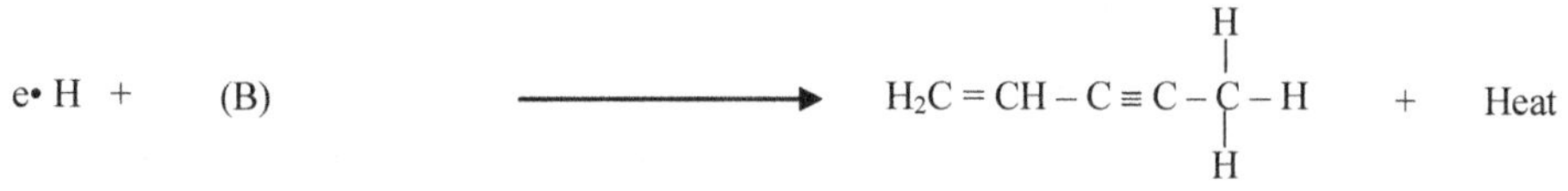

(C) Vinyl methyl acetylene   10.147a

Overall equation: Vinyl cyclopropene $\xrightarrow{>400^0 C}$ Vinyl methyl acetylene   10.147b

The methyl acetylene group can be observed to be a group far more radical-pushing than H, and the resonance stabilization provider. Note that the product above cannot be obtained by molecular rearrangement, but by decomposition via Equilibrium mechanism. Therefore, the methyl acetylenes obtained does mean that it is more nucleophilic than cyclopropene. While Molecular rearrangement phenomena involve movement from less nucleophilic to more nucleophilic, Electroradicalization and Activated/Equilibrium States of existence rearrangement phenomena reverse the movement. The rearrangement above is via Electroradicalization mechanism as already explained.

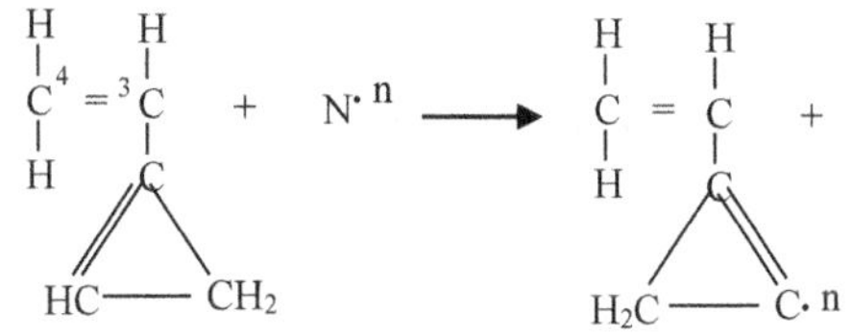

(II)   10.148a

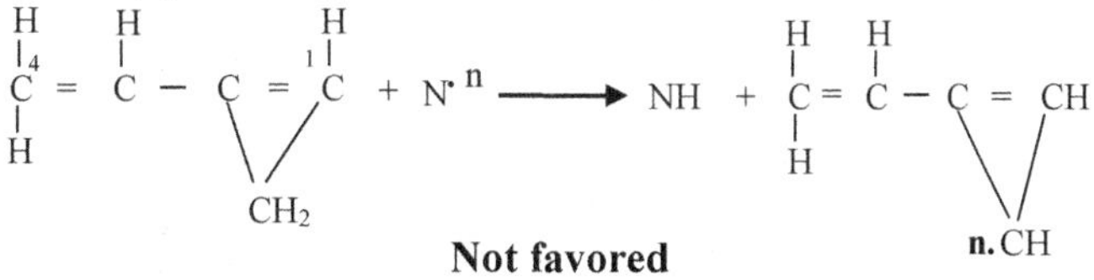

Not favored   10.148b

This is only possible free-radically and never chargedly, noting that one H atom may need to be replaced if not suppressed. The H atom abstracted in the first equation is not a transfer species. When suppressed, the monomer will favor both free-radical routes as will shortly become obvious.

All the mono-forms after replacement are nucleophiles, since only the (positively charged for four or five membered rings and above) electro-free-radical route can be favored. Thus, it can be observed that the group shown below (I) is a radical-pushing resonance stabilization group of very great capacity.

59

$$RHC = CH- \quad \geq \quad RHC = C=CH - \quad > \quad RC \equiv C - \quad > \quad \overset{|}{\underset{(I)}{RC \overset{C}{\diagdown} CH_2}} \quad \geq \quad \mathbf{R} \qquad 10.149$$

Order of Radical-pushing capacities of Resonance stabilization groups

Note that the capacity of the ringed group has been greatly reduced, because the CH_2 group in the ring seems to be fully resonance stabilized. Hence, the need for (I) to carry R does arise unlike the order shown in Equations 10.96b and 10.119c where they are not resonance stabilized. This order above is not the order of their resonance stabilization capabilities or nucleophilic capacities as a monomer.

When two external resonance stabilization groups are symmetrically placed on cyclopropene, the followings are obtained.

$$10.150$$

[NOT FAVORED-Charges cannot be removed from their carriers as has been shown] 10.151

From the above, only the radical state is favored, for which both routes are favored, since the internally located $>CH_2$ group is fully resonance stabilized. Since the monomer is symmetric, the following relationship can be obtained.

60

$$(CH_2 = CH-) \quad > \quad [-H] \quad > \quad (HC \equiv C-) \quad > \quad \left(\begin{array}{c} -C \\ H_2C - C \\ | \\ CH \\ || \\ CH_2 \end{array} \right)$$

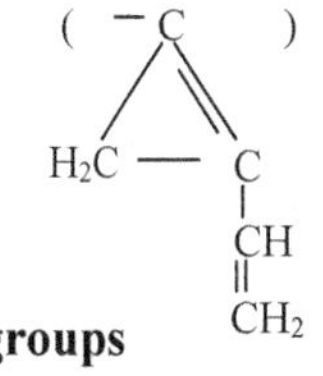

Order of Radical-pushing capacity of resonance stabilization groups　　　10.152

Another way by which transfer species of the first kind of the ninth type can be shielded for cycloalkenes, is to provide double resonance stabilization groups on both sides. This will be particularly useful for larger sized rings, not for the three-membered one as will shortly become obvious.

When the H atom in (I) of Equation 10.146 is replaced with CH_3 groups, the followings are obtained.

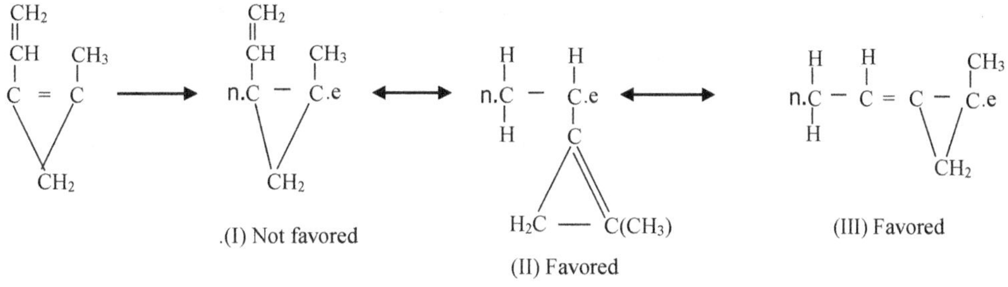

.(I) Not favored　　　(II) Favored　　　(III) Favored

NUCLEOPHILES　　　10.153

When it is said not to be favored, such mono-forms never exist at any point in time, as has always been thought to be the case. Only two mono-forms can exist for Diene, three for trienes and so on. Transfer species is provided by CH_3 group free-radically. Nevertheless, it is believed that the resonance-stabilized forms undergo molecular rearrangement by decomposition mechanism as follows.

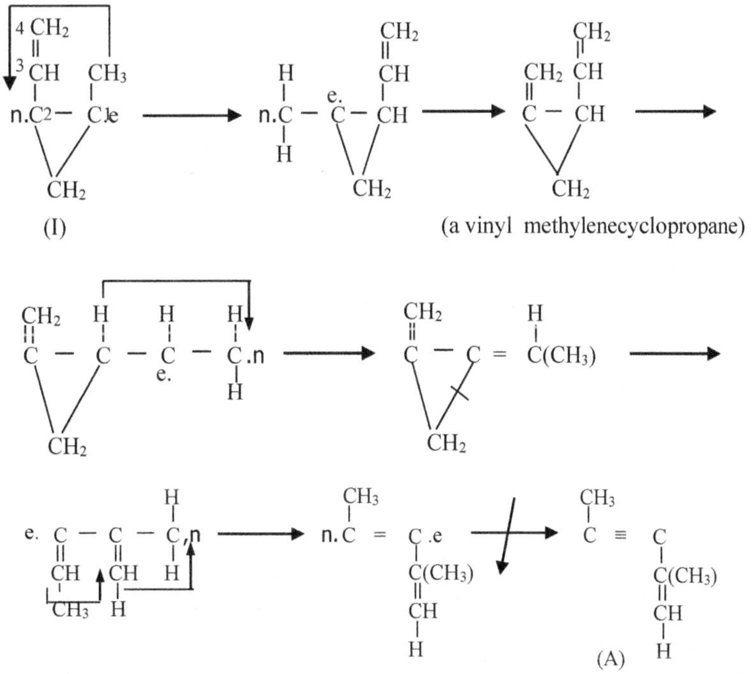

(I)　　　(a vinyl methylenecyclopropane)

(A)　　　10.154

(I) above is not one of the mono-forms. It has only been used for exploratory purposes

61

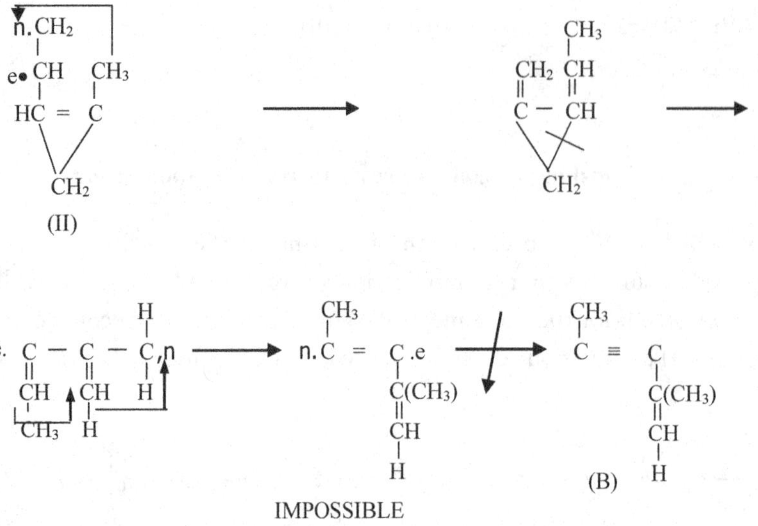

(II)

IMPOSSIBLE   (B)   10.155

(III)

IMPOSSIBLE   (C)   10.156

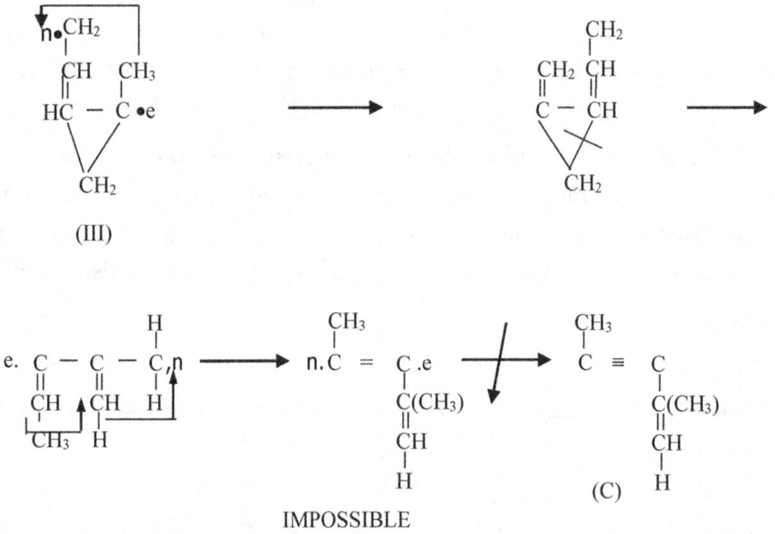

Based on the mono-forms of Equation 10.153 [(II) and (III)], no molecular rearrangement of the third kind can be favored to produce the same acetylenic monomer (A), (B), and (C), by application of the laws and concepts which have so far been developed. It is important to note the transfer species involved during the two different types of molecular rearrangements. The acetylenic monomer could not be obtained, because $-CH_3$ group is of greater radical-pushing capacity than $-C(CH_3) = CH_2$ group. The laws of Conservation of transfer of transfer species cannot be broken. It is obvious that $H_2C =$ is less radical-pushing than $= CH(CH_3)$ group as already stated into law. Secondly, the point of scission is that with the highest radical-pushing potential difference. It is also important to note that, in general, the 1, 4- addition mono-form is the most stable, in view of the reduced number of steps involved during rearrangements. ***Therefore, for this three-membered ring, one believes that the transfer species involved above should not be from CH_2 group in the ring, but from CH_3 group as used all above. The CH_2 group in the three-membered ring is fully resonance stabilized, i.e., well shielded. For larger sized rings, the situation is slightly different as will shortly become obvious.***

The case shown below can undergo molecular rearrangements.

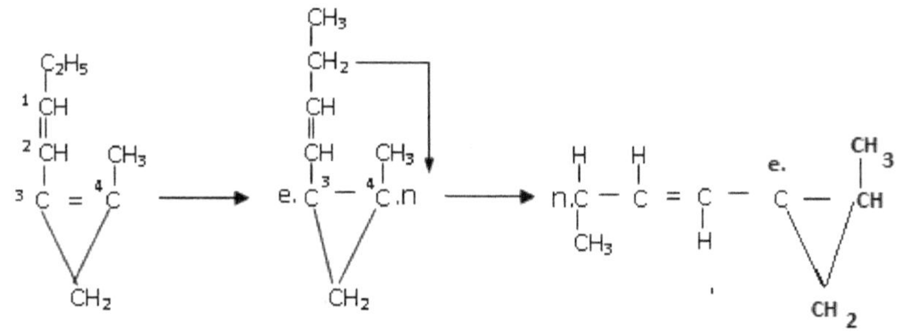

10.157

(I)

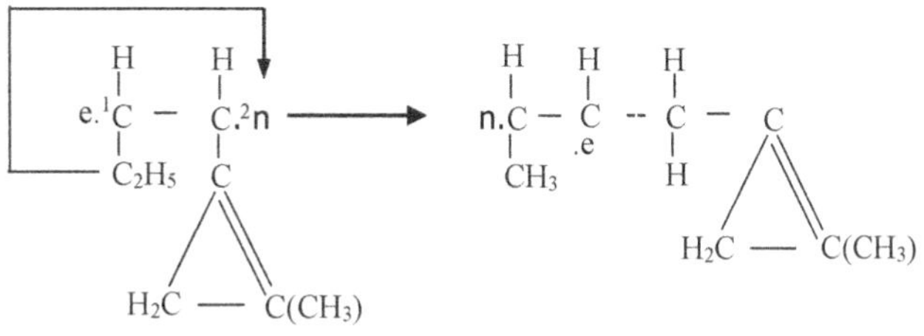

Right Activated state

(II)

Favored

10.158

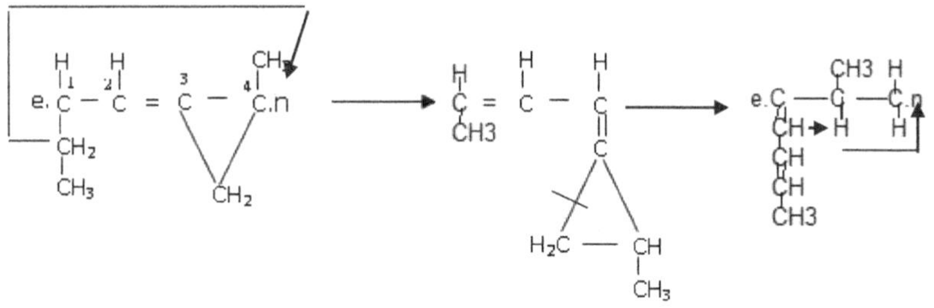

(II)

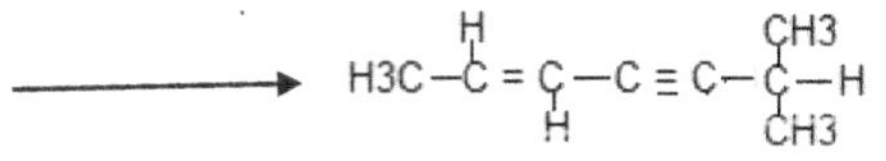

A vinyl acetylene

(III)

10.159

Based on Equation 10.149, the rearrangement above is fully favored. However, there is limit to the size of C_2H_5 carried on the alkene side. The maximum size of R is C_4H_9. For molecular rearrange-ment to be favored the R group on the alkene side must be greater than the R group on the ring side. Only one mono-form was used above. The first case above is not a mono-form. However, all of them undergo molecular rearrangement of the third kind as reflected in the last equation to give vinyl acetylene, (III). Based on the fact that the 3,4 center is the most nucleophilc center which can never be activated, clearly indicates that the following is valid.

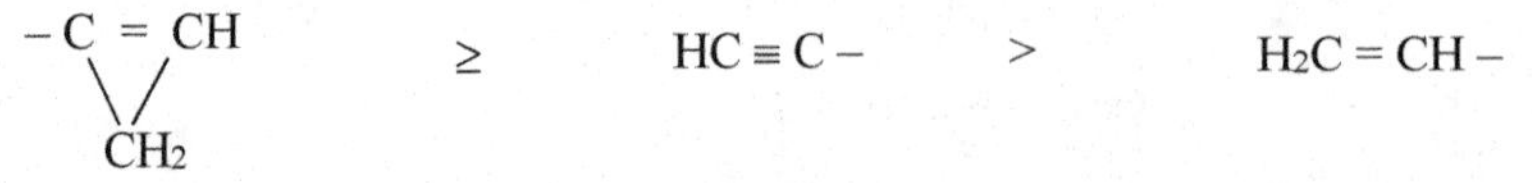

$$-C = CH \qquad \geq \qquad HC \equiv C - \qquad > \qquad H_2C = CH -$$
$$\backslash \ /$$
$$CH_2$$

Order of Resonance Stabilization capabilities and Nucleophilicity of their monomers   10.160

Unlike 1,3-pentadiene and propene, the corresponding 1,4- mono-forms of vinyl cyclo-propenes undergo molecular rearrangement to give different monomer. Since the three-membered ring alone cannot be opened, it rearranges back to its original self. When one compares the order above with those of Equation 10.149 and the like, one can observe and begin to understand the great wonderful logical world of CHEMISTRY.

Now, consider the case of a cyclopropane ring fused to a diene with radical-pushing group(s) as shown below.

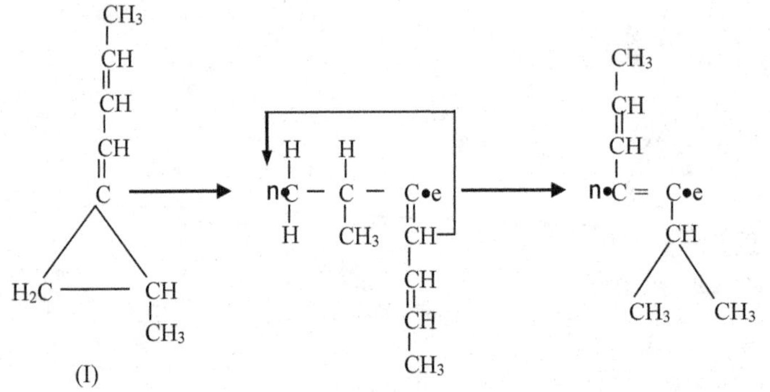

(I)   10.161

When the ring is opened as shown above, molecular rearrangement of the third kind takes place to give what is shown above. When the above is decomposed by Equilibrium mechanism, the same product as above is obtained.

When **_the radicals are not well placed,_** the followings are obtained.

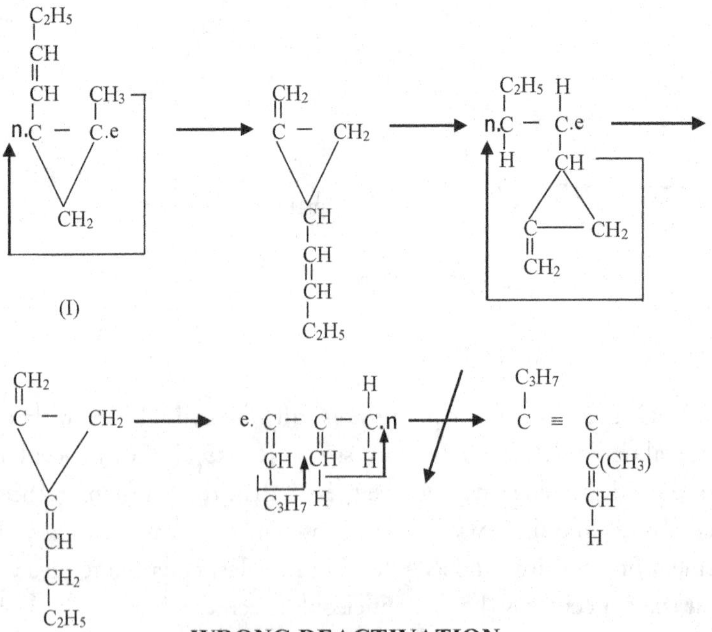

(I)

WRONG DEACTIVATION   10.162

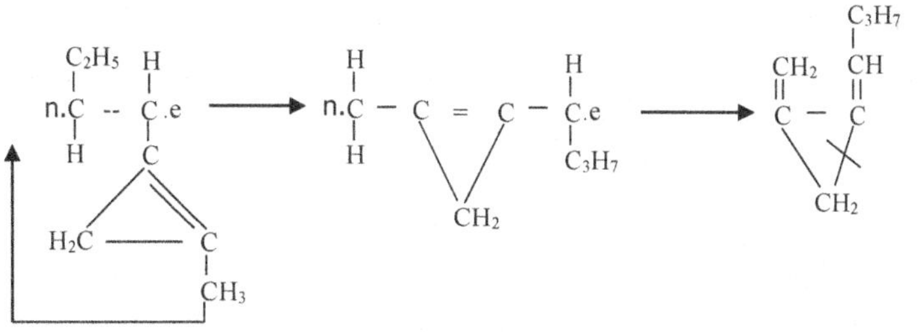

$$\text{(II)}$$

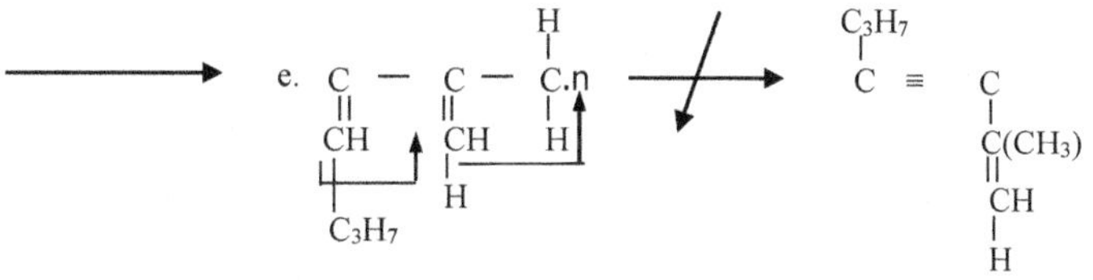

WRONG DEACTIVATION   10.163

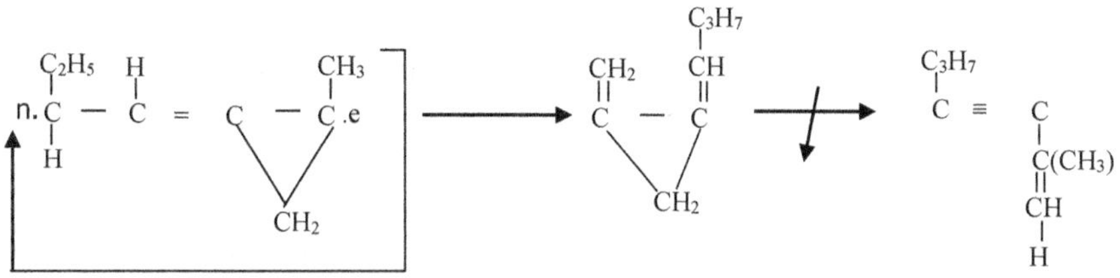

WRONG DEACTIVATION   10.164

Thus, it can be observed that the wrong activated states of the monomers are as shown above, not like those shown in Equations 10.157 and 159. Even if the CH_3 group is replaced with H, the activated state as in Equations 10.157 to 10.159 remains the same.

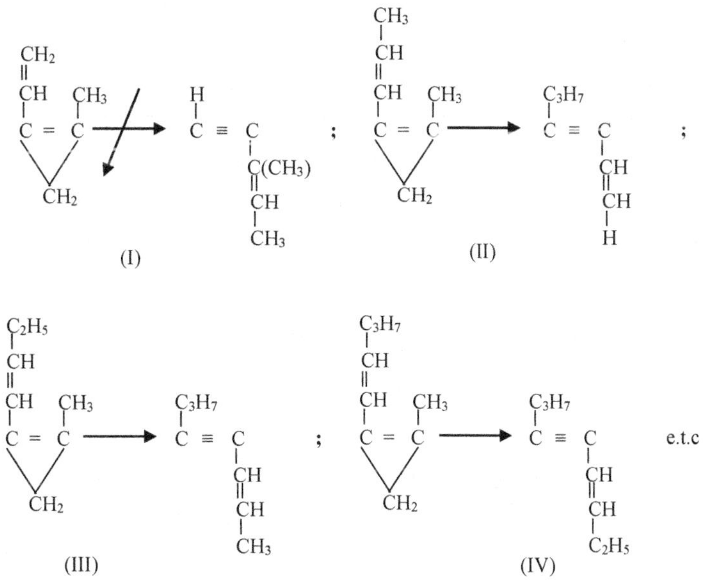

$$\text{(I)} \qquad\qquad \text{(II)}$$

$$\text{(III)} \qquad\qquad \text{(IV)} \qquad\qquad 10.165$$

From analysis so far above, the followings are beginning to be inherently clear –

65

i) That the presence of $H_2C=$ group cumulatively placed to the ring, does not make the three membered rings to be so strained not to exist. It exists with point of scission under Decomposition mechanism and Equilibrium mechanism.

ii) That the consideration of the three-membered ring as has been done so far was necessitated by the fact that, with it, so much was going to be revealed about the way Nature operates, since as shown below, there are marked differences between three-membered rings and all the other members, wherein it is only in the three-membered ring, one carbon center is connected to the two centers carrying double bond.

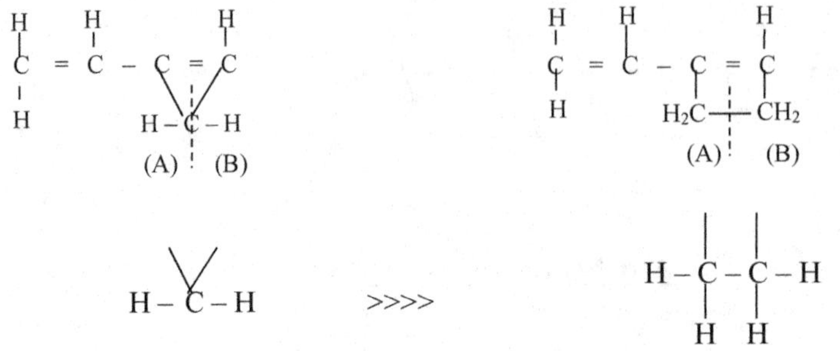

PUSHING CAPACITY FROM *ONE* CENTER TO *TWO* CENTERS   PUSHING CAPACITY FROM *TWO* CENTERS TO *TWO* CENTERS

10.166

iii) That as the size of the ring increases from four upwards, the influence of resonance stabilization diminishes. In Equation 10.166 above, notice the dotted line of demarcation between the resonance stabilized group in the ring marked as (A)-shielded and the non-resonance stabilized region of the ring marked as (B). For the odd-membered rings like the three-membered ring, the C center seems to be split into half, while for even-membered rings, it is the bond in the middle that is the splitting point depending on what the carbon centers are carrying. If it is all H, it will be in the middle.

iv) That in view of the fact that the C center cannot be split into two parts when odd, and the fact that there is no point of scission in the three membered ring unlike the others, hence the only CH_2 in the three-membered ring is fully resonance stabilized and not partly. For the other odd –membered rings, splitting takes place just next to a bond connected the Central C atom.

v) Thus, while the resonance stabilization capacity decreases with increasing size of the ring, the radical pushing capacity of the closed groups shown in Equation 10.166 increases with increasing size of the rings as shown below only for four-membered rings and above.

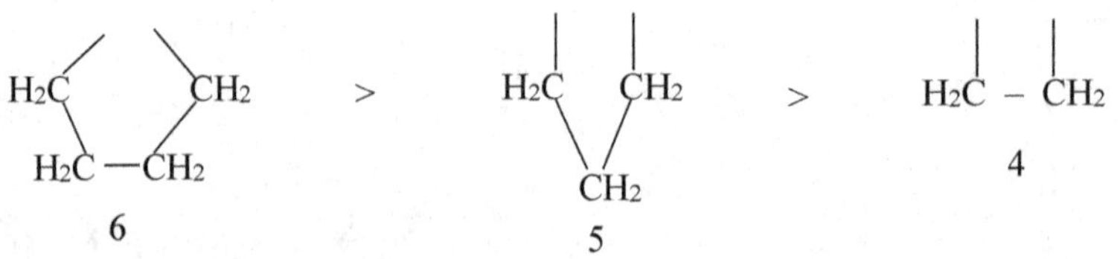

Order of Radical-pushing capacities of Closed groups   10.167a

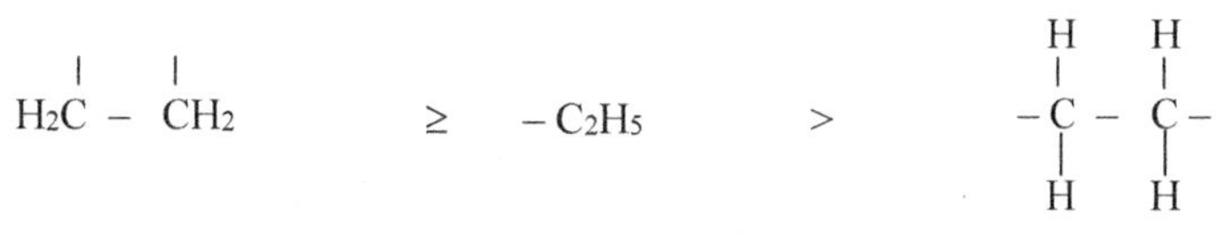

$$H_2C - CH_2 \qquad \geq \qquad - C_2H_5 \qquad > \qquad -\overset{\overset{\displaystyle H}{|}}{\underset{\underset{\displaystyle H}{|}}{C}} - \overset{\overset{\displaystyle H}{|}}{\underset{\underset{\displaystyle H}{|}}{C}} -$$

CLOSED   OPENED

Order of Radical-pushing capacities of Opened and Closed groups 10.167b

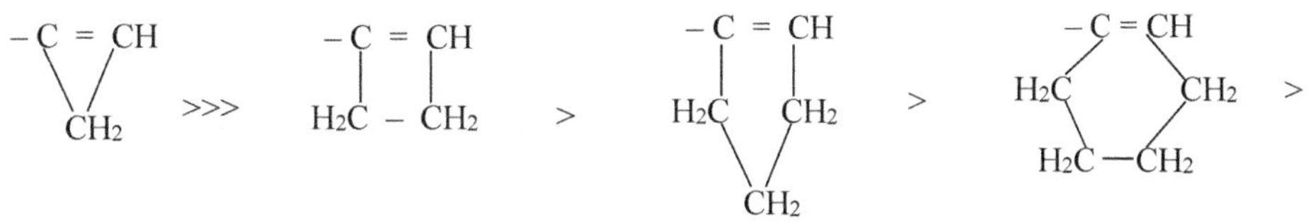

Order of Resonance stabilization capabilities of Resonance stabilization groups 10.168

vi) That for three-membered ringed vinyl cyclopropene and members that carry radical-pushing groups on the double bond of the ring, molecular rearrangement to give another compound can take place only when the R group carried by the alkene is greater in capacity than the R group on the ring with however some limitations. This will almost apply to larger sized rings. Depending on the capacities of the groups carried, it will take place for some of them. For the first member of vinyl cyclobutenes, molecular rearrange-ment can take place as shown below since the capacity of H in the ring is not greater than the capacity of H outside the ring unlike the case of cyclopropene.

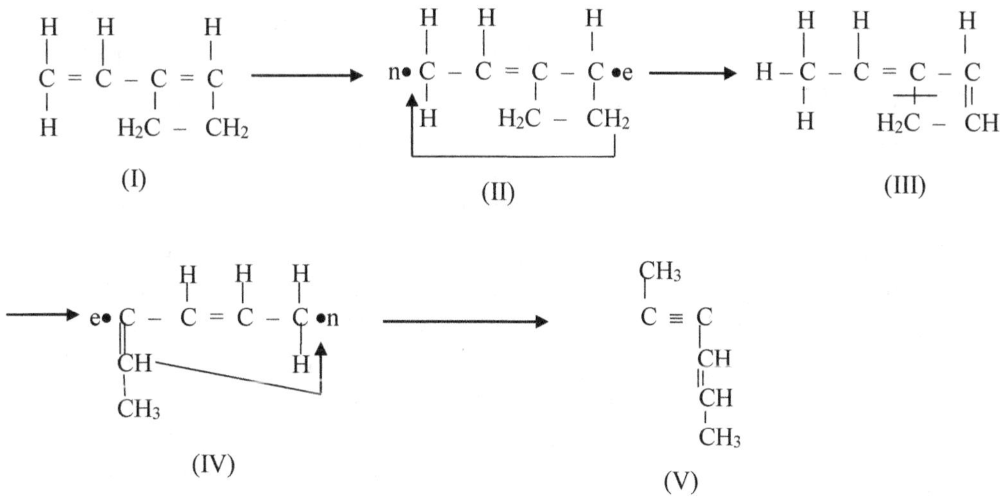

10.169

As a Nucleophile, no matter what the vinyl center is carrying, the ring will always be the provider of resonance stabilization.

Nevertheless, the exploratory journey on cyclopropenes continues. When the CH_3 group on cyclopropene is replaced with C_2H_5, the followings cannot take place.

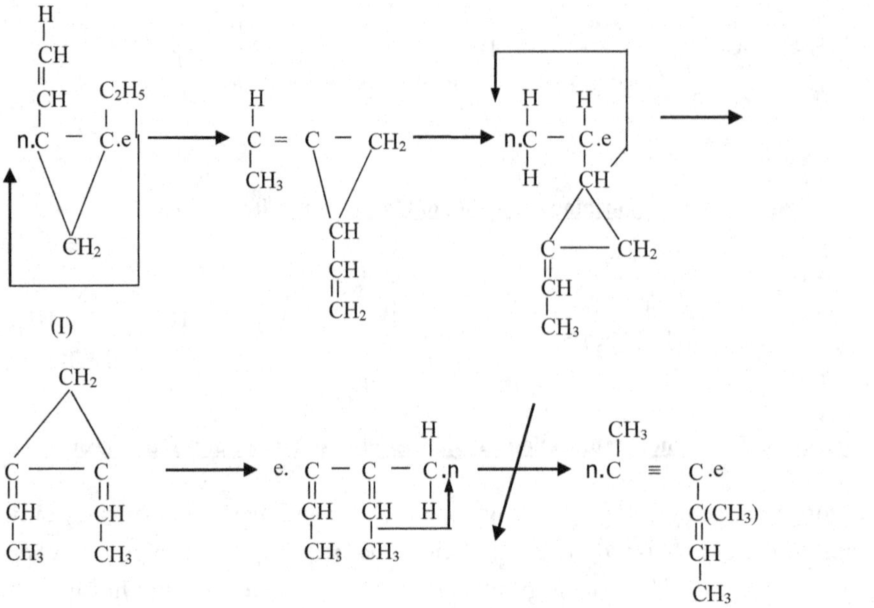

(Not favoured) (I)a

Impossible activated state

IMPOSSIBLE REACTIONS 10.170

All these possible and impossible reactions are being shown using complex molecular compounds which up to date have never been known to exist, because we including the author do not know anything; for the more we know, the more we begin to realize that we know nothing. CHEMISTRY IS A COMPLEX ORDERED NETWORK SYSTEM which requires the use of EYE OF THE NEEDLE. CHEMISTRY is very LOGICAL IN CHARACTER. CHEMISTRY does not discriminate, but integrates. CHEMISTRY has the ability to accomplish all things in HUMANITY all depending ON THE OPERATING CONDITIONS. CHEMISTRY IS THE FOUNDATION OF LIFE AND TRANSITION (known in present-day world as DEATH). For indeed, based on CHEMISTRY, there is NOTHING CALLED DEATH, but death of the physical body. It is a cycle where there is nothing called WASTE. The SO-CALLED WASTES UNLIKE HOMO-SAPIENCE MUST be recycled to give another product, whether useful or not, all depending on the OPERATING CONDITIONS, noting that we live in a complex world where the GOOD side of life and the BAD side must co-exist for any to exist. These have been stated into laws.

Part- hydrocarbon radical-pushing groups are very unique in the sense that all of them are more radical-pushing than the hydrocarbon ones. This has been a unique observation. Interestingly enough, all of them contain H or R group. Such groups include, the amines-NH_2, NHR, NR_2, OH, OR, SH, SR, and so on. When NH_2 is involved, the followings are obtained only radically.

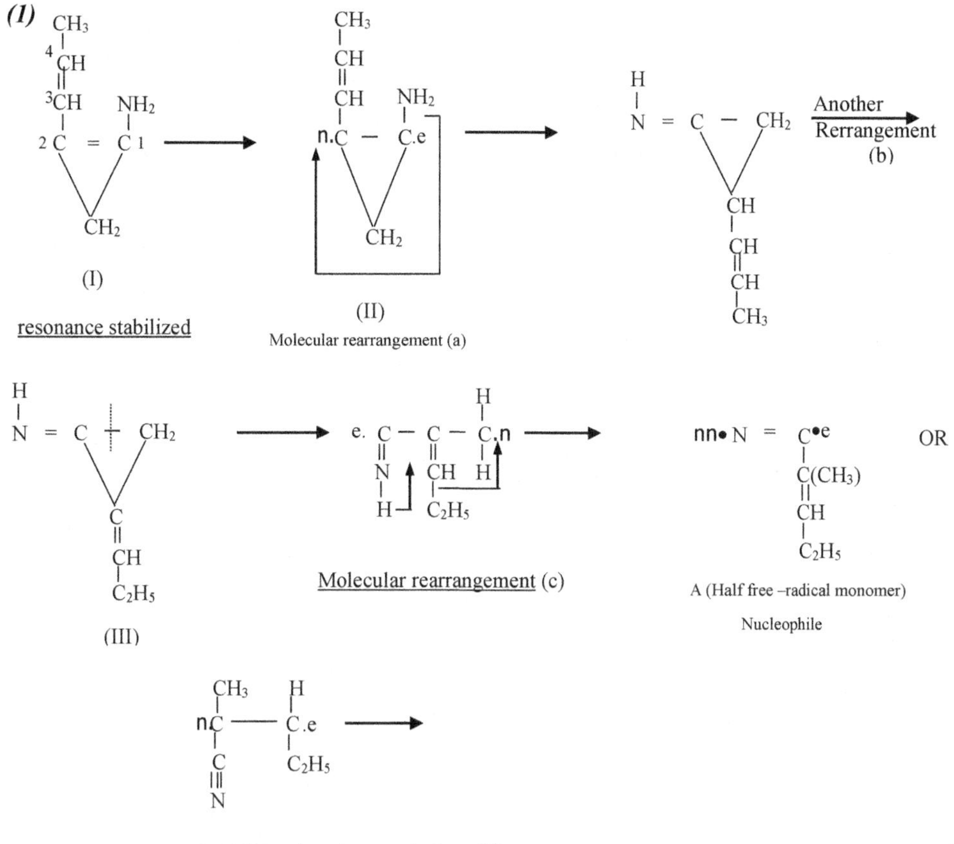

(1)

$$CH_3 - \overset{4}{C}H = \overset{3}{C}H - \overset{2}{C} = \overset{1}{C} \quad NH_2 \;/\; CH_2$$

(I)
resonance stabilized

$\longrightarrow$

$$CH_3 - CH = CH - n.C - C.e \quad NH_2 \;/\; CH_2$$

(II)
Molecular rearrangement (a)

$\longrightarrow$

$$\overset{H}{\underset{}{N}} = C - CH_2 \;/\; CH \quad CH = CH - CH_3$$

Another
Rerrangement
(b)

$$\overset{H}{N} = C \,|\, CH_2 \;/\; C = CH - C_2H_5$$

(III)

$\longrightarrow$

$$e. \, C - C - C, n$$
$$\overset{\shortparallel}{N} \quad CH \quad H$$
$$\underset{H}{|} \quad C_2H_5$$

Molecular rearrangement (c)

$\longrightarrow$

$$nn\bullet N = C^{\bullet e}$$
$$C(CH_3)$$
$$CH$$
$$C_2H_5$$

A (Half free –radical monomer)
Nucleophile

OR

$$n\overset{CH_3}{\underset{C}{\overset{|}{C}}} - \overset{H}{\underset{C_2H_5}{\overset{|}{C}}.e$$
$$\overset{\shortmid}{\underset{N}{C}}$$

$\longrightarrow$

B   (Full free charged monomer) Electrophile

10.171a

Note the point of scission in the ring, where a double bond is conjugatedly placed to C – C single bond in the ring. If the double bond was in the ring or conjugatedly placed to a C – N single bond in the ring, then point(s) of scission will exist in the ring as shown below. Recall also that the radical-pushing capacity of H–N= group is greater than that of H_2C= group. Hence, the point of scission is as indicated above.

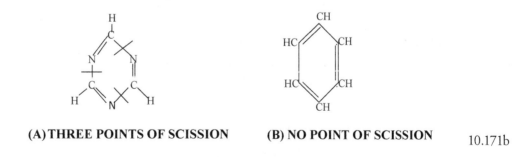

(A) THREE POINTS OF SCISSION   **(B) NO POINT OF SCISSION**   10.171b

Examples of cases like this will be seen as we continue downstream based on the use of universal data.

(2)

10.172

Just like above, the scission here is in order. Despite the fact that the wrong center was activated in **(1)**, the same acrylonitrile type of monomer was obtained in **(1)** and **(2)**. It is said to be wrong because it is not one of the mono-forms when resonance stabilized. The original monomer (I), a Female, molecularly rearranged to give a Male. This is incomprehensible though not new from current developments in NEW FRONTIERS. The same as above will apply to larger sized rings.

FEMALE   MALE   10.173

The acrylonitrile obtained will not favor its natural route- the nucleo-free-radical route. It will only favor the natural route chargedly when the CN group is trans-placed to C_2H_5 group above using negatively charged paired initiator. ***Despite all the above, it has been an exploration, if based on the law of conservation of transfer of transfer species, the last step of molecular rearrangement of the third kind is not favored.***

One shocking revelation as has been said before was finding out that NH_2 and OH types of groups which are anionic in character are radical-pushing groups of very great capacity instead of being radical-pulling groups, unlike Cl, $OCOCH_3$, OSO_3H, etc. types of groups. Based on what has been observed just above and based on what has been seen in Equations 10.96b, 10.119c, 10.132 to 10.137 for OH and NH_2, there is no doubt that the followings are valid.

$$= NH \quad \gg \quad NH_2 \quad > \quad OH \quad > \quad H_2C \quad > \quad H_2C= \quad \gg \quad H_3C\text{- (Any R)}$$

Order of Radical-pushing capacity of groups.

10.174

Hence, the transfers of H from NH_2, OH groups, as transfer species shown in those equations are valid. The reason why these orders and concern raised above are in order, is because of the marked difference between the Hydrocarbon families and the Non-hydrocarbon families. The reactions above (Equations 10.172 and 10.173) will also not be favored by NHR and NR_2 types of groups. With OH group, no rearrangement is possible. If any exists, a carbene is obtained.

With radical-pulling groups replacing CH_3, the ring may not be opened, after undergoing molecular rearrangement of the first kind. The opening of the ring will depend on the capacity of for example =C=O group relative to other groups. Based on the considerations so far, one should know what to expect with the following groups.

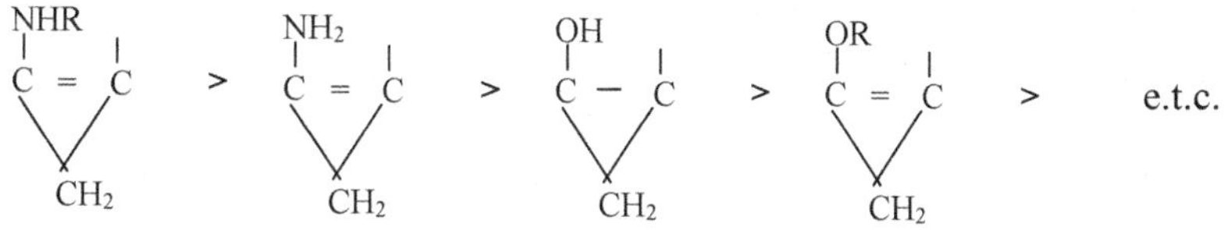

$$
\begin{array}{ccccccccc}
\underset{CH_2}{\overset{\overset{\displaystyle NHR}{|}}{C}} = C & > & \underset{CH_2}{\overset{\overset{\displaystyle NH_2}{|}}{C}} = C & > & \underset{CH_2}{\overset{\overset{\displaystyle OH}{|}}{C}} - C & > & \underset{CH_2}{\overset{\overset{\displaystyle OR}{|}}{C}} = C & > & e.t.c.
\end{array}
$$

10.175

Order of Radical-pushing capacity of Resonance Stabilization groups

With radical-pulling groups such as Cl, COOR, $CONH_2$, OCOR, in place of pushing NH_2, OH types of groups, the followings should be expected using Cl which has no transfer species.

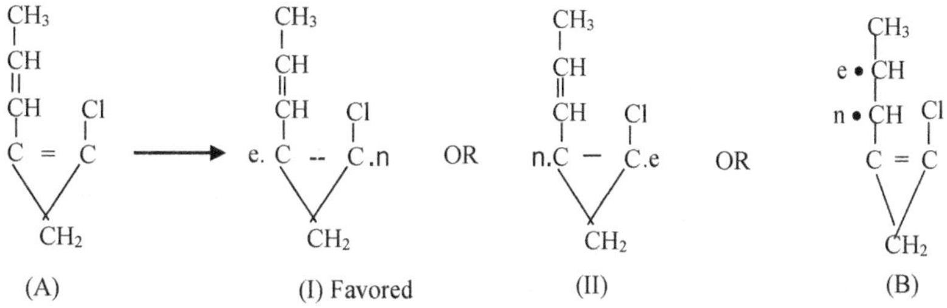

$$\text{(A)} \longrightarrow \text{(I) Favored} \quad OR \quad \text{(II)} \quad OR \quad \text{(B)}$$

10.176

It can be misleading to use the wrong activation center. (I) is the true activated state. That is the less nucleophilic center. With (I), rearrangement is favored to give a vinyl acetylene, since the law of Conservation of transfer of transfer species is not broken. For the other groups, with transfer species, one knows what to expect. While for COR, there is no rearrangement, with the others, there may be rearrangement with no favored opening of the ring.

With radical-pushing groups of lesser capacity than H, such as CF_3, C_2F_5, etc., the situation is different as we already know. When CF_3 is put in place of Cl above, if (I) above is the true activated state, the followings are to be expected.

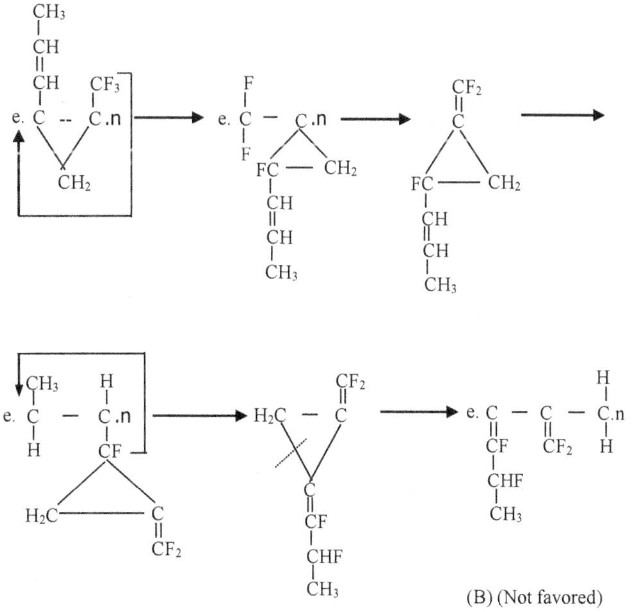

(B) (Not favored)

10.177

The reaction above is not favored, because the transfer species used is wrong. H should have been the transfer species which cannot be transferred because it is more radical-pushing than CF_3 group. One can see the very great need of establishing the order of these groups, knowing what they are, why they are what they are, how they operate, and so much more. However, notice how the fictitious ring above has been scissioned. The point of scission above is in order, since $F_2C=$ is far less radical pushing than H. However, whatever the size of CF_3 group, no rearrangement is possible.

$$\text{(II) Nucleophile} \qquad \text{No molecular rearrangement} \qquad \qquad 10.178$$

No transfer species can be transferred for molecular rearrangement and no route is favored free-radically. "Assuming" that it could be activated chargedly, no route will be favored, noting that CF_3 is not a radical-pulling group, but a "charge"-pulling group. Note that, while charges cannot be removed from their carriers, leaving their carriers behind, they can however be pushed or pulled from their carriers when about to be created. **Therefore, we have charge-pulling or pushing groups just as radical-pulling and pushing groups, as one has been using so far.** What make centers to carry charges are the types of initiators present in its environment. It is the initiator that makes the radical-pushing or pulling group push or pull radically. When one radical is pushed, it is radically; when two radicals are pushed, it is chargedly. Like the case above fully recalled below, it can partly be activated chargedly, and fully radically. In the presence of a charge-like initiator, only the alkene center can be activated. If the alkene center was acetylene center, nothing happens, i.e., no activation.

$$\underline{\text{Activated state}} \qquad \qquad \underline{\text{Activated state}}$$

$$\text{(Favors no free- routes)}$$

$$10.179$$

$$\underline{\text{Possible activated state}} \qquad \qquad \underline{\text{Possible activated state}}$$

$$\text{(Favors no resonance stabilization)} \qquad \qquad 10.180$$

The charges carried by the first center in the charged case is in order, but it's being resonance stabilized chargedly is impossible. The 1, 4-mono-form shown above chargedly does not exist. Even then, for the two cases above, the ring is the first center to be activated and never the alkene as used above. These are some of the forbidden things, all of which we do in our world to achieve goals! *With THE ALMIGHTY INFINITE GOD, ALL THINGS THAT WORK ACCORDING TO HIS LAWS- THE LAWS OF NATURE ARE POSSIBLE, but with HUMANS all things are possible.*

From all considerations so far, the following relationship is valid.

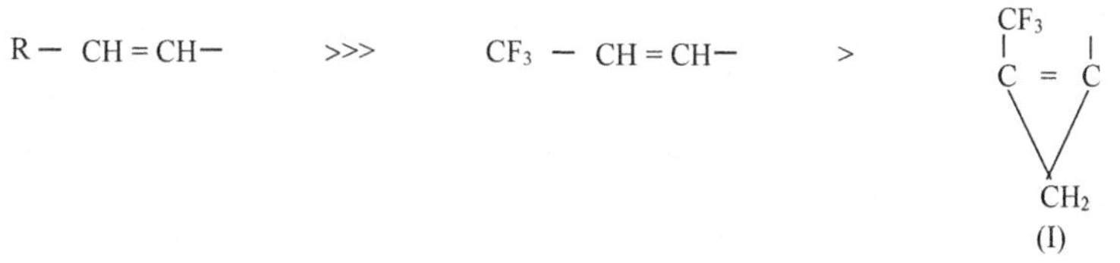

$$R - CH=CH- \quad >>> \quad CF_3 - CH=CH- \quad > \quad (I)$$

Order of Radical-pushing capacity of Resonance stabilization groups　　　10.181

Regardless the size of CF_3, C_2F_5 and higher groups, the followings can never take place.

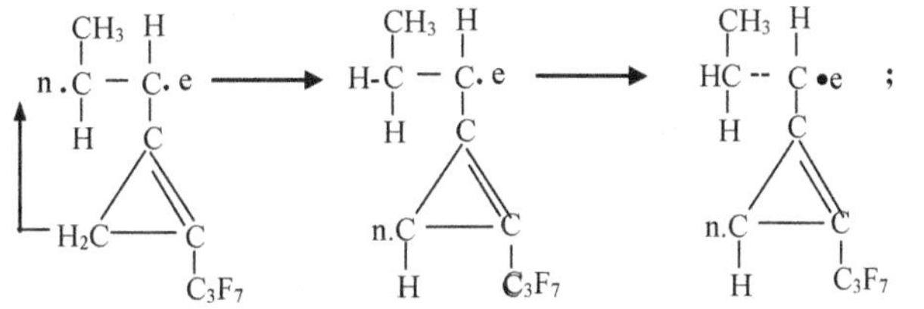

Impossible Transfer　　　10.182

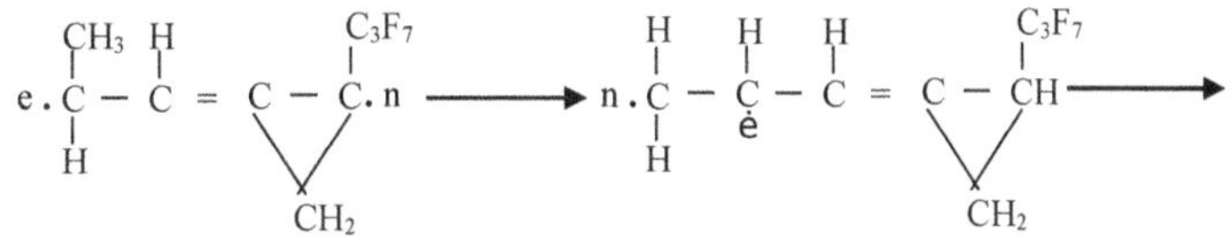

Impossible Transfer　　　10.183

The first case above is impossible because, the CH_2 group in the ring is resonance stabilized and the center has not been properly activated. In the second case, the radicals have been properly placed on the centers, except that H cannot be transferred, because H is more radical-pushing than C_3F_7 group. When the radicals are well placed, CF_3 or C_2F_5 or higher groups can never be a transfer species here. Thus, regardless the capacity of primary fluoro-alkylane groups with respect to CH_3 group, higher groups, or even H, the monomers though resonance stabilized will not favor molecular rearrange-ment of any kind. The same applies to secondary and tertiary fluoro-alkylane groups.

10.2.2.2 Acetylenic cyclopropenes (Triynes)

Based on what has been seen so far, particularly with respect to Equilibrium mechanism decomposition of cyclopropene, methyl cyclopropene and methylene cyclopropane in Equations 10.81, 10.91b and 10.93a respectively and Decomposition mechanism decomposition of methylene cyclopropane in Equations 10.105 and 10.108a, it is believed that the following relationships are valid under the conditions of resonance stabilization.

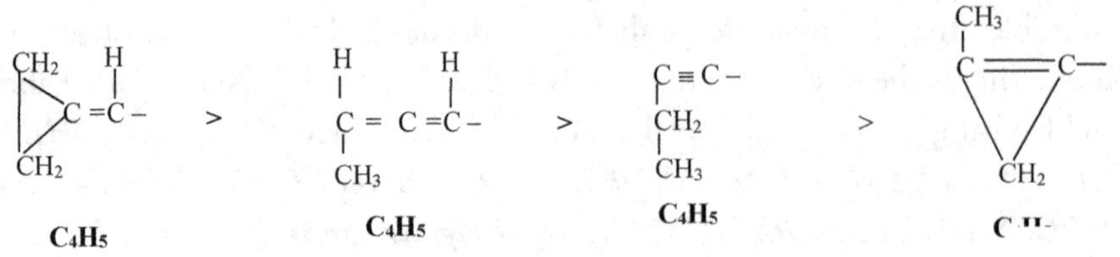

$$\begin{array}{ccccccc}
\underset{\underset{CH_2}{|}}{\overset{\overset{CH_2}{|}}{}}\!\!\!>\!\!\!C=\underset{}{\overset{H}{\underset{|}{C}}}- & > & \underset{\underset{CH_3}{|}}{\overset{H}{\underset{|}{C}}}=C=\overset{H}{\underset{|}{C}}- & > & \overset{C\equiv C-}{\underset{\underset{CH_3}{|}}{\underset{CH_2}{|}}} & > & \underset{\underset{CH_2}{|}}{\overset{\overset{CH_3}{\underset{|}{C}}}{}}\!\!\!>\!\!\!C=C- \\
\mathbf{C_4H_5} & & \mathbf{C_4H_5} & & \mathbf{C_4H_5} & & \mathbf{C\,H_{-}}
\end{array}$$

<u>**Order of Radical-pushing capacity**</u>   10.184

The first two are oddly cumulenic, while the last three or all of them are resonance stabilization groups. Indeed, whether they are isomeric or not, the same order virtually remains, particularly with respect to cyclopropene, noting that the CH_2 group in the ring is no longer a part of the substituent groups carried by the ring, since it is resonance stabilized. When not resonance stabilized, the CH_2 group in the ring becomes a part of the substituent groups carried by the ring like in the first case. In vinyl acetylene, the resonance stabilization provider is the acetylene group which is less radical-pushing than the ethene group or H. Yet acetylene is more nucleophilic than ethene. Since cyclopropane is more nucleophilic than ethene, one should also expect cyclopropene to be more nucleophilic than acetylene. Hence the order above is valid. It should also be noted that cumulene becomes more nucleophilic than acetylene when it rearranges molecularly. As far as hydrocarbon cyclopropenes are concerned it has been shown that none of them can molecularly rearrange to a more stable compound such as substituted acetylenes. Shown below are two isomers of acetylene and cyclopropene.

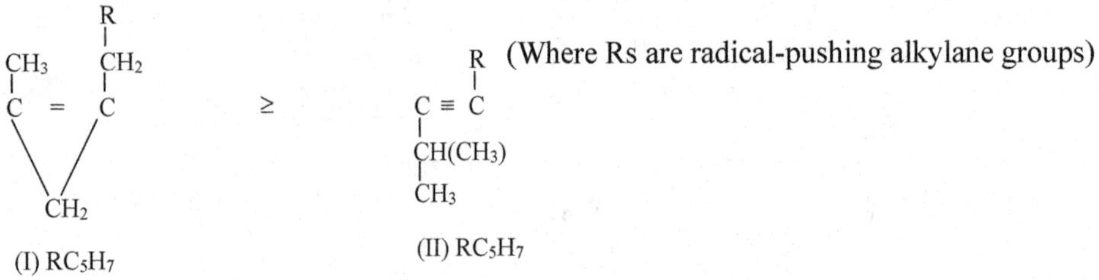

$$\underset{\text{(I) } RC_5H_7}{\underset{\underset{CH_2}{|}}{\overset{\overset{R}{\underset{|}{CH_2}}}{\underset{CH_3}{|}\,C}=\underset{}{C}}} \qquad \geq \qquad \underset{\text{(II) } RC_5H_7}{\underset{\underset{CH_3}{|}}{\underset{CH(CH_3)}{|}}\overset{\overset{R}{\underset{|}{C}}}{C}\equiv C}$$

(Where Rs are radical-pushing alkylane groups)

<u>**Order of Nucleophilicity (R ≤ C₃H₇)**</u>   10.185

The Alkylane groups are of the primary types. The relationships reflect in part the observations which have been made so far. As will be shown downstream, the difference between their total bond energies is the SE in the ring. When R is greater than C_3H_7, (I) cannot ***artificially*** rearrange to form acetylenes, unless when the CH_3 group is changed. The word *"artificially"* is meant to imply the decomposition of the cumulenic ringed isomer of (I), i.e., an alkylenemethylcyclopropane (A) below

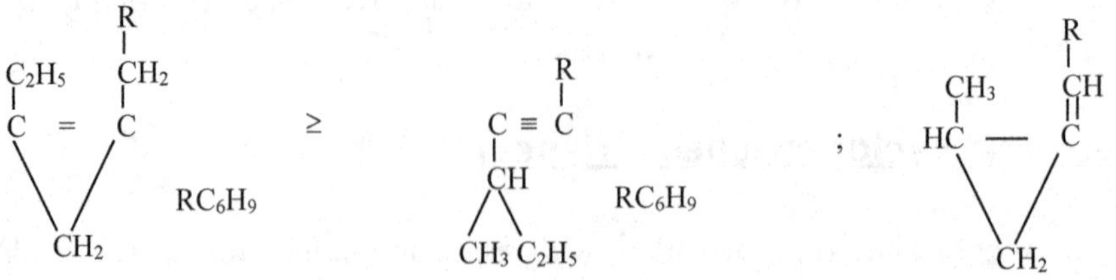

$$\underset{RC_6H_9}{\underset{\underset{CH_2}{|}}{\overset{\overset{R}{\underset{|}{CH_2}}}{\underset{C_2H_5}{|}\,C}=\underset{}{C}}} \quad \geq \quad \underset{RC_6H_9}{\underset{\underset{CH_3\;C_2H_5}{}}{\underset{CH}{|}}\overset{\overset{R}{\underset{|}{C}}}{C}\equiv C} \quad ; \quad \underset{\text{(A) } RC_5H_7}{\underset{\underset{CH_2}{|}}{\overset{CH_3}{\underset{|}{HC}}-\overset{\overset{R}{\underset{|}{CH}}}{\underset{}{C}}}}$$

<u>**Order of Nucleophilicity (R ≤ C₄H₉)**</u>   10.186

The relationship in Equation 10.184 can further be ascertained as follows.

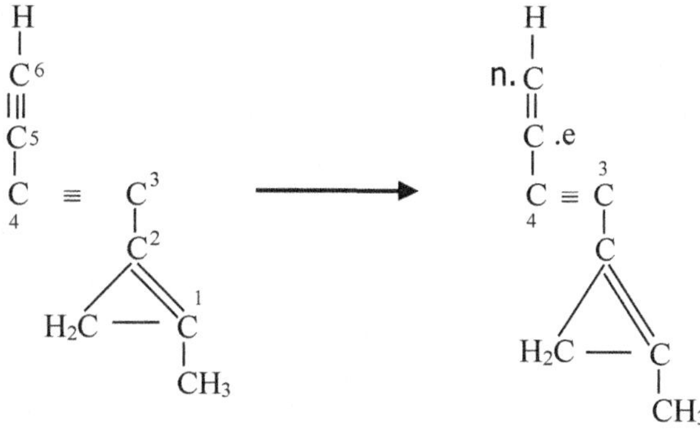

$$
\begin{array}{ccc}
\text{H} & & \text{H} \\
| & & | \\
\text{C}^6 & & \text{n. C} \\
||| & & || \\
\text{C}^5 & & \text{C .e} \\
| & & | \quad _3 \\
\text{C} \equiv \text{C}^3 & \longrightarrow & \text{C} \equiv \text{C} \\
_4 \quad | & & _4 \quad | \\
\text{C}^2 & & \text{C} \\
H_2\text{C} - \text{C}^1 & & H_2\text{C} - \text{C} \\
| & & | \\
\text{CH}_3 & & \text{CH}_3 \\
(I) & & (II)
\end{array}
\qquad 10.187
$$

(I) is analogous to a nucleophilic 1,3,5- hexatriyne. It favors only the electro-free-radical route, because, of the type of transfer species carried. It cannot be activated chargedly.

$$
(III) \longleftrightarrow (IV) \longleftrightarrow (V)
$$

$$
(VI) \longleftrightarrow (VII) \longleftrightarrow (VIII)
$$

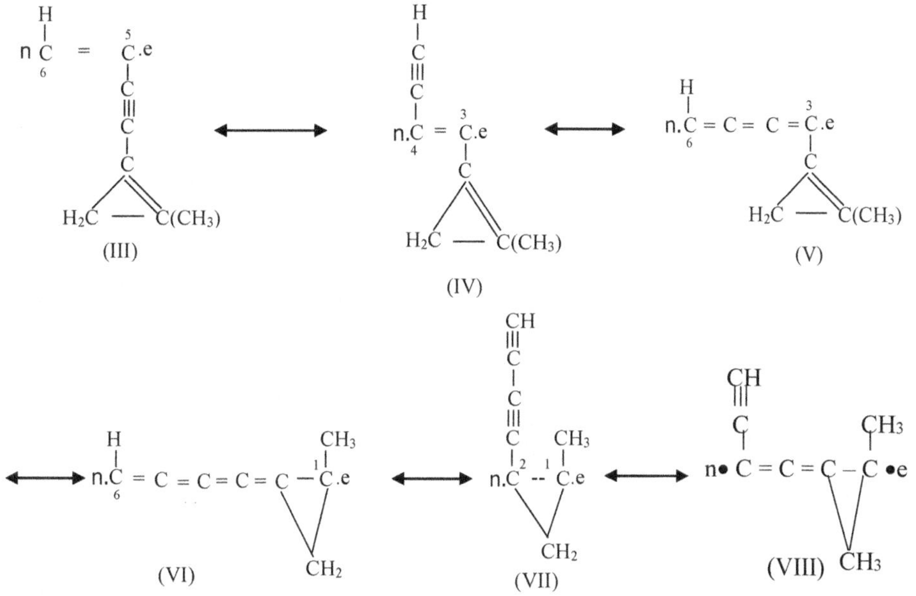

The mono-forms are (III), (V) and (VI) 10.188

Activation begins with the least Nucleophilic center and that is the center shown in (II) or (III). From (III), it moves to (V) and from there finally to (VI). Never are the other centers involved. Only larger sized rings can sustain these well unsaturated systems. On the other hand, the first member of the family or monomer if it exists, will always be in Equilibrium state of existence. For it to be used, it must either be suppressed or replacement of one of the externally located H atoms with a group as shown above with CH_3. In fact, just like methyl cyclopropene, the two have to be replaced.

Nevertheless, only low molecular weight polymers or telomers can be produced, when the right initiators and right environment are not used and provided. The initiators required to activate and polymerize them if resonance stabilization is to be allowed must be electro-free-radical initiators. Though it cannot be activated chargedly, for larger membered rings, no resonance stabilization can take place and therefore only 6,5-monoform can be obtained, for which without doubt, the system will be sterically hindered.in the absence of molecular rearrangement. When weak initiators are used, molecular rearrangement will take place where possible when suppressed.

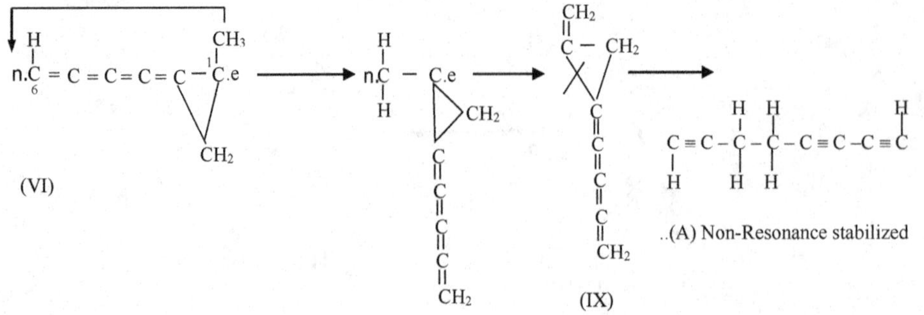

(VI)

NOT FAVORED

..(A) Non-Resonance stabilized

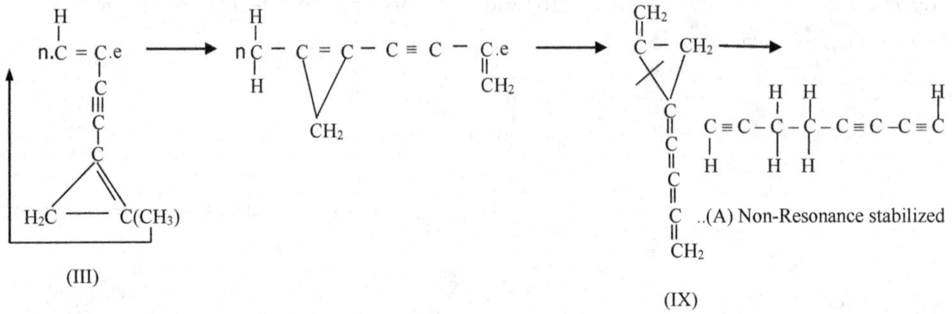

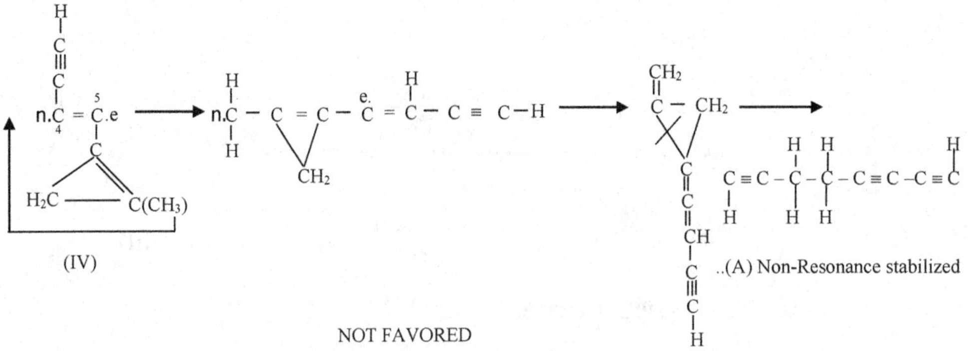

(IX)

10.189

(III)

NOT FAVORED

..(A) Non-Resonance stabilized

(IX)

10.190

(IV)

NOT FAVORED

..(A) Non-Resonance stabilized

10.191

Notice that all the mono-forms molecularly rearrange to give the same (A) through series of molecular rearrangements. For the three-membered ring, the (A) obtained is a non-resonance stabilized compound, clear indication that the point of scission in the ring is wrong because the followings are valid.

$$HO- \;>\; H_2C=C=C=C=C=C= \;>\; H_2C=C=C=C=C= \;>\; H_2C=C=C=C= \;>$$

(EVEN) (ODD) (EVEN)

$$H_2C=C=C= \;>\; H_2C=C= \;>\; H_2C= \;>>\; R$$

(ODD) (EVEN) (ODD)

Order of Radical-pushing capacities of Radical-pushing groups . 10.192

Based on the laws which have been stated so far, *whether the groups are odd or even in the absence of hetero atom terminally located, they are all radical-pushing groups with capacities as shown*

above. When the real point of scission is applied, as an exercise, show that the monomer rearranges back to itself, with the ring reformed after a carbene is formed as an intermediate.

Replacing the H atom on the acetylenic end of (I) of Equation 10.187, the followings are obtained.

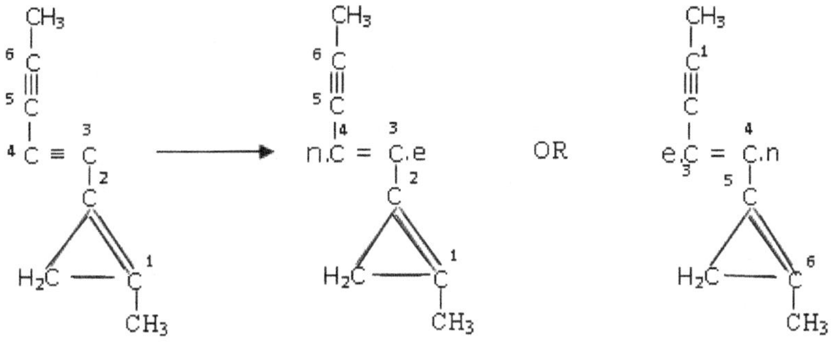

$$
\begin{array}{ccc}
\text{(A)} & \text{(I)} & \text{(II)}
\end{array}
$$

$$\text{(A)} \longrightarrow \text{(I)} \quad OR \quad \text{(II)} \tag{10.193}$$

Which is the true activated state (Not activated center), on the basis of Equation 10.184? Now considering the mono-forms of (I), the followings are to be expected.

FOR *(I)*

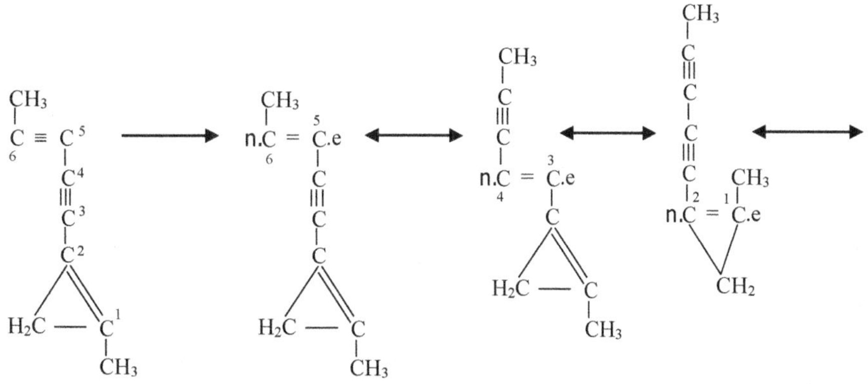

$$
\begin{array}{cccc}
\text{(A)} & \text{(III) Favored} & \text{.(I) Not Favored} & \text{(IV) Not Favored}
\end{array}
$$

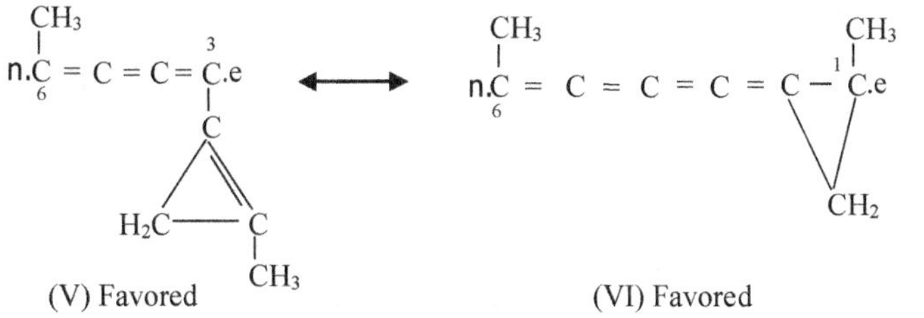

$$
\begin{array}{cc}
\text{(V) Favored} & \text{(VI) Favored}
\end{array} \tag{10.194}
$$

They seem to favor molecular rearrangements to produce resonance stabilized hybrid of acetylene and ethenes as shown below for some of the mono-forms.

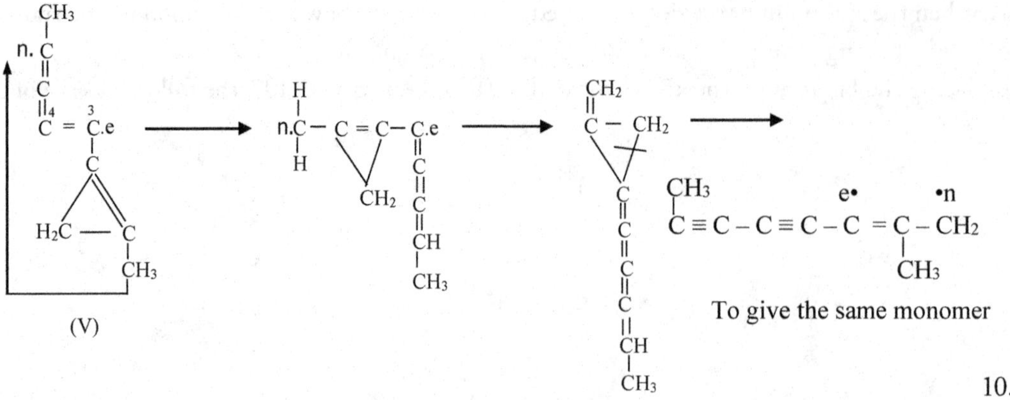

$$
\text{(V)} \longrightarrow \longrightarrow \longrightarrow \qquad 10.195
$$

To give the same monomer

$$
\text{C} \equiv \text{C} - \text{C} \equiv \text{C} - \text{C} = \text{C} - \text{CH}_2
$$

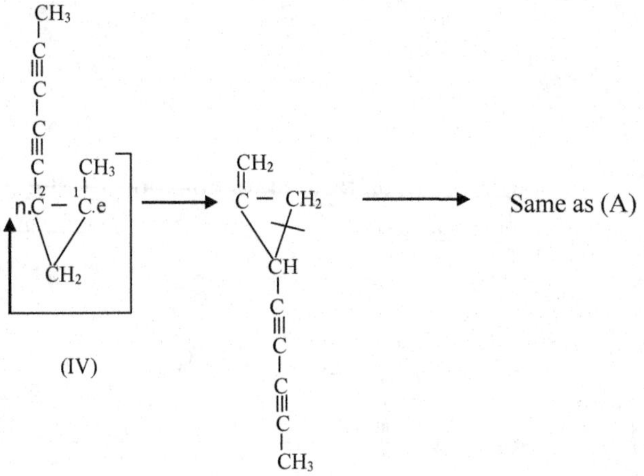

$$
\text{(IV)} \longrightarrow \longrightarrow \text{Same as (A)} \qquad 10.196
$$

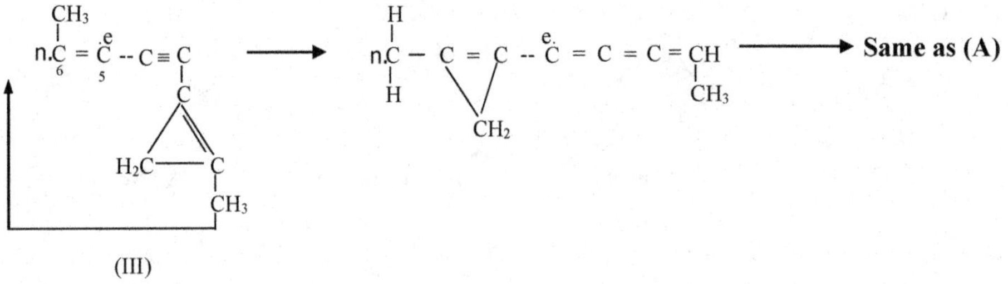

$$
\text{(III)} \longrightarrow \longrightarrow \textbf{Same as (A)} \qquad 10.197
$$

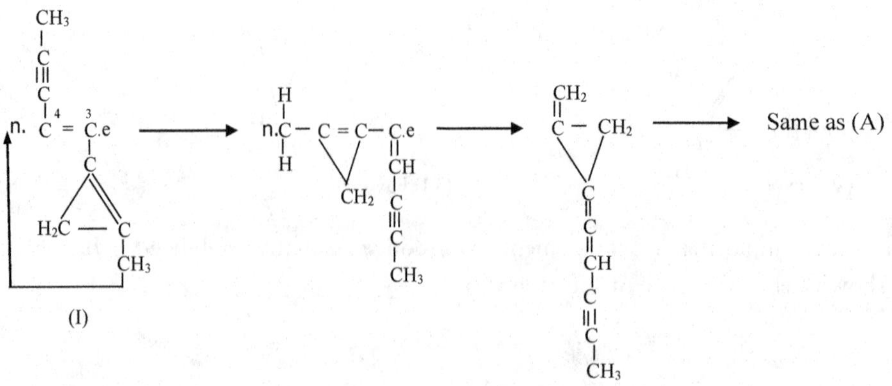

$$
\text{(I)} \longrightarrow \longrightarrow \longrightarrow \text{Same as (A)} \qquad 10.198
$$

For (II) of Equation 10.193, the followings are obtained.

FOR *(II)*

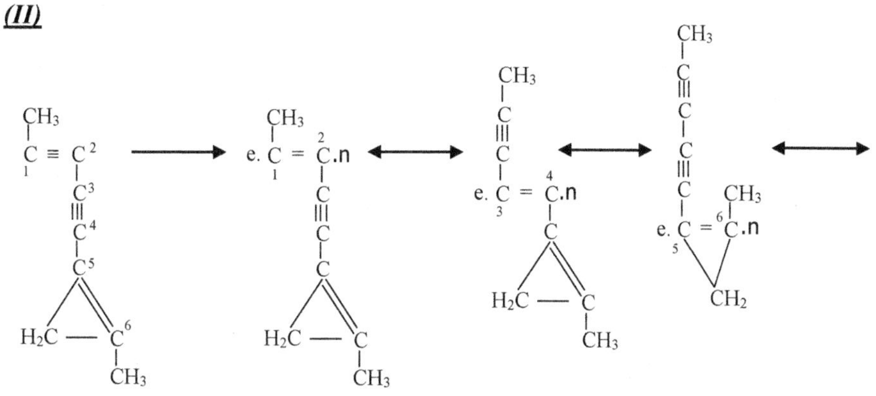

(A)　　　　　　(VII)　　　　　　(II)　　　　　　(VIII)

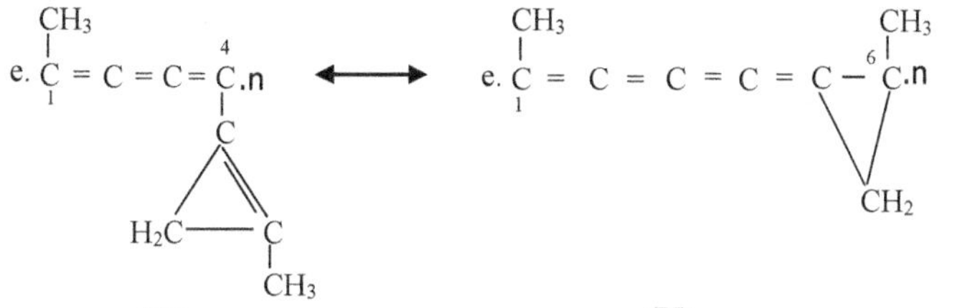

(IX)　　　　　　　　　　(X)　　　　　　　　　　　　　　10.199

They all seem to favor molecular rearrangements to produce the same mono-form (C) as shown below for them. (VII), (IX) and (X) are the mono-forms.

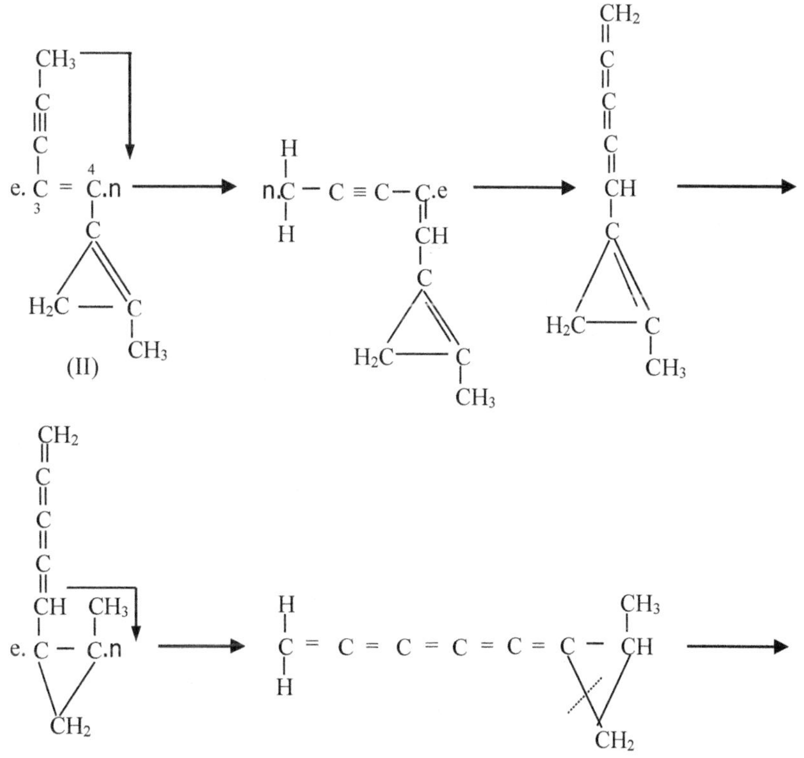

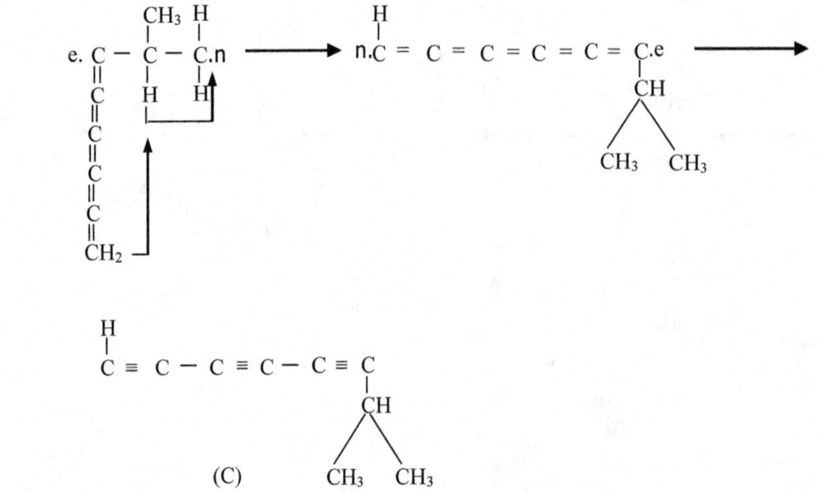

The point of scission is in order, and that which has been obtained is that which is to be expected

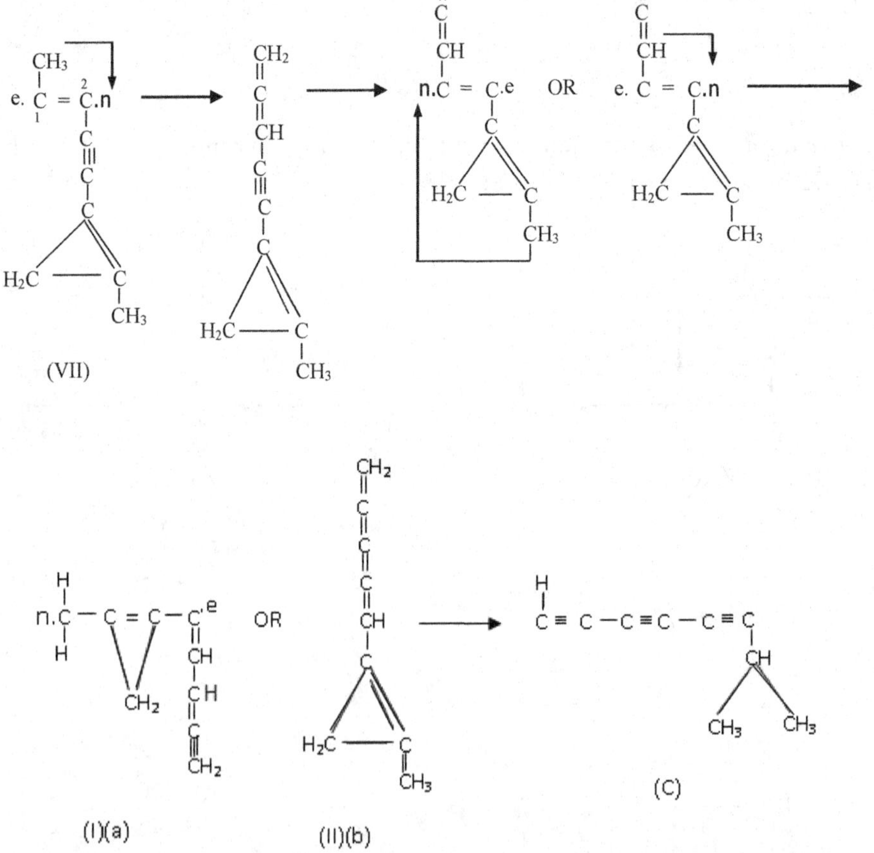

(C)   10.200

(VII)   10.201

(I)(a)   (II)(b)   (C)

80

$$
\begin{array}{c}
CH_3 \\
| \\
C \\
||| \\
C \\
| \\
C \\
||| \\
C \\
| \quad CH_3 \\
e.\ C - C.n \\
5 \quad 6 \\
| \\
CH_2 \\
(VIII)
\end{array}
\longrightarrow
\begin{array}{c}
H \\
| \\
n.C - C \equiv C - C \equiv C - C.e \\
| \\
H \\
H_2C - CH \\
| \\
CH_3
\end{array}
\longrightarrow
\begin{array}{c}
CH_2 \\
|| \\
C \\
|| \\
C \\
|| \\
C \\
|| \\
C \\
H_2C - CH \\
| \\
CH_3
\end{array}
\longrightarrow
$$

$$
\begin{array}{c}
H \\
| \\
C \equiv C - C \equiv C - C \equiv C \\
| \\
CH \\
(C) \quad CH_3 \quad CH_3
\end{array}
\qquad 10.202
$$

$$
\begin{array}{c}
CH_3 \\
| \\
e.\ C = C = C = C.n \\
1 \qquad 4 \\
C \\
H_2C - C \\
| \\
CH_3 \\
(IX)
\end{array}
\longrightarrow
\begin{array}{c}
CH_2 \\
|| \\
C \\
|| \\
C \\
| \\
C \\
| \\
CH \\
| \\
C \\
H_2C - C \\
| \\
CH_3
\end{array}
\longrightarrow
\begin{array}{c}
H \\
| \\
C \equiv C - C \equiv C - C \equiv C \\
| \\
CH \\
CH_3 \quad CH_3 \\
(C)
\end{array}
\qquad 10.203
$$

$$
\begin{array}{c}
CH_3 \qquad CH_3 \\
| \qquad | \\
e.\ C = C = C = C = C - C.n \\
1 \qquad\qquad 6 \\
| \\
CH_2 \\
(X)
\end{array}
\longrightarrow
\begin{array}{c}
CH_3 \\
| \\
HC - CH_2 \\
| \\
C \\
|| \\
C \\
| \\
C \\
|| \\
C \\
| \\
CH_2
\end{array}
\longrightarrow
\begin{array}{c}
H \\
| \\
C \equiv C - C \equiv C - C \equiv C \\
| \\
CH \\
CH_3 \quad CH_3 \\
(C)
\end{array}
\qquad 10.204
$$

As has been established, $H_2C=$ or $H_2C=C=C=$ or $H_2C=C=$ or $H_2C=C=C=C=$ groups are strong radical-pushing groups of greater capacity than alkylane groups. The 1,3-cumulenic center is more nucleophilic than the acetylene center. However, based on the fact that the centers for (II) have been rightly activated on the basis of Equation 10.149, it is *(II)* that is favored. In *(I)*, the resonance stabilization provider is the ring which is more radical-pushing than the acetylenic side. In (I), the products which are the same obtained from all the mono-forms are the same as the monomer, that which is not to be expected. In *(II)*, the resonance stabilization provider is the ring which is less radical pushing than the acetylenic side. In order words, as has already been explained differently above, cyclopropene is more nucleophilic than acetylene as partly reflected in Equations 10.185 and 10.186 for isomers of them.

$$H \quad H$$
$$| \quad |$$
$$C = C$$
$$\diagdown \diagup$$
$$CH_2$$
$$\qquad > \qquad$$
$$H$$
$$|$$
$$C \equiv C$$
$$|$$
$$H$$

Order of Nucleophilicity 10.205

A resonance stabilized monomer in which the activation centers are of the same nucleophilicity and favor both routes radically, must be symmetric with no transfer species. Examples include butadiene, hexatriene, butadiyne and hexa-triyne.

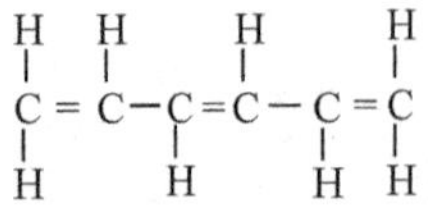

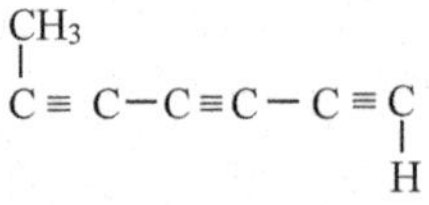

$$H \quad H \qquad H \qquad H$$
$$| \quad | \qquad | \qquad |$$
$$C = C - C = C - C = C \qquad ; \qquad C \equiv C - C \equiv C - C \equiv C$$
$$| \quad | \qquad | \quad | \qquad |$$
$$H \quad H \qquad H \quad H \qquad H$$

with CH_3 above the first left carbon of the second structure

Hexa-triene   Methyl Hexa-triyne 10.206

Only the first case is symmetric. The second becomes symmetric only when the CH_3 group is changed to H or H is changed to CH_3 group and this unlike the first case cannot be activated chargedly. Now comparing the two look-alike monomers, one symmetric and the second non-symmetric shown below, the followings are obvious.

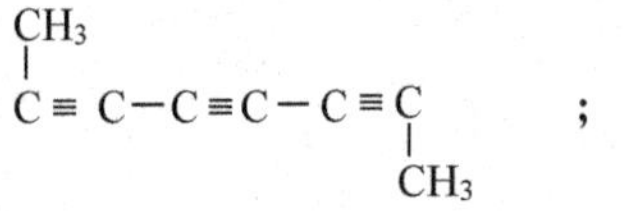

$$CH_3$$
$$|$$
$$C \equiv C - C \equiv C - C \equiv C \qquad ; \qquad C \equiv C - C \equiv C - C = C$$
$$|$$
$$CH_3$$

with CH_3 groups and CH_2 below for the second structure

(I) Resonance stabilized 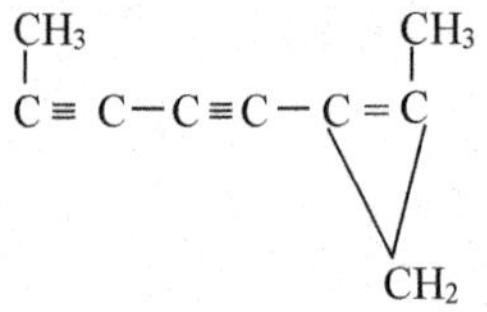   (II) Resonance stabilized

	(I)	Three activation centers are of same nucleophilicity.	(I)	Three activation centers are of different nucleophilicity.
	(II)	Cannot be activated chargedly.	(II)	Cannot be activated chargedly.
	(III)	Favors only electro-free-radical route.	(III)	Favors only electro-free-radical route.

10.207

*In general, the followings as have already been shown are valid **when the groups are on their own or used alone**.*

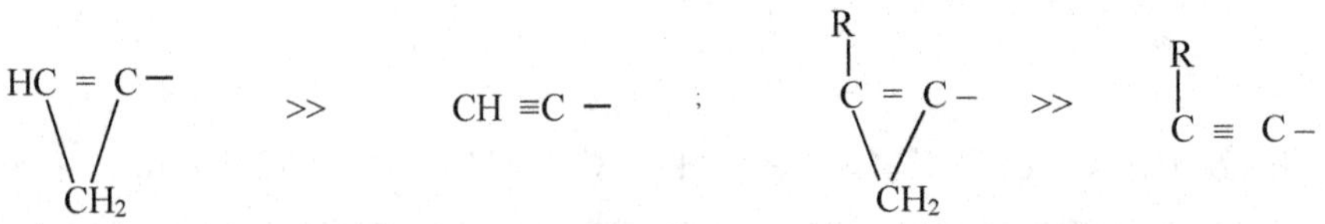

$$HC = C - \qquad >> \qquad CH \equiv C - \qquad ; \qquad C = C - \qquad >> \qquad C \equiv C -$$
$$\diagdown \diagup \qquad\qquad\qquad\qquad\qquad\qquad \diagdown \diagup$$
$$CH_2 \qquad\qquad\qquad\qquad\qquad\qquad\qquad CH_2$$

with R above for the last two structures

Order of radical-pushing capacity in the absence of resonance stabilization 10.208

But when used as resonance stabilization groups, the capacities of the groups are reduced. This is one of unique characteristic qualities of resonance stabilization phenomena.

Simplifying further, the followings seem also to be valid, since molecular rearrangements were involved.

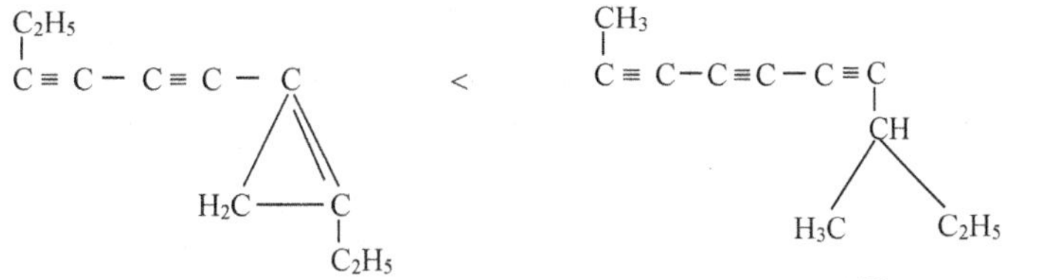

$$C_2H_5 - C \equiv C - C \equiv C - C \quad < \quad CH_3 - C \equiv C - C \equiv C - C \equiv C$$

(with ring structure $H_2C - C - C_2H_5$) and (C) structure with CH, H_3C, C_2H_5

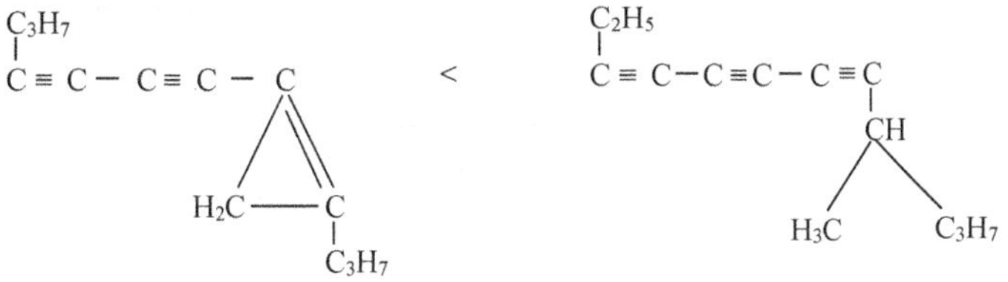

Order of Nucleophilicty (C)   10.209a

$$C_3H_7 - C \equiv C - C \equiv C - C \quad < \quad C_2H_5 - C \equiv C - C \equiv C - C \equiv C$$

(with ring structure $H_2C - C - C_3H_7$) and (D) structure with CH, H_3C, C_3H_7

Order of Nucleophilicty (D)   10.209b

When the groups externally located are different, they all still molecularly rearrange on decomposi-tion to give triynes as shown below.

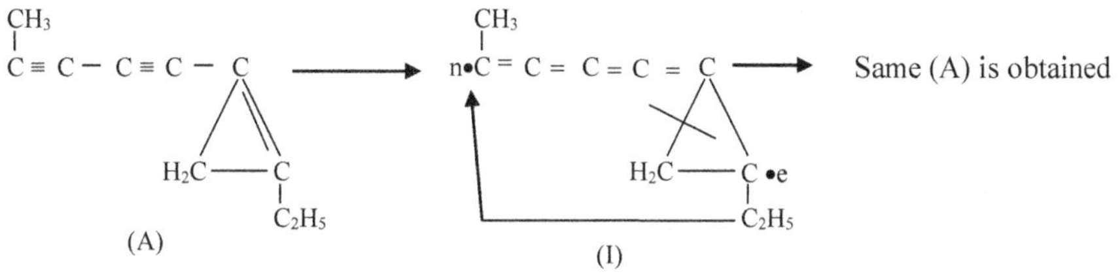

$$CH_3 \text{ structure (A)} \longrightarrow n \cdot C = C = C = C = C \text{ (I)} \longrightarrow \text{ Same (A) is obtained}$$

(A) with ring $H_2C - C - C_2H_5$

(I) with ring $H_2C - C \cdot e - C_2H_5$

                                    10.210a

The case above is different from the case shown below.

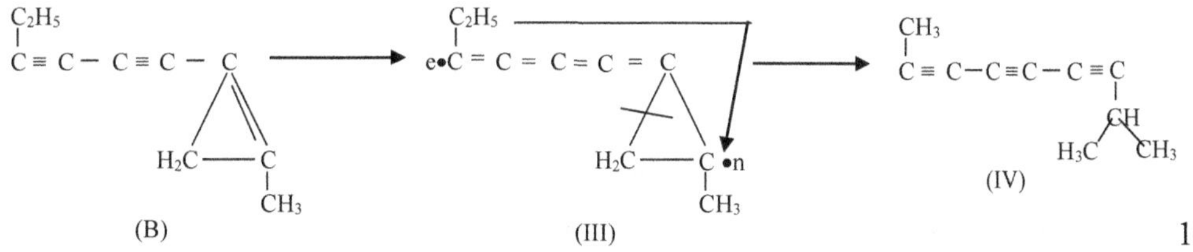

$$C_2H_5 \text{ structure (B)} \longrightarrow e \cdot C = C = C = C = C \text{ (III)} \longrightarrow CH_3 - C \equiv C - C \equiv C - C \equiv C \text{ (IV)}$$

(B) with ring $H_2C - C - CH_3$

(III) with ring $H_2C - C \cdot n - CH_3$

(IV) with CH, H_3C, CH_3

                                    10.210b

Two different groups-CH_3 and C_2H_5 have been used above for (A) and (B). In both, the resonance provider is the ring. In (A), the ring is more radical pushing than the last acetylenic end. In (B), the reverse is the case. When molecular rearrangement of the first kind of the first type takes place, via Decomposition mechanism, the ring is forced to open. This is then followed by molecular rearrange-ment of the third kind of the first type to give conjugatedly placed acetylenes. It was easier for (B) than for (A), because the ring which is far less radical-pushing than acetylene when same groups are carried, was now made to be more radical-pushing than acetylene, in view the larger capacity of the group placed on her. In fact, the product from (A) remains (A), that which is not to be expected. This is like the case of Equation 10.187 where the triene rearranges back to itself. Based on the considerations above, it was observed that in general, the followings are valid

$$HO- \; > \; R_2C=C=C=C=C=C= \; (Even) \; > \; R_2C=C=C=C=C= \; (Odd) \; > \; R_2C=C=C=C= \; (Even)$$

$$> \; R_2C=C=C= \; (Odd) \; > \; R_2C=C= \; (Even) \; > \; R_2C= \; (Odd) \; >> \; R \; (Of \; any \; size)$$

Order of Radical-pushing capacities of Radical-pushing groups

                                    10.192

Note that the R group carried could be H or any alkylane group. The R group are alkylane groups such as CH_3. *Hence in general for only C and H containing cyclopropenes, these cannot molecularly rearrange to acetylenic types of monomers; for C and H containing vinyl cyclo-propenes, these undergo molecular rearrangement only where the R in the vinyl group is greater than the R in the cyclopropene via Decomposition mechanism to give resonance stabilized vinyl acetylenes with limitations; while for C and H containing di-acetylenic cyclopropenes (Triynes), these undergo molecular rearrangement to give resonance stabilized triynes only where the R in the acetylene group is equal to or greater than the R in the cyclopropene. The same applies to Diynes.*

It is important to note that, (A) of Equation 10.210a is less Nucleophilic than (B) of Equation 10.210b. Nevertheless, it is important to note that the following is valid.

$$R - C \equiv C - C \equiv C - \qquad \geq \qquad R - C \equiv C - \qquad \text{(Where R is an alkylane group)}$$

Order of Radical-pushing capacity
$$10.211$$

(C) and (D) of Equations 10.209a and 209b, and (IV) of Equation 10.210b are all resonance stabilized since the radicals (Not charges) carried by the different mono-forms (Not activation centers) are as follows for (C).

centers) are as follows for (C).

$$10.212$$

(I) is 6, 5-mono-form. (II) is 6, 3- mono-form and (III) is 6, 1- mono-form. They have been so-called because they are all nucleophiles, with same transfer species. (I) of Equation 10.187 like the others, was noted to undergo molecular rearrangement of the third kind. When the CH_3 group in (I) of Equation 10.187 is replaced with higher groups, molecular rearrangements still remain favored, but back to itself. When R is C_2H_5, the following is obtained. When R is greater than C_2H_5, rearrangement ceases. This is clear indication of the fact that when R equals C_2H_5 the nucleophilicity of the ring with the group equals the nucleophilicity of Diyne (i.e. two acetylenes).

$$10213a$$

Just as with $R = CH_3$, the same resonance stabilized compound is also produced here, *clear indication that these compounds cannot produce a different product when mildly decomposed*.

However, in general, the following is valid.

$$C_3H_7CH = CCH_3\text{-} \quad > \quad C_2H_5CH = CCH_3\text{-} \quad > \quad \textbf{(H)} \quad \geq \quad CH_3CH = CCH_3\text{-}$$

Order of Radical-pushing and resonance stabilization capacity 10.213b

Finally, consider the case shown below, wherein the ring is carrying H while the acetylenic end is carrying CH_3 group.

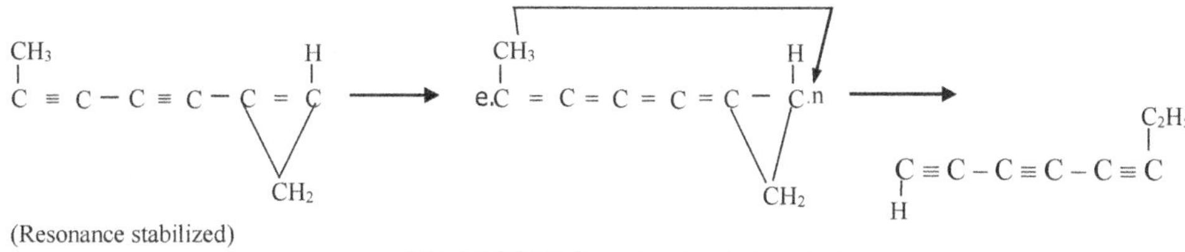

(Resonance stabilized)

VIA DECOMPOSITION MECHANISM

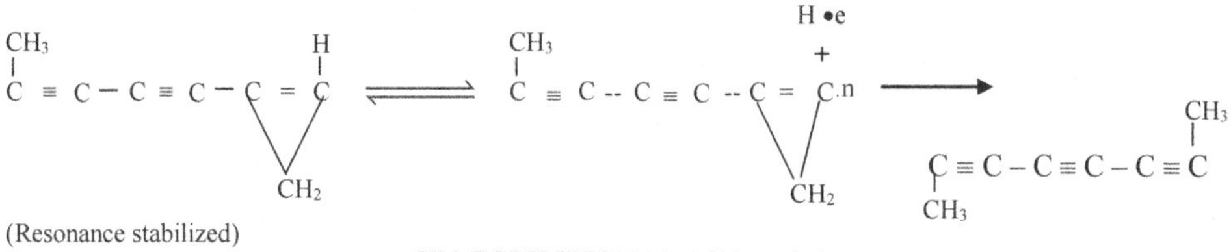

10.214a

(Resonance stabilized)

VIA EQUILIBRIUM MECHANISM

10.214b

While there is limitation via Decomposition mechanism, there is no limitation via Equilibrium mechanism, because different but similar products are obtained from both routes for this case. Via Decomposition mechanism, the CH_3 group cannot exceed C_3H_7. When this limitation is compared with the limitation of Equation 10,210b, there is no doubt that cyclopropene is more nucleophilic than acetylene. One can see that CHEMISTRY is a complete logical network system.

As an exercise to readers, what happens when the externally located radical-pushing group carried by the ring is C_2F_5, while the one carried by the externally located pushing groups at the other end is CH_3? What also happens when the H atoms in the ring are also changed? One cannot escape from asking questions, because these are some of the major origins of our problems in humanity. Little things we seem to **neglect** or **ignore** is the source of ***NEGLIGENCE or IGNORANCE*** in Humanity.

Since, N is very important in Living systems, there is need to look at compounds that carry them in different ways. We have seen what happens in general when for example NH_2 group is put in place of CH_3 on the ring or acetylene alone. Now, consider NH_2, NHR, NR_2, OH and OR types of groups. For NH_2, the followings should be expected.

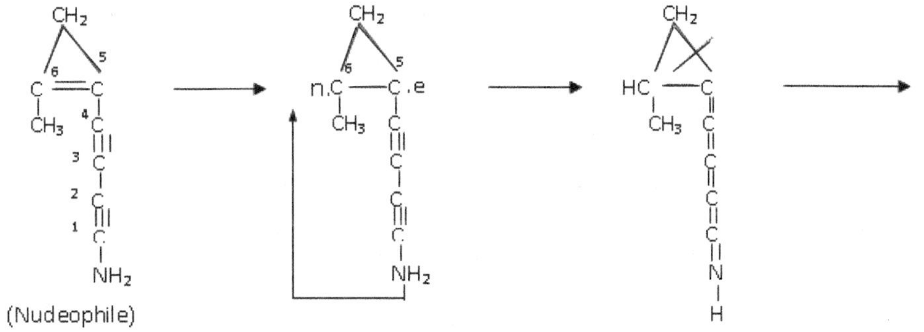

(Nucleophile)

Full-free resonance
stabilized mono-form

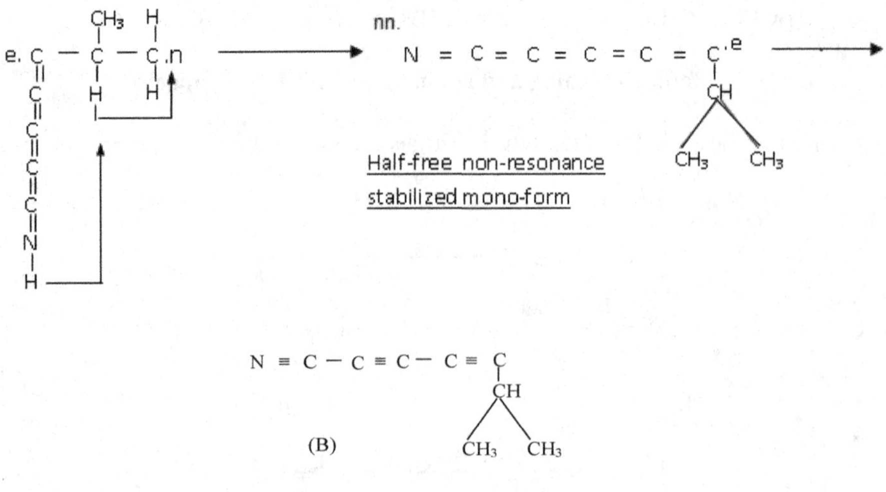

$$\text{e. } \overset{\displaystyle CH_3 \quad H}{\underset{\displaystyle \underset{H}{\overset{\displaystyle\|}{C}}}{C} - \underset{H}{\overset{\displaystyle |}{C}} - \underset{H}{\overset{\displaystyle |}{C}}.n} \longrightarrow$$

$$\overset{nn.}{N = C = C = C = C = C^{.e}} \longrightarrow$$

Half-free non-resonance stabilized mono-form

$$N \equiv C - C \equiv C - C \equiv C$$

(B)

(Electrophile)

10.215

(B) above is partially resonance stabilize only in the diene section. The reaction above is favored under certain operating conditions. It will apply to larger sized rings where MSRE can be attained. The case above has not been scissioned artificially in order to show what to expect when a non-ringed isomer is being searched for in order to determine the SE in the ring. Usually, the isomer chosen or used for such purposes, must be such that bears the same character and property as the ring as will be shown downstream. For example, acetaldehyde and vinyl alcohol are isomers of three-membered cyclic ether (Epoxide). But the real isomer to use in determining the SE in the ring is the vinyl alcohol and not the acetaldehyde which bears no semblance with the epoxide.

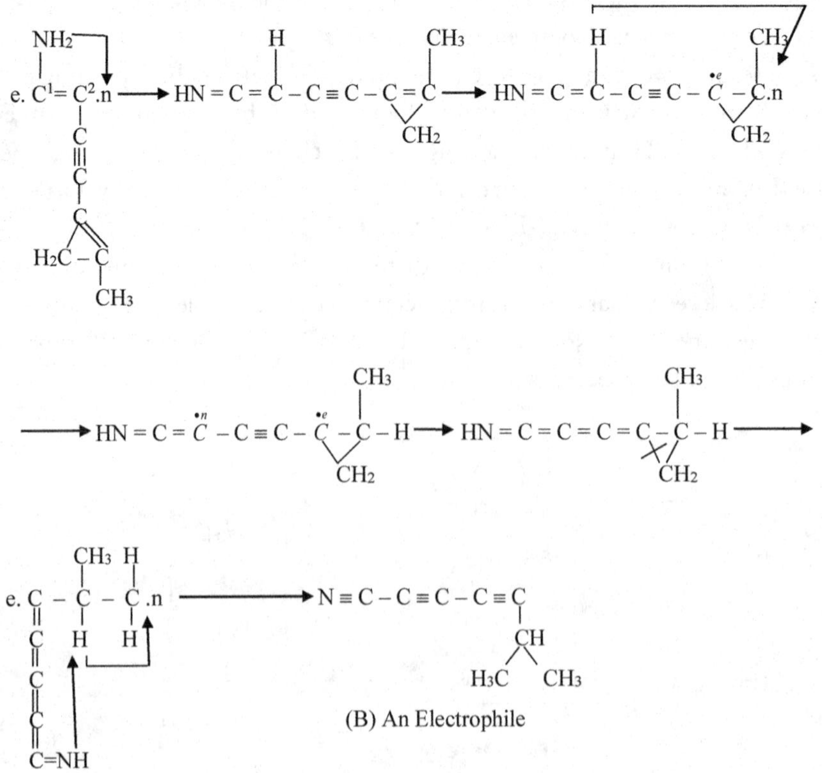

$$\text{e. } C^1 = C^2.n \longrightarrow HN = C = C - C \equiv C - C = C \longrightarrow HN = C = C - C \equiv C - \overset{\bullet e}{C} - C.n$$

$$\longrightarrow HN = C = \overset{\bullet n}{C} - C \equiv C - \overset{\bullet e}{C} - C - H \longrightarrow HN = C = C = C = C - C - H \longrightarrow$$

$$\text{e. } C - C - C.n \longrightarrow N \equiv C - C \equiv C - C \equiv C$$

(B) An Electrophile

10.216

86

This monomer, originally a Nucleophile is resonance stabilized as shown below. From the three different mono-forms, molecular rearrangements based on the operating conditions can be made to readily take place to give the same monomer which is an Electrophile.

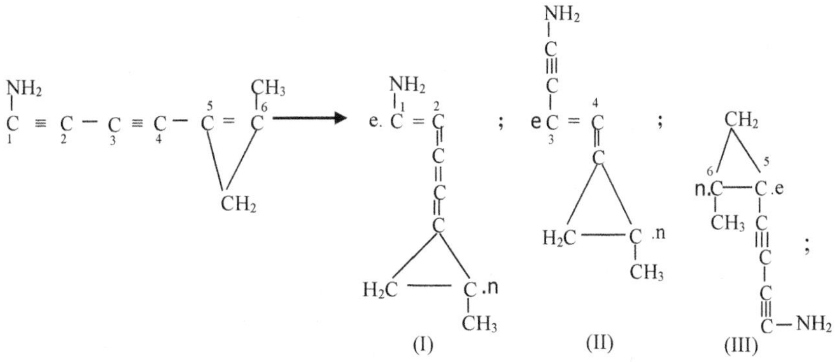

10.217

The least nucleophilic center is 6, 5 –mono-form (III) which resonance stabilizes to 6, 3 –mono-form (II), and finally to 6, 1-mono-form (I). All these mono-forms artificially rearrange to give an ELECTROPHILE. Based on the mechanisms seen so far, one can observe that NHR (when R$\geq$CH$_3$) and NR$_2$ types of groups cannot produce ELECTROPHILES. This has been the same observation so far with other families of compounds where they exist.

Now the next question to ask is what happens when the NH$_2$ group is reversely placed, that is, on the ring side?

10.218

i) Nucleophile

ii) Resonance stabilized

iii) Least Nucleophilic center (LNC)-6, 5.

iv) Molecularly rearranges to Males

i) Nucleophile

ii) Resonance stabilized

iii) LNC- 6, 5.

iv) Molecularly "rearranges" to Nitriles

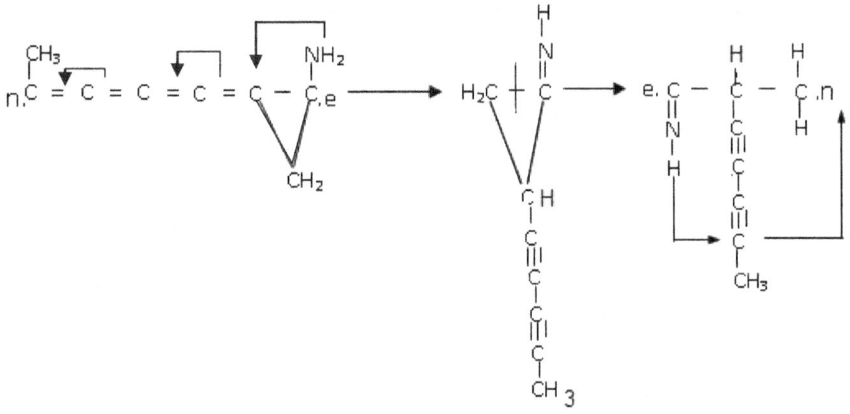

For the case of (B) as a simple exercise, consider the 1, 2-mono-form

nn.

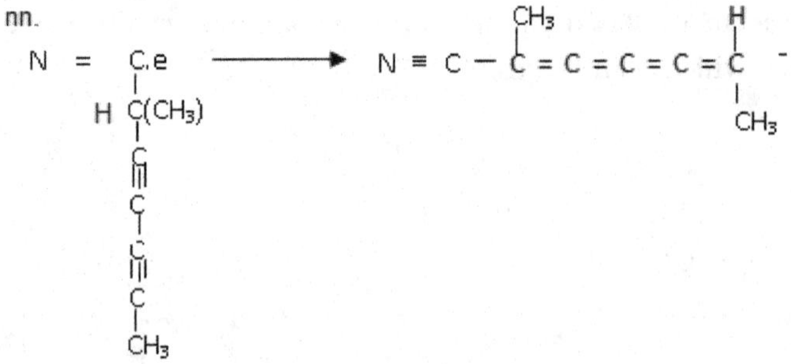

$$N = \overset{|}{\underset{|}{C}}e \longrightarrow N \equiv C - \overset{CH_3}{\underset{|}{C}} = C = C = C = \overset{H}{\underset{|}{C}} \cdot$$

(with the side groups H, $\overset{|}{C}(CH_3)$, $\overset{|}{C}$, $\overset{|}{C}$, CH_3 and CH_3)

10.219

The same Nitrile is obtained for all the mono-forms. The Nucleophile was transformed to a Nucleophile. However, note that the activation center used above can never be activated alone being not the least nucleophilic center. It is in fact the most nucleophilic center, the resonance provider.

Shown below is what to expect when NHR is used for only R equal to CH_3 and no higher group.

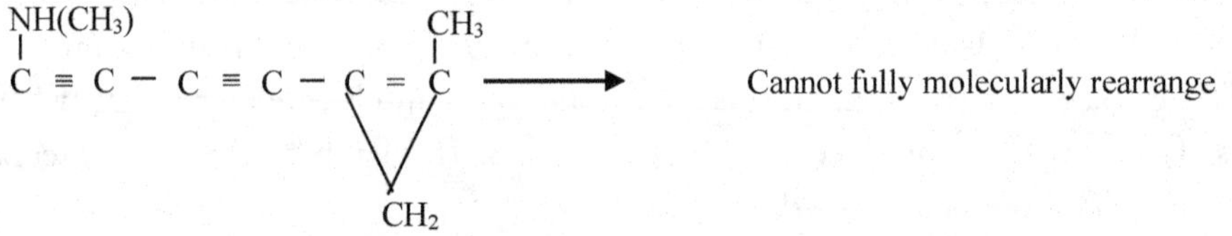

$$\overset{NH(CH_3)}{\underset{}{C}} \equiv C - C \equiv C - \overset{CH_3}{\underset{}{C}} = \overset{}{\underset{}{C}} \longrightarrow \quad \text{Cannot fully molecularly rearrange}$$

(with CH_2 bridging below)

10.220

When the ring is opened, transfer species can be moved sequentially to give a stable product, like the case already shown and shown below with $N(CH_3)_2$.

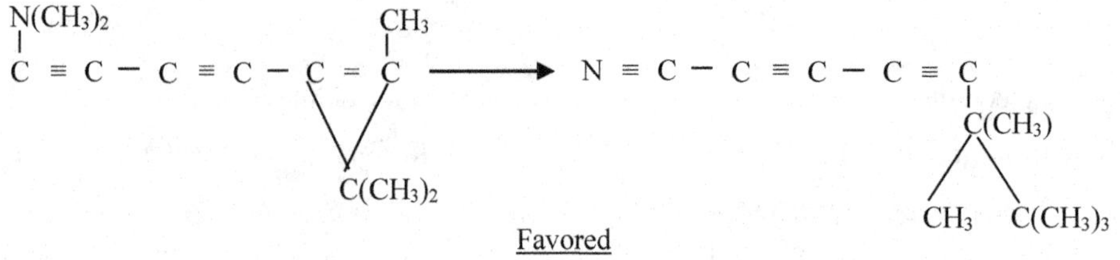

$$\overset{N(CH_3)_2}{\underset{}{C}} \equiv C - C \equiv C - \overset{CH_3}{\underset{}{C}} = \overset{}{\underset{}{C}} \longrightarrow N \equiv C - C \equiv C - \overset{}{\underset{}{C}} \equiv C$$

(with $C(CH_3)_2$ bridging below; product with $C(CH_3)$, CH_3, $C(CH_3)_3$)

Favored

10.221

With OR groups which has no transfer species, no molecular rearrangement of the first kind can take place. The original monomers are still resonance stabilized. With OH group, molecular rearrangements can take place when decomposed only via Equilibrium mechanism to give ketenes.

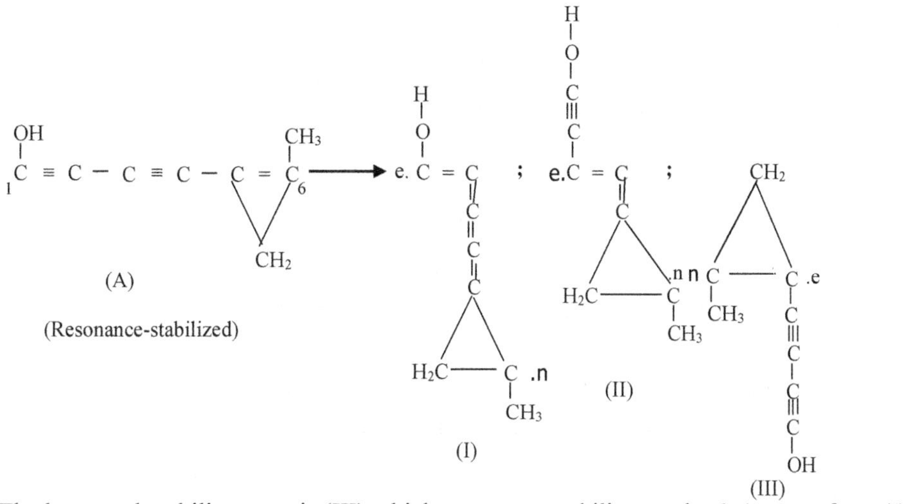

10.222

The least nucleophilic center is (III) which resonance stabilizes to the 6, 1-mono-form (I).

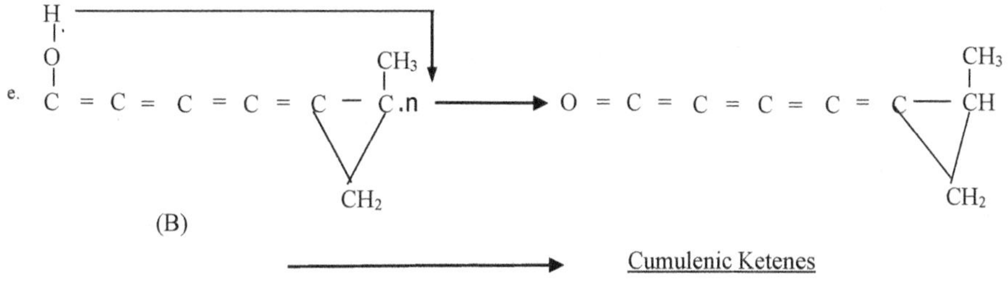

10.223

For any of the mono-forms, where rearrangements take place, cumulenic ketenes can be produced.

Now consider the case where the OH and CH_3 groups are reversely placed.

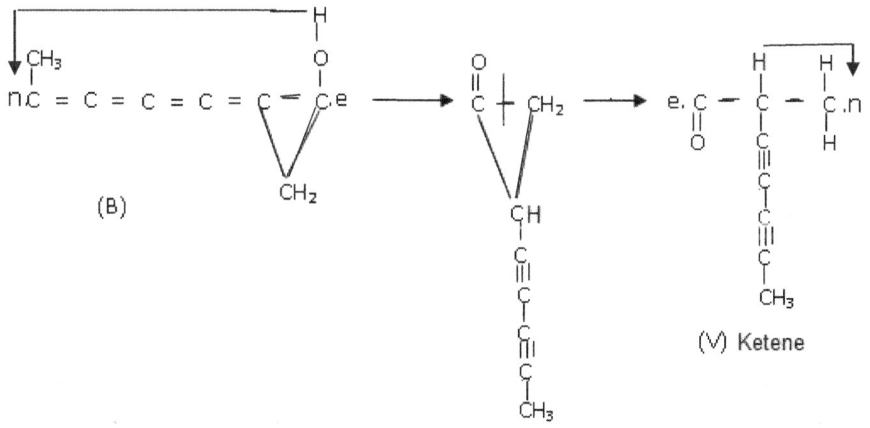

10.224

Though the ring can be opened, it cannot fully molecularly rearrange due to the absence of second transfer species on the O center. The center activated above is the most nucleophilic center which is not to be expected. The situation is limited to OH group. When R > CH_3, rearrangement can no longer take place. One can imagine all along the great significance of HYDROGEN and METHYL (CH_3) groups – e•CH_3 and n•CH_3. A look at the families of 1,3- ($H_2C=C=O$) and 1,4- ($H_2C=C=C=O$) Ketenes and the families of the corresponding nitrogen containing one (1, 3- $H_2C=C=NH$, and 1, 4- $H_2C=C=C=NH$) as has been done in the past will reveal that the followings are valid.

$$[HN= < HN=C=C= < HN=C=C=C=C=] > [O= < O=C=C= < O=C=C=C=C=]$$
Radical-pushing groups

$$H_2C=C=C=C=C= > H_2C=C=C=C= > H_2C=C=C= \gg H_2C= \gg R > H$$

ORDER OF RADICAL-PUSHING CAPACITIES OF PUSHING GROUPS

$$[HN=C=C=C= > HN=C=] > O=C=C=C=C=C= > O=C=C=C= > O=C= > N\equiv C- \ ;$$
Radical-pulling groups

ORDER OF RADICAL-PULLING CAPACITIES OF PULLING GROUPS   10.225

$N\equiv C$- is one of the strongest primary radical-pulling groups if not the strongest, while $- C\equiv CH$ is one of the weakest primary C/H radical-pushing groups, the weakest probably being cyclopropene group (Cyclopropyne) in that family of singlets, i.e., $- X$. With duplets, i.e., $=X$, notice the difference between HN= (EVEN) and HN=C= (ODD), and between O= (EVEN) and O=C= (ODD). There is ODD or EVEN for $H_2C=$, $H_2C=C=$, $H_2C=C=C=$, etc., types of groups.

With radical-pulling groups, the rings cannot readily be opened, in view of the fact that so much energy or force will be required to attain the MRSE for their rings.

10.226

Only three activation centers are resonance stabilized radically, with none of them favoring charged activation. However, chargedly, only the C=O center can be activated. The transfer species for the monomer is OCH_3 group, being an Electrophile. The least Nucleophilic center is the 2,1--mono-form. Even after molecular rearrangement of the first kind, the rings cannot easily be opened. In the rearrangement, the only transfer species involved is the OCH_3 group and not H. If H is a transfer species for a similar monomer as used above ($(CH_3)CH=CH(COOCH_3)$), an allylic Nucleophile is obtained, that which is not to be expected, being an Electrophile. If the right transfer species is used and the product becomes a Nucleophile, then that is acceptable. CH_3 is the transfer species for the C=O center which rearranges back to itself.

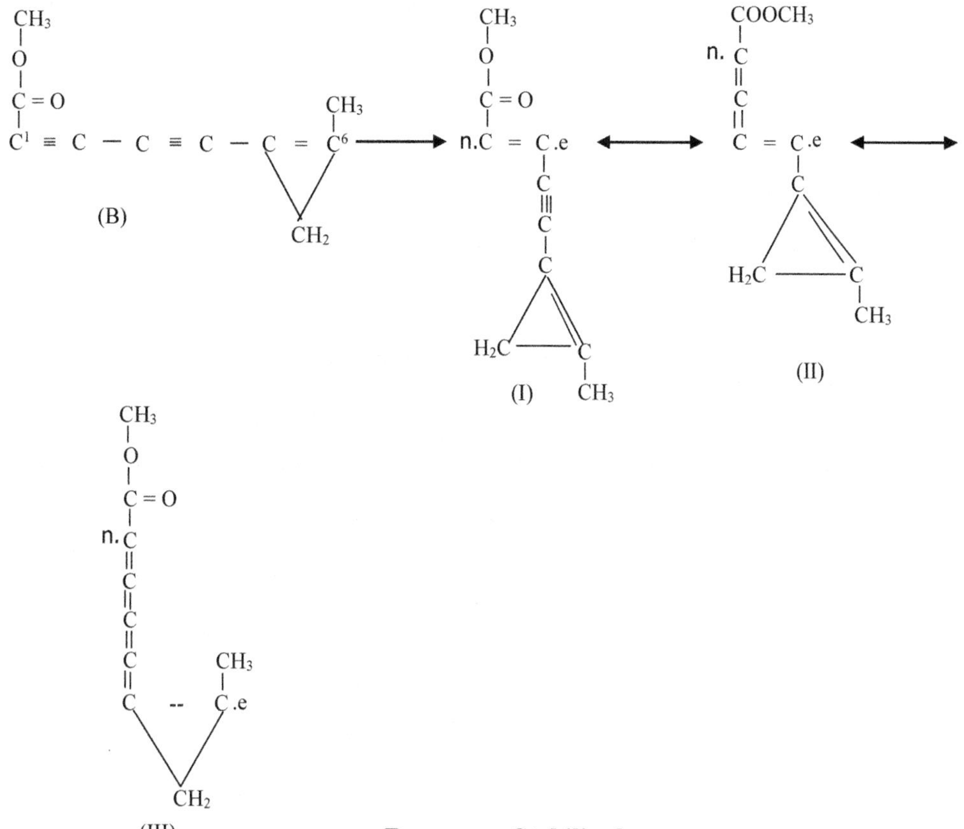

$$
\begin{array}{c}
CH_3 \\
| \\
O \\
| \\
C=O \\
| \\
C^1 \equiv C - C \equiv C - C = C^6 \\
\end{array}
\qquad (B)
\longrightarrow
\qquad (I) \qquad (II)
$$

(III)

Resonance Stabilized

10.227

Like the case above, only three activation centers are resonance stabilized radically, with none of them favoring charged activation. The monomer is an Electrophile. The least Nucleophilic center is the 2,1--mono-form. Even after molecular rearrangement of the first kind, the rings may easily be opened with no possibility to molecularly rearrange.

Like symmetrical nucleophilic triynes which can be fully resonance stabilized, the sym-metrical electrophilic triynes cannot be fully resonance stabilized.

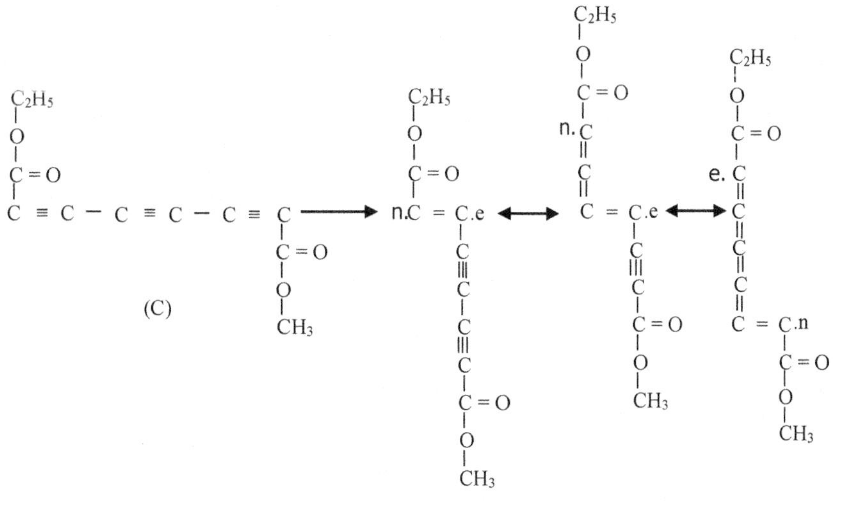

10.228

Based on the laws stated, these cannot fully be resonance stabilized as shown above, since the C=O center is not involved. Free radically, only the nucleo-free-radical route is favored via the $C \equiv C$ centers

91

or electro-free-radical route via the C = O center. Based on the developments so far, it is obvious that the following relationship is valid.

Order of Radical-pulling capacity

10.229

The unique features of cyclopropenes can thus be observed. Indirectly, triynes have been considered. While cyclopropene group has different capacity from that corresponding to acetylenic group, structurally the followings are valid.

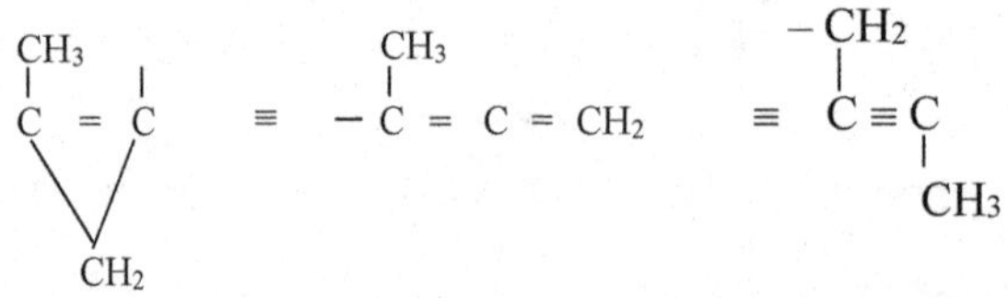

Structural equivalence (isomeric groups)

10.230

A network system seems to exist between cyclopropenes, cumulenes, and acetylenes. For the first time, it has been observed that all nucleophilic triynes are resonance stabilized.

Having considered vinylcyclopropenes, butadiyne cyclopropene, there is need to consider cyclopropene carrying acetylene, that which should have been presented before this sub-section. It has been so done for specific reasons.

10.2.2.3 Acetylenic Cyclopropene (Di-yne)

It has been shown above in many ways that cyclopropene is more nucleophilic than acetylene

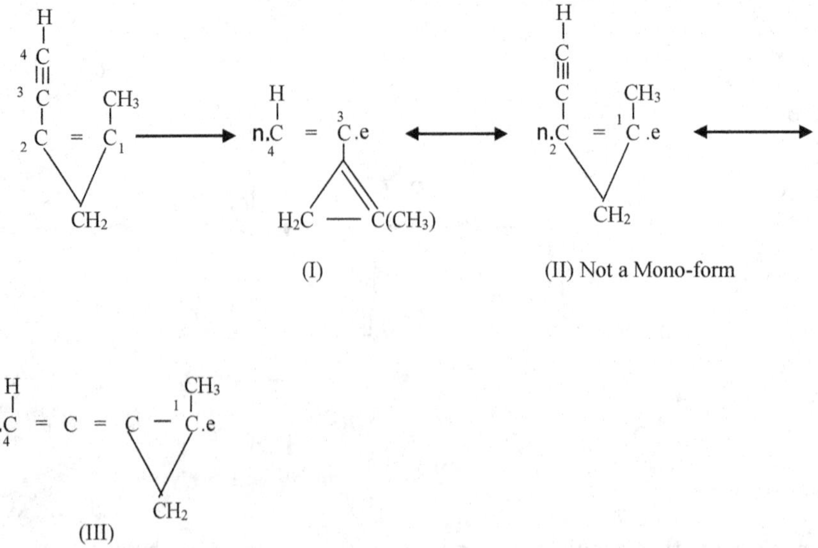

10.231

The (I) and (III) mono-forms are the resonance stabilized mono-forms and are nucleophiles, with the ring providing the resonance and the CH_3 group providing the transfer species. The less nucleophilic center here is 4,3-mono-form. They cannot favor molecular rearrangement. Presence of a passive catalyst, i.e., suppressing agent may be required to keep the H on the acetylene stable, otherwise the H atom may have to be replaced if activation is to take place.

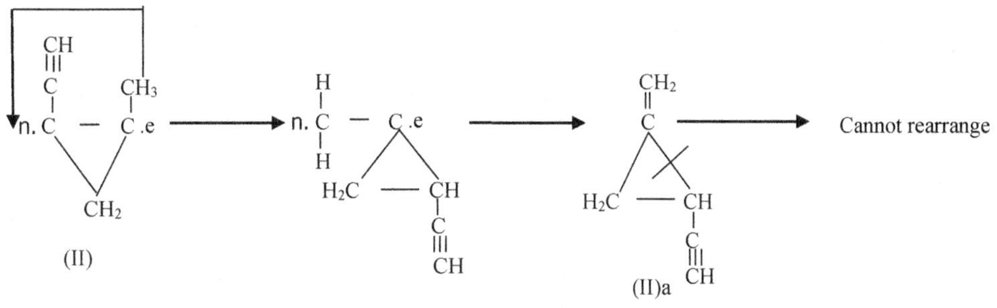

10.232

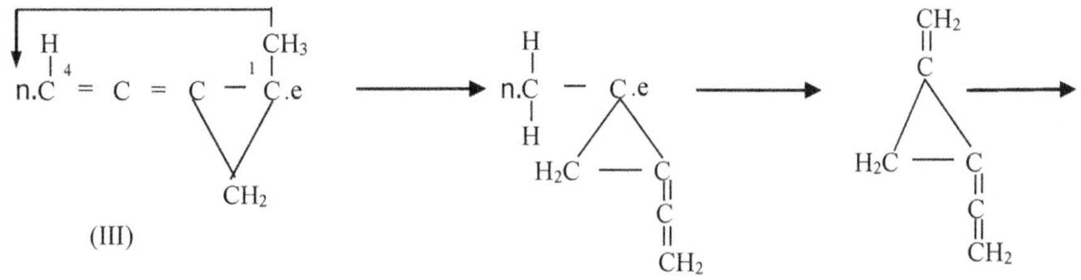

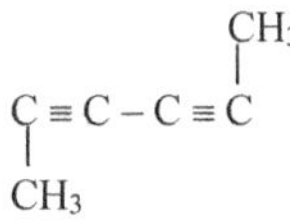

(IV) Not favored

(V) Same as monomer [FAVORED]

10.233a

(II) is not one of the mono-forms. (I) will rearrange to give the same monomer, i.e., back to itself. All the mono-forms molecularly rearrange to give the same resonance stabilized compound with no opening of ring. This is to be expected, since a resonance stabilized monomer or compound does not have to molecularly re-arrange to give a resonance stabilized monomer. The monomer (IV) above, its linear isomer can be observed to be a strongly symmetric nucleophilic Di-yne which indeed is two conjugatedly placed cumulenes as shown below. (IV) above cannot be obtained by rearrangement.

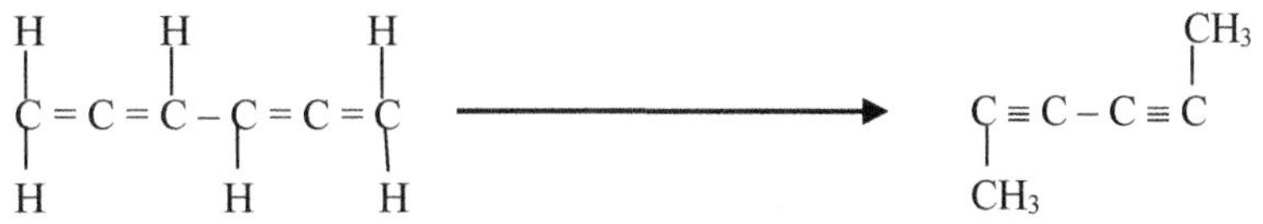

(IV)

10.233b

When H on the acetylenic end is replaced with CH_3 group, making it more nucleophilic than above, the followings are possibilities. In the first possibility below, the radicals have been well placed.

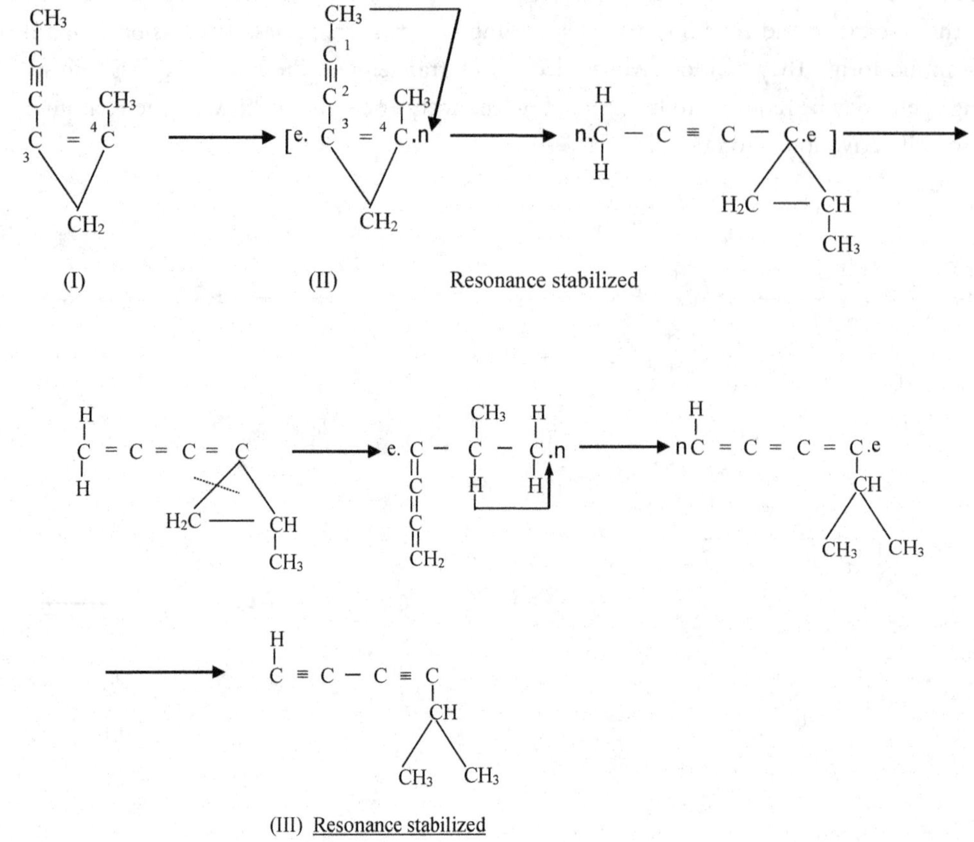

(I)   (II)   Resonance stabilized

(III) <u>Resonance stabilized</u>   10.234

When the radicals on the active centers are now wrongly placed on activation, then the followings are obtained.

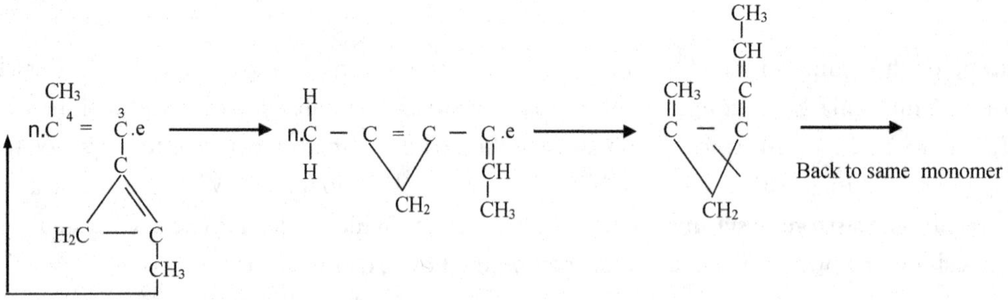

Back to same monomer   10.235

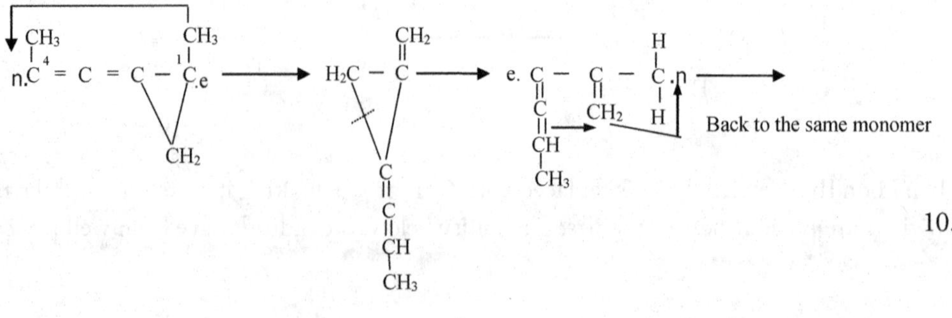

Back to the same monomer   10.236a

Just like the case above and the case of triynes (See Equation 10.195, the ring can be opened to give the same monomer. This clearly shows that the monomer has either been wrongly activated or the H atom on the acetylenic end has not been replaced with a radical-pushing group. ***Note that while $H_2C=$ is a non-resonance stabilization group, $H_2C=C=$ is a resonance stabilization group with a difference, such as in 1,4- cumulene. These can be called SECONDARY type of resonance stabilization groups, while all the others encountered so far (e.g. $-CH=CH_2, - C\circ CH$) are PRIMARY resonance stabilization groups.***

(A)   (B)   (C)   (D)   (E)   (F)

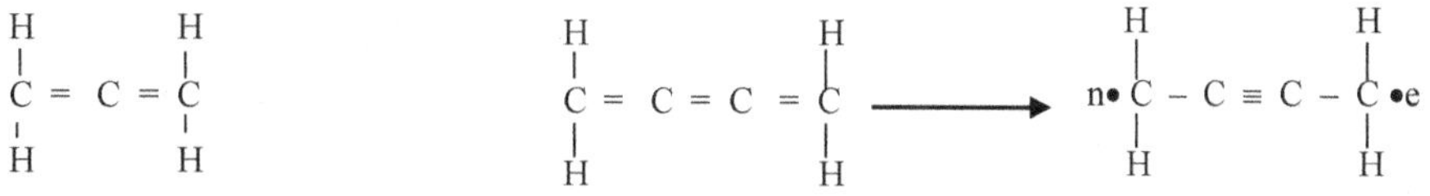

1, 3- Cummulene   1, 4- Cummulene

NON-RESONANCE STABILIZED   **RESONANCE STABILIZED**

10.236b

While the active center in (A), (B), (C), (E) and (F) are carrying resonance stabilization groups, the reverse is the case for the group in (D). While (C) has no transfer species for rearrangement, the others and (D) which is not a resonance stabilization group, have transfer species. The transfer species on them can only be abstracted nucleo-free-radically. One can observe how NATURE operates. As shown above, one can now see why 1, 3-Cumulene is not resonance stabilized, but 1, 4-cumulene is. Though the 2,3-mono-form of 1, 4-cumulene can undergo molecular rearrangement to give vinyl acetylene, it can never be obtained because that center is never activated. The 1,2- or 3,4-mono-form of 1, 4-cummulene cannot undergo molecular rearrangement.

With C_2H_5 in place of CH_3, the followings are obtained.

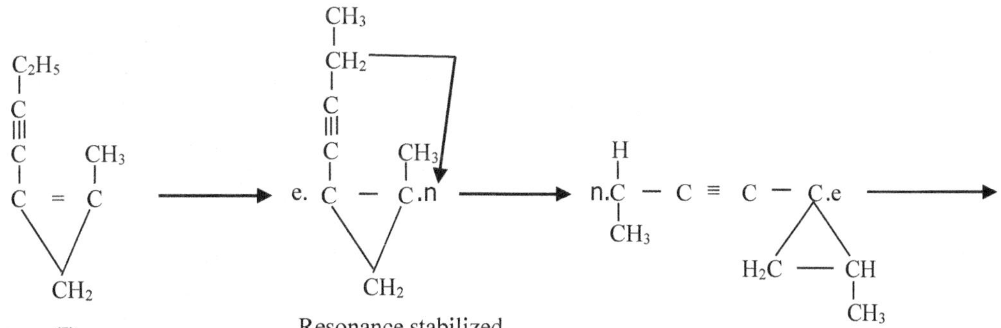

(I)   Resonance stabilized

95

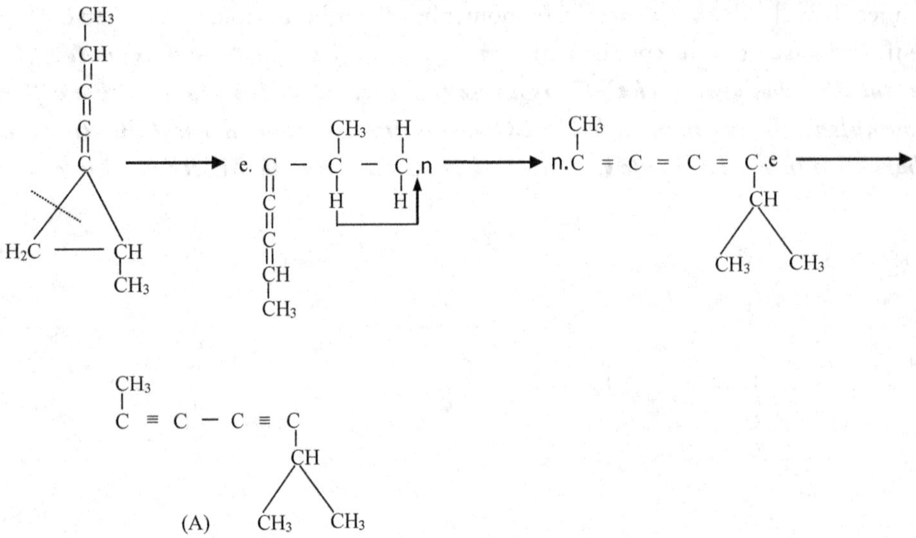

$$
\text{(A)} \quad CH_3\text{-}C \equiv C \text{-} C \equiv C\text{(}CH(CH_3)_2\text{)}
$$

10.237

At first, one will think that the wrong center has been activated, that which is true. This is the center as we already know, can never get activated in the presence of the other center which is less nucleophilic. Secondly, one will think that the radicals have been wrongly placed, that which is not true. One can see the very great importance in establishing the order of capacities of components, for without all that is being done so far, we will continue to live in a world difficult to describe. Though whatever center is used, the same products are obtained, that center above should never be used in establishing the mechanism of a particular reaction. Note that the product above is similar to the case of triynes in Equation 10 210b.

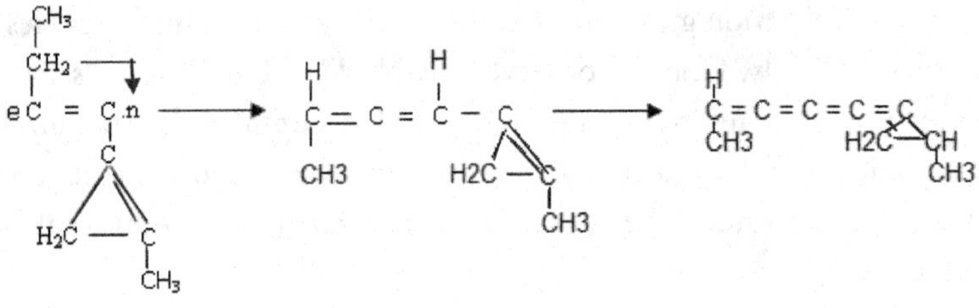

(II) resonance stabilized

$\longrightarrow$ (A)

10.238

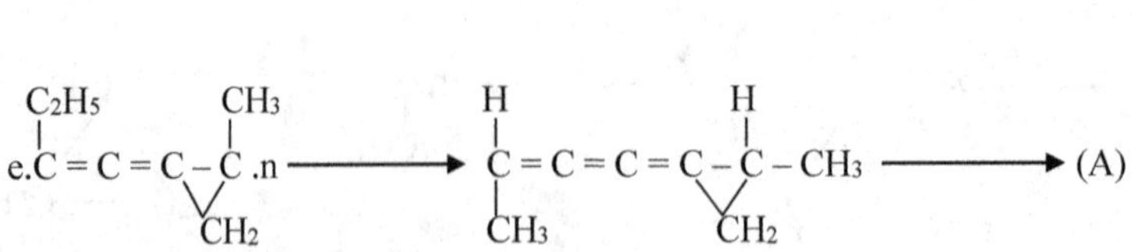

(III) Resonance stabilized

10.239

As shown above, all three mono-forms, two of which are resonance stabilized, rearrange to give the same mono-form, (A), a substituted butadiyne.

Shown below are three monomers to be used for the purpose of identification and extensive study.

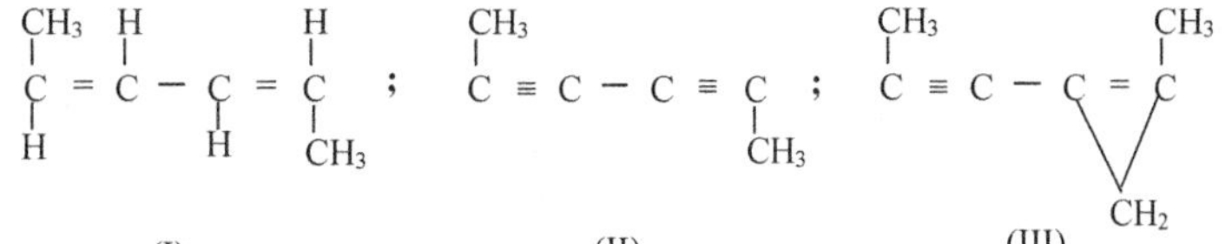

$$
\begin{array}{ccccc}
\text{CH}_3 \quad \text{H} \qquad \text{H} & & \text{CH}_3 & & \text{CH}_3 \qquad \text{CH}_3 \\
| \qquad | \qquad | & & | & & | \qquad\qquad | \\
\text{C} = \text{C} - \text{C} = \text{C} & ; & \text{C} \equiv \text{C} - \text{C} \equiv \text{C} & ; & \text{C} \equiv \text{C} - \text{C} = \text{C} \\
| \qquad\quad | \qquad | & & | & & \bigtriangledown \\
\text{H} \qquad\quad \text{H} \quad \text{CH}_3 & & \text{CH}_3 & & \text{CH}_2 \\
(\text{I}) & & (\text{II}) & & (\text{III})
\end{array}
$$

$$10.240$$

(II) and (III) were identified with respect to nucleophilic capacity. (I) "looks" symmetric, implying that any of the active centers of the monomers can carry any charge or radical, but not the case. When resonance stabilized, this is only possible free-radically starting from the 3,4-mono-form.

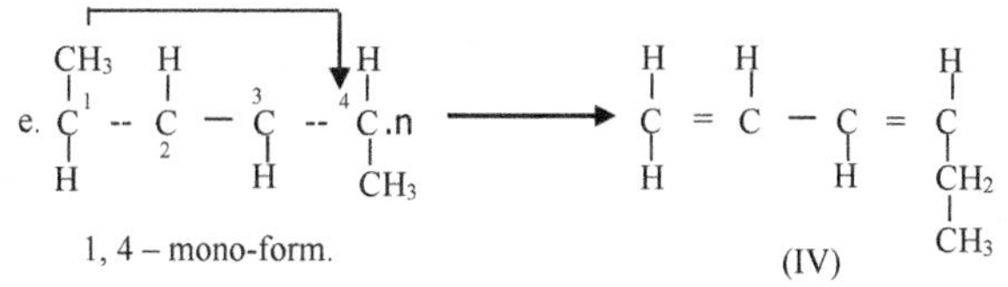

$$10.241$$

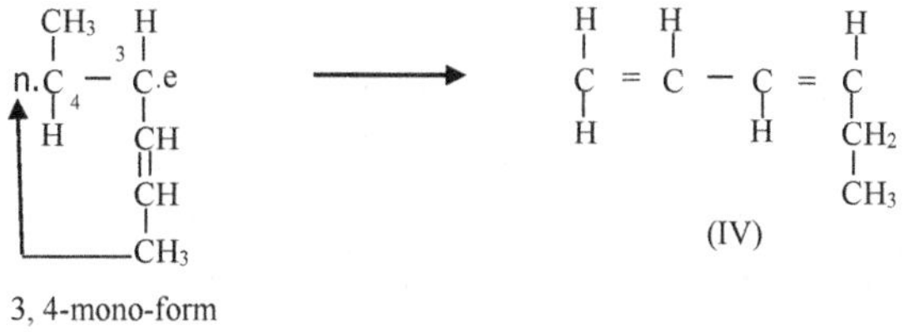

$$10.242a$$

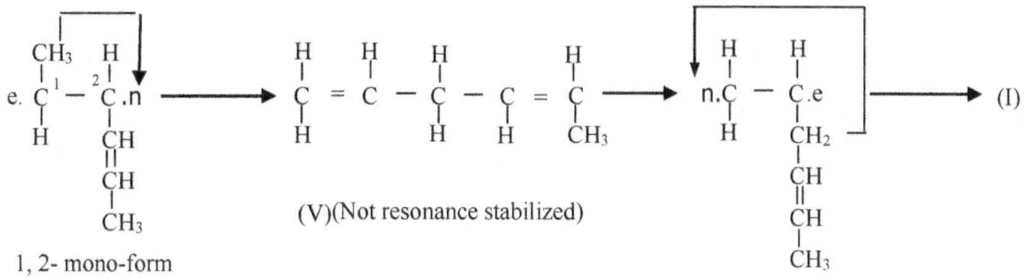

$$10.242b$$

Since for 3,4- mono-form, the radicals are fixed based on the last two equations above (i.e., the externally located C center must always carry a nucleo-free-radical being the least nucleophilic center), hence, only the 1, 4-mono-form of (I) is truly symmetric, since the radicals carried by the active centers are not fixed. Being externally located is not the general rule, since for the diyne analogue of the case above ($H_3C - C \equiv C - C \equiv C - CH_3$), the same monomer is obtained for 1,2- or 3,4- mono-form regardless the placement of the radicals. In fact, all of them rearrange back to itself unlike the case above. Based on the above, the following is valid.

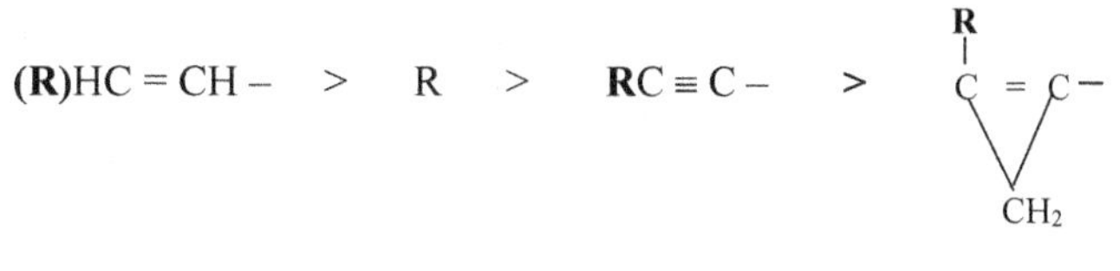

$$
\textbf{(R)}\text{HC} = \text{CH} - \quad > \quad \text{R} \quad > \quad \text{RC} \equiv \text{C} - \quad > \quad \overset{\textstyle \text{R} \atop |}{\underset{\textstyle \text{CH}_2}{\text{C} \bigtriangledown \text{C}}} -
$$

ORDER OF RADICAL-PUSHING CAPACITIES $\qquad$ 10.243

The structure of the 1,2- or 3,4- mono-form clearly shows that it is non-symmetric. The case above is to be expected, since 2-butenes (Cis- and Trans-) molecularly rearrange to 1-butene.

The 1,4-mono-form of (II) of Equation 10.240 is symmetric, unlike its other mono-forms. However, all the mono-forms molecularly rearrange to give the same mono-form as shown below.

$$\left\{ \begin{array}{c} \overset{CH_3}{\underset{C}{\overset{|}{{}^1C}}} \equiv \overset{}{\underset{C}{C^2}} \longrightarrow n.\ \overset{CH_3}{\underset{C}{\overset{|}{C}}} = \underset{CH_3}{\overset{|}{C}}.e \\ \overset{|}{\underset{CH_3}{C}} \end{array} \right\} ; \quad \left\{ \begin{array}{c} \overset{CH_3}{\underset{C}{\overset{|}{{}^3C}}} \equiv \overset{}{\underset{C}{C^4}} \longrightarrow n.\ \overset{CH_3}{\underset{C}{\overset{|}{C}}} = \underset{CH_3}{\overset{|}{C}}.e \end{array} \right\} ;$$

1,2-mono-form ; 3,4-mono-form

$$\left\{ \overset{CH_3}{\underset{CH_3}{\overset{|}{C}}} \equiv C - C \equiv C \longrightarrow e.\ \overset{CH_3}{\underset{CH_3}{\overset{|}{C}}} = C = C = C.n \right\} \longrightarrow \overset{H}{\underset{H}{\overset{|}{C}}} = C = C = C = \overset{H}{\underset{CH_3}{\overset{|}{C}}}$$

1,4-mono-form

Via molecular rearrangements   10.244

There is no third kind of rearrangement here, since a ring is not involved. Secondly, all depend on the types of groups carried by the centers. Thirdly, all the mono-forms can be observed to be resonance stabilized as the level of unsaturation increases from two to three. That is the essence of Equation 10.244 above. Fourthly, the 1,5- cumulene obtained above undergoes molecular rear-rangement back to the diyne. This is not a reversible reaction, but that based on operating conditions. It like $H_2C=CH-C\equiv C-CH=CH_2$ are linear isomers of benzene.

None of the mono-forms of (III) of Equation 10.240 is symmetric. Like (II), all the mono-forms of (III) can undergo molecular rearrangement as already shown. Only (I) can carry charges, and when it does it cannot be resonance stabilized. Apart from the symmetric cases in (I) and (II), the radicals carried by the centers of activation of all of them are fixed.

$$\overset{CH_3}{\underset{CH_3}{\overset{|}{C}}} \equiv C - C \equiv \overset{}{\underset{CH_3}{C}} \longrightarrow e.\ \overset{CH_3}{\underset{CH_3}{\overset{|}{C}}} = C = C = C.n \longleftrightarrow n\ \overset{CH_3}{\underset{C}{\overset{|}{C}}} = \underset{CH_3}{\overset{|}{C}}.e$$

(II)   (Symmetric)

10.245

$$\overset{CH_3}{\underset{}{\overset{|}{C}}} \equiv C - \overset{CH_3}{\underset{}{\overset{|}{C}}} = \overset{CH_3}{\underset{CH_2}{\overset{|}{C}}} \longrightarrow e.\overset{CH_3}{\underset{}{\overset{|}{C}}} = C = \overset{CH_3}{\underset{CH_2}{\overset{|}{C}}} - \overset{CH_3}{\underset{}{\overset{|}{C}}}.n \longleftrightarrow e.\overset{CH_3}{\underset{}{\overset{|}{C}}} = \underset{H_2C - \overset{|}{C}CH_3}{\overset{|}{C}}.n$$

(III)   (Non-symmetric)

10.246

While (I) has one symmetric mono-form with limitation, (II) has one symmetric mono-forms with no limitation, (III) has two non-symmetric mono-forms. The limitation is stated in Equation 10.243. As has become obvious based on the mechanism of resonance stabilization provided so far, (II) is a fully resonance-stabilized symmetric monomer, while (I) is not.

For those monomers carrying C_2H_5 and above as substituents groups, the followings are obtained with C_2H_5.

98

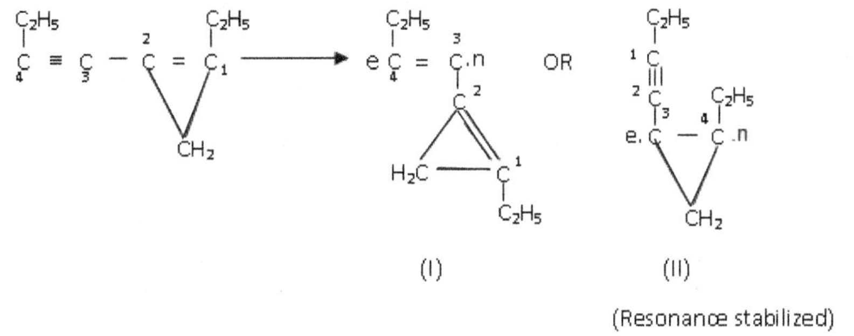

$$
\underset{4}{C} \equiv \underset{3}{C} - C = \underset{1}{C} \qquad \longrightarrow \qquad e\,\underset{4}{C} = \underset{2}{C}.n \qquad OR \qquad \underset{2}{C}
$$

(I) (II)

(Resonance stabilized) 10.247

Since they are resonance stabilized, the 1,4-mono-form should exist. All the mono-forms favor only one route radically. The 1,4-monoform, (I) and (II) can molecularly rearrange as shown below beginning with the forbidden center.

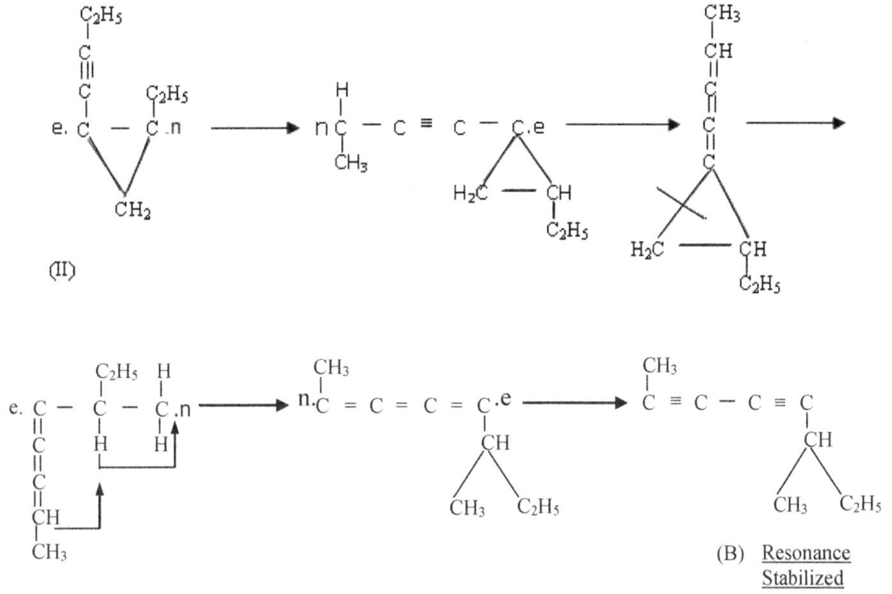

(II)

10.248

$$(B) \quad \underline{\text{Resonance Stabilized}}$$

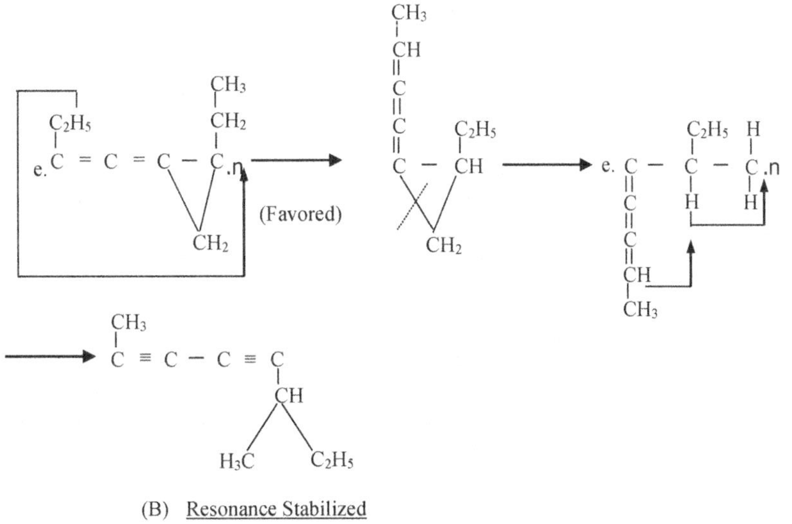

(Favored)

$$(B) \quad \underline{\text{Resonance Stabilized}} \qquad 10.249$$

99

Like similar cases encountered so far, the mono-forms rearrange to produce (B) above with radicals well placed.

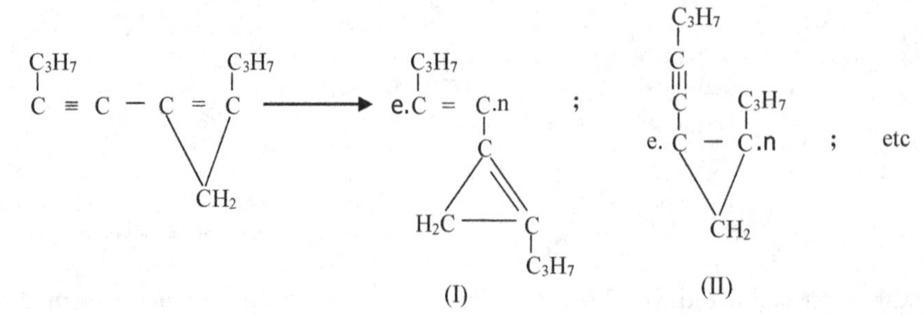

$$
\underset{\substack{|\\ CH_2}}{C \equiv C - C = C} \; \longrightarrow \; e.C = C.n \quad ; \quad e. \; C - C.n \quad ; \quad etc
$$

(I)   (II)   10.250

(II)

$$
e. \; C - C.n \; \longrightarrow \; n\overset{H}{\underset{C_2H_5}{C}} - C \equiv C - C - CH \; \longrightarrow \; C - CH \; \longrightarrow
$$

$$
e. \; C - C.n \; \longrightarrow \; C \equiv C - C \equiv C
$$

(B)   10.251

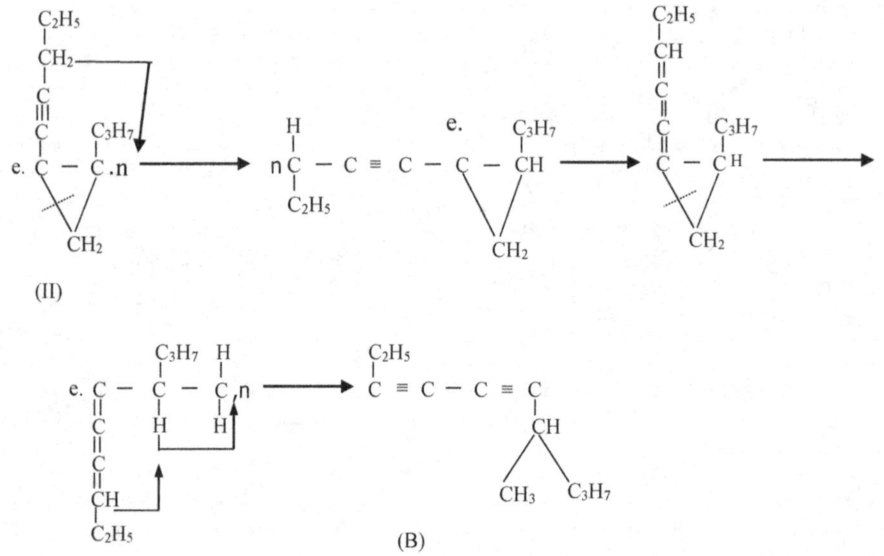

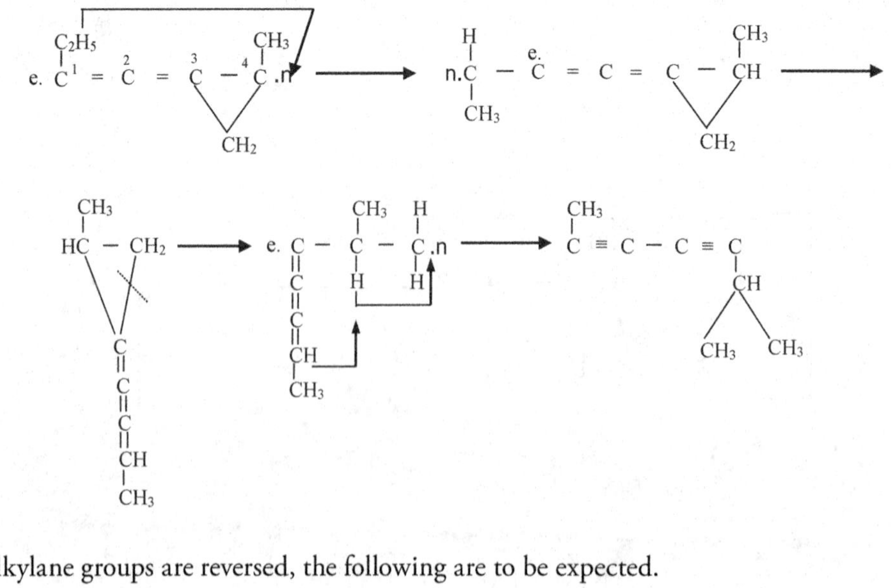

10.252

When the alkylane groups are reversed, the following are to be expected.

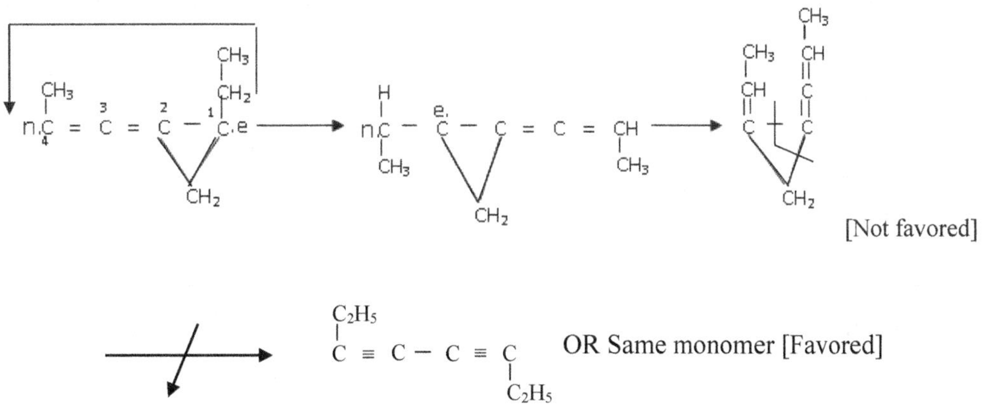

$$\text{n.}\underset{4}{C} = \underset{3}{C} = \underset{2}{C} - \underset{1}{C.e} \quad \longrightarrow \quad \text{n.}C - C - C = C = CH \quad \longrightarrow \quad \text{[Not favored]}$$

$$\longrightarrow\!\!\!/ \qquad \underset{C_2H_5}{\overset{C_2H_5}{C}} \equiv C - C \equiv C \qquad \text{OR Same monomer [Favored]}$$

10.253

From all the reactions above and so far, the following are conclusively obvious for acetylenic cyclopropene.

(i) Acetylenic cyclopropenes are all unsymmetrical and resonance stabilized.

(ii) All their mono-forms where the R group on acetylene side is equal to or greater than the R group on cyclopropene side, molecularly rearrange to produce same resonance-stabilized butadiyne monomer.

(iii) The radicals carried by the active centers of the 1,4-, 1,2- or 3,4- addition mono-forms are fixed.

(iv) That $R - C = C-$ is less than $RCH_2 - C \equiv C-$ in radical-pushing as already confirmed.

(v) That it is $H_2C = C = CH-$ type of group that first molecularly rearranges to

$H_3C\,C \circ C$ as follows in general, depending on what is on the other side-

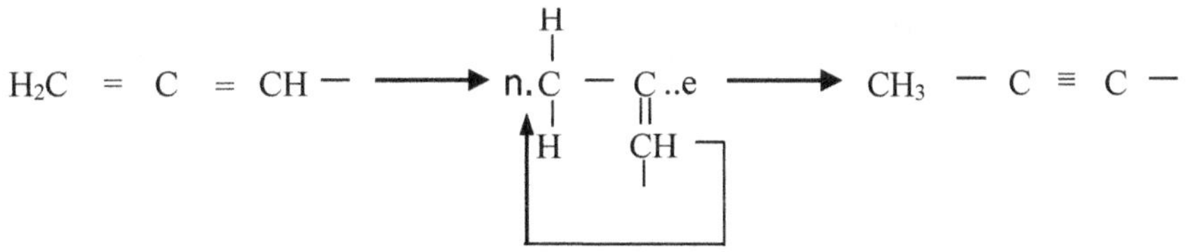

$$H_2C = C = CH - \longrightarrow \text{n.}C - C..e \longrightarrow CH_3 - C \equiv C -$$

10.254

during the steps of molecular rearrangements.

$$(CH_3)CH = C = CH - \longrightarrow n\,C - C.e \longrightarrow C_2H_5 - C \equiv C -$$

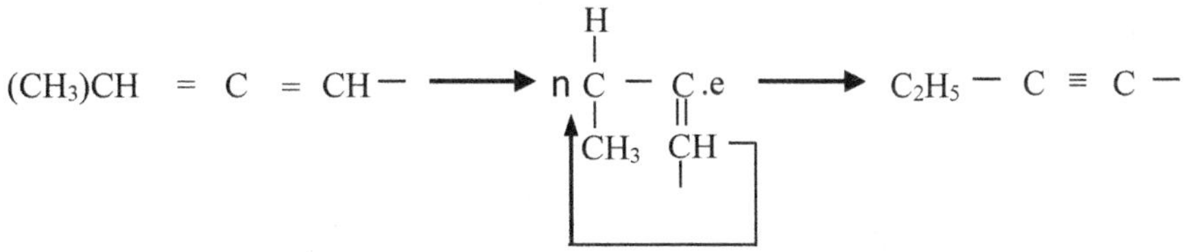

10.255

(vi) That with radical-pushing groups, the acetylenic activation center or the ring of these monomers can be involved in molecular rearrangements, depending on which center is carrying the larger group.

(vii) That only electro-free-radical route will favor the polymerization of these resonance stabilized monomers.

(viii) That there are limitations for molecular rearrangement with respect to the size of R groups.

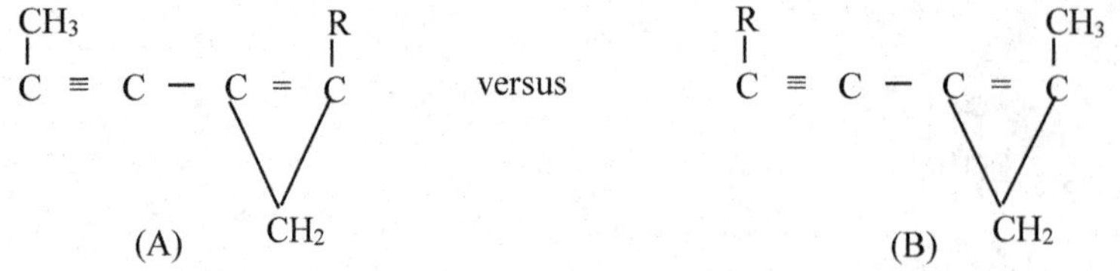

$$
\begin{array}{c}
\overset{CH_3}{\underset{|}{C}} \equiv C - \overset{R}{\underset{|}{C}} = C \\
\diagdown \diagup \\
CH_2
\end{array}
\qquad \text{versus} \qquad
\begin{array}{c}
\overset{R}{\underset{|}{C}} \equiv C - \overset{CH_3}{\underset{|}{C}} = C \\
\diagdown \diagup \\
CH_2
\end{array}
$$

(A)   (B)   10.256

But, when the CH_2 group internally located in the ring begins to change, the situation changes drastically as shown below.

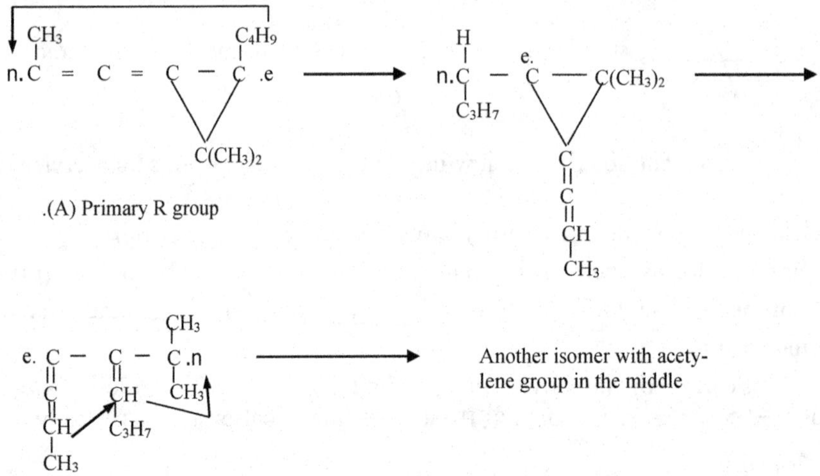

.(A) Primary R group

Another isomer with acetylene group in the middle

10.257

.(A) Secondary R group

No rearrangement possible

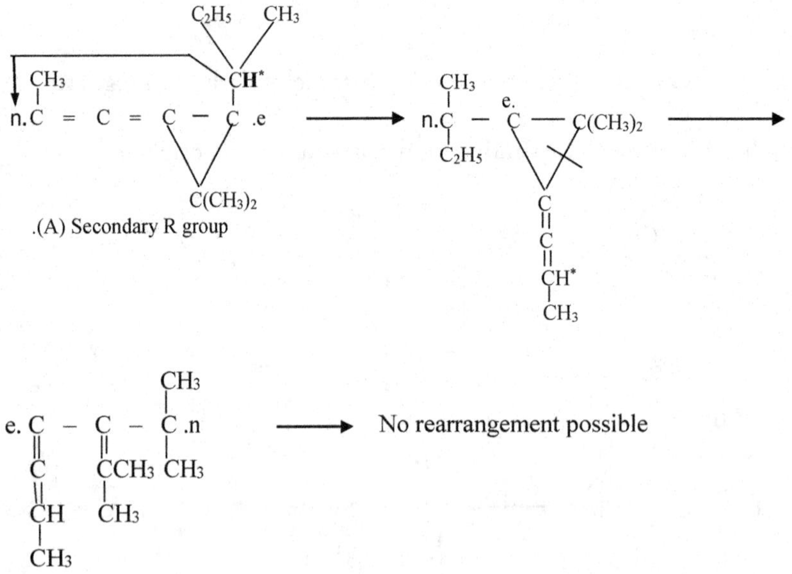

10.258

But when the H^* in (A) above is changed to CH_3, the followings are obtained.

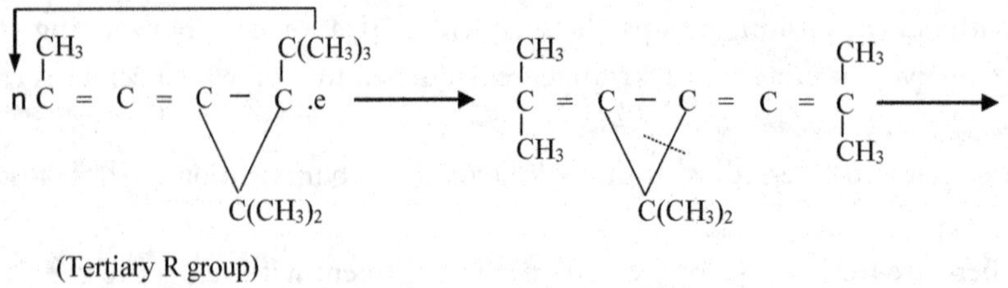

(Tertiary R group)

e.
$$
\underset{\underset{\underset{CH_3}{|}}{\overset{\overset{\overset{\overset{CH_3}{|}}{CCH_3}}{||}}{C}}}{C} - \underset{\underset{CH_3}{|}}{\overset{\overset{CCH_3}{|}}{C}} - \underset{\underset{CH_3}{|}}{\overset{\overset{CH_3}{|}}{C}} .n \longrightarrow \text{Same as the monomer}
$$

10.259

One can observe a distinction between primary, secondary and tertiary alkanes and the importance of these groups when the CH_2 group in the ring begins to carry alkane groups.

e.
$$
C = C = \underset{\underset{C(CH_3)_2}{|}}{\overset{\overset{CH_3}{|}}{C}} - \underset{\overset{C(CH_3)_3}{|}}{\overset{}{C}} .n \longrightarrow n. \underset{\underset{CH_3}{|}}{\overset{\overset{CH_3}{|}}{C}} - \overset{\bullet e}{C} = C = \underset{\underset{C(CH_3)_2}{|}}{\overset{}{C}} - \underset{\underset{CH_3}{|}}{\overset{\overset{CH_3}{|}}{C}} - CH_3 \longrightarrow
$$

e.
$$
\underset{\underset{\underset{C(CH_3)_2}{|}}{\overset{\overset{C}{||}}{C}}}{C} - \underset{\underset{CH_3}{|}}{\overset{\overset{CH_3}{|}}{C}} - \underset{\underset{CH_3}{|}}{\overset{\overset{CH_3}{|}}{C}} .n \longrightarrow \underset{\underset{\underset{C(CH_3)_3}{|}}{C(CH_3)_2}}{\overset{\overset{CH_3}{|}}{C}} \equiv C - C \equiv C
$$

FAVORED

10.260

From all the considerations above, the following is valid as has already been stated into law.

$$
- C(CH_3)_3 \quad > \quad - CH(CH_3)C_2H_5 \quad > \quad - C_4H_9
$$

<u>Tertiary</u>   <u>Secondary</u>   <u>Primary</u>

ORDER OF RADICAL-PUSHING CAPACITIES   10.261

It should not be forgotten that all the monomers above are resonance stabilized.
From all the considerations so far, the followings are obvious.

$$
\underset{\underset{CH_2}{\diagdown\diagup}}{\overset{\overset{\overset{\overset{CH_3}{|}}{C}}{\underset{\underset{C}{|||}}{}}}{C} = \underset{CH_3}{C}} \quad < \quad \underset{\underset{CH_2}{\diagdown\diagup}}{\overset{\overset{\overset{\overset{C_2H_5}{|}}{C}}{\underset{\underset{C}{|||}}{}}}{C} = \underset{C_2H_5}{C}} \quad < \quad \underset{\underset{CH_2}{\diagdown\diagup}}{\overset{\overset{\overset{\overset{C_3H_7}{|}}{C}}{\underset{\underset{C}{|||}}{}}}{C} = \underset{C_3H_7}{C}} \quad < \quad \underset{\underset{CH_2}{\diagdown\diagup}}{\overset{\overset{\overset{\overset{C_4H_9}{|}}{C}}{\underset{\underset{C}{|||}}{}}}{C} = \underset{C_4H_9}{C}}
$$

ORDER OF NUCLEOPHILICITY   10.262

OR

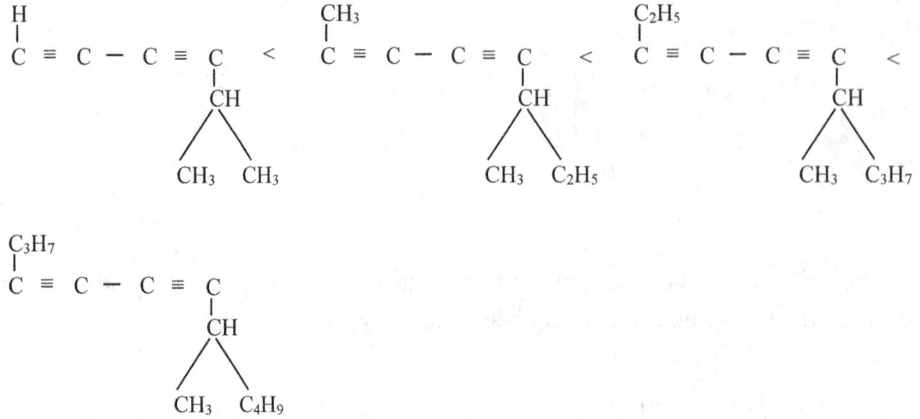

$$\underset{\substack{\text{CH}\\|\\\text{CH}_3\quad\text{CH}_3}}{\text{C}\equiv\text{C}-\text{C}\equiv\underset{|}{\text{C}}}\overset{\text{H}}{\underset{}{}} \quad < \quad \underset{\substack{\text{CH}\\|\\\text{CH}_3\quad\text{C}_2\text{H}_5}}{\text{C}\equiv\text{C}-\text{C}\equiv\underset{|}{\text{C}}}\overset{\text{CH}_3}{} \quad < \quad \underset{\substack{\text{CH}\\|\\\text{CH}_3\quad\text{C}_3\text{H}_7}}{\text{C}\equiv\text{C}-\text{C}\equiv\underset{|}{\text{C}}}\overset{\text{C}_2\text{H}_5}{} \quad <$$

$$\underset{\substack{\text{CH}\\|\\\text{CH}_3\quad\text{C}_4\text{H}_9}}{\text{C}\equiv\text{C}-\text{C}\equiv\underset{|}{\text{C}}}\overset{\text{C}_3\text{H}_7}{}$$

ORDER OF NUCLEOPHILICITY

[Structures showing four cyclic/branched alkyne-alkene compounds with substituents CH$_3$, C$_2$H$_5$, C$_3$H$_7$, C$_4$H$_9$ connected to CH$_2$ rings, separated by < signs]

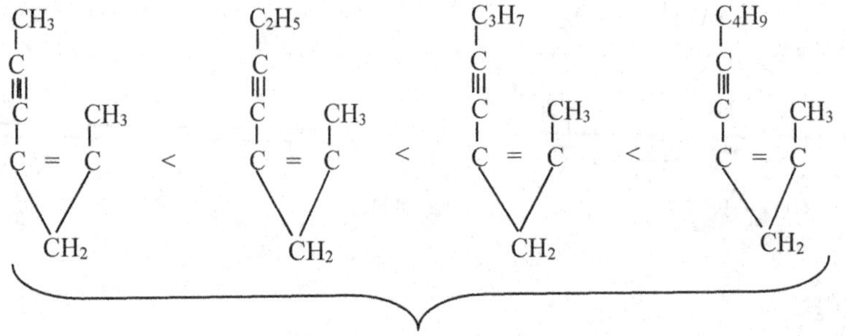

10.263

ORDER OF NUCLEOPHILICITY

OR

10.264

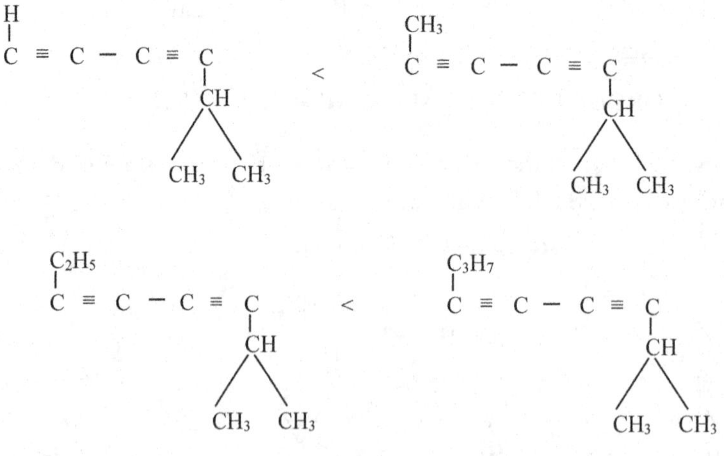

$$\underset{\substack{\text{CH}\\|\\\text{CH}_3\quad\text{CH}_3}}{\text{C}\equiv\text{C}-\text{C}\equiv\underset{|}{\text{C}}}\overset{\text{H}}{} \quad < \quad \underset{\substack{\text{CH}\\|\\\text{CH}_3\quad\text{CH}_3}}{\text{C}\equiv\text{C}-\text{C}\equiv\underset{|}{\text{C}}}\overset{\text{CH}_3}{}$$

$$\underset{\substack{\text{CH}\\|\\\text{CH}_3\quad\text{CH}_3}}{\text{C}\equiv\text{C}-\text{C}\equiv\underset{|}{\text{C}}}\overset{\text{C}_2\text{H}_5}{} \quad < \quad \underset{\substack{\text{CH}\\|\\\text{CH}_3\quad\text{CH}_3}}{\text{C}\equiv\text{C}-\text{C}\equiv\underset{|}{\text{C}}}\overset{\text{C}_3\text{H}_7}{}$$

Order of Nucleophilicity

10.265

Now, with radical-pushing groups such as OH, OR, NH$_2$ and NR$_2$, the situation is different. Their monomers are resonance stabilized. When OH is placed on the ring, the followings are to be expected.

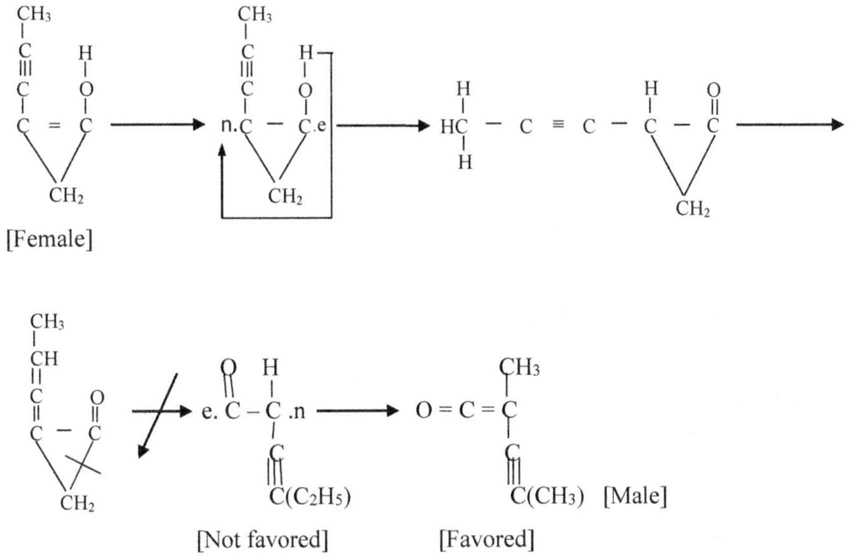

[Female]

[Not favored]   [Favored]   10.266

Free-radically, rearrangement to give a Ketene by decomposition mechanism is possible despite the fact that the laws of Conservation of transfer of transfer species does not apply, being an Electrophile..

With OH group on the acetylene center, the followings are to be expected.

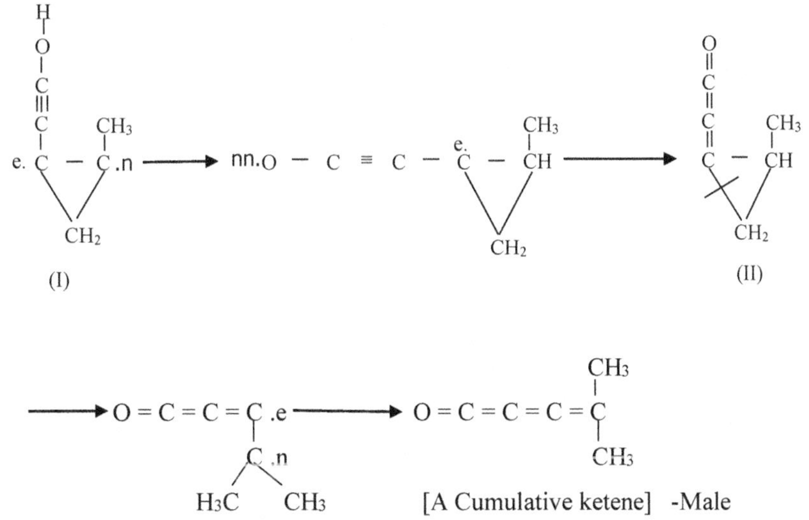

(I)   (II)

H_3C CH_3 [A Cumulative ketene] -Male   10.267

(II) above can only be opened via Equilibrium decomposition mechanism to give another more cumulative ketene. Otherwise, the ketene cannot be formed via Decomposition mechanism. So are the other mono-forms.

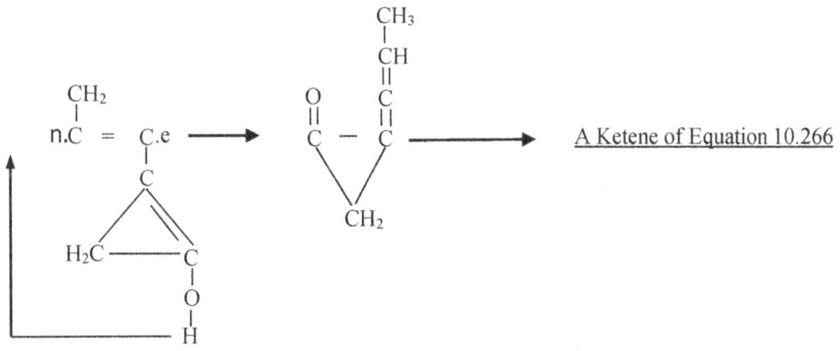

A Ketene of Equation 10.266

10.268

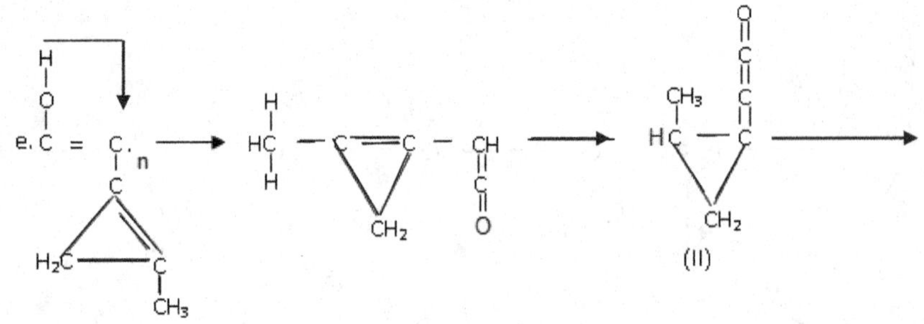

A cumulative Ketene of Equation 10,267   10.269

Polymerization will readily be favored only free-radically with OH and not with OR groups.

With NH_2 group in place of H on acetylene, the followings are expected.

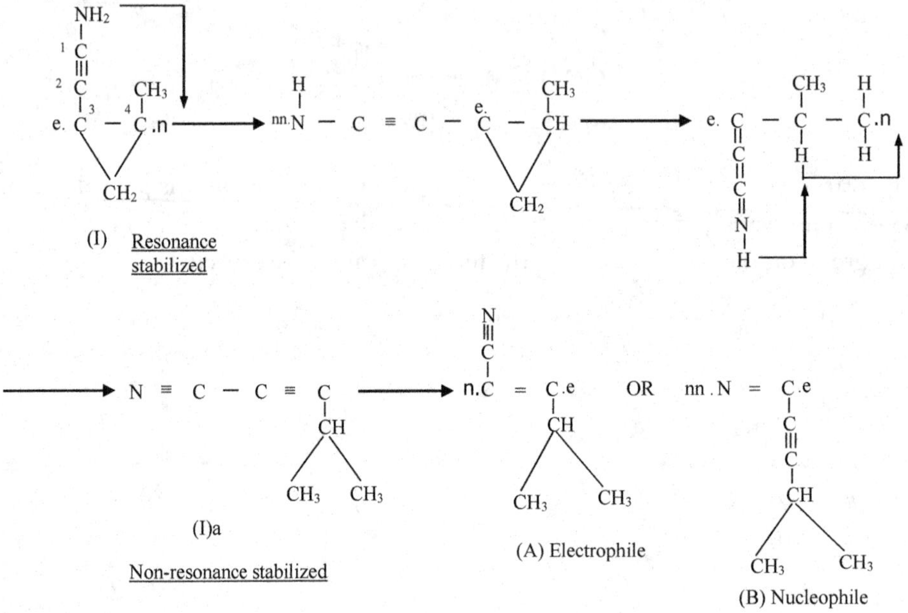

(I)   Resonance stabilized

(I)a

Non-resonance stabilized   (A) Electrophile   10.270

(B) Nucleophile

With NH_2 group on the ring, the followings are also expected.

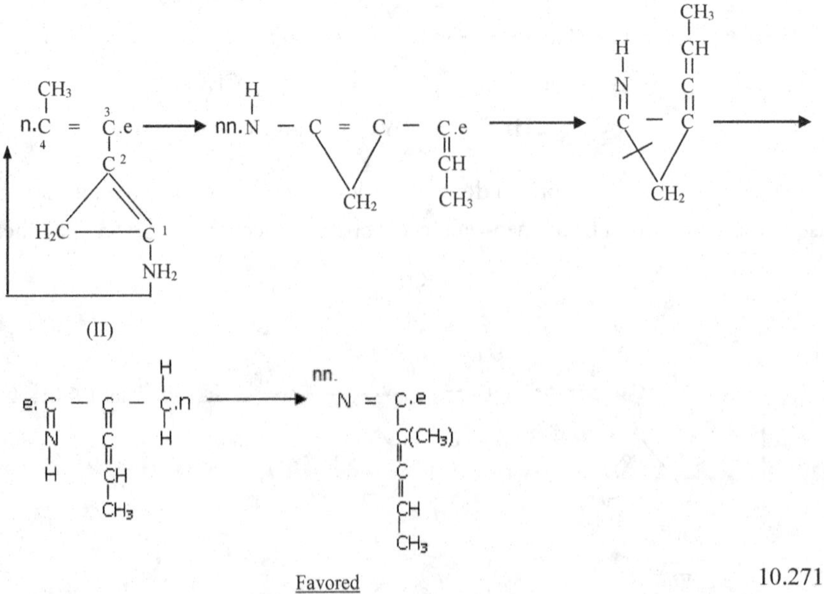

(II)

Favored   10.271   10.271

$$
\begin{array}{c}
CH_3 \\
| \\
NH \\
| \\
^1C \\
||| \\
^2C \quad CH_3 \\
| \quad | \\
e.\,C - ^4C.n \\
| \\
CH_2 \\
(III)
\end{array}
\longrightarrow
\begin{array}{c}
CH_3 \\
| \\
nn\,.N - C \equiv C - \overset{e.}{C} - CH \\
| \\
CH_2
\end{array}
\longrightarrow
\begin{array}{c}
CH_3 \quad H \\
| \quad | \\
e.\,C - C - C.n \\
|| \quad | \quad | \\
C \quad H \quad H \\
|| \\
C \\
|| \\
N \\
| \\
CH_3
\end{array}
$$

$$
\longrightarrow
\begin{array}{c}
N \equiv C - C \equiv C \\
| \\
C(CH_3) \\
/ \quad \backslash \\
CH_3 \quad CH_3 \\
(II)a
\end{array}
\longrightarrow
\begin{array}{c}
N \\
||| \\
C \\
| \\
C = C.e \\
n. \quad | \\
C(CH_3) \\
/ \quad \backslash \\
CH_3 \quad CH_3 \\
(C)
\end{array}
\quad OR \quad
\begin{array}{c}
nn. \\
N = C.e \\
| \\
C \\
||| \\
C \\
| \\
C(CH_3) \\
/ \quad \backslash \\
CH_3 \quad CH_3 \\
(D)
\end{array}
$$

Electrophile   Nucleophile   10.272

Compare this with that of Equation 10.270. All these are being used to illustrate fundamental principles.

$$
\begin{array}{c}
CH_3 \\
| \\
NCH_3 \\
| \\
^1C \\
||| \\
^2C \quad CH_3 \\
| \quad | \\
e.\,C - ^4C.n \\
\backslash \quad / \\
C(CH_3)_2 \\
(IV)
\end{array}
\longrightarrow
\begin{array}{c}
CH_3 \\
| \\
nn\,.N - C \equiv C - \overset{e.}{C} - C(CH_3) \\
\backslash \quad / \\
C(CH_3)_2
\end{array}
\longrightarrow
\begin{array}{c}
CH_3 \quad CH_3 \\
| \quad | \\
e.\,C - C - C.n \\
|| \quad | \quad | \\
C \quad CH_3 \quad CH_3 \\
|| \\
C \\
|| \\
N \\
| \\
CH_3
\end{array}
$$

$$
\longrightarrow
\begin{array}{c}
N \equiv C - C \equiv C \\
| \\
C(CH_3) \\
/ \quad \backslash \\
CH_3 \quad C(CH_3)_3 \\
(IV)b
\end{array}
\longrightarrow
\begin{array}{c}
N \\
||| \\
C \\
| \\
n\,.C = C.e \\
| \\
C(CH_3) \\
/ \quad \backslash \\
CH_3 \quad C(CH_3)_3 \\
(E) \text{ Electrophile}
\end{array}
\quad OR \quad
\begin{array}{c}
nn\,.N = C\,.e \\
| \\
C \\
||| \\
C \\
| \\
C(CH_3) \\
/ \quad \backslash \\
CH_3 \quad C(CH_3)_3 \\
(F) \text{ Nucleophile}
\end{array}
$$

__FAVORED__   10.273

However, from the products obtained, (A), (C) and (E), there is no doubt that the following is valid, noting that for $N(CH_3)_2$, the internal hydrogen atoms had to be changed to allow for the unquestionable rearrangement to be favored.

107

$$— N(CH_3)_2 \quad > \quad — NHCH_3 \quad > \quad — NH_2$$

ORDER OF RADICAL-PUSHING CAPACITIES 10.274

From the order the following relationship is also valid,

$$H_2N — C \equiv C— \quad < \quad HCH_3N — C \equiv C— \quad < \quad (CH_3)_2N— C \equiv C—$$

ORDER OF RADICAL-PUSHING CAPACITIES 10.275

With radical-pulling groups, one will begin with F, CF_3, C_2F_5 types of groups.

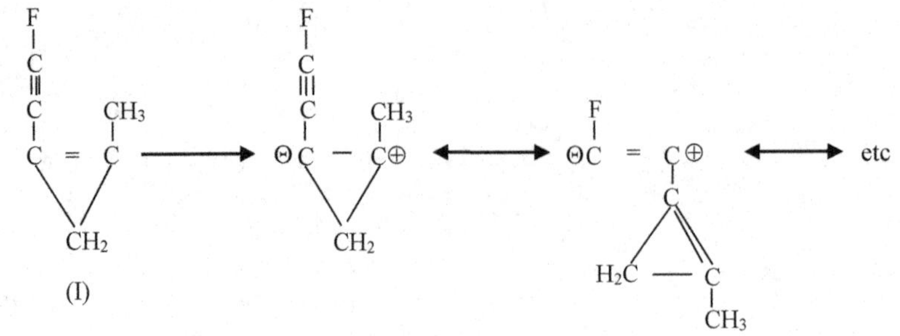

Impossible activated state 10.276

(I) is resonance stabilized and cannot be activated chargedly due to electrostatic forces of repulsion both from the ring and F atom. Free-radically, the following is obtained.

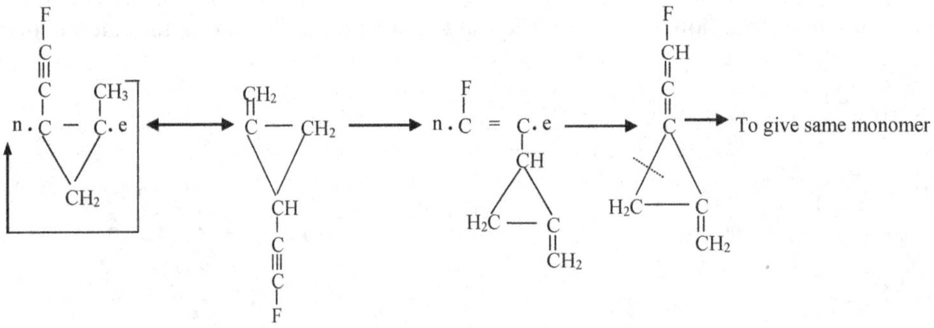

10.277

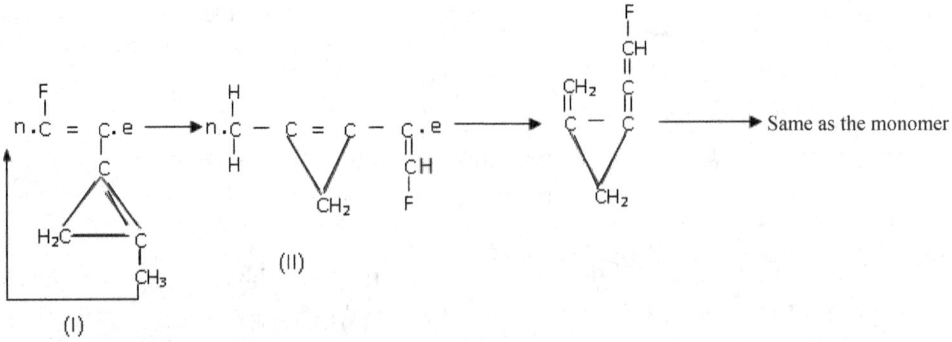

10.278

As can be observed so far, when molecular rearrangements take place, transformation of a compound from Female to Male and vice visa and transformation of resonance stabilized compound to non-resonance stabilized compound and vice visa are not hindered. All these depend on what the compound is carrying and the operating conditions.

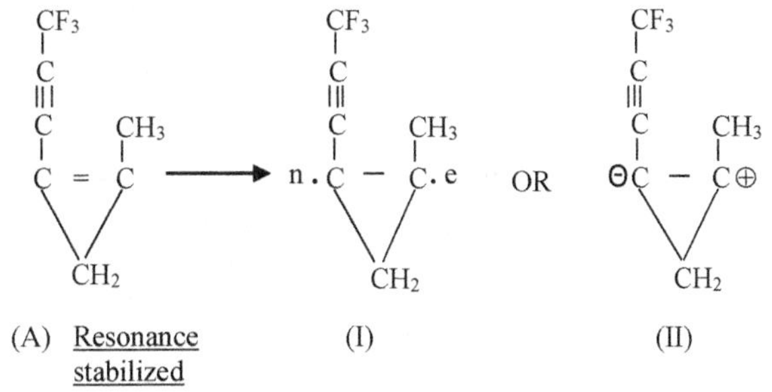

$$10.279$$

(I) will favor molecular rearrangement to produce the same molecule. (II) is not supposed to be shown. It has been placed there for exploratory purposes.

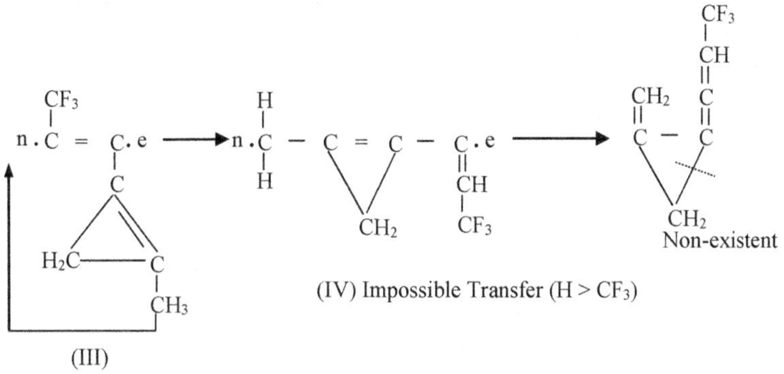

$$10.280$$

(IV) cannot exist, because H is more radical-pushing than CF_3. Radically the monomer above is resonance stabilized. Only the acetylenic center will favor being used for polymerization in the absence of steric limitations. In fact, as it is, it cannot favor any free-radical route.

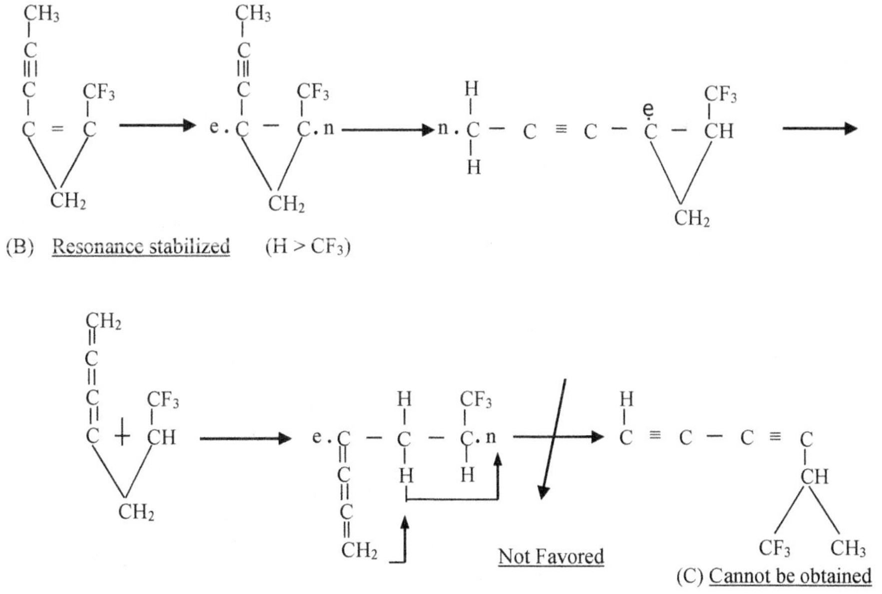

$$10.281$$

No rearrangement could take place here, because of the nature of transfer species involved, that which does not obey the Laws of Conservation of transfer of transfer species.

Chargedly, the following is obtained for (B).

$$
\begin{array}{c}
CH_3 \\
| \\
C \\
\lll \\
C \quad CF_3 \\
| \quad | \\
C = C \\
\diagdown\diagup \\
CH_2
\end{array}
\longrightarrow
\begin{array}{c}
CH_3 \\
| \\
C \\
\lll \\
C \quad CF_3 \\
| \quad | \\
\oplus C - C\ominus \\
\diagdown\diagup \\
CH_2
\end{array}
\longrightarrow
\begin{array}{c}
F \\
| \\
\oplus C - C\ominus - CH_2 \\
| \\
F \\
\qquad CF \\
\qquad | \\
\qquad C \\
\qquad \lll \\
\qquad C \\
\qquad | \\
\qquad CH_3
\end{array}
$$

Impossible Transfer
<u>Impossible activated state</u>

10.282

Transfer of F is impossible, the monomer being a Nucleophile. Radically, there is no rearrangement of any type. Considering the second activation center of (B), the followings are to be expected. This is the more nucleophilic center which indeed should not be the one to be activated.

$$
\begin{array}{c}
CH_3 \\
| \\
e \cdot C = C \cdot n \\
| \\
C \\
\diagup\diagdown \\
H_2C \quad\quad C \\
| \\
CF_3
\end{array}
\quad OR \quad
\begin{array}{c}
CH_3 \\
| \\
\oplus C = C\ominus \\
| \\
C \\
\diagup\diagdown \\
H_2C \quad\quad C \\
| \\
CF_3
\end{array}
$$

<u>No molecular rearrangements</u>   <u>Cannot exist</u>

10.283

It should be noted that (A) and (B) are resonance stabilized monomers, noting that (C) of Equation 10.281 cannot be obtained.

$$
\begin{array}{c}
C(CF_3)_3 \\
| \\
C \\
\lll \\
C \quad CF_3 \\
| \quad | \\
C = C \\
\diagdown\diagup \\
C(CF_3)_2
\end{array}
\longrightarrow
\begin{array}{c}
C(CF_3)_3 \\
| \\
C \\
\lll \\
C \quad CF_3 \\
| \quad | \\
e \cdot C - C \cdot n \\
\diagdown\diagup \\
C(CF_3)_2
\end{array}
\longrightarrow
\begin{array}{c}
CF_3 \\
| \\
n \cdot C - C \equiv C - \quad\quad\quad CF_3 \\
| \qquad\qquad\qquad\qquad\quad \overset{e}{C} - C(CF_3) \\
CF_3 \qquad\qquad\qquad\qquad \diagdown\diagup \\
\qquad\qquad\qquad\qquad\qquad C(CF_3)_2
\end{array}
\longrightarrow
$$

(I)   <u>Resonance stabilized</u>

$$
\begin{array}{c}
CF_3 \\
| \\
F_3CC - C(CF_3)_2 \\
\diagdown\diagup \\
C \\
\| \\
C \\
\| \\
C \\
| \\
C(CF_3) \\
| \\
CF_3
\end{array}
\longrightarrow
\begin{array}{c}
CF_3 \quad CF_3 \\
| \quad\quad | \\
e \cdot C - C - C \cdot n \\
\| \quad CF_3 \quad CF_3 \\
C \\
\| \\
C \\
| \\
C(CF_3) \\
| \\
CF_3
\end{array}
\quad \underline{Favored} \quad
\begin{array}{c}
CF_3 \\
| \\
\overset{}{C} \equiv \underset{3}{C} - \overset{2}{C} \equiv \underset{1}{C} \\
\underset{4}{} \qquad\qquad | \\
\qquad\qquad C(CF_3) \\
\qquad\qquad \diagup\diagdown \\
\qquad F_3C \quad C(CF_3)_3
\end{array}
$$

(II)
<u>Resonance stabilized</u>

10.284

(II) above will not favor any free-radical polymerization. Electro-free-radically, F is the transfer species, while nucleo-free-radically CF_3 is the transfer species. When only one center is activated during activation, the center is the 4,3- center which was 1,2- center in the original monomer.

110

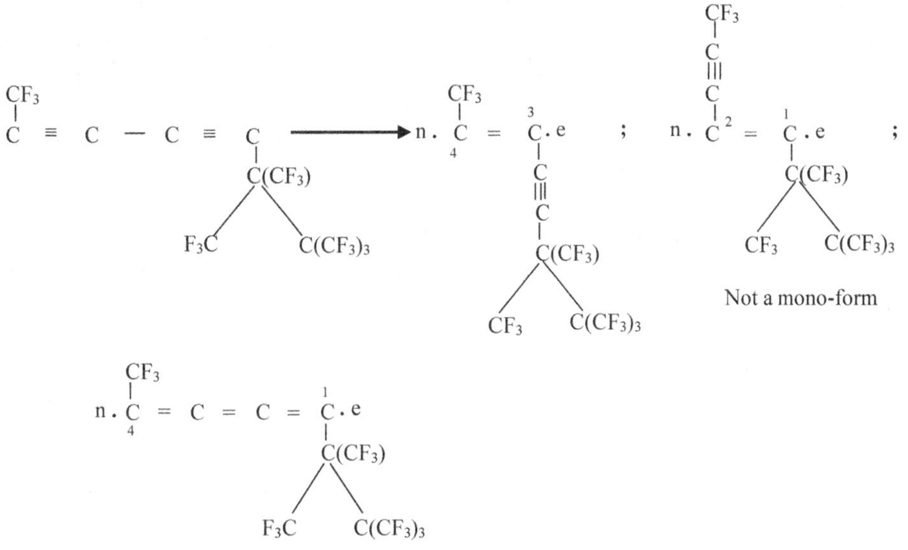

$$
\begin{array}{c}
CF_3 \\
| \\
C \equiv C - C \equiv C \\
\qquad\quad | \\
\qquad\quad C(CF_3) \\
\qquad\quad / \ \backslash \\
\quad F_3C \quad\ C(CF_3)_3
\end{array}
\longrightarrow
\begin{array}{c}
CF_3 \\
| \\
n \cdot \underset{4}{C} = \overset{3}{C} \cdot e \\
\quad\quad | \\
\quad\quad C \\
\quad\quad ||| \\
\quad\quad C \\
\quad\quad | \\
\quad\quad C(CF_3) \\
\quad\quad / \ \backslash \\
\ CF_3 \quad C(CF_3)_3
\end{array}
\ ; \
\begin{array}{c}
CF_3 \\
| \\
C \\
||| \\
C \\
| \\
n \cdot \overset{2}{C} = \overset{1}{C} \cdot e \\
\qquad | \\
\qquad C(CF_3) \\
\qquad / \ \backslash \\
\ CF_3 \quad C(CF_3)_3 \\
\text{Not a mono-form}
\end{array}
\ ;
$$

$$
\begin{array}{c}
CF_3 \\
| \\
n \cdot \underset{4}{C} = C = C = \overset{1}{C} \cdot e \\
\qquad\qquad\qquad | \\
\qquad\qquad\qquad C(CF_3) \\
\qquad\qquad\qquad / \ \backslash \\
\qquad\quad F_3C \quad\ C(CF_3)_3
\end{array}
\qquad\qquad\qquad\qquad 10.285
$$

With - $C(C_2F_5)(CF_3)_2$ group replacing $-C(CF_3)_3$ above, molecular rearrangements still remains favored.

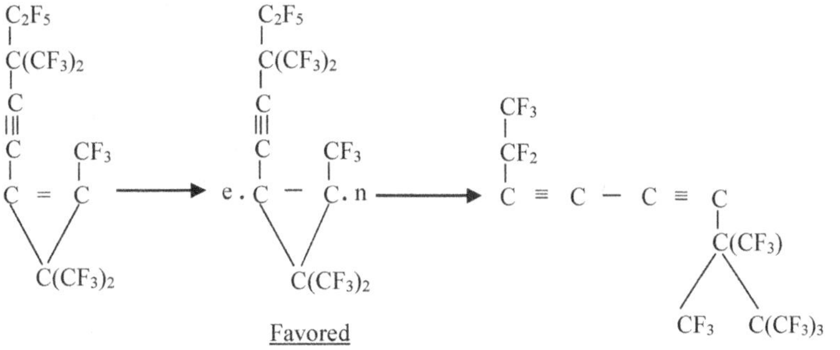

$$
\begin{array}{c}
C_2F_5 \\
| \\
C(CF_3)_2 \\
| \\
C \\
||| \\
C \quad\ CF_3 \\
| \ / \\
C = C \\
\ \backslash \\
\ C(CF_3)_2
\end{array}
\longrightarrow
\begin{array}{c}
C_2F_5 \\
| \\
C(CF_3)_2 \\
| \\
C \\
||| \\
C \quad\ CF_3 \\
| \ / \\
e \cdot C - C \cdot n \\
\ \backslash \\
\ C(CF_3)_2 \\
\underline{\text{Favored}}
\end{array}
\longrightarrow
\begin{array}{c}
CF_3 \\
| \\
CF_2 \\
| \\
C \equiv C - C \equiv C \\
\qquad\qquad\quad | \\
\qquad\qquad\quad C(CF_3) \\
\qquad\qquad\quad / \ \backslash \\
\qquad\quad CF_3 \ C(CF_3)_3
\end{array}
\quad 10.286
$$

Fluorinated acetylenes are different free-radically. Shown below are some of them.

$$
\begin{array}{cccc}
\begin{array}{c} F_3C \\ | \\ C \equiv C \\ \qquad | \\ \qquad F \end{array}
&
\begin{array}{c} F_5C_2 \\ | \\ C \equiv C \\ \qquad | \\ \qquad F \end{array}
&
\begin{array}{c} F_5C_2 \\ | \\ C \equiv C \\ \qquad | \\ \qquad CF_3 \end{array}
&
\begin{array}{c} F_5C_2 \\ | \\ C \equiv C \\ \qquad | \\ \qquad C_3F_7 \end{array} \\
(A) & (B) & (C) & (D)
\end{array}
\qquad 10.287a
$$

(A), (B), (C), and (D) are Nucleophiles. While (A) has no transfer species when activated, (B), (C) and (D) have. The transfer species in (B) is CF_3 nucleo-free-radically the route which is not natural to the monomer. The transfer species in (C) and (D) are F electro-free- radically and CF_3 nucleo-free

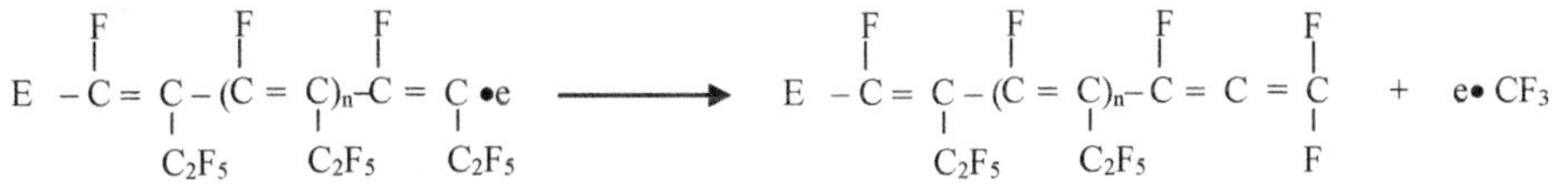

$$
\begin{array}{c}
\quad\ F \quad\quad F \quad\quad F \\
\quad\ | \qquad | \qquad | \\
E - C = C - (C = C)_n - C = C \bullet e \\
\qquad\ | \qquad\quad | \qquad\quad | \\
\qquad C_2F_5 \quad C_2F_5 \quad C_2F_5
\end{array}
\longrightarrow
\begin{array}{c}
\quad\ F \quad\quad F \quad\quad F \quad\quad F \\
\quad\ | \qquad | \qquad | \qquad | \\
E - C = C - (C = C)_n - C = C = C \\
\qquad\ | \qquad\quad | \qquad\qquad\quad | \\
\qquad C_2F_5 \quad C_2F_5 \qquad\quad F
\end{array}
+ \ e \bullet CF_3
$$

CHARACTER OF A NUCLEOPHILE IN (B)   10.287b

-radically. Therefore, (C) and (D) cannot be polymerized. If C_3F_7 was primary, then the transfer species nucleo-free-radically is C_2H_5. Free-radically C_nF_{2n+1} groups are radical-pushing groups of lower capacity than H. Chargedly, they are radical-pulling groups, like CN, COOH, COOR, COR, OCOR types of groups. ***One can observe that C_nF_{2n+1} types of groups have dual characters unlike C_nH_{2n+1} types of groups which have only one transfer species, H.*** This is also very much unlike allylic groups of the types – CH_2F, -CH_2CONH_2, -CH_2COOH, --CH_2CN, -CH_2COOCH_2 etc. which are radical-pushing groups all of greater capacity than H, but of lesser capacity than CH_3. Though –CF_2H is not allylic, it is still radical-pushing of greater capacity than H. As far as the author is concerned, in C/F family, CF_2H is fluoro-allylic.

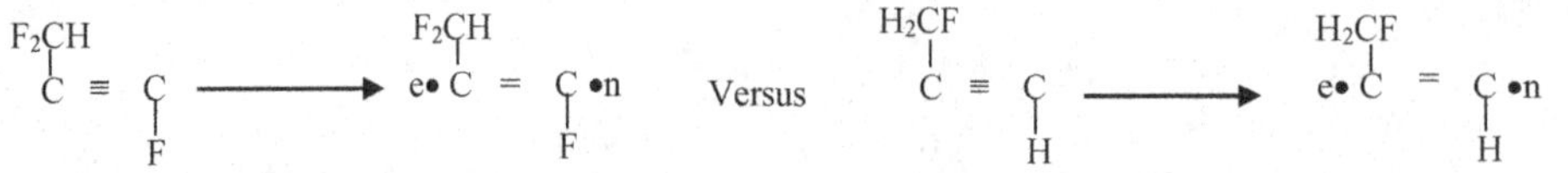

(A) C/F-Allylic group-CF₂H   **(B) C/H-Allylic group-CH₂F**   10.287c

Now consider, cases where allylic groups are involved.

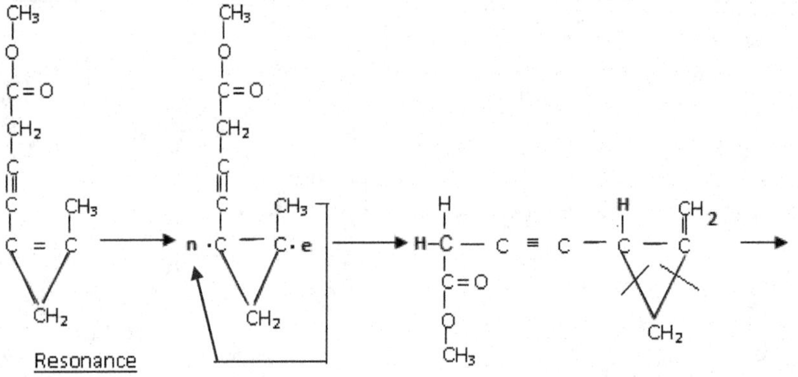

(B) The same monomer   10.288

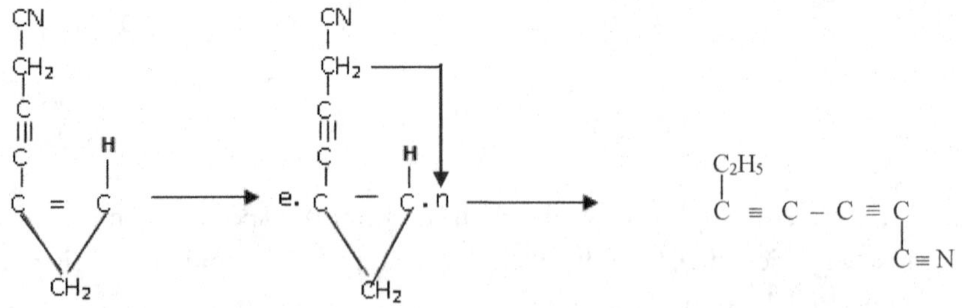

(B) Resonance stabilized   10.289

Just as $HOOC-CH_2C\equiv C-$ is a radical-pushing resonance stabilization group, so also is $NC-CH_2C\equiv C-$ group. It should be recalled that all the butadiynes obtained above are resonance stabilized. The last one is an electrophile since the $C\equiv N$ or $C=O$ centers are never involved. Only the Y centers are involved.

With COOR, $CONH_2$ and COR types of groups, the analysis of the past remain the same, that is the ring cannot readily be opened. As Electrophiles, the transfer species can never be H.

Cannot easily be opened

10.290

Free-radically, the rings cannot easily be opened. When opened at the point indicated above, it breaks down to give an Electrophile- $(OCH_3)CH_3C=CH(HC=C=C=O)$. When the ring carries the COOR group, after the first rearrangement, the ring cannot easily be opened. However, when opened, a resonance stabilized product may still be obtained.

10.3. Proposition of rules of Chemistry and concluding remarks

So far with only the considerations of cycloalkanes and cyclopropenes, one has begun to identify the driving forces favoring the opening of rings. In view of the unique character of cyclopropenes, it was the only member of cycloalkenes considered for study in this chapter. For the first time, one has identified what the Strain energy in a ringed compound is. It is an invisible π-bond nucleophilic in character. For the first time, using the concept of the mechanics of springs, one has identified what is called or referred to as minimum required strain energy (MRSE) which any ring must possess before it can be opened when external forces in form of energy are applied. There is also maximum required strain energy ($M_{ax}RSE$), yet to be fully identified. It is indeed the energy which makes a ring which is supposed to exist to never exist as ring such as cyclobutadiene.

For the first time, one has started to identify points of scissions in rings. When a ring has point of scission, it can be opened at that point when the MRSE is attained for the ring. For the first time also, the influence of types of substituent or substituted groups carried by the connecting centers in rings have begun to be identified. Why cyclopropenes are not known to be useful monomers have been explained. How they can be used as monomers have also been provided. In the process, new phenomena were identified with respect to molecular rearrangements and resonance stabilizations. What is particularly unique is that these phenomena are limited only to nucleophiles (female monomers). In the process, the analysis of cyclopropenes, diynes, triynes, vinyl cyclopropene, methylene cyclopropane and vinyl cyclopropanes were considered. For the first time, it was noted that not all conjugatedly placed triple bonds (e.g., H_3C-CºC-CºC-CºN) are fully resonance sta-bilized monomers. This was also however noted to be limited to electrophiles and not nucleophiles. New resonance stabilization groups were identified. One has begun to move slowly and steadily using the three basic mechanisms for all Systems-Combination, Equilibrium and Decomposition mechanisms. Any mechanism which cannot be interpreted into a mathematical form using mathematical languages without the need of making necessary and unnecessary assumptions, is no mechanism, Most of the rings were opened via Decomposition mechanism while some were opened via Equilibrium mechanism. High operating conditions are usually required for opening of these rings.

Rule 583: This rule of Chemistry for **Ringed compounds,** states that, the strain energy (SE) is that energy which a ring must possess before it can exist as a ring similar to that which a spring must possess before it can be called a spring in Physics and which in Chemistry is said to be *the invisible π-bond of different capacities; in order words no ring with a SE of zero can exist.*
(Laws of Creations in Chemistry as applied to Physics)

Rule 584: This rule of Chemistry for **Ringed compounds,** states that, the SE is the same for a family of ringed compounds, for which if a family has many sizes, the same energy is carried by each member, thus making the smallest size to be the most strained member of the family, followed by the next size and so on in decreasing order; thus making the invisible π-bond weaker as we move from the smallest size to the next and so on in the imaginary world, but stronger in the real world.
(Laws of Creations in Chemistry as applied to Physics and Mathematics)

Rule 585: This rule of Chemistry for **Ringed compounds,** states that, the SE for every family of ringed compounds, is carried by the first member of the family which is the smallest size of ring and it is the value of the SE that eventually determines what the size of the last member of that family will be, the point where the SE is close to zero but never zero in the absence of any molecular rearrangements.
(Laws of Creations for Limitations in Ringed compounds)

Rule 586: This rule of Chemistry for **Ringed compounds,** states that, since every ring belongs to a particular family and every family has its own value of SE as an identity carried by the first member of the family, therefore the SE of a family is a function of the atoms that form the connecting centers of the ring and the types of groups carried by the connecting centers; for which the order of SE between different families cannot be used as a measure of the order of nucleophilicity between families, but only as a measure for a particular family.
(Laws of Creations for Ringed compounds)

Rule 587: This rule of Chemistry for **Ringed compounds,** states that, for a particular family of rings, the size of a ring is a *measure of the order of nucleophilicity of the ring,* whether the ring is a Nucleophile or an Electrophile; for which the smaller the size of the ring, the easier it is to open the ring and therefore the less nucleophilic is the ring as shown below for cycloalkanes.

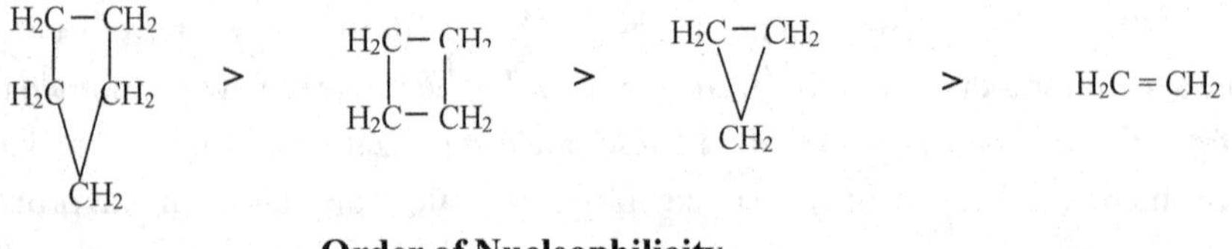

$$H_2C - CH_2 \quad > \quad H_2C - CH_2 \quad > \quad H_2C - CH_2 \quad > \quad H_2C = CH_2$$

Order of Nucleophilicity

(Laws of Creations for Ringed compounds)

Rule 588: This rule of Chemistry for **Ringed compounds,** states that, before any ring can be opened, the ring must have *a point of scission,* for which if none exists (such as in Benzene), the ring cannot be opened no matter how much energy from different means is put into the ring.
(Laws of Creations for Ringed compounds)

Rule 589: This rule of Chemistry for **Ringed compounds,** states that, the point of scission in a ring is always a single σ- bond between two centers with *the largest radical potential difference,* which is measured or determined by the capacities of the groups carried by the two centers forming the bond; for which when the centers are the same and they carry same groups the potential difference is zero and any bond can be a point of scission, as shown below for some of them.

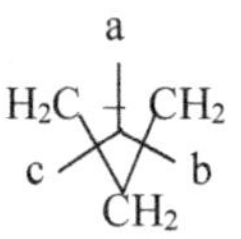

(I) (3) 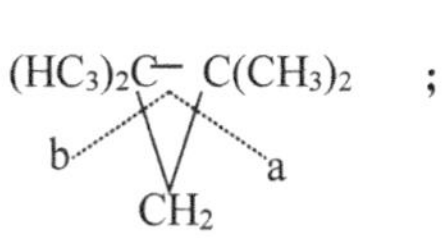   (II) (2) 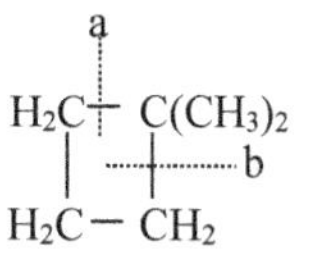   (III) (2) 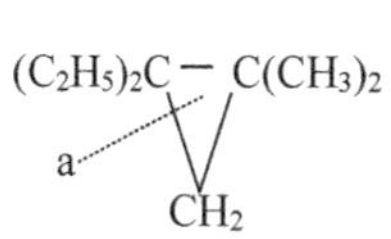   (IV) (1)

(Laws of Creations for Ringed compounds)

Rule 590: This rule of Chemistry for **Ringed compounds,** states that, when a family of ringed compounds exists and some members are not present, this can only be from the smallest size upwards, in view of the presence of Max.RSE in the rings, such as the case of three-membered cyclic ketone not known to exist.

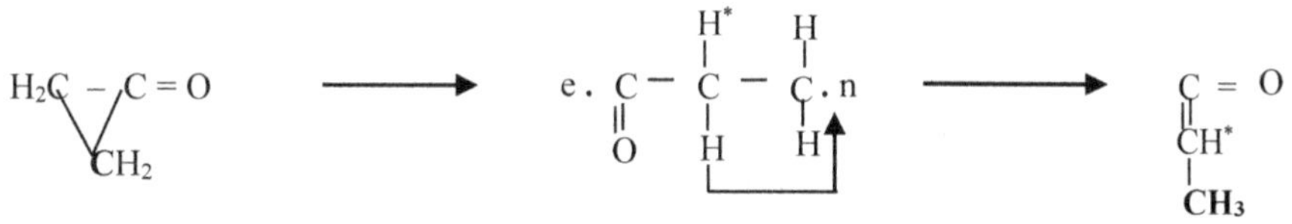

Molecular rearrangement of 3rd kind   (Electrophile)

(Laws of Creations for Ringed compounds)

Rule 591: This rule of Chemistry for **Ringed compounds,** states that, due to electrostatic forces of repulsion chargedly or strong radical-pushing capacity of $H_2C=$, $RHC=$, $R_2C=$ types of groups, scission of a single bond between two center adjacently located to a $C = C$ double bond inside any ring is impossible; such is the case with benzene, cyclopropenes and more shown below.

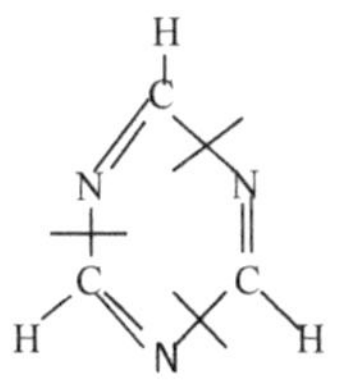

(B) THREE POINTS OF SCISSION 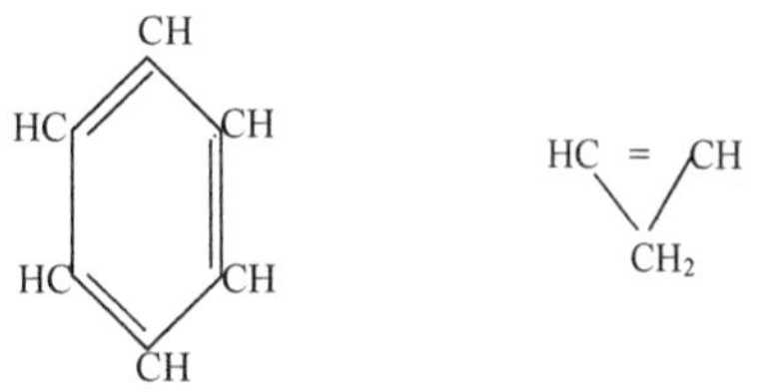   **(B) NO POINT OF SCISSION**

CAN EXIST    CANNOT EXIST

(Laws of Creations for π-bonded Rings)

Rule 592: This rule of Chemistry for **Ringed compounds,** states that, the amount of energy required to be added to the SE to open a ring is called the ***Minimum required strain energy (MRSE) which implies that Total Energy in Ring <u>at time of opening</u> equals SE plus MRSE.***
(Laws of Creations for Ringed compounds)

Rule 593: This rule of Chemistry for **Ringed compounds,** states that, the MRSE for a ring is a measure of the degree of unsaturation of the ring for which the smaller it is, the more invisibly unsaturated is the ring for a family and the less nucleophilic is the ring; for which in general all rings can be said to be ***invisibly unsaturated*** whether a visible π-bond is present in the ring or not.
(Laws of Creations for Ringed compounds)

Rule 594: This rule of Chemistry for **Ringed compounds,** states that, ***the first factor determining the ability of a ring to possess the MRSE***, is the size of the ring; for which all three-membered rings for all families where they exist require the smallest MRSE, being the most strained.
(Laws of Creations for Ringed compounds)

Rule 595: This rule of Chemistry for **Ringed compounds,** states that, ***the second factor determining the ability to possess the MRSE***, is the number of double, polar, electrostatic, or triple bonds present in a ring; for which when a triple bond is present, there is far more strain energy than when a polar or electrostatic or double bond is present in the same ring size, which in turn is more strained than when saturated; the strain energy increasing with increasing presence or numbers of these bonds.
(Laws of Creations for Ringed compounds)

Rule 596: This rule of Chemistry for **Ringed compounds,** states that, ***the third factor determining the ability of a ring to possess the MRSE***, is the types of groups carried by the connecting centers of a ring; for which when the groups are all radical-pushing groups, the SE of the ring is increased, and when one, two or more radical-pulling groups are present, the SE in decreased; for example, while cyclopropane can be opened, fluorinated cyclopropane may not or cannot be opened.
(Laws of Creations for Ringed compounds)

Rule 597: This rule of Chemistry for **Ringed compounds,** states that, ***the fourth factor determining the ability of a ring to possess the MRSE***, is the presence of paired unbonded radicals on the con-necting centers of the ring; for which the more their presence in a ring, the more strained is the ring.
(Laws of Creations for Ringed compounds)

Rule 598: This rule of Chemistry for **Ringed compounds,** states that, ***the fifth factor determining the ability of a ring to possess the MRSE***, is presence of hetero atoms in the connecting centers of a ring, for which the more their presence, the more strained is the ring.
(Laws of Creations for Ringed compounds)

Rule 599: This rule of Chemistry for **Ringed compounds,** states that, *the first driving force* favoring the opening of a ring **instantaneously**, is provision of MRSE for the ring, via application of heat without the use of functional centers in the ring when present or application of electro-dynamic forces provided by centers carrying paired unbonded radicals or paired radical or charged coordination centers or irradiation and much more.
(Laws of Creations for Ringed compounds)

Rule 600: This rule of Chemistry for **Ringed-compounds,** states that, *the second driving force* favoring the opening of a ring **instantaneously,** is the size of the ring, for which the larger the size, the more difficult it is to attain the MRSE; for example, cyclohexane cannot readily be opened.
(Laws of Creations for Ringed compounds)

Rule 601: This rule of Chemistry for **Ringed compounds,** states that, when the first member of a family which is supposed to exist does not exist at STP (e.g. three-membered cyclic amide, three-membered cyclic ketone), the reason is because the ring is carrying equal to or far more than the *Maximum required strain energy (MaxRSE)* for the family and this energy is more than the SE of existing first member; for which for a particular family, while the SE and MaxRSE are fixed, the MRSE is not fixed, but varies from size to size.
(Laws of Creations for Ringed compounds)

Rule 602: This rule of Chemistry for **Ringed compounds,** states that, Molecular *rearrangement of the third kind* is that which involves movement of transfer species of *the first kind of third type* from one center to another center adjacently located continuously without breaking the laws of Conservation of transfer of transfer species, after *opening of the ring instantaneously* to form a stable product as shown below for a ring with transient existence –

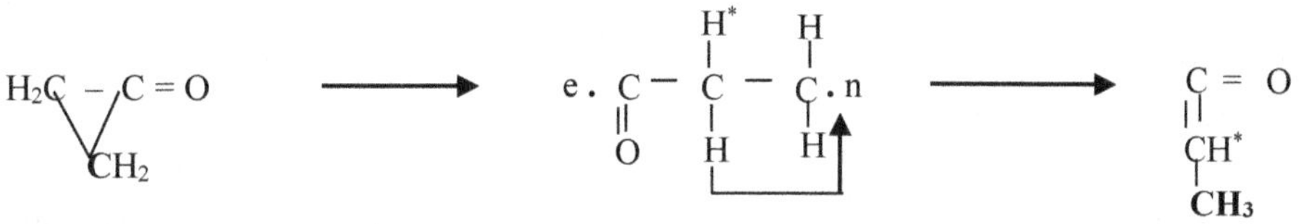

Molecular rearrangement of 3rd kind   (Electrophile)

for which a strained three-membered Electrophilic ring has been converted to a ketene also an Electrophile.
(Laws of Creations for Molecular Rearrangements of the Third kind)

Rule 603: This rule of Chemistry for **Ringed compounds,** states that, when a ring is opened instantaneously, this can only be done either via Decomposition or Equilibrium mechanisms where possible or both as shown below for three-membered cyclic amide which is known not to exist.

Stage 1:   **EQUILIBRIUM MECHANISM**

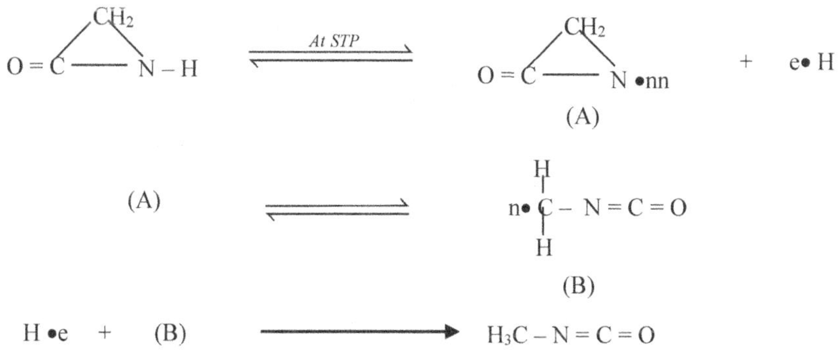

Overall Equation: Three-membered Cyclic Amide ⟶ **Methyl Isocyanate**

117

Stage 1:   <u>**DECOMPOSITION MECHANISM**</u>

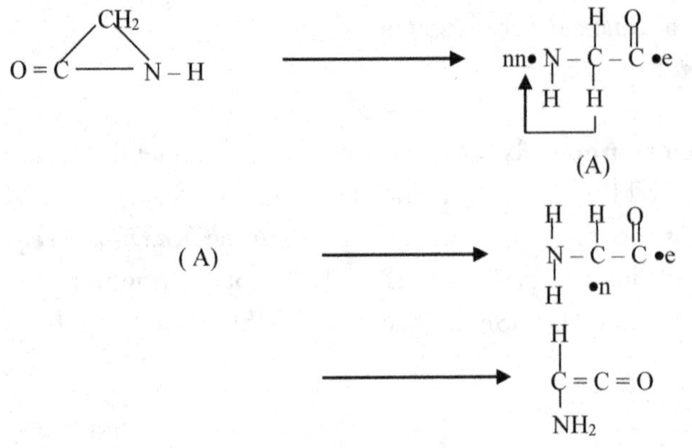

Overall Equation: Three-membered Cyclic Amide ——————→ A Ketene

noting how the point of scission has differed based on State of existences.

(Laws of Creations for Ringed compounds)

<u>**Rule 604:**</u> This rule of Chemistry for **Cyclopropane,** states that, this three-membered ring is the first member of its own family-the cycloalkanes family and as such, the most strained and the easiest to open instantaneously, since these families carry no functional center and visible activation center.

(Laws of Creations for Cyclopropane)

<u>**Rule 605:**</u> This rule of Chemistry for **Cyclopropane,** states that, when opened instantaneously in the absence of an initiator, no products can be obtained, since Molecular rearrangement of the third kind is not favored by it; while in the presence of an initiator, both free-radical and charged routes are favored by it to give poly (cyclopropane) as shown below using paired-media positive charges from BF_3/ROR combination.

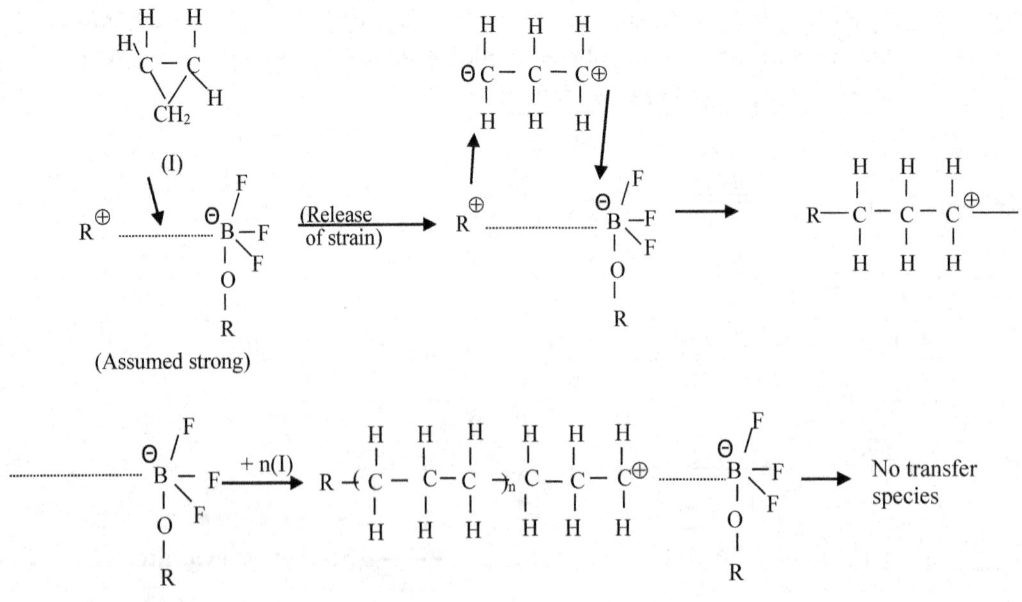

(Laws of Creations for Cyclopropane)

Rule 606: This rule of Chemistry for **Cyclopropane,** states that, when made to exist in Equilibrium state of Existence using a passive catalyst, it decomposes via Equilibrium mechanism as shown below, to give propene.

Stage 1:

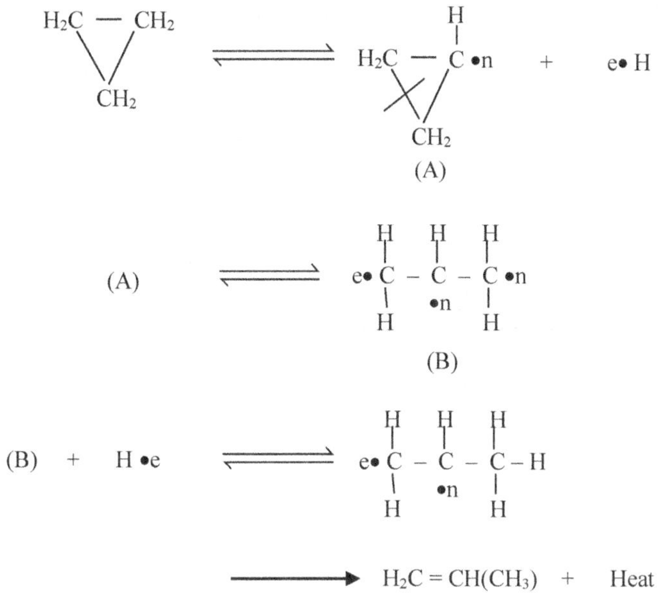

Overall Equation: Cyclopropane ⟶   Propene +   Heat

(Laws of Creations for Cyclopropane)

Rule 607: This rule of Chemistry for **Cyclopropane,** states that, when made to carry *one radical-pushing alkylane group,* the number of point of scission is reduced from 3 to 2 and the monomer becomes a stronger nucleophile and more strained, the degree of added strain depending on the capacity of the group carried; for which they can be made to undergo Molecular rearrangement of the third kind to give alkenes and the only routes favored by all of them are the natural routes- the electro-free- radical and positively charged routes.
(Laws of Creations for Cyclopropanes)

Rule 608: This rule of Chemistry for **Cyclopropane,** states that, when made to carry *two radical-pushing alkylane groups of the same capacity,* the number of points of scission is reduced from 3 to 2 and the monomer becomes a stronger nucleophile and more stained; for which Molecular rearrangement of the third kind cannot take place and the only routes favored are the electro-free-radical and positively charged routes.
(Laws of Creations for Cyclopropanes)

Rule 609: This rule of Chemistry for **Cyclopropane,** states that, when made to carry *two radical-pushing alkylane groups of different capacities,* the number of points of scission is reduced from 3 to 1 and the monomer becomes a stronger nucleophile and more stained; for which they can undergo molecular rearrangement of the third kind to give alkenes and the routes favored by them are the electro-free-radical and positively charged routes.
(Laws of Creations for Cyclopropanes)

Rule 610: This rule of Chemistry for **Cyclopropane,** states that, when made to carry radical-pushing alkylane groups, the followings are to be expected for the order of strain energy and nucleophilicity.

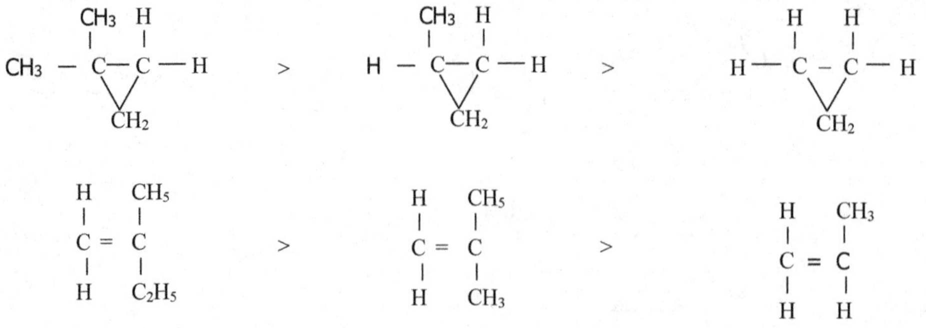

$$CH_3 - \underset{\underset{CH_2}{\overset{|}{\bigvee}}}{\overset{\overset{CH_3}{|}}{C}} - \overset{\overset{H}{|}}{\underset{|}{C}} - H \quad > \quad H - \underset{\underset{CH_2}{\overset{|}{\bigvee}}}{\overset{\overset{CH_3}{|}}{C}} - \overset{\overset{H}{|}}{\underset{|}{C}} - H \quad > \quad H - \underset{\underset{CH_2}{\overset{|}{\bigvee}}}{\overset{\overset{H}{|}}{C}} - \overset{\overset{H}{|}}{\underset{|}{C}} - H$$

$$\underset{\underset{C_2H_5}{|}}{\overset{\overset{H}{|}}{C}} = \underset{\underset{C_2H_5}{|}}{\overset{\overset{CH_5}{|}}{C}} \quad > \quad \underset{\underset{CH_3}{|}}{\overset{\overset{H}{|}}{C}} = \underset{\underset{CH_3}{|}}{\overset{\overset{CH_5}{|}}{C}} \quad > \quad \underset{\underset{H}{|}}{\overset{\overset{H}{|}}{C}} = \underset{\underset{H}{|}}{\overset{\overset{CH_3}{|}}{C}}$$

ORDER OF STRAIN ENERGY AND NUCLEOPHILICITY

(Laws of Creations for Cyclopropanes)

Rule 611: This rule of Chemistry for **Cyclopropane,** states that, when made to carry resonance stabilization group of the vinyl type (Vinyl cyclopropane), despite the fact that the ring is little more strained than when H was there, the monomer is more nucleophilic than cyclopropane and the ring is more nucleophilic than the vinyl group-

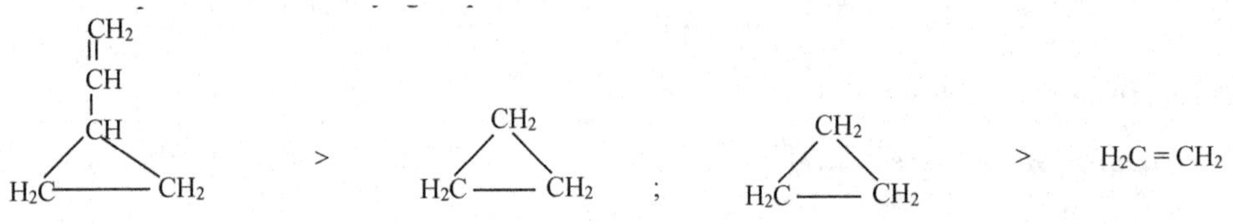

(I)   (II)
Order of Nucleophilicity   **Order of Nucleophilicity**

for which the followings are to be expected electro-free-radically,

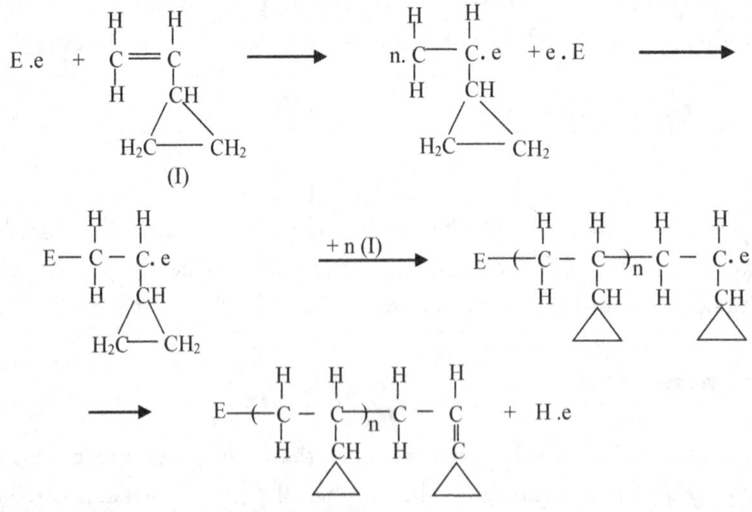

Transfer species of the 1ˢᵗ kind of the ninth type.

with the presence of transfer species of the first kind of the ninth type.

(Laws of Creations for Vinyl cyclopropane)

120

Rule 612: This rule of Chemistry for **Vinyl cyclopropane**, states that, the radical-pushing capacity of cyclopropane ring can be obtained as follows when R group is placed on the vinyl group, *via instantaneous opening of the ring artificially, since this is impossible without first activating the vinyl center when R(ºAlkylane groups) is less than C_2H_5-*

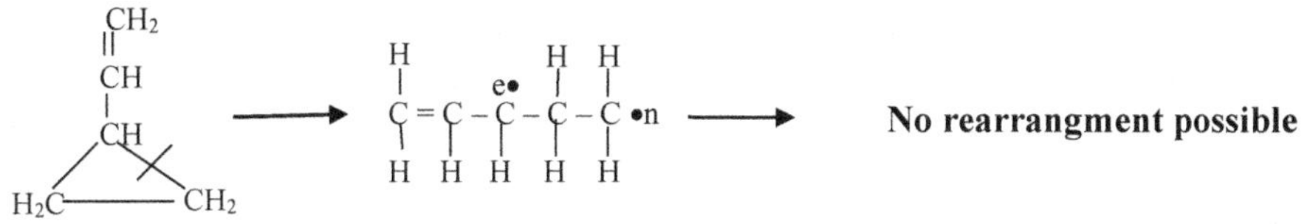

(I) NOT FAVORED

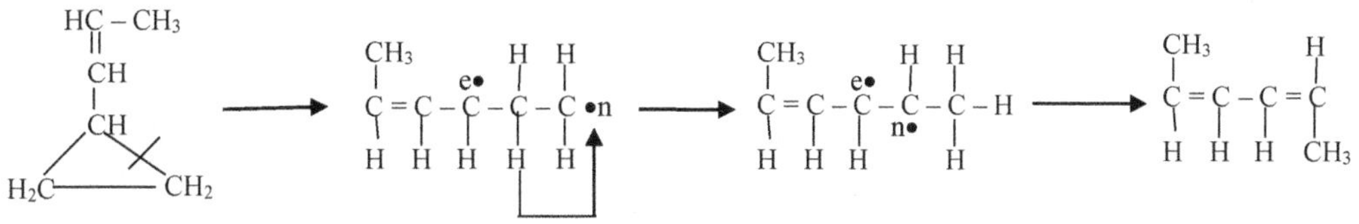

(II) FAVORED

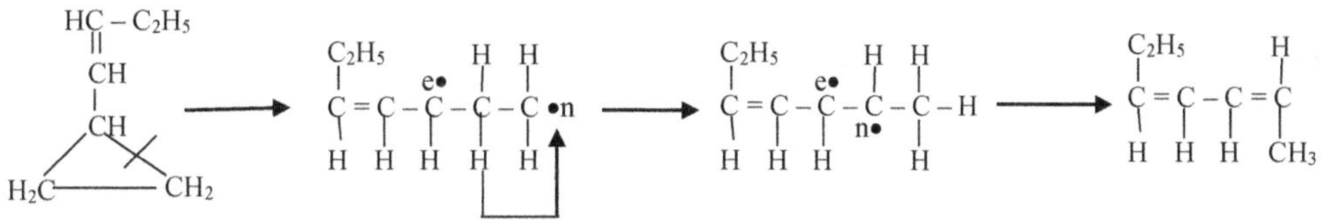

(III) FAVORED

from which the following is valid-

ORDER OF NUCLEOPHILICITY   ORDER OF RADICAL-PUSHING CAPACITY

(Laws of Creations for Vinyl cyclopropane)

Rule 613: This rule of Chemistry for **the use of $AlBr_3$ as initiator,** states that, just like BF_3, $SnCl_4$, and BF_3, it can only be used as Electrostatically positively charged-paired initiator as shown below-

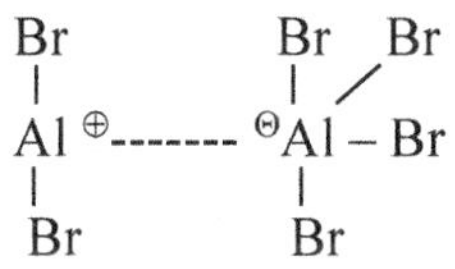

ELECTROSTATICALLY POSITIVELY CHARGED-PAIRED INITIATOR
(Laws of Creation for Initiators)

Rule 614: This rule of Chemistry for **Vinyl cyclopropanes,** states that, when the ring is made to carry one or two radical-pushing alkylane group specially located, under certain operating condi-tions, the ring can be made to readily decompose via Equilibrium mechanism to give a cyclo-pentadiene without any rearrangement; noting that this could be done not by instantaneously opening of the ring, but by activation of the externally located activation center followed by opening of the ring as shown below, for which heat must be applied-

Stage 1:

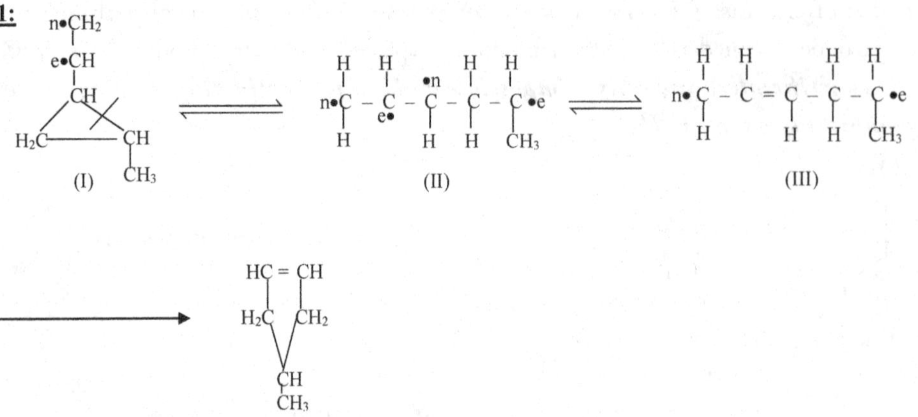

$$n\bullet CH_2$$
$$e\bullet CH$$
$$CH$$
$$H_2C \quad CH$$
$$(I) \quad CH_3$$

$$\rightleftharpoons \quad n\bullet\overset{H}{\underset{H}{C}} - \overset{H}{\underset{e\bullet}{C}} - \overset{\bullet n}{\underset{H}{C}} - \overset{H}{\underset{H}{C}} - \overset{H}{\underset{CH_3}{C}}\bullet e$$

$$(II)$$

$$\rightleftharpoons \quad n\bullet\overset{H}{\underset{H}{C}} - \overset{H}{\underset{H}{C}} = \overset{}{\underset{H}{C}} - \overset{H}{\underset{H}{C}} - \overset{H}{\underset{CH_3}{C}}\bullet e$$

$$(III)$$

$$\longrightarrow$$

$$HC = CH$$
$$H_2C \quad CH_2$$
$$CH$$
$$CH_3$$

4-Methyl cyclopentadiene

noting herein how a three-membered ring has been expanded to a five-meebered ring using a resonance stabilization group where possible.

(Laws of Creations for Vinyl cyclopropane)

Rule 615: This rule of Chemistry for **Vinyl cyclopropanes,** states that, for the two monomers shown below, since $H_2C = C(CH_3)$ – group is less radical-pushing than H, the followings are to be expected when $SnCl_4$ is used as source of initiator; for which (I) may not readily favor opening of the ring, while (II) will as shown below, noting that the route is the positively charged route-

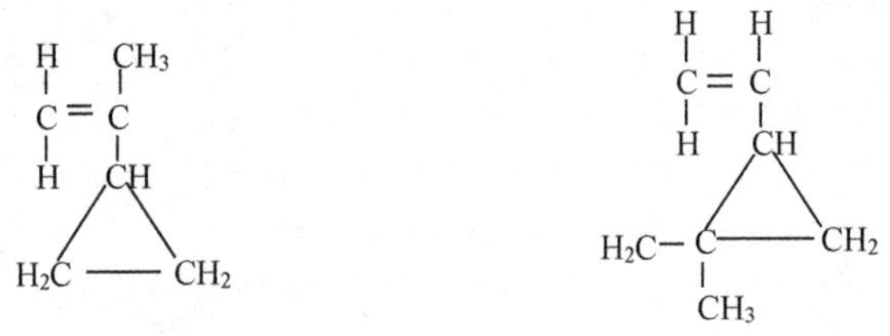

$$\overset{H}{\underset{H}{C}} = \overset{CH_3}{\underset{CH}{C}}$$
$$H_2C \underline{\quad\quad} CH_2$$

(I) C_6H_{10}

$$\overset{H}{\underset{H}{C}} = \overset{H}{\underset{CH}{C}}$$
$$H_2C - \overset{}{\underset{CH_3}{C}} \underline{\quad} CH_2$$

(II) C_7H_{11}

Derivatives of vinylcyclopropane

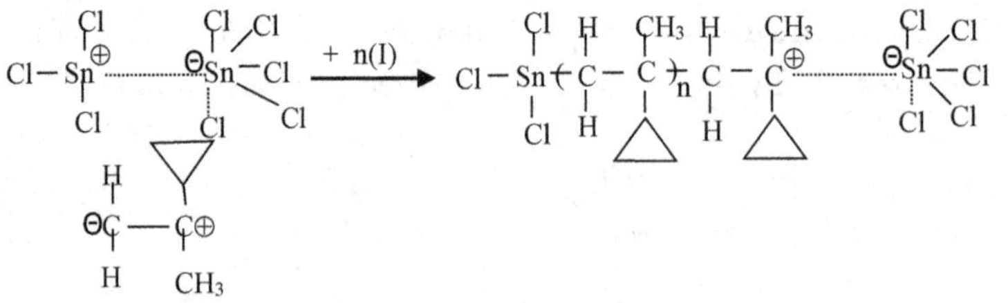

$$Cl - \overset{Cl}{\underset{Cl}{\overset{|}{Sn}}}\oplus \cdots \ominus\overset{Cl}{\underset{Cl}{\overset{|}{Sn}}} \diagup Cl \quad + n(I) \longrightarrow Cl - \overset{Cl}{\underset{Cl}{\overset{|}{Sn}}} \left(\overset{H}{\underset{H}{C}} - \overset{CH_3}{\underset{}{C}} \right)_n \overset{CH_3}{\underset{H}{C}} - \overset{\oplus}{\underset{}{C}} \cdots \ominus\overset{Cl}{\underset{Cl}{\overset{|}{Sn}}}\diagup Cl$$

$$\ominus\overset{H}{\underset{H}{C}} - \overset{}{\underset{CH_3}{C}}\oplus$$

(I)

$$Cl - \overset{Cl}{\underset{Cl}{\overset{|}{Sn}}}\oplus \cdots \ominus\overset{Cl}{\underset{Cl}{\overset{|}{Sn}}}\diagup Cl \quad + n(II) \longrightarrow Cl - \overset{Cl}{\underset{Cl}{\overset{|}{Sn}}} \left(\overset{H}{\underset{H}{C}} - \overset{H}{\underset{H}{C}} = \overset{}{\underset{R}{C}} - \overset{H}{\underset{R}{C}} - \overset{H}{\underset{H}{C}} \right)_n \overset{H}{\underset{R}{C}} - \overset{H}{\underset{H}{C}} = \overset{}{\underset{R}{C}} - \overset{H}{\underset{R}{C}} - \overset{\oplus}{\underset{}{C}} \cdots \ominus\overset{Cl}{\underset{Cl}{\overset{|}{Sn}}}\diagup Cl$$

$$H_2C \underline{\quad\quad} CR_2$$
$$\overset{H}{\underset{}{C}H} \quad CH$$
$$\ominus\overset{}{\underset{H}{C}} - \overset{}{\underset{H}{C}}\oplus$$

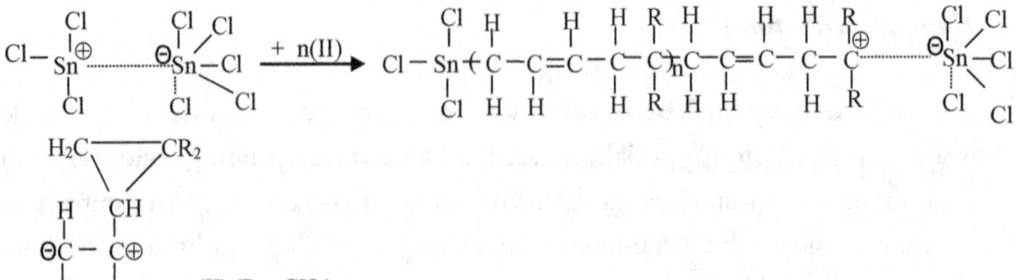

(II) $(R \equiv CH_3)$

noting that the paired unbonded radicals on the Sn centers have been datively blocked using $SnCl_4$.
(Laws of Creation for Vinyl cyclopropanes)

Rule 616: This rule of Chemistry for **Cyclopropane,** states that, when made to carry resonance stabilization group of the ***acetylenic type,*** despite the fact that the ring is now less strained than when H was there, the monomer is still more nucleophilic than cyclopropane in view of the fact that the following is valid-

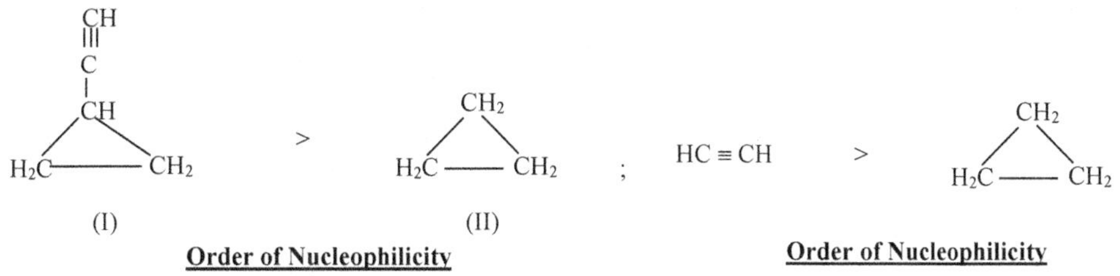

for which only the ring is the center of action and based on the capacity of the acetylenic group compared to H, the ring cannot be opened, and there may be need to replace the H atom on acetylenic side with radical-pushing group if the ring is to be opened.
(Laws of Creations for Acetylenic cyclopropane)

Rule 617: This rule of Chemistry for **Cyclopropane,** states that, when made to carry one radical-pushing group of the type OH, NH_2 and NHR, the number of point of scission is reduced from 3 to 2 and the monomer becomes more nucleophilic and more strained than when an alkylane group is carried; for which they can be made to undergo Molecular rearrangement of the third and first kinds to give a ketone for OH and aldimines for NH_2 and NHR as shown below and the only routes favored by them are the electro-free-radical and positively charged routes.

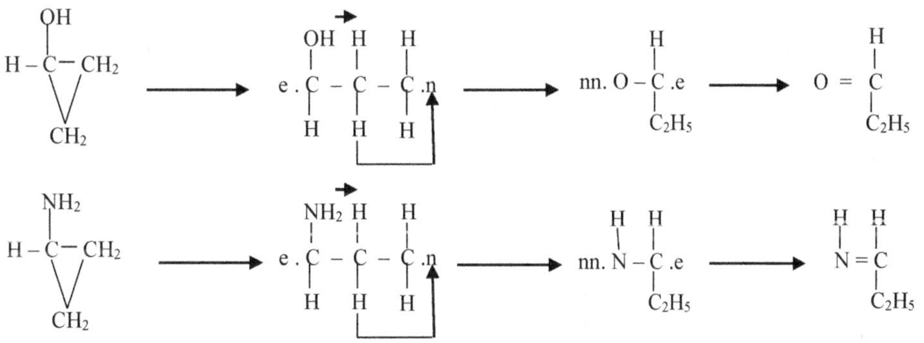

(Laws of Creations for Cyclopropanes)

Rule 618: This rule of Chemistry for **Cyclopropane,** states that, when made to carry one radical-pushing group of the type OR, NR_2, the number of point of scission is reduced from 3 to 2 and the monomer becomes more nucleophilic and more strained than when an alkylane group is carried; for which they cannot undergo Molecular rearrangement of the third kind (to give vinyl alcohol for OR and vinyl amine for NR_2) and the only routes favored by them are the electro-free-radical and positively charged routes when the ring is opened.
(Laws of Creations for Cyclopropanes))

Rule 619: This rule of Chemistry for **Cyclopropane,** states that, when made to carry one radical-pushing group of the type CF_3, C_2F_5, though two points of scission exist in them, the rings can rarely be opened, because the ring is less strained; for which if any attempt is made to open the ring, the following are to be expected for this example, with no possibility of Molecular rearrangement of the third kind taking place, noting that, $:NR_3$ has been used here to open the ring instantaneously.

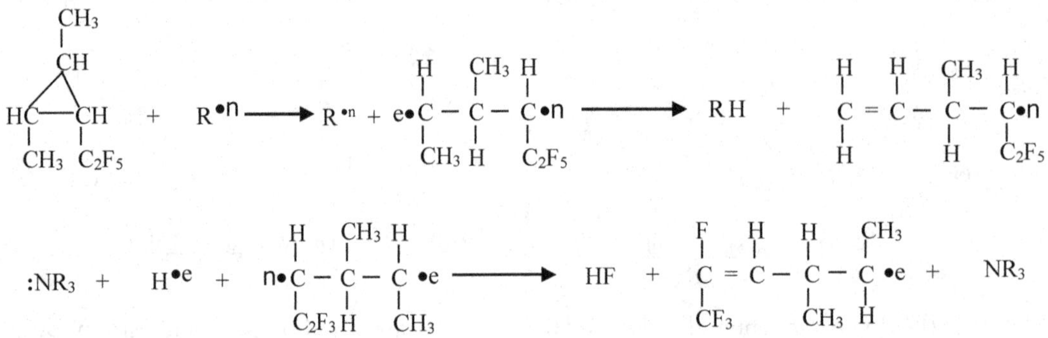

(Laws of Creations for Cyclopropanes)

Rule 620: This rule of Chemistry for **Cyclopropane,** states that, when made to carry one or more radical-pulling groups of the type COOH, COOR, $CONH_2$, OCOR, Cl, CN, though points of scission exist in the rings, they cannot be opened even when radical-pushing groups are present, because the ring is far less strained than when H was in place; for which nothing can be done, and even the externally located activation centers cannot be used because the following is valid.

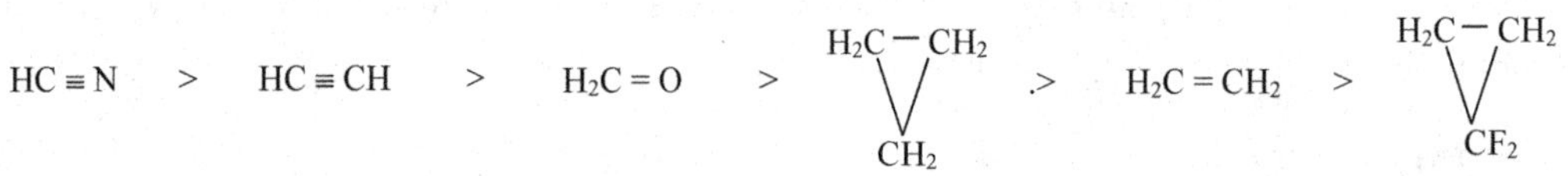

ORDER OF NUCLEOPHILICITY OF SOME ACTIVATION CENTERS

noting that those with COOH, $CONH_2$, CN, types of groups are indeed Electrophiles, wherein the invisible π-bond are the Y centers and the C=O, C≡N centers are the X centers; hence they cannot readily be opened

(Laws of Creations for Cyclopropanes)

Rule 621: This rule of Chemistry for **Cyclobutane,** states that, being the first member of its own family and the second member of the Cycloalkane family with SE of the order of ¾ of the SE of three-membered ring, it is still strained to favor instantaneous opening of the ring which cannot undergo Molecular rearrangement of the third kind, but can still be used as monomer to give poly(cyclobutane) via all free-radical and free-charged routes as shown below electro-free-radically using $TiCl_3$ the route natural to it.

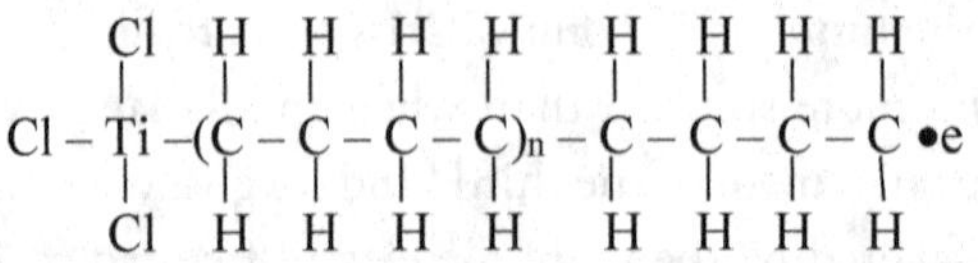

(Laws of Creations for Cyclobutane)

Rule 622: This rule of Chemistry for **Cyclobutane,** states that, when it can be made to exist in Equilibrium state of existence using a passive catalyst, it decomposes to 1-butene via Equilibrium mechanism as shown below.

Stage 1:

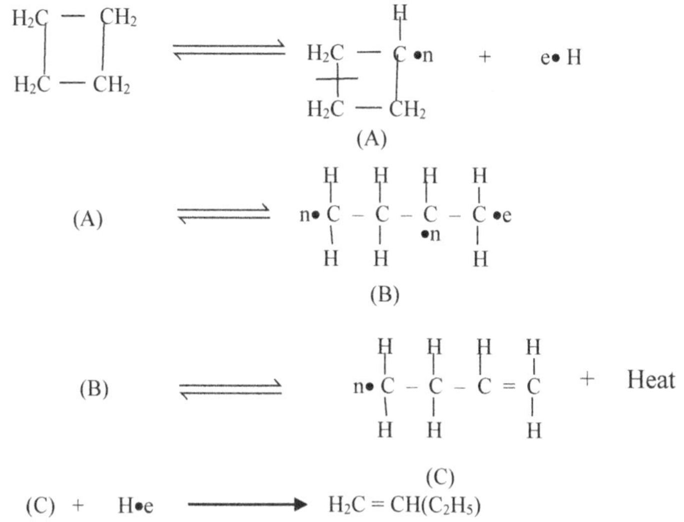

Overall Equation: Cyclobutane ⟶ 1-Butene + Heat

(Laws of Creations for Cyclobutane) [See Rule 606]

Rule 623: This rule of Chemistry for **Cyclobutane,** states that, when made to carry radical- pushing alkylane types of groups, it becomes easier to open the ring, because it is now more strained and more nucleophilic than cyclobutane and routes favored by them whether they are able to undergo molecular rearrangement of the third kind or not, are only the electro-free-radical or positively charged routes as shown below for one of them using an Electrostatically positively charged-paired initiator from $AlBr_3$.

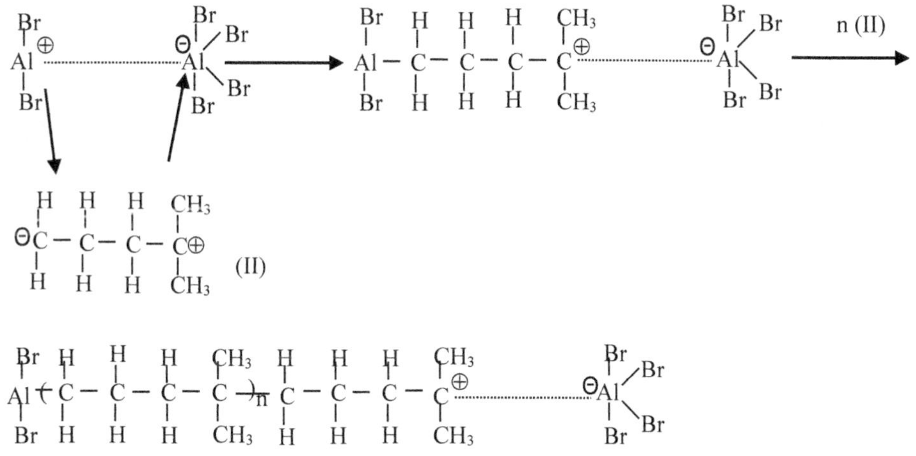

(Laws of Creations for Cyclobutanes)

Rule 624: This rule of Chemistry for **Cyclobutane,** states that, when made to carry radical-pushing alkylane groups, the followings are to be expected of the order of strain energy and nucleophilicity.

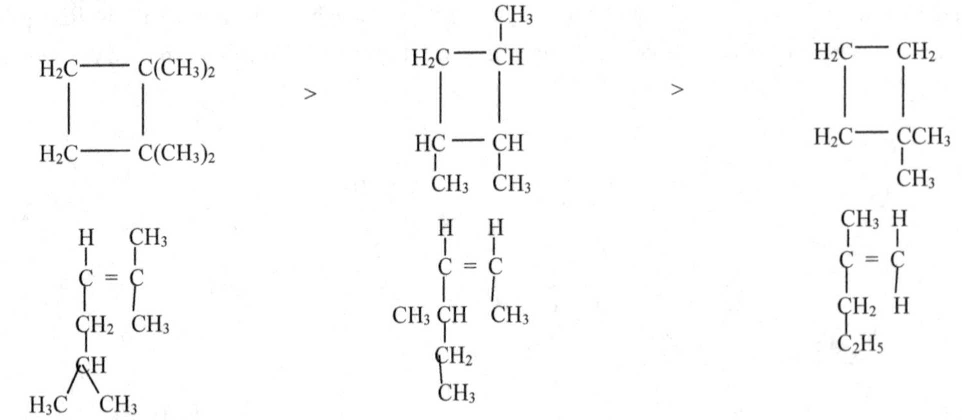

Substituted cyclobutanes – Order of Strain Energy and Nucleophilicity

(Laws of Creations for Cyclobutanes)

Rule 625: This rule of Chemistry for **Cyclobutane,** states that, when made to carry resonance stabilization group of the vinyl type (Vinyl cyclobutane), despite the fact that the ring is less strain than when H was there, the monomer is still more nucleophilic than cyclobutane and the ring is more nucleophilic than the vinyl group up to a limit as shown below;

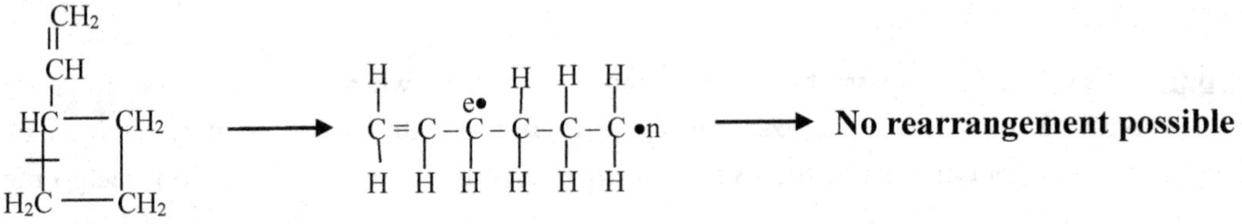

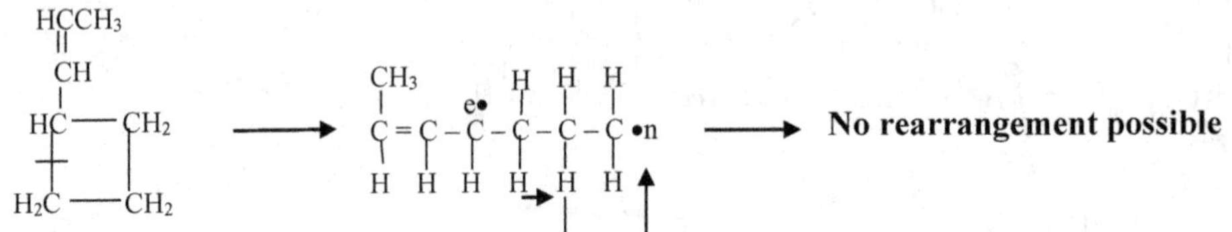

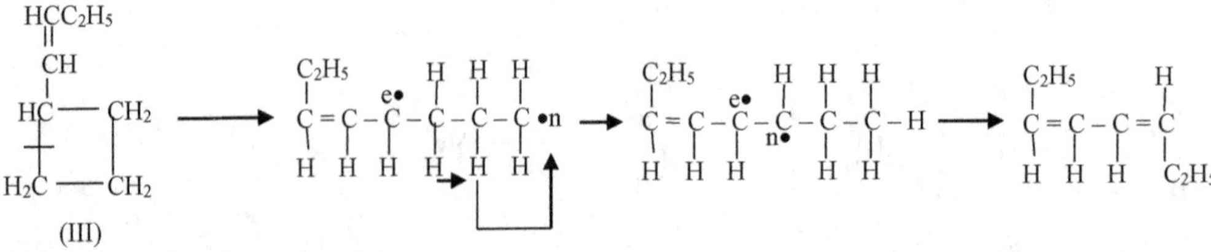

for which the followings are valid-

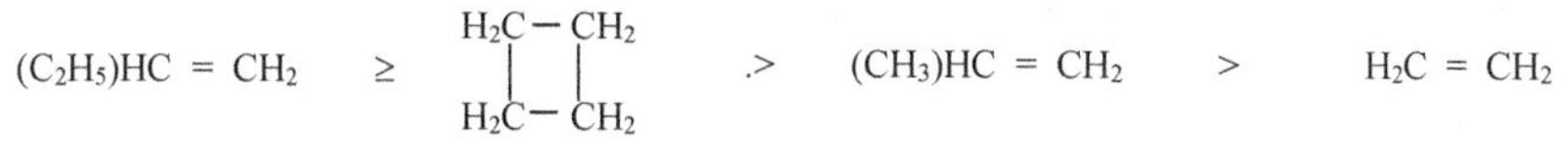

$$(C_2H_5)HC = CH_2 \quad \geq \quad \begin{array}{c} H_2C - CH_2 \\ | \quad\quad | \\ H_2C - CH_2 \end{array} \quad .> \quad (CH_3)HC = CH_2 \quad > \quad H_2C = CH_2$$

ORDER OF NUCLEOPHILICITY

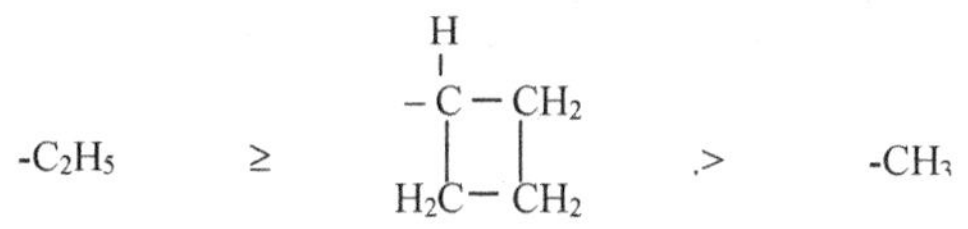

$$-C_2H_5 \quad \geq \quad \begin{array}{c} H \\ | \\ -C - CH_2 \\ | \quad\quad | \\ H_2C - CH_2 \end{array} \quad .> \quad -CH_3$$

ORDER OF RADICAL-PUSHING CAPACITY

(Laws of Creations for Cyclobutanes)

Rule 626: This rule of Chemistry for **Unsaturated groups of the type C_nH_{2n-1},** states that, when n equals 1 (CH), it does not belong to the family of groups in question herein, but when n equals 2 and above, a family of groups exist as shown below; for which all but the first (n=2) can be called **CYCLOALKYLANE** groups.

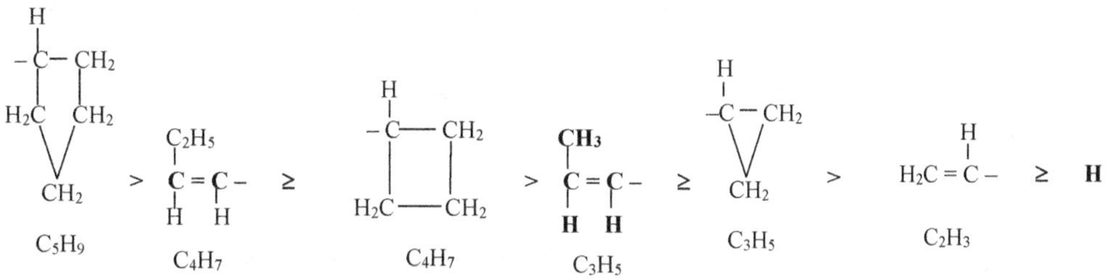

ORDER OF RADICAL-PUSHING CAPACITIES OF SPECIAL GROUPS

(Laws of Creations for Cycloalkylane groups)

Rule 627: This rule of Chemistry for **Vinyl Cyclobutane,** states that, when the ring is made to carry one or two radical-pushing alkylane group specially located, under certain operating conditions, the ring can be made to readily decompose via Equilibrium mechanism to give a cyclo-hexadiene without any rearrangement; noting that this could be done not by instantaneously opening of the ring, but by activation of the externally located activation center followed by opening of the ring as shown below, for which heat may not be required-

Stage 1:

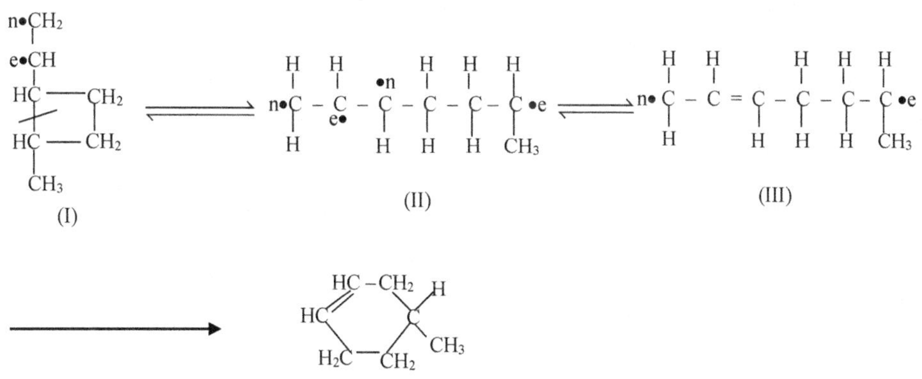

4-Methyl cyclohexadiene

127

noting herein how a four-membered ring has been expanded to a six-meebered ring using a resonance stabilization group where possible.
(Laws of Creations for Vinyl cyclobutane) [See Rule 614]

Rule 628: This rule of Chemistry for **Cyclobutane,** states that, when made to carry radical-pulling groups, the rings will be far less strained and difficult to open instantaneously.
(Laws of Creations for Cyclobutanes)

Rule 629: This rule of Chemistry for **Cyclopentane,** states that, this is the first member of its own family and the third member of the Cycloalkane family whose SE is far less than 3/5 the SE in the first member because of more conformational instability due to its size, and being more nucleophilic and less strained, the ring cannot be opened instantaneously, unless when made to carry radical pushing groups and special resonance stabilization group. [See Rule 625]
(Laws of Creations for Cyclopentane)

Rule 630: This rule of Chemistry for **Cyclohexane and higher members of the Cycloalkane family tree,** states that, because they are far more nucleophilic and less strained than the first three members, their rings cannot be opened instantaneously unless when their rings are saturated with radical-pushing groups and/or special resonance stabilization groups. [See Rule 625]
(Laws of Creations for Cycloalkanes)

Rule 631: This rule of Chemistry for **Cycloalkenes [Cyclopropene in particular],** states that, these are special springs which cannot be broken anywhere inside and when stress is applied, it begins to expand and stretch proportionately (like Hooke's Law) until a point is reached when it looks linearly placed and when stress is released, it comes back to its original state.
(Laws of Creations for Cyclopropenes)

Rule 632: This rule of Chemistry for **Cycloalkenes,** states that, like acetylenes and even-membered cumulene, the members of this family favor only free-radical existence and not charged existence, because when activated chargedly, there is electrostatic forces of repulsion between a negative charge and invisible π-bond which ***is imaginary but real,*** adjacently located to it.
(Laws of Creations for Cycloalkenes)

Rule 633: This rule of Chemistry for **Cycloalkenes,** states that, in view of the presence of a double bond in their rings, the following is valid for them.

Cycloalkenes > Cycloalkanes
Order of Nucleophilicity and Strain Energy
(Laws of Physics in Chemistry)

Rule 634: This rule of Chemistry for **Cyclopropene,** states that, in view of the size of the ring and its high degree of unsaturation (carbon to hydrogen ratio equals 1:4/3) favored by the presence of the ring,

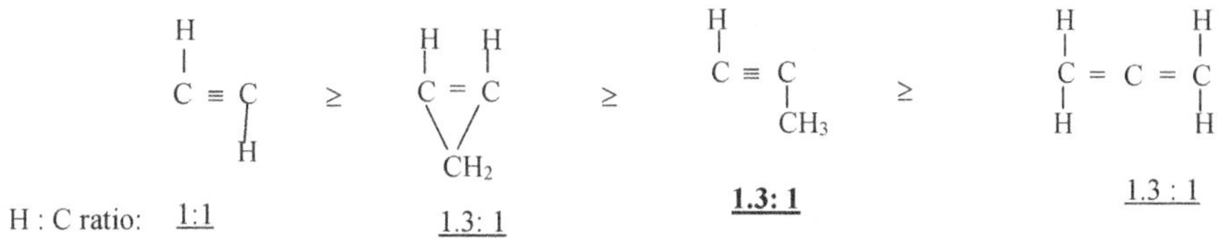

$$
\begin{array}{ccccccc}
\overset{\displaystyle H}{\underset{\displaystyle |}{C}} \equiv \underset{\displaystyle |}{\underset{\displaystyle H}{C}} & \geq & \overset{\displaystyle H}{\underset{\displaystyle |}{C}} = \overset{\displaystyle H}{\underset{\displaystyle |}{C}} & \geq & \overset{\displaystyle H}{\underset{\displaystyle |}{C}} \equiv \underset{\displaystyle CH_3}{\underset{\displaystyle |}{C}} & \geq & \overset{\displaystyle H}{\underset{\displaystyle |}{C}} = C = \overset{\displaystyle H}{\underset{\displaystyle |}{C}}
\end{array}
$$

H : C ratio: 1:1   1.3 : 1   **1.3 : 1**   1.3 : 1

ORDER OF UNSATURATION

hence cyclopropene exists in Equilibrium state of existence when not suppressed all the time as shown below-

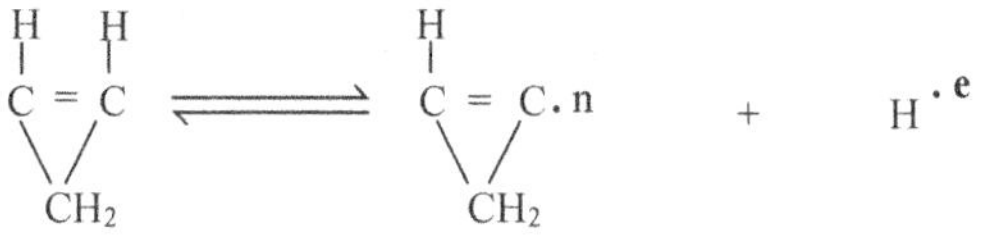

$$
\underset{CH_2}{\overset{H\ \ H}{C=C}} \rightleftharpoons \underset{CH_2}{\overset{H}{C=C \cdot n}} \quad + \quad H \cdot {}^e
$$

for which one or two H atoms on the carbon centers carrying double bond must be replaced with substituent groups before it can be used as a monomer.
(Laws of Creations for Cyclopropene)

Rule 635: This rule of Chemistry for **Cyclopropene**, states that, this is the first member of its own family and the Cycloalkene family, which unlike others in the family has no point of scission and therefore cannot be opened instantaneously no matter what the operating conditions are. [See Rule 631]
(Laws of Creations for Cyclopropene)

Rule 636: This rule of Chemistry for **Cyclopropene,** states that, when exposed to harsh operating conditions, i.e. temperature of the order of 425^0C, the followings take place via Equilibrium mechanism in one stage-

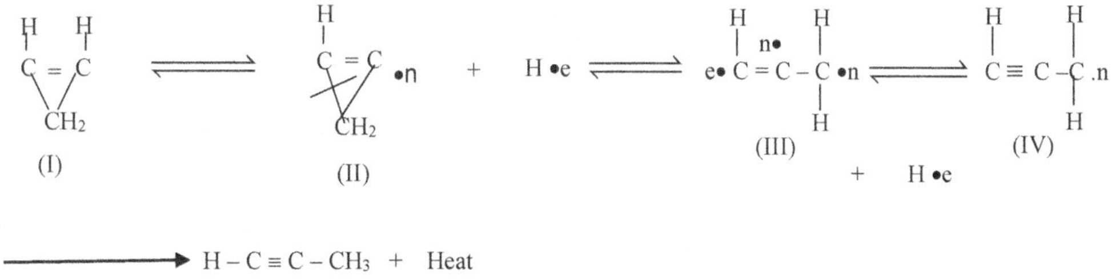

$$
\underset{CH_2}{\overset{H\ \ H}{C=C}} \rightleftharpoons \underset{CH_2}{\overset{H}{C=C}} \cdot n \quad + \quad H \cdot e \rightleftharpoons e \cdot C = C - \overset{H}{\underset{H}{C}} \cdot n \rightleftharpoons C \equiv C - \overset{H}{\underset{H}{C}} \cdot n
$$

(I)   (II)   (III)   (IV)

+   H •e

$$
\longrightarrow H - C \equiv C - CH_3 \quad + \quad Heat
$$

Propyne or Methyl acetylene

ELECTRORADICALIZATION PHENOMENON

wherein (II) obtained from (I) in Equilibrium state of existence is scissioned to form a triple bond (IV), obtained as a result of the fact that *the invisible electro-free-radical did not grabbed the visible nucleo -free-radical to form a bond, clearly indicating that the invisible π-bond is imaginary, but that in which* the H held in Equilibrium state of existence finally added to the new center, after deactivation to give propyne.
(Laws of Creations for Cyclopropene)

Rule 637: This rule of Chemistry for **Cyclopropene,** states that, when suppressed (that which may not be possible), it can be polymerized only electro-free-radically, because of presence of transfer species of the first kind of the tenth type as shown below; the transfer species said to be

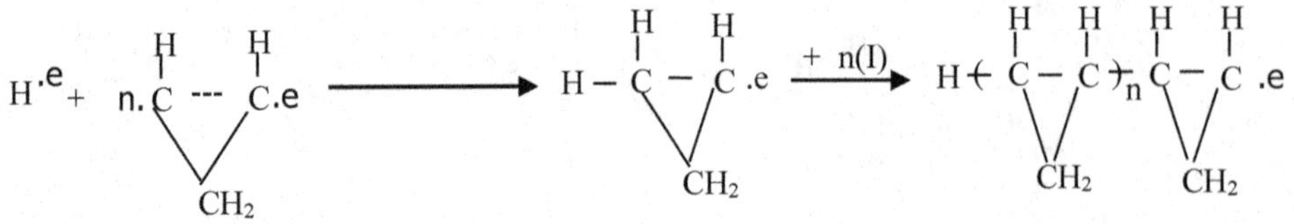

(I) SUPPRESSED   (II) a

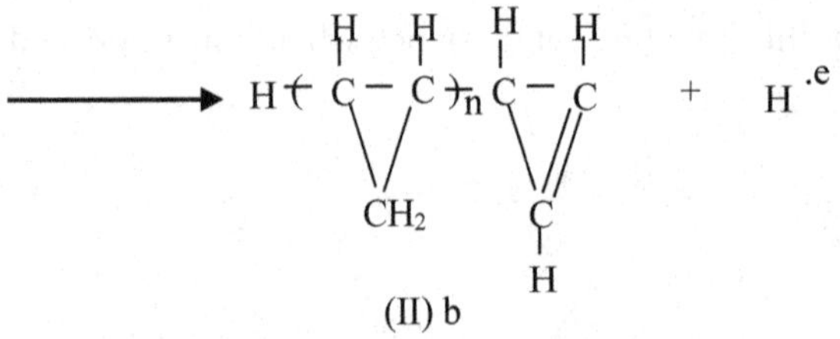

(II) b

Transfer species of first kind of the tenth type

of the tenth type because it is located inside an unsaturated ring and as such shared by two active centers as opposed to the other types of transfer species of the first kind, noting that the laws of conservation of transfer of transfer species has not been broken.
(Laws of Creations for Cyclopropene)

Rule 638: This rule of Chemistry for **Cyclopropene,** states that, when made to carry one radical pushing alkylane group, since H to C ratio a measure of degree of unsaturation is still below 2.0, some of them are still unstable as shown below-.

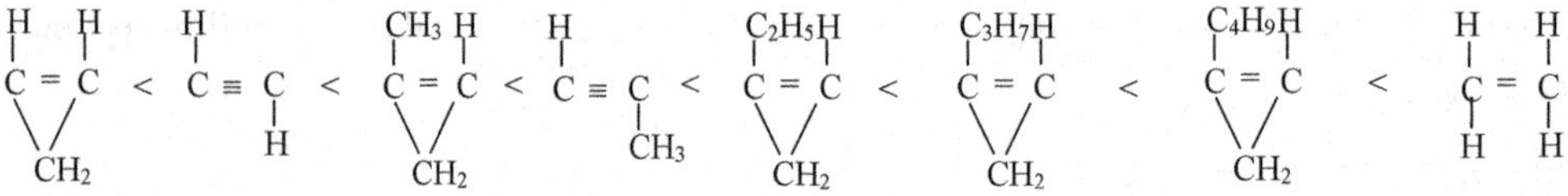

| 1.3: 1 | 1:1 | 1.5: 1 | 1.3: 1 | 1.6: 1 | 1.67: 1 | 1.7: 1 | 2:1 |

← —— UNSTABLE ZONE ——→   ← —— STABLE ZONE ——→

ORDER OF INSTABILITY OF SOME SPECIAL MONOMERS

for which it can be observed that the most unstable of all of them are the first members – acetylene, cyclopropene, followed by Methyl acetylene and then followed by Methyl cyclopropene.
(Laws of Creations for Cyclopropenes)

Rule 639: This rule of Chemistry for **Methyl cyclopropene,** states that, when suppressed, it cannot molecularly rearrange to give methylene cyclopropane as shown below,

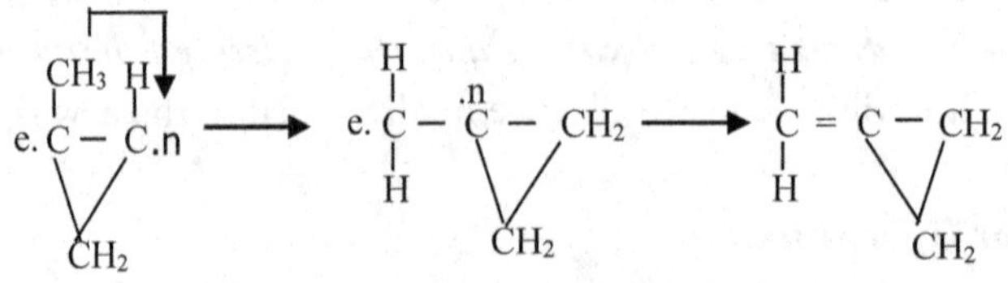

(I)   (II)   (III)

Non-favored Rearrangement

$$\text{e.}\underset{\overset{|}{CH_2}}{\overset{\overset{CH_3}{|}}{C}} - \overset{\overset{H}{|}}{C}.n \longrightarrow \text{e.}\underset{\overset{|}{CH_2}}{\overset{\overset{CH_3}{|}}{C}} - \overset{\overset{H}{|}}{C}.n$$

Favored Rearrangement

because the following is valid-

$$H_2C\langle \qquad > \qquad H_2C = \qquad >>> \qquad -CH_3 \text{ [Any R]}$$

$$\text{(Closed)} \qquad\qquad \text{(Opened)} \qquad\qquad \text{(Opened)}$$

Order of Radical-pushing capacity

for which it is the second reaction above that is favored; that wherein the monomer rearranges back to itself via molecular rearrangement of the first kind of the tenth type.

(Laws of Creations for Methyl cyclopropene)

Rule 640: This rule of Chemistry for **Methylene cyclopropane,** states that, though it cannot be obtained from methyl cyclopropene by rearrangement, it is a direct product of methylene and allene as shown below in a one stage Equilibrium mechanism system.

Stage 1:

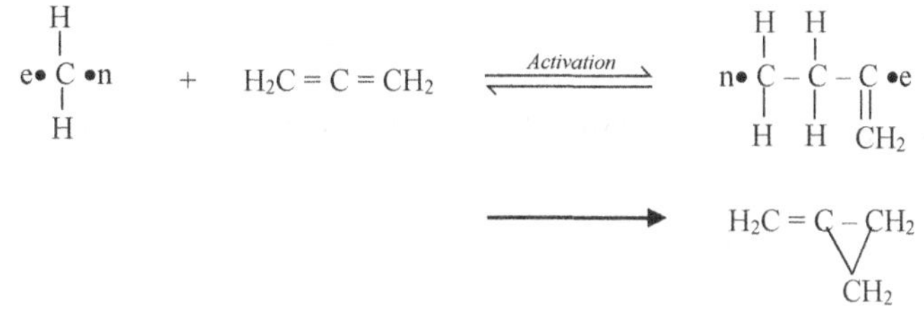

$$e\bullet \overset{\overset{H}{|}}{\underset{\overset{|}{H}}{C}} \bullet n \quad + \quad H_2C = C = CH_2 \xrightarrow[]{Activation} n\bullet \overset{\overset{H}{|}}{\underset{\overset{|}{H}}{C}} - \overset{\overset{H}{|}}{\underset{\overset{|}{H}}{C}} - \overset{\overset{CH_2}{||}}{C} \bullet e$$

$$\longrightarrow H_2C = \underset{\overset{|}{CH_2}}{C} - CH_2$$

Overall equation: H_2C + $H_2C=C=CH_2 \longrightarrow$ Methylene cyclopropene

noting the close relatives of this compound.

(Laws of Creations for Methylene cyclopropane)

Rule 641: This rule of Chemistry for **Methyl cyclopropene,** states that, in view of the high degree of unsaturation (1.5) compared to that of ethene (2.0) compared to that of acetylene (1.0), methyl cyclopropene is unstable, for which its Equilibrium state of existence even at STP is as follows-

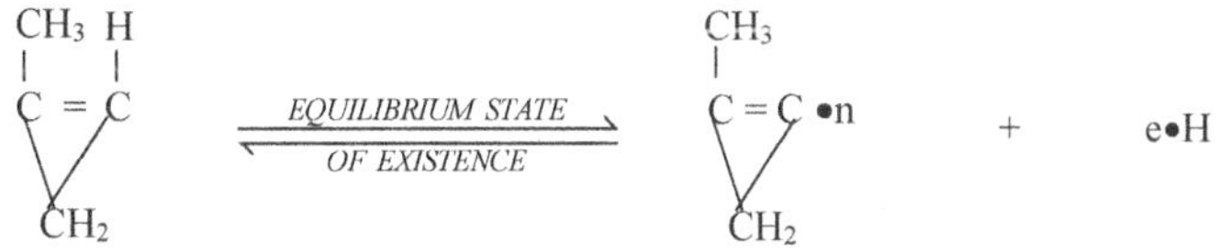

$$\underset{\overset{|}{CH_2}}{\overset{\overset{CH_3}{|}}{C}} = \overset{\overset{H}{|}}{C} \xrightleftharpoons[\text{OF EXISTENCE}]{\text{EQUILIBRIUM STATE}} \underset{\overset{|}{CH_2}}{\overset{\overset{CH_3}{|}}{C}} = C \bullet n \quad + \quad e\bullet H$$

EQUILIBRIUM STATE OF EXISTENCE OF
METHYL CYCLOPROPENE

(Laws of Creations for Methyl cyclopropene)

Rule 642: This rule of Chemistry for **Methyl cyclopropene,** states that, when kept in Activated/ Equilibrium state of existence, the followings are obtained-

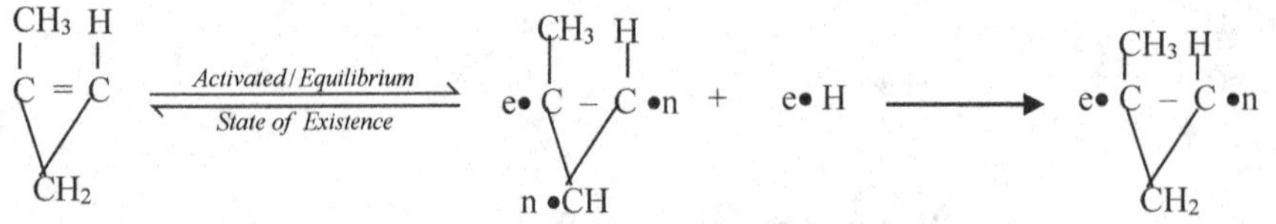

Methyl cyclopropene

for which the H held is that which is involved as transfer species, noting that the same methyl cyclopropene is obtained.

(Laws of Creations for Methyl cyclopropene)

Rule 643: This rule of Chemistry for **Methyl cyclopropene,** states that, when heated <u>above 400^0C,</u> it decomposes to give $H_3CC \equiv CCH_3$ as shown below via Equilibrium mechanism.

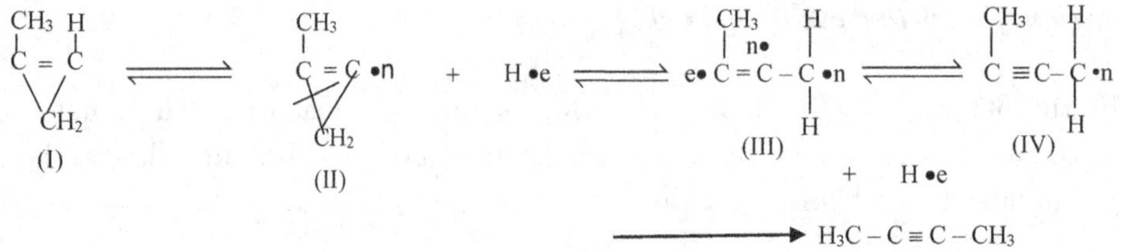

Butyne or Di-methyl acetylene

ELECTRORADICALIZATION PHENOMENON [2nd Type]

[See Rule 636]
(Laws of Creations for Methyl cyclopropene)

Rule 644: This rule of Chemistry for **Methyl cyclopropene,** states that, when used as an agent *(1-MCP)* for extraction of some components such as ethene, this it does as follows-

Stage1:

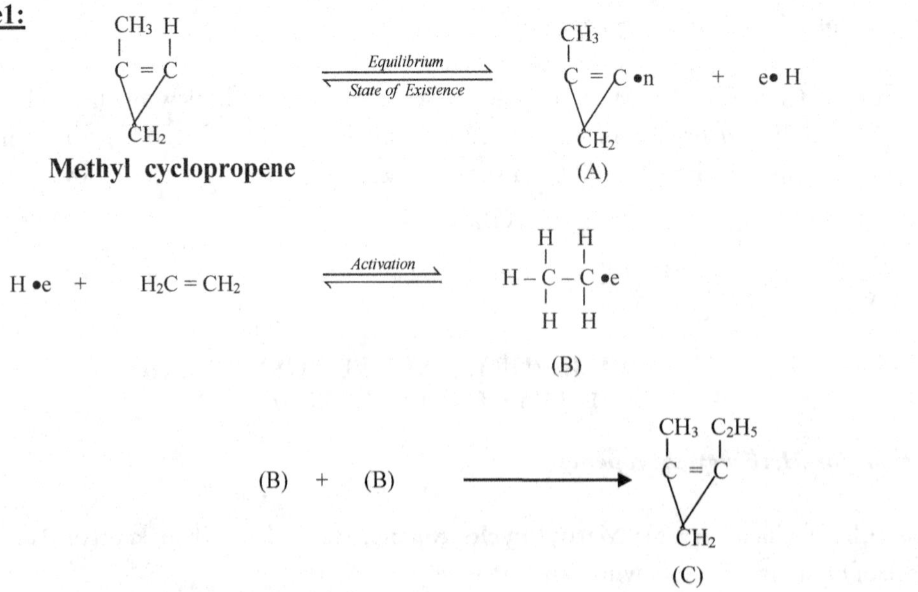

Methyl cyclopropene (A)

(B)

(C)

Overall Equation: Methyl cyclopropene + Ethylene ⟶ (C)

for which an unstable fully substituted cyclopropene (C) is obtained.

(Laws of Creations for Methyl cyclopropene)

Rule 645: This rule of Chemistry for **Cyclopropenes** where the two H atoms are replaced with radical-pushing alkylane groups, states that, these are still unstable compounds, since they undergo molecular-rearrangement to give an unstable more nucleophilic cyclopropene as shown below-

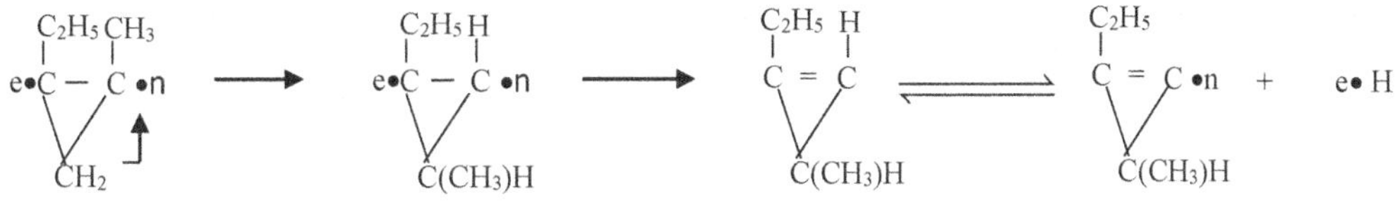

FAVORED REARRANGEMENT

(A)Unstable

for which the (A) above can be used to grab another ethene for example via Equilibrium mechanism to form (B) shown below, bringing the total number of molecules of ethene removed to two-

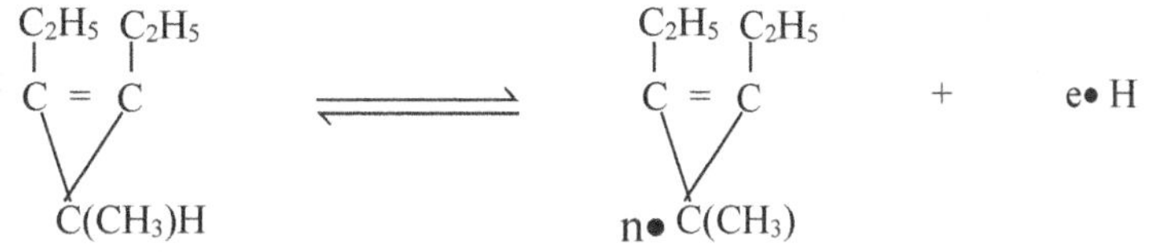

(B)Unstable

and the (B) formed, exists in Equilibrium state of existence to grab another molecule of ethene to form (C) shown below-

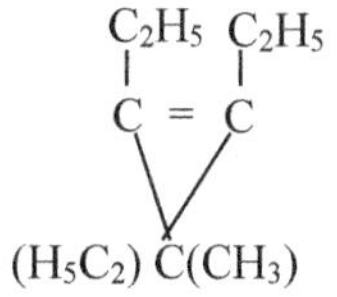

(C)

thus, bringing the total number of molecules of ethene grabbed to three, noting that the (C) formed above can no longer exist in Equilibrium state of existence at STP.
(Laws of Creations for Cyclopropenes)

Rule 646: This rule of Chemistry for **Cyclopropene**, states that, when made to carry two-radical-pushing groups on the activation center, these cannot be used as monomers, since they will readily undergo *Molecular rearrangement of the first kind of the tenth type,* wherein transfer species of the tenth type is involved; unless the initiator used is so strong as to prevent the rearrangement from taking place radically and when polymerization is possible, this can only be done electro-free-radically as is the case for all the Nucleophilic members of cyclopropene families when suppressed.
(Laws of Creations for Cyclopropenes)

Rule 647: This rule of Chemistry for **Cyclopropene,** states that, when made to carry radical-pushing alkylane groups in place of all the H atoms such as shown below, the rings become more strained, more nucleophilic and only molecular rearrangement of the first kind is favored by some to produce most of time the same monomer, all of which can only be polymerized electro-free-radically in the absence of steric limitations;

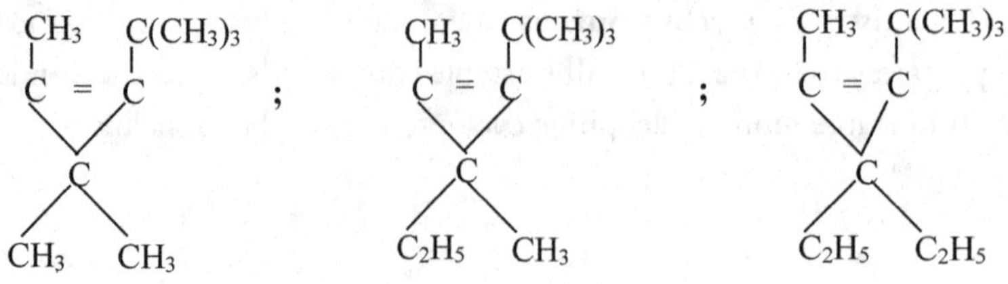

$$\begin{array}{ccc} \underset{\substack{| \\ C \,=\, C \\ | \\ \underset{CH_3 \quad CH_3}{C}}}{CH_3 \quad C(CH_3)_3} & ; & \underset{\substack{| \\ C \,=\, C \\ | \\ \underset{C_2H_5 \quad CH_3}{C}}}{CH_3 \quad C(CH_3)_3} & ; & \underset{\substack{| \\ C \,=\, C \\ | \\ \underset{C_2H_5 \quad C_2H_5}{C}}}{CH_3 \quad C(CH_3)_3} \\ (I) & & (II) & & (III) \end{array}$$

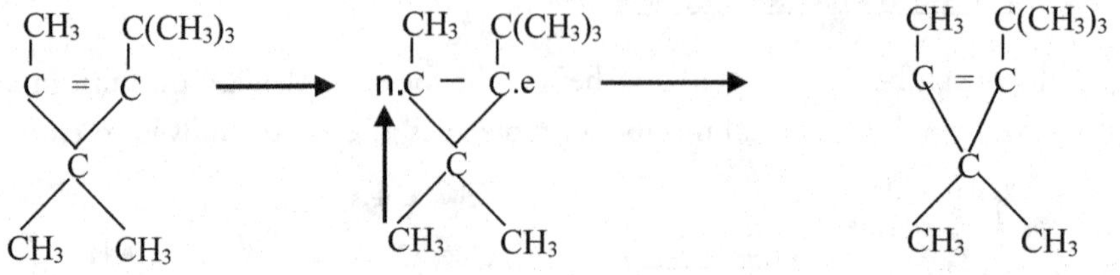

$$\underset{(I)}{\underset{\substack{| \\ C \,=\, C \\ | \\ \underset{CH_3 \quad CH_3}{C}}}{CH_3 \quad C(CH_3)_3}} \longrightarrow \underset{\substack{| \\ n.C \,-\, C.e \\ | \\ \underset{CH_3 \quad CH_3}{C}}}{CH_3 \quad C(CH_3)_3} \longrightarrow \underset{\substack{| \\ C \,=\, C \\ | \\ \underset{CH_3 \quad CH_3}{C}}}{CH_3 \quad C(CH_3)_3}$$

noting that only (I) above will favor any movement of transfer species (CH_3), to give same monomer. *(Laws of Creations of Cyclopropenes)*

Rule 648: This rule of Chemistry for **Methylene cyclopropane,** states that, even at STP, this compound an isomer of methyl cyclopropene is unstable, for which its Equilibrium state of existence is as follows-

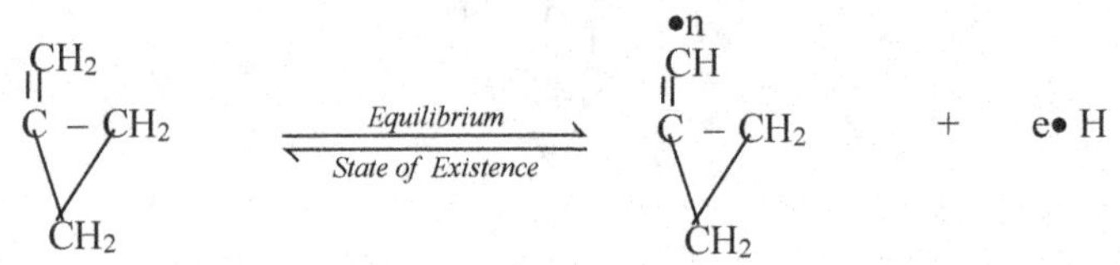

$$\underset{\substack{| \\ C \,-\, CH_2 \\ | \\ CH_2}}{\overset{CH_2}{\|}} \underset{\substack{\textit{State of Existence}}}{\overset{\textit{Equilibrium}}{\rightleftharpoons}} \underset{\substack{| \\ C \,-\, CH_2 \\ | \\ CH_2}}{\overset{\overset{\bullet n}{CH}}{\|}} \quad + \quad e\bullet H$$

EQUILIBRIUM STATE OF EXISTENCE OF
METHYLENE CYCLOPROPANE

for which when compared with methyl cyclopropene (See Rule 641), the following is valid in general-

$$\text{H in Ring} \quad > \quad \text{H outside the same ring}$$

Radical-pushing capacity of H inside and outside a threemembered ring only
(Laws of Creations for Methylene cyclopropane/Methyl cyclopropene)

Rule 649: This rule of Chemistry for **Methylene cyclopropane,** states that, ***this compound which does not have Activated/Equilibrium state of existence,*** decomposes when heated <u>at or above 400ºC</u> to give acetylene and ethene via Equilibrium mechanism as shown below-

Stage 1:

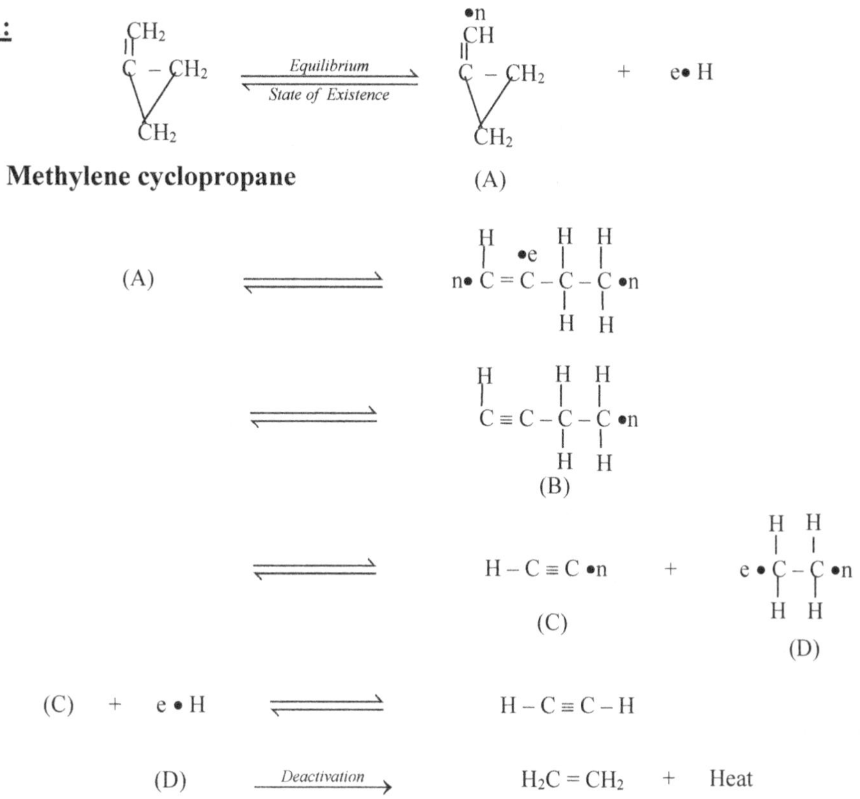

$$
\underset{\text{Methylene cyclopropane}}{
\begin{array}{c}
CH_2 \\
\| \\
C - CH_2 \\
\diagdown \diagup \\
CH_2
\end{array}}
\quad \underset{\text{State of Existence}}{\overset{\text{Equilibrium}}{\rightleftharpoons}} \quad
\underset{(A)}{
\begin{array}{c}
\bullet n \\
CH \\
\| \\
C - CH_2 \\
\diagdown \diagup \\
CH_2
\end{array}}
\quad + \quad e \bullet H
$$

$$
(A) \quad \rightleftharpoons \quad
n \bullet \overset{\displaystyle H}{\underset{}{C}} = \overset{\bullet e}{C} - \overset{\displaystyle H}{\underset{\displaystyle H}{C}} - \overset{\displaystyle H}{\underset{\displaystyle H}{C}} \bullet n
$$

$$
\rightleftharpoons \quad
\underset{(B)}{C \equiv C - \overset{\displaystyle H}{\underset{\displaystyle H}{C}} - \overset{\displaystyle H}{\underset{\displaystyle H}{C}} \bullet n} \quad \text{(with } H \text{ on first C)}
$$

$$
\rightleftharpoons \quad \underset{(C)}{H - C \equiv C \bullet n} \quad + \quad \underset{(D)}{e \bullet \overset{\displaystyle H}{\underset{\displaystyle H}{C}} - \overset{\displaystyle H}{\underset{\displaystyle H}{C}} \bullet n}
$$

$$
(C) \quad + \quad e \bullet H \quad \rightleftharpoons \quad H - C \equiv C - H
$$

$$
(D) \quad \overset{\text{Deactivation}}{\longrightarrow} \quad H_2C = CH_2 \quad + \quad Heat
$$

Overall Equation: Methylene cyclopropane $\longrightarrow$ $HC\equiv CH$ + $H_2C=CH_2$

for which, if the operating condition is lower than above, ethyl acetylene is formed.
(Laws of Creations for Methylene cyclopropane)

Rule 650: This rule of Chemistry for **Methylene cyclopropane,** states that, apart from the Laws of Conservation of transfer of transfer species, methylene cyclopropane cannot molecularly rearrange to methyl cyclopropane based on Rule 648, for which the following is valid-

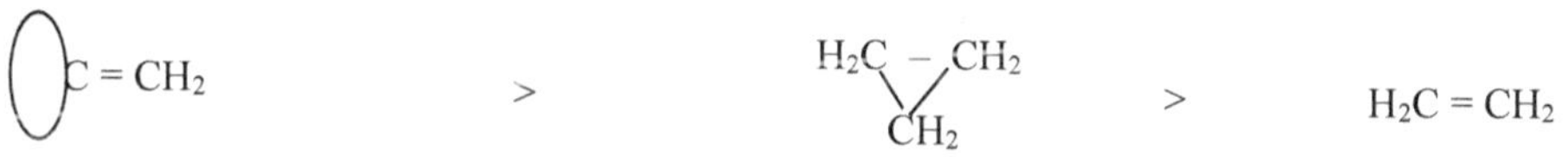

$$
\begin{array}{c}
\bigcirc C = CH_2
\end{array}
\quad > \quad
\begin{array}{c}
H_2C - CH_2 \\
\diagdown \diagup \\
CH_2
\end{array}
\quad > \quad
H_2C = CH_2
$$

ORDER OF NUCLEOPHIICITY FOR THREE-MEMBERED RINGS

(Laws of Creations for Methylene cyclopropane)

Rule 651: This rule of Chemistry for **Methylene cyclopropane,** states that, being cumulenic in character, when suppressed and heated at and above 400^0C, it decomposes to give more nucleophilic acetylenes via Decomposition mechanism as shown below- [See Rule 649]

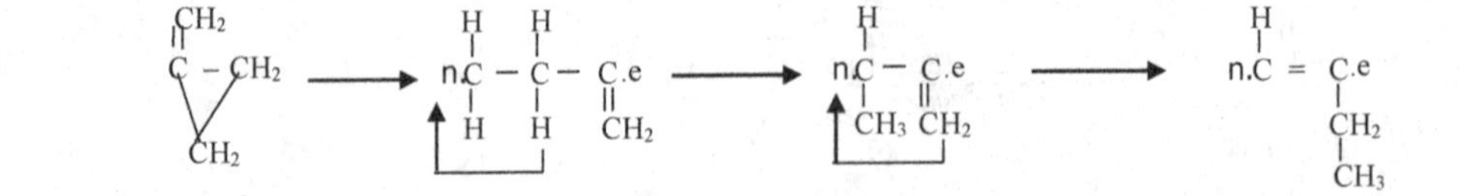

Methylene cyclopropane   (A) Less Stable   (B) More Stable

(FAVORED)

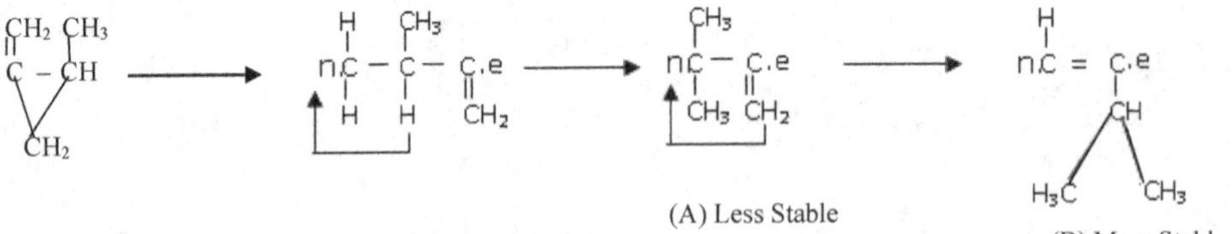

(A) Less Stable   (B) More Stable

(FAVORED)

(Laws of Creations for Methylene cyclopropane)

Rule 652: This rule of Chemistry for **Molecular rearrangement of the Third kind,** states that, without opening of a ring instantaneously, this phenomenon does not exist.
(Laws of creations for Molecular rearrangement of the Third kind)

Rule 653: This rule of Chemistry for **1,3- Cumulenes,** states that, when $H_2C=C=CH_2$, $F_2C=C=CH_2$ and $F_2C=C=CF_2$ are compared, the followings are obtained.

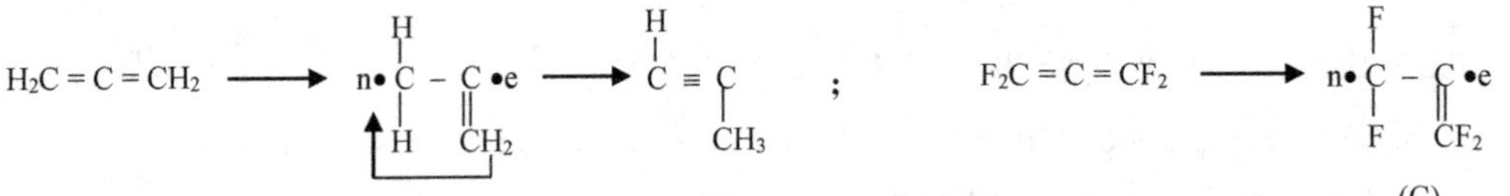

NUCLEOPHILE   "BOTH"

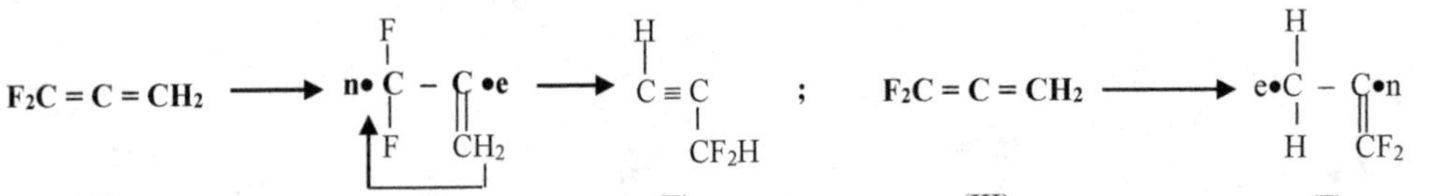

X-center   Y-center

AN ELECTROPHILE

from which in general, the followings are valid.

$$R_2C = \; > \; RHC= \; > \; H_2C= \; > \; RR_FC= \; > \; HR_FC= \; > \; (R_F)_2C= \; > \; RFC= \; > \; FHC= \; > \; R_FFC= \; >>$$

$$-R \; > \; -H \; > \; F_2C= \; > \; -R_F$$

Where $R \equiv C_nH_{2n+1}$; $R_F \equiv C_nF_{2n+1}$

ORDER OF RADICAL-PUSHING CAPACITIES OF RADICAL-PUSHING GROUPS

noting that, if R_F could also be a radical-pulling group such as COOR, $CONH_2$, $C \equiv N$ and the likes, it's capacity will be different [H > $F_2C=$ > $(R_F)_2C=$] .
(Laws of Creations for 1,3-Cumulenes) [See Rule 422]

Rule 654: This rule of Chemistry for **the type of 1,3-cumulene shown below,** states that, this is a unique type of electrophile, wherein the X and Y centers are cumulatively placed; which unlike the linear case, the X center is the center that

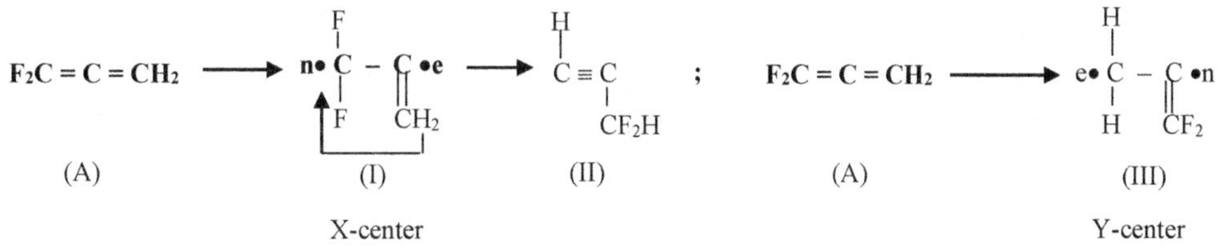

AN ELECTROPHILE

undergoes molecular rearrangement of the first kind to give acetylenes, while the Y center cannot rearrange, because H_2FC is radical-pushing while F is radical-pulling.
(Laws of Creations for 1,3-cumulenes)

Rule 655: This rule of Chemistry for **Cyclopropenes,** states that, in order to make them cumulenic in character, the CH_2 group in the ring must be changed to F_2C type of group, since the group is less radical pushing than H as shown below-

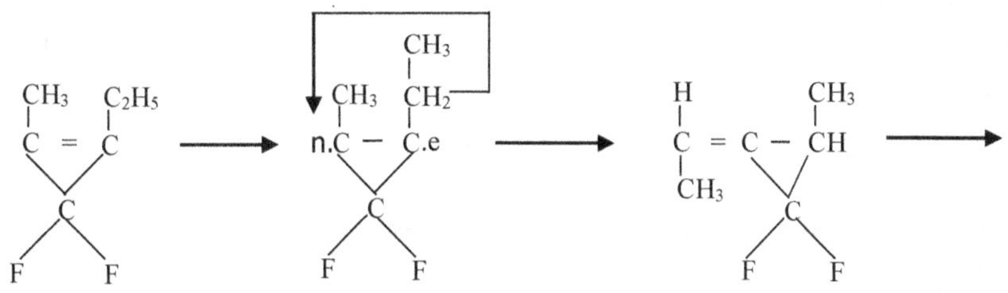

CUMULENE FORMATION

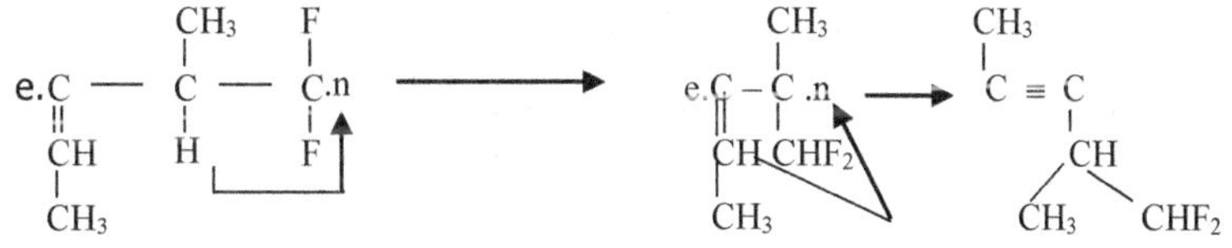

FAVORED

at the end of which on decomposition triple bonded monomers (acetylenic in character) are formed with limitations, that wherein the size of CH_3 the non-provider of transfer species cannot exceed C_2H_5, unless C_2H_5 the transfer species provider is changed.
(Laws of Creation for Molecular rearrangements of the First and Third kinds for Cyclopropenes)

Rule 656: This rule of Chemistry for **Cyclopropene wherein the CH_2 group is changed to CF_2 group,** states that, despite the fact that the ring is far less strained, the compound is still unstable, since it can readily exist in Equilibrium state of existence as shown below-

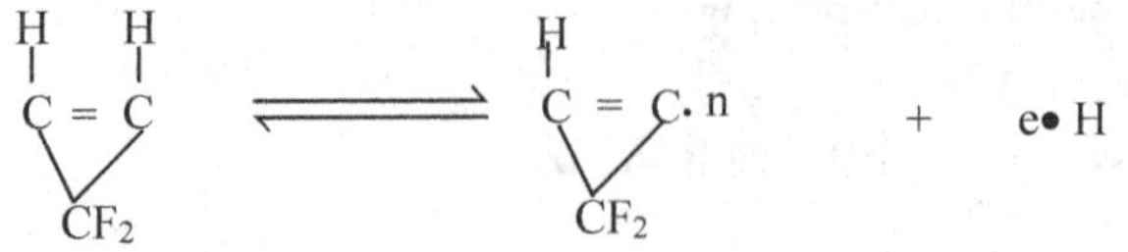

$$\underset{CF_2}{\overset{\overset{\displaystyle H \quad H}{|\quad\ |}}{C = C}} \quad \rightleftharpoons \quad \underset{CF_2}{\overset{\overset{\displaystyle H}{|}}{C = C \cdot n}} \quad + \quad e \bullet H$$

noting that it is not the F atom that is held, but H atom, *while if F was replaced with Cl or Br, the reverse will be the case.*
(Laws of Creations for Perfluoro cyclopropene)

Rule 657: This rule of Chemistry for **Cyclopropenes shown below,** states that, these are nucleo-philes with the following characteristics in terms of order of Nucleophilicity, the radical- pushing capacities of the groups carried internally;

$$\underset{CH_2}{\overset{\overset{\displaystyle H \quad H}{|\quad\ |}}{C = C}} \quad >> \quad \underset{CFH}{\overset{\overset{\displaystyle H \quad H}{|\quad\ |}}{C = C}} \quad >> \quad \underset{CF_2}{\overset{\overset{\displaystyle H \quad H}{|\quad\ |}}{C = C}} \quad >> \quad \underset{CF_2}{\overset{\overset{\displaystyle F \quad F}{|\quad\ |}}{C = C}}$$

ORDER OF NUCLEOPHILICITY

$$H_2C\big< \quad >> \quad HFC\big< \quad >> \quad R > H > F_2C\big<$$

Order of Radical-pushing capacities of closed groups

for which while the first will favor only its natural route when suppressed-electro-free-radical route, the second will favor no route when suppressed, the third will favor the route not natural to it- the nucleo-free-radical route when suppressed and the fourth will favor also only the route not natural to it- the nucleo-free-radical route without being suppressed.
(Laws of Creations for Per fluoro cyclopropenes)

Rule 658: This rule of Chemistry for **Cyclopropenes,** states that, when groups such as OH, OR, NH_2, NHR, NR_2, =NH, and the likes are placed on the ring, they have the ability of suppressing the influence of the CH_2 group in the ring, because the following is valid-

$$= NH \quad > \quad - NH_2 \quad > \quad - OH \quad > \quad H_2C\big< \quad > \quad H_2C= \quad >> \quad H_3C- (Any\ R)$$

Order of Radical-pushing capacity of groups.

(Laws of Creations for Cyclopropenes)

Rule 659: This rule of Chemistry for **Cycloketones,** states that, if the ring is opened instantaneously under very mild operating conditions or even at STP, it can undergo molecular rearrangement of the third kind, to produce ketenes, because the Laws of Conservation of transfer of transfer species will not be broken.
(Laws of Creations for Cycloketones)

Rule 660: This rule of Chemistry for **Cyclopropanone,** states that, the reasons why this ring is not commonly known to exist, is because when it readily exists in Equilibrium state of existence it also decomposes to give methyl ketene as shown below for cyclopropanone via Equilibrium decomposition mechanism-

Stage 1:

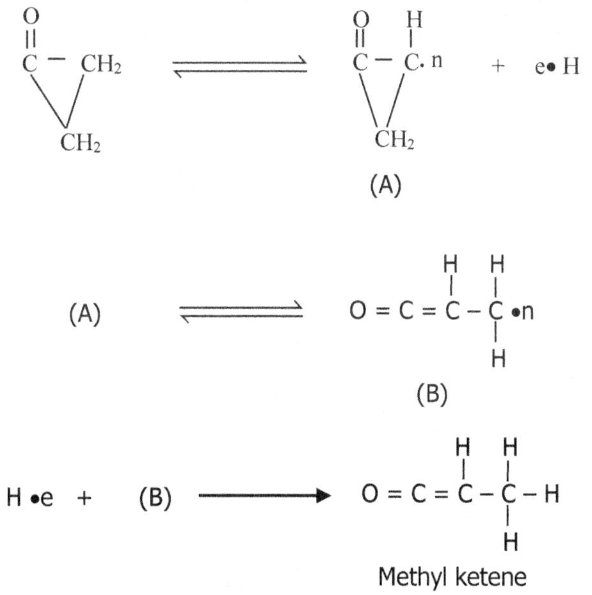

$$\text{Cyclopropanone (O=C–CH}_2\text{, CH}_2) \rightleftharpoons (A) + e\bullet H$$

(A)

$$(A) \rightleftharpoons O=C=C(H)(H)-C(H)(H)\bullet n$$

(B)

$$H\bullet e + (B) \longrightarrow O=C=C(H)(H)-C(H)(H)-H$$

Methyl ketene

Overall Equation: Cyclopropanone $\longrightarrow$ Methyl ketene

(Laws of Creations for Cyclopropanone) [See Rule 602]

Rule 661: This rule of Chemistry for **Cyclopropenes**, states that, when made to carry one radical-pushing group of the type OH or NHR, this will favor molecular rearrangement of the first kind only when CF_2 is put in place of CH_2, since with CH_2 the cyclic ketone or aldimine obtained are too strained to exist as shown below for OH;

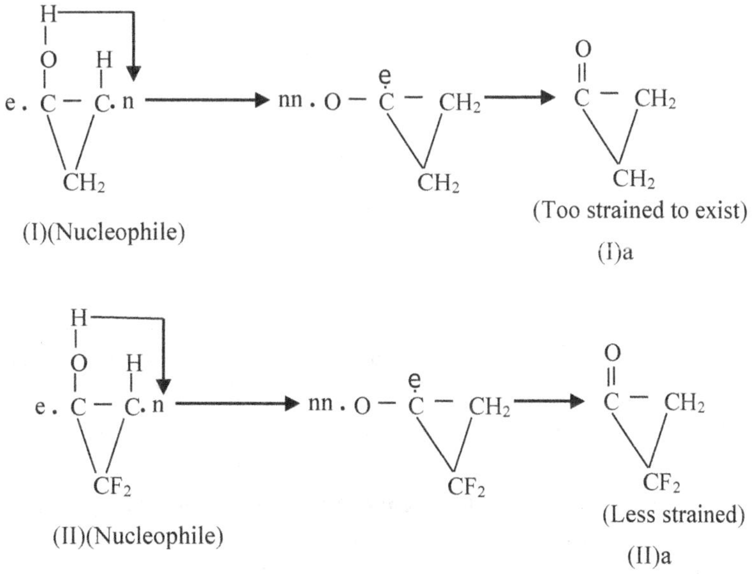

(I)(Nucleophile) $\longrightarrow nn\bullet O-\dots \longrightarrow$ (I)a (Too strained to exist)

(II)(Nucleophile) $\longrightarrow nn\bullet O-\dots \longrightarrow$ (II)a (Less strained)

for which with (II)a if it exists, the ring may be opened, with the possibility of favoring molecular rearrangement of the third kind and this implies that (I) cannot exist or has only transient existence and same applies to NHR.

(Laws of Creations for Cyclopropenes)

Rule 662: This rule of Chemistry for **Cyclopropene,** states that, when made to carry radical-pushing group of the type NH_2, this compound unlike OH group can undergo molecular rearrangement of the first and third kinds when ring is opened instantaneously to produce Nitriles as shown below-

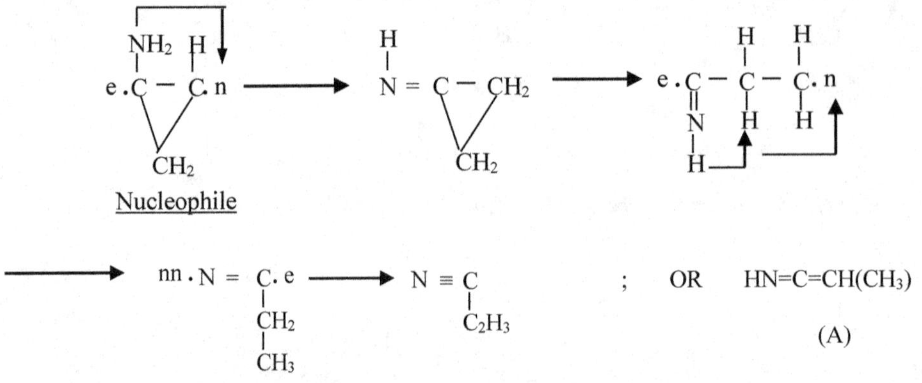

Nucleophile

Nucleophile (More)

noting that if the last movement is not favored (that which is believed to be favored for these types of cumulenic Electrophiles), (A) above will be obtained.
(Laws of Creations for Cyclopropenes)

Rule 663: This rule of Chemistry for **Cycloalkenes of the type shown below,** states that, the use of NHR as a group for rearrangement, is a function of what is carried by the other center, for which the followings are to be expected when decomposed instantaneously,

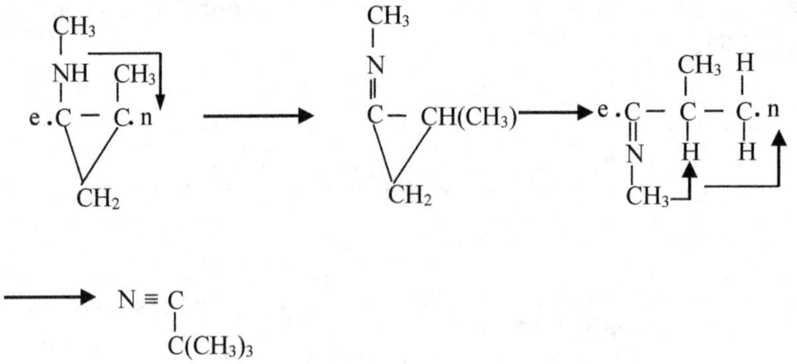

FAVORED

noting that a nitrile is obtained and when the group is changed to NHC_2H_5 group, the rearrangement of the third kind ceases with movement of C_2H_5 to produce $C_2H_5N=C=C(CH_3)_2$.
(Laws of Molecular rearrangement of the third kind)

Rule 664: This rule of Chemistry for **Cycloalkanes of the type shown below,** states that, the use of NR_2 as group for rearrangement, is a function of what is carried by the other center, for which the followings are to be expected when decomposed instantaneously,

140

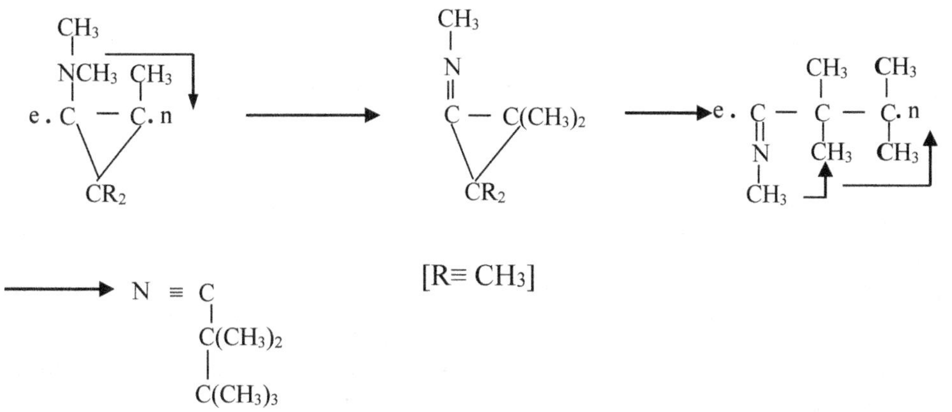

FAVORED

noting that a nitrile or a cumulenic aldimine is obtained and when the group above is changed to $N(CH_3)$ C_2H_5 and higher, rearrangement ceases with movement of C_2H_5 to produce a cumulenic aldimine $(H_5C_2N=C=CCH_3(C(CH_3)_3)$.
(Laws of Molecular rearrangement of the third kind)

Rule 665: This rule of Chemistry for **Cyclopropene,** states that, when made to carry radical-pushing group of the type OR or NR_2, no form of molecular rearrangement is possible, for which they cannot be opened.
(Laws of Creations for Cyclopropenes)

Rule 666: This rule of Chemistry for **Cyclopropene,** states that, when made to carry one radical-pushing group of lower capacity than H, such as CF_3, C_2F_5, *no molecular rearrangement is favored by them and no route is favored by them*, as shown below for one of them.

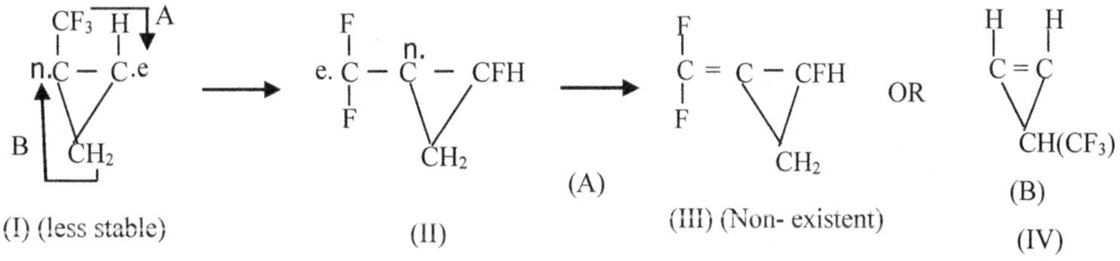

NUCLEOPHILE

wherein H (for B) the transfer species nucleo-free-radically cannot be transferred, H being more radical-pushing than CF_3 and F (for A) the transfer species electro-free-radically cannot be transferred, the monomer being a Nucleophile.
(Laws of Creations for Cyclopropenes)

Rule 667: This rule of Chemistry for **Cyclopropene,** states that, when the Hs are changed to CF_3 types of groups, the monomer becomes more stable, less strained and less nucleophilic and only molecular rearrangement of the first kind is favored by some to produce most of time the same monomer, some of which can only be polymerized electro-free-radically in the absence of steric limitations.

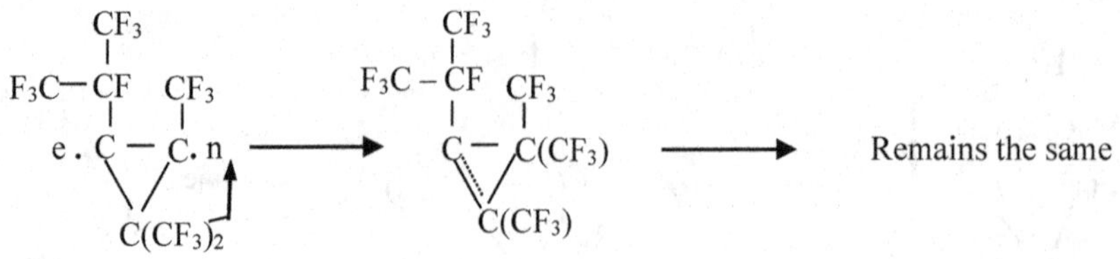

(Laws of Creations for Cyclopropenes)

Rule 668: This rule of Chemistry for **Cyclopropene,** states that, when made to carry radical-pulling groups such as COOR, $CONH_2$, COR groups, though the first two can undergo molecular rearrangement of the first kind to give cyclopropanes, the rings cannot be opened, because of the **strong radical-pulling capacity of =C=O group;** for which if attempts are made to open the ring instantaneously, molecular rearrangement of the third kind will not be favored by it, because

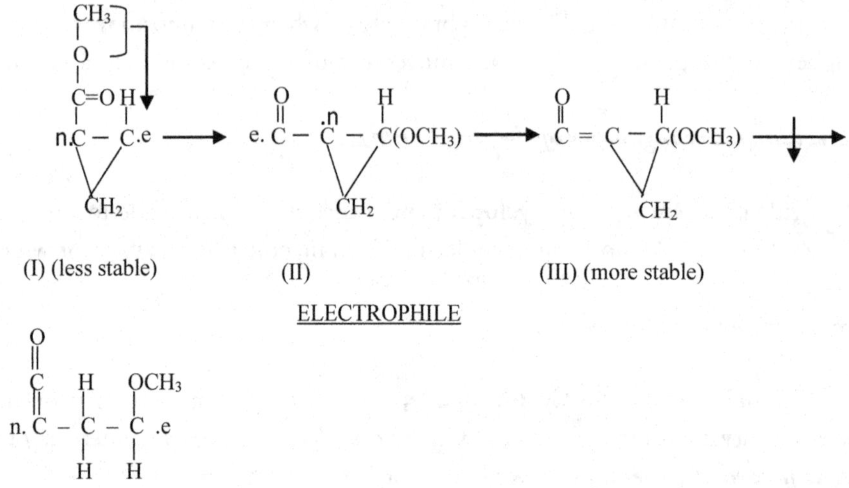

(I) (less stable) (II) (III) (more stable)

ELECTROPHILE

(Living telomer)

= C = O is a strong radical-pulling group of greater capacity than –HC=O group and that
–HC=C=O group is a strong radical-pushing group, noting that the monomer is an Electro-phile.
(Laws of Creations for Cyclopropenes)

Rule 669: This rule of Chemistry for **Vinyl cyclopropene,** states that, this is an unsymmetrical resonance stabilized monomer which can only be activated free-radically as shown below-

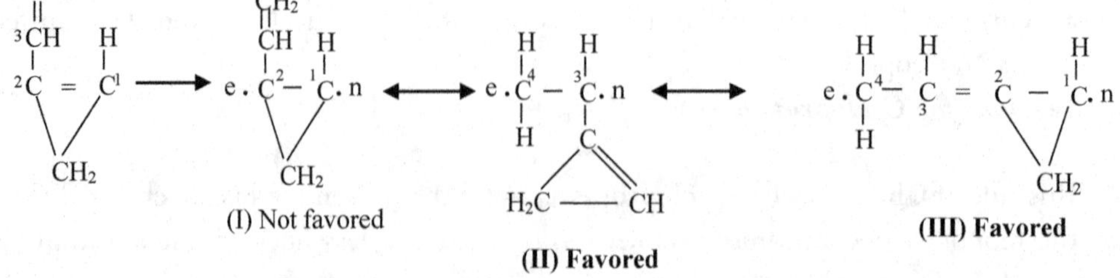

(I) Not favored **(II) Favored** **(III) Favored**

for which being non-symmetric, the radicals shown on the active centers of (II) and (III), are fixed, noting that the >CH_2 group is fully resonance stabilized, with (II) being the less nucleophilic center and the

142

group providing resonance stabilization, the ring being more nucleophilic and (I) favoring no existence; yet, the monomer favors both free-radical routes when suppressed.
(Laws of Creations for Vinyl cyclopropene)

Rule 670: This rule of Chemistry for **Monomers or compounds such as vinyl acetylene, vinyl cyclopropene and the likes,** states that, it is the more nucleophilic center that provides one H atom loosely bonded all the time when not suppressed as shown below for vinylcyclopropene

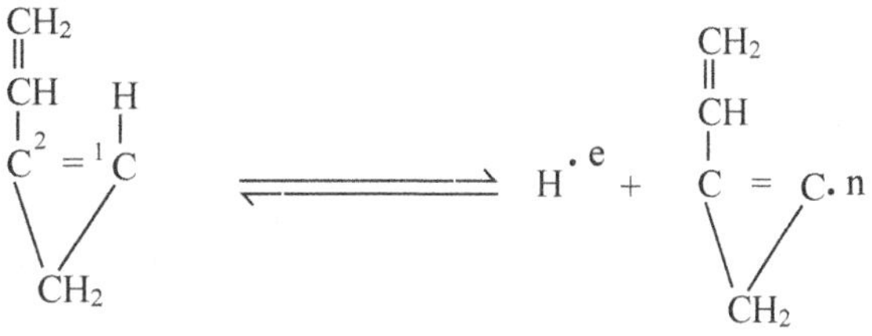

(I) (Non-symmetric)

when in Equilibrium state of existence.
(Laws of Creations for Vinyl cyclopropene)

Rule 671: This rule of Chemistry for **Vinyl cyclopropene,** states that, when not suppressed, the followings take place free-radically-

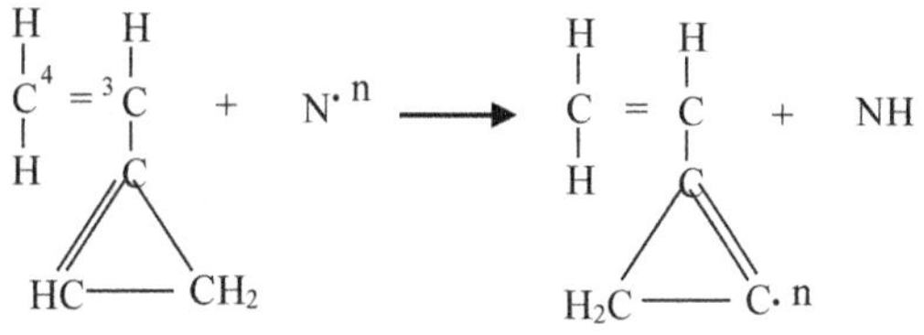

(II)

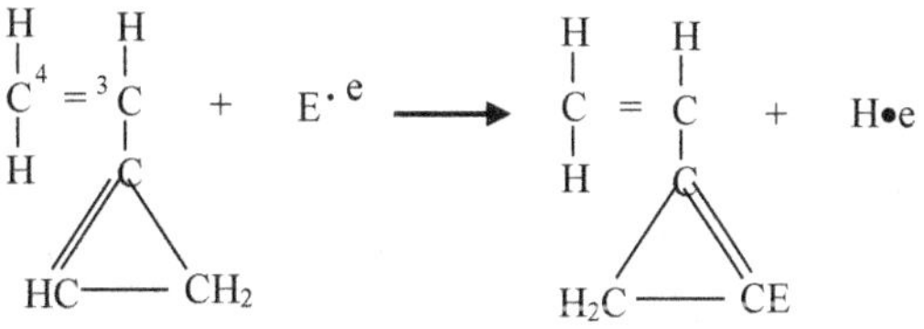

(II)

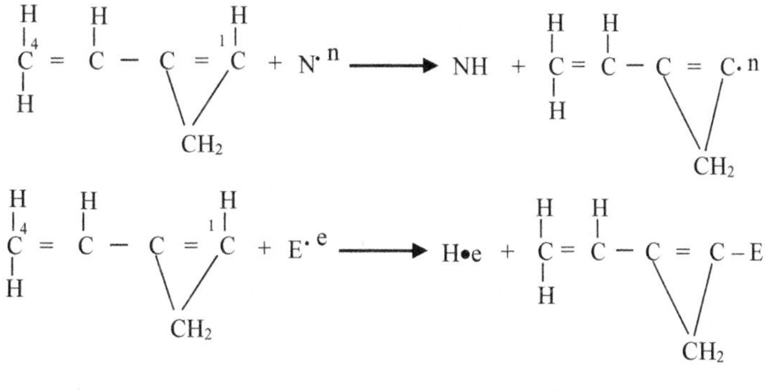

for which that one H atom may need to be replaced if not suppressed, noting that the H atom abstracted or replaced is not a transfer species.
(Laws of Creations for Vinyl cyclopropene)

143

Rule 672: This rule of Chemistry for **Vinyl cyclopropene,** states that, when heated at or above 400^0C, it can undergo Electro-radicalization phenomenon to give Vinyl methyl acetylene or methyl 1, 4- cumulene as shown below.

Stage 1:

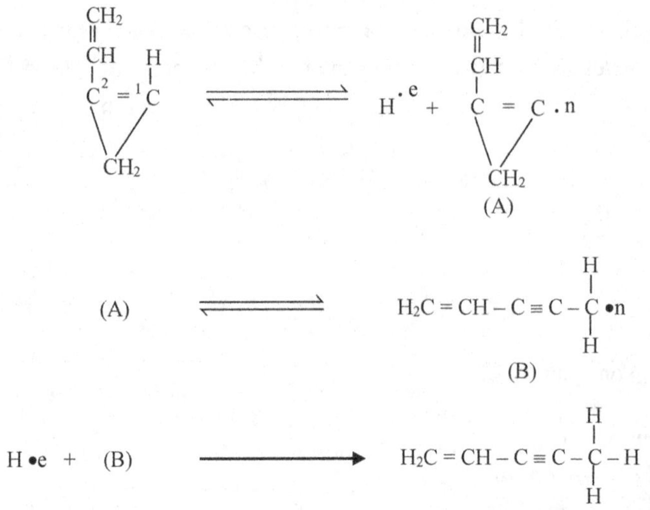

(A)

(B)

(C) Vinyl methyl acetylene

Overall equation: Vinyl cyclopropene $\xrightarrow{>400^0C}$ Vinyl methyl acetylene

(Laws of Creations for Vinyl cyclopropene)

Rule 673: This rule of Chemistry for **Cyclopropene,** states that, when used as a resonance stabilization group, the followings are valid-

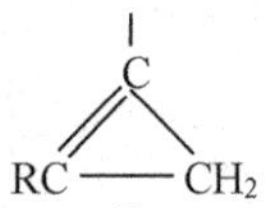

(I)

Order of Radical-pushing capacities of Resonance stabilization groups

noting that the capacity of the ringed group has been greatly reduced, because the CH_2 group in the ring is now fully resonance stabilized for which the need for (I) to carry R arises.

(Laws of Creations for Cyclopropene)

Rule 674: This rule of Chemistry for **Vinyl cyclopropene of the type shown below,** states that, just like isoprene, the R group in the compound is well shielded and so also is the $>CH_2$ group in the ring; for which both free-radical routes remain favored when suppressed.

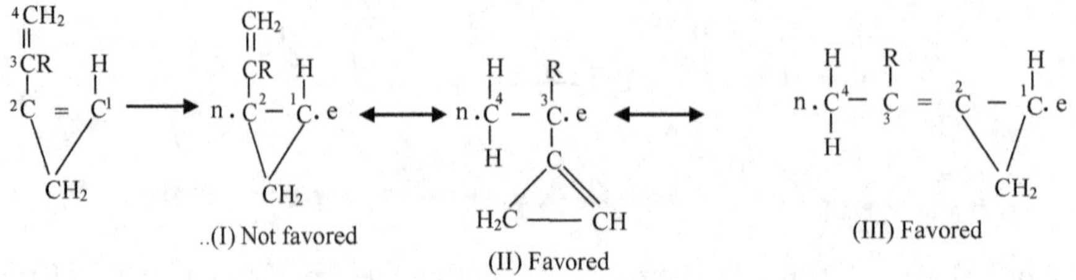

..(I) Not favored   (II) Favored   (III) Favored

(Laws of Creations for Vinyl cyclopropenes)

144

Rule 675: This rule of Chemistry for **Vinyl cyclopropenes of the type shown below,** states that, almost like pentadiene, the R group in the compound is not resonance stabilized as shown below-

.(I) Not favored   (II) Favored   (III) Favored

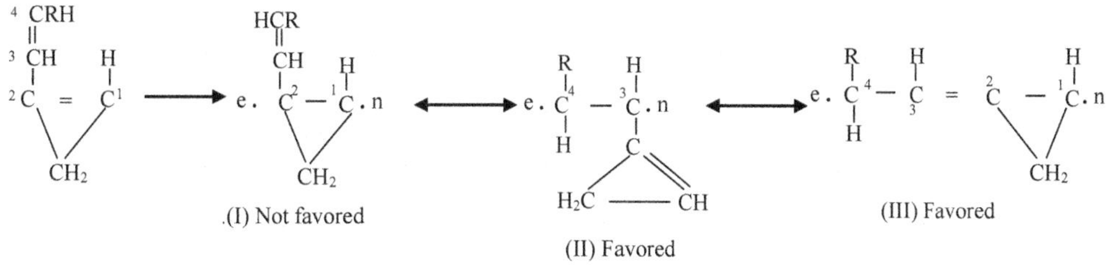

since the monomer is now more nucleophilic favoring only the electro-free-radical route.
(Laws of Creations for Vinyl cyclopropene)

Rule 676: This rule of Chemistry for **Vinyl cyclopropenes of the type shown below,** states that, the R group in the compound is not resonance stabilized as shown below-

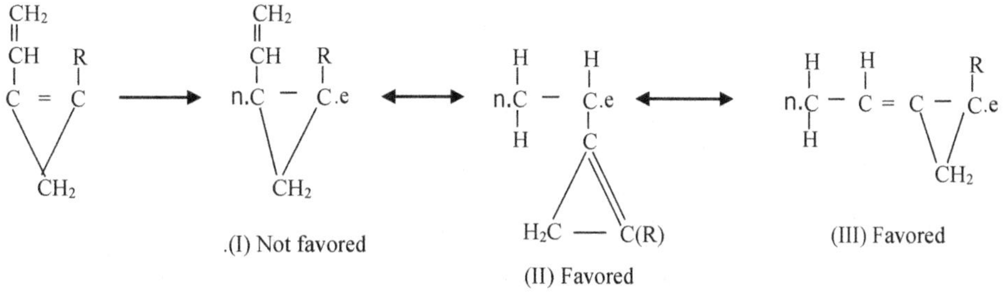

.(I) Not favored   (II) Favored   (III) Favored

since the compound or monomer is now more nucleophilic favoring only the electro-free-radical route.
(Laws of Creations for Vinyl cyclopropenes)

Rule 677: This rule of Chemistry for **Vinyl cyclopropenes of the type shown below,** states that, the R and R^1 groups in the compound are not resonance stabilized as shown below, (where $R > R^1$), since the compound or monomer is more nucleophilic favoring only the electro-free-radical route;

.(I) Not favored   (II) Favored   (III) Favored

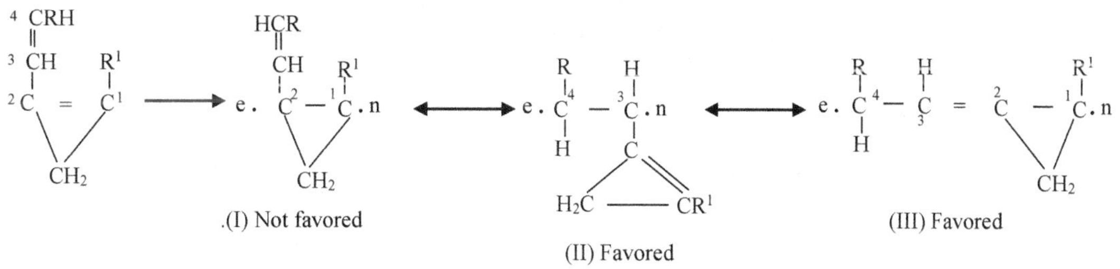

for which when $R^1 > R$, no molecular rearrangement of the third kind can take place since the same mono-form is obtained, while when $R^1 < R$, molecular rearrangement of the third kind can take place with limitations as shown below-

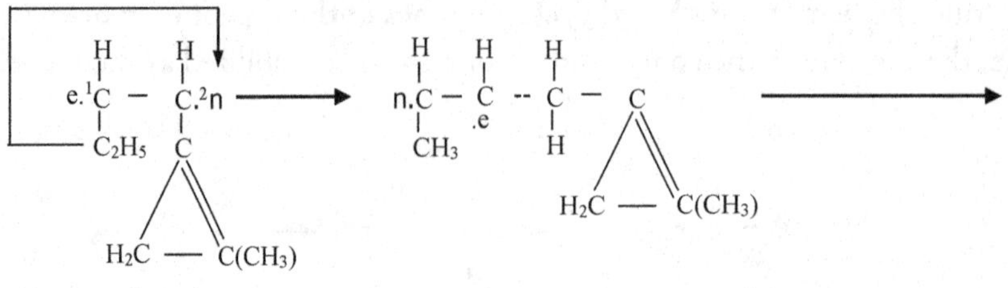

$$
\begin{array}{c}
\text{e.}^1\text{C} - \text{C.}^2\text{n} \\
\end{array}
\quad\longrightarrow\quad
\text{n.C} - \text{C} \cdots \text{C} - \text{C}
\quad\longrightarrow
$$

Right Activated state

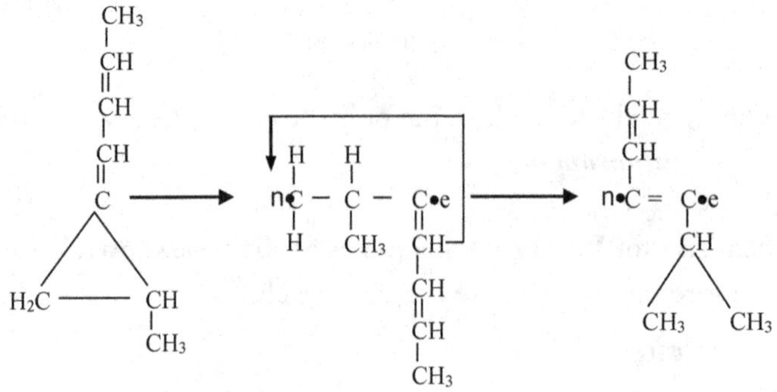

$$
\begin{array}{c}
CH_3 \\
CH \\
CH \\
CH \\
C
\end{array}
\quad\longrightarrow\quad
\begin{array}{c}
n\dot{C} - C - C\bullet e
\end{array}
\quad\longrightarrow\quad
n\bullet C = C\bullet e
$$

noting the limitations when C_2H_5 is replaced with C_5H_{11}.
(Laws of Creations for Vinyl cyclopropene)

Rule 678: This rule of Chemistry for **The first members of vinyl cyclopropene, acetylene, and alkenes in terms of their resonance stabilization capabilities,** states that, the following is valid for them-

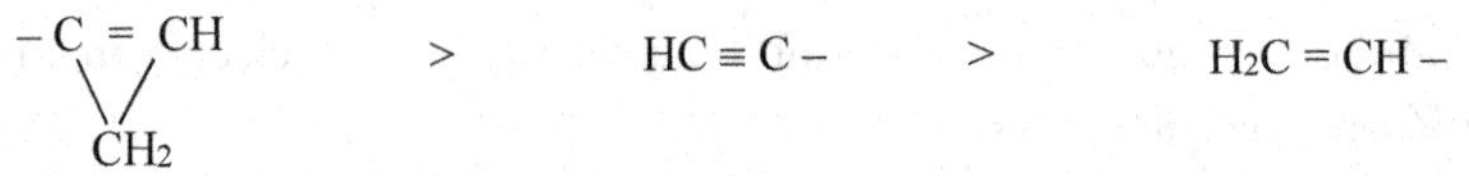

$$
\underset{CH_2}{-C = CH} \quad > \quad HC \equiv C - \quad > \quad H_2C = CH -
$$

Order of Resonance Stabilization capabilities and Nucleophilicity of their monomers

(Laws of Creations for Cyclopropene)

Rule 679: This rule of Chemistry for **Cyclopropene,** states that, when two resonance stabilization groups are symmetrically placed on the ring, the followings are obtained;

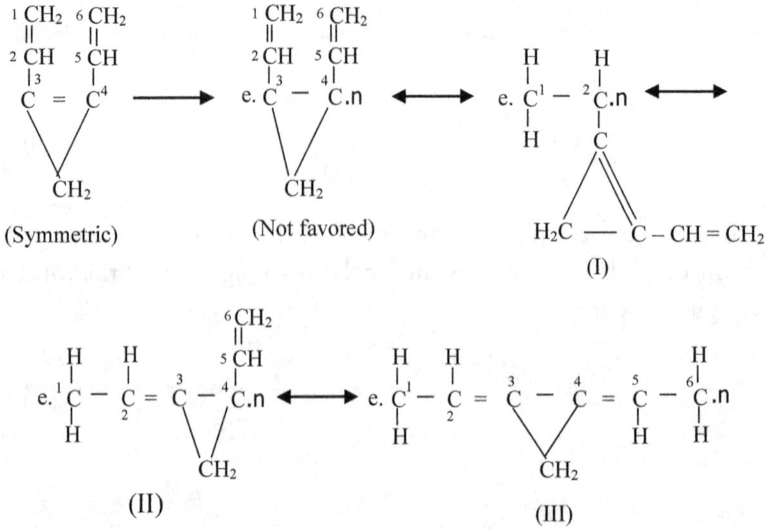

(Symmetric) → (Not favored) ↔ (I)

(II) ↔ (III)

146

for which the center providing resonance stabilization is the ring, and the resonance stabilized mono-forms (I), (II), and (III) favor both free-radical routes more so easily electro-free-radically being natural to it, noting that these cannot take place chargedly.
(Laws of Creations for Resonance stabilization)

Rule 680: This rule of Chemistry for **Resonance stabilization groups,** states that, in view of the symmetric character of di-vinyl cyclopropene, the following relationships are valid-

$$(CH_2 = CH -) \quad > \quad [-H] \quad > \quad (HC \equiv C–) \quad > \quad (\; ^-C \;)$$

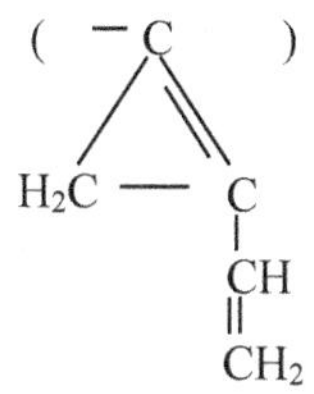

Order of Radical-pushing capacity of resonance stabilization groups

(Laws of Creations for Resonance stabilization groups)

Rule 681: This rule of Chemistry for **Cycloalkenes,** states that, there are marked differences between three-membered rings and all the other members, wherein it is only in the three-membered ring, one carbon center is connected to the two centers carrying double bond.

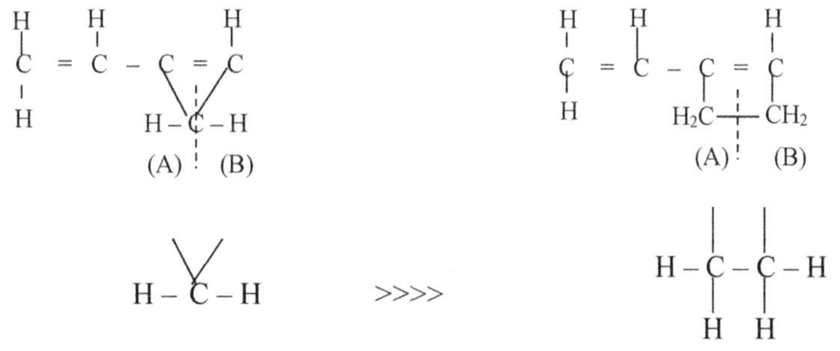

PUSHING CAPACITY FROM *ONE* CENTER TO *TWO* CENTERS

PUSHING CAPACITY FROM *TWO* CENTERS TO *TWO* CENTERS

(Laws of Creations for Cycloalkenes)

Rule 682: This rule of Chemistry for **Cycloalkenes,** states that, as the size of the ring increases from four upwards, the influence of resonance stabilization decreases with increasing size of the rings and the radical pushing capacity of the closed groups increases with increasing size of the rings as shown below only for four-membered rings and above.

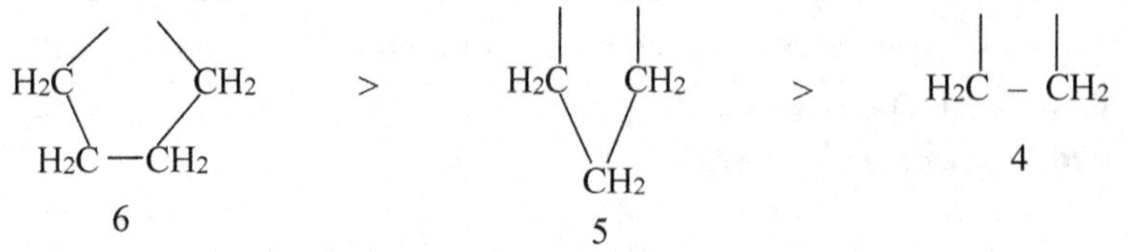

Order of Radical-pushing capacities of Closed groups

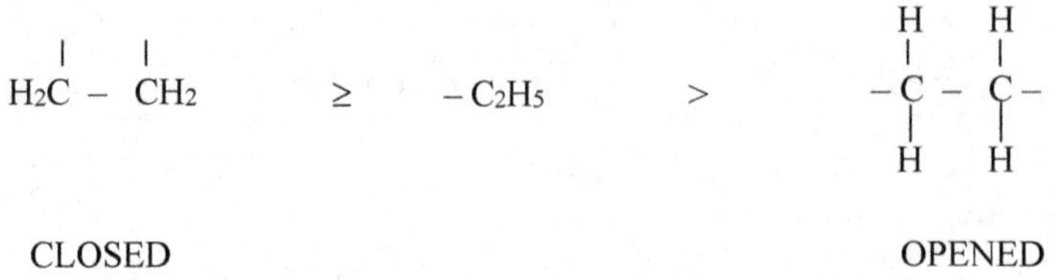

CLOSED   OPENED

Order of Radical-pushing capacities of Opened and Closed groups

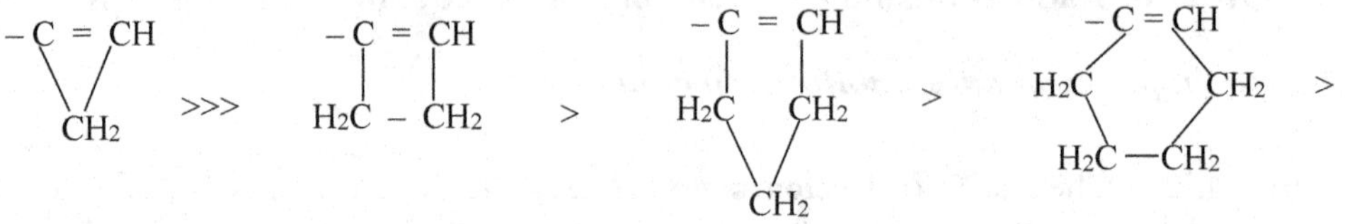

Order of Resonance stabilization capabilities of Resonance stabilization groups

(Laws of Creations for Cycloalkenes)

Rule 683: This rule of Chemistry for **Vinyl Cycloalkenes,** states that, while for cyclopropene, molecular rearrangement of the third kind can never take place, for cyclobutene and higher members, they can readily undergo molecular rearrangement of the third kind as shown below for cyclobutene-

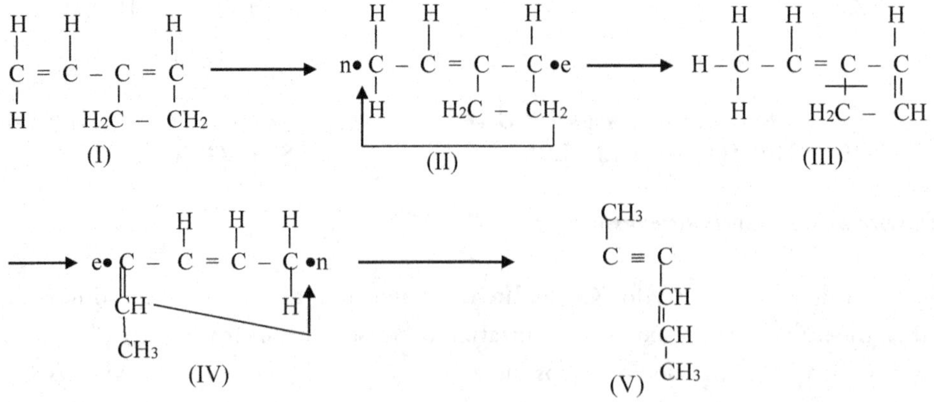

since the capacity of H atom in the ring is not greater than the capacity of H atom externally located to the ring unlike in cyclopropene, noting that as Nucleophiles, only the electro-free-radical route is favored by them.

(Laws of Creations for Cycloalkenes)

Rule 684: This rule of Chemistry for **Resonance stabilized cyclopropene wherein NH_2 is the radical-pushing group in place of H as shown below,** states that, though it may seem that these undergo molecular rearrangement of the third kind only radically as shown below to give a vinyl nitrile or preferably an acrylonitrile which is an electrophile, from all the mono-forms, it is still favored, despite the fact that the final transfer species involved (H), is not the same transfer species involved nucleo-free-radically for the X center; being an Electrophile-

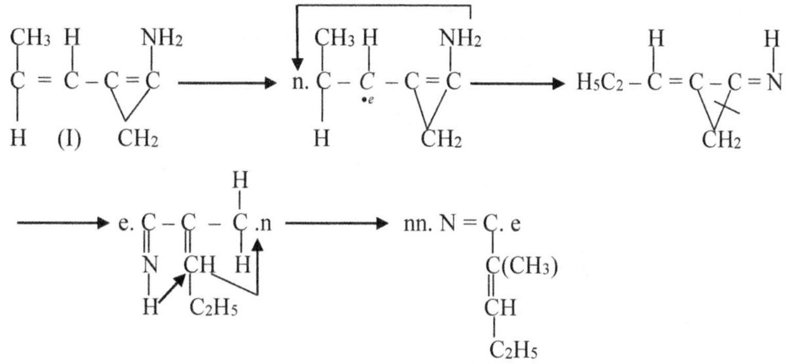

noting that these cannot also take place with NHR and NR$_2$ groups.
(Laws of Creations for Vinyl cyclopropenes)

Rule 685: This rule of Chemistry for **Resonance stabilized cyclopropene wherein OH is the radical-pushing group in place of H as shown below,** states that, these cannot undergo molecular rearrangement of the third kind radically as shown below for one of the mono-forms.

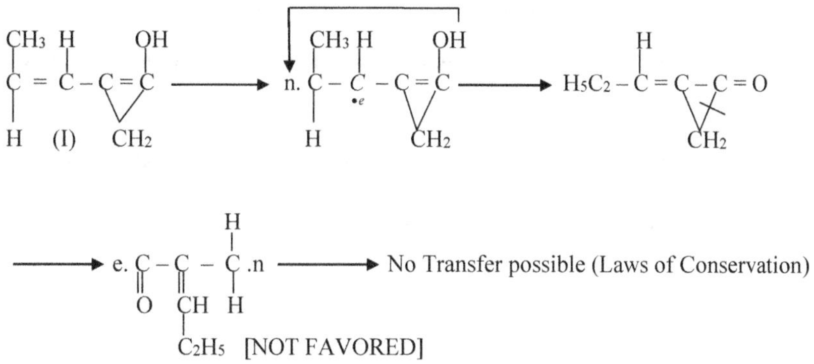

(Laws of Creations for Vinyl cyclopropenes)

Rule 686: This rule of Chemistry for **Resonance stabilized cyclopropene wherein NH$_2$ is the radical-pushing group in place of H as shown below,** states that, these undergo molecular rearrangement of the third kind only radically when the ring is opened instantaneously as shown below to give an electrophile from all the mono-forms,

149

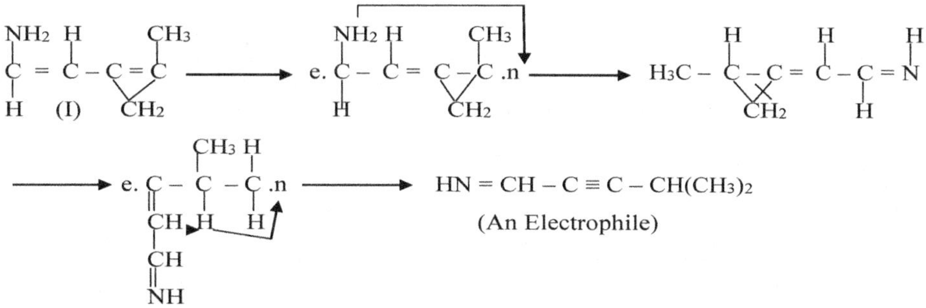

noting that these can take place with NHR but not with NR_2 groups where their use will depend on the size of R group with respect to the opposite alkylane group (e.g. CH_3 above)
(Laws of Creations for Vinyl cyclopropenes)

Rule 687: This rule of Chemistry for **Resonance stabilized cyclopropene wherein OH is the radical-pushing group in place of H as shown below,** states that, these undergo molecular rearrangement of the third kind only radically as shown below to give an electrophile from all the mono-forms,

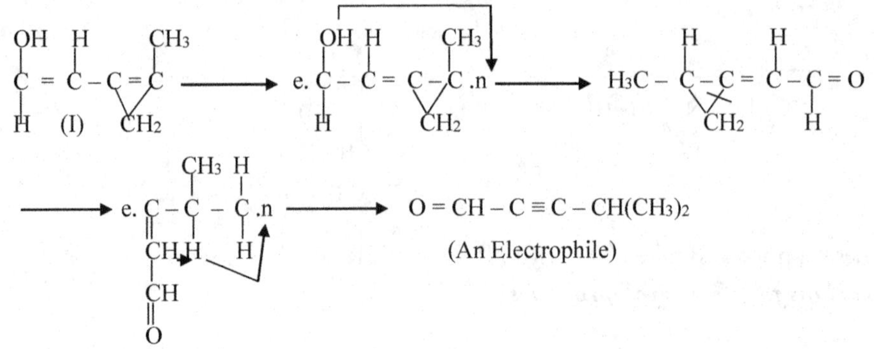

$$O = CH - C \equiv C - CH(CH_3)_2$$

(An Electrophile)

noting that these cannot take place with OR groups where their use will depend on the size of R group with respect to the opposite alkylane group (e.g. CH_3 above)
(Laws of Creations for Vinyl cyclopropenes)

Rule 688: This rule of Chemistry for **Radical-pushing groups of the opened and closed types,** states that, the following is valid for some of them-

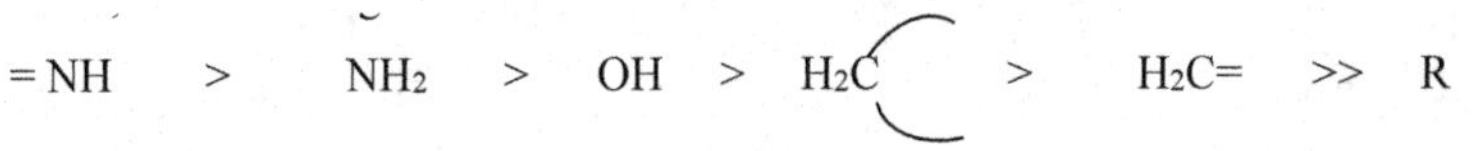

$$= NH \quad > \quad NH_2 \quad > \quad OH \quad > \quad H_2C \quad > \quad H_2C= \quad >> \quad R$$

Order of Radical-pushing capacity of groups.

noting the hierarchy of appearance of these groups and the marked difference between the Hydrocarbon and the Non-hydrocarbon families.
(Laws of Creations for Radical-pushing groups)

Rule 689: This rule of Chemistry for **Resonance stabilized cyclopropene groups with special radical pushing groups,** states that, some of the followings are the simple resonance stabilization providers with the following order of radical-pushing capacity-

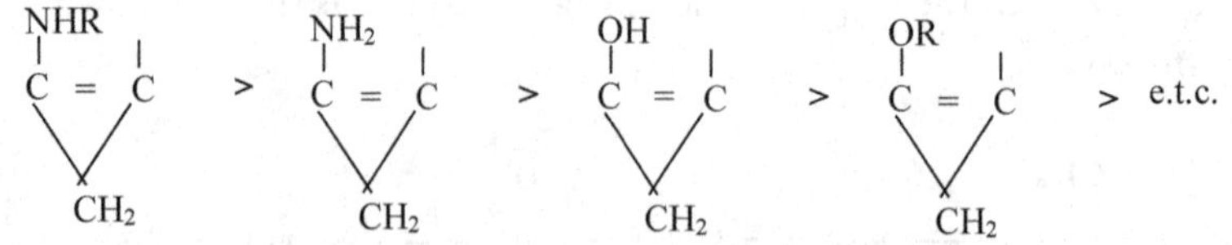

Radical-pushing capacities of resonance stabilization groups

(Laws of Creations Cyclopropene groups)

Rule 690: This rule of Chemistry for **Resonance stabilized cyclopropene wherein $C_{2n}F_{2n+1}$ is the radical-pushing group in place of H as shown below,** states that, these cannot undergo molecular rearrangement of the third kind radically as shown below to give a stable monomer, since molecular rearrangement of the first kind is not possible, because H is more radical-pushing than $C_{2n}F_{2n+1}$,

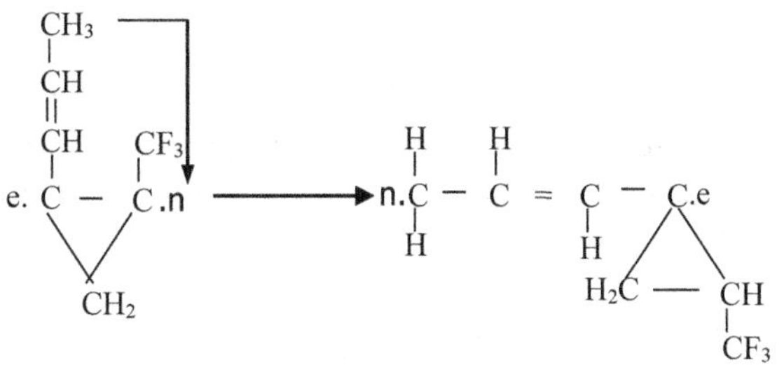

(As a nucleophile)   Impossible Transfer

clear indication that the following is valid-

Order of Radical-pushing capacity during resonance stabilization

(Laws of Creations for Vinyl cyclopropenes)

Rule 691: This rule of Chemistry for **Resonance stabilization groups,** states that, the following is their order of radical-pushing capacities, based on what they are carrying-.

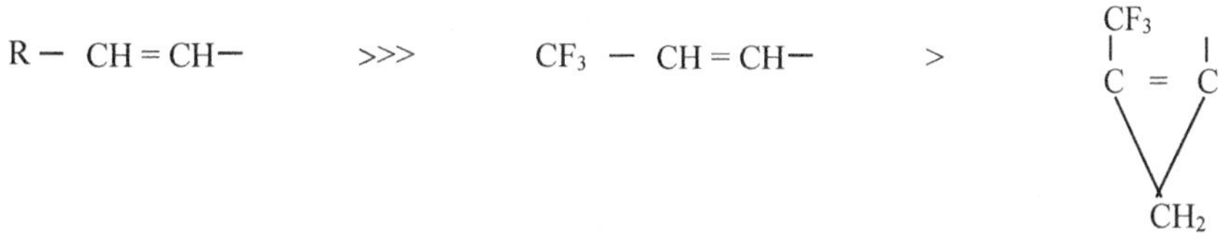

Order of Radical-pushing capacity of Resonance stabilization groups

(Laws of Creations for Resonance stabilizations)

Rule 692: This rule of Chemistry for **Vinyl cyclopropene with radical-pulling groups such as Cl, COOR, CONH$_2$, OCOR, in place of H on the ring,** states that, it is only for Cl, the possibility of opening of the ring exists for which the followings are to be expected using Cl which has no transfer species, when activated-.

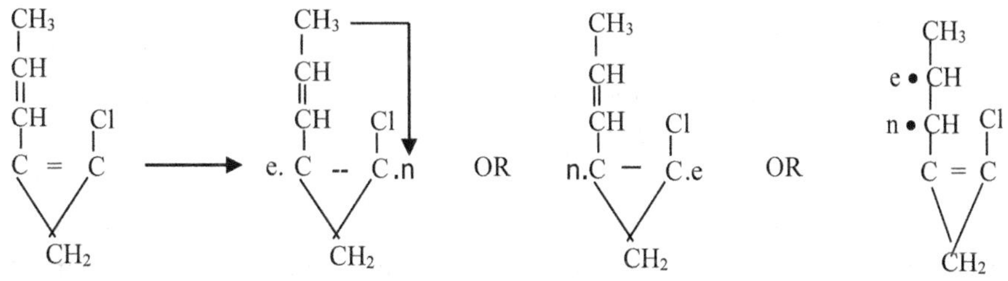

(A)   (I) Favored   (II) Wrongly activated   (B)

for which, (I) is the true activated state, i.e., the less nucleophilic center where possibility of the ring opening to form a vinyl acetylene may exist; while for the other groups, which make the vinyl cyclopropene become Electrophiles, their rings can never be opened, particularly with respect to COR groups which has no transfer species.

(Laws of Creations for Vinyl cyclopropenes)

Rule 693: This rule of Chemistry for **Members of groups in families of cyclopropenes and their isomers from methylene cyclopropane, cumulenes and acetylenes,** states that, the followings are their orders of radical-pushing capacities when used to provide resonance stabilization-

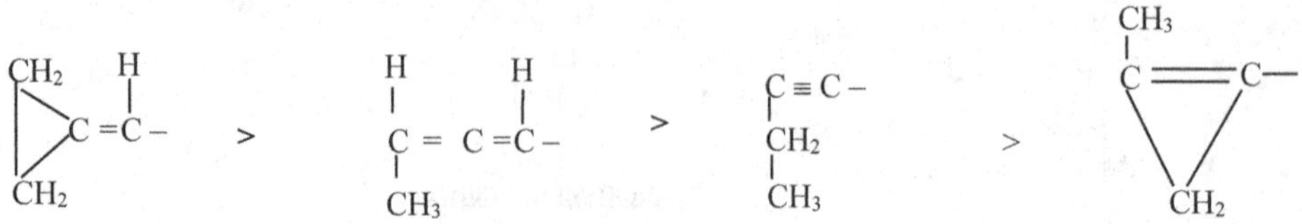

C_4H_5 **C_4H_5** **C_4H_5** **C_4H_5**

Order of Radical-pushing capacity

(Laws of Creations for Resonance stabilization groups)

Rule 694: This rule of Chemistry for **Cyclopropenes and Acetylenes,** states that, the followings are valid for their isomers-

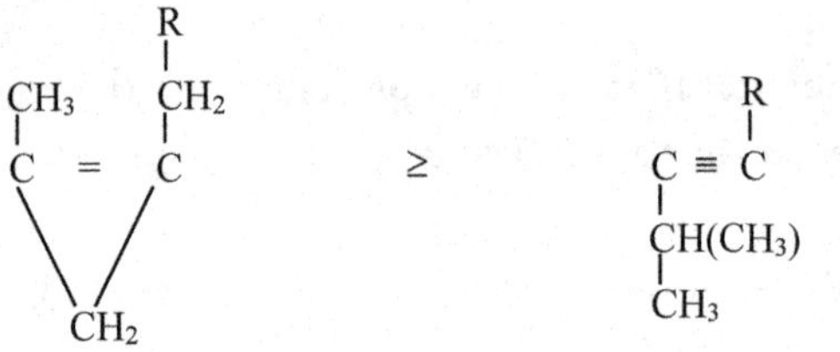

(Where Rs are radical-pushing alkylane groups)

(I) RC_5H_7 (II) RC_5H_7

Order of Nucleophilicity ($R \leq C_3H_7$)

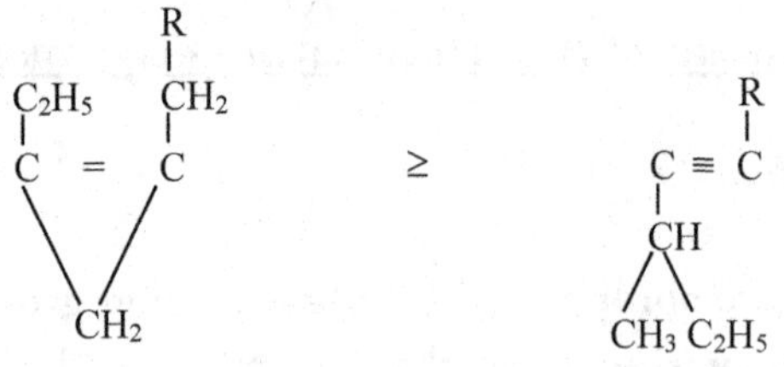

(III) RC_6H_9 (IV) RC_6H_9

Order of Nucleophilicity ($R \leq C_4H_9$)

for which, when R is greater than C_3H_7 for (I) and C_4H_9 for (III), the order of nucleophilicity is reversed, with acetylene becoming more nucleophilic.

(Laws of Creations for Acetylenes and cyclopropenes)

Rule 695: This rule of Chemistry for **The first member of Butadiyne cyclopropene whose structure is shown below,** states that, this is a resonance stabilized non-symmetric monomer or compound which when used as a monomer when suppressed favors both free-radical routes-

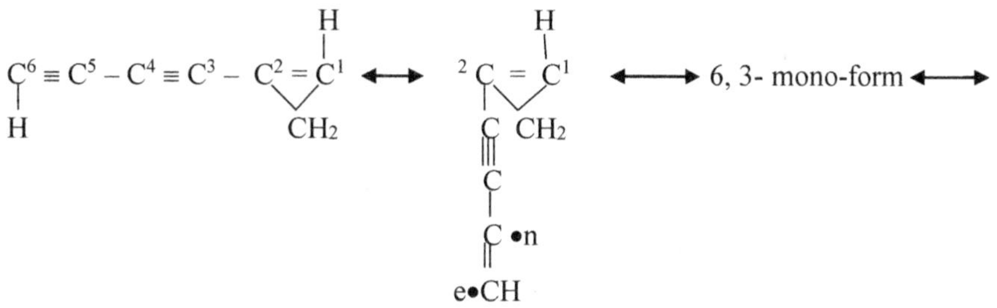

$$C^6 \equiv C^5 - C^4 \equiv C^3 - C^2 = C^1 \quad \longleftrightarrow \quad {}^2C = C^1 \quad \longleftrightarrow \quad 6,3\text{- mono-form} \quad \longleftrightarrow$$

[Least Nucleophilic center]

for which no form of molecular rearrangement can take place and the ring can never be opened; noting that the most nucleophilic center is the ring, the resonance stabilization provider and activa-tion commences at the 6,5 center to give the 6,3- and 6,1- mono-forms.

(Laws of Creations for Butadiyne cyclopropene)

Rule 696: This rule of Chemistry for **The first member of butadiyne cyclopropene,** states that, when not suppressed, one of the externally located H atoms is still loosely bonded electro-free-radically and the hydrogen atom is as shown below-

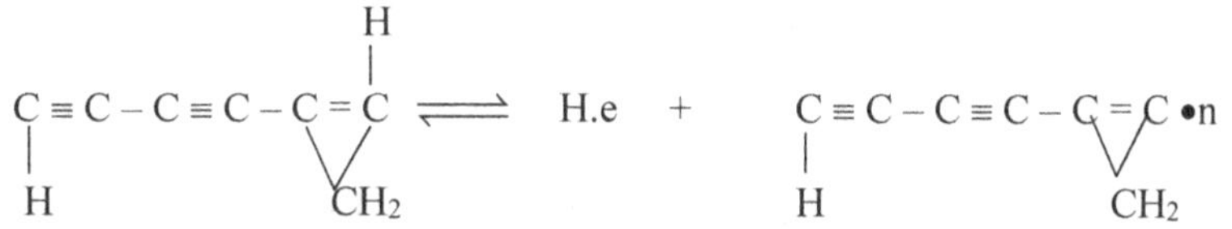

$$C \equiv C - C \equiv C - C = C \quad \rightleftharpoons \quad H.e \quad + \quad C \equiv C - C \equiv C - C = C \bullet n$$

EQUILIBRIUM STATE OF EXISTENCE

for which there may be need to replace it, with radical-pushing group.

(Laws of Creations for Butadiyne cyclopropene)

Rule 697: This rule of Chemistry for **Butadiyne cyclopropene,** states that, when the ring is made to carry radical-pushing alkylane group (R), the followings are the mono-forms-

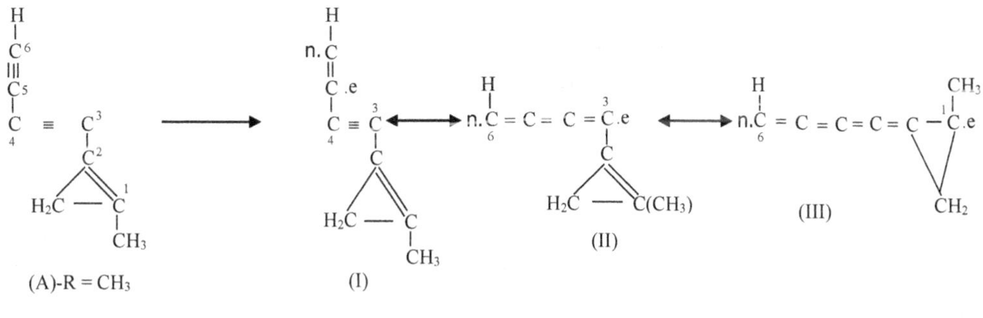

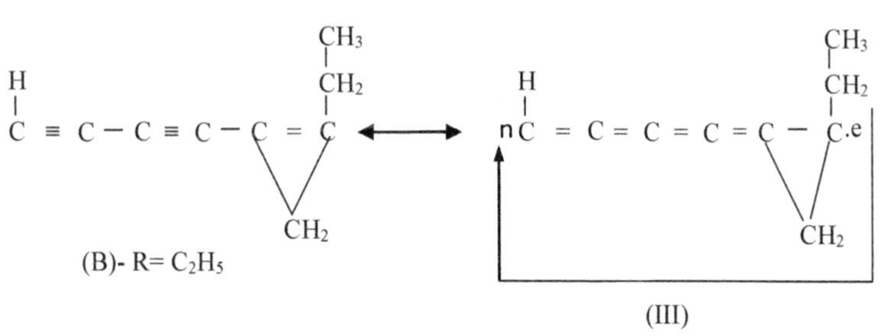

(A)-R = CH₃   (I)   (II)   (III)

(B)- R= C₂H₅   (III)

for which, when one of the mono-forms is decomposed, molecular rearrangement of the third kind is favored to give the products shown below-

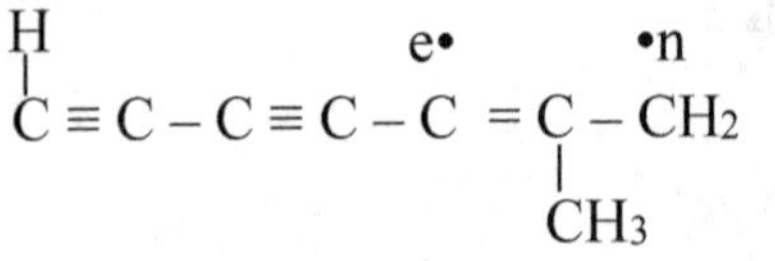

$$
\underset{\overset{|}{H}}{C} \equiv C - C \equiv C - \overset{e\bullet}{C} = \overset{\bullet n}{C} - CH_2
$$
$$
\underset{CH_3}{|}
$$

$$
\underset{\overset{|}{H}}{C} \equiv C - C \equiv C - \overset{e\bullet}{C} = \overset{\bullet n}{C} - CH_2
$$
$$
\underset{C_2H_5}{|}
$$

To give the same monomer   To give the same monomer

From (A)   From (B)

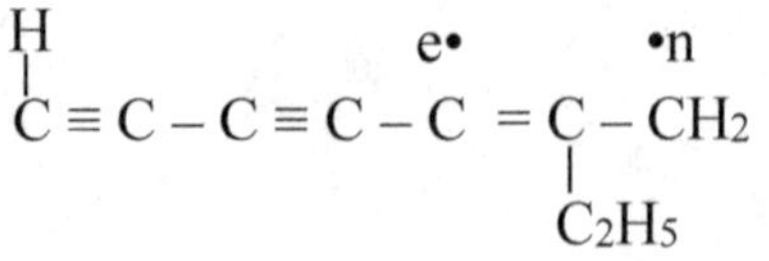

noting that the same compound is produced.
(Laws of Creations for Butadiyne cyclopropenes)

Rule 698: This rule of Chemistry for **Radical-pushing groups,** states that, based on considerations of Butadiyne cyclopropenes, the followings are valid- .

$H_2C=C=C=C=C=C= >$ $H_2C=C=C=C=C= >$ $H_2C=C=C=C= >$ $H_2C=C=C= >$

(Even)   (Odd)   (Even)   (Odd)

$H_2C=C= >$ $H_2C= >> R$

(Even)   (Odd)

Order of Radical-pushing capacities of Radical-pushing groups
(Laws of Creations for Radical-pushing groups) [See Rule 428]

Rule 699: This rule of Chemistry for **Radical-pushing groups,** states that, while $-RCH = CH_2$ which are radical pushing resonance stabilization groups, *are radical-pulling in character,* $- H_2C = CHR$ which also are radical-pushing resonance stabilization groups, *are radical-pushing in character;* the former is of lesser capacity than H, while the latter is of greater capacity than H, as shown below-

$C_3H_7CH = CCH_3 - > C_2H_5CH = CCH_3 - >$ **(H)** $\geq CH_3CH = CCH_3 - > H_2C = CCH_3 -$

Order of Radical-pushing and resonance stabilization capacity
for which the same seem to apply to ODD and EVEN cumulenyl groups.
(Laws of Creations for Radical-pushing groups)

Rule 700: This rule of chemistry for **Butadiyne cyclopropenes which carry one alkylane group on the butadiyne center,** states that, these resonance stabilized monomers cannot only be made to undergo molecular rearrangement of the third kind via Decomposition mechanism, but can also via Equilibrium mechanism be made to give the different isomeric tri-ynes as shown below-

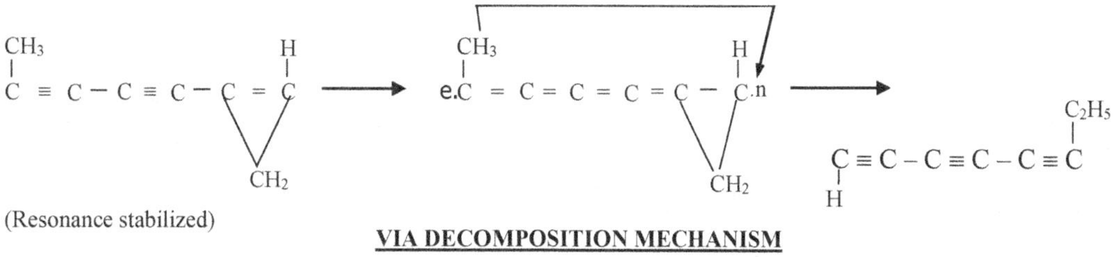

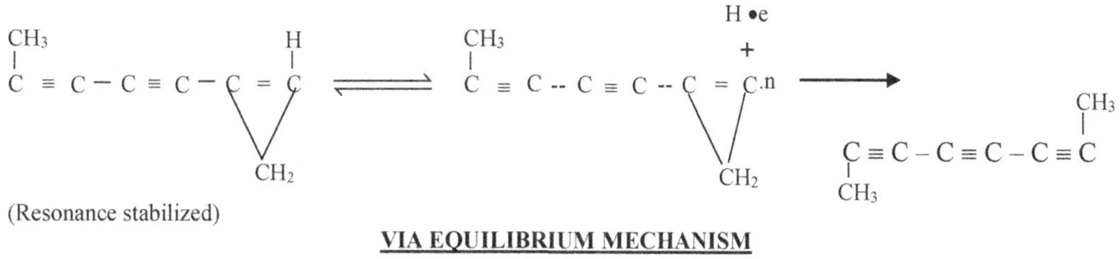

(Resonance stabilized)

VIA DECOMPOSITION MECHANISM

(Resonance stabilized)

VIA EQUILIBRIUM MECHANISM

for which via decomposition mechanism, rearrangement stops when the CH_3 (R) becomes C_3H_7. as shown below.

When $R \equiv C_2H_5$ When $R \equiv C_3H_7$

(Laws of Creations for Butadiyne cyclopropene)

<u>**Rule 701:**</u> This rule of Chemistry for **Butadiyne cyclopropenes which carry alkylane groups (R) of equal capacity,** states that, these resonance stabilized monomers can undergo molecular rearrangement of the third kind to give tri-ynes as shown below-

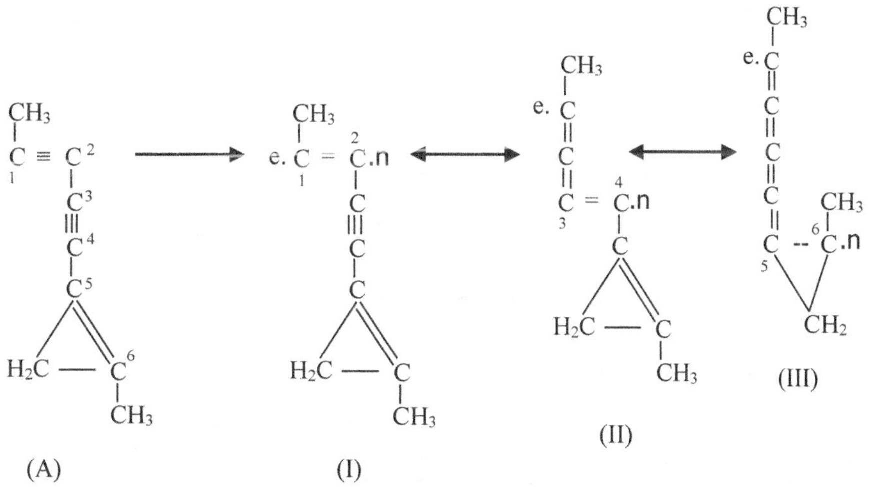

(A)   (I)   (II)   (III)

155

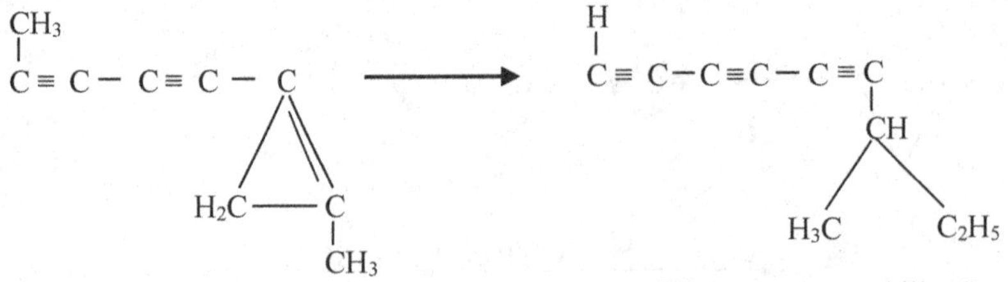

$$CH_3 - C \equiv C - C \equiv C - C \overset{H_2C - C}{\underset{CH_3}{\triangle}} \longrightarrow H - C \equiv C - C \equiv C - C \equiv C - \underset{H_3C \quad C_2H_5}{CH}$$

(I) (resonance stabilized)   (A) (resonance stabilized)

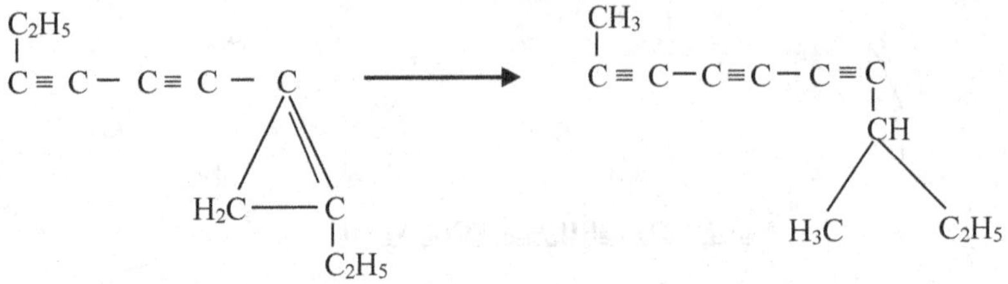

$$C_2H_5 - C \equiv C - C \equiv C - C \overset{H_2C - C}{\underset{C_2H_5}{\triangle}} \longrightarrow CH_3 - C \equiv C - C \equiv C - C \equiv C - \underset{H_3C \quad C_2H_5}{CH}$$

(I) (resonance stabilized)   (C) (resonance stabilized)

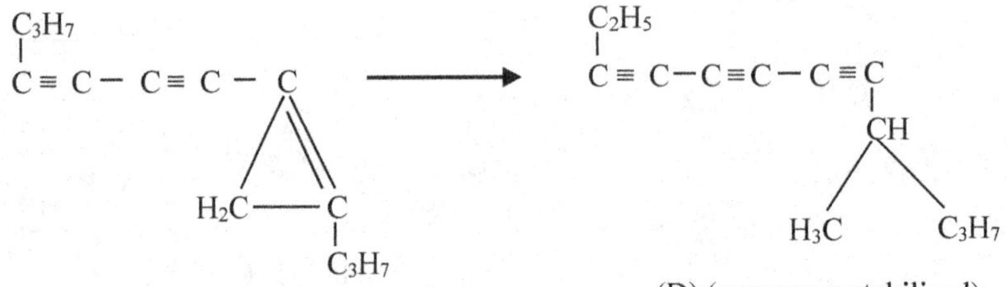

$$C_3H_7 - C \equiv C - C \equiv C - C \overset{H_2C - C}{\underset{C_3H_7}{\triangle}} \longrightarrow C_2H_5 - C \equiv C - C \equiv C - C \equiv C - \underset{H_3C \quad C_3H_7}{CH}$$

(II) (resonance stabilized)   (D) (resonance stabilized)

noting that, there is no limitation placed on the size of the group, but the acetylene side; the only route favored by them being the electro-free-radical route.
(Laws of Creations for Butadiyne cyclopropene)

Rule 702: This rule of Chemistry for **Acetylene and Cyclopropene,** states that, based on all considerations of their families, the followings are valid for them-

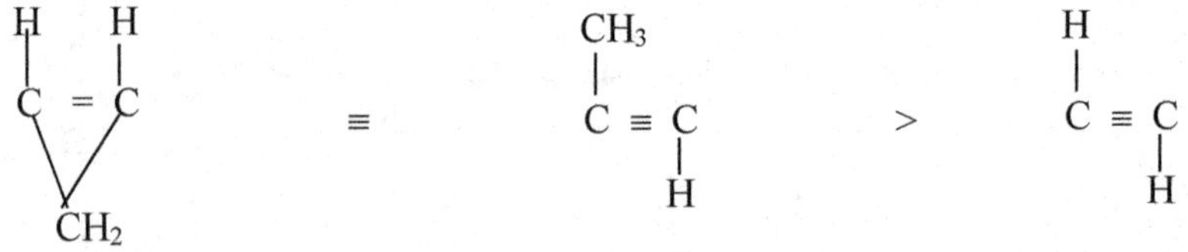

$$\underset{CH_2}{\overset{H \quad H}{\underset{}{C = C}}} \equiv \underset{H}{\overset{CH_3}{C \equiv C}} > \underset{H}{\overset{H}{C \equiv C}}$$

Order of Nucleophilicity

(Laws of Creations for Cyclopropene/Acetylene)

Rule 703: This rule of Chemistry for **Cyclopropene and Acetylene,** states that, in general, the followings are valid when the groups from them are on their own or used alone (See Rules 678 and 694) -

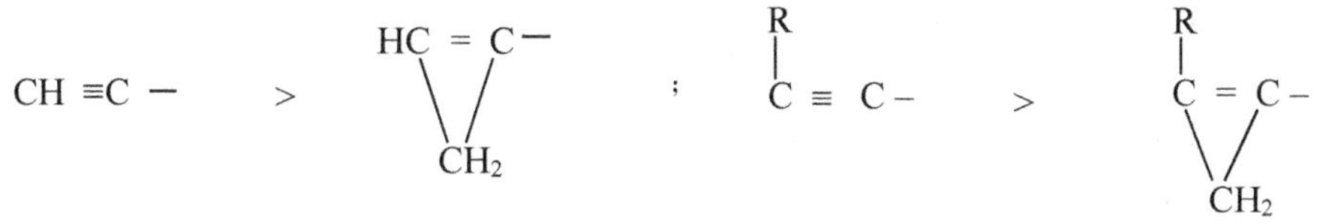

$$CH \equiv C - \quad > \quad \underset{CH_2}{\overset{HC\,=\,C-}{\diagdown\diagup}} \qquad ; \qquad \overset{R}{\underset{}{\overset{|}{C}}} \equiv C - \quad > \quad \overset{R}{\underset{CH_2}{\overset{|}{C}\,=\,C-}}$$

Order of radical-pushing capacity in the presence of resonance stabilization

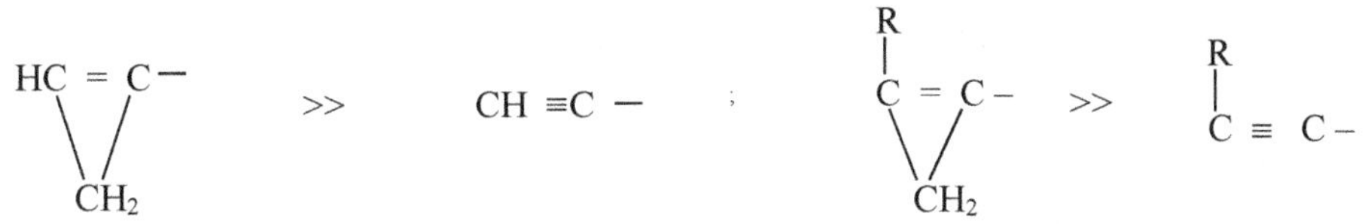

$$\underset{CH_2}{\overset{HC\,=\,C-}{\diagdown\diagup}} \quad >> \quad CH \equiv C - \qquad ; \qquad \overset{R}{\underset{CH_2}{\overset{|}{C}\,=\,C-}} \quad >> \quad \overset{R}{\underset{}{\overset{|}{C}}} \equiv C -$$

<u>Order of radical-pushing capacity in the absence of resonance stabilization</u>

for which in general, the last equation above is valid when the groups are on their own or used alone, but when resonance stabilization is provided by them, the reverse is the case, because the CH_2 group is resonance stabilized.

(Laws of Creations for Cyclopropene/Acetylene)

<u>**Rule 704:**</u> This rule of Chemistry for **Rearrangements by decomposition,** states that, when a ringed compound is decomposed via Equilibrium mechanism *where possible,* the product or sum of the products obtained is of the same nucleophilicity as the ringed compound, while when decomposed via Decomposition mechanism wherein molecular rearrangements are involved, the product obtained is of stronger nucleophilicity than the ringed compound as shown below-

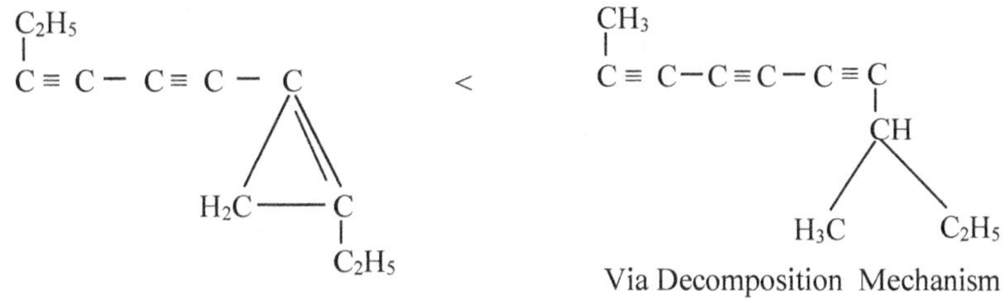

Order of Nucleophilicity

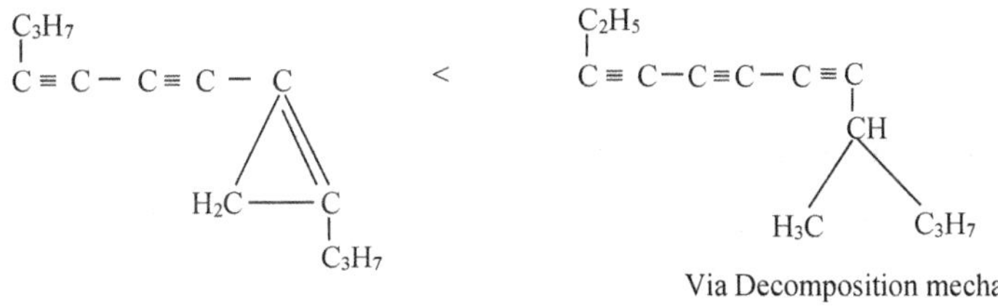

Order of Nucleophilicty

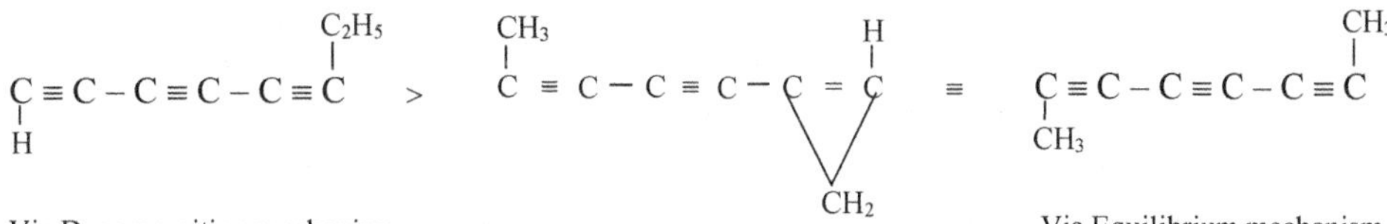

ORDER OF NUCLEOPHILICITY

157

noting that Equilibrium mechanism takes place only when there is one H atom terminally located on the ring.

(Laws of Creations for Rearrangement by decomposition)

Rule 705: This rule of Chemistry for **Butadiyne cyclopropene,** states that, when radical-pushing alkylane groups externally located are different, only when the R group carried by the Acetylene is greater than or equal to that on the cyclopropene will molecularly rearrangement take place on decomposition to give triynes as shown below-

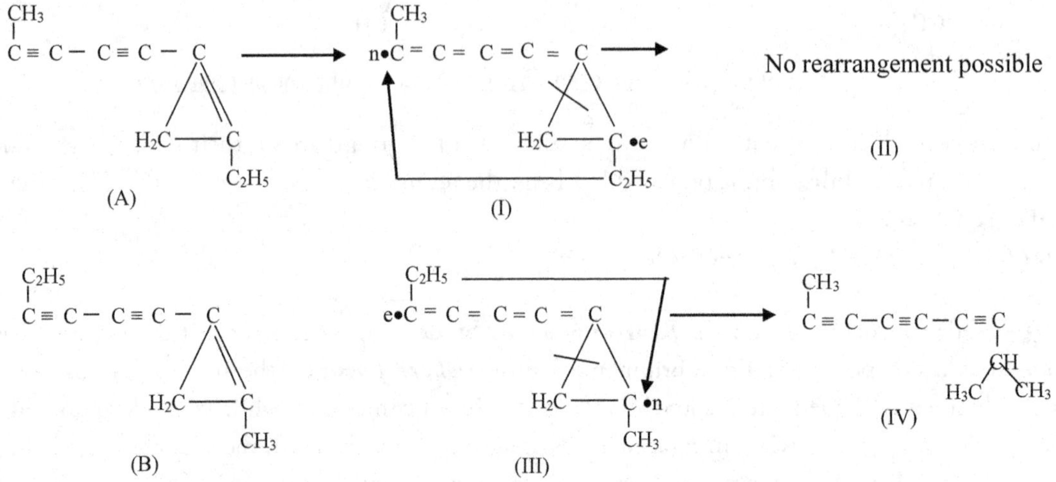

for which in both cases above, the resonance stabilization provider is the ring, noting that in (A), the ring is more radical pushing than the last acetylenic end and no product could readily be obtained, while in (B), a triyne is obtained.

(Laws of Creations for Butadiyne cyclopropenes)

Rule 706: This rule of Chemistry for **Resonance stabilization groups,** states that, in general, the followings are valid-

$$R_F - C \equiv C - C \equiv C - \quad \geq \quad R_F - C \equiv C - \quad \text{(Where } R_F \text{ is for example} \\ \text{CN, COOH, Cl, etc.)}$$

Order of Radical-pulling capacity

$$R - C \equiv C - C \equiv C - \quad \geq \quad R - C \equiv C - \quad \text{(Where R is an alkylane group)}$$

Order of Radical-pushing capacity

(Laws of Creations for Resonance stabilization)

Rule 707: This rule of Chemistry for **Butadiyne cyclopropenes which carry one radical pushing group of the type NH$_2$ on the butadiyne center,** states that, these monomers which are Nucleophiles and resonance stabilized, have 6,5-, 6,3- and 6,1- mono-forms all of which undergo molecular rearrangement of the third kind, based on the operating conditions to give the same monomer which is an Electrophile that is partly resonance stabilized.

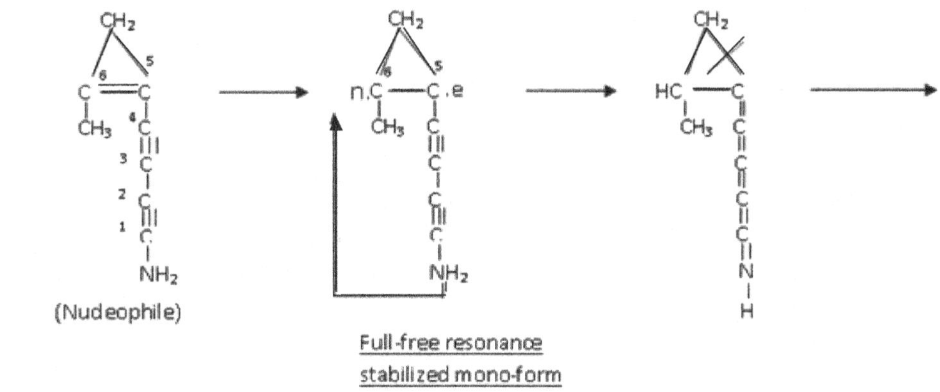

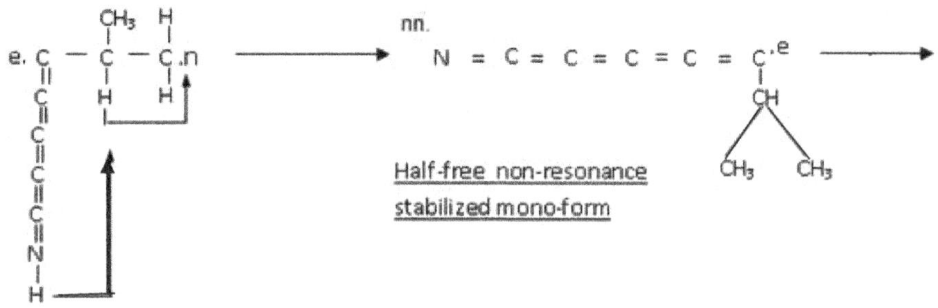

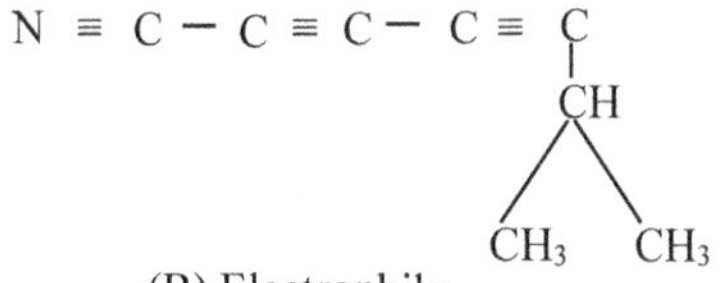

(Nucleophile)

Full-free resonance
stabilized mono-form

Half-free non-resonance
stabilized mono-form

$$N \equiv C - C \equiv C - C \equiv C$$

(B) Electrophile

(Laws of Creations for Butadiyne cyclopropenes)

Rule 708: This rule of Chemistry for **Butadiyne cyclopropenes which carry one radical pushing group of the types NHR, NR$_2$ on the butadiyne center,** states that, these monomers, originally resonance stabilized Nucleophiles, the mono-forms of which are 6,5-, 6,3- and 6,1- mono-forms undergo molecular rearrangement of the third kind only when R is CH$_3$, to give the same monomer which is an Electrophile.

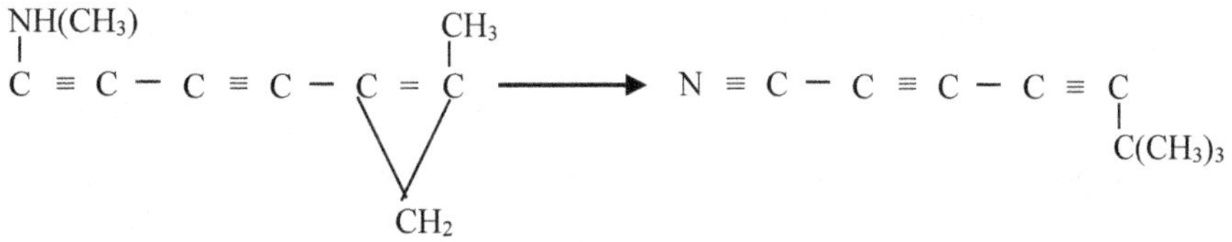

Favored

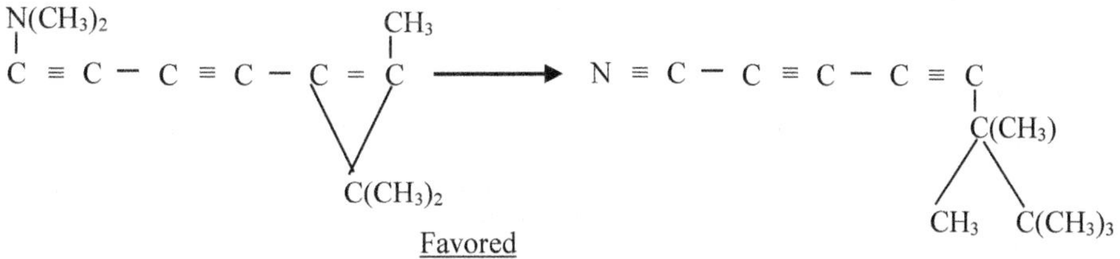

Favored

159

noting the replacement of H in the CH_2 group to CH_3 group for $N(CH_3)_2$, i.e., NR_2, and the two types of transfer species used for $NHCH_3$, H and CH_3 without breaking the Laws of Molecular rearrange-ment and Conservation of transfer of transfer species.
(Laws of Creations for Butadiyne cyclopropenes)

Rule 709: This rule of Chemistry for **Butadiyne cyclopropenes which carry NH_2 radical-pushing group on the ring center and an alkylane group (R) at the other end,** states that, the three mono-forms which are resonance stabilized favor molecular rearrangement of the third kind to give Nitriles which are Nucleophiles as shown below for one of the mono-forms

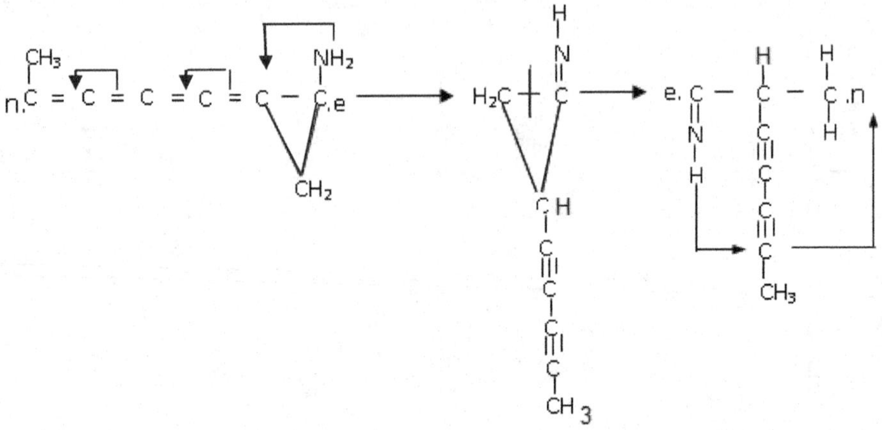

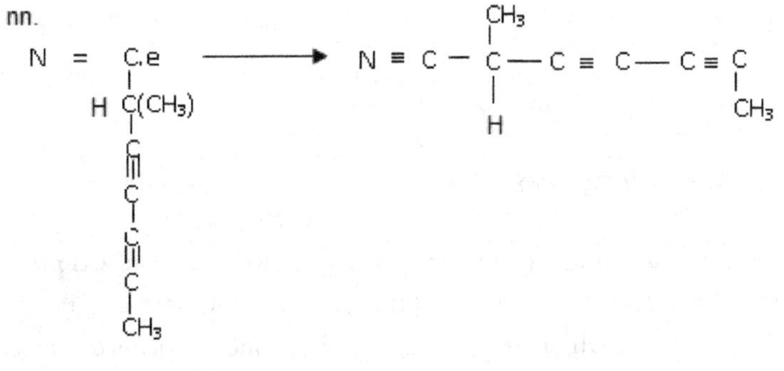

for which a Nucleophile is produced after a rearrangement.
(Laws of Creations for Butadiyne cyclopropenes)

Rule 710: This rule of Chemistry for **Butadiyne cyclopropenes which carry NHR or NR_2 radical -pushing groups,** states that, only the NHR will favor molecular rearrangement of the third kind only for R equal to CH_3; while NR_2 will do only when all the Hs are replaced with CH_3 or higher as shown below-

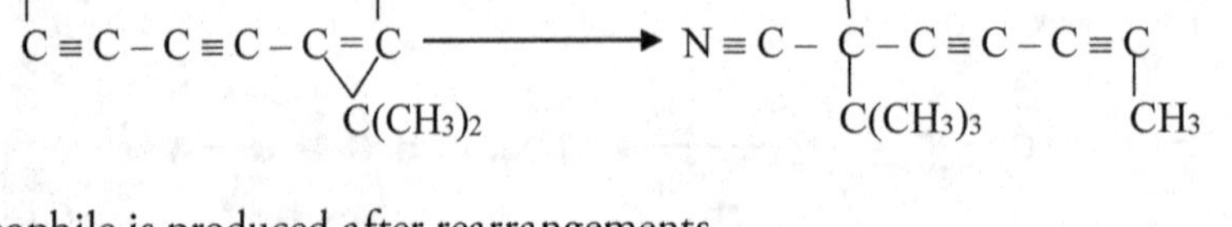

for which a Nucleophile is produced after rearrangements.
(Laws of Creations for Butadiyne cyclopropenes)

Rule 711: This rule of Chemistry for **Butadiyne cyclopropenes carrying OH radical-pushing groups on the butadiyne center,** states that, though resonance stabilized, with possibility of molecular rearrangements of the first and third kinds taking place to first form an extended cyclopropanone followed by decomposition to give 1, 6-Ketene via Decomposition mechanism.

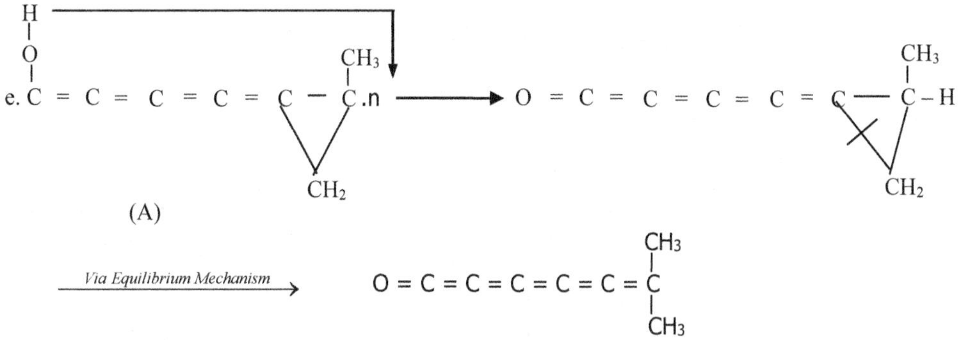

$$
\text{e. } C = C = C = C = C - C.n \quad \longrightarrow \quad O = C = C = C = C = C - C-H
$$

(A)

$$
\xrightarrow{\textit{Via Equilibrium Mechanism}} \quad O = C = C = C = C = C = C
$$

1, 6-Dimethyl ketene

(Laws of Creations for Butadiyne cyclopropenes)

Rule 712: This rule of Chemistry for **Butadiyne cyclopropenes which carry OH group on the ring,** states that, while their rings can be opened, they can undergo molecular rearrangement of the third kind to give a Ketene.

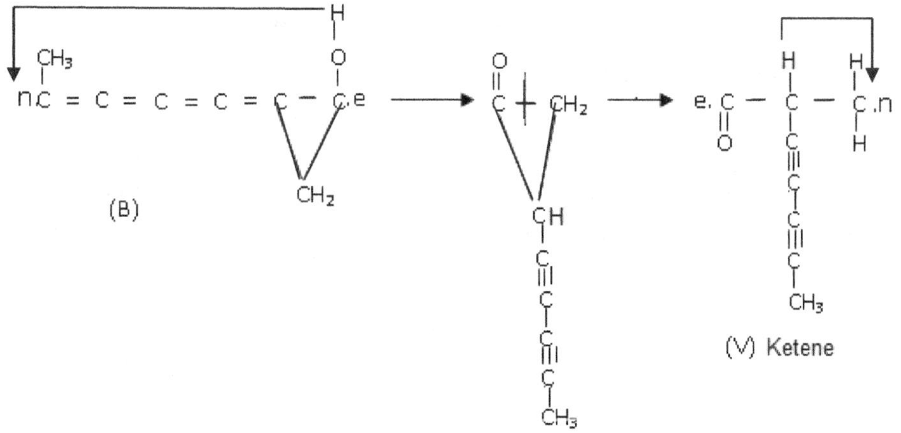

$$
\text{n } C = C = C = C = C - C.e \quad \longrightarrow \quad \quad \longrightarrow \quad \text{e. } C - C - C.n
$$

(B)

(V) Ketene

(Laws of Creations for Butadiyne cyclopropenes)

Rule 713: This rule of Chemistry for **Butadiyne cyclopropene carrying radical-pulling group,** states that, the rings cannot be opened, in view of the absence of any means of attaining MRSE in the rings, for which groups such as $HOOC - C \equiv C -$, $CH_3OOC - C \equiv C -$, $H_2NOC - C \equiv C -$, etc. cannot be readily transformed as follows via molecular rearrangement, since $C \equiv C$ is more nucleophilic than $C = O$.

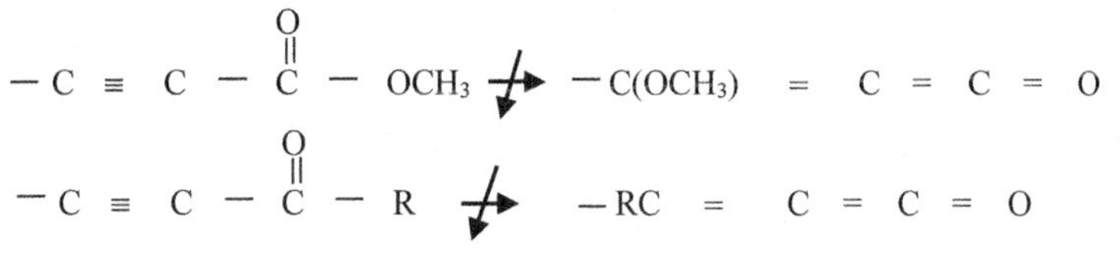

$$
- C \equiv C - \overset{\overset{\text{O}}{\|}}{C} - OCH_3 \quad \nrightarrow \quad - C(OCH_3) = C = C = O
$$

$$
- C \equiv C - \overset{\overset{\text{O}}{\|}}{C} - R \quad \nrightarrow \quad - RC = C = C = O
$$

161

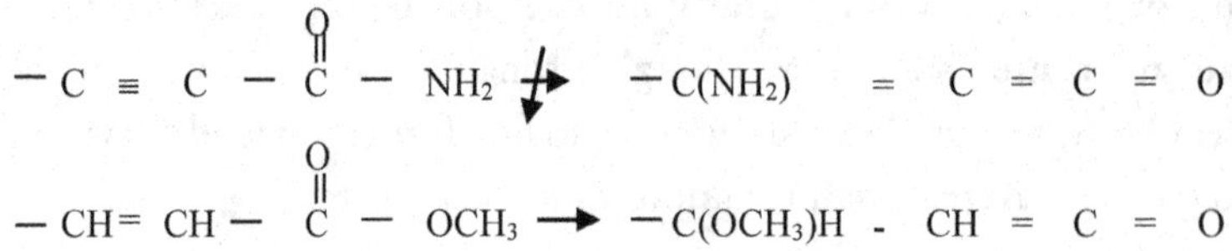

$$-C \equiv C - \overset{\overset{O}{\|}}{C} - NH_2 \;\not\!\!\rightarrow\; -C(NH_2) \;=\; C \;=\; C \;=\; O$$

$$- CH = CH - \overset{\overset{O}{\|}}{C} - OCH_3 \longrightarrow -C(OCH_3)H - CH = C = O$$

Radical-pulling resonance stabilization groups

(Laws of Creations for Radical-pulling groups)

Rule 714: This rule of Chemistry for **the families of 1,3- ($H_2C=C=O$) and 1,4- ($H_2C=C =C=O$) Ketenes and the families of the corresponding nitrogen containing ones (1,3- $H_2C=C=NH$, and 1,4- $H_2C=C=C=NH$) as has been done in the past**, states that, the followings are valid.

1,4- $H_2C=C=C=NH$) as has been done in the past, states that, the followings are valid.

$$[HN=> HN=C=C= > HN=C=C=C=C=] \;>\; [O= < O=C=C= < O=C=C=C=C=]$$

[EVEN] {See Rule 532}   [EVEN] {See Rule 493}

ORDER OF RADICAL-PUSHING CAPACITIES OF PUSHING GROUPS

$$[HN=C=C=C= > HN=C=]> O=C=C=C=C=C= > O=C=C=C= > O=C= > N\equiv C - \; ;$$
$$[ODD] \qquad\qquad\qquad\qquad [ODD]$$

Radical-pulling groups

ORDER OF RADICAL- PULLING CAPACITIES

(Laws of Creations for Special Substituent groups)

Rule 715: This rule of chemistry for **Butadiyne cyclopropene carrying radical-pulling groups on the ring center,** states that, only three mono-forms exist when resonance stabilized being an electrophile with the least nucleophilic center being 1,2- mono-form as shown below; for which their rings cannot be opened and even when opened, molecular rearrangement of the third kind can never take place.

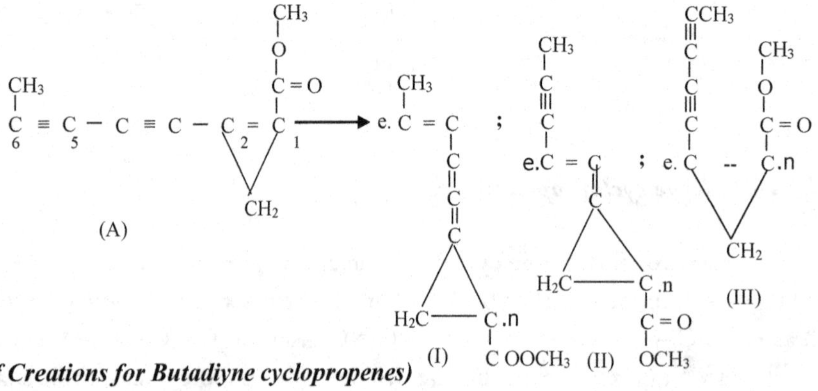

(Laws of Creations for Butadiyne cyclopropenes)

Rule 716: This rule of Chemistry for **Butadiyne cyclopropenes carrying radical-pulling groups on the butadiyne center,** states that, these are electrophiles which favor limited resonance stabilization with three mono-forms as shown below-

162

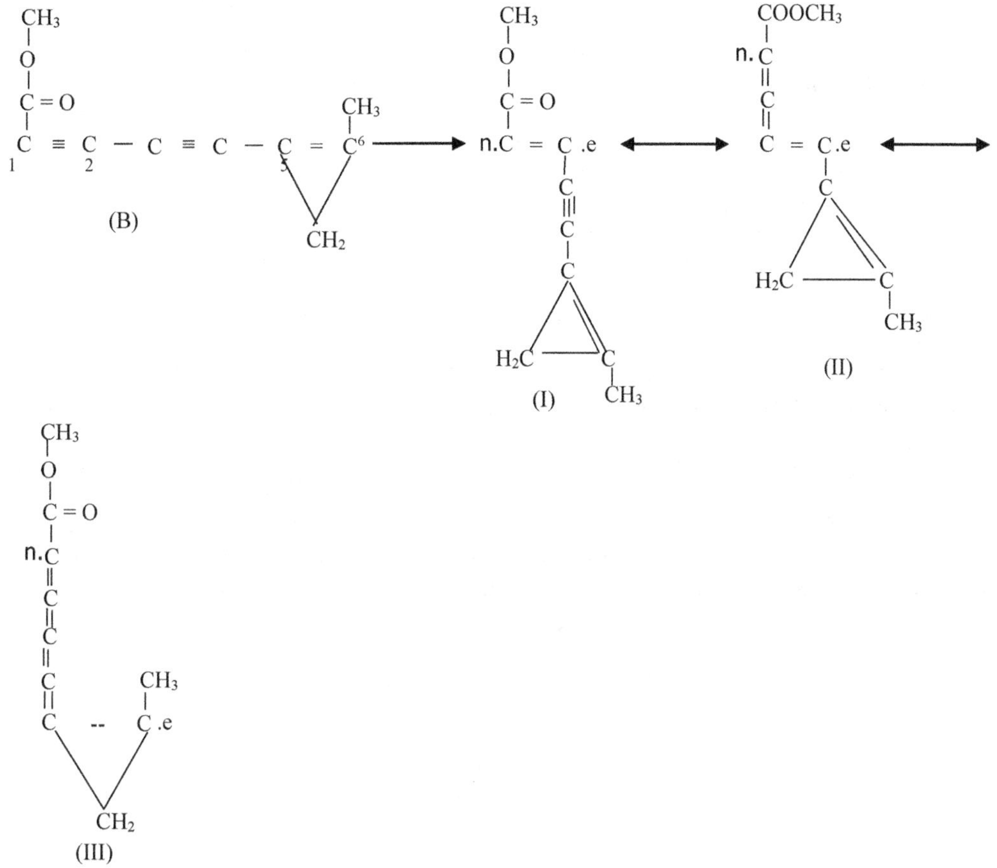

(B)   (I)   (II)

(III)

for which the 1,2-- mono-form is the least Nucleophilic center and when the ring which cannot be opened is opened, no molecular rearrangement of the third kind can take place.

(Laws of Creations for Butadiyne cyclopropenes)

<u>**Rule 717:**</u> This rule of Chemistry for **Cyclopropenes and Acetylenes carrying radical-pulling groups of the types COOH, COOR, CONH$_2$,** states that, when used as resonance stabilization group, the following relationship is valid-.

(I)   >   (II)

<u>**Order of Radical-pulling capacity**</u>

while when not used as resonance stabilization groups, the reverse is the case.

(Laws of Creations for Cyclopropenes)

Rule 718: This rule of Chemistry for **Cyclopropenes, Acetylenes and Cumulenes,** states that, while cyclopropene groups have different capacity from that corresponding to acetylenic and cumulenic groups, structurally the followings are valid-

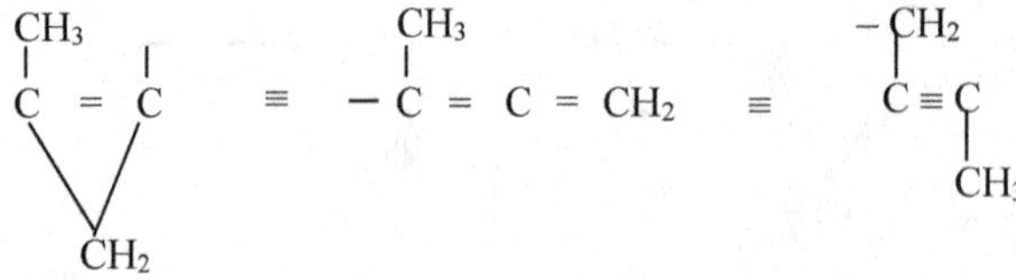

Structural equivalence (isomeric groups)

for which a network system seems to exist between cyclopropenes, cumulenes, and acetylenes.
(Laws of Creations for Isomeric alkyl-type of groups)

Rule 719: This rule of Chemistry for **The first member of acetylenic cyclopropene,** states that, this is a resonance stabilized nucleophilic monomer/compound (a Diyne) which cannot undergo molecular rearrangement of any kind, and the least nucleophilic center is 1,2- mono-form with all the mono-forms having the ability to favor both free-radical routes when suppressed, and when not suppressed, exists in Equilibrium state of existence with H held coming from the ring side, that highlighted asteriskically, noting that the 3,4- center is never activated alone

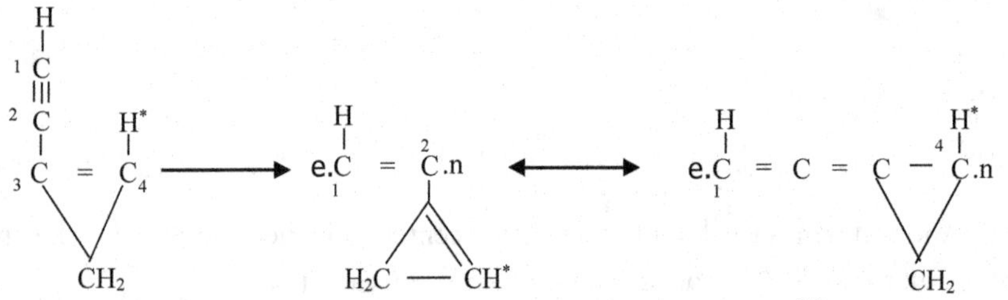

(I)   (II)

(Laws of Creations for Acetylenic cyclopropenes)

Rule 720: This rule of chemistry for **Acetylenic cyclopropenes carrying radical-pushing group on the ring,** states that, these resonance stabilized monomers/compounds, with the ring providing resonance stabilization and the CH_3 group providing the transfer species; the least nucleophilic center being the acetylene center as shown below-

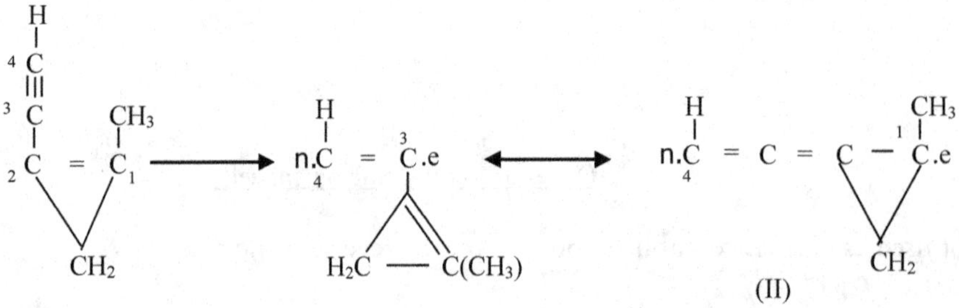

(I)   (II)

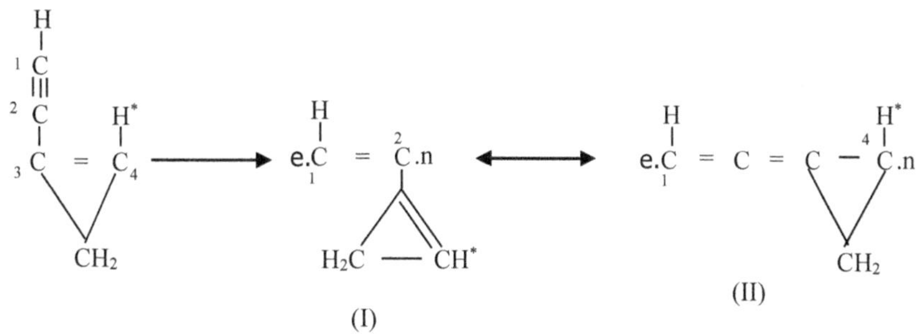

$$
\begin{array}{c}
H \\
| \\
{}^1C \\
||| \\
{}^2C \quad H^* \\
| \quad | \\
{}_3C = {}^4C \\
\diagdown\diagup \\
CH_2
\end{array}
\quad\longrightarrow\quad
\begin{array}{c}
H \\
| \\
e.C = {}^2C.n \\
{}_1 \\
\diagdown\diagup \\
H_2C - CH^*
\end{array}
\quad\longleftrightarrow\quad
\begin{array}{c}
H \quad\quad H^* \\
| \quad\quad | \\
e.C = C = C - {}^4C.n \\
{}_1 \\
\diagdown\diagup \\
CH_2
\end{array}
$$

(I) (II)

all favor molecular rearrangement of the first kind to produce the same monomer in the presence of a passive catalyst, i.e., suppressing agent, otherwise, the H on the acetylene may have to be replaced if activation is to take place, and for which the ring can be opened to allow for molecular rearrangement of the third type taking place as shown below-

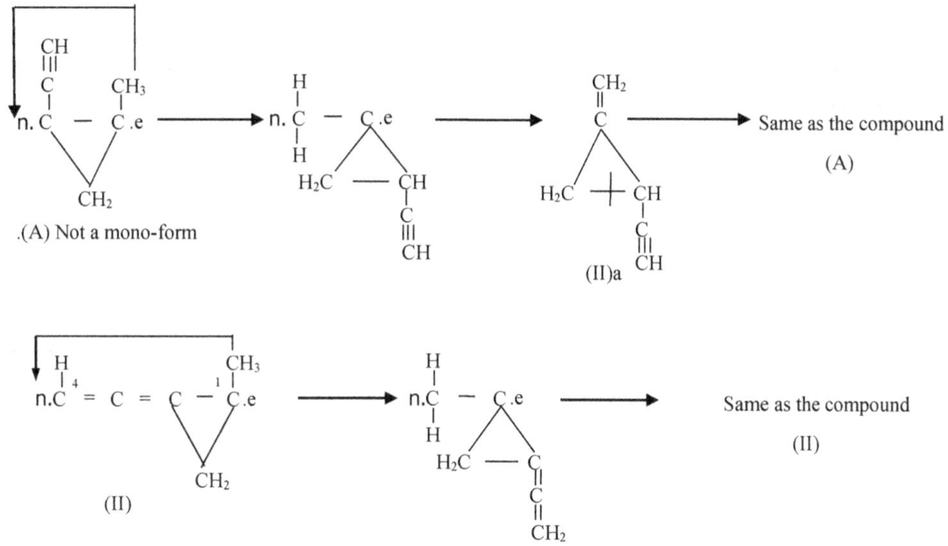

$$
\begin{array}{c}
CH \\
||| \\
C \quad CH_3 \\
| \quad | \\
n.C - C.e \\
\diagdown\diagup \\
CH_2
\end{array}
\quad\longrightarrow\quad
\begin{array}{c}
H \\
| \\
n.C - C.e \\
| \\
H \\
\diagdown\diagup \\
H_2C - CH \\
\quad\quad | \\
\quad\quad C \\
\quad\quad ||| \\
\quad\quad CH
\end{array}
\quad\longrightarrow\quad
\begin{array}{c}
CH_2 \\
|| \\
C \\
\diagdown\diagup \\
H_2C + CH \\
\quad\quad | \\
\quad\quad C \\
\quad\quad ||| \\
\quad\quad CH
\end{array}
\quad\longrightarrow\quad
\text{Same as the compound}
$$

.(A) Not a mono-form   (II)a   (A)

$$
\begin{array}{c}
H \quad\quad CH_3 \\
| \; {}^4 \quad\quad | \\
n.C = C = C - {}^1C.e \\
\diagdown\diagup \\
CH_2
\end{array}
\quad\longrightarrow\quad
\begin{array}{c}
H \\
| \\
n.C - C.e \\
| \\
H \\
\diagdown\diagup \\
H_2C - C \\
\quad\quad || \\
\quad\quad C \\
\quad\quad || \\
\quad\quad CH_2
\end{array}
\quad\longrightarrow\quad
\text{Same as the compound}
$$

(II)   (II)

to produce the same monomer or compound.

(Laws of Creations Acetylenic cyclopropenes)

Rule 721: This rule of Chemistry for **Acetylenic Cyclopropenes carrying radical-pushing alkylane groups (R) on the acetylene center,** states that, these resonance stabilized monomers/ compounds, with resonance stabilization provided by the ring center, can undergo molecular rearrangement of the third kind to give diynes as shown below when suppressed,

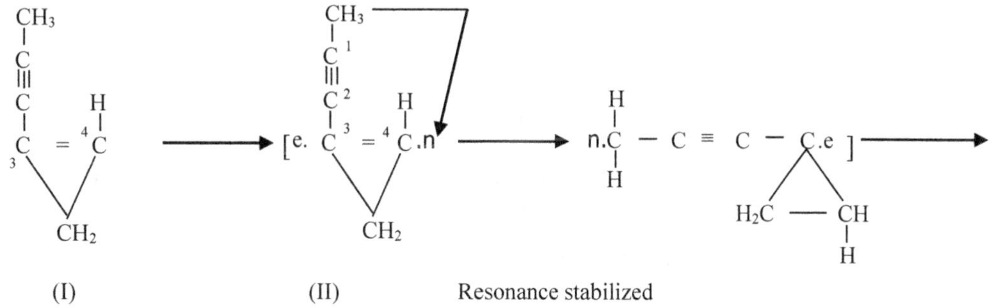

$$
\begin{array}{c}
CH_3 \\
| \\
C \\
||| \\
C \\
| \quad H \\
| \quad | \\
{}_3C = {}^4C \\
\diagdown\diagup \\
CH_2
\end{array}
\quad\longrightarrow\quad
\begin{array}{c}
CH_3 \\
| \\
{}^1C \\
||| \\
{}^2C \quad H \\
| \quad | \\
[e. \; {}^3C = {}^4C.n \\
\diagdown\diagup \\
CH_2
\end{array}
\quad\longrightarrow\quad
\begin{array}{c}
H \\
| \\
n.C - C \equiv C - C.e \;] \\
| \\
H \\
\diagdown\diagup \\
H_2C - CH \\
\quad\quad\quad | \\
\quad\quad\quad H
\end{array}
\quad\longrightarrow
$$

(I)   (II)   Resonance stabilized

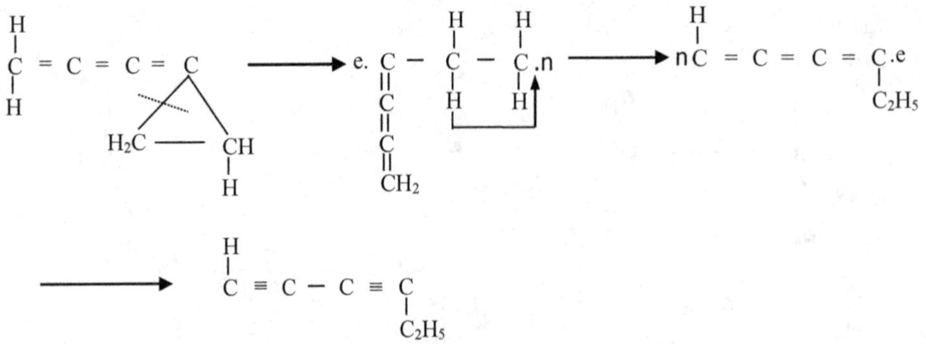

$$
\begin{array}{c}
\overset{H}{\underset{H}{|}}C = C = C = C \quad \longrightarrow \quad e.\ C - C - C.n \quad \longrightarrow \quad nC = C = C = C.e
\end{array}
$$

(with the cyclopropene ring $H_2C - CH$ / H below the first structure, CH_2 chain below the second, and C_2H_5 on the last)

$$
\overset{H}{\underset{C_2H_5}{|}}C \equiv C - C \equiv C
$$

(III)   Resonance stabilized

for which the R group cannot exceed C_3H_7 for the case above and only electro-free-radical route is favored by them.

(Laws of Creations for Acetylenic cyclopropenes))

Rule 722: This rule of Chemistry of **Acetylenic Cyclopropenes carrying equal radical-pushing alkylane groups (R) on both centers,** states that, these resonance stabilized monomers/compounds, with resonance stabilization provided by the ring center, can undergo molecular rearrangement of the third kind to give diynes as shown below,

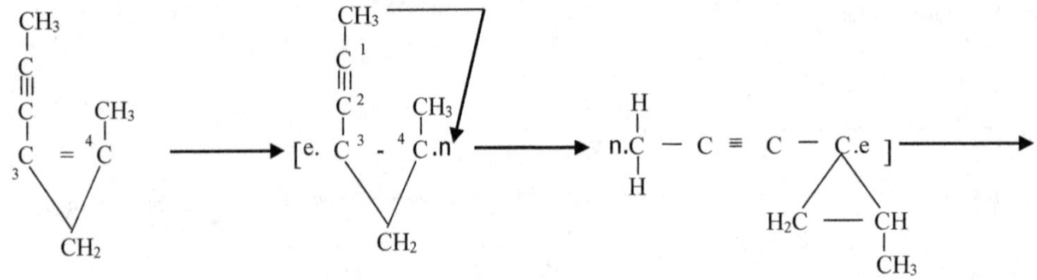

(I)   (II)   Resonance stabilized

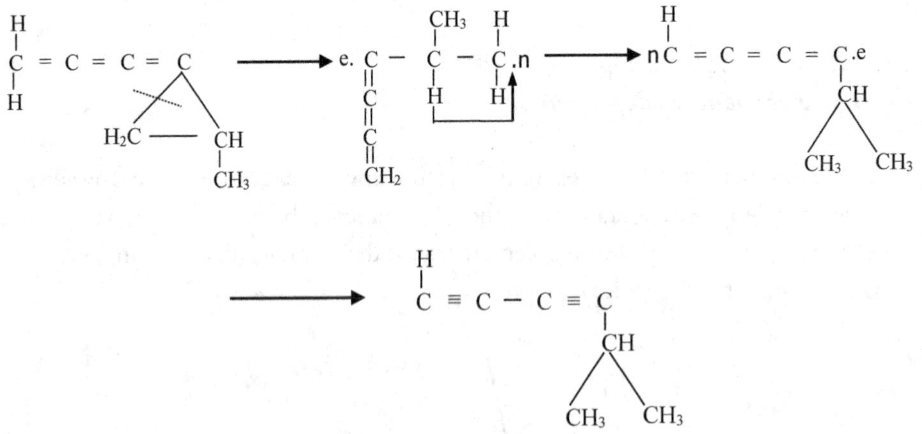

(III) Resonance stabilized

for which the R group on acetylene side cannot exceed C_4H_9 for the case above and only the electro-free-radical route is favored by them.

(Laws of Creations for Acetylenic cyclopropenes))

166

Rule 723: This rule of Chemistry of **Acetylenic Cyclopropenes carrying different radical-pushing alkylane groups (R),** states that, these resonance stabilized monomers/compounds, with resonance stabilization provided by the ring center when acetylene is the carrier of the larger group, can undergo molecular rearrangement of the third kind to give diynes as shown below-

$$
\begin{array}{c}
CH_3 \\
| \\
CH_2 \\
| \\
C \\
\equiv \\
\end{array}
$$

(I)

(B)

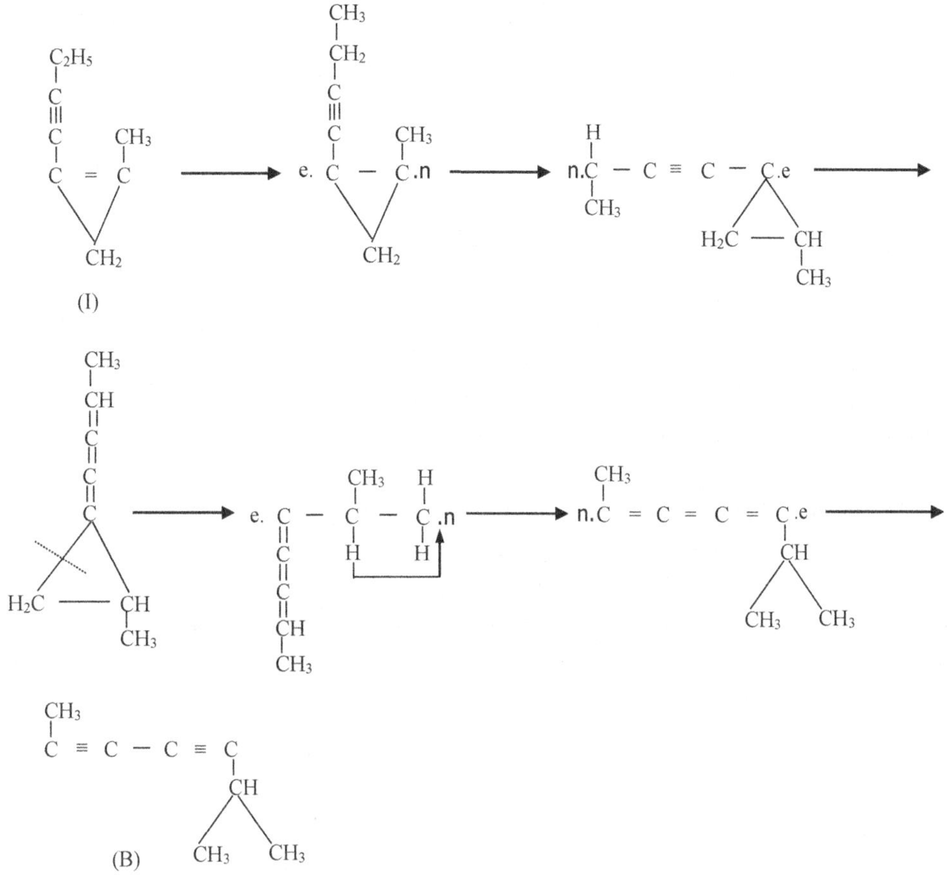

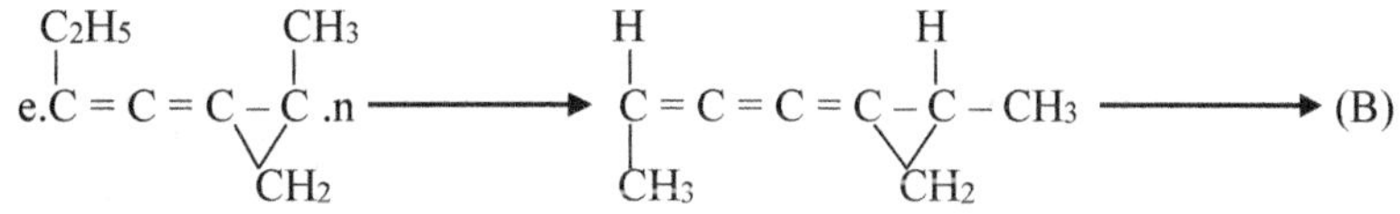

(III) Resonance stabilized

for which as shown above, the two mono-forms rearrange to give the same monomer, (B), a substituted butadiyne, noting that, once the R group on the acetylenic side exceeds C_4H_9, the rearrangement of the third kind ceases.

(Laws of Creations for Acetylenic cyclopropenes)

Rule 724: This rule of Chemistry of **Acetylenic Cyclopropenes carrying different radical-pushing alkylane groups,** states that, these resonance stabilized monomers/compounds, with resonance stabilization provided by the ring center where ring is the carrier of the larger group, cannot undergo molecular rearrangement of the third kind to give diynes as shown below-

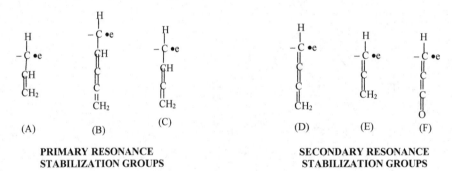

for which as shown above, the two mono-forms rearrange to give the same monomer, which is a resonance stabilized, nucleophilic substituted acetylenic compound with transfer species against the Conservation laws; clear indication that this monomer cannot be decomposed.

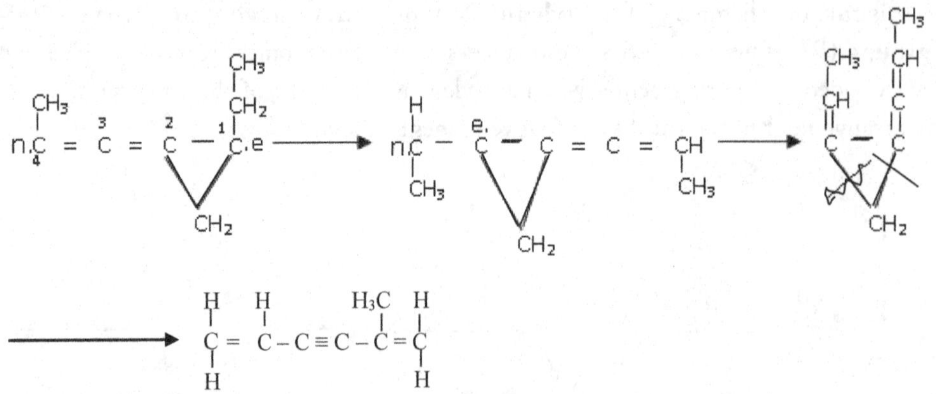

(Laws of Creations for Acetylenic cyclopropenes)

Rule 725: This rule of Chemistry for **Resonance stabilization groups**, states that, there are two kinds of resonance stabilization groups- the Primary kind and the Secondary kind- *for which while $H_2C=$ is a non-resonance stabilization group, $H_2C=C=$ is a resonance stabilization group with a difference, such as used in 1,4- cumulene and these are called* **the SECONDARY kind of resonance stabilization groups,** *while all the others encountered so far (e.g. $-CH=CH_2$, $-C \equiv CH$) are* **PRIMARY resonance stabilization groups.**

(A)   (B)   (C)   (D)   (E)   (F)

PRIMARY RESONANCE
STABILIZATION GROUPS

SECONDARY RESONANCE
STABILIZATION GROUPS

(Laws of Creations for Resonance Stabilization groups)

Rule 726: This rule of Chemistry for **1,3- and 1,4- Cumulenes,** states that, while 1,3- Cumulene is not resonance stabilized, 1,4-cumulene is and though the 2,3-mono-form of 1,4-cumulene can undergo molecular rearrangement to give vinyl acetylene, this can never be obtained, because that center is never activated,

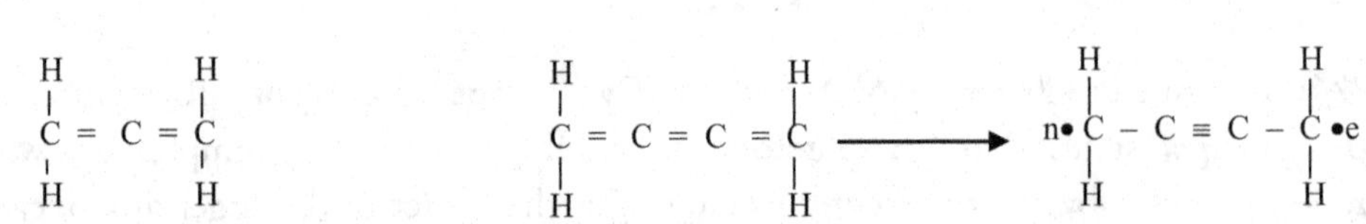

1, 3- Cummulene

NON-RESONANCE STABILIZED

1, 4- Cummulene

RESONANCE STABILIZED

noting that the 1,2- or 3,4-mono-form of 1,4-cummulene cannot undergo molecular rearrangement.
(Laws of Creations for Cumulenes)

Rule 727: This rule of Chemistry for **Acetylenic Cyclopropene and Butadiyne**, states that, from all the considerations so far, the followings are obvious.

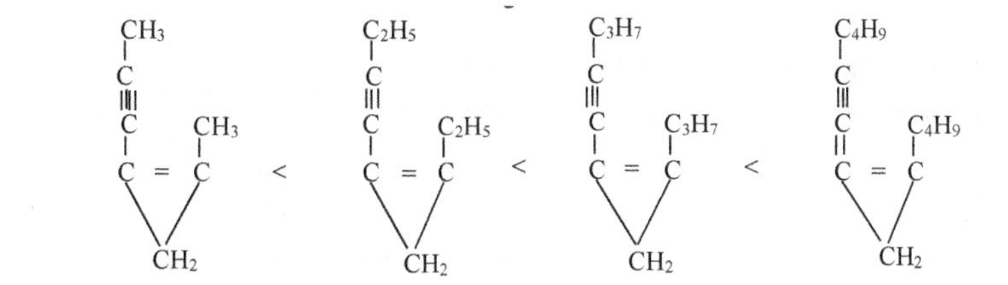

Order of nucleophilicity

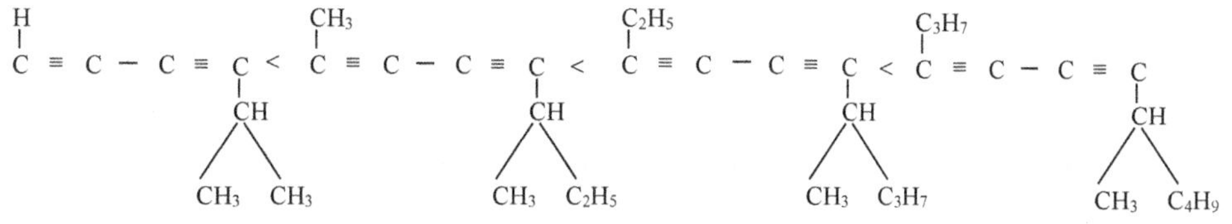

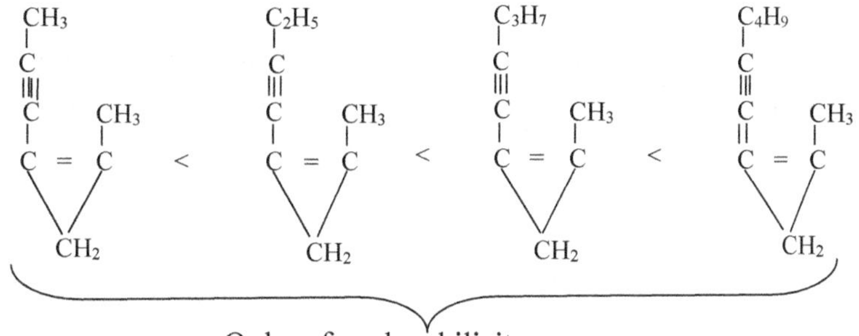

Order of nucleophilicity

OR

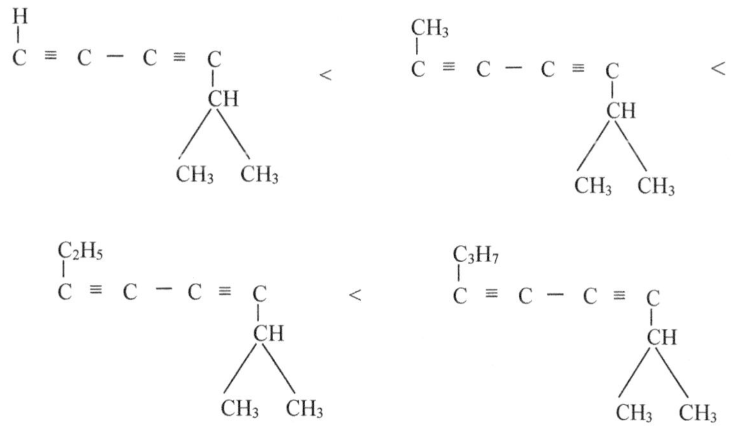

Order of Nucleophilicity

for which the latter is more nucleophilic than the former.
(Laws of Creations for Butadiynes/Acetylenic cyclopropenes)

Rule 728: This rule of Chemistry for **the three families of monomers shown below**, states that,

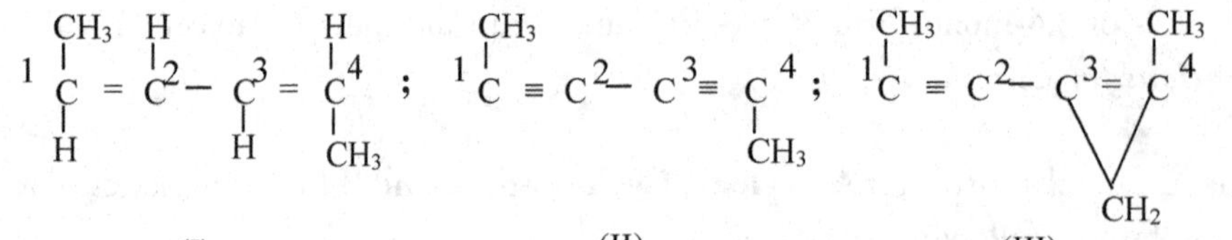

$$1 \begin{array}{c} CH_3 \\ | \\ C \\ | \\ H \end{array} = \begin{array}{c} H \\ | \\ C^2 - \end{array} \begin{array}{c} H \\ | \\ C^3 \\ | \\ H \end{array} = \begin{array}{c} H \\ | \\ C^4 \\ | \\ CH_3 \end{array} \quad ; \quad 1 \begin{array}{c} CH_3 \\ | \\ C \end{array} \equiv C^2 - C^3 \equiv \begin{array}{c} CH_3 \\ | \\ C^4 \\ | \\ CH_3 \end{array} \quad ; \quad 1 \begin{array}{c} CH_3 \\ | \\ C \end{array} \equiv C^2 - \begin{array}{c} CH_3 \\ | \\ C^3 = C^4 \\ \diagdown \diagup \\ CH_2 \end{array}$$

(I)   (II)   (III)

while (I) has two structurally non-symmetric mono-forms (3,4- and 1,4- mono-forms), (II) has two structurally non-symmetric mono-forms (1,2- and 1,4- mono-forms), (III) has two structurally non-symmetric mono-forms (1,2- and 1,4-mono-forms) and based on the mechanism of resonance stabilization provided so far, only (II) is a fully resonance-stabilized symmetric monomer, while (III) is not and because of (I) where only 1,4- mono-form is symmetric, the followings are valid for the limitations-

$$(\textbf{R})HC = CH - \quad > \quad R \quad > \quad RC \equiv C - \quad > \quad \begin{array}{c} \textbf{R} \\ | \\ C = C - \\ \diagdown \diagup \\ CH_2 \end{array}$$

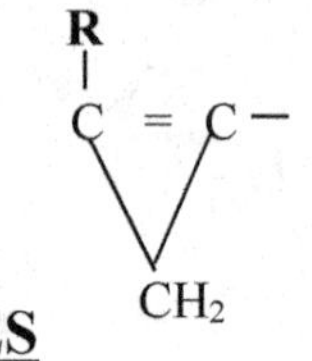

ORDER OF RADICAL-PUSHING CAPACITIES

(Laws of Creations for Symmetry)

Rule 729: This rule of Chemistry for **Acetylenic Cyclopropenes,** states that, when the CH_2 group internally located in the ring begins to change, the situation changes drastically as shown below.

Primary C_4H_2:

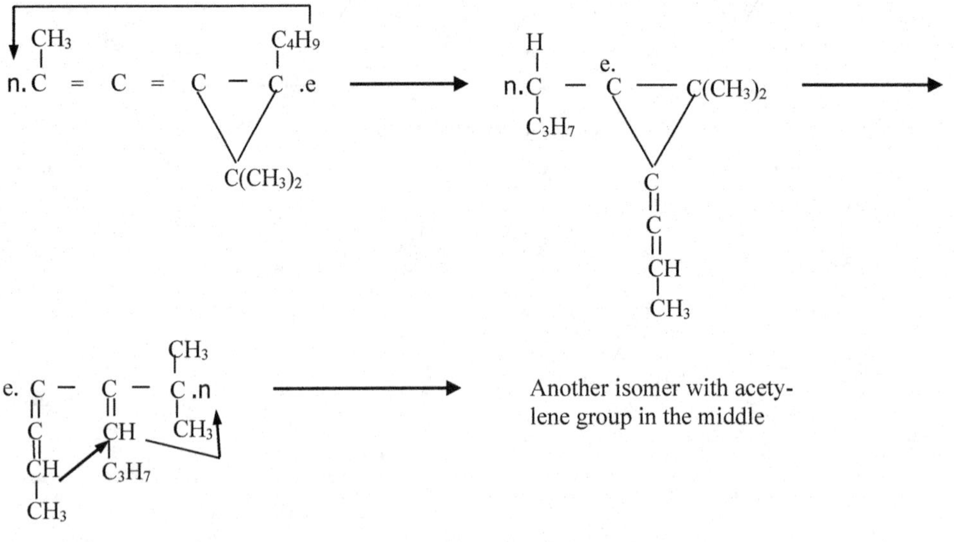

noting that this rearrangement is not favored based on the Conservation law.

Secondary C_4H_9:

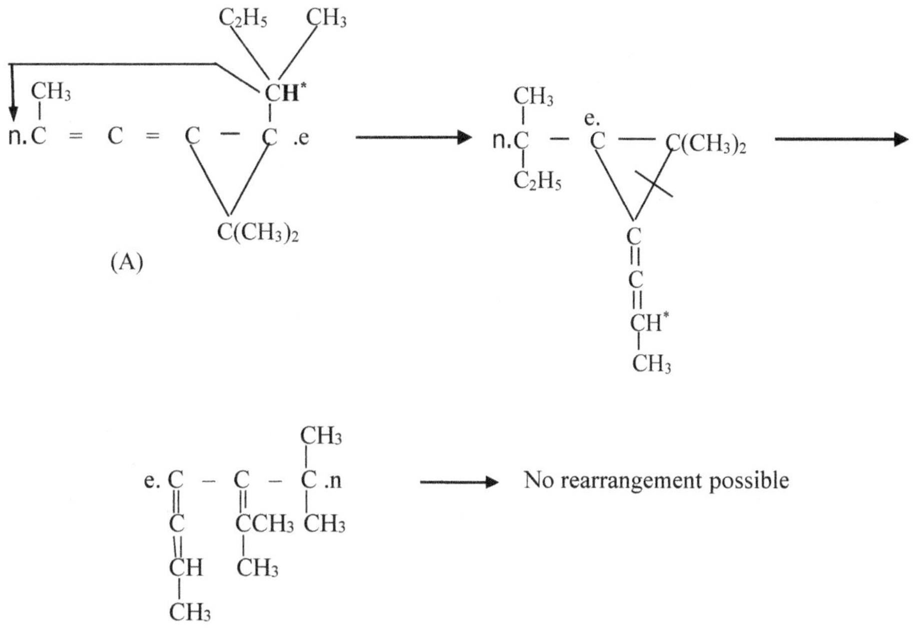

(A)

ii)

noting that, no rearrangement can take place when the ring is opened;

Tertiary C_4H_9:

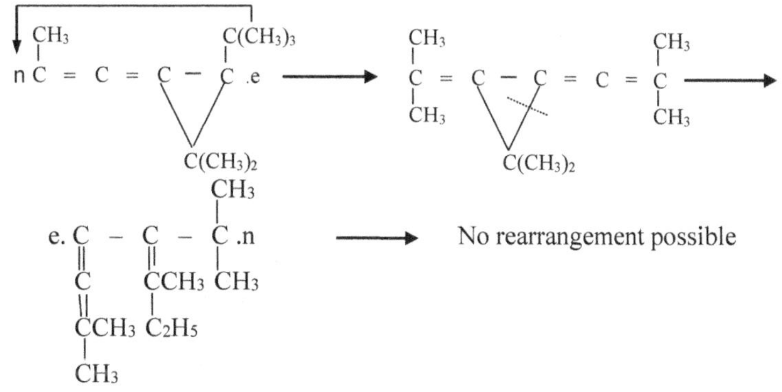

iii)

from which nevertheless when the groups are reversedly placed, rearrangement is favored and the followings can be said to be generally valid-

$$- C(CH_3)_3 \quad > \quad - CH(CH_3)C_2H_5 \quad > \quad - C_4H_9$$

Tertiary   Secondary   Primary

ORDER OF RADICAL-PUSHING CAPACITIES

(Laws of Creations for Alkylane or Alkanyl groups)

Rule 730: This rule of Chemistry for **Acetylenic Cyclopropenes carrying radical-pushing groups of the type OH on the ring center,** states that, their monomers are resonance stabilized for which the followings are to be expected, that is, the monomer can be made to undergo molecular rearrange-ment of

171

the third kind to produce ketenes, since the Laws of Conservation of transfer of transfer species are still indirectly obeyed after opening of the ring, in view of transformation from Female to Male.

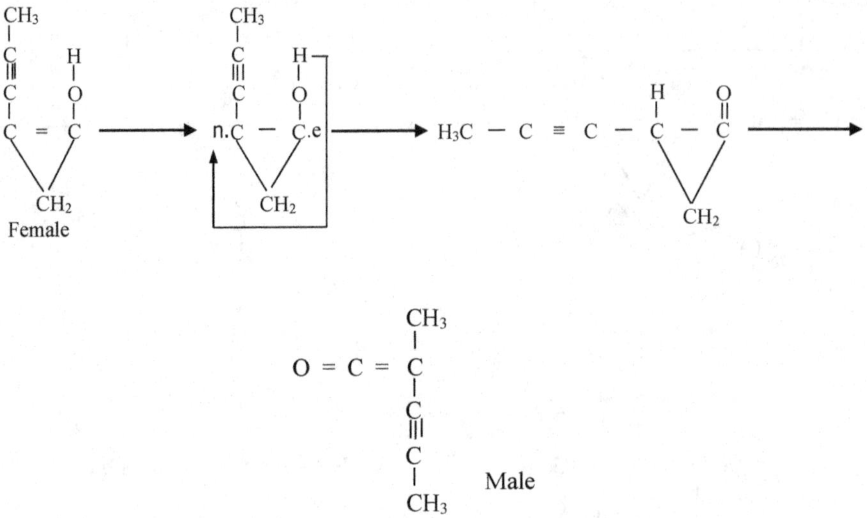

Female → Male

(Laws of Creations for Acetylenic cyclopropenes)

Rule 731: This rule of Chemistry for **Acetylene Cyclopropenes carrying radical pushing groups of the type OH on the acetylenic center,** states that, their monomers are resonance stabilized for which the followings are to be expected, that is, the monomer can be made to undergo molecular

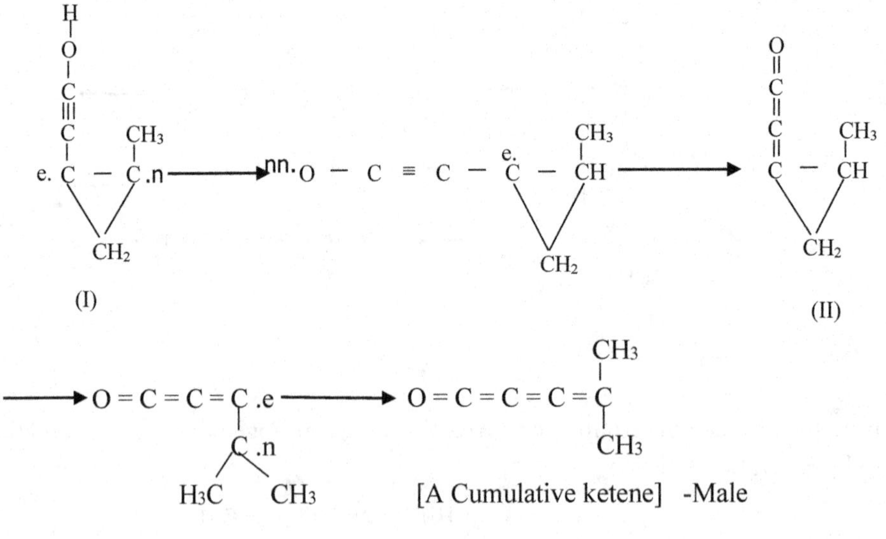

(I)   (II)

$$O = C = C = C = C \overset{CH_3}{\underset{CH_3}{|}}$$

[A Cumulative ketene] -Male

rearrangement of the third kind to give a more cumulative Ketene, since the Laws of Conservation of transfer of transfer species will not be broken, being now an Electrophile from a resonance stabilized Nucleophile.

(Laws of Creations for Acetylenic cyclopropenes))

Rule 732: This rule of Chemistry for **Acetylenic Cyclopropenes with NH$_2$ group in place of H on acetylene side,** states that, the followings are expected- ·

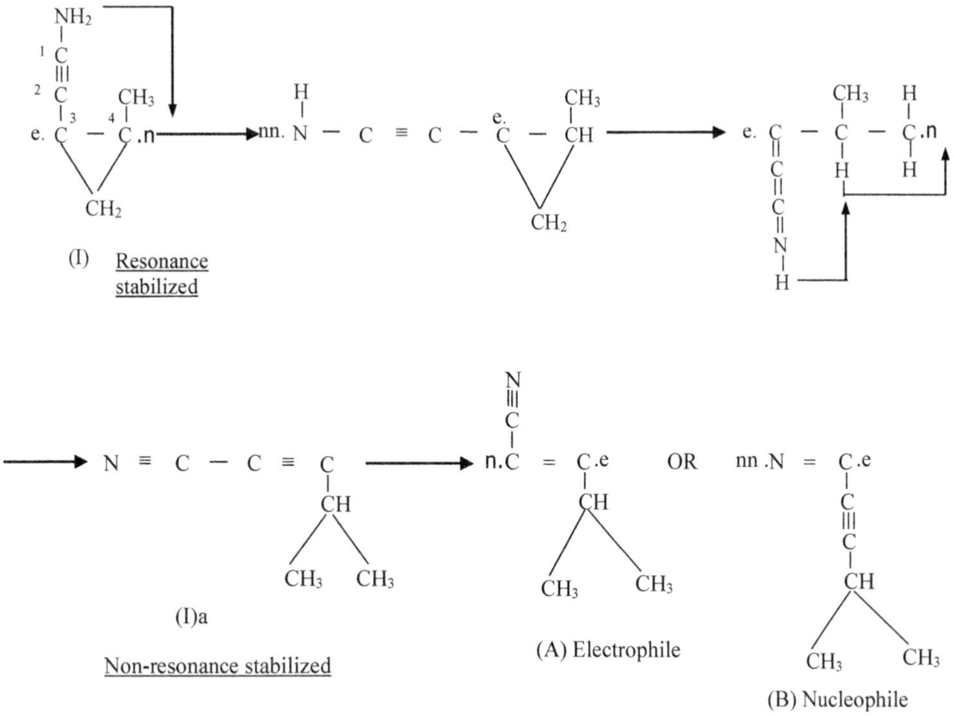

$$NH_2-C\equiv C-\underset{\underset{CH_2}{|}}{\underset{e.\ C}{}}-\underset{CH_3}{\underset{|}{C}}.n \longrightarrow nn.\ N-C\equiv C-\underset{\underset{CH_2}{|}}{C}-\underset{CH_3}{CH} \longrightarrow e.\ C-\underset{H}{\underset{|}{C}}-\underset{H}{\underset{|}{C}}.n$$

(I) Resonance
      stabilized

$$\longrightarrow N\equiv C-C\equiv \underset{\underset{CH_3\ \ CH_3}{CH}}{C} \longrightarrow n.C=\underset{\underset{CH_3\ \ CH_3}{CH}}{C}.e \qquad OR \qquad nn.\ N=\underset{\underset{CH_3\ \ CH_3}{CH}}{C}.e$$

(I)a   (A) Electrophile   (B) Nucleophile

Non-resonance stabilized

for which the resonance stabilized Nucleophilic monomer is transformed to a non-resonance stabilized Electrophile via molecular rearrangement of the third kind.

(Laws of Creations for Acetylenic cyclopropenes)

Rule 733: This rule of Chemistry for **Acetylenic Cyclopropenes carrying NHR type of groups on the acetylenic center,** states that, these resonance stabilized monomers, favor molecular rearrangement of the third kind only when R is CH_3 as shown below for this case -

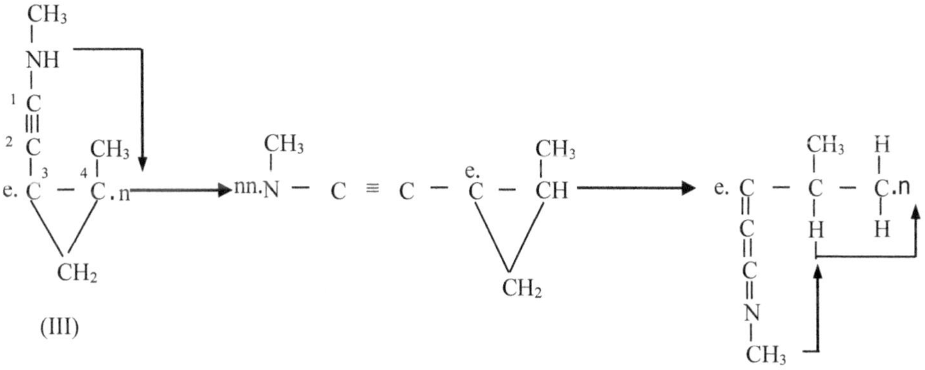

$$\underset{CH_3}{\overset{CH_3}{\underset{|}{NH}}}-C\equiv C-\underset{\underset{CH_2}{|}}{\underset{e.\ C}{}}-\underset{CH_3}{\underset{|}{C}}.n \longrightarrow nn.N-C\equiv C-\underset{\underset{CH_2}{|}}{C}-\underset{CH_3}{CH} \longrightarrow e.\ C-\underset{H}{\underset{|}{C}}-\underset{H}{\underset{|}{C}}.n$$

(III)

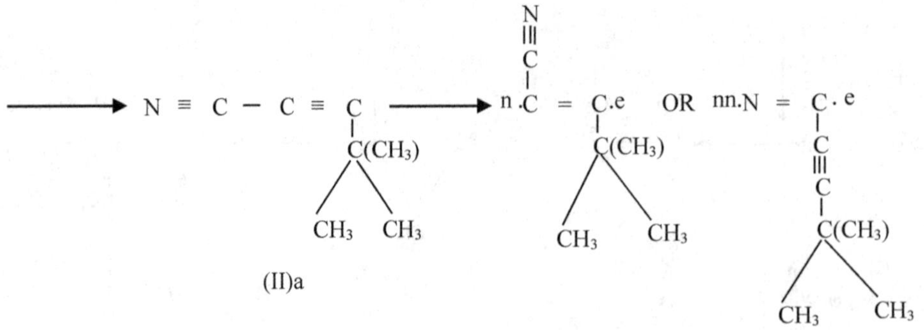

$$N \equiv C - C \equiv \underset{\underset{\underset{CH_3 \quad CH_3}{\diagdown \diagup}}{\overset{|}{C(CH_3)}}}{C} \quad \longrightarrow \quad n.\underset{\underset{\underset{CH_3 \quad CH_3}{\diagdown \diagup}}{\overset{|}{C(CH_3)}}}{C} = \overset{e}{C} \quad OR \quad nn.N = \underset{\underset{\underset{CH_3 \quad CH_3}{\diagdown \diagup}}{\overset{|}{\overset{|}{\underset{C(CH_3)}{C}}}}}{\overset{\overset{N}{\overset{|||}{C}}}{C}} \cdot e$$

(II)a

(C) Electrophile

(D) Nucleophile

<u>FAVORED</u>

(Laws of Creations for Acetylenic cyclopropenes)

<u>Rule 734:</u> This rule of Chemistry for **Acetylenic Cyclopropenes carrying NR_2 types of groups on the acetylenic center,** states that, these resonance stabilized monomers, favor molecular rearrangement of the third kind only when R is CH_3 as shown below for the case –

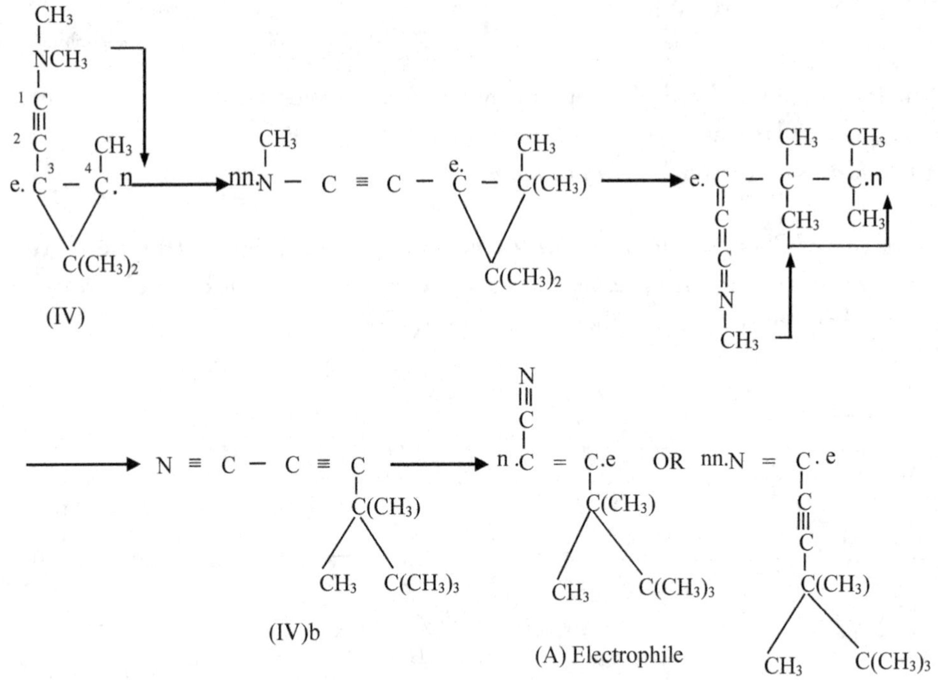

(IV)

$$\longrightarrow \quad N \equiv C - C \equiv \underset{\underset{\underset{CH_3 \quad C(CH_3)_3}{\diagdown \diagup}}{\overset{|}{C(CH_3)}}}{C} \quad \longrightarrow \quad n.\underset{\underset{\underset{CH_3 \quad C(CH_3)_3}{\diagdown \diagup}}{\overset{|}{C(CH_3)}}}{C} = \overset{e}{C} \quad OR \quad nn.N = \underset{\underset{\underset{CH_3 \quad C(CH_3)_3}{\diagdown \diagup}}{\overset{|}{C(CH_3)}}}{\overset{\overset{N}{\overset{|||}{C}}}{C}} \cdot e$$

(IV)b

(A) Electrophile

(B) Nucleophile

<u>FAVORED</u>

noting that for $N(CH_3)_2$, the internal hydrogen atoms on CH_2 had to be changed to allow for the unquestionable rearrangement to be favored for which the followings can be seen to be valid-

$$— N(CH_3)_2 \quad > \quad — NHCH_3 \quad > \quad — NH_2$$

<u>**ORDER OF RADICAL-PUSHING CAPACITIES**</u>   [See Rule 729]

and from the order the following relationship is also valid,

$$(CH_3)_2N — C \equiv C — \quad > \quad HCH_3N — C \equiv C — \quad > \quad H_2N — C \equiv C —$$

<u>**ORDER OF RADICAL-PUSHING CAPACITIES**</u>

(Laws of Creations for Acetylenic cyclopropenes)

<u>**Rule 735:**</u> This rule of Chemistry for **Acetylenic Cyclopropenes carrying NH₂ group on the ring,** states that, the followings are expected,

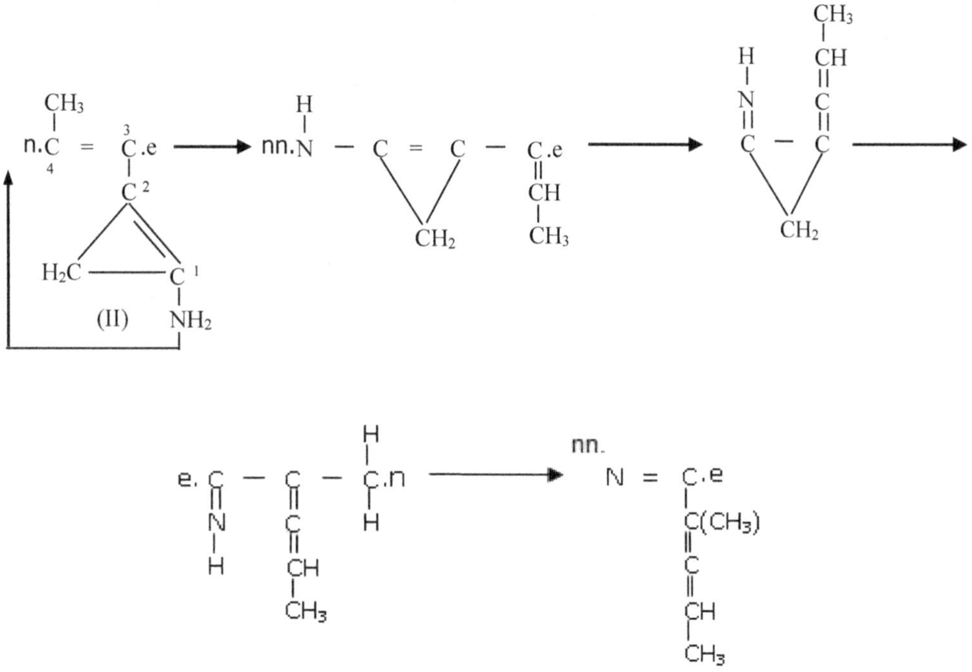

Electrophile

for which molecular rearrangement of the third kind can take place, based on the laws of molecular rearrangement and the laws of conservation of transfer of transfer species, to give an Electrophile.
(Laws of Creations for Acetylenic cyclopropenes)

<u>**Rule 736:**</u> This rule of Chemistry for **Acetylene Cyclopropene carrying F on the acetylene center,** states that, though resonance stabilized, they cannot undergo molecular rearrangement of the third kind when the ring is opened instantaneously as shown below for one of the mono-forms-

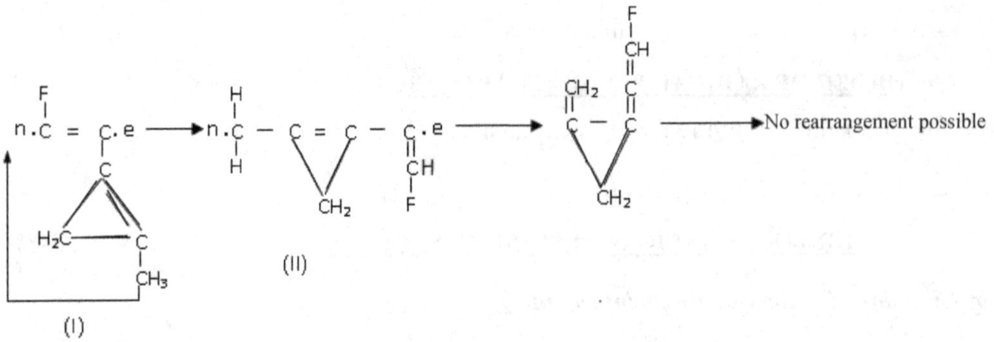

(Laws of creations for Acetylenic cyclopropenes)

<u>**Rule 737:**</u> This rule of Chemistry for **Acetylenic Cyclopropenes, which carry C_nF_{2n+1} types of groups in place of one H atom,** states that, these compounds or monomers cannot undergo any form of molecular rearrangements, regardless the operating conditions, since the group is lower in radical pushing capacity than H.

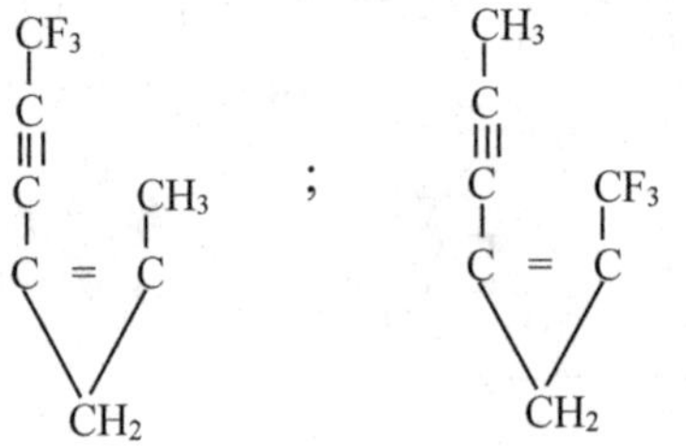

(Laws of Creations for Acetylenic cyclopropenes)

<u>**Rule 738:**</u> This rule of Chemistry for **the type of Acetylenic Cyclopropene carrying CF_3 tertiarily on the acetylene center shown below,** states that, this monomer is resonance stabilized and can also undergo molecular rearrangement of the third kind to give a butadyne.

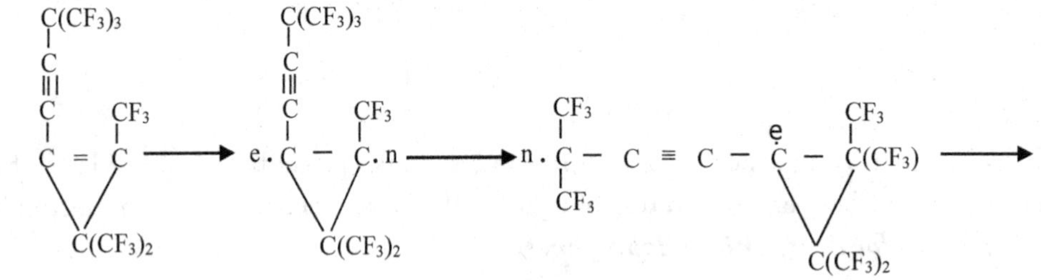

(I) <u>Resonance stabilized</u>

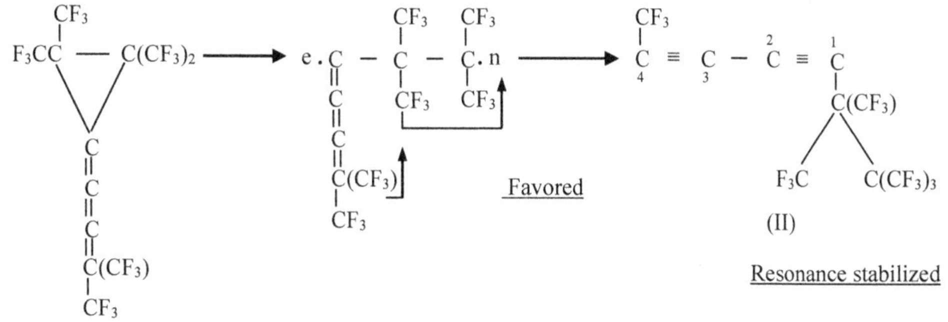

Favored

(II)

Resonance stabilized

(Laws of Creations for Acetylenic cyclopropenes)

Rule 739: This rule of Chemistry for **Fluorinated Acetylenes shown below,** states that, these are Nucleophiles, for which (A) has no transfer species when activated (i.e., it will favor both routes)

(A) (B) (C) (D)

while (B), (C) and (D) have; the transfer species for (B) is CF_3 nucleo-free-radically the route not natural to the monomer and the transfer species for (C) and (D) are F electro-free- radically and CF_3 nucleo-free-free-radically and shown below is the case of (B) electro-free-radically noting that (C)

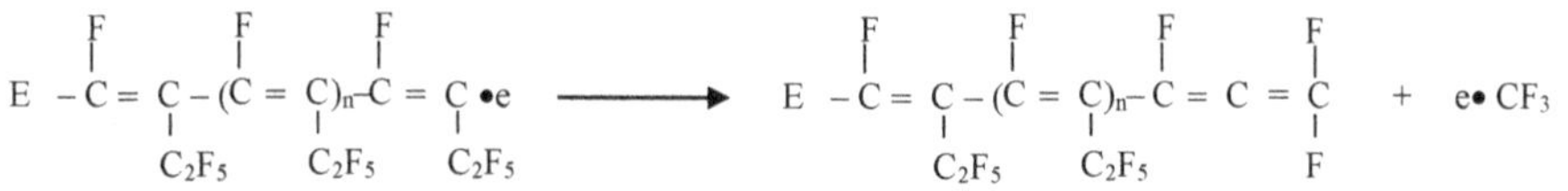

CHARACTER OF A NUCLEOPHILE IN (B)

and (D) cannot be polymerized, all these clearly indicating *that C_nF_{2n+1} types of groups have dual characters unlike C_nH_{2n+1} types of groups which have only one transfer species, H.*
(Laws of Creation of Duality of C/F alkylane types of groups)

Rule 740: This rule of Chemistry for **The dualities of C_nF_{2n+1} groups,** states that, based on the spectrum of radical pushing and pulling groups some of which are shown below,

Pulling Groups: -C=NH, -COH, -CN, -COOH, -COOR, -CONH$_2$, -COR, **-CF$_3$**, **-F,**
Anionic Charge: No No No No No No No No Yes

Pushing Groups: -NH$_2$, -OH, -SH, -CH$_3$, -CH=CH$_2$, =CH$_2$, =O, =C=C=CH$_2$, **-CF$_3$**, **-H**
Cationic Charge: No No No No No No No No No Yes
only F and H can carry ionic charges in the classifications above, noting that common to both Radical-pushing and -pulling groups is CF_3 which chargedly it is radical-pulling, while radically, it is radical-pushing but of lower capacity than H and also noting that it is not the only one with such unique character.
(Laws of Creations for Duality of C_nF_{2n+1} groups)

Rule 741: This rule of Chemistry for **Allylic groups of the types – CH$_2$F, -CH$_2$CONH$_2$, -CH$_2$COOH, --CH$_2$CN, -CH$_2$COOCH$_2$ etc. which are radical-pushing groups all of greater capacity than H, but**

of lesser capacity than CH$_3$, states that, though –CF$_2$H is not C/H allylic in character, it is still radical-pushing of greater capacity than H, but lesser capacity than CH$_2$F, and in C/F family, it is fluoro-allylic as shown below.

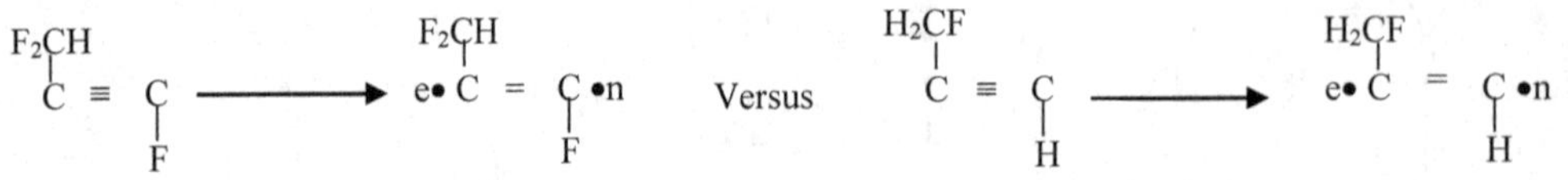

(A) C/F-Allylic group-CF$_2$H · (B) C/H-Allylic group-CH$_2$F

(Laws of Creations for Fluoro Allylic groups)

Rule 742: This rule of Chemistry for **Acetylenic Cyclopropenes carrying allylic groups of the types –CH$_2$F, -CH$_2$CF$_3$, -CH$_2$COOH, --CH$_2$CN, -CH$_2$COOCH$_2$, etc. which are radical-pushing groups,** states that, these are resonance stabilized to a limited extent and undergo molecular rearrangement of the third kind to give resonance stabilized cumulenic acetylene-

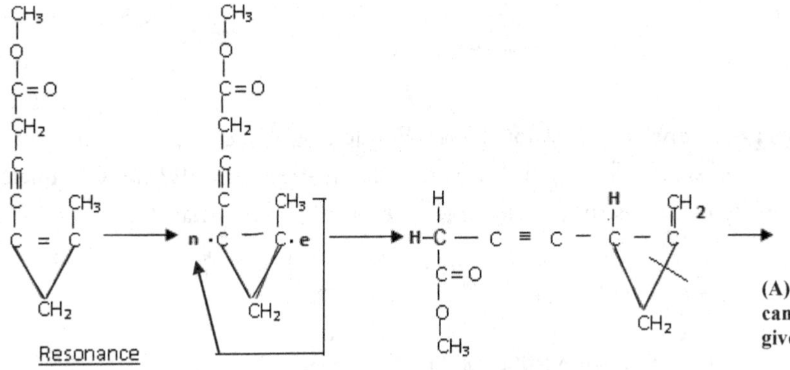

(A) The mono-forms cannot rearrange to give the same monomer.

i)

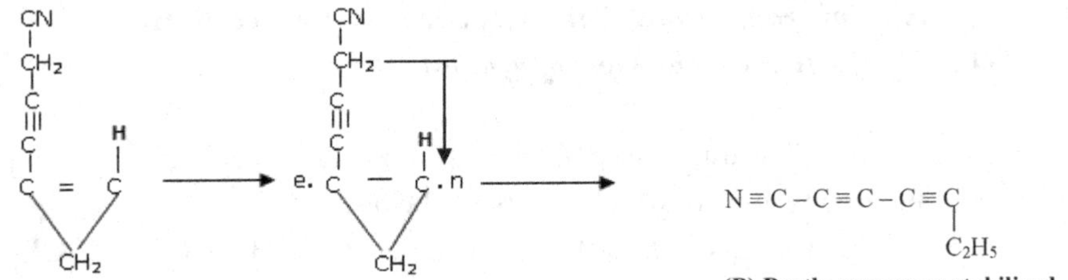

(B) Partly resonance stabilized

ii)

for which for the first case there will be no rearrangement and for the last case there may be need to provide a suppressing agent, noting that (B) is an Electrophile.
(Laws of Creations for Acetylenic cyclopropenes)

Rule 743: This rule of Chemistry for **Acetylenic Cyclopropenes**, states that, when they carry COOR, CONH$_2$ and COR types of groups, the rings cannot readily be opened, noting that as Electrophiles, the transfer species can never be H, but OCH$_3$ group; for if opened, transfer species

Underline: Cannot easily be opened

becomes H to give a resonance stabilized ketene [$(CH_3O)(CH_3)HC - HC = C = C = C = C = O$].
(Laws of Creations for Acetylenic cyclopropenes)

One hundred and sixty-one foundation rules have been proposed for cycloalkanes, cyclo-propenes, triynes, diynes, vinyl cyclopropanes and more. From these rules, other more complex cases can readily be handled. By these rules, the characters of the different members of the families have clearly been identified. Based on the new chemistry of atoms, molecules, ions, radicals and charges and more which have been established so far, one is beginning to explain most of the observations of so many years from different points of view. Even new ones have been highlighted.

Candidly speaking, there is nothing like S_N1, E1, S_N2, "John Doe" reactions, etc., to explain the mechanisms of reactions which were discovered by accident without understanding how the reactions were obtained. That has been the case in general, just like the Ziegler and Natta case. How syndiotactic and isotactic placements are obtained are unknown to them. As can be observed, only three mechanisms cut across chemical systems, all based on the applications of the LAWS OF NATURE. All these will become obvious when inorganic chemistry of compounds have been considered. As a mathematician, physicist, chemist, philosopher, and an engineer, for so many years, one has never believed in the methods of trials and errors, hypothetical rules, rules of the thumb, semblances, coping, necessary and unnecessary assumptions, empirical rules, accidental creations, exclusive claim to knowledge or anything, and so on. One has always believed that the reason why all the above exist is because the eyes of the needle have never been used, in order words, there has been no order yet established in any discipline. To know what Orderliness is, one has to know how NATURE operates.

References

1. E. L. Eliel, "Stereochemistry of carbon compounds:, Chap. 7, McGraw-Hill Book company. Inc., New York, 1962.

2. P. J. Flory, "Principles of Polymer Chemistry", Chap. 11, Cornell University Press, Ithaca, New York, 1953.

3. W. H. Carothers and J. W. Hill, J. Am. Chem. Sec., 55: 5043 (1933).

4. C. R. Noller, "Textbook of Organic Chemistry," W. B. Saunders Company, (1966), pgs. 638 658.

5. G. Odian, "Principles of Polymer Systems". McGraw-Hill Book company, (1970), pgs. 63 – 69.

6. J. Brandrup and E. H. Immergut (eds.), "Polymer handbook," pp. 11 – 363, Interscience Publishers, John Wiley & Sons, Inc., New York, 1966.

7. P. H. Plesh and P. H. Westermann, Polymer, 10:105 (1969).

8. R. M. Joshi and B. J. Zwolinski, heats of Polymerization and Their Structural and Mechanistic Implementations, in G. E. Ham (ed.), "Vinyl Polymerization", Vol. 1. Part I, chap. 8, Marcel Dekker, Inc., New York, 1967.

9. C. R. Noller, "Textbook of Organic Chemistry", W. B. Saunders Company, (1966), pg. 601.

10. R. A. Patsiga, J. Macromol Sci., -Revs. Macromol. Chem. CI(2) : 223 (1967).

11. G. Odian, "Principles of Polymer Systems". McGraw-Hill Book company, (1970), pg. 503.

12. I. L. Finar. "Organic Chemistry", Volume 2, Longmans, Green and Co Ltd, (1964) pg. 287.

13. H. E. Gunning, E. W. R. Steacic, (1949), Journal of Chemical Physics, 17, 351-7.

14. R. J. Scott, H. E. Gunning, (1952), Journal of Chemical Physics, 17, 151-7.

15. Handbook of ring-opening polymerization, Philippe Dubois, Olivier Coulembier, Jean-Marie Raquez-2009.

16. A. D. Kelley, A. J. Berlin, L. P. Fisher, "Rearrangements of the propagating chain end in the cationic polymerization of vinylcyclopropene and related compounds" Journal of Polymer Science, Part a-1: Polymer Chemistry, Volume 5, Issue 1, pages 227-230, Jan. 1967.

17. T. Takashashi. I. Yamashita, "1,5-Polymerization of vinylcyclopropene", Journal of Polymer Science, Part B: Polymer Letters, Volume 3, Issue 4, pages 251-255, Apr. 1965.

18. T. Takashashi, "Polymerization of vinylcyclopropanes. II", Journal of Polymer Science, Part A-1: Polymer Chemistry, Volume 6, Issue 2, pages 403-414,Feb. 1968.

19. F.. Lautenschlaeger, H. Schnecko, "Polymerization of unsaturated episulfides", Journal of Polymer Science, Part A-1: Polymer Chemistry, Volume 8, Issue 9, pages 2579-2594. Sept. 1970.

20. F. L. Carter, V. L. Frampton, (1964), "Review of the Chemistry of Cyclopropene compounds", Chemical Review, 64: 497-525.

21. G. Snavely, S. Hassoon, D. Snavely, "Vibrational overtone activation of methylcyclopropene", Office of Naval Research, N00014-88-K-4130. R & T Code 4131063, Tech.Report # 12, 1993.

22. K. B. Wilberg, R. A. Fenoglio, J. Am. Chem. Soc., 90, 3395 (1968).

23. K. B. Wilberg, W. J. Bartley, F. D. Lossing, J. Am. Chem. Soc., 84, (1962).

24. J. W. Knowlton, F. D. Rossani, J. Res. Natl. Bur.. Stand., 43, 113 (1949).

25. P. Davison, H. Monty Frey, R. Walsh, "Vinylidene: probable intermediacy in methylene cyclopropane decomposition and heat of formation", Chemical Physics Letters., Vol. 120, Issue 2, (4 Oct. 1986) pg. 227-228.

26. B. Chow, P. McCourt, (2006), "Plant hormone receptors: perception is everything", Genes Dev. 20 (15), 1998-2088.

27. E. C. Sisler, M. Serek, (2003), " Compounds interacting with the ethylene receptor in plants", Plant Biology 5: 473-80.

28. M. Serek, G. Tamari, E. C. Sisler, A. Borochov, (1995), " Inhibition of ethylene- induced cellular senescence symptoms by 1-methylcyclopropene, a new inhibitor of ethylene action", Physiol. Plant 94 : 229-232.

29. E. C. Sisler, E. Dupille, M. Serek, " Effect of 1-methycyclopropene and methylene-cyclopropane on ethylene binding and ethylene action on cut carnations", Plant Growth Regulation, Vol. 18, Numbers 1-2, 79-86. 1996.

Problems

10.1   (a)   List five of the established observations for cyclic monomers made over the years.

        (b)   What are the factors that determine the ease of polymerization of cyclic monomers.

        (c)   Based on current developments, what are the factors that determine the ease of polymerizations of cycloalkanes?

10.2   (a)   Do cycloalkanes such as cyclopropane have functional or activation centers in the ring? Explain.

        (b)   What is a minimum required strain energy? In what ways can it be acquired for cycloalkanes?

        (c)   What are initiation centers with electrostatic forces for opening of rings instantaneously? Give examples of such activation forces.

10.3.   (a)   What is a radical-pushing potential difference between two connecting centers? When does it apply? Can there be a very large radical-pushing potential difference, but yet the ring cannot be opened? Explained if the answer is yes or no.

        (b)   Identify the single bond with the largest radical-pushing potential difference for the following rings.

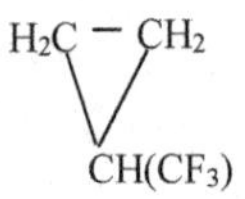

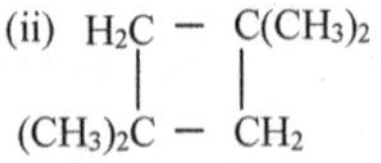

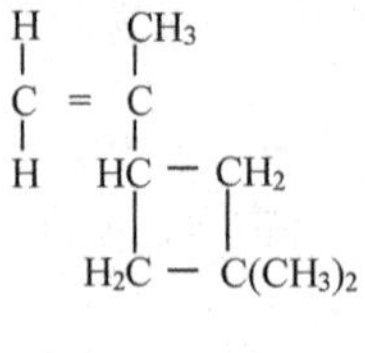

182

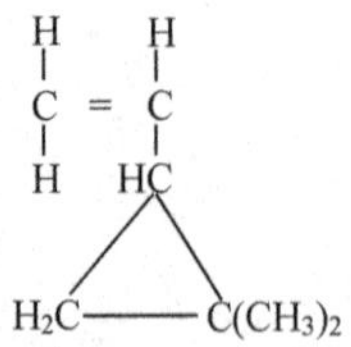

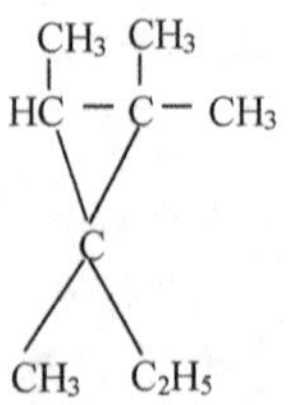

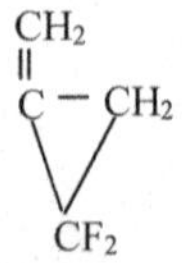

(vi) 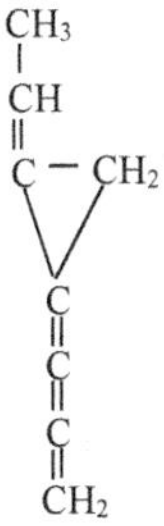 [structure with CH₃, CH, C—CH₂, etc.]   (viii) [cyclobutane structure]   (ix) 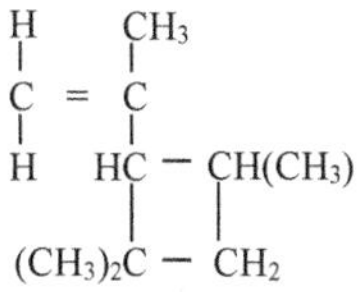 [structure]   (x) 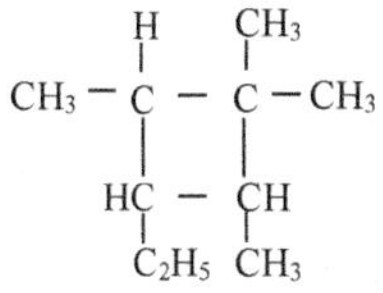 [structure]

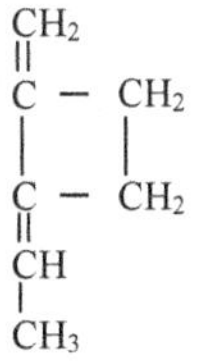

(c)   Which of the rings above cannot be opened? Explain.

10.4.   (a)   What are the driving forces favoring the existence of molecular rearrangement of the third kind?

(b)   Show that cyclopropanes are less nucleophilic than cyclobutanes of same characters.

(c)   Show that cyclopropane is more nucleophilic than ethene. Can it be generally said that cycloalkanes are more nucleophilic than alkenes? Explain.

10.5.   (a)   Which of the monomers or compounds shown below will favor molecular rearrangement of the third kind when instantaneously opened?

(i) 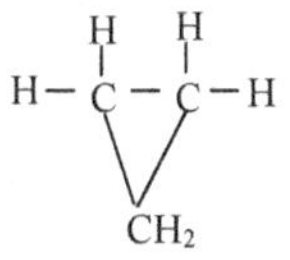   H—C—C—H / CH₂   (ii) 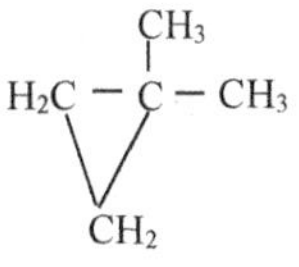   H₂C—C—CH₃ with CH₃ / CH₂   (iii) 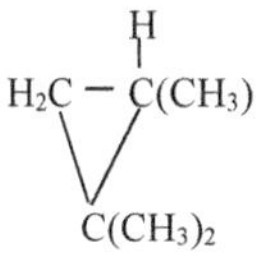   H₂C—C(CH₃) with H / C(CH₃)₂

(iv) 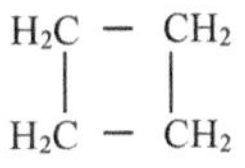   H₂C—CH₂ / H₂C—CH₂   (v) 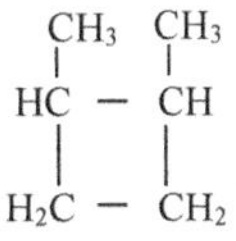   HC—CH with CH₃ CH₃ / H₂C—CH₂   (vi) 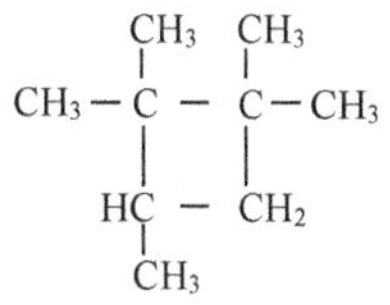   CH₃—C—C—CH₃ with CH₃ CH₃ / HC—CH₂ / CH₃

(vii) 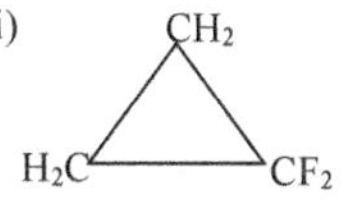 [triangle: CH₂ top, H₂C and CF₂ bottom]   (viii) [triangle: C(CF₃)₂ top, H₂C and C(CF₃)₂ bottom]   (ix) 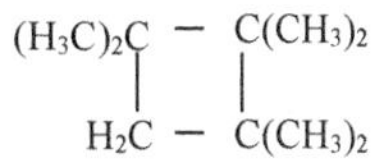   (H₃C)₂C—C(CH₃)₂ / H₂C—C(CH₃)₂   (x) 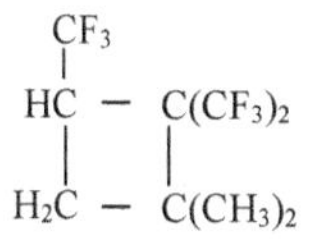   HC—C(CF₃)₂ with CF₃ / H₂C—C(CH₃)₂

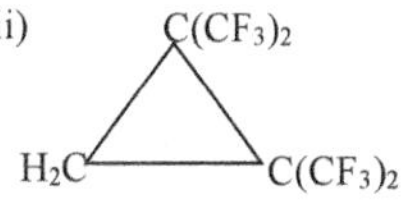

(b)   For those that favor it, show the order of nucleophilicity of the monomers obtained.

(c)   How is the order of strain energy in a ring a measure of the nucleophilicity of the monomers?

10.6.   (a)   Show the order of strain energy in the monomers of Q 10.5. (a).

(b)   Shown below are some vinyl cyclopropanes and vinyl cyclobutanes.

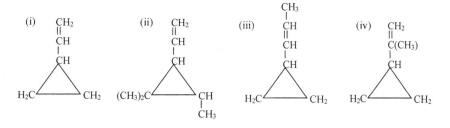

(i)   CH₂=CH—CH / triangle H₂C—CH₂   (ii)   CH₂=CH—CH / triangle (CH₃)₂C—CH—CH₃   (iii) 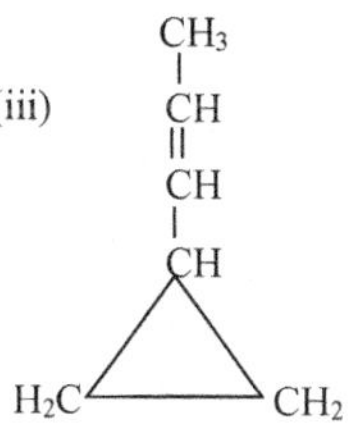   CH₃ / CH₂=CH—CH / triangle H₂C—CH₂   (iv)   CH₂=C(CH₃)—CH / triangle H₂C—CH₂

183

$$\underset{\underset{H_2C \diagdown\underline{\hspace{1.2cm}}\diagup CH_2}{\text{(v)}\quad \overset{\displaystyle \overset{H}{\underset{|}{C(CH_3)}}}{\underset{\overset{\parallel}{\underset{|}{C(CH_3)}}}{\underset{CH}{}}}}{}$$

(v) H / C(CH₃) ‖ C(CH₃) / CH — triangle H₂C—CH₂

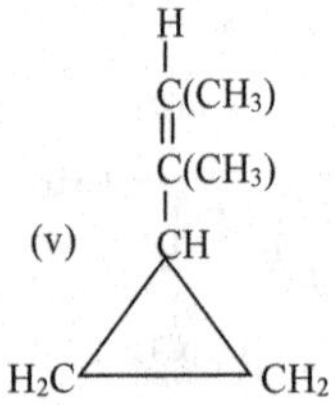

(vi) 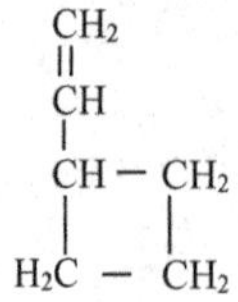
CH₂
‖
CH
|
CH — CH₂
| |
H₂C — CH₂

(vii) 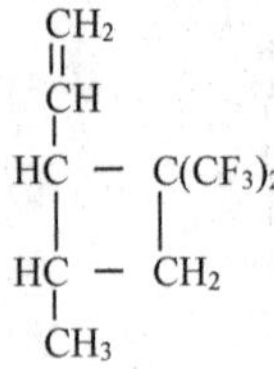
CH₂
‖
CH
|
HC — C(CF₃)₂
| |
HC — CH₂
|
CH₃

(viii)
CH₃
|
CH
‖
CH
|
HC — C(CH₃)₃
| |
H₂C — CH₂

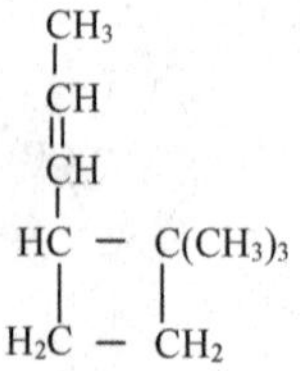

(ix) 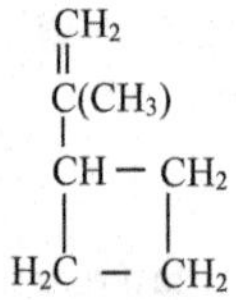
CH₂
‖
C(CH₃)
|
CH — CH₂
| |
H₂C — CH₂

(x) 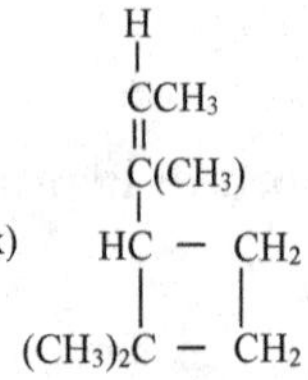
H
|
CCH₃
‖
C(CH₃)
|
HC — CH₂
| |
(CH₃)₂C — CH₂

Show the order of the strain energy for the family members. Which of the rings can be opened? Show how they can be opened.

10.7. (a)   Why can cyclopropene ring not be opened instantaneously?

(b)   How can the activation center be used as a growing center?

(c)   What are unique about cyclopropene?

10.8. (a)   What is molecular rearrangement of the third kind?

(b)   What are the driving forces favoring the existence of this type of molecular rearrangement?

(d)   Distinguish between molecular rearrangements of first and third kinds.

10.9. (a)   Distinguish between the two monomers or compounds shown below.

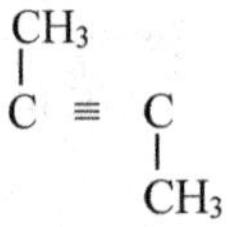

$$\underset{\underset{CH_3}{|}}{\overset{\overset{CH_3}{|}}{C}} \equiv C \qquad \text{and} \qquad \overset{\overset{CH_3 \quad CH_3}{|\qquad\ |}}{C = C}$$

(with CH₂ below the C=C, forming a ring)

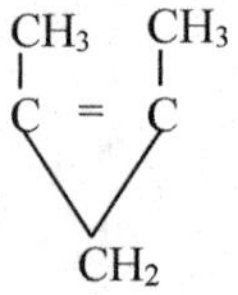

(b)   Show that NH_2 is less radical-pushing than $NHCH_3$ which in turn is less radical-pushing than $N(CH_3)_2$ using a cyclopropene ring.

10.10 (a)   Using the series of monomers shown below,

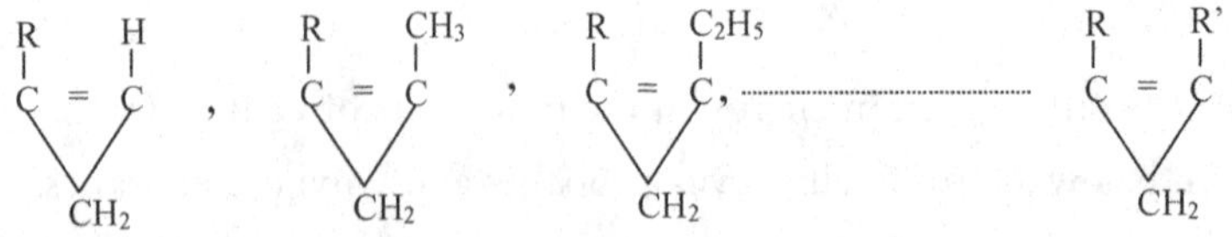

R   H   R   CH₃   R   C₂H₅   R   R'
| | | | | | | |
C = C   ,   C = C   ,   C = C, ⋯⋯⋯⋯⋯⋯⋯⋯   C = C
  \ /   \ /   \ /   \ /
  CH₂   CH₂   CH₂   CH₂

Provide a network system for R groups when the ring is opened instantaneously. R groups are placed on groups such as RCH=CH −, RC≡C-.

(b)   What are the conditions favoring the non-stable existence of some of the monomers shown below-

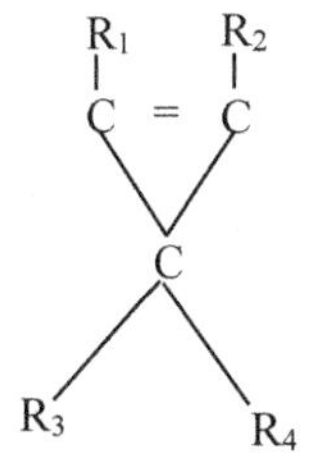

$$
\begin{array}{ccc}
R_1 & & R_2 \\
| & & | \\
C & = & C \\
& \diagdown \ \diagup & \\
& C & \\
& \diagup \ \diagdown & \\
R_3 & & R_4
\end{array}
$$

(where R_1, R_2, R_3, and R_4 are radical-pushing alkylane groups including H)

10.11   (a)   What are the relationships between cyclopropenic groups and alkenylic groups?

(b)   Shown below are two monomers-

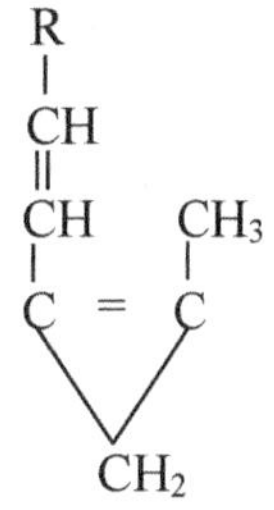

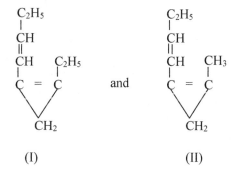

(I)   (II)

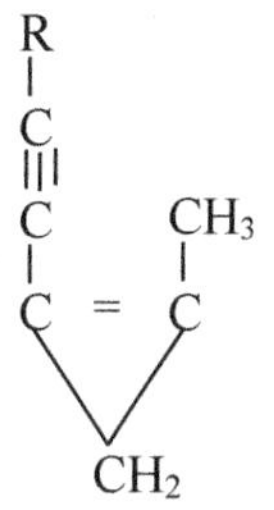

(i)   Distinguish between the two of them.

(ii)   Under what conditions do they favor molecular rearrangement of the third kind?

(iii)   When can the R group provide transfer species?

10.12.   (a)   Shown below are two monomers-

$$
\begin{array}{cc}
C_2H_5 & \qquad\qquad C_2H_5 \\
| & \qquad\qquad | \\
CH & \qquad\qquad CH \\
\| & \qquad\qquad \| \\
CH \quad C_2H_5 & \qquad CH \quad CH_3 \\
| \qquad | & \qquad\quad | \qquad | \\
C = C & \qquad\quad C = C \\
\diagdown\diagup & \qquad\qquad \diagdown\diagup \\
CH_2 & \qquad\qquad CH_2
\end{array}
$$

(I)   and   (II)

Can they undergo molecular rearrangement of the third kind? Explain

(b)   Show how the nucleophilicities of (I) of Q. 10.11 (b) are affected by varying the size of R.

(c)   What is the order of nucleophilicity of the monomers in (a)?

10.13.   (a)   Shown below are two compounds-

185

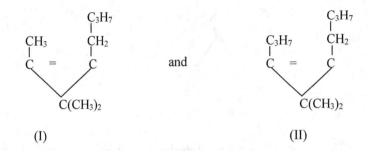

(I)   (II)

(i)   Which one will undergo molecular rearrangement of the third kind? If none, explain.

(ii)   What are the characters of the compounds?

(b)   What is common and different between cyclopropenes and allene?

(c)   Under what conditions can fluorinated alkylane groups be used to favor the existence of an unstable cyclopropene?

10.14.   Prove the validity of the following relationships in terms of radical-pushing capacities.

(i) $CH_3 - C \equiv C -$ $\leq$ CH_3

(ii) $H - C \equiv C -$ $\leq$ $H -$

(iii) $F - C \equiv C -$ $<$ $H -$

(iv) $R - C \equiv C - C \equiv C -$ $\leq$ $R - C \equiv C -$

(v) $CH_3 - C \equiv C -$ $\leq$ $C_2H_5 - C \equiv C -$

10.15   (a)   Distinguish between the following sets of paired groups.

(i) $CH_2 = C = CH -$ and $CH_2 = C = C(CH_3) -$

(ii) $O = C = CH -$ and $O = C = C(CH_3) -$

(iii) $HN = C = CH -$ and $HN = C = C(CH_3) -$

(b)   Find the equivalence of the following groups-

(i) $CH_3 O \overset{O}{\overset{\|}{C}} - C \equiv C -$ (ii) $H_2N - \overset{O}{\overset{\|}{C}} - C \equiv C -$ (iii) $R - \overset{O}{\overset{\|}{C}} - C \equiv C -$

(c)   Why are $=CF_2$, $=CH (COOCH_3)$, $=CH (CN)$, etc. said to be radical-pushing groups?

10.16   (a)   Show the equivalence of the following groups free-radically where possible.

(i) $F_3 C - C \equiv C -$,   (ii) $F_5C_2 - C \equiv C -$,   (iii) $(F_3C)_2FC - C \equiv C -$,

(iv) $(CH_3)_3C - C \equiv C -$,   (v) $CH_2F - C \equiv C -$,   (vi) $CH_3 O CH_2C - \overset{O}{\overset{\|}{C}} - C \equiv C -$

(b)   Show that tertiary alkylane groups are more radical-pushing than secondary alkylane groups which in turn are more radical-pushing than primary alkylane groups, using cyclopropene rings.

10.17   (a)   Shown below are two compounds or monomers-

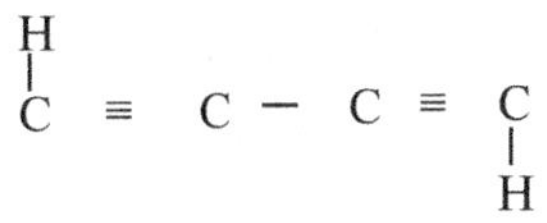

$$H-C \equiv C - C \equiv C{-H}$$

$$\text{and}$$

$$H-C \equiv C - C = C{-H} \quad (CH_2)$$

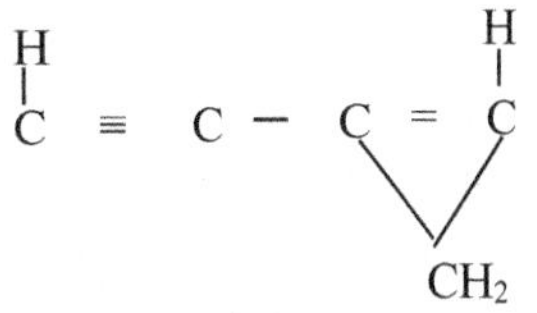

(I)   (II)

(i)   Are the two monomers resonance stabilized? Explain

(ii)   What are their mono-forms?

(iii)   Are all their mono-forms symmetric? Explain

(iv)   What is the character of each of the monomers?

(b)   Shown below are also two compounds or monomers

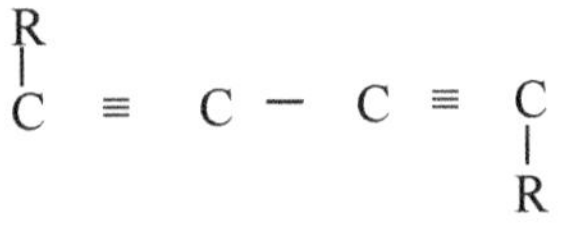

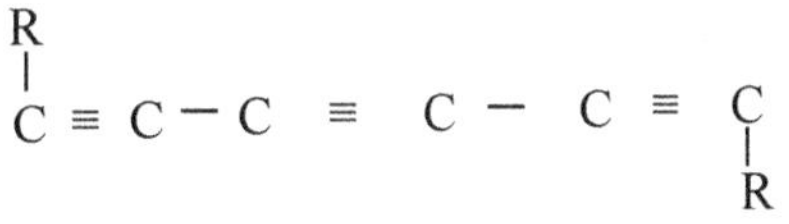

$$R-C \equiv C - C \equiv C{-R}$$

$$R-C \equiv C - C \equiv C - C \equiv C{-R}$$

(I)   (II)

(where R is a radical-pushing group with capacity $\geq CH_3$)

(i)   Are the two monomers stable? Explain.

(ii)   Are all their mono-forms symmetric? Explain.

(iii)   What is the character of the mono-forms?

10.18.   (a)   Shown below are two symmetrically placed diynes and triynes.

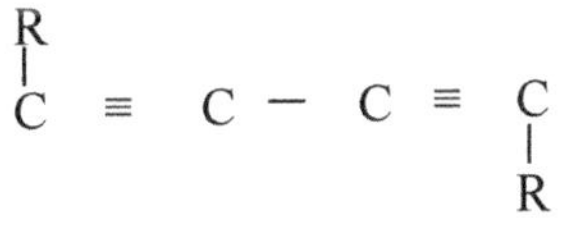

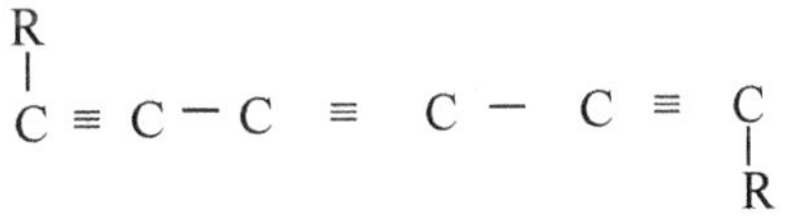

$$R-C \equiv C - C \equiv C{-R}$$

$$R-C \equiv C - C \equiv C - C \equiv C{-R}$$

(I)   (II)

(where R is a radical-pushing alkylane group)

(i)   Show how resonance stabilization takes place in them.

(ii)   Which of the mono-forms are symmetric in character?

(iii)   What happens when the two Rs become different (R_1 and R_2)?

(iv)   Shown below are two butadiyne cyclopropene-

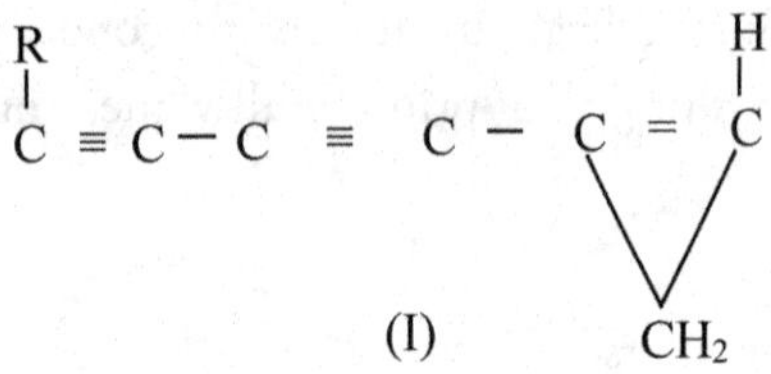

$$
\underset{(I)}{\overset{R}{\underset{|}{C}} \equiv C - C \equiv C - \overset{H}{\underset{|}{C}} = C}\diagdown_{CH_2}
\quad and \quad
\overset{H}{\underset{|}{C}} \equiv C - C \equiv C - \overset{R}{\underset{|}{C}} = C\diagdown_{CH_2}
\quad (II)
$$

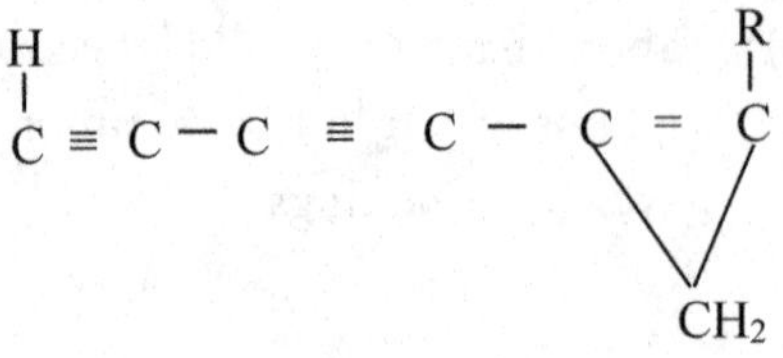

(i)   For what sizes of R as alkylane groups, will the monomers favor molecular rearrangement of the third kind when the ring is opened instantaneously?

(ii)   Distinguish between (I) and (II) in terms of their abilities to favor stable existences.

(iii)   What is an inherently symmetric monomer? Use methyl cyclopropene to show it.

10.19.   Of what significance is the CH_3 group in monomeric systems?

10.20 (a)   List three factors so far identified which determine the ability of a ring to possess the MRSE.

(b)   List five driving forces so far identified favoring the instantaneous opening of ring.

(c)   Of what significances have the study of cyclopropenes been to us in the current developments?

CHAPTER ELEVEN

TRANSFER OF TRANSFER SPECIES IN CYCLOALKENES, CYCLODIENES, CYCLOTRIENES AND CYCLOTETRAENES

11.0. Introduction

Though cyclopropenes is the first member of cycloalkenes, it has separately been considered on its own merits in the last chapter, in view of its higher SE compared to other members of the family, its unique features which have never been known before and most importantly, establishment of New Frontiers for the other members in the family and for other families. By its consideration, it has helped to validate most of the rules (laws of nature) which have been proposed. For instance, without the law of conservation of transfer of transfer species, it would not have been possible to identify molecular rearrangement of different kinds in particular the third kind. Obviously, with respects to the size of cyclopropane and cyclopropene rings, one must be forced to ask so many questions, which in most cases are overlooked, because the more you look, the less you see.

Just as cycloalkanes have fully been considered, so also will all polymerizable members of cycloalkenes be considered in this chapter. The remaining members are still uniquely important, particularly with even-membered ones such as cyclobutenes, cyclohexenes and so on. *With increasing size of their rings, the less unsaturated they become, and the less strained they are.*

With increasing presence of double bonds in the rings, the less the presence of small membered rings. With cyclodienes, there is no three-membered ring. With cyclotrienes, there are no three, four and even five-membered rings, and so on. Apart from the increased unsaturation of the rings, increased strain energy in the rings, there is need to question their non-existences. Just as there exists the minimum required strain energy (MRSE), so also there exists a maximum required strain energy (MaxRSE), the energy above which no ring can exist, That energy can be measured for the cyclodiene family using cyclo-butadiene (unsubstituted), a monomer or compound that is known to favor only transient existence as a highly reactive gas.

So far, two types of initiations have been identified, based on the types of initiators involved – free-media and coordination or paired-media initiators. With coordination initiators, there is always a re-orientation of the activated monomers, due to influence of electrostatic forces. Two additional new types of initiations using coordination initiators will also be identified with the families of monomers to be considered herein. Special coordination initiators are required to achieve these initiations.

With increasing unsaturation of rings, it gets to a point where no unsaturation can take place. Cyclopropene is fully unsaturated but not resonance stabilized. Cyclobutadiene is fully unsaturated, fully resonance stabilized but favors transient existence. Cyclohexatriene is fully unsaturated, fully resonance stabilized and favors full existence. Cyclooctatetraene is fully unsaturated, partially resonance stabilized, and favors dual existence depending on the operating conditions. The same applies to other fully unsaturated larger membered rings. Of all these members which seem to belong to a unique family, the most unique is cyclohexatriene called Benzene.

1.1 Cycloalkenes

1.1.1 Cyclobutenes

Compared to cyclopropene (H:C ratio), cyclobutene are far less unsaturated. Unlike cyclopropenes, two modes of polymerizations are possible -

(i) Via opening of the ring
(ii) Via the double bond

Cyclobutane cannot be opened instantaneously using initiators, since they cannot provide the minimum required strain energy. However, with heat, it can be opened instantaneously. Also with the presence of a double bond in the ring, little additional energy will be required to attain the minimum required strain energy than that that will be required for cyclobutane. Releasing the strain energy will depend on what is desired to be done. Is our mission to use it as a monomer for polymer production or just as a compound to produce another compound for other applications? If the mission is for polymerization, then for the ring to be opened instantaneously, the followings are important.

(i) the type of initiator to be used,
(ii) the strength of the initiator,
(iii) the other ancillary very important operating conditions such as Temperature, Pressure, Concentrations, Molar ratios, and type of reactor.

If the strength of the initiator is very weak and temperature is low to unzip the ring instantaneously, then the internally located double bond can be used. In general, mostly metallic types have been used for the polymerization of cyclobutene[1]. This activation can only take place if the nucleophilicity of the invisible π-bond in the ring is greater than the nucleophilicity of the visible π-bond in ring. If equal or less, activation cannot take place chargedly or radically. This is for all of them including cyclopropene. *If a ring cannot be opened, whether there is or are points of scission in the ring or not, the reason is because the nucleophilicity of the invisible π-bond in the ring is greater than the nucleophilicity of the visible π-bond(s) in the ring.*

Anionically, before the ring is opened using its (Anionic) electrostatic forces, the double bond is already activated, so that the followings occur.

$$R:^{\ominus} \;+\; \underset{H_2C \rule{1cm}{0.4pt} CH_2}{\overset{H \quad H}{C = C}} \longrightarrow R:^{\ominus} \;+\; \underset{H_2C \rule{1cm}{0.4pt} CH_2}{\overset{H \quad H}{\oplus C - C \ominus}} \rightleftharpoons$$

Free negative
charge (WEAK)

$$RH \;+\; \underset{HC = CH}{\overset{H \quad H}{\ominus C \rule{1cm}{0.4pt} C - H}}$$

<u>Transfer species of 1st kind of 10th type</u> 11.1

$$R:^{\ominus} \;+\; \underset{H \qquad\quad H \qquad\; H}{\overset{H \quad H \qquad\qquad H}{\oplus C - C = C - C\ominus}} \longrightarrow$$

(Strong)

$$R - \underset{H \qquad\quad H \quad\; H}{\overset{H \quad H \qquad\qquad H}{C - C = C - C\ominus}}$$

POSSIBLE CHARGEDLY 11.2

Between the two reactions above, any can be favored, depending on the strength of the initiators radically or chargedly. When the initiator is weak, it is the double bond that is activated if and only if the nucleophilicity of the double bond is less than that of the invisible π-bond. If not, the ring cannot be opened. ***When the conditions for opening of the ring exist, that is, that in which the nucleophilicity of the invisible π-bond is now made to be less than the nucleophilicty of the visible π-bond, the visible double bond is the last to be activated.*** The ring is now opened instantaneously chargedly or radically. ***If the nucleophilicty of one visible π-bond in a ring with one or more points of scission had been greater than that of the invisible π-bond, then the ring will not exist as a ring.*** When the ring is opened, the double bond remains unactivated. But chargedly, no matter the operating conditions, when the double bond is the first to be activated for some of them where the conditions exist, then the ring cannot be opened instantaneously. Chargedly therefore, either only the 1,2-disubstituted ethenes will appear along the chain with no ring opened or the 1,4-mono-form will appear along the chain when opened, only via positively charged route for many of them. This has in the past been so reported[1]. Free-radically, not only the same as above will apply electro-free-radically, but also when the ring is opened instantaneously, both nucleo-free- and electro-free-radical routes are favored. The transfer species in the first reaction above is of the first kind of the tenth type. With negative charges and anionically, polymerization of the cyclobutene is not favored whether the initiator is weak or strong, the monomer being Nucleophilic.

When weak or strong positively-charged-paired initiators are involved, the followings are obtained -

(I) OR (II) 11.3

As has already been said, when the initiator is weak, it is the double bond that is first activated if its nucleophilicity is less than or equal to that inside the ring. Chargedly, the ring here carrying only H can still be opened if the initiator is strong.

Before the ring is used as a monomer by the weak coordination center, it is first trans-placed in view of the influence of electrostatic and electrodynamic forces of repulsion resulting from equal placements of the folded segment of the ring along the horizontal axis after activation. Hence, it is (II) above that is favored. With a weak negatively-charged -paired initiator, the followings are to be expected. No growing of the chain negatively is possible being Nucleophilic whether trans-placed or not. However, based on (II) of Equation 11.3, note should be taken of a new form of placement along the chain as was shown in the first Volume, because of the even character of the ring.

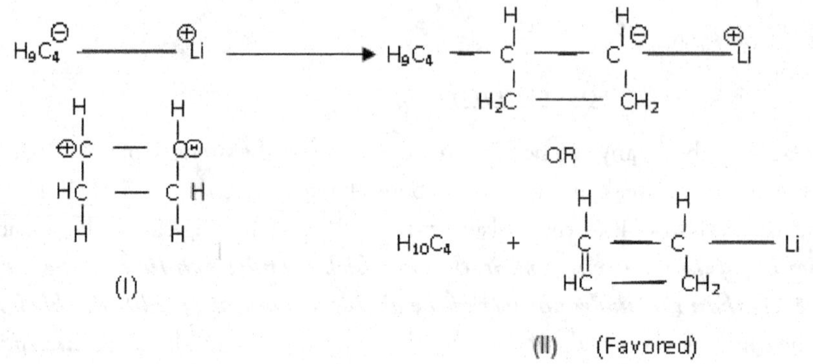

$$11.4$$

This is what is expect unlike the case of Equation 11.1, because to have free negatively charged center without pairing is impossible. However, it is being shown for the purpose of simplicity. Non-free-negatively charged center can be paired and can be free (i.e. isolated).

From all indications, when the initiator is weak, trans-existence of the cyclic activated cyclobutene are more favored when coordination types of initiators are used.

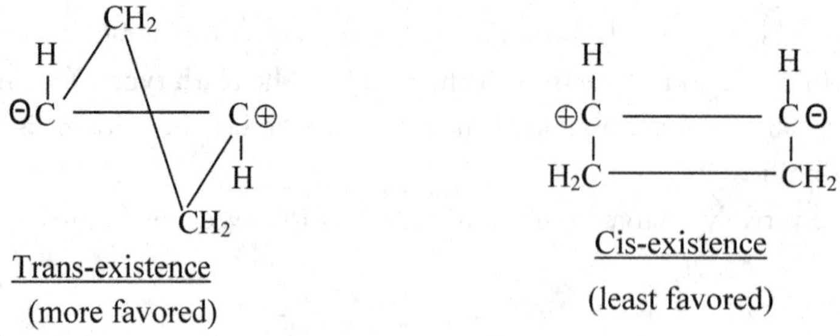

Trans-existence
(more favored)

Cis-existence
(least favored)

$$11.5$$

When weak/moderate electrostatically positively charged coordination initiators with reservoirs for the monomers are involved, the followings are therefore to be expected.

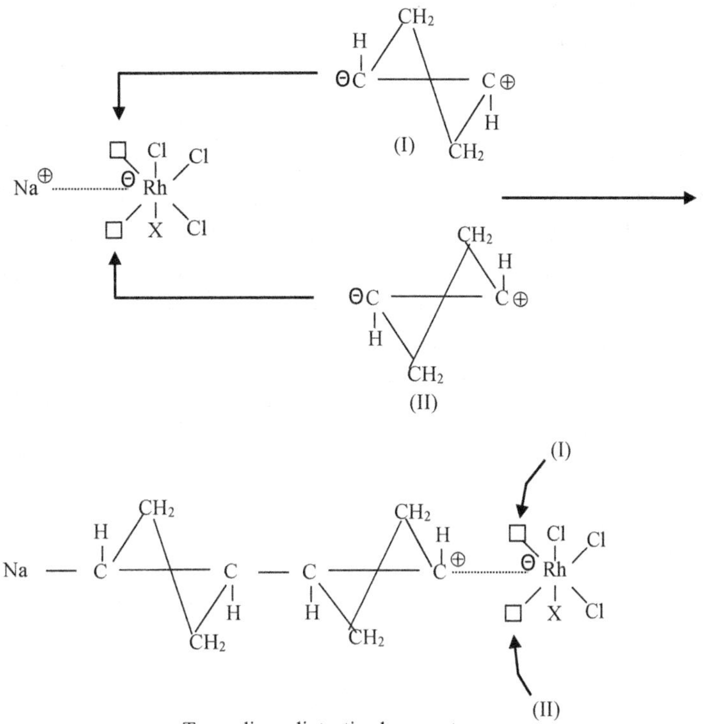

Trans-di-syndiotactic placement.   11.6

The real positively charged-paired initiator from NaX/RhCl$_3$ combination has not yet been shown. Though what has been shown above exist, it cannot actually be used, because the equation will not be chargedly balanced. A cation can never be used to carry a chain where the charge is covalent and not ionic. However, it will continue to be used until the time for correction arises after introducing new concepts. Note the orientation of the monomers in the vicinity of the coordination centers and the type of placement when one or two reservoirs are present.

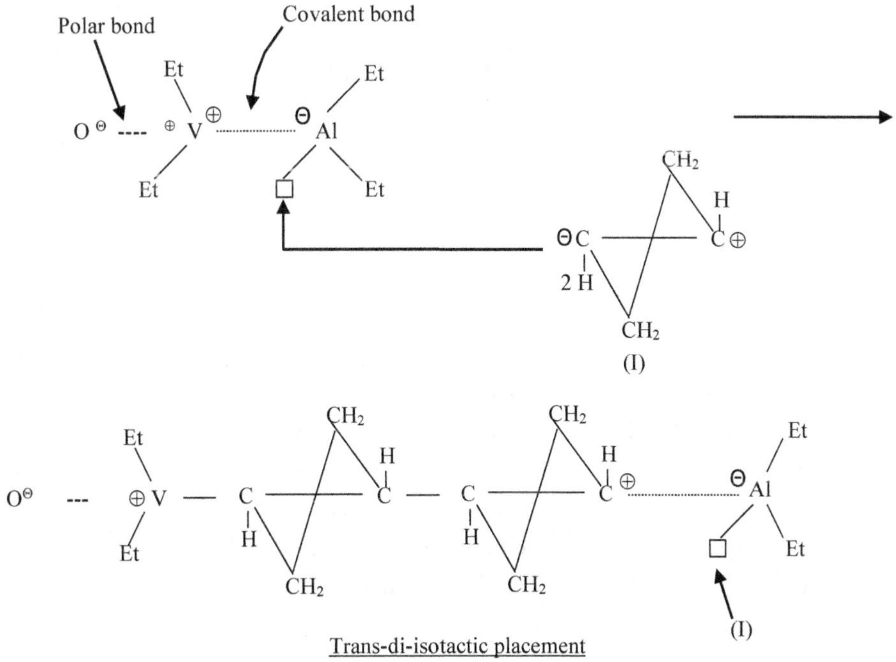

Trans-di-isotactic placement   11.7

Polycyclobutene is the polymer obtained above, because of its origin. The first initiator is obtained from Na/Rh combination, while the second is from V/Al combination. When only the double bond is activated, it is then trans-placed due to electrostatic forces, followed by addition to obtain different types of placements depending on the number of vacant orbitals on the counter-charged center. Worthy of note is the polar bond between V and O and the way the CH_2 groups have been placed along the chain. They seem to look the same, but not, because these types of placements, can lead to different functions in systems.

When a familiar type of strong positively-charged-paired coordination initiator is used as if free-radically paired in a condition where the nucleophilicity of the invisible π-bond becomes less than or equal to that of the visible π-bond in the ring, the followings are obtained[1].

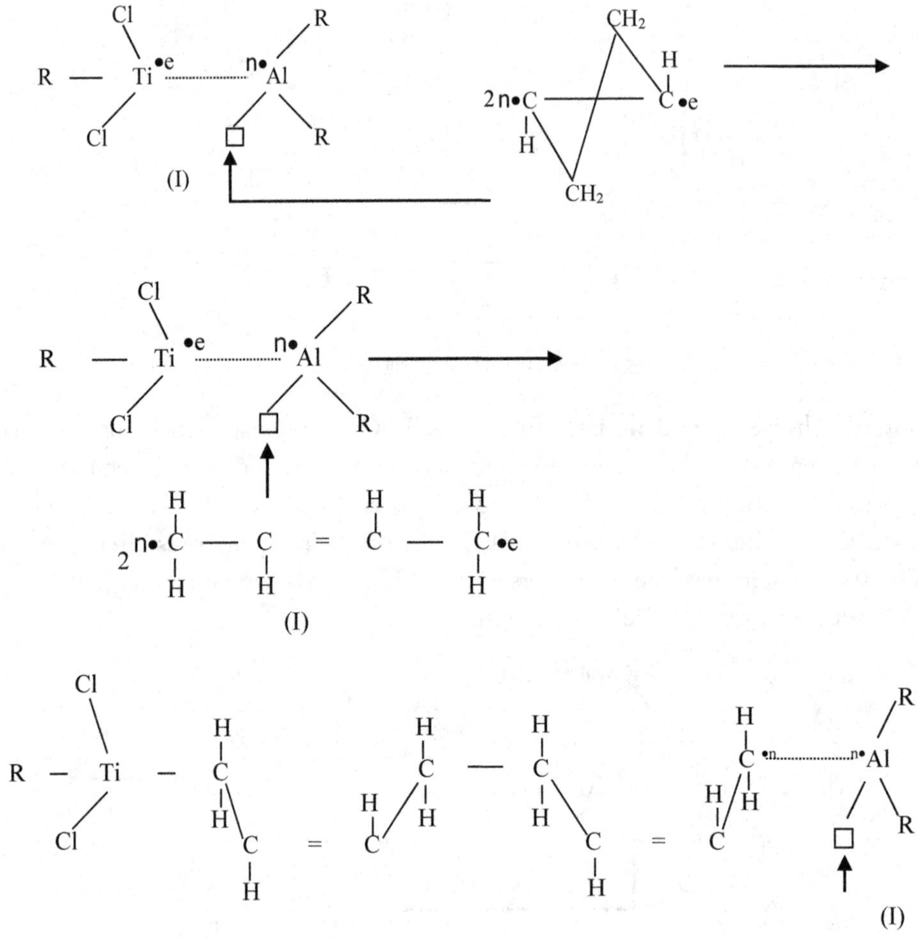

(I)

(I)

(Cis - placement)

11.8

In view of the strength of the coordination center and the fact that minimum required strain energy has been made to exist in the ring, it unzips instantaneously at the point of scission indicated above with adequate orientation before addition. Since only one vacant orbital is present on the counter-charged center, cis-polybutadiene is obtained, just like when butadiene is involved only free-radically and not chargedly. As shown below, the ring can only be opened in the 1-4- Carbon-carbon single bond and never on a bond adjacently located to a double bond only when the double bond is not activated. It has been shown chargedly below when the double bond has already been activated.

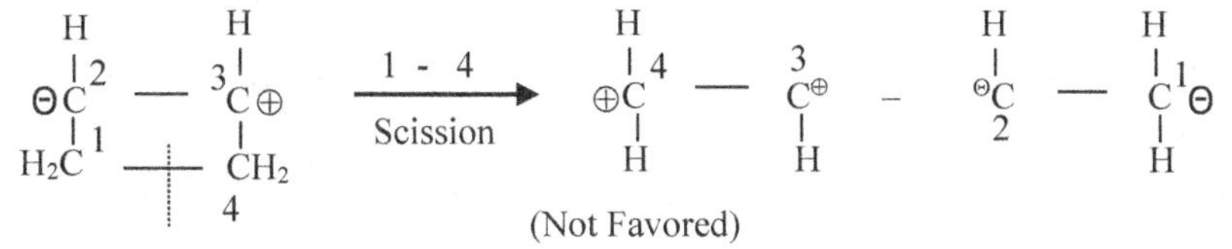

$$\text{(Eq. 11.9 — chemical structures with "1 - 4 Scission" and "(Not Favored)")}$$

(Not Favored)   11.9

(Not Favored)   11.10

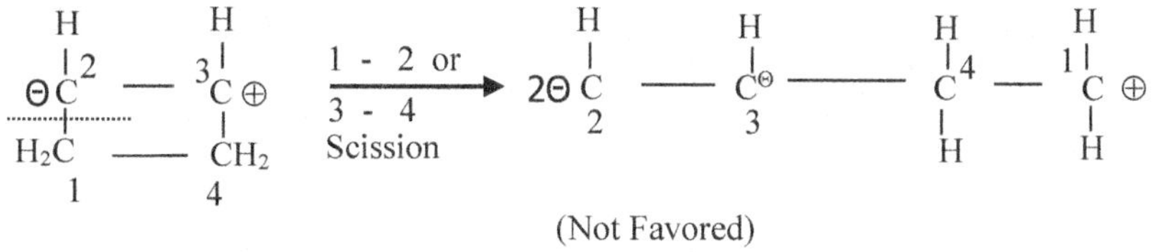

But however, as has already been stated into law in the last chapter, one can see another reason why scission of the type shown in last equation can never take place. When an activation center is activated, scission can never take place next door to it or anywhere in the ring, radically or chargedly.

Free-radically in the absence of electrostatic forces, activation can only take place via the double bond as follows.

11.11

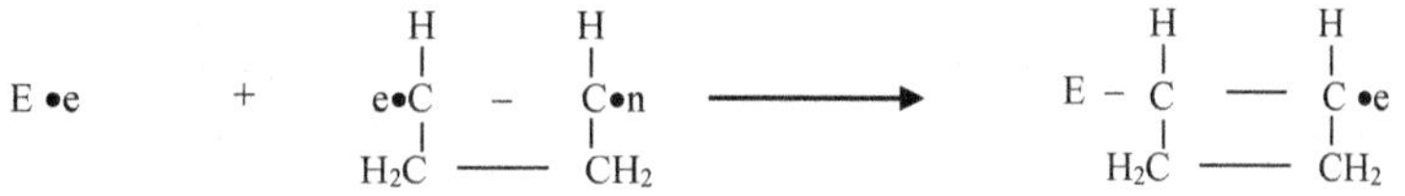

11.12

Due to the presence of transfer species of the first kind of the tenth type, no initiation is favored nucleo-free-radically. Electro-free-radically, the route is favored with the possibility of having trans-placement of the ring, but with much difficulty, in view of absence of coordination.

For alkylane substituted cyclobutene, consider **cyclomethyl-butene (3-methyl-l-cyclobutene)[1].** Note that the numbering in parenthesis is not questionable.

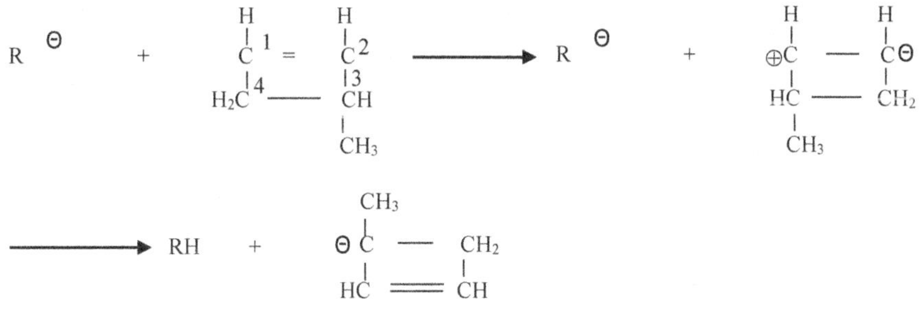

Transfer species of first kind of tenth type   11.13

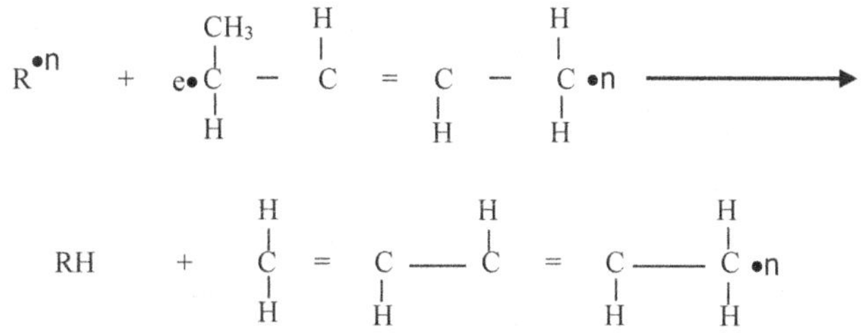

Transfer species of first kind of first type   11.14

In the last equation above, the ring can also be opened instantaneously chargedly if the nucleo-philicity of the invisible π-bond can be made less than that of the visible π-bond. Though negatively charged initiation will not be favored, two different types of transfer species can be involved, depending on the operating conditions. Once the double bond is activated, the transfer species is instantaneously abstracted before the ring opens, when the initiator is weak. This is valid only when the double bond is less nucleophilic than the invisible π-bond in the strained ring. ***Strong initiators or heat decrease the nucleophilicity of the invisible π-bond.*** Therefore, the opening of a ring can be seen to be dependent on what the ring is carrying and the operating conditions.

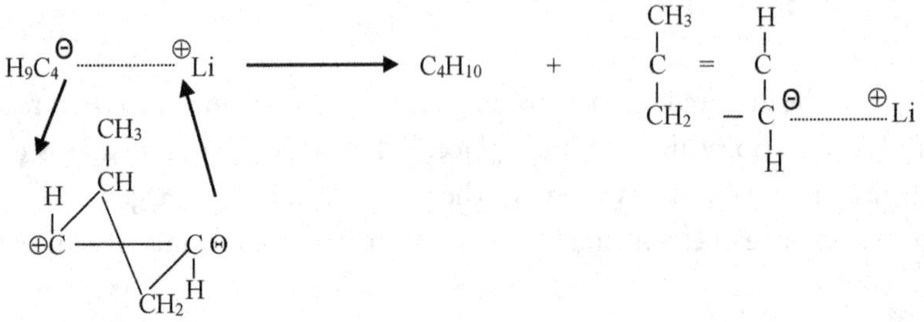

11.15

It is important to note the trans-placement and orientation of the activated monomer above. The transfer species of the tenth type is involved, when the ring is not opened. It is H that is the transfer species because H is less-radical-pushing than CH_3 group. As already said, the initiator used above is dual in character only radically. It has been so used above wrongly as universally used. The monomer being a Nucleophile, will be carried by Li with Initiation Step being favored only radically. If the initiator was such that could not open the ring, when the nucleophilicity of the visible π bond is less than or equal to that in the ring as shown below, then the followings are obtained.

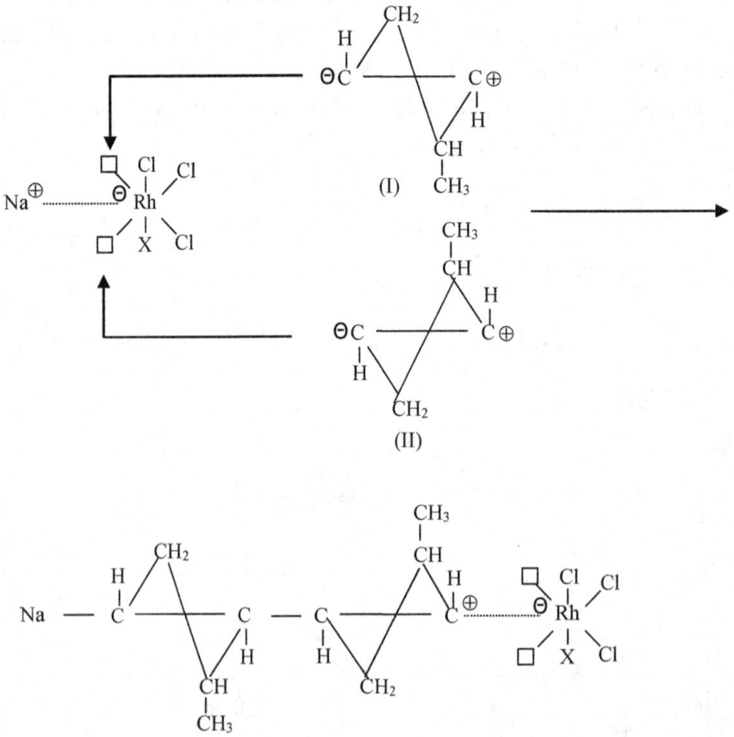

Trans-di-syndiotactic placement. (tds)

11.16

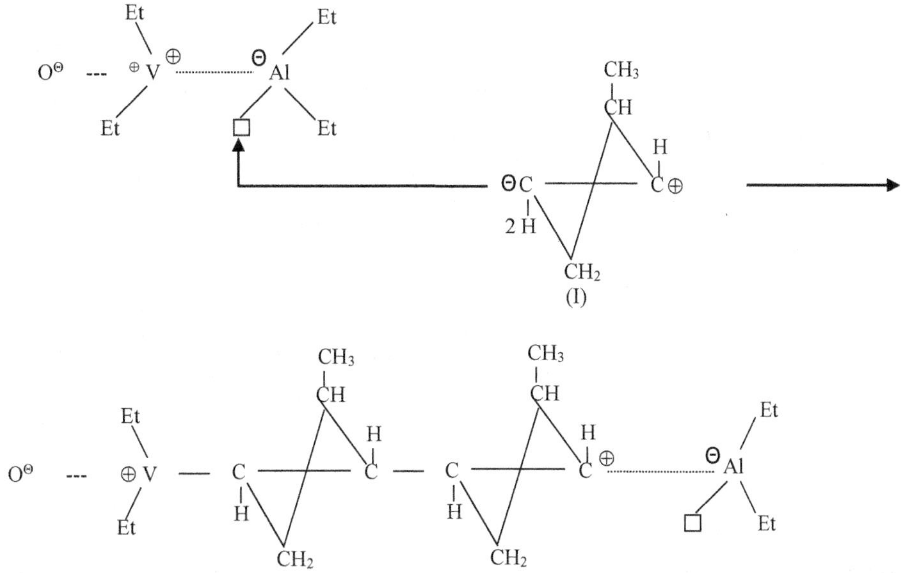

Trans-di-isotactic placement (tdi)

11.17

Substituted polycyclobutenes are obtained in the reactions above. With very strong initiators, the ring is unzipped at the only point of scission. As will be shown downstream, the strength of the active center of Paired-media initiator is partly dependent on the electropositivities of the metallic centers relative to that of the counter centers.

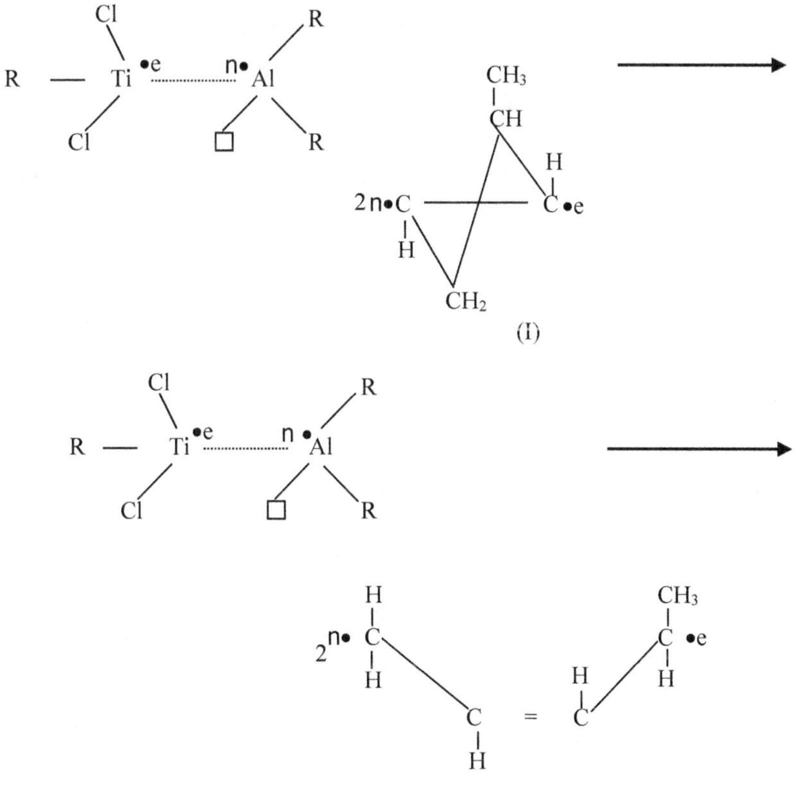

(Favored)

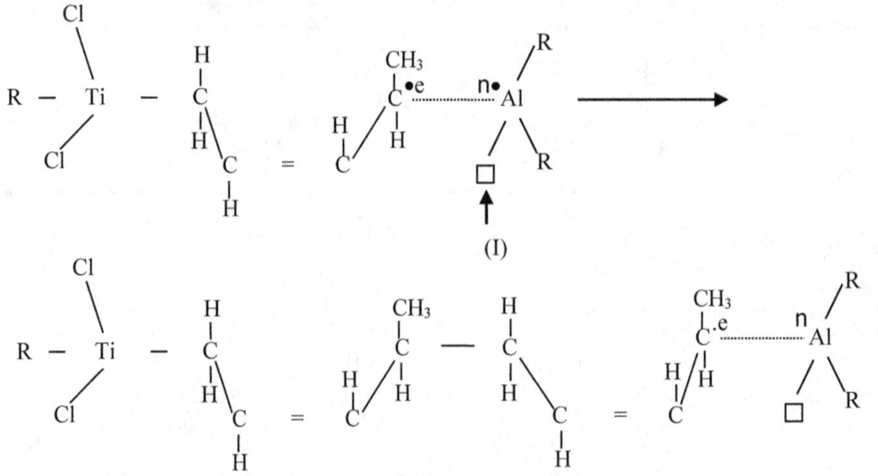

Cis - 1, - methyl butadiene polymer

11.18

The first monomer addition above is the Initiation Step, followed by propagation with one monomer unit. As it can be noticed, only poly (1- methyl butadiene) like polypentadiene from pentadiene, can be obtained when the ring is opened. Poly(isoprene) can never be obtained from the ring above, since the monomer cannot molecularly rearrange when not opened. When opened and allowed to rearrange the same monomer is obtained. If the ring is opened, the product above is the product that which can readily be obtained from the non-ringed so-called pentadiene. *If opened as above, then the nucleophilicity of the strain in the ring is less than that of the visible π-bond in the ring.* This will largely be the case when the ring is carrying radical-pushing groups in place of H, since the ring will be more strained to favor less MRSE. For cases like this, the double bond is never used, unless the initiator is too weak to unzip the ring radically. Note that far lower temperature will be required to unzip cyclobutene than cyclobutane, because cyclobutene is more strained than cyclobutane.

For the growing polymer chains of Equations 11.6 and 11.16, the following are obtained.

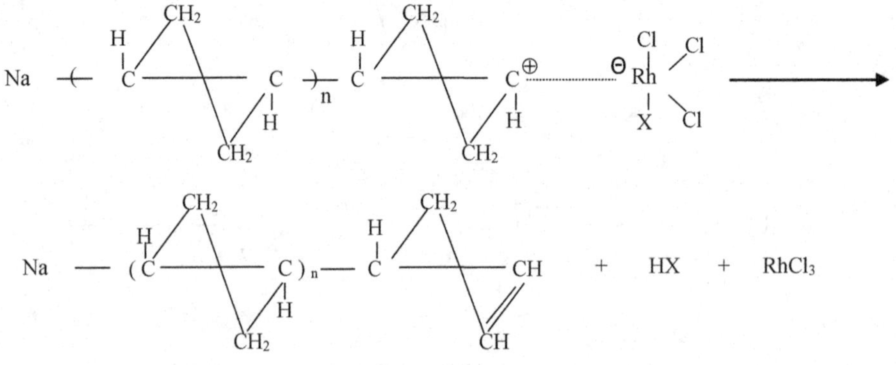

Transfer species of first kind of the tenth type

11.19

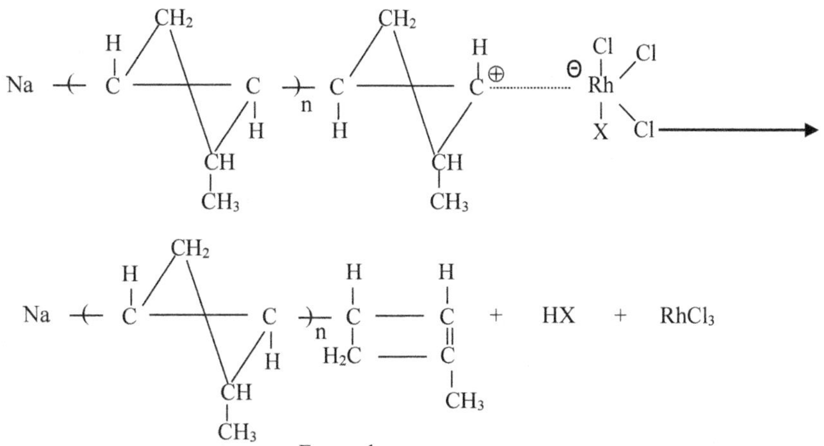

Transfer species of the first kind of the tenth type 11.20

Transfer species of tenth type is released for both reactions to give dead terminal double bond polymers. All these are based on the applications of the Laws of conservation of transfer of transfer species. ***Unlike cyclopropene which has repulsive forces due to its size, cyclobutene does not have it. Hence, it can be activated chargedly.***

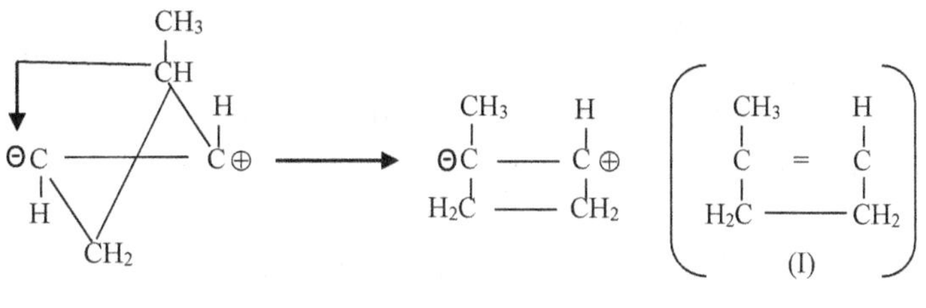

Impossible activated state

11.21

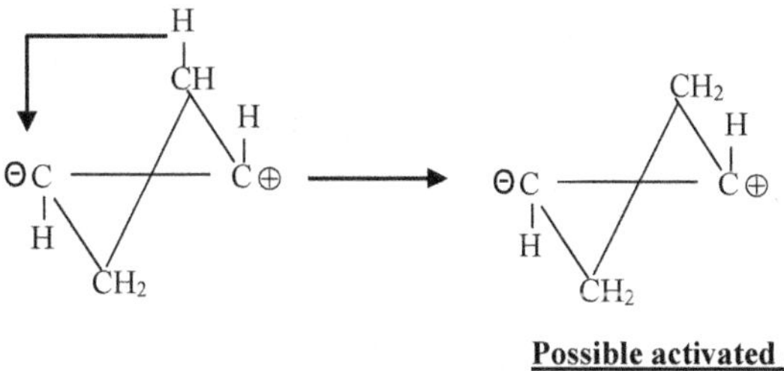

Possible activated state 11.22

For cyclobutene, transfer of transfer species is favored to produce the same monomer. The same applies free-radically. Hence, one should expect slower rate of polymerization. For 1-methyl-cyclobutene, molecular rearrangement is not favored. It should be noted that the transfer species involved are of the first kind of the tenth type. Note the similarity of (I) of Equation 11.21 and the terminal of the dead terminal cyclic double bond polymer of Equation 11.20. Free- radically, like cyclobutene, only the electro-free-radical route can be favored being a stronger Nucleophile.

Considering replacing one H atom on the double bond with an alkylane group (R), the followings should be expected beginning with CH_3.

(Free Charge)

$$\text{(I)} \qquad\qquad \text{(II) FAVORED} \qquad\qquad 11.23a$$

Nucleo-free-radically

$$\text{(III)} \qquad\qquad 11.23b$$

It should be noted that though the ring has been opened radically, the case above can also be opened chargedly, under same operating conditions, but with charged initiators. The substituent group in the ring is more radical-pushing than the CH_3 group on top of the double bond. If it was not, then methylene cyclobutane would have been obtained. *Methylene cyclobutane is well known to exist and used as a monomer to produce isopolyisoprene as well as cyclized polyisoprene, uncyclized polyenes and polymeric chain with cyclobutane structures[2].* Free-radically with methyl cyclo-butene, when instantaneously opened, polyisoprene is produced. But with methylene cyclobutane, polyisoprene is not one of the products. This could be due to the operating conditions. However, when used as a monomer, there is no doubt that when the initiator is weak, a fraction of it (3-methyl cyclobutene) molecularly rearranges to give 1-methyl cyclobutene from which cyclized polyisoprene was obtained. From the remaining fractions of the methylene cyclobutane when present, the other polymers were obtained via instantaneous opening of its ring with and without molecular rearrangement of the third kind. When no rearrangement is allowed to take place, isopolyisoprene (that in which the C=C double bonds are in the exo position with respect to the chain) is produced, and when allowed to take place, polyenes are produced. Where polymeric chains were said to be obtained "cationically" (i.e., with positively charged-paired initiators), this was obtained by direct activation of the methylene cyclobutane without opening of its ring which obviously will be sterically hindered, unless when coordination centers with two vacant orbitals are used. One can see the great importance of strength of initiators used in polymerization systems. Hence, coming back to Equation 11.23a, (II) is favored, when the initiator is weak chargedly or radically. If the initiator is strong enough to unzip the ring, this can be done radically as shown by favored (III) of Equation 11.23b. This is that in which polyisoprene is obtained in the route not natural to it. It can also be done chargedly only with few types of initiators such as $H_9C_4{}^{\ominus}.....^{\oplus}Li$, where the $Li^{\oplus}$ center cannot be used. The chain will not grow as long as that obtained electro-free-radically or with positively charged-paired initiators, their natural routes.

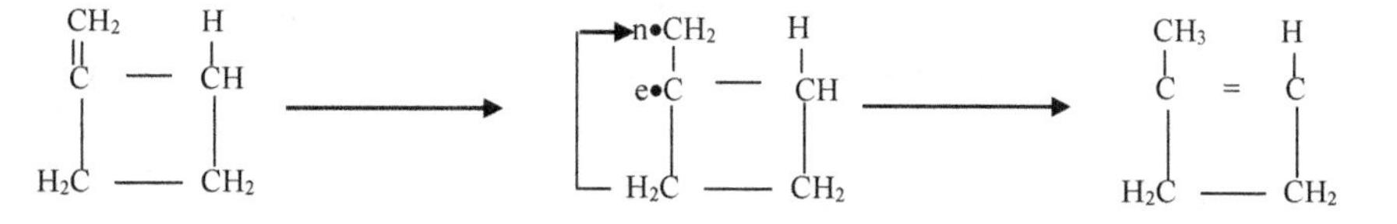

Methylene cyclobutane   Methyl cyclobutene

11.24

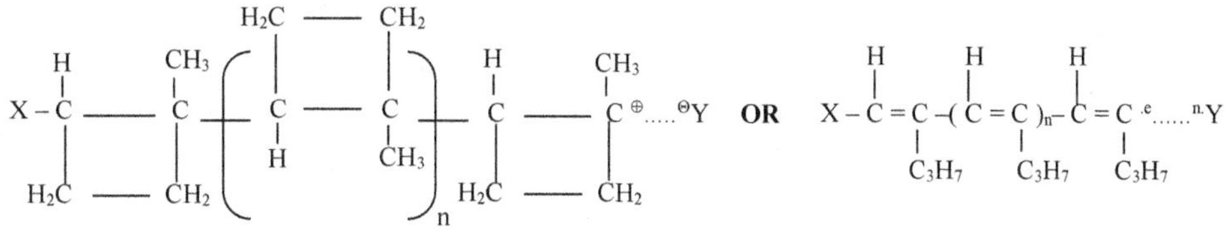

Cyclized polyisoprene from "Z/N" initiators   Uncyclized polylenes from Z/N initiators

11.25

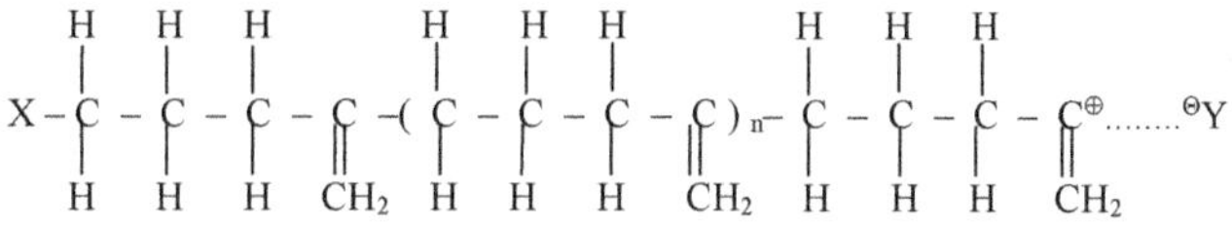

Isopolyisoprene from Gps IV, V, VI & especially V and Cr catalysts

11.26

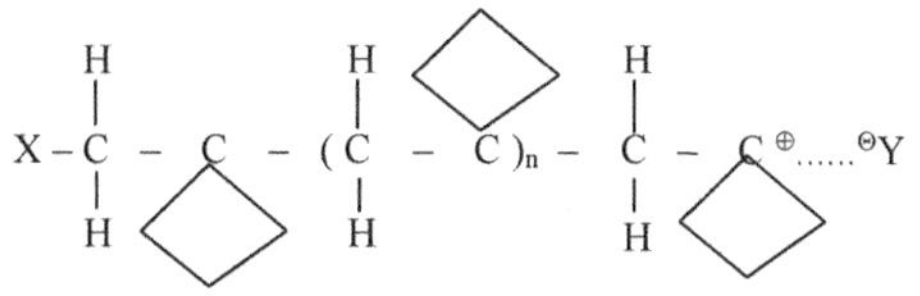

Chain with Cyclobutane structure – Obtained "Cationically"

11.27

One can imagine the different types of polymers produced from methylene cyclobutane an isomer of 2-methyl cyclobutene. When the types of initiators used and other operating conditions are properly put in place, very unique polymeric products as opposed to mixtures of two of them can readily be obtained. Most of the initiator are chargedly paired while few are radically paired or free. Not all the placements have been shown above. All the growing chains can be killed by Starvation.

Considering molecular rearrangement for the activated ringed cyclo-1-methyl-butene, the following is to be expected.

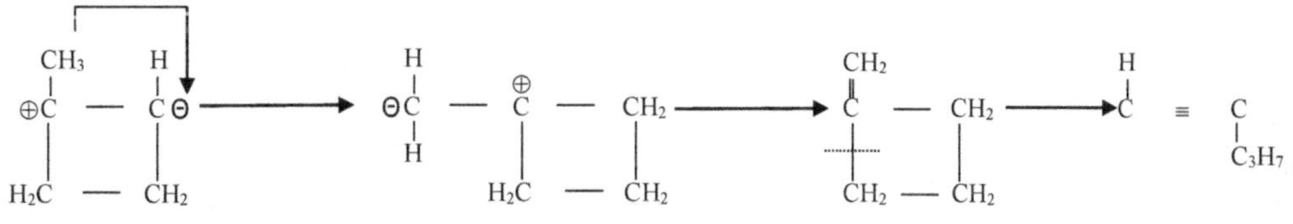

(I)   **Not favored**

11.28a

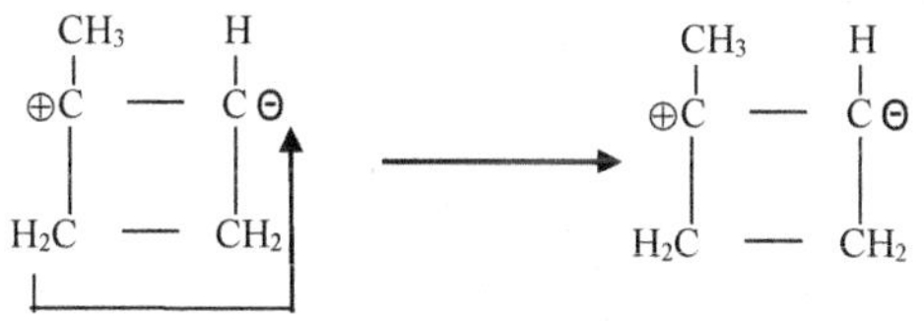

(II)   **Favored**

11.28a

Since substituent group in the ring is more radical-pushing than the CH_3 group, CH_3 group cannot therefore provide the transfer species. Hence methylene cyclobutane cannot be obtained from 1-methyl cyclobutene via rearrangement. It is (II) above that is favored. This is the case where it rearranges back to itself. When a larger group is used in place of CH_3, larger in fact than C_2H_5, since the group on each center of the ring is $-CH_2-CH_2-$ the followings are to be expected.

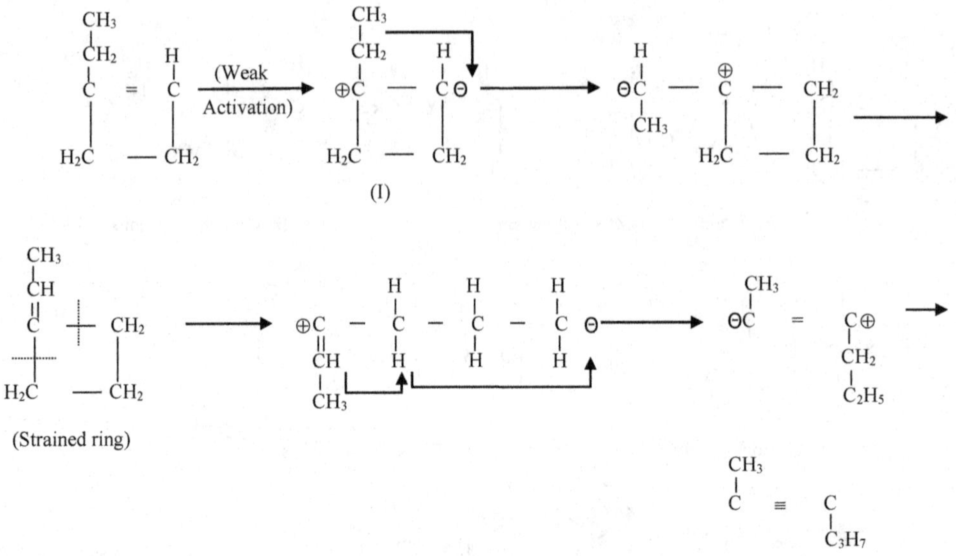

(I)

$$11.29$$

It is important to note that acetylenic monomer is obtained via molecular rearrangements of the first and third kinds. The last equation above unlike (I) of Equation 11.28a is favored. They have been used to find the limitations placed on making the rearrangement possible. It is only when R group is greater than or equal to C_2H_5 that molecular rearrangement from outside the ring commences and ends with R equals C_4H_9 as shown below.

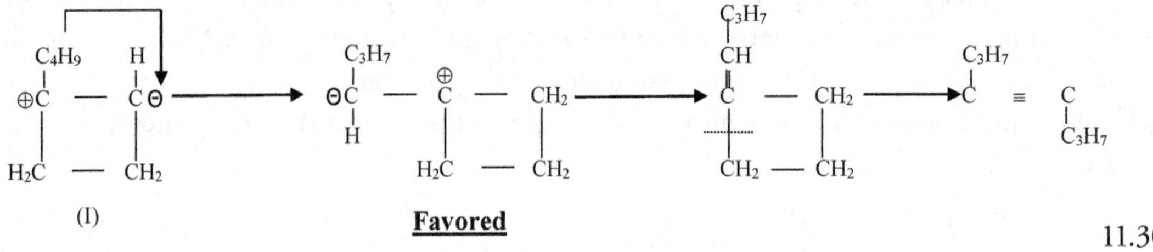

(I)   **Favored**

$$11.30a$$

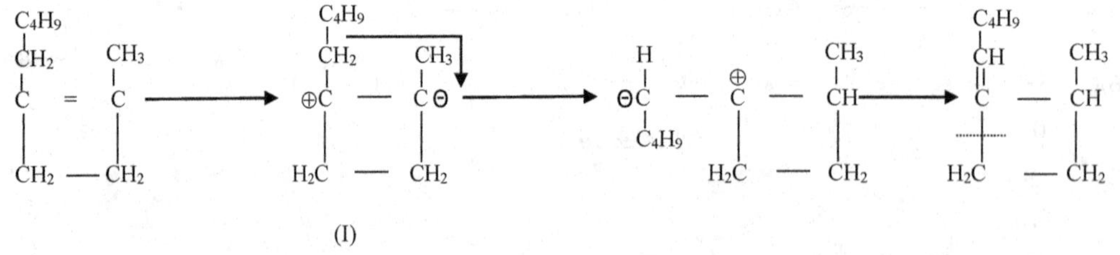

(I)

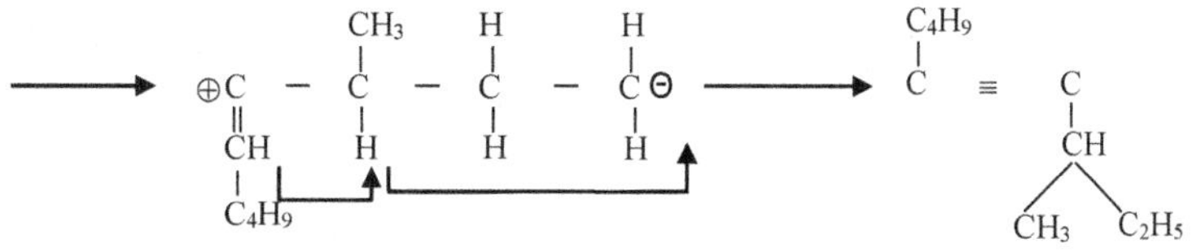

(I)   11.30b

Note that though the reactions above have been shown chargedly, they all indeed take place radically, because never can resonance stabilization phenonmena take place chargedly. All these are being shown chargedly exploratively.

With the cyclobutene shown below, this will undergo molecular rearrangement as shown below to give the same monomer. Because the 1,2,3-trimethylcyclobutene is strained, it can readily be decomposed to give only 2,3-Dimethyl pentadiene.[3]

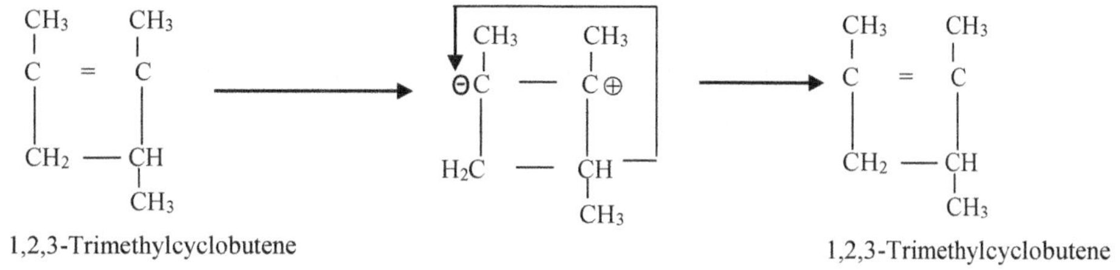

11.31 a

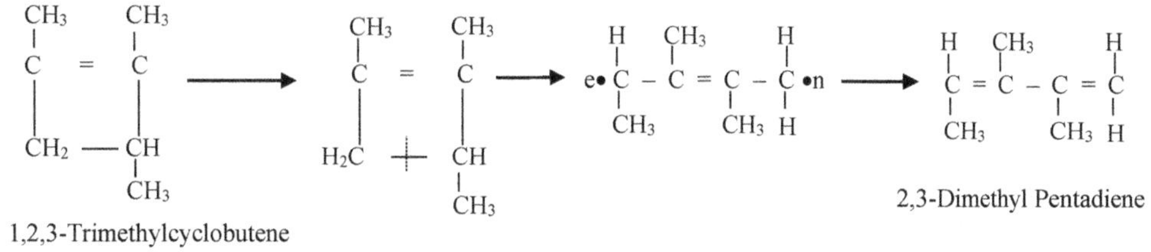

11.31b

Notice that in the first equation, transfer species is coming from a group in the ring, being of greater capacity than that outside the ring. Notice also the point of scission in the second equation and the charges or radicals carried by the centers. The ring can only be opened instantaneously radically and chargedly. Radically or chargedly, only the electro-free-radical or positively charged route is favored by the monomer after the ring is instantaneously opened. Before the double bond is activated and used, the ring is already opened via the only point of scission.

One can observe marked differences between cyclopropene which is always in Equilibrium state of existence and cyclobutene which is quite stable at STP. Based on the considerations above, there is no doubt that the following is valid.

ORDER OF RADICAL-PUSHING CAPACITY   11.32a

On the basis of the above considerations, one cannot on a one to one basis compare the two cases shown below. They belong to two very different families, though they both belong to the cycloalkenes family. That is one of the countless ways it is generally in our world with in particular Homo Sapience.

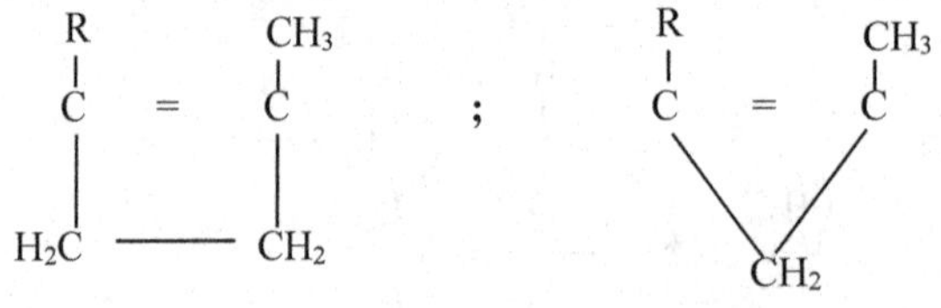

$$
\begin{array}{ccc}
\overset{\displaystyle R}{\underset{\displaystyle H_2C - CH_2}{\overset{|}{C}} = \overset{\overset{\displaystyle CH_3}{|}}{\underset{|}{C}}} & ; & \overset{\displaystyle R}{\underset{\displaystyle CH_2}{\overset{|}{C}} = \overset{\overset{\displaystyle CH_3}{|}}{C}}
\end{array}
$$

(I)   (II)   11.32b

While (I) will favor existence of acetylenic monomers only for $R \geq C_5H_{11}$, (II) will not favor such rearrangements. Yet it is closer to acetylene than cyclobutene (which gives more nucleophilic acetylene) is. When R is H, while (II) will decompose via Equilibrium mechanism to highly nucleo-philic acetylene, (I) cannot; yet it is more nucleophilic than methyl cyclopropene.

Coming back to (I) of Equation 11.29, the following should then be expected when transfer species of the tenth type, is involved. This is what to expect for the size of group and for the use of greater than/equal to sign ($\geq$) in Equation 11.32a. The same monomer is obtained. Like the case of Equation 11.31a, the ring here (i.e., (I) of Equation 11.29) can also be instantaneously opened to give a butadiene wherein, the C_2H_5 group is shielded. However, unlike the case of Equation 11.31a, both routes are favored when the ring is opened, if and only if R is less than C_2H_5 in capacity.

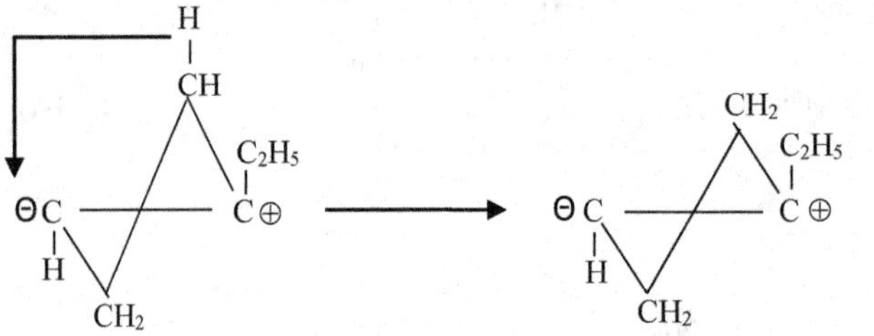

$$11.33$$

For the corresponding case of chloroprene, the followings are obtained.

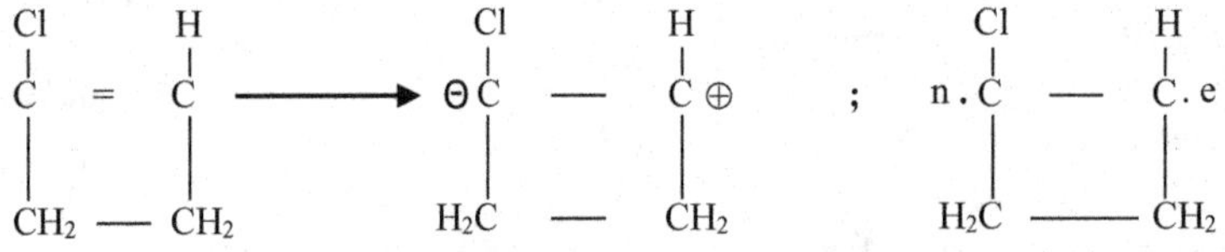

$$
\begin{array}{ccc}
\overset{\displaystyle Cl}{\underset{\displaystyle CH_2 - CH_2}{\overset{|}{C}} = \overset{\overset{\displaystyle H}{|}}{\underset{|}{C}}} \longrightarrow \overset{\displaystyle Cl}{\underset{\displaystyle H_2C - CH_2}{\overset{|}{\ominus C}} - \overset{\overset{\displaystyle H}{|}}{\underset{|}{C \oplus}}} & ; & \overset{\displaystyle Cl}{\underset{\displaystyle H_2C - CH_2}{n \cdot \overset{|}{C}} - \overset{\overset{\displaystyle H}{|}}{\underset{|}{C \cdot e}}}
\end{array}
$$

(I) (Existence not possible)   (II)   11.34

Radically, polymerization is favored only electro-free-radically, while chargedly the monomer cannot be activated due to electrostatic forces of repulsion. It can be opened free-radically as shown below. It can also be opened instantaneously chargedly, if the double bond is not activated.

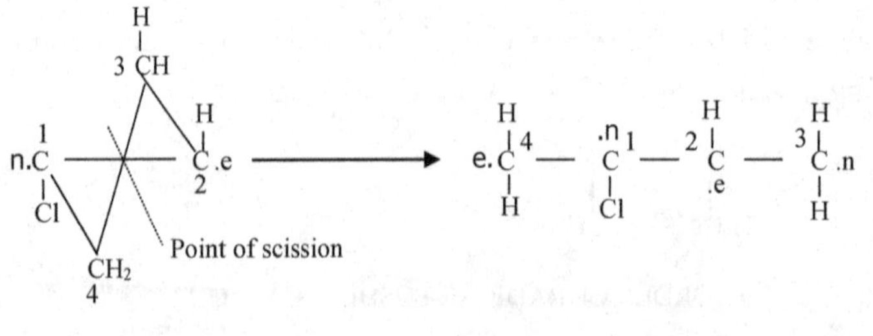

$$11.35$$

When the ring is instantaneously opened where possible, poly(chloroprene) can be observed to be produced. It can be polymerized both electro- and nucleo-free-radically, unlike when the ring is not opened. No molecular rearrangement can be favored by the monomer radically.

When the Cl is replaced with a radical-pushing halogenated substituent group, the followings are obvious.

$$R:^{\ominus} \;+\; \underset{\underset{H_2C \;-\; CH_2}{\big|}}{\overset{\overset{CF_3}{\big|}}{C}} = \underset{\big|}{\overset{\overset{H}{\big|}}{C}} \longrightarrow R:^{\ominus} \;+\; \underset{\underset{H_2C \;-\; CH_2}{\big|}}{\overset{\overset{H}{\big|}}{\oplus C}} - \underset{\big|}{\overset{\overset{CF_3}{\big|}}{C \ominus}} \longrightarrow$$

(Nucleophile)

$$RH \;+\; \underset{\underset{\overset{C}{\big|}}{\big|}}{\overset{\overset{H}{\big|}}{\ominus C}} - CH_2$$

11.36a

$$R^{\oplus} \;+\; \underset{\underset{H_2C \;-\; CH_2}{\big|}}{\overset{\overset{CF_3}{\big|}}{\ominus C}} - \underset{\big|}{\overset{\overset{H}{\big|}}{C\oplus}} \longrightarrow RF \;+\; \underset{\underset{H_2C \;-\; CH_2}{\big|}}{\overset{\overset{F}{\big|}}{\underset{F}{C}}} = C - \overset{\overset{H}{\big|}}{C\oplus}$$

(weak)

11.36b

Chargedly, no route is favored. The same applies free-radically when not opened. It will be more difficult to open the ring here than the last one free-radically. However, when opened, both routes will be favored free-radically. Molecular rearrangement can never be favored by it no matter the size of CF_3 type of group.

With COOR, $CONH_2$ types of groups with transfer species, (but not COR or CN types), we now have electrophiles whose rings will be very difficult to open, because they are less strained. They will not favor any route. They can be made to undergo molecular rearrangement to give a ketonic ring which when opened cannot readily rearrange to produce a stable product. With COR and CN types of groups with no transfer species, their rings cannot be opened and no molecular rearrangement can take place. For these cases with no transfer species via the less nucleophilic center (C=C), the CN or CO centers can readily be used only electro-free-radically or positively but with steric limitations if it is to be used as a monomer.

Now considering vinylcyclobutenes, one will begin with the first member.

$$\overset{1}{C}H_2 = \overset{2}{C}H - \overset{3}{C} = \overset{4}{C} \longrightarrow n.\overset{1}{C} - \overset{2}{C}.e \longleftrightarrow \longrightarrow$$

$$n.\overset{1}{C} - C = C - \overset{4}{C}.e \longleftrightarrow \longrightarrow n.\overset{3}{C} - \overset{4}{C}.e$$

NOT A MONO-FORM

11.37

It can be observed that the monomer is resonance stabilized only radically with the following valid relationships, noting that the resonance stabilization provider is the ring, being more nucleophilic and the last mono-form above can only exist as an artificial mono-form, i.e., cannot exist.

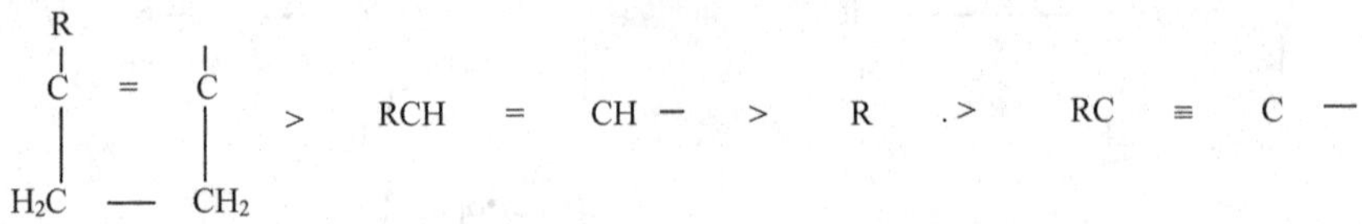

$$
\begin{array}{c}
\overset{R}{\underset{|}{\overset{||}{C}}} = \overset{|}{\underset{|}{C}} \\
\underset{H_2C}{|} \;-\; \underset{CH_2}{|}
\end{array}
\;>\; RCH = CH- \;>\; R \cdot \;>\; RC \equiv C-
$$

Order of radical-pushing capacity of resonance stabilization groups　　　11.38

The mono-forms of Equation 11.37 can be observed to favor only the electro-free-radical route. Its polymerization will be limited with propagation due to steric limitations, unless when syndiotatically placed. 1, 4- mono-form will only exist along the chain, being the more stable mono-form of the two mono-forms, unless when the initiator is weak where the 1,2- mono-form will largely exist from close to the beginning to the middle of the chain. In the last chapter (See Rule 683), its decomposition to give $CH_3CH=CH-C\equiv CCH_3$ was shown to be possible via molecular rearrange-ments because the capacity of H in the ring is not greater than the capacity of H externally located to the ring unlike with cyclopropene. It's decomposition to give $H_2C=CH-C\equiv CC_2H_5$ is also possible only if it can be made to exist in Equilibrium state of existence.

When the H on the C=C activation center on the ring in the presence of the resonance stabilization group is replaced with CH_3 and higher groups, no molecular rearrangements are favored, unlike the corresponding case of cyclopropene.

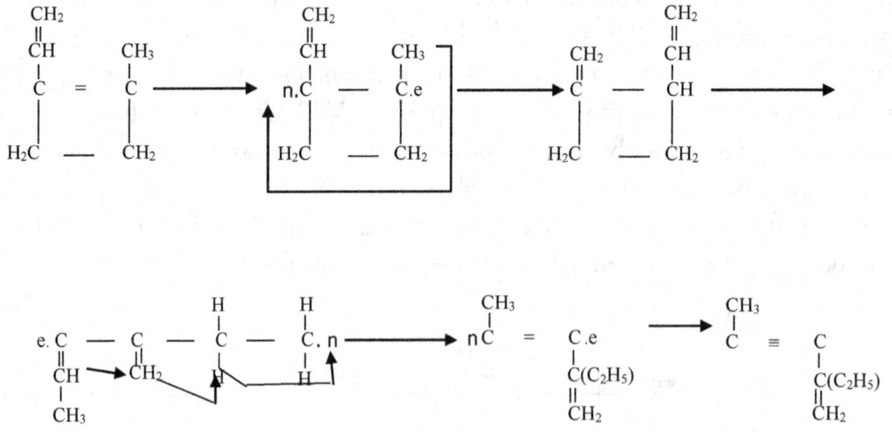

(I) Impossible State　　　11. 39

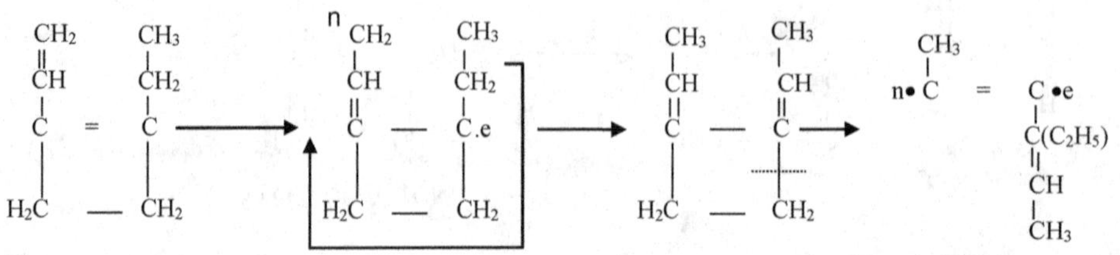

(II) Impossible existence　　11.40a

206

When molecular rearrangement of the third kind was allowed to take place, nothing could be obtained. This does not mean that the original monomer is not resonance stabilized. In the first equation, the wrong mono-form was used, i.e., the more nucleophilic center. The same product will however still be obtained.

When however, the 1-ethyl-2 vinylcyclobutene of the last equation above is heated above 100^0C, it was reported to give 2 ethyl-3-vinylbuta-1,3-diene[3]. This can be explained as follows.

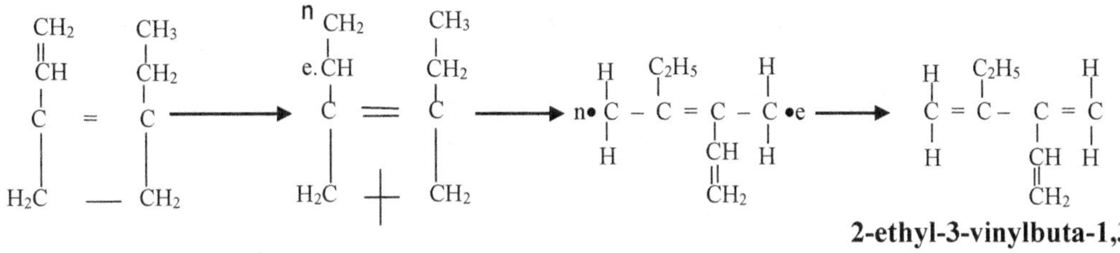

2-ethyl-3-vinylbuta-1,3-diene

11.40b

Indeed, the activation in the first step above takes place, since the vinyl center is least nucleophilic.

It is a one stage Decomposition mechanism system in which the ring is never activated. Chargedly, the last product cannot be obtained, but the activated product can only be used positively. Worthy of note is the type of unique product obtained – a double–headed resonance stabilized diene, observed to favor both routes. Downstream, the product above will be recalled when Resonance stabilization phenomena are to be fully classified. Also note that, what were not shielded before ring opening are now shielded when the ring is opened. Even the alkene group is shielded. It was also only the ring that broke down to a diene. *Invariably, what these observations imply is that the ring has remained the resonance stabilization provider both before and after opening of the ring.* The ring with radical-pushing groups either on the double bond or inside the ring, is strained enough to favor its instantaneous opening. Note that the center activated above is the less nucleocleophilic center. The same will apply to the first case above with CH_3 group and even the first member.

All the types shown below will favor decomposition via molecular rearrangements when $R^3 R^1$. These are cases carrying radical-pushing groups at the terminals. As it seems, these will not readily favor instantaneous opening of their rings to give double-headed dienes.

11.41

When R^1 and R equal to CH_3, the followings are obtained *when the centers are wrongly activated.*

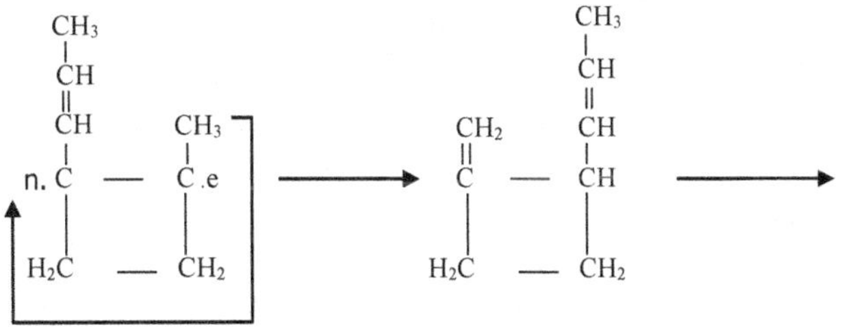

Resonance stabilized

$$\text{e. C} - \text{C} - \overset{\overset{\textstyle H}{|}}{\text{C}} - \overset{\overset{\textstyle H}{|}}{\text{C.n}} \longrightarrow \text{n. C} = \text{C .e}$$

(with substituents: $\overset{||}{\text{CH}} \rightarrow \overset{||}{\text{CH}_2}$, C_2H_5; $C(C_2H_5)$, CH_2)

Impossible Existence

11.42a

It should be noted that the CH_3 group on the ring is not alone by itself as the only radical-pushing group on the C center. Even when the CH_3 group on the ring side is changed to C_4H_9, molecular rearrangement of the third kind is never favored (i.e. for $R^1 > R$). For these cases, they can readily be opened instantaneously to give double-headed dienes. For example, when the last case above is opened instantaneously, the followings are obtained.

Resonance stabilized

Instantaneous opening is not possible

Resonance stabilized

11.42b

With instantaneous opening of the ring, only the ring is broken down to a diene like the others. Also like the others, the former alkene group is now the resonance stabilization provider instead of the ring, unlike the case where R^1 is C_2H_5 and R is H shown in Equation 11.40b. *In general as it seems, when R^1 is less than or equal to R, the rings cannot be opened instantaneously, but can be opened via molecular rearrangement.* When R^1 is greater than R, the ring can be opened instantaneously and rearrangement cannot take place. The case above rearranges to a vinyl acetylene ($H_2C=CH-C\equiv C-CH(CH_3)(C_2H_5)$), but cannot be opened instantaneously.

When only the alkene group is allowed to carry radical-pushing groups, the followings are to be expected.

11.43a

11.43b

Molecular rearrangement ceases when R group on the alkene side becomes greater than C_4H_9. The products obtained are vinyl acetylenes which are strongly Nucleophilic. The component now left to be shielded becomes H. For as long as R is greater than R^1, molecular rearrangement will always take place with limitations.

When opened instantaneously, the followings are obtained.

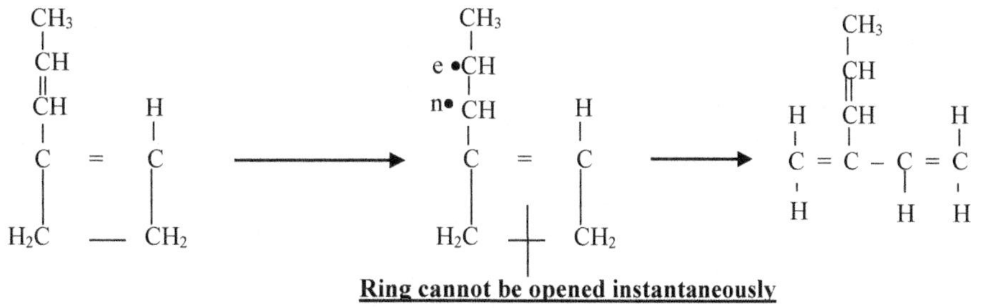

Ring cannot be opened instantaneously   11.43c

acetelinic cyclobutene

Though the vinyl group still remains not involved in the decomposition, it is now the resonance stabilization provider in the butadiene, but not the ring. Hence, it is believed that for such cases as has been said above, their rings cannot be opened instantaneously, since the ring is no-longer the resonance stabilization provider. The invisible π-bond in the ring now becomes the least nucleophilic center via the group which was originally resonance stabilized, while the group which was not resonance stabilized when the ring was not opened is now resonance stabilized after opening of the ring. One can observe how NATURE operates. It is too much to comprehend.

With acetylenic resonance stabilization groups replacing alkenic groups, the followings are to be expected for the first member when suppressed.

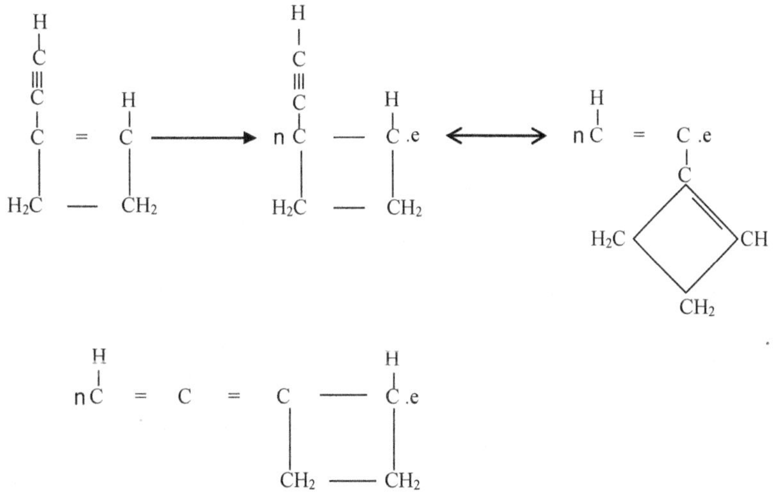

11.44

The monomer is resonance stabilized, favoring only the electro-free-radical route. Note that the second case above is not one of the mono-forms, with the acetylene being the resonance stabilization provider. From the considerations so far, there is no doubt that the following relationships are valid.

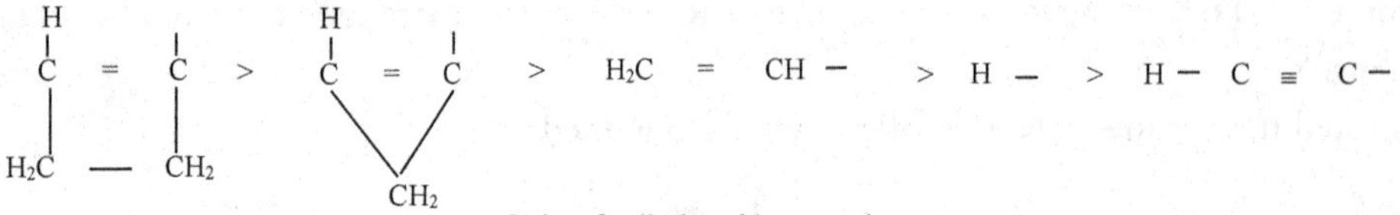

$$\begin{array}{c} H \\ | \\ C \\ | \end{array} = \begin{array}{c} | \\ C \\ | \end{array} \quad > \quad \begin{array}{c} H \\ | \\ C \\ \end{array} = \begin{array}{c} | \\ C \\ \end{array} \quad > \quad H_2C = CH - \quad > \quad H - \quad > \quad H - C \equiv C -$$

Order of radical-pushing capacity   11.45a

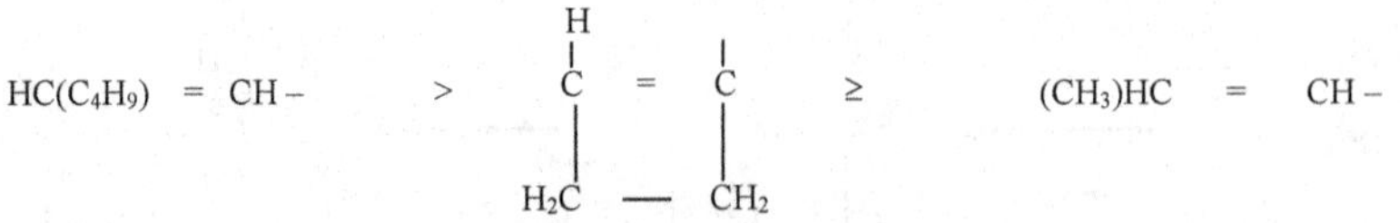

$$HC(C_4H_9) = CH - \quad > \quad \begin{array}{c} H \\ | \\ C \\ | \end{array} = \begin{array}{c} | \\ C \\ | \end{array} \quad \geq \quad (CH_3)HC = CH -$$

Transition point for Molecular rearrangement of the 3rd kind   11.45b

Replacing the externally located H atoms one at a time, the followings are obtained.

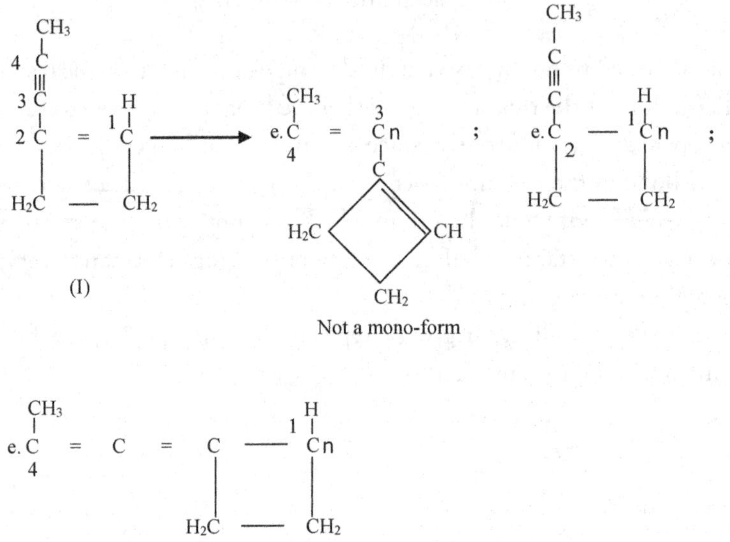

(I)

Not a mono-form

Resonance stabilized   11.46

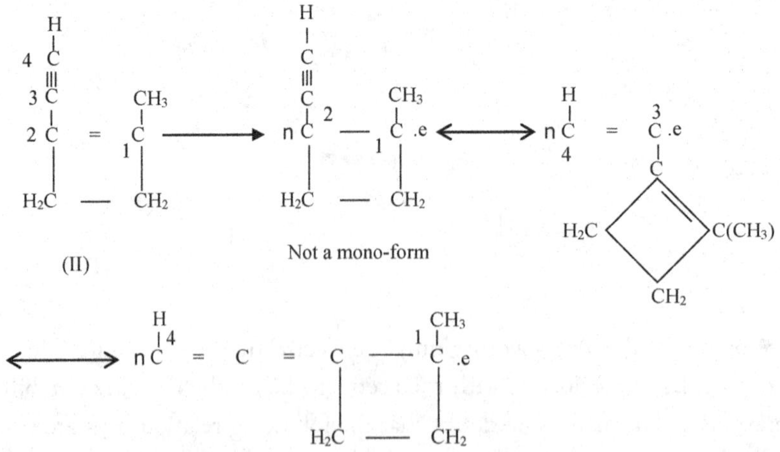

(II)

Not a mono-form

Resonance stabilized   11.47

210

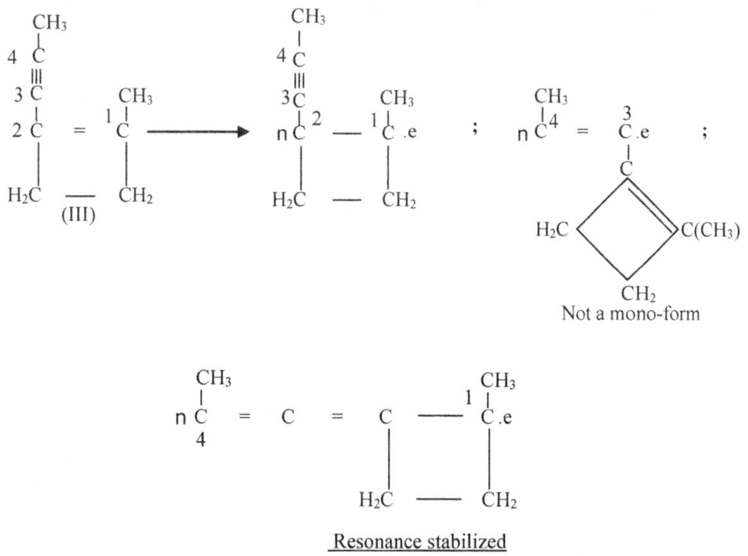

$$\begin{array}{c}
CH_3 \\
| \\
4\ C \\
||| \\
3\ C \\
| \quad CH_3 \\
2\ C = \overset{1}{C} \\
| \quad | \\
H_2C — CH_2 \\
(III)
\end{array}
\longrightarrow
\begin{array}{c}
CH_3 \\
| \\
4\ C \\
||| \\
3\ C \\
n\ C\overset{2}{} — \overset{1}{C}.e \\
| \quad | \\
H_2C — CH_2
\end{array}
\ ;\
\begin{array}{c}
CH_3 \\
n\ \overset{4}{C} = \overset{3}{C}.e
\end{array}
\ ;$$

Not a mono-form

$$n\ \overset{CH_3}{\underset{4}{C}} = C = C — \overset{CH_3}{\underset{}{\overset{1}{C}}.e \\
| \\
H_2C — CH_2$$

<u>Resonance stabilized</u>   11.48

For all of them, the radicals have been well placed. (I), (II) and (III) can be made to undergo molecular rearrangements, to give more nucleophilc monomers. With (I), the followings are obtained.

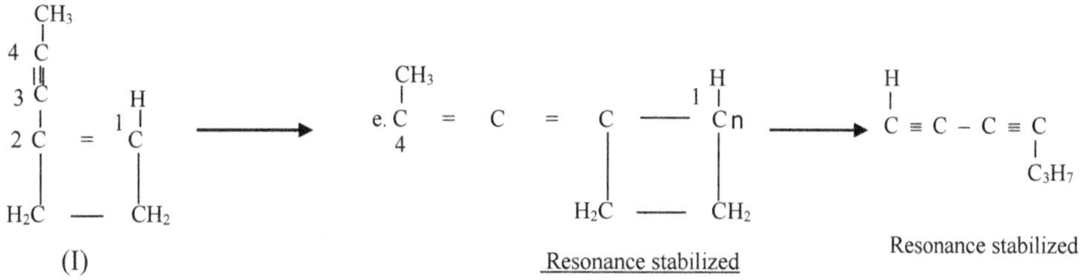

$$\begin{array}{c}
CH_3 \\
4\ C \\
|| \\
3\ C \quad H \\
2\ C = \overset{1}{C} \\
| \quad | \\
H_2C — CH_2 \\
(I)
\end{array}
\longrightarrow
\begin{array}{c}
CH_3 \\
e.\ C = C = C — \overset{H}{\overset{1}{C}}n \\
4 \\
| \\
H_2C — CH_2
\end{array}
\longrightarrow
\begin{array}{c}
H \\
| \\
C \equiv C — C \equiv C \\
| \\
C_3H_7
\end{array}$$

(I)   <u>Resonance stabilized</u>   Resonance stabilized   11.49a

The monomer here can be observed to be resonance stabilized. With (II), the situation is different as shown below.

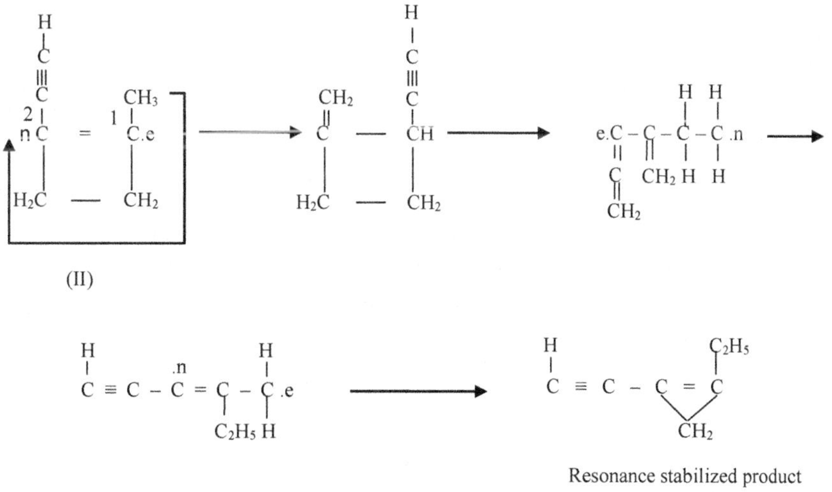

$$\begin{array}{c}
H \\
C \\
||| \\
C \\
2\ | \quad CH_3 \\
n\ C = \overset{1}{C}.e \\
| \quad | \\
H_2C — CH_2 \\
(II)
\end{array}
\longrightarrow
\begin{array}{c}
CH_2 \\
|| \\
C \\
| \\
CH \\
| \\
H_2C — CH_2
\end{array}
\longrightarrow
\begin{array}{c}
H \quad H \\
e.C — C — C — C.n \\
|| \quad || \quad | \quad | \\
C \quad CH_2\ H\ H \\
|| \\
CH_2
\end{array}
\longrightarrow$$

$$\begin{array}{c}
H \quad\quad H \\
| \quad .n \quad | \\
C \equiv C — C = C — C.e \\
| \quad | \\
C_2H_5\ H
\end{array}
\longrightarrow
\begin{array}{c}
H \quad\quad C_2H_5 \\
| \quad\quad | \\
C \equiv C — C = C \\
| \\
CH_2
\end{array}$$

Resonance stabilized product   11.49b

Unlike the case with ethene, molecular rearrangement of the third kind is favored here to give a very different type of product, wherein the four membered ring was reduced to a three membered ring, bringing the acetylene group back as if it was never involved. Note that the center activated above is the wrong center, being the most nucleophilic center. However, the same product obtained when the right centers are used, still remains obtained. The same as above will also apply to (III). When the CH_3 group in (I) is replaced with higher groups, rearrangements remains favored with limitations.

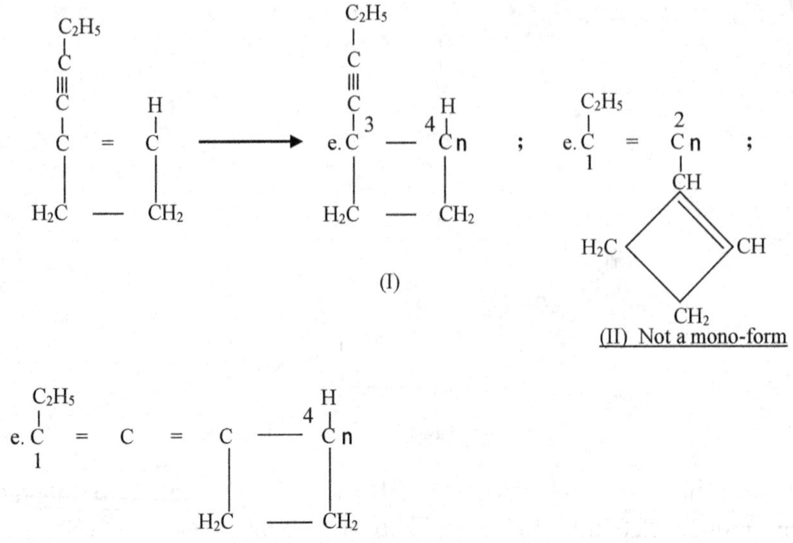

(I)

(II) Not a mono-form

(III)

11.50

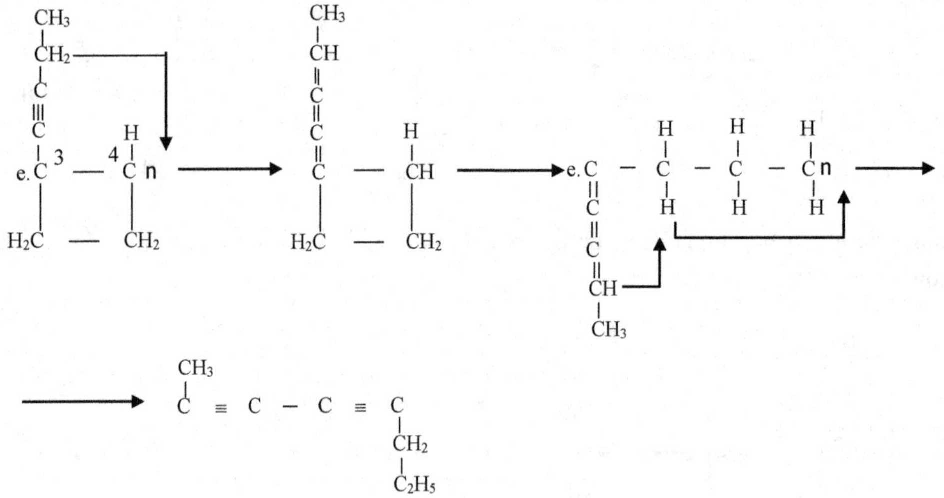

Resonance stabilized

11.51a

Like (I) of Equation 11.49a, all the mono-forms can molecularly rearrange, since the laws are fully obeyed. The last reaction above has been used to illustrate the limit to which R group can get to, for which the followings are valid.

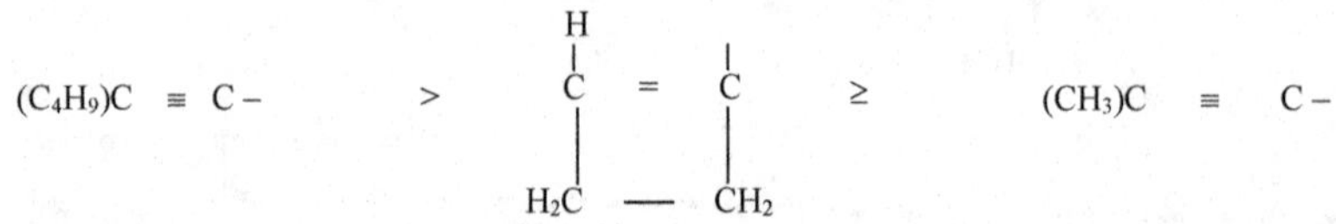

$$(C_4H_9)C \equiv C - \quad > \quad \underset{H_2C - CH_2}{\overset{H}{C}} = C \quad \geq \quad (CH_3)C \equiv C -$$

Transition point for Rearrangement for 3rd kind

11.51b

One can observe that the alkenes and acetylenes have identical transition point for rearrangement. They differ only in terms of the products obtained.

When R is less than R^1 in capacity, rearrangement is not possible as shown below.

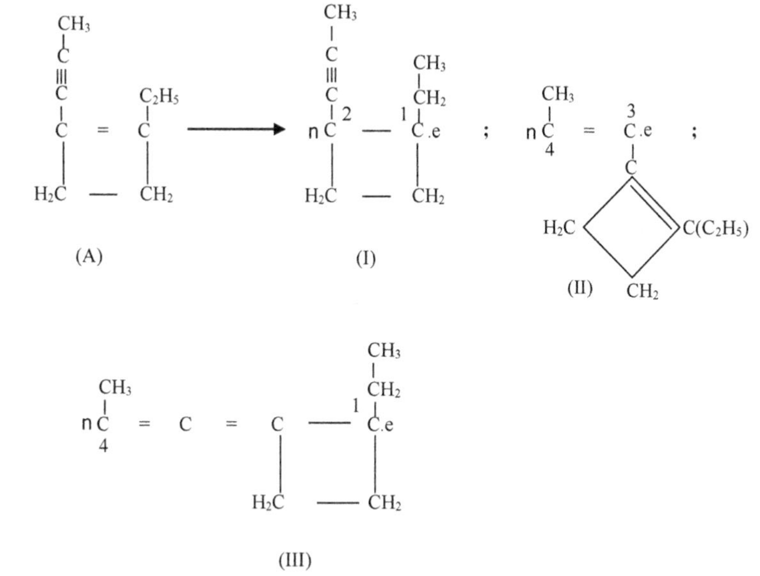

CH₃
|
C
|||
C
|
C = C — C.e   ;   nC = C.e   ;
| | | |
H₂C — CH₂   CH₃

(A)   (I)   (II)

11.52

(III)

When R is greater than R^1, the followings are obtained.

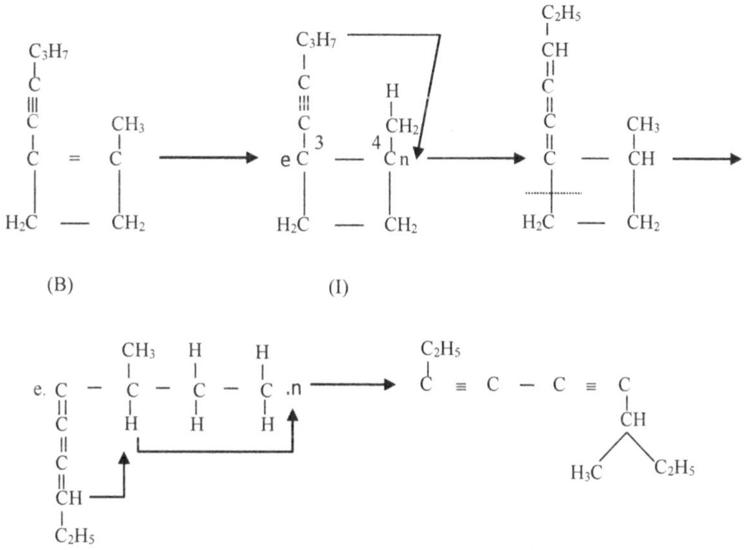

11.53

(B)   (I)

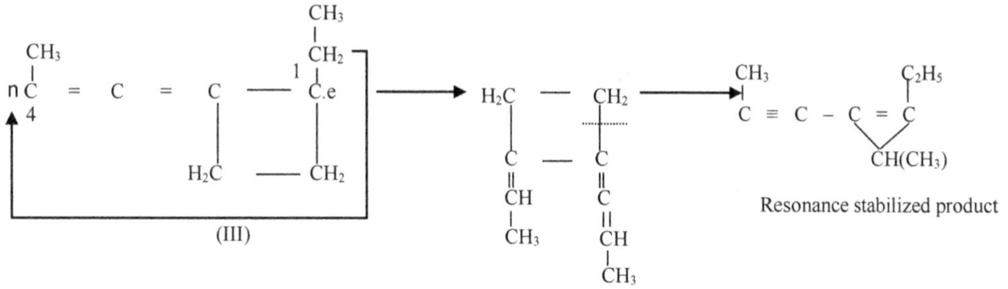

Resonance stabilized product

11.54

213

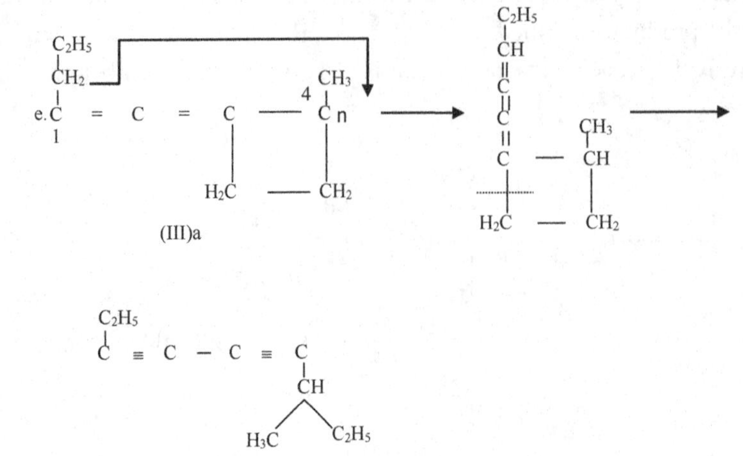

$$
\begin{array}{c}
C_2H_5 \\
| \\
CH_2 \\
| \\
e.C = C = C - \overset{4}{\underset{|}{C}}n \\
1
\end{array}
\quad
\begin{array}{c}
CH_3 \\
\end{array}
\longrightarrow
\qquad \longrightarrow
$$

(III)a

$$
\begin{array}{c}
C_2H_5 \\
| \\
C \equiv C - C \equiv C \\
| \\
CH \\
H_3C \qquad C_2H_5
\end{array}
\qquad\qquad 11.55
$$

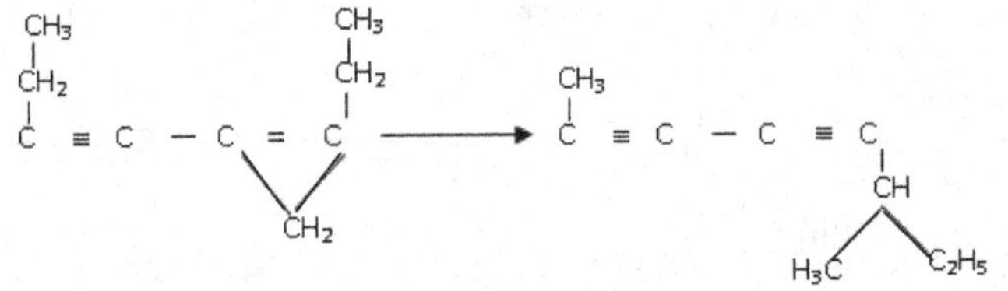

$$
\begin{array}{c}
CH_3 \qquad\qquad CH_3 \\
| \qquad\qquad\quad | \\
CH_2 \qquad\qquad CH_2 \\
| \qquad\qquad\quad | \\
C \equiv C - C = C \\
| \\
CH_2
\end{array}
\longrightarrow
\begin{array}{c}
CH_3 \\
| \\
C \equiv C - C \equiv C \\
| \\
CH \\
H_3C \qquad C_2H_5
\end{array}
\qquad 11.56
$$

Based on the last two equations above, one can observe the order of nucleophilicity between the three-membered and four-membered rings and how same product can be obtained from these logical network systems.

Imagine if the radicals are not properly placed, then in place of what was shown in Equation 11.48, would have been the followings-

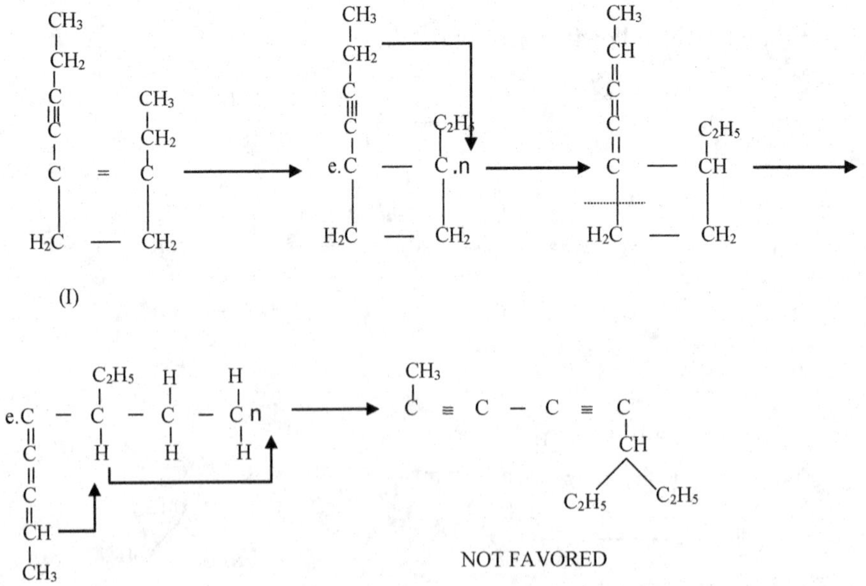

$$
\begin{array}{ccc}
CH_3 & CH_3 & CH_3 \\
| & | & | \\
CH_2 & CH_2 & CH \\
| & | & || \\
C & C & C \\
||| & ||| & || \\
C & C & C \\
| & | & | \\
C = C & e.C - C.n & C - CH \\
| & | & | \qquad | \\
H_2C - CH_2 & H_2C - CH_2 & H_2C - CH_2
\end{array}
\longrightarrow
$$

(I)

$$
\begin{array}{c}
C_2H_5 \quad H \quad H \\
| \qquad | \quad | \\
e.C - C - C - Cn \\
|| \quad | \quad | \quad | \\
C \quad H \quad H \quad H \\
|| \\
C \\
|| \\
CH \\
| \\
CH_3
\end{array}
\longrightarrow
\begin{array}{c}
CH_3 \\
| \\
C \equiv C - C \equiv C \\
| \\
CH \\
C_2H_5 \qquad C_2H_5
\end{array}
$$

NOT FAVORED

11.57

Instead of a resonance stabilized acetylene cyclopropene compound similar to that from Equation 11.53, what is shown above is obtained as will shortly be shown.

Even when the ring is instantaneously opened, the radicals shown above are not carried by the active centers. Shown below is a case where the radicals are ***now well placed,*** for which the followings are indeed obtained.

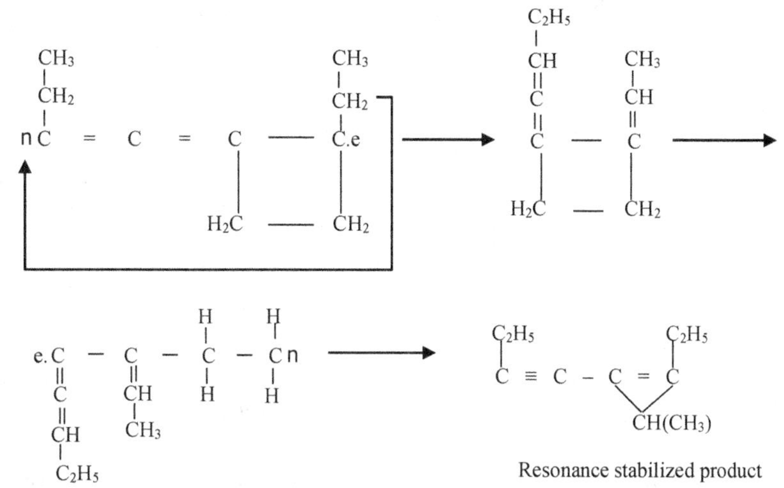

Resonance stabilized product   11.58

When the acetylenic monomer obtained here (Equation 11.58) is compared with that of Equation 11.56, it is unquestionable that cyclobutenes are more nucleophilic than cyclopropenes when resonance stabilized with acetylene.

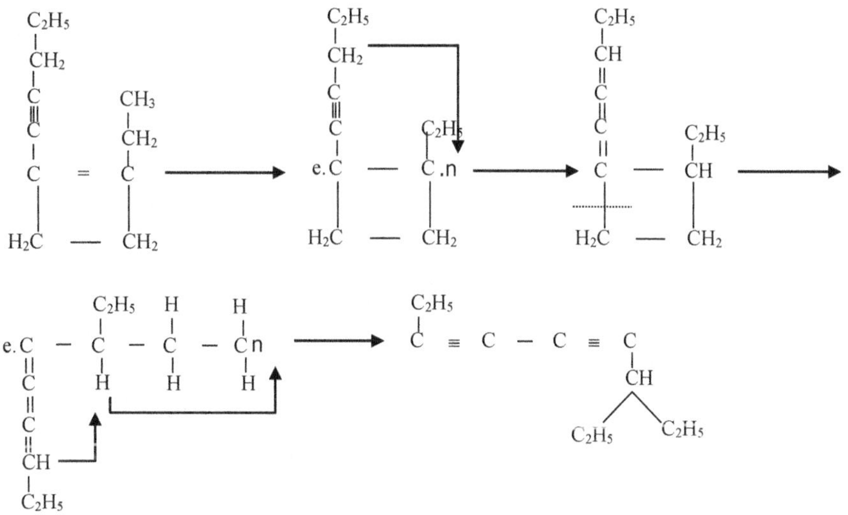

11.59

What we have been seeing so far are laws of rearrangements and progressions in Mathematics. One can observe how closely related cyclobutenes can be to acetylenes despite the reduced level of unsaturation. ***In order words though increased or decreased level of unsaturation has a part to play with respect to the nucleophilicity of a monomer, it has more to do with its favored existence.***

Having to some extent considered alkylane groups, there is need to look at other radical-pushing groups. But before doing that, it is important to note that one has not considered the instantaneous opening of the rings which carry acetylenes conjugatedly placed as was done with ethenes and others. Can the ring in the last equation above not be opened instantaneously when attempted, since one point of scission exists in the ring and the ring is carrying more radical-pushing groups than the case in Equation 11.40b? As it seems, such rings when resonance stabilized from outside or inside, can never be opened instantaneously, because acetylene will always be the resonance stabilization provider when the ring is opened. The acetylene carrying a radical-pushing group greater than C_2H_5, is naturally the resonance stabilization provider when the ring is carrying H being more nucleophilic than the cyclobutene. It has been reported that 1-phenyl-1,4-dichloro-1-butene is converted upon reaction with Mg in ether, into a complex mixture containing 1-phenylcyclobutene (A), trans-1-chloro-1-phenyl-1-butene (B), trans-1-phenyl-1-butene (C), cis-1-phenyl-1-butene (D), and 1-phenyl-1,3-butadiene (E)[4].

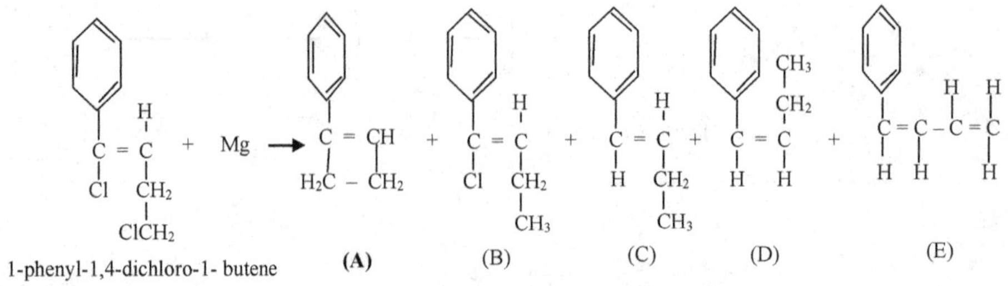

1-phenyl-1,4-dichloro-1- butene   (A)   (B)   (C)   (D)   (E)   11.60

Obviously, this is a series/parallel Equilibrium reaction mechanism system in which a fraction of the reactant is in Equilibrium state of existence, while the remaining fraction is in Stable state of existence. Hydrolysis were definitely involved, because Grignard reagents were formed in some of the stages. It is a very unique reaction with many stages, the mechanisms of which will be provided downstream. However, it should be noted that none of the other products can be obtained from (A) via opening of its ring. If the ring had been opened instantaneously, the product shown below would have been obtained.

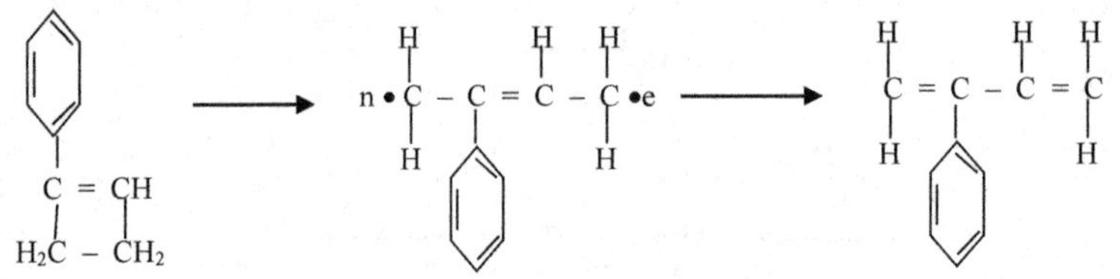

11.61

This is a case where the resonance stabilization provider is the benzene ring when not opened, and when opened the benzene ring remains the resonance stabilization provider. This is different from the condition already identified for opening of the ring. It is indeed the reverse in that if a ring is to be opened, the resonance stabilization provider should always be the ring before and after opening.

A situation where such ring was reportedly opened was when the ring was carrying radical-pulling groups such C°N, a resonance stabilization group and a metallic complex as shown below[5].

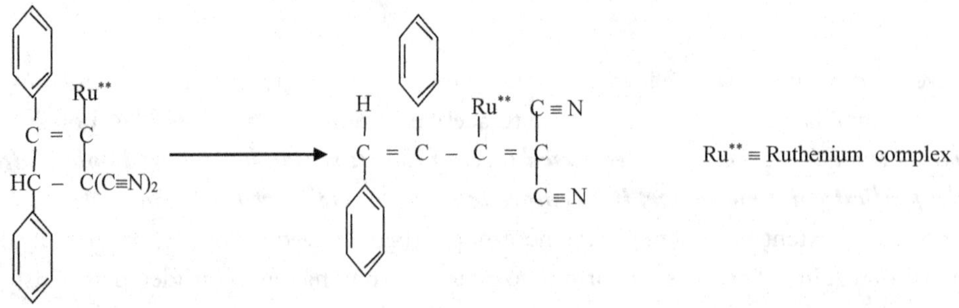

$Ru^{**} \equiv$ Ruthenium complex

11.62

The ring above was opened instantaneously, because (apart from the reduced strain energy of the ring provided by the presence of two very strong radical-pulling groups $N \equiv C -$), increased strain energy provided by the presence of a transition metallic center on the double bond and most importantly the presence of benzene ring in place of H inside the ring made it sufficient enough to make the ring less nucleophilic than the benzene ring conjugatedly placed to it. Notice that, it is that benzene ring that replaced H in CH_2 group that is now the resonance stabilization provider. Without that benzene ring, the four-membered ring can never be opened. It is only when the ring is the resonance stabilization provider, that the ring can be instantaneously opened under certain conditions. On the basis of all the considerations so far for these members of families-Cycloalkenes, di-alkenes and alkynes, the followings are valid, using cyclobutene.

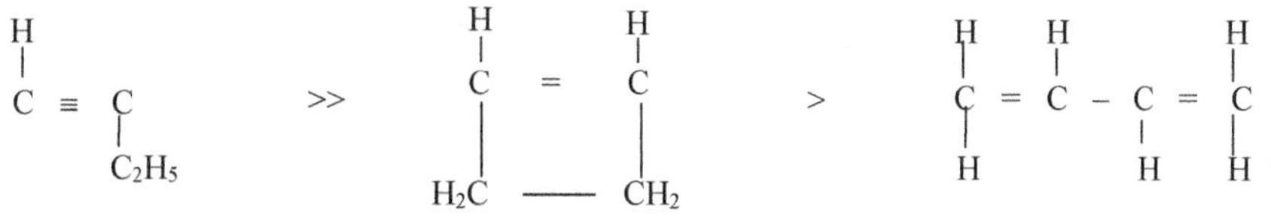

$$
\begin{array}{ccc}
\underset{\underset{C_2H_5}{|}}{\overset{\overset{H}{|}}{C}} \equiv C & >> & \text{(vinyl cyclobutene structure)} & > & \text{(diene structure)}
\end{array}
$$

ORDER OF NUCLEOPHILICITY   11.63

This very important equation above sends so many messages. With vinyl cyclobutene, the resonance stabilization provider is the ring, while with acetylenic cyclobutene, the provider is acetylene. From the equation above, one can clearly see the equivalence of alkenes and alkynes for cyclopropene, cyclobutene, cyclopentene, cyclohexene and so on. It is not for all cases where the nucleophilicity of the ring is greater than the nucleophilicity of the alkene or alkyne that will be opened instantaneous- ly. When the nucleophilicity of the alkene or alkyne in the compound is greater than the nucleophilicity of ring, then the ring will not be opened. For example, the ring shown in Equation 11.59 cannot be opened as shown below, because the ring is less nucleophilic than the acetylene

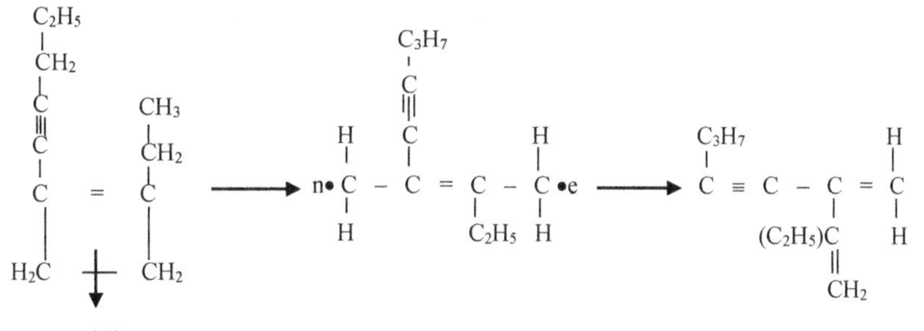

(A)   **NOT FAVORED**   (B)   11.64

The product formed is (B), that in which the resonance stabilization provider is the acetylene.

With OH group in cyclobutene, the followings are obtained.

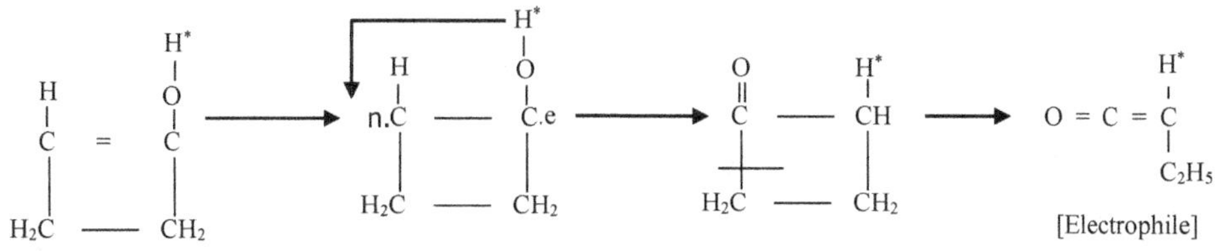

(C) (strained)   11.65

(C) can be made to fully undergo molecular rearrangement of the third kind as already stated into law. One should recall the case of vinyl alcohol where Molecular rearrangement and Enolization are reversible depending on the operating conditions. ***With this monomer, only molecular rearrange-ment of the first kind of the sixth type takes place based on Rule 332, since the OH group like some alkylane groups is more radical-pushing than the group carried by the ring.*** After the first rearrangement, in the presence of heat, it decomposes to Ethyl ketene, When the OH group is replaced with HO–C≡C – group, the same (C) type of compound is obtained.

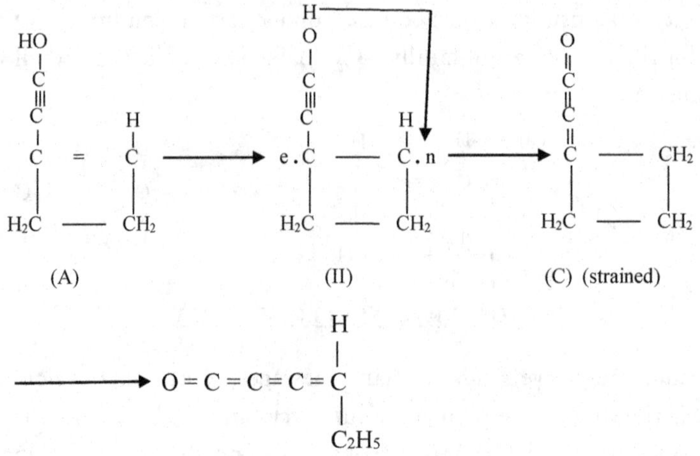

$$O = C = C = C = \underset{\underset{C_2H_5}{|}}{\overset{\overset{H}{|}}{C}}$$

[Electrophile]   11.66

Worthy of note is the fact that the two nucleophilic monomers above were converted or rearranged to give electrophiles. Recalled below are the followings-

$$O = \quad ; \quad O = C = C = \quad ; \quad O = C = C = C = C =$$

Radical - pushing groups [EVEN]

$$O = C = \quad ; \quad O = C = C = C = \quad ; \quad O = C = C = C = C = C =$$

Radical – pulling groups [ODD]   11.67

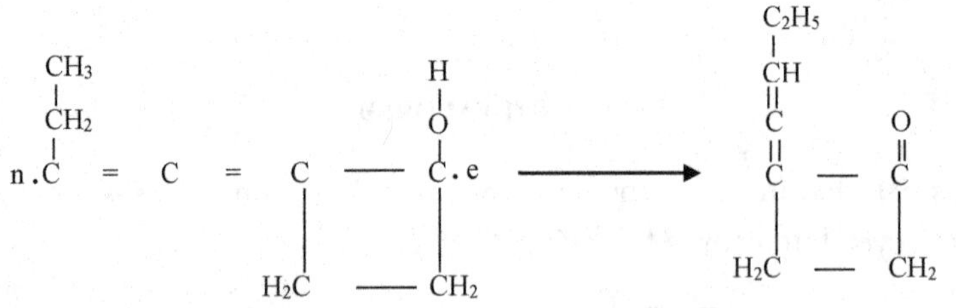

11.68

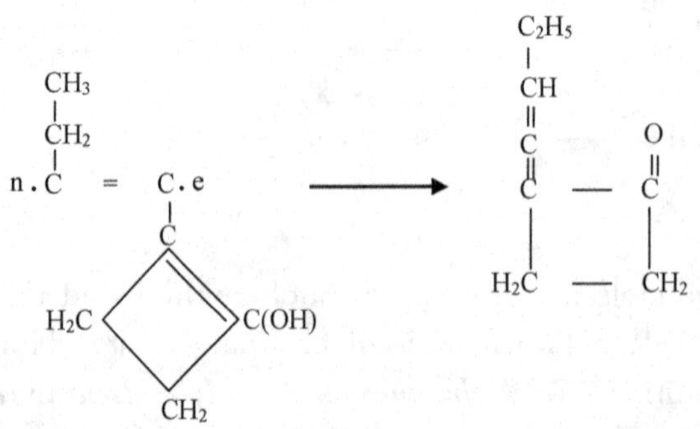

11.69

The last two reactions are favored and the products are ketones. They cannot undergo molecular rearrangement of the third kind to produce ketene, because the transfer species will no longer be the same as against the Laws of Conservation of transfer of transfer species. The transfer species above are of the first kind of the sixth type.

With NH_2, NHR, NR_2 types of groups, one should expect the followings with the groups placed on the acetylene center.

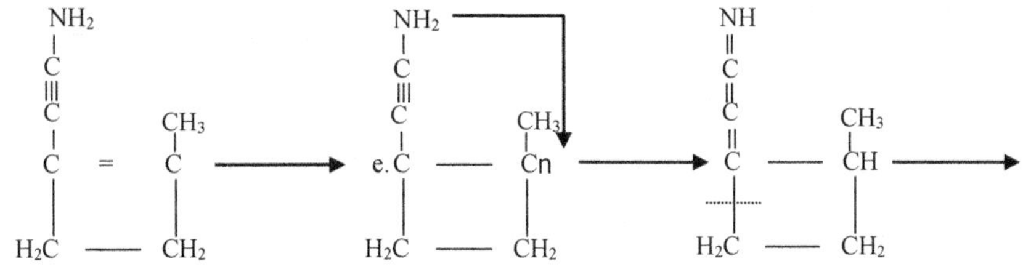

$$
\begin{array}{ccc}
\overset{NH_2}{\underset{\underset{\underset{H_2C \,-\, CH_2}{|}}{\overset{|}{C}}}{\underset{\|}{\overset{\|}{C}}}} = \overset{CH_3}{\underset{\underset{}{|}}{C}} & \longrightarrow & e.C \cdots \\
\end{array}
$$

(A) (Resonance stabilized) - <u>Nucleophile</u>

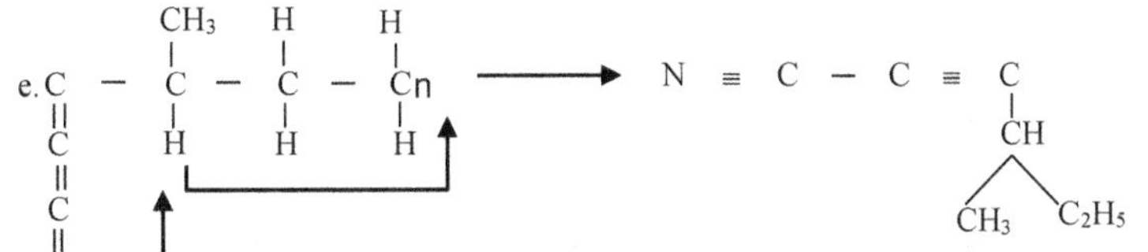

$$
e.C \;-\; \overset{CH_3}{\underset{H}{C}} \;-\; \overset{H}{\underset{H}{C}} \;-\; \overset{H}{\underset{H}{Cn}} \longrightarrow N \equiv C \;-\; C \equiv C
$$

(Non-resonance stabilized)
<u>Electrophile</u>   11.70

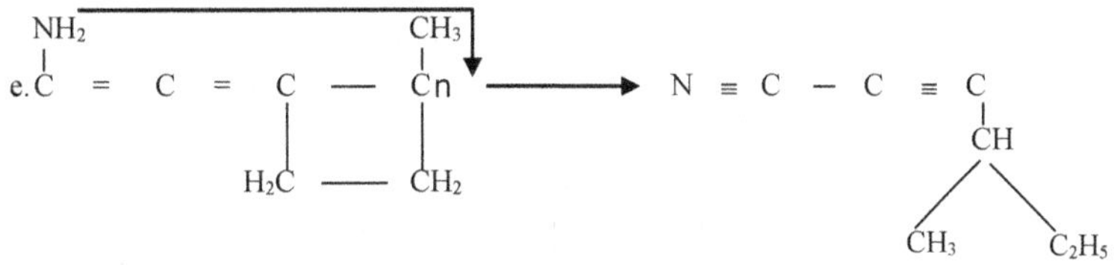

$$
e.\overset{NH_2}{C} = C = \overset{CH_3}{C} \;-\; Cn \longrightarrow N \equiv C \;-\; C \equiv C
$$

11.71

Like (A) of Equation 11.64, the two mono-forms here undergo molecular rearrangement of the first and the third types to produce non-resonance-stabilized acetylenic monomers of same characters, unlike the case shown below

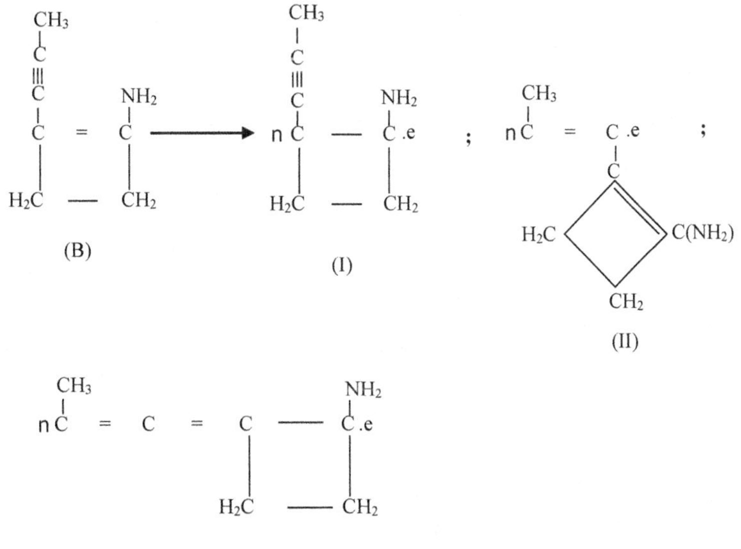

(B)   (I)   (II)

(III)   11.72

These will not molecularly rearrange. Shown below, are examples of possible and impossible cases.

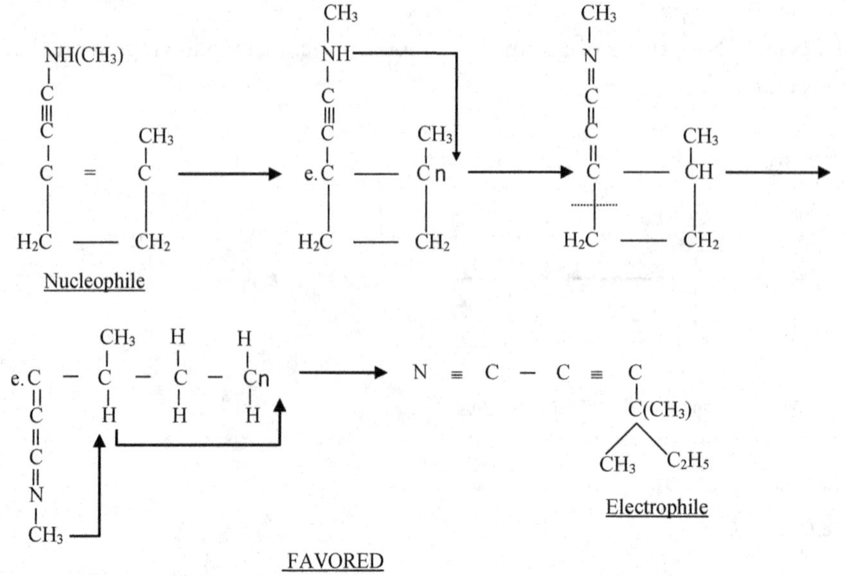

Nucleophile

FAVORED

Electrophile

11.73

When the $NH(CH_3)$ is replaced with $NH(C_2H_5)$, the same also applies only when the CH_3 group is replaced with C_2H_5 group. This is the limit. Thus, all the cases shown below will favor this rearrangement to give nitriles.

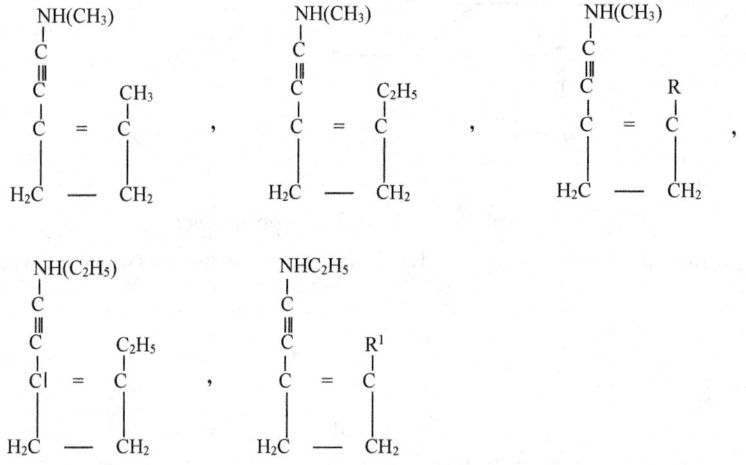

Where $R \leq CH_3$, $R^1 \leq C_2H_5$

11.74

With NR_2 types of groups, the followings are obtained.

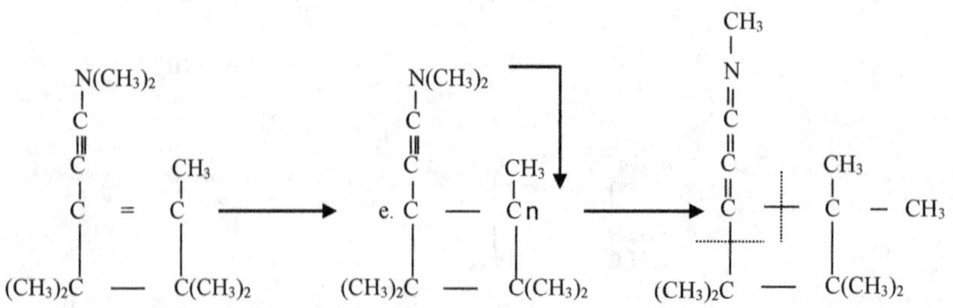

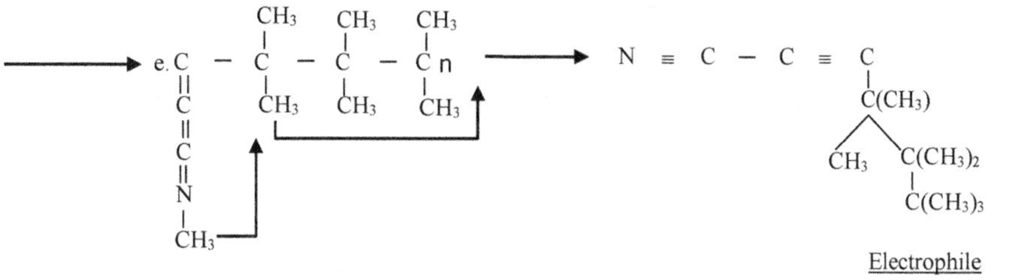

FAVORED

11.75

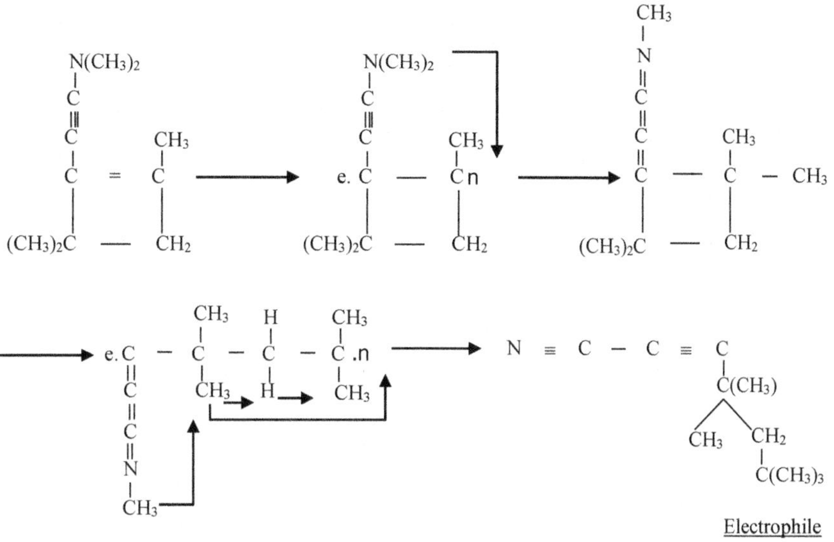

NOT FAVORED

11.76

It is said not to be favored because, CH_3 group cannot move to a center carrying H after the first movement. The conditions favoring the existence of molecular rearrangements of the third kind can be observed for higher cycloalkanes where possible.

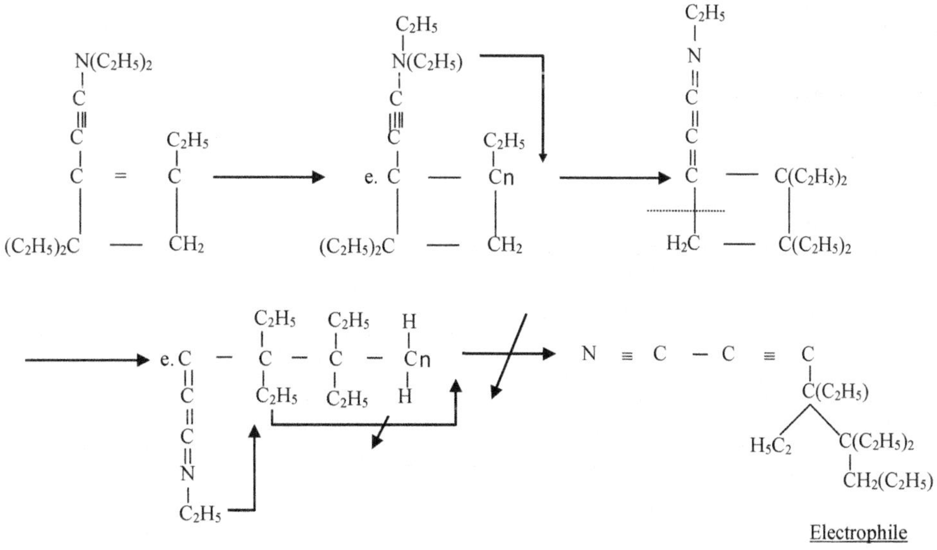

NOT FAVORED

11.77

More examples of cases which cannot favor molecular rearrangement of the third kind are shown below.

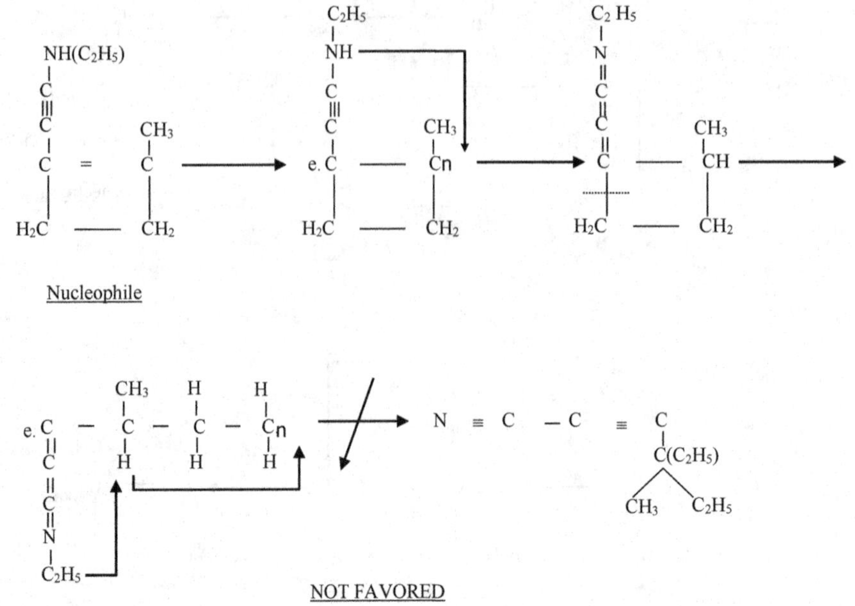

Nucleophile

NOT FAVORED

11.78

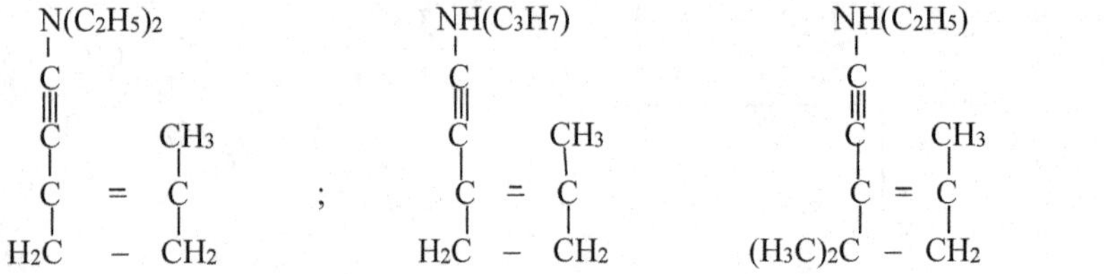

11.79

None of the examples will favor rearrangement, because as shown in the second but last equation C_2H_5 will not be the transfer species abstracted during initiation, but CH_3. In such cases, the ring cannot be opened at all.

In general, one can observe the strong relationship between cyclopropene and cyclobutene and the same will almost apply to the members of the family where the MRSE can be provided. All the analysis so far only apply radically.

11.1.2. Cyclopentenes

Compared to cyclobutene, cyclopentenes are for less unsaturated and less strained. Therefore, it seems that cyclopentene cannot possess the minimum required strain energy, to favor its opening instantaneously. Unlike cyclobutenes, it seems that there is only one major mode of polymerization, and that is

(i)   Via the double bond (difficult)

Like cyclopropenes, it seems that there is no way by which the first membered ring can be opened instantaneously. Instantaneous opening of the ring in the presence of strong coordination initiators is impossible, though there are at least two points of scission in the ring as shown below.

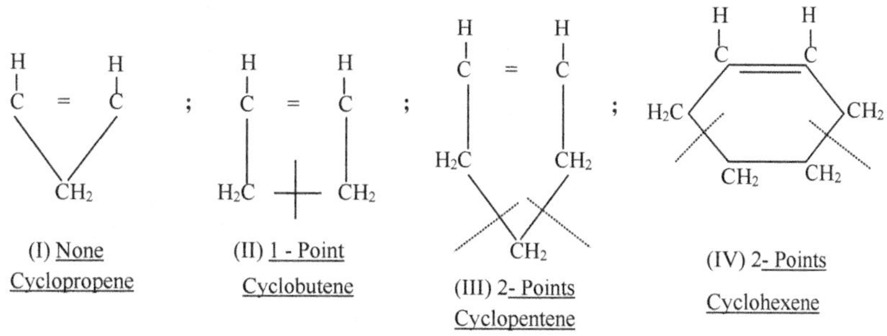

(I) None
Cyclopropene

(II) 1 - Point
Cyclobutene

(III) 2 - Points
Cyclopentene

(IV) 2 - Points
Cyclohexene

Points of Scission   11.80

The number of points of scission can be reduced by the presence and location of substituents groups as shown below. Their presence makes it easier for the ring to be opened instantaneously.

(I) 2 - Points

(II) 1 - Point

(III) 1 - Point

(IV) 1 - Point   etc

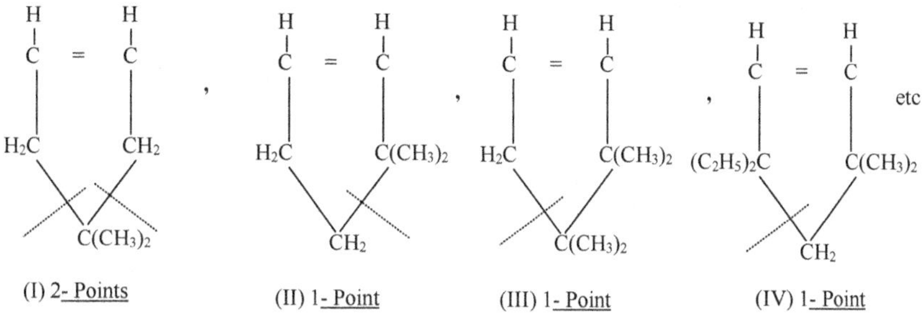

Points of scission   11.81

In the absence of symmetry, the point of scission should be determined by the single bond with the largest radical-pushing potential difference. Nevertheless, even where many of them are opened instantaneously, the best that can be obtained are living isolated mono-enes (III), (IV) and (I) or butadienes (II).

When the ring is activated, the followings are obtained, using a weak initiator.

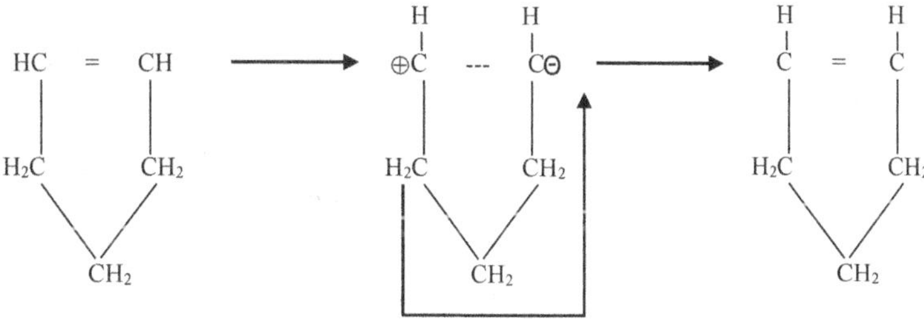

(I) Molecular rearrangement of the tenth type   11.82

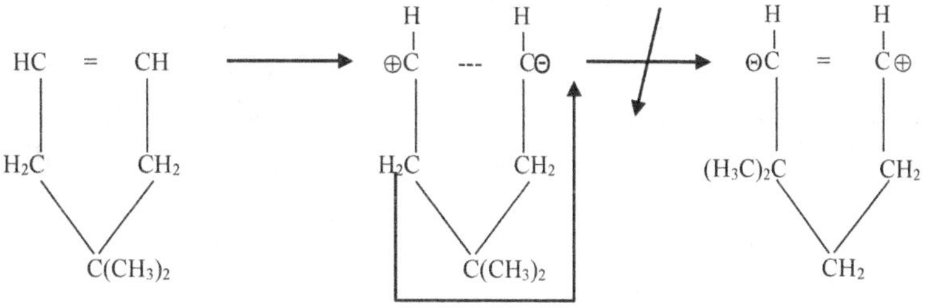

(II) Impossible Molecular rearrangement   11.83a

223

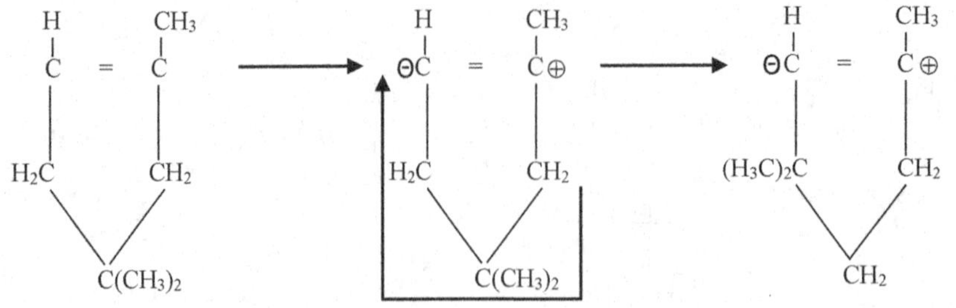

(III) Impossible Molecular rearrangement

11.83b

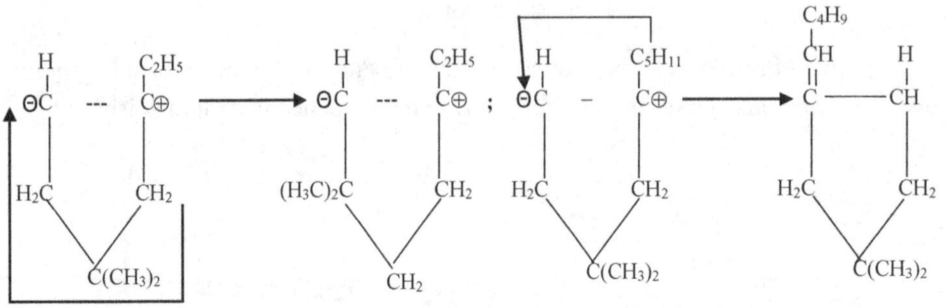

(IV) Impossible Molecular
rearrangement

(V) Molecular rearrangement
of the sixth type

11.84

Molecular rearrangement of the tenth type is rarely favored. It is the sixth type that predorrminates with limitations. For example, when R in (V) is larger than what has been used above, molecular rearrangement of the sixth type will not take place.

In the presence of weak or strong initiators, particularly of the Ziegler-Natta or coordination types, all of them cannot be readily used as monomers, due to the odd-numbered character of the monomers. They, being odd-numbered, unlike cyclobutene, cannot be trans-placed as shown below.

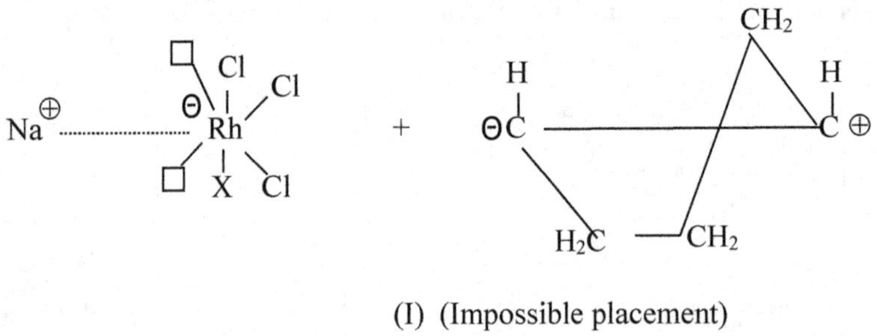

(I) (Impossible placement)

11.85

Since (I) cannot be symmetrically placed across the activation centers, trans-placements of odd-membered rings are impossible, since the conformation of the activated monomer can take so many shapes, noting that there are countless numbers of conformations but only one fixed configuration. The possibility of having cis- or isotactic placement of the monomer units along a chain is also impossible due to steric limitations. Trans- or Syndiotactic placement may be possible at very low polymerization temperatures and in particular when there are vacant orbitals or reservoirs with less or smaller pendant groups on the counter center.

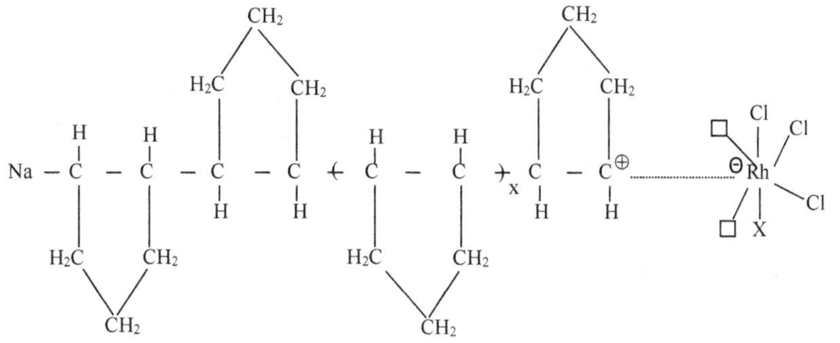

Syndiotactic polycyclopentene growing chain 11.86

Thus, just as cis-placement is not favored for even-membered cycloalkenes, it is also not favored for odd-membered cycloalkenes. For the growing polymer chain above, transfer species of the first kind of the tenth type can be released to kill the chain. This is the same transfer species that will prevent the initiation of the monomer with negative charges and nucleo-free-radically. Electro-free-radically, it can also be polymerized, but with more difficulty, unless where pairing radically is found to be possible.

(IV) of Equation 11.82 will be visited shortly. In the presence of strong initiators, imagine if the five-membered rings of Equations 11.81 and 11.83a are instantaneously opened as follows.

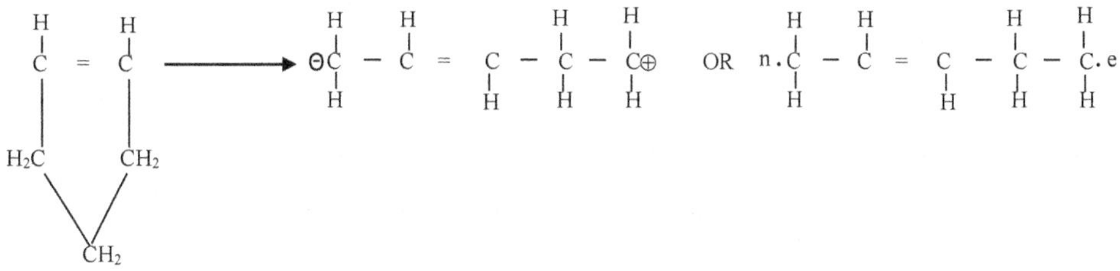

11.87

Opening of the rings here chargedly is possible in the absence of electrodynamic forces of repulsion, since the double bond cannot be the first to be activated when the ring is about to be opened. For this case, when opened radically double bonds will appear along the chain in the absence of rearrange-ment. Note that H and CH_3 types of groups **cannot be moved** as nucleo-free-radicals or negative charges. When H is moved with electro-free-radical or positive charge, only in few cases will stable products be produced. It could not be obtained with the case above, not even vinyl cyclopropane.

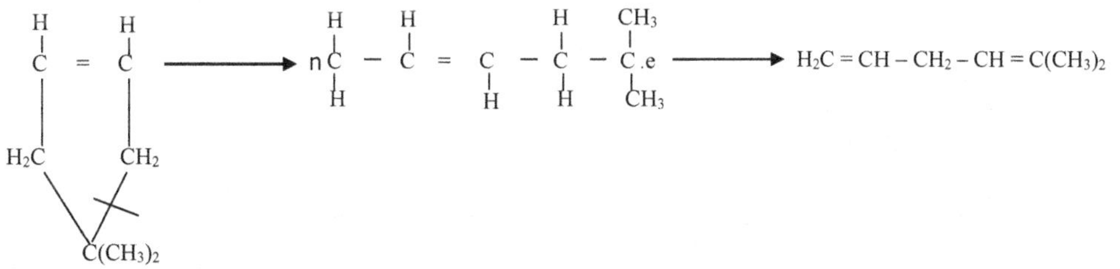

11.88

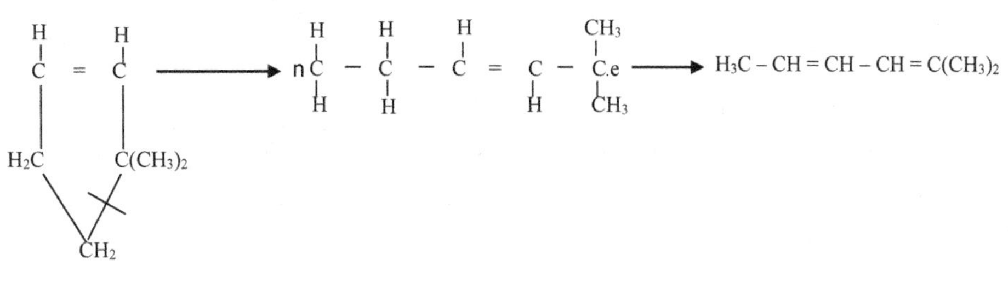

11.89

Opening of the rings here like the case above is meaningful only radically, since stable substituted conjugatedly and isolatedly placed dienes are the products, via movement of electro-free-radicals and H.e. Notice that the products are not the same. The more the presence of radical-pushing groups, the greater the possibility of opening of the ring. Note that while the product from the first case above is not resonance stabilized, the product from the last case is resonance stabilized.

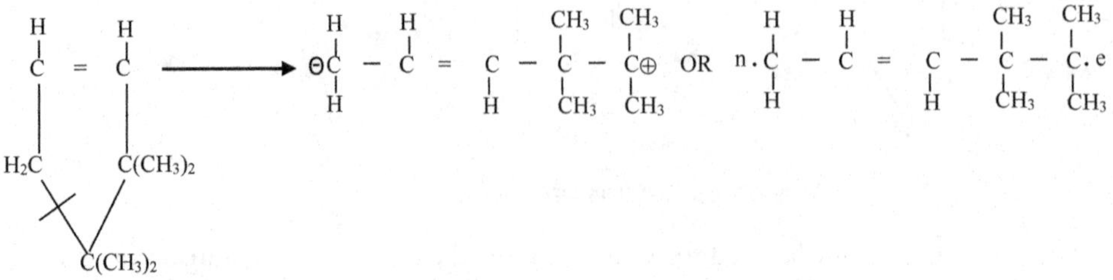

11.90

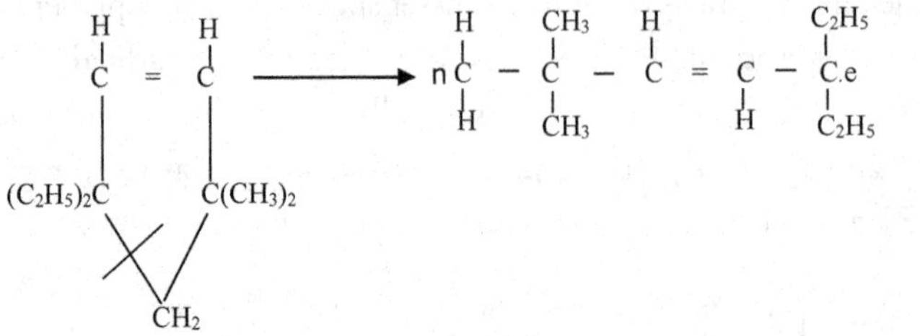

11.91

In some of them like the case of Equation 11.90 above, when opened instantaneously, movement of radicals can take place to produce a vinyl cyclopropane, though the ring may be too strained to exist as a ring. One can observe what it takes to open the ring instantaneously as shown below for cyclopentene that which cannot readily be opened instantaneously.

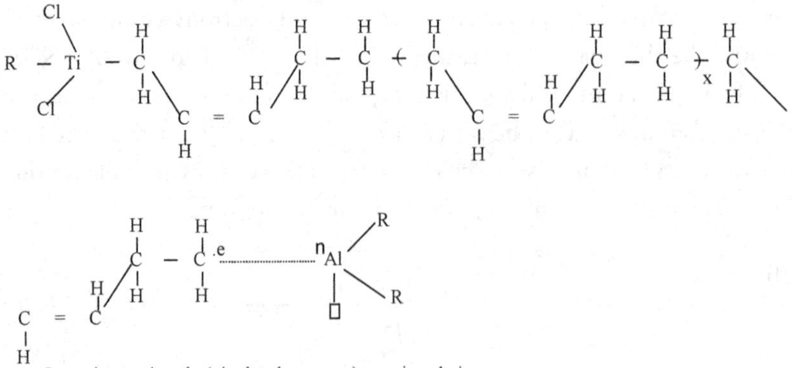

Iso - cis - tactic poly (vinylcyclopropene) growing chain

11.92

Instantaneous opening of such a ring with a double bond and its polymerization may be possible using very strong electro-free-radical coordination initiators. Strong nucleo-free-radical initiators can also polymerize the monomer above as long as the ring can be opened instantaneously. The monomers of Equations 11.88 to 11.91, will only favor the use of electro-free-radical coordination initiators. When cyclobutene ring is opened, the polymer therein obtained is polybutadiene. When the ring is not opened, the polymer obtained is poly(cyclobutene). Thus, one can observe how the production of polybutadienes, vinyl cyclopropenes, poly(cyclo-pentenes), poly(alkenes) and more from cyclopentene can be made possible. With radical-pushing groups in the cyclopentene ring, existence of MRSE in the ring is far more favored than when radical-pulling groups are present.

The second means by which the rings can be opened instantaneously is via the use of a modified form of (V) of Equation 11.84. The ring obtained from the modified form of (V) of that equation may be strained enough to be opened instantaneously. It is said to be modified, because the middle placed $C(CH_3)_2$ group has been changed to CH_2 for specific reasons. If it has the MRSE under higher operating conditions, then the following is obtained.

$$\begin{array}{c} C_4H_9 \\ | \\ CH \\ \| \\ C \ + \ CH_2 \\ \| \\ H_2C \quad CH_2 \\ \diagdown \diagup \\ CH_2 \end{array} \longrightarrow \ e.C - \overset{H}{\underset{CH_2}{\overset{|}{C}}} - \overset{H}{\underset{H}{\overset{|}{C}}} - \overset{H}{\underset{H}{\overset{|}{C}}} - \overset{H}{\underset{H}{\overset{|}{C}}}n \longrightarrow \ \overset{C_4H_9}{\underset{}{C}} \equiv \overset{}{\underset{C_4H_9}{C}}$$

$$\text{(with } CH_2 \text{ and } C_4H_9 \text{ substituents)} \qquad 11.93$$

Molecular rearrangement begins when the single R group on one of the active carbon centers is greater than half the capacity in the ring (i.e. C_3H_7) and stops when the single R group is C_5H_{11} for the case above.

$$\left(\begin{array}{c} R \quad CH_3 \\ | \quad\ | \\ C = C \\ | \quad\ | \\ H_2C \quad CH_2 \\ \diagdown \diagup \\ CH_2 \end{array} \right) \quad \begin{array}{c} C_3H_7 \quad CH_3 \\ | \quad\ | \\ C = C \\ | \quad\ | \\ H_2C \quad CH_2 \\ \diagdown \diagup \\ CH_2 \end{array} \longrightarrow \begin{array}{c} C_2H_5 \\ | \\ CH_2 \quad CH_3 \\ | \quad\ | \\ e.C - C .n \\ | \quad\ | \\ H_2C \quad CH_2 \\ \diagdown \diagup \\ CH_2 \end{array} \longrightarrow$$

$$\begin{array}{c} C_2H_5 \\ | \\ CH \quad CH_3 \\ \| \quad\ | \\ C - CH \\ | \quad\ | \\ H_2C \quad CH_2 \\ \diagdown \diagup \\ CH_2 \end{array} \longrightarrow e.C - \overset{CH_3}{\underset{CH}{\overset{|}{C}}} - \overset{H}{\underset{H}{\overset{|}{C}}} - \overset{H}{\underset{H}{\overset{|}{C}}} - \overset{H}{\underset{H}{\overset{|}{C}}}n \longrightarrow \overset{C_2H_5}{C} \equiv \overset{}{\underset{CH}{C}}$$

$$\text{(with } C_2H_5 \text{ and } CH_3, C_3H_7 \text{ substituents)} \qquad 11.94$$

The rearrangement stops here when the R group is C_6H_{13}.

$$\begin{array}{c} C_8H_{17} \quad CH_3 \\ | \quad\ | \\ C = C \\ | \quad\ | \\ H_2C \quad HC(CH_3) \\ \diagdown \diagup \\ HC(CH_3) \end{array} \longrightarrow \begin{array}{c} C_7H_{15} \\ | \\ CH_2 \quad CH_3 \\ | \quad\ | \\ e.C - C .n \\ | \quad\ | \\ H_2C \quad HC(CH_3) \\ \diagdown \diagup \\ HC(CH_3) \end{array} \longrightarrow \begin{array}{c} C_7H_{15} \\ | \\ CH \quad CH_3 \\ \| \quad\ | \\ C - CH \\ | \quad\ | \\ H_2C \quad HC(CH_3) \\ \diagdown \diagup \\ HC(CH_3) \end{array} \longrightarrow$$

$$e.C - \overset{CH_3}{\underset{CH}{\overset{|}{C}}} - \overset{CH_3}{\underset{H}{\overset{|}{C}}} - \overset{CH_3}{\underset{H}{\overset{|}{C}}} - \overset{H}{\underset{H}{\overset{|}{C}}}n \longrightarrow \overset{C_7H_{15}}{C} \equiv \overset{}{\underset{CH}{C}}$$

$$\text{(with } C_7H_{15} \text{ and } CH_3, HC(CH_3); HC(CH_3)_2 \text{ substituents)} \qquad 11.95$$

By the reactions above, how the strain energy in such rings can be increased to favor its instantaneous opening, have been demonstrated without breaking the driving forces favoring the existence of molecular rearrangement of the third kind. The rearrangement stops when R is C_8H_{17}.

Following the same pattern of analysis for cycloalkanes, the followings are obvious for cycloalkenes.

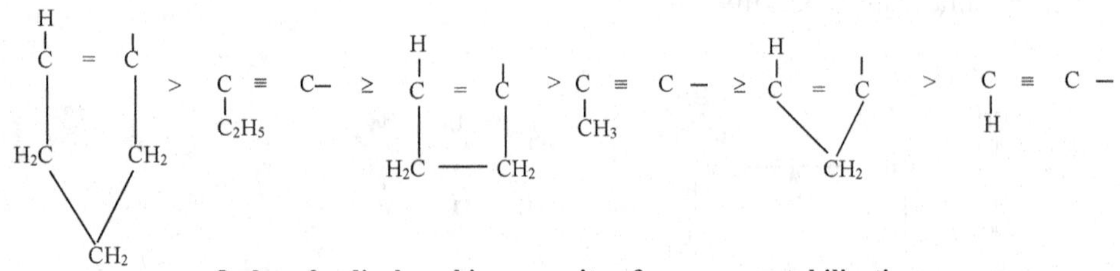

Order of radical-pushing capacity of resonance stabilization groups   11.96

With vinyl cyclopentenes, consider the followings when resonance stabilization takes place, that which can only be done free-radically.

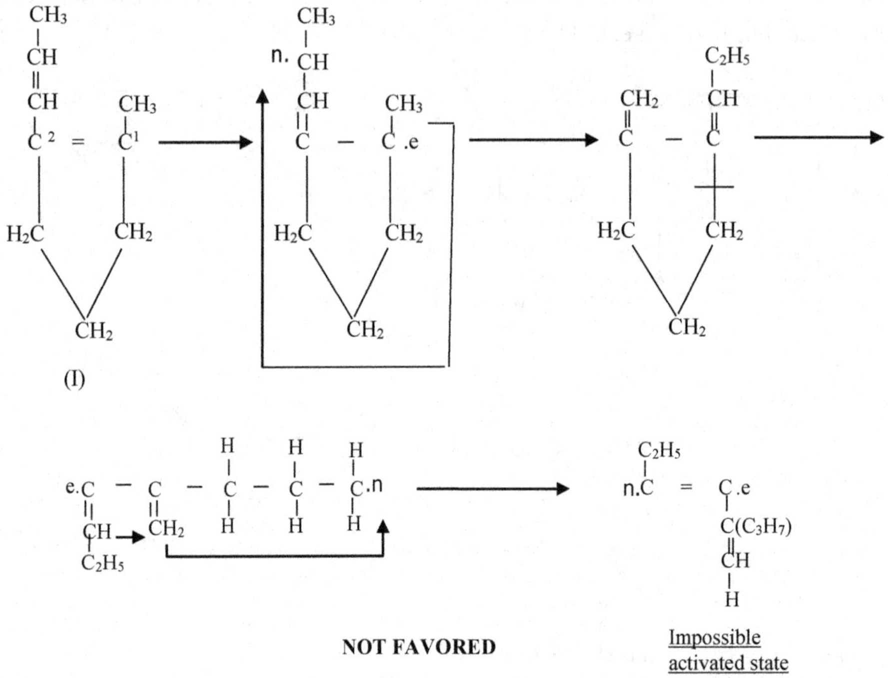

(I)

NOT FAVORED   Impossible activated state

11.97

In the first case, CH_3 group as used above, cannot be the provider of transfer species. When the transfer species comes from the ring, no rearrangement is possible. When the radicals are now well placed, the followings are obtained.

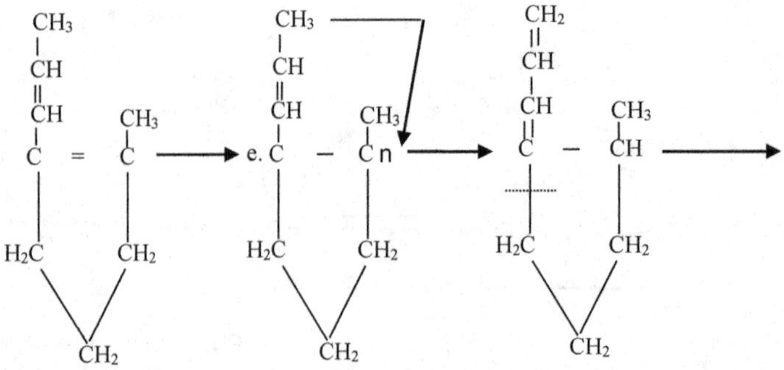

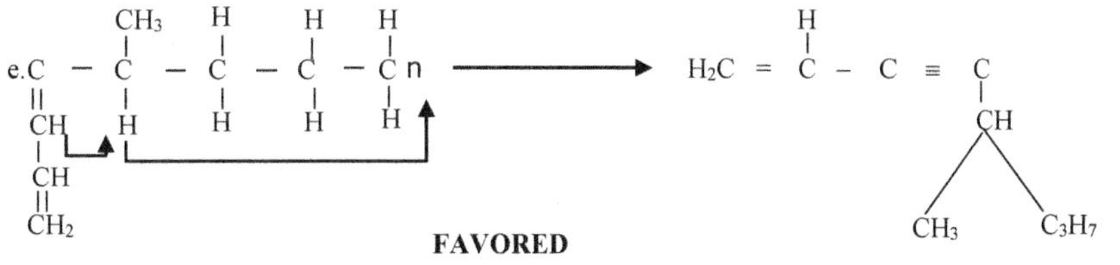

FAVORED

11.98

Though the center activated is the wrong center, the same product will still be obtained. A vinyl acetylene is obtained, clear indication of the strong nucleophilic character of cyclopentene. As the ring size increases, the possibility of opening can only be done when the groups carried are strongly radical-pushing in character. When the monomer of Equation 11.90 is made to carry an alkene, the ring cannot be opened instantaneously as shown below, since the product obtained (A) in the absence of further decomposition is that in which the ring whose size has been reduced is not the resonance stabilization provider just as what it was before the ring was opened. The alkene group has now become the provider in (A).

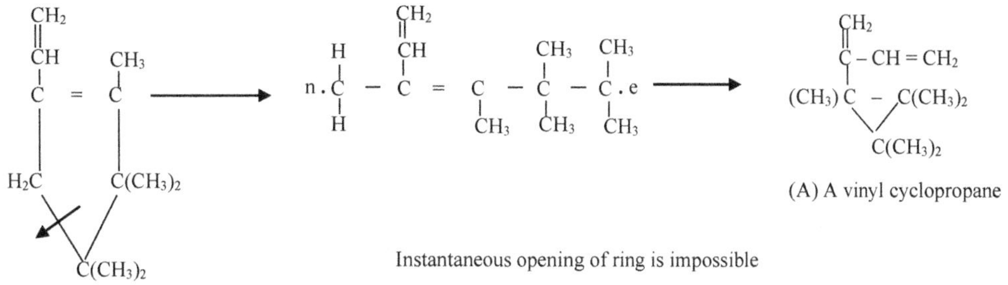

(A) A vinyl cyclopropane

Instantaneous opening of ring is impossible

11.99

When the R group carried by the alkene center is now made larger than what the ring is carrying, the situation remains the same. It cannot be opened instantaneously, but can undergo molecular rearrangements of the first and third kinds.

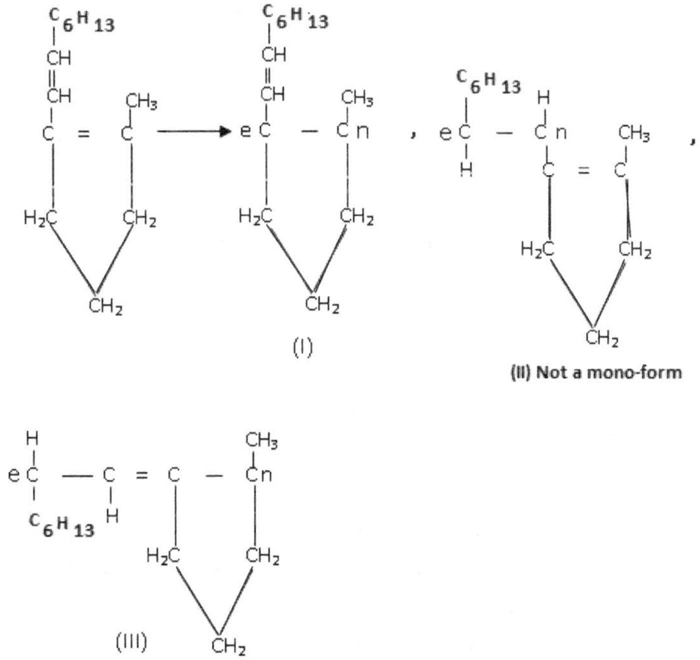

(I)

(II) Not a mono-form

(III)

11.100

229

When acetylenes are carried, the followings are to be expected

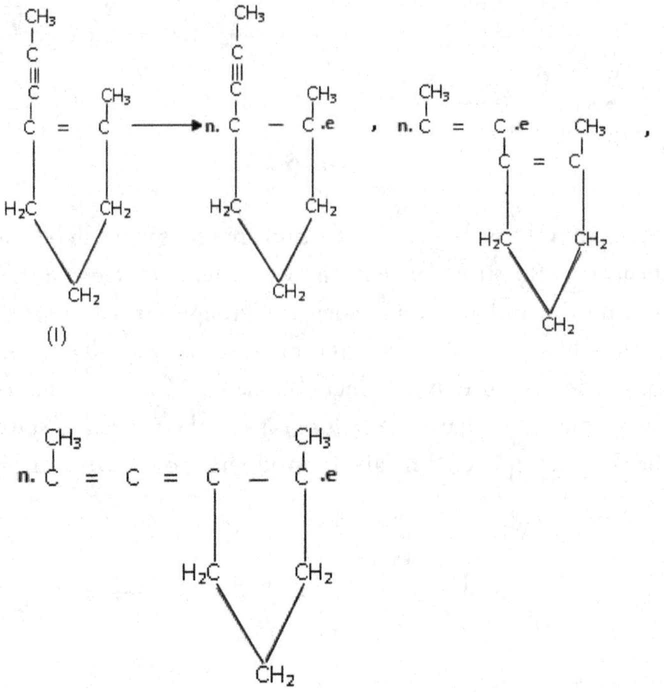

$$
\begin{array}{c}
\text{CH}_3 \\
| \\
\text{C} \\
\parallel\!\parallel \\
\text{C} \\
| \\
\text{C} = \text{C} \quad \text{CH}_3 \\
\end{array}
\longrightarrow
\text{n. C} - \text{C .e} \quad , \quad \text{n. C} = \text{C .e} \quad \text{CH}_3 \quad ,
$$

(I)

$$
\text{n. C} = \text{C} = \text{C} - \text{C .e}
$$

<u>Rresonance stabilized</u>   11.101a

The second is not a mono-form, because the resonance stabilization provider is the ring.

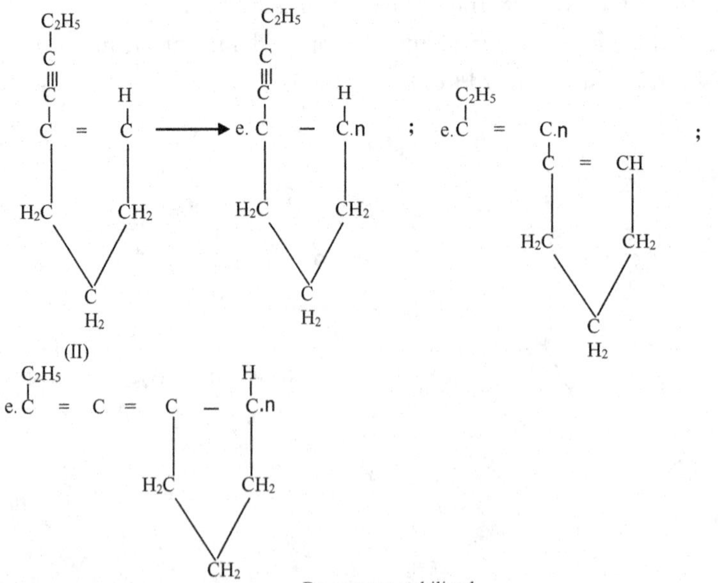

$$
\begin{array}{c}
\text{C}_2\text{H}_5 \\
| \\
\text{C} \\
\parallel\!\parallel \\
\text{C}
\end{array}
\longrightarrow
\text{e. C} - \text{C.n} \quad ; \quad \text{e. C} = \text{C.n} \quad ;
$$

(II)

$$
\text{e. C} = \text{C} = \text{C} - \text{C.n}
$$

<u>Rresonance stabilized</u>   11.101b

The second one above is also not a mono-form, because it is the resonance stabilization provider. ***No alkenic group can provide resonance stabilization for the ring opened or not. More acetylenic groups can be the providers of resonance stabilization.*** The acetylene is the provider of resonance stabilization for the two cases above, because the followings are valid.

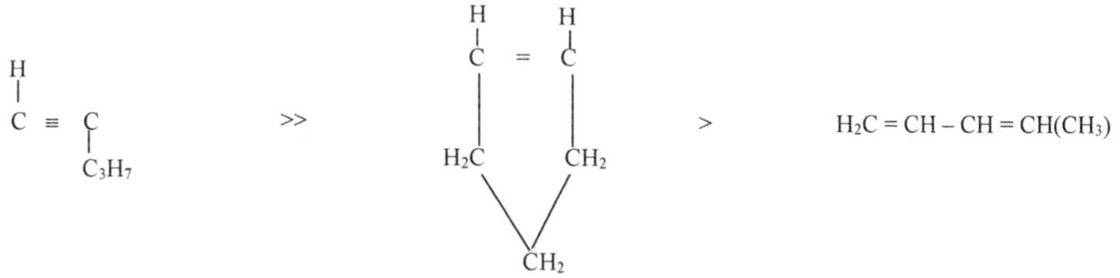

$$
\underset{\overset{|}{C_3H_7}}{\overset{\overset{H}{|}}{C}} \equiv C \qquad >> \qquad \underset{CH_2}{\overset{\overset{\overset{H}{|}}{C} = \overset{\overset{H}{|}}{C}}{\underset{H_2C \qquad CH_2}{\diagdown\diagup}}} \qquad > \qquad H_2C = CH - CH = CH(CH_3)
$$

<u>ORDER OF NUCLEOPHILICITY</u>

11.102a

In the last equation above unlike in Equation 11.96, we did not identify with the acetylene and alkene, because of the capacity of the double bond in the ring as well as the invisible π-bond. Of the monomers in Equations 11.101a and 11.101b, only the last can be made to undergo molecular rearrangement of the third kind to give resonance stabilized acetylenic monomer.

For the case below, molecular rearrangement is possible.

$$
- C_3H_7 \qquad > \qquad \underset{CH_2}{\overset{\overset{\overset{H}{|}}{C} = \overset{\overset{H}{|}}{C}}{\underset{H_2C \qquad CH_2}{\diagdown\diagup}}} \qquad > \qquad \underset{H_2C \!-\! CH_2}{\overset{\overset{\overset{H}{|}}{C} = C-}{}}
$$

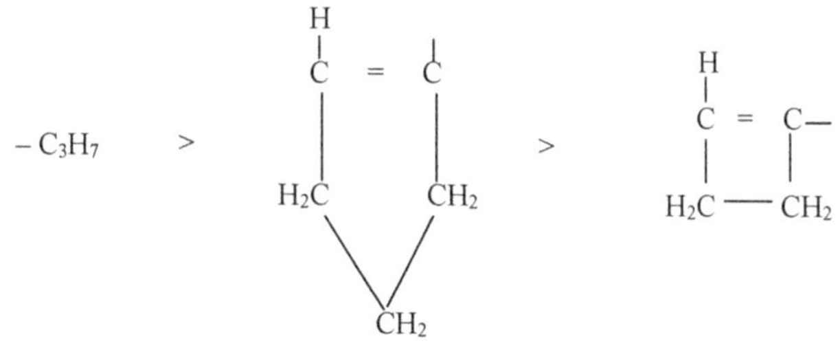

ORDER OF RADICAL-PUSHING CAPACITY

11.102b

In the last equation above unlike in Equation 11.96, we did not identify with the acetylene and alkene, because of the capacity of the double bond in the ring as well as the invisible π-bond. Of the monomers in Equations 11.101a and 11.101b, only the last can be made to undergo molecular rearrangement of the third kind to give resonance stabilized acetylenic monomer.

For the case below, molecular rearrangement is possible.

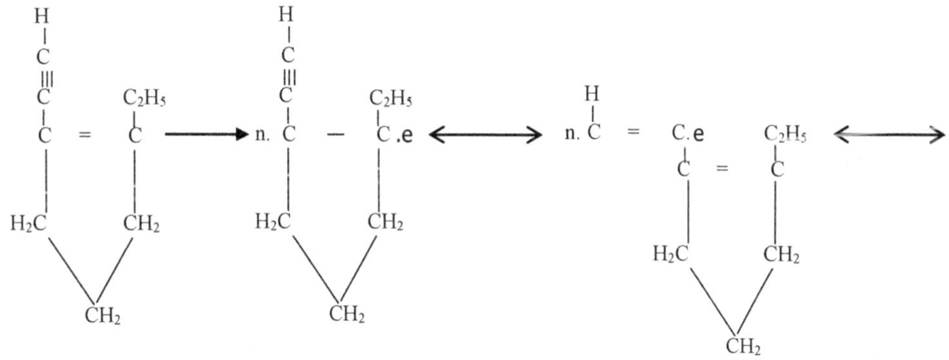

(III)   <u>Resonance stabilized</u>

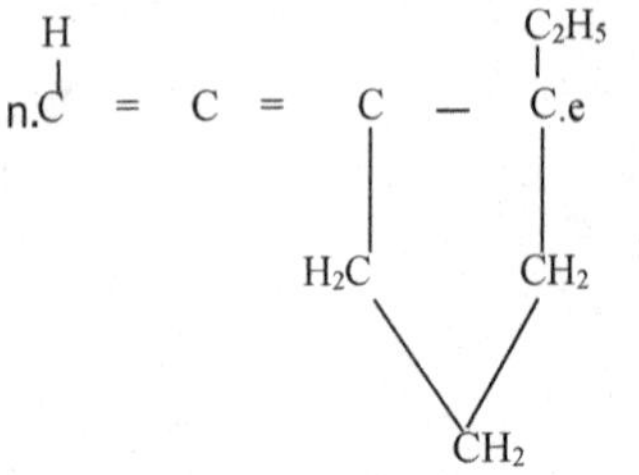

$$n.\overset{\overset{\displaystyle H}{|}}{C} = C = C - \overset{\overset{\displaystyle C_2H_5}{|}}{C}.e$$

with H_2C and CH_2 bridging to CH_2

11.103

(I) cannot rearrange to a resonance stabilized form. So also is (III). Only (II) as already said, will undergo molecular rearrangement as shown below.

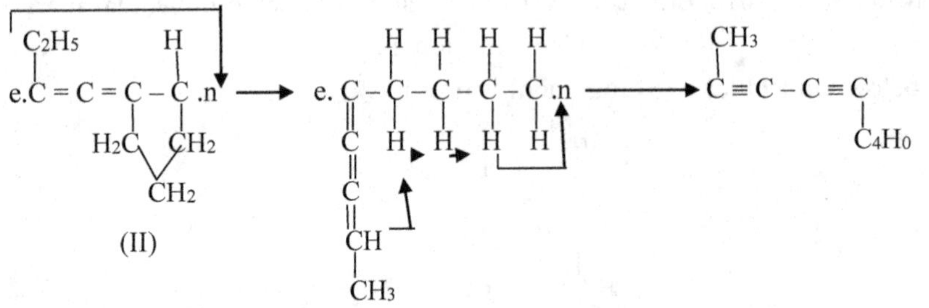

$$e.C = C = C - C .n \longrightarrow e. C - C - C - C - C .n \longrightarrow C \equiv C - C \equiv C$$

(II)   [FAVORED]

11.104a

When C_6H_{13} is put in place of C_2H_5, rearrangement cannot take place. It ceases at C_5H_{11}.

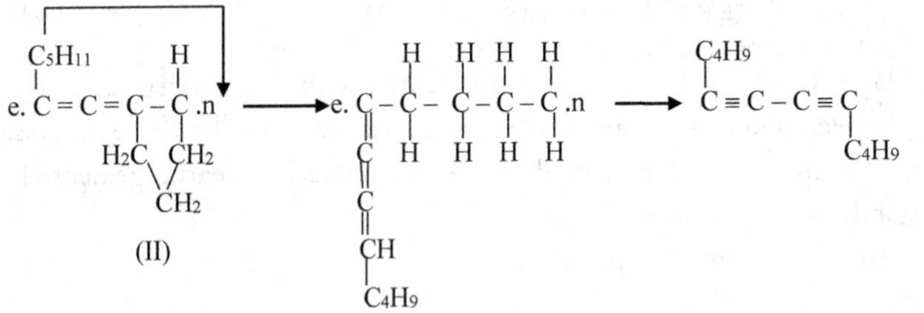

$$e. C = C = C - C .n \longrightarrow e. C - C - C - C .n \longrightarrow C \equiv C - C \equiv C$$

(II)   [FAVORED]

11.104b

For (I) of Equation 11.101a, the followings are to be expected.

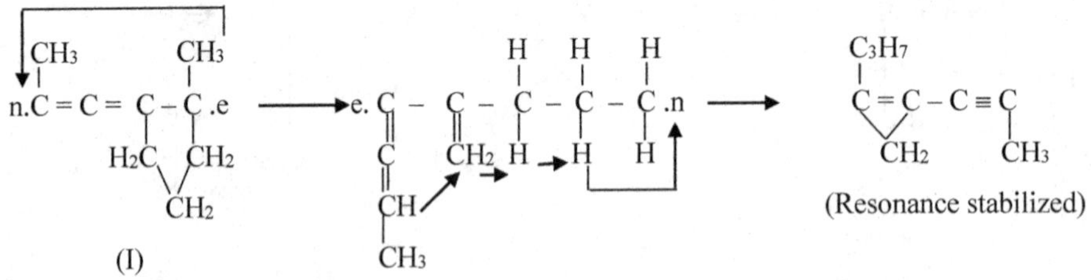

$$n.C = C = C - C .e \longrightarrow e. C - C - C - C - C .n \longrightarrow C = C - C \equiv C$$

(I)   [FAVORED]   (Resonance stabilized)

11.105

The rearrangement is favored. When CH_3 on the ring side is changed to C_4H_9, rearrangement ceases. Anything in life whether for positive or negative development, has its own limitations with its domains and boundaries.

For (III) of Equation 11.103, the followings are obtained.

11.106

Like the others, when C_2H_5 exceeds C_4H_9, rearrangement ceases. Note that all these can only take place radically.

From the analysis so far, the order of the characters of members of cycloalkenes is obvious. The nucleophilicity of the members increases, with increasing size of the rings when same groups are carried.

Cyclopropenes   >   Cyclobutenes   >   Cyclopentenes   >   Cyclohexenes

<u>Order of nucleophilicity</u>

11.107

Occurrence of molecular rearrangement of the third kind decreases with increasing size of the ring, due to decreasing size of strain energy in the ring with increasing size, increasing size of R group needed for molecular rearrangement of the first kind to take place and increasing difficulty in transfer of transfer species of the first type (e.g. H) to the active centers after the ring is opened.

Presence of resonance stabilization phenomena with these nucleophiles seems to increase with increasing ring size, while its effect on the ring seem to decrease with increasing ring size. Not all nucleophilic diynes and triynes, all resonance stabilized, can undergo molecular rearrangements. All electrophilic diynes, triynes and vinyl cycloalkanes are worthy of note, in terms of their applications to other natural phenomena.

While some electrophilic cyclopropenes, cyclobutenes will favor polymerization via the double bonds, electrophilic cyclopentenes, and higher members may not. The groups carried by the internal carbon centers may be selectively chosen and only negatively charged or nucleo-free-radical coordination initiators can be used.

<u>Possibke Polymerizable Electrophiles via double bond</u>

11.108

(III) will be more difficult to polymerize than (II) which in turn will be more than (I), due to steric limitations and countless numbers of conformations that can be possible, increasing with increasing size of the ring. It is largely for this reason, cycloheptene (seven-membered ring) cannot be polymerized via the double bond. The active centers cannot be properly externally located, in order to minimize steric effects. All the above will favor the nucleo-free-radical route, while only (II) and (III) will in addition favor only the negatively charged route.

11.1.3. Cyclohexenes

Like cyclobutenes, this is an even-numbered ring, in which only trans-placement of the molecular structure is favored when activated using coordination initiators. Unlike cyclobutenes, the ring cannot easily be opened instantaneously, since the six-membered unsaturated rings may require so much energy to attain the minimum required strain energy needed to unzip the ring instantaneously. To open the ring, it must be heavily loaded with the right radical-pushing groups well placed. Therefore, mostly activation of the p - bond will be considered.

However, cyclohexene, 4-methylcyclohexene, and 4-vinylcyclohexene have been decom-posed in a single-pulse shock tube to give ethene and 1,3-butadiene, propene and 1,3-butadiene, and only 1,3-butadiene respectively[6]. Based on the New Frontiers, these can be explained as follows.

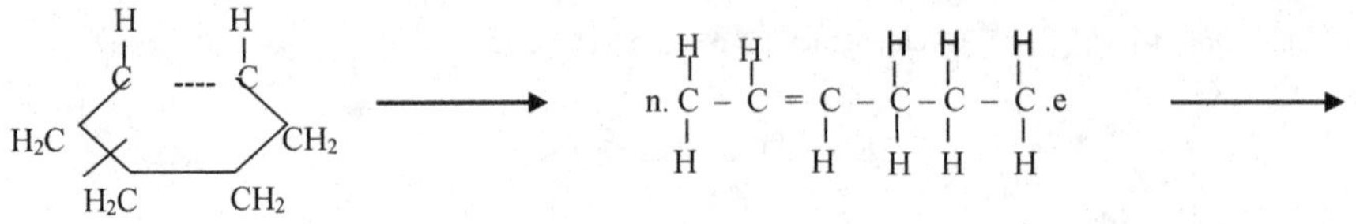

(I) Cyclohexene

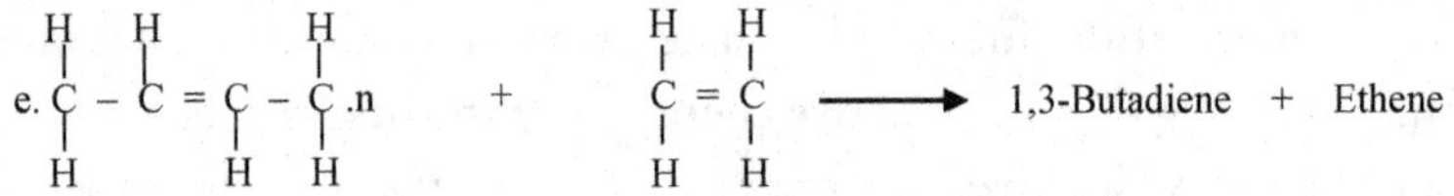

$$11.109$$

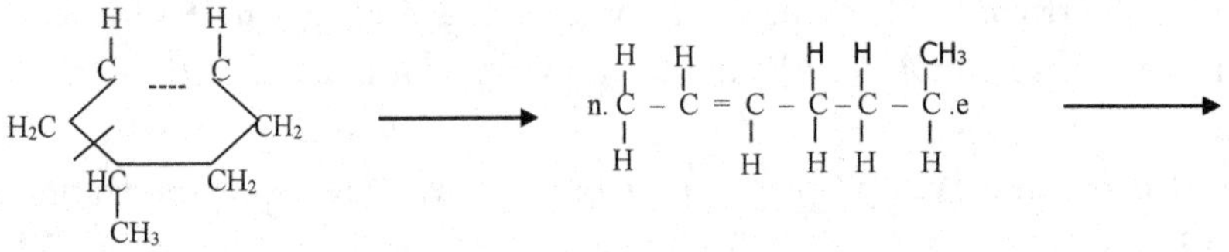

(II) 4-Methylcyclohexene

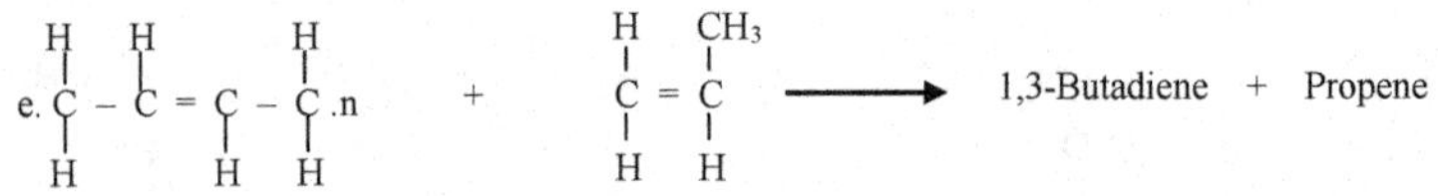

$$11.110$$

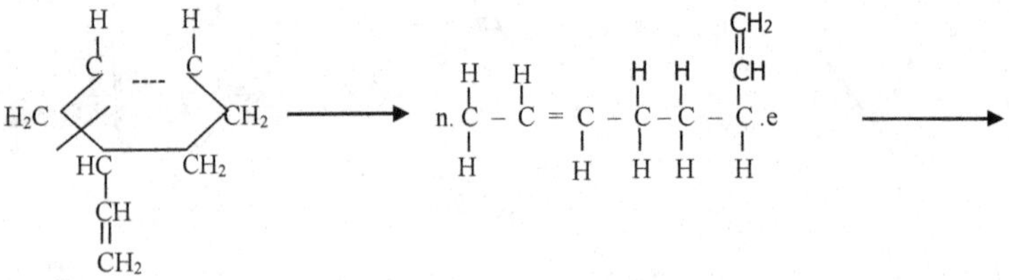

(III) 4-vinylcyclohexene

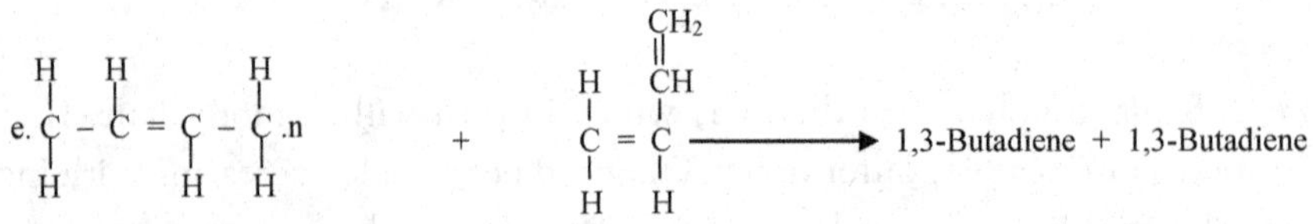

$$11.111$$

All these are Equilibrium or Decomposition mechanisms, each in a single stage. The possibility of getting other products other than above is impossible. Based on the operating conditions cyclobutene could not be a product. It is easier to break a ring than to form a ring! It is very important to note the point of scission. It is the point with the maximum radical potential difference. For the cyclohexene for example, the radical potential difference in the middle is zero. In the first case, the ring has three points of scission for which only one or two can be used depending on the types of groups carried.

Note that all first members of the cycloalkene family have a common finger-print, i.e. Equilibrium state of existence, that in which one particular H is loosely held. The intensity at which it is held differs from size to size. Same applies to all families of compounds which can exist as such. Therefore, if operating conditions can be found to keep the cyclohexene in Equilibrium state of existence, then the followings will be obtained.

Stage 1:

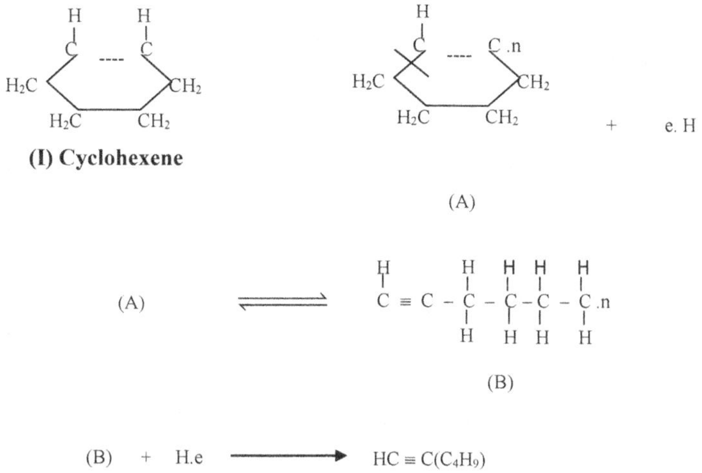

(I) Cyclohexene

(A)

(B)

$$(B) \ + \ H.e \longrightarrow HC \equiv C(C_4H_9)$$

11.112a

Overall equation: Cyclopentene $\longrightarrow$ Butyl acetylene

11.112b

Just as cyclopentene will decompose via Equilibrium mechanism to propyl acetylene, so also cyclohexene will decompose to give butyl acetylene.

Like the other members, nucleo-free-radically, no polymerization can be favored, in view of the presence of transfer species of the first kind of the sixth type. Chargedly, only the positively-charged route will be favored just as only the electro-free-radical route will be favored radically. The characters of these monomers can only be altered by changing or varying the sources of transfer species of the first and sixth types of the first kind. Whether the ring can be opened or not, the use of positively-charged paired initiators is possible.

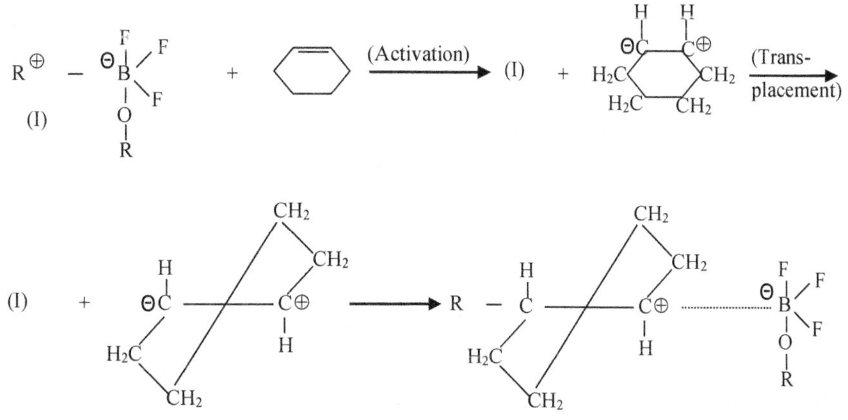

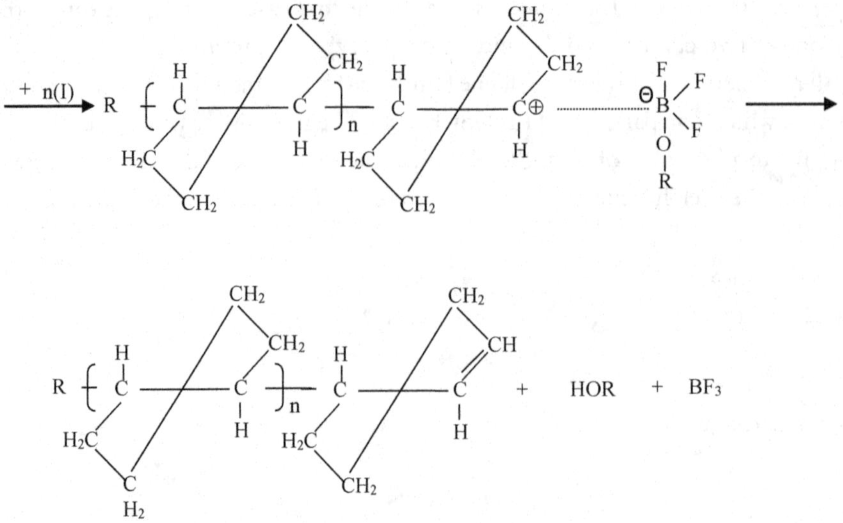

Transfer species of 1st kind of sixth type
11.113

When positively-charged coordination initiators with vacant orbitals are involved, trans-di-isotactic and trans-di-syndiotactic placements can be obtained for one and two vacant orbitals respectively. These would be fully explained during considerations of propagation steps, where the micro-structures are built. The location of the single double bond in the ring does not alter the reactions and analysis above. For the first member above, molecular rearrangement of the sixth type readily takes place to produce same activated monomer, taking longer times in the presence of weak initiators.

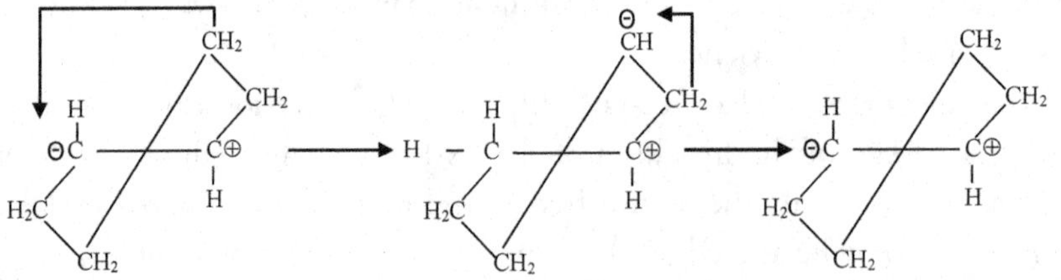

Molecular rearrangement of the sixth type
11.114

When one of the H atoms located on the double bond is replaced with alkylane groups, the followings are obtained beginning with CH_3.

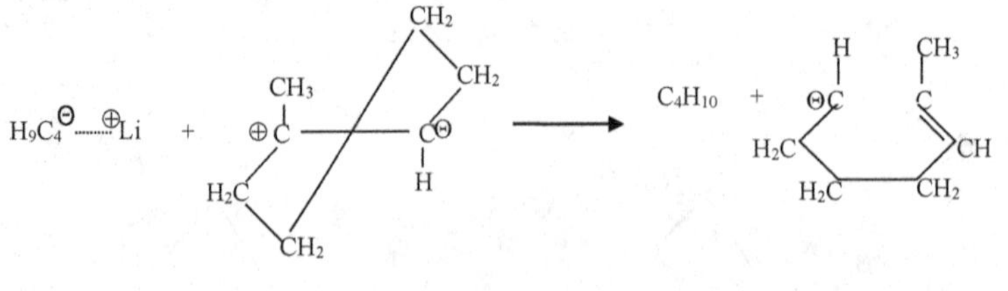

11.115

As has been said and will be fully shown downstream, the initiator above mistakenly called "Anionic ion-paired initiator" is not. It has a dual character only radically and has the ability to identify if the monomer is a Male or a Female. Indeed, it will initiate the monomer above with the Li center as the carrier.

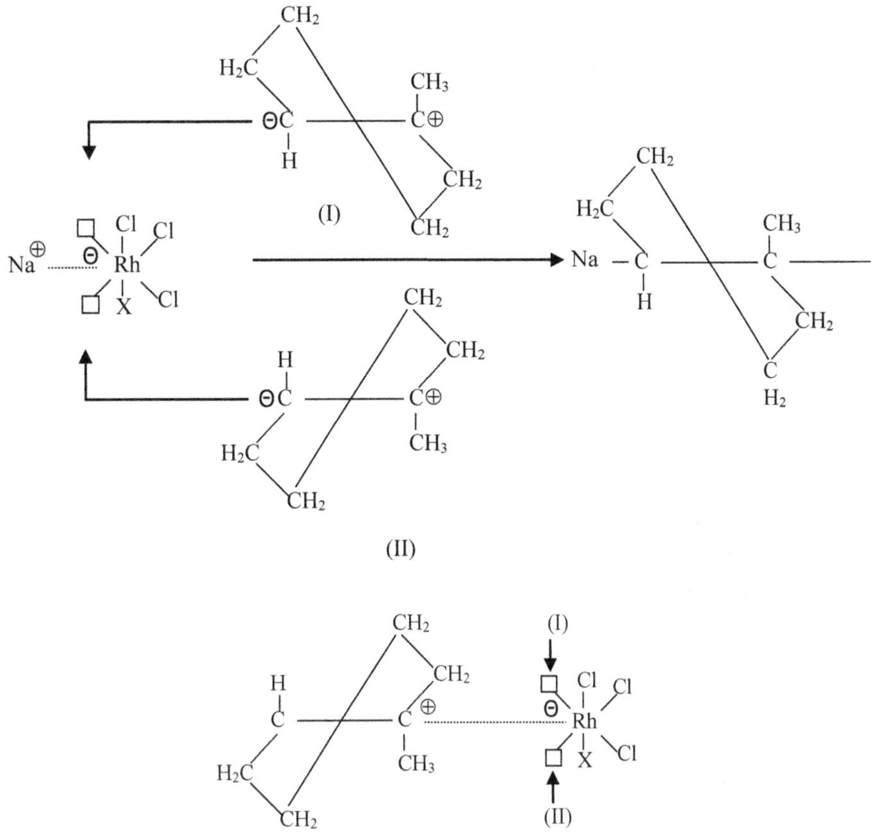

11.116

From the orientation of the activated monomer obtained when a substituent group of less radical-pushing capacity than those in the ring is involved, it is transfer species of the sixth type that is involved. It will need a group such as C_5H_{11} to alter the orientation as shown below

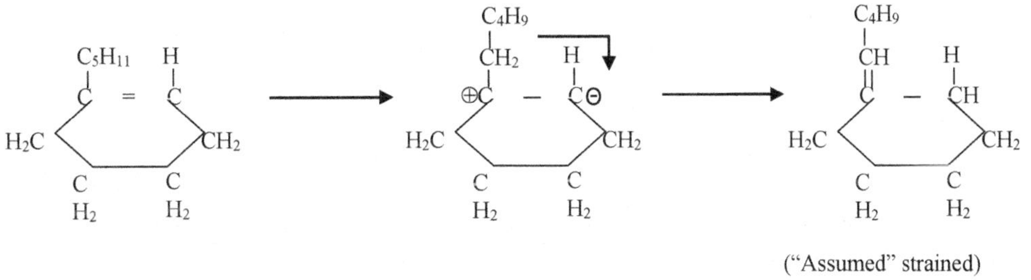

("Assumed" strained)

11.117

In view of the size of the groups, full propagation may not be favored. Nevertheless, it is important to take note of the influence of electrostatic forces in these systems. With negatively charged-paired initiators for the specific cases just as above in Equation 11.115, transfer species of the first type will be involved for C_5H_{11} group.

Now, if the ring is strained enough, then after molecular rearrangement of the first kind of the first type, it can be opened as follows.

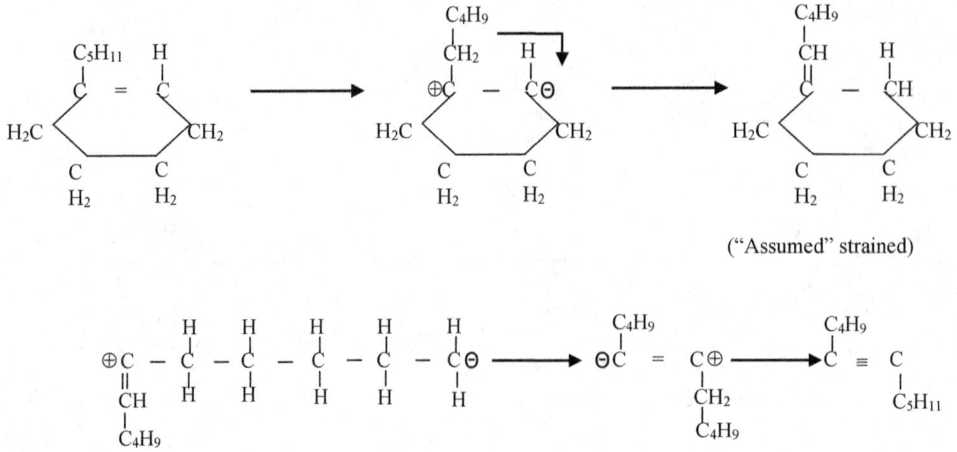

$$C_5H_{11}\text{—}C=C\text{—}H \quad \cdots \quad \text{("Assumed" strained)}$$

(structures for 11.118)

$$\underset{\underset{C_4H_9}{\overset{|}{CH}}}{\overset{||}{\underset{}{\oplus C}}}-\overset{H}{\underset{H}{C}}-\overset{H}{\underset{H}{C}}-\overset{H}{\underset{H}{C}}-\overset{H}{\underset{H}{C}}-\overset{H}{\underset{H}{\underset{}{C\ominus}}} \longrightarrow \ominus C = \underset{\underset{C_4H_9}{\overset{|}{CH_2}}}{\overset{C_4H_9}{\overset{|}{C\oplus}}} \longrightarrow \underset{\underset{C_5H_{11}}{\overset{|}{}}}{\overset{C_4H_9}{C}} \equiv C$$

$$11.118$$

The rearrangement is favored up C_6H_{13}. In life, there are limitations to everything with their domains and boundaries. When radical-pushing groups are introduced into the ring, depending on the capacity of the external radical-pushing group, the MRSE can be easily introduced into the ring.

(structures for 11.119)

[IMPOSSIBLE TRANSFER]

$$N \equiv C\text{—}CH_2\text{—}C(CH_3)_2\text{—}C(CH_3)_2\text{—}C(CH_3)_2\text{—}CH_3$$

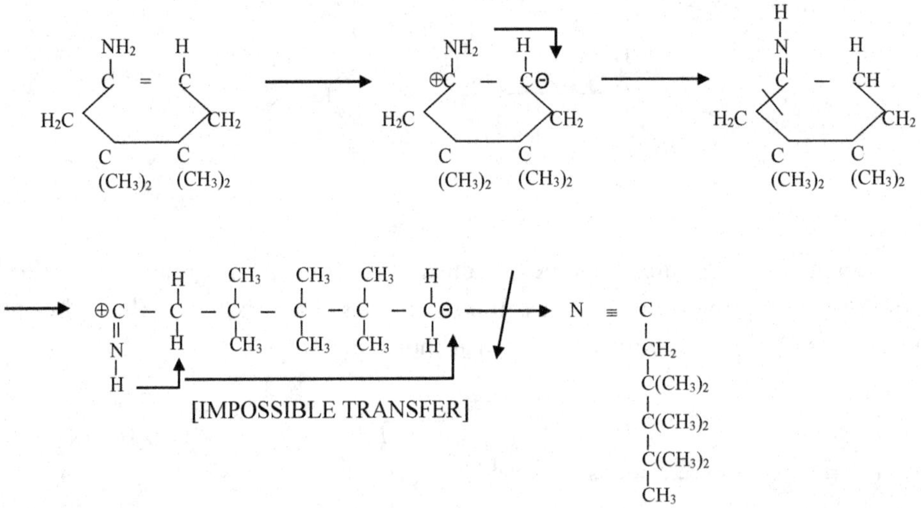

$$11.119$$

Molecular rearrangement of the third kind is not favored here based on the laws already put in place. CH_3 cannot be transferred. In attempting to make the ring have the MRSE, one can observe how important the choice of groups and their placements in the rings are.

In the presence of resonance stabilization groups, based on the foundations which have been laid so far, one knows what to expect in these complex logical network systems guided by wonderful laws which have no exception. Shown below are some first members.

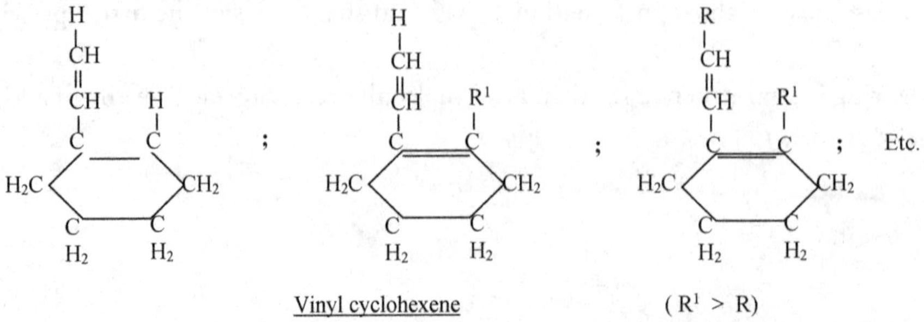

(structures for 11.120a)

Vinyl cyclohexene   (R¹ > R)   11.120a

Cyclohexene diyne ; ; Cyclohexene Triyne ; Etc.

Examples of some resonance stabilized cyclohexenes.   11.120b

Just as the CH_3 group was identified with three- membered ring, C_2H_5 with four-membered ring, C_3H_7 with five –membered ring, so also is C_4H_9 identified with cyclohexene. Apart from the three-membered ring, where all the groups in the ring are resonance stabilized and the ring is unique in character, with other larger membered rings where part of the groups in the ring are resonance stabilized leaving the remaining part non-resonance stabilized, the situation is different as has been clearly indicated so far. Their rings can only be opened instantaneously as long as the ring remains the resonance stabilization provider before and after opening of the ring.

When radical-pulling groups such as COOR, $CONR_2$, CF_3 etc. types are involved, no polymerization route will be favored by the monomers, due to steric limitations. The pattern of analysis with other higher cycloalkenes should be clearly obvious.

11.2. Cyclodienes

These are cases with two double bonds in the ring. In view of the very large strain energy that will exist in a three membered ring with two double bonds, existence of cycloallenes is impossible.

Cycloallene - Non existent

Order of strain energy in three-membered   11.121

The same will similarly apply to cyclobutadiene, which though is strongly closed-loop resonance stabilized, favors only transient existence at STP and far below. Like 1,3-butadiene, it has no transfer species. One cannot say that though it has the same C to H ratio with benzene, it cannot readily exist because of its size, since cyclooctatetraene exists but rearranges very readily to styrene. It is the size which gives it such a large SE, *since the radical-pushing capacity of H_2C in cyclopropene is less than that of HC=CH in cyclobutadiene.*

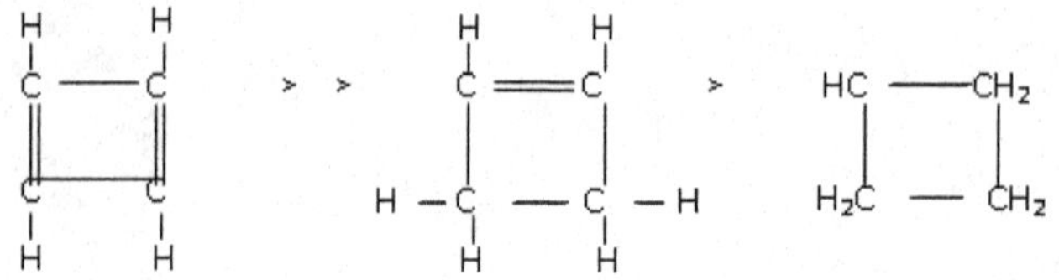

Highly Strained

Order of strain energy in four-membered rings

11.122

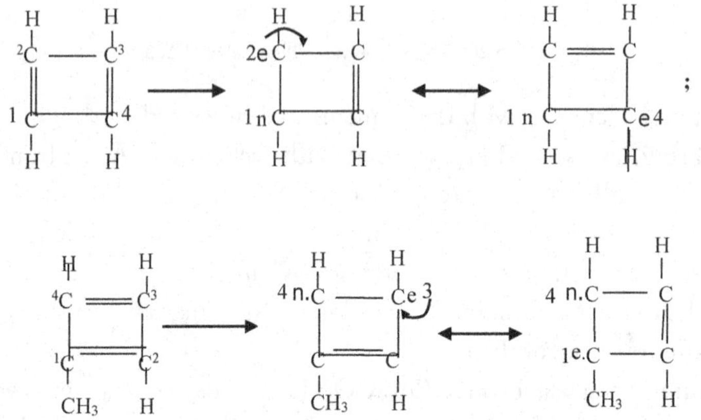

11.123

Limited Continuous closed -loop resonance stabilization

Only cyclopentadiene upwards will favor stable unactivated existence in view of the absence of excessive strain energy in the rings (maximum required strain energy), which the cases above have.

1.1.1 Cyclopentadiene

This is a cyclic conjugated diene, whose ring cannot be opened, since like cyclopropene, there is no point of scission. There is no doubt that the monomer will have the minimum required strain energy but not the maximum.

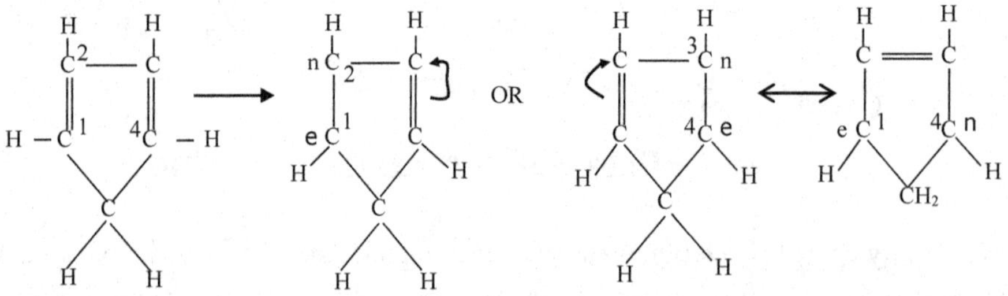

(I) 1,2- addition   (II) 3,4- addition   (III) 1,4- addition

Discrete closed loop resonance stabilized

11.124

As shown above, the monomer favors closed-loop discrete resonance stabilization. For this case, 1,2 - and 3,4 - addition mono-forms are the same. Cyclopentadiene readily undergoes 1,4 - addition to the double bonds. With maleic anhydride for example, the following is obtained. Though the reaction has been thought to be ionic for so many years[7], it can only take place free-radically.

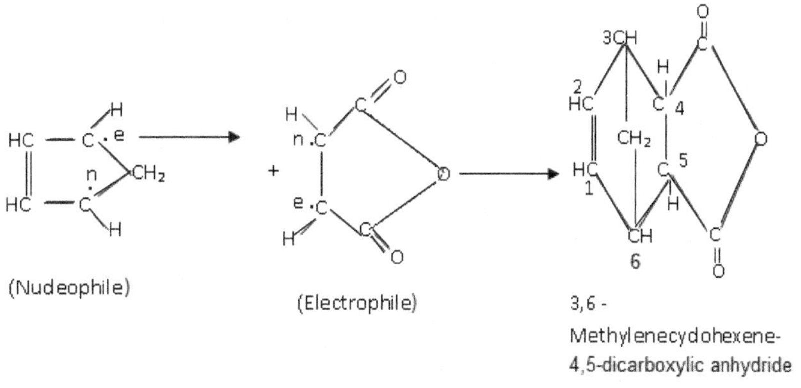

(Nucleophile)   (Electrophile)   3,6 -
Methylenecyclohexene-
4,5-dicarboxylic anhydride

11.125

Though the monomer is resonance stabilized like 1,3-pentadiene and 1-methyl-vinyl acetylene, unlike 1,3-pentadiene and 1-methyl-vinyl acetylene which have transfer species of the first kind of first type, this has transfer species of the first kind of the tenth type on substituent group equally shared by two active centers in the 1,4 -mono-form as shown below.

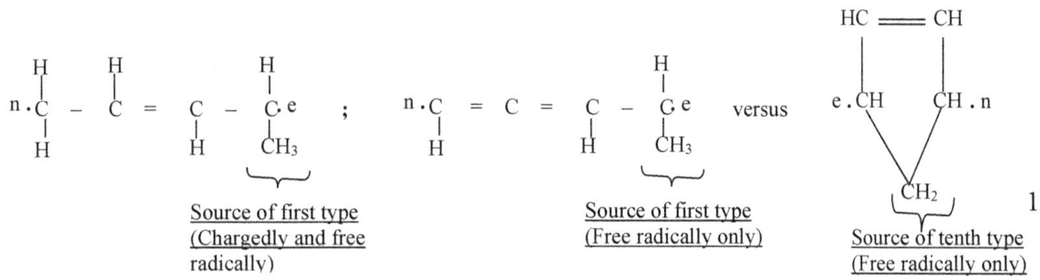

11.126

The resonance stabilized forms in Equation 11.124 however, can favor only electro-free-radical homopolymerization route and not even the positively charged routes, because like cyclopropene, it cannot be activated chargedly due to the fact that charges cannot be removed from their carriers.

They readily undergo molecular rearrangement of the first kind as shown below, only free-radically. Chargedly, like cyclopropene, it is not possible. The transfer species involved is of the first kind of the tenth type. Chargedly, however, one double bond can be activated

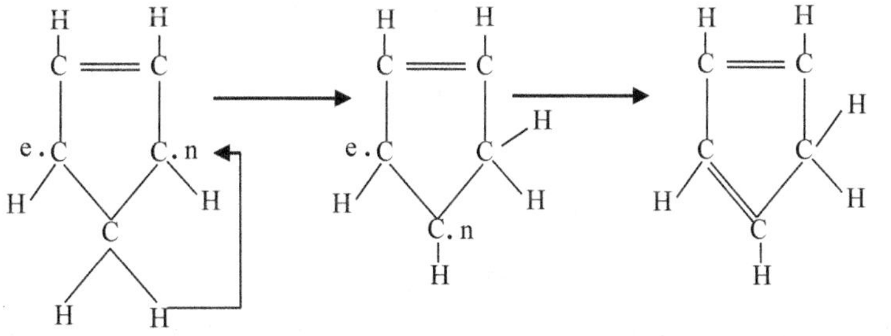

11.127

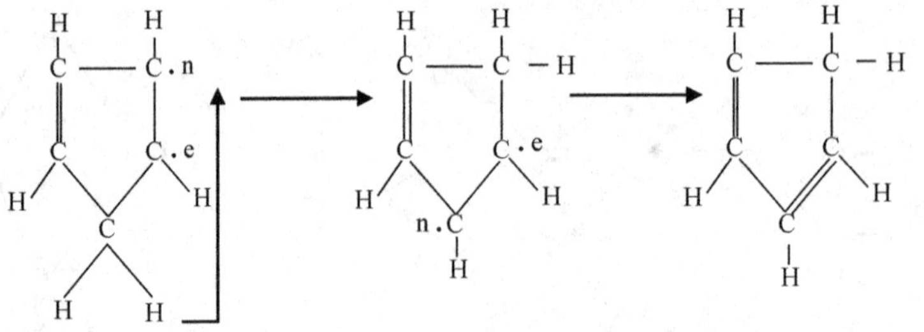

11.128

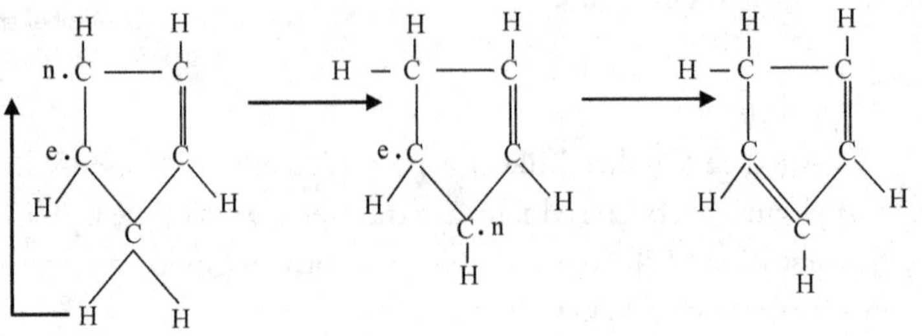

11.129

Cyclopentadiene is known to dimerize slowly on standing to dicyclopentadiene which can dissociate at its boiling point to the monomer[8]. The reaction can only be radical in character, noting that there is no transfer species.

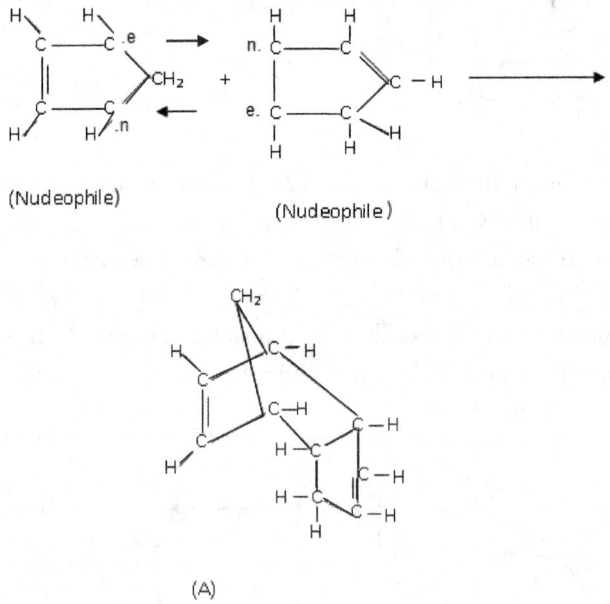

(Nucleophile)   (Nucleophile)

(A)

11.130

It should be noted that it is the 1,2– mono-form that diffuses to another 1,2- mono-form to give the 1,4– mono-form and add to it to favor the existence of (A), under the control of not exceeding the conditions required for attaining the most stable SE for stable existence. It is partly for this reason (MRSE & MaxRSE) that there is no 1,2- to 1,2- addition or 1,4- to 1,4- addition (i.e. addition between two centers of same nucleophilicity) as will be fully explained downstream. Even then, it is only the electro-radical that diffuses all the time in the presence or absence of nucleo-radical, while nucleo-radical diffuses only when the electro-radical is not in the environment as is the case in all Initiation steps in polymerization systems. More importantly is that the driving force favoring the existence of (A), is the formation of a six-membered hexene which is more stable. With electro-free-radical initiator, only the use of special coordination initiators will favor the polymerization of the monomer-cyclopentadiene via only 1,4-addition mono-form as shown below, due to steric limitations. In fact, even in the absence of steric limitations, never does 1, 2- addition mono-form exist along the growing chain. It only exists barely at the beginning of the chain growth when a weak initiator was initially used.

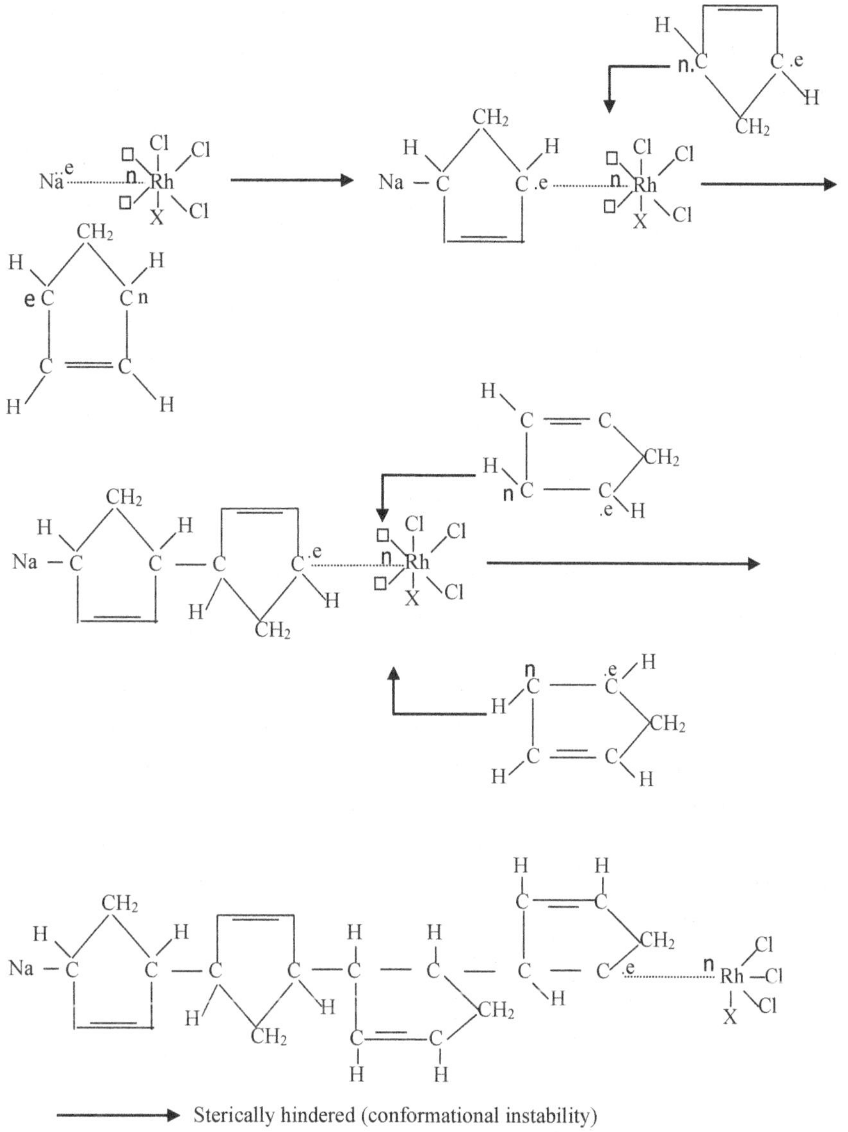

11.131

⟶ Sterically hindered (conformational instability)

Note that the NaX/RhCl₃ combination used above radically is impossible, since pairing cannot take place radically when electrostatically bonded. It was shown to identify with this fact and for exploratory purposes. Pairing radically can only take place with covalent and ionic bond only between specific centers. The addition above should be more favored as the strength of active center increases in the route natural to it. There should be no 1, 2- addition after 1,4- addition. It should be the reverse. It should largely be 1,4- addition. Obviously pairing can take place chargedly-ionically, covalently, and electrostatically. It is important to note that some electrostatically paired centers can be made to carry radicals while chargedly paired (largely via Backward addition as will be shown downstream). The radicals are usually from the counter-charged center when in Equilibrium state of existence. Can an electrostatically paired center be made to carry radicals when paired? Can a covalently paired center such as the Z/N cases (Between two metals) carry radicals when paired? Can a covalently paired center such as LiC_4H_9 (Between a metal and a free non-metal) carry radicals when paired? Can an ionically paired center such as $NaOCH_3$ (Between an ionic metal and a non-free non-metal) carry radicals when paired? The case above is between an ionic metal and a non-ionic metal and the bond is electrostatic. The Rh center has three paired unbonded radicals and seven vacant orbitals. As will be shown downstream, the paired unbonded radicals must be shielded to make the center free, before the center can be used. It is well known that they can carry charges. The $Na^\oplus$ can never carry a chain cationically. On the other hand, when free-media initiators are used, due to electrodynamic forces of repulsion resulting from the π-bonds in the rings, syndiotactic placement may be obtained as is the case with vinyl chloride.

For so many years since 1984, there has been many reports over the cationic and anionic polymerizations[9-11] of 1,3-pentadiene, cyclopentadiene, even to the point of awarding patents for their anionic polymerizations[12,13]!!! All these types of information in the literature too numerous to list have been big plugs to progress in humanity. How can 1,3-pentadiene with transfer species of the first kind of the first type be polymerized when nucleo-free-radicals or negatively charged initiators are used[12]? When organo-metallic initiators are used, the route for 1,3-pentadiene can only be positively or electro-free-radically. With an organo-metallic initiator such as LiC_4H_9, the carrier of the polymer chain is Li only radically, the monomer being a Female, i.e. a Nucleophile, and not C_4H_9. In addition positively, only 1,2- mono-form can exist along the chain, since resonance stabilization cannot take place chargedly to give the 1,4- mono-form. Radically both can exist along the chain with the 1,2- monomer first to appear if the initiator is weak in strength, then followed by 1,4- mono-form all along the chain to the end. The same initiator LiC_4H_9 which as has been said is dual in character, can also be used free-radically with the carrier being Li in the absence of the counter center. If the counter center is present, it will prevent the chain from growing. The chain obtained does not need a terminating agent since it can be killed by starvation to give a dead terminal double bond polymer. All the claims in the literature and in patents are absolute nonsense, creating so much confusions because cyclopentadiene cannot unlike 1,3-pentadiene be activated chargedly due to electrostatic forces of repulsion resulting from the negative charge and the very strong invisible π-bond inside the ring. It is strong, because the C to H ratio in cyclopentadiene is 1 to 1.2 smaller than 1 to 1.3 for cyclopropene. The CH_2 group in the cyclopentadiene as shown in Equation 11.126, is externally located to the conjugatedly placed double bonds. So also are the groups at the terminals of the two double bonds. These and the CH_2 group are not resonance stabilized. Only the groups carried inside the double bond (Internally located groups) are resonance stabilized.

<table>
<tr><td>

R'C ══ CR'   R' ≡ RESONANCE STABILIZED

e.CR   CR.n

R & CH₂ NOT RESONANCE STABILIZED

CH₂

**FAVORS ONLY ELECTRO-
FREE-RADICAL ROUTE.**
</td><td>

HC ══ CH

e.CH   CH.n

C
H F

FAVORS NO ROUTE   11.132
</td></tr>
</table>

When the CH_2 group whose radical-pushing capacity is less than that in cyclopropene is partly changed to CHF group, one can see the wonders of NATURE. Notice how the radicals have been placed above, due to the fact that there is one and only one configuration. Such cyclopentadiene will favor no free-radical route. Electro-free-radically F is the transfer species and nucleo-free-radically H is the transfer species.

So many weird types of initiators with weird solvents have over the years been used as additives, promoters, activators and more, without understanding how NATURE operates, i.e. understanding the mechanisms of systems and so much more. Initiator such as HCl adducts of cyclopentadiene (CPD) or a vinyl ether, $SnCl_4$ (so-called Lewis acid catalyst), nBu_4NCl (as an additive) in dichloromethane at -78^{0}C[9,10], have in the past been used to polymerize cyclopentadiene "cationically". The first initiator is an electro-free-radical electrostatic initiator as shown below using HCl and vinyl ether.

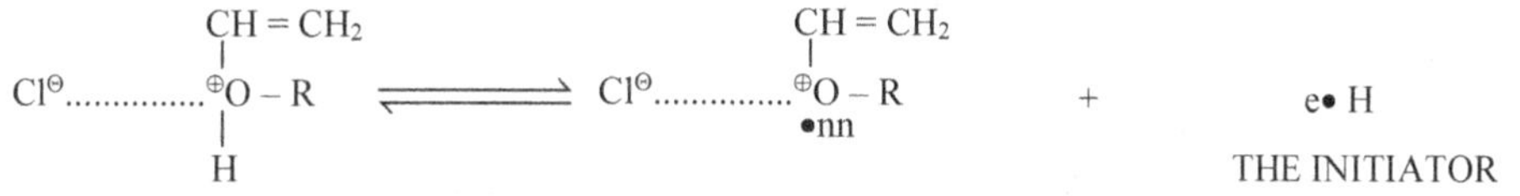

$$Cl^{\ominus}............\overset{\overset{\textstyle CH=CH_2}{|}}{\underset{\underset{\textstyle H}{|}}{\overset{\oplus}{O}-R}} \quad\rightleftharpoons\quad Cl^{\ominus}............\overset{\overset{\textstyle CH=CH_2}{|}}{\underset{\bullet nn}{\overset{\oplus}{O}-R}} \quad + \quad e\bullet H$$

$$\text{THE INITIATOR}$$

$$11.133$$

This is just like the use of fifty percent dilute HCl. The initiator is the H atom. This is what the monomer carries, later to be terminated by Cl when the optimum chain length has been reached, leaving the vinyl ether behind after the job or by starvation to form a dead terminal double bonded ring, leaving behind HCl and the vinyl ether.

For the second initiator, the followings are obtained.

Stage 1: $\quad SnCl_4 \quad\rightleftharpoons\quad Cl_3Sn\bullet e \quad + \quad Cl\bullet nn$

$$Cl_3Sn\bullet e \rightleftharpoons Cl_3Sn^{\oplus} \quad + \quad \bullet e$$

$$e\bullet \quad + \quad Cl\bullet nn \quad\rightleftharpoons\quad Cl^{\ominus}$$

$$Cl^{\ominus} \quad + \quad SnCl_4 \quad\rightleftharpoons\quad {}^{\ominus}SnCl_5$$

$$Cl_3Sn^{\oplus} \quad + \quad {}^{\ominus}SnCl_5 \quad\longrightarrow\quad Cl_3Sn^{\oplus}........{}^{\ominus}SnCl_5 \text{ [THE INITIATOR]} \qquad 11.134a$$

Overall equation: $\quad 2SnCl_4 \longrightarrow Cl_3Sn^{\oplus}.......{}^{\ominus}SnCl_5 \qquad 11.134b$

Chargedly, the initiator (the so-called Lewis acid catalyst) can be used as it is with the active center being the positively charged side. What has been shown above is incomplete as will be shown downstream, because the paired unbonded radicals on the Sn centers have not yet been shielded. Its Equilibrium state of existence is as follows.

$$Cl_3Sn^{\oplus}........{}^{\theta}SnCl_5 \quad\rightleftharpoons\quad Cl_3Sn^{\oplus}........\underset{\bullet e}{{}^{\theta}SnCl_4} \quad + \quad Cl\bullet nn$$

$$\textbf{THE INITIATOR} \qquad 11.134c$$

$$H^{\oplus}........{}^{\theta}SnCl_5 \quad\rightleftharpoons\quad H^{\oplus}........\underset{\bullet e}{{}^{\theta}SnCl_4} \quad + \quad Cl\bullet nn$$

$$\textbf{THE INITIATOR}^{\text{''}} \qquad 11.134d$$

The real initiator has been identified above. It is the electro-free-radical on the counter-charged center. ***The Cl•nn cannot close the chain since it cannot diffuse or move.*** The initiator was prepared via Equilibrium mechanism. It is then used via Combination mechanism to polymerize the monomer. ***However, the existence of the last initiator above is questionable, because the presence of $SnCl_4$ will not allow HCl to exist in Equilibrium state of existence.*** The existence of 1,2- mono-form along the chain will only depend on the strength of the initiator initially used and its appearance will only be from the beginning of the chain. The carrier of the chain is $Cl_3Sn^{\oplus}....{}^{\theta}SnCl_4$, leaving behind HCl after the growing chain has killed itself via release of H to form a dead terminal double bonded ringed polymer.

For the third initiator-nBu_4NCl (said to be used as an additive) in dichloromethane (CH_2Cl_2) at -78^{0}C, the followings are obtained.

THE INITIATOR   11.135

The initiator here can be observed to be n-Butyl carrying the electro-free-radical for the polymerization of the monomer. The counter-charged center carrying the nucleo-free-radical on the N center cannot close the chain since she cannot diffuse and the mechanism is not Equilibrium mechanism, but Combination mechanism. ***It is the CH_2Cl_2 that is the additive, the presence of which made it possible to operate at such a low temperature.*** While the carrier of the chain here is $n\text{-}C_4H_9$, when LiC_4H_9 is used as initiator, the carrier of the chain is Li, that which cannot be used for this of monomer if $t\text{-}C_4H_9$ is present in the system. One can see the wonders of Nature, that which is incomprehensible.

Cyclopentadiene like cyclohexadiene which was reported to favor anionic polymerization with the use of Alkyllithium/N,N,N',N'-Tetramethylethylenediamine $(CH_3)_2NCH_2CH_2N(CH_3)_2$ as initiator[14], can only be polymerized via electro-free-radical route, and no other route.

IMPOSSIBLE MONO-FORM ADDITIONS
11.136

Though free-media initiator from R.e was used in the reactions above, the placements shown cannot take place. One cannot begin with 1,4-addition and then latter have 1,2-addition along the chain. Secondly, one cannot have isotactic placement with such ring, due to steric limitations. Invariably, one cannot use a coordination center that gives only isotactic placement. With paired-media or indeed free-media initiator, the followings are to be expected.

1,4 - addition -polymerization
11.137

1,4 - addition mono-forms being the most stable of the two mono-forms, will largely predominate. The placement above is syndiotactic placement (see Equation 11.86). Higher polymerization temperatures will be required to achieve the 1,4 - addition polymerization above radically only. The lower the temperatures, the more the favored existence of 1,2-monomer along the chain.

Anionic polymerization of acrylonitrile (A Male) by cyclopentadienyl sodium has in the past been

246

reported[15]. Indeed, this is the route (as well as nucleo-free-radicals) natural to the monomer. Of particular interest is the initiator used.

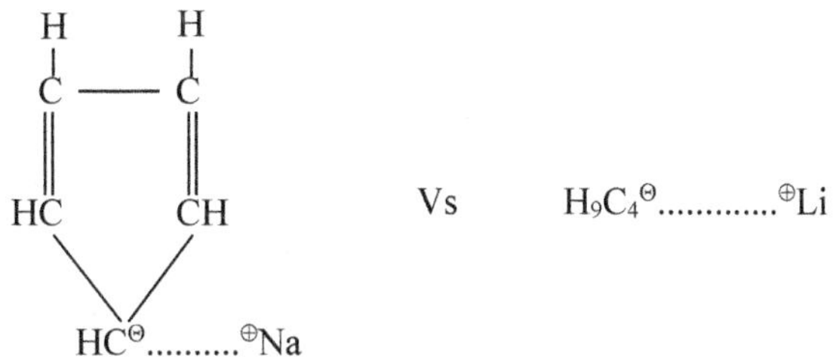

COVALENTLY CHARGED PAIRED INITIATORS 11.138

These like the Z/N types are all Covalently charged-paired initiators. They have dual characters, i.e. two active centers- one for a Male (Electrophiles) and the other for a Female (Nucleophiles, except that the $Na^{\oplus}$ cannot be used for females). Only the positive center is ionic, while the negative center is non-ionic, but covalent in character. It is non-ionic, because it is non-polar. It is free. When the initiator above is used on acrylonitrile, the carrier of the chain is the ring and not Na. Based on what one is seeing above there is no doubt that the cyclopendiene is a unique hydrocarbon. It is heavily unsaturated and strongly Nucleophilic in character. The Equilibrium state of existence of cyclopentadiene is as follows.

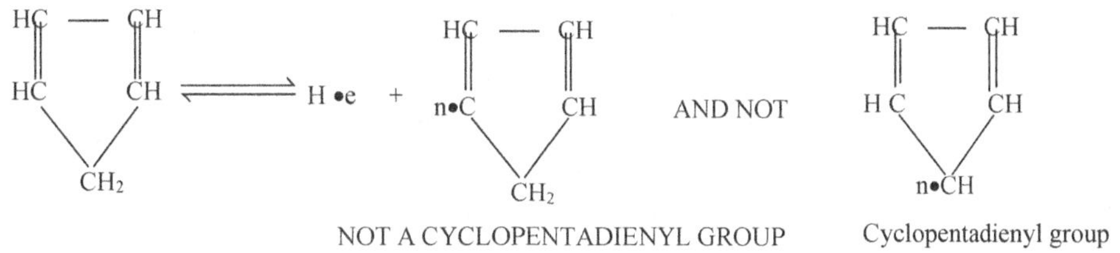

NOT A CYCLOPENTADIENYL GROUP   Cyclopentadienyl group   11.139

When I_2 like some others are made to react with the cyclopentadiene when kept in Stable state of existence, the H removed is not that held in Equilibrium state of existence, but that in CH_2 and ***that is what makes it cyclopentadienyl in character.*** When heated, above 1000K, many kinds of products have been reportedly obtained[16-21]. The products include acetylene and H-absorption profiles at **1260 to 1600K** with pressures between 0.7 and 5.6 bar in a single pulse shock tube[16]. Another found benzene, acetylene and ethylene at **1300 to 1700K** in a small flow tube reactor[17]. In another case, in order of abundance, the products obtained were acetylene, ethylene, methane, allene, propyne, butadiene, propylene and benzene at **1080 to 1550K** and pressures behind the shock was between 1.7 to 9.6 atm.[18]. Thermal decomposition of cyclopentadiene to $c\text{-}C_5H_5$ (cyclopentadienyl radical) and H and the reverse, were studied quantum chemically at the G2M level of theory[19]. The ring opening of the cyclopentadienyl radical was found to be the crucial step in the mechanism[18, 19,21]. Even Indene, benzene and naphthalene were found to be the major products of decomposition in a laminar flow reactor operating in the temperature range of **600 to 950°C**[21]. Obviously, the decomposition of the cyclopentadiene starting via Decomposition mechanism is impossible, since it has no point of scission to start with, and it cannot undergo any rearrangement to give another product. Its decomposition can only take place starting with Equilibrium mechanism. When held in Equilibrium state of existence, as shown in the last equation above, the group held is not a cyclopentadienyl group as mistakenly universally thought to be, but something else.

i) Shown below is the mechanism of decomposition of cyclopentadiene ***when all the fractions are held in Equilibrium state of existence.***

EQUILIBRIUM MECHANISM

Stage 1:

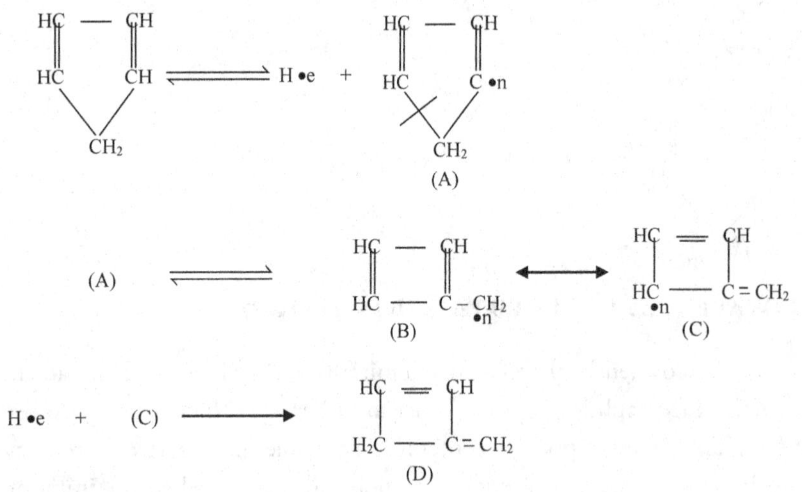

(A)

(A) ⇌ (B) ↔ (C)

H•e + (C) ⟶ (D)

Overall equation: 2Cyclopentadiene ⟶ 2(D)   11.140a

DECOMPOSITION MECHANISM   11.140b

Stage 2:

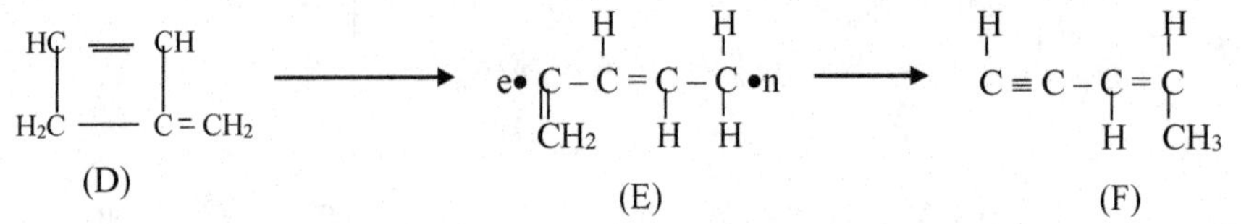

(D)   (E)   (F)

                                                        11.141a

Overall equation: 2Cyclopentadiene ⟶ (E) + (F)   11.141b

Stage 3a:

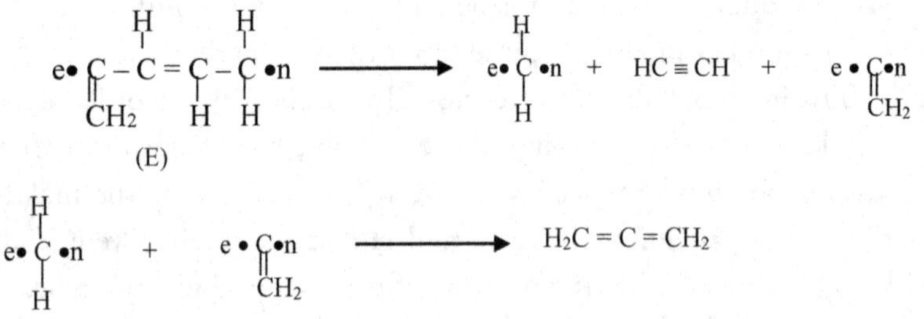

(E)

$$e\bullet \overset{H}{\underset{H}{C}}\bullet n \;+\; e\bullet \overset{}{\underset{CH_2}{C}}\bullet n \longrightarrow H_2C = C = CH_2$$

                                                        11.142a

Overall equation: 2Cyclopentadiene ⟶ Acetylene + Propyne + (F)   11.142b

Stage 3b:

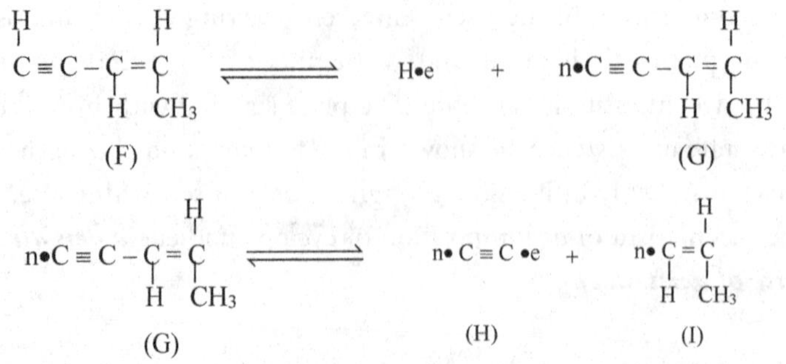

(F)   (G)

(G)   (H)   (I)

248

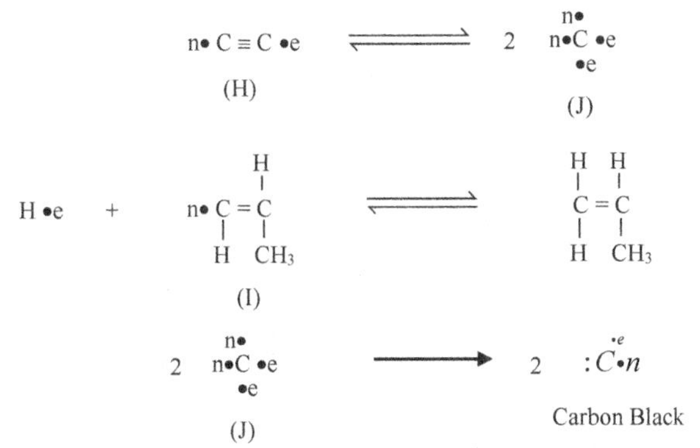

$$n\bullet C \equiv C \bullet e \quad \rightleftharpoons \quad 2 \quad n\bullet C \bullet e$$

(H)   (J)

$$H \bullet e \quad + \quad n\bullet C = C \quad \rightleftharpoons \quad C = C$$

(I)

$$2 \quad n\bullet C \bullet e \quad \longrightarrow \quad 2 \quad :C\bullet n$$

(J)   Carbon Black   11.143a

Overall equation: 2Cyclopentadiene ⟶ Acetylene + Propyne + Propene +

2Carbon Black   11.143b

One had expected to see acetylene or cumulenic units when decomposed via Equilibrium mechanism. However, based the operating conditions close to those used in the Petroleum industries for Cracking of hydrocarbons (above 1000^0C to 1750^0C), one is not surprised at the products obtained above based on the applications of the current developments in the NEW FRONTIERS. *When the operating conditions are such that all the cyclopentadienes are held in Equilibrium state of existence, the products obtained above are essentially the main products.* The system above is a series/parallel network system. In the first stage, methylene cyclobutene (D) was first obtained. This was what could not be obtained from cyclopropene, because of the size of the ring. Instead, an allene is obtained for cyclopropene which then rearranges to methyl acetylene (See Chapter 10). In stage 2, the (D) decomposed into (E). Half of (E) rearranged to Methyl vinyl acetylene (F), leaving the remaining half or fraction which could not rearrange behind in its activated state to decompose. In stages 3a and 3b, both in parallel, i.e. at the same time, the fraction of (E) left and (F) formed decompose to give the products shown. The stages above could indeed be four or more, because for example, the allene produced in Stage 3a, molecularly rearranges to methyl acetylene (Propyne). Where some observed products such as benzene and butadiene for example, these were formed based on their operating conditions-temperature, pressure, solvents used, presence of unwanted components, types of reactors, and so on. As will be shown shortly with the case of Toluene, the presence of benzene is a result of large number of moles of cyclopentadiene decomposed, for which three moles of acetylene were used to produce benzene as shown below.

Stage 4 or5:

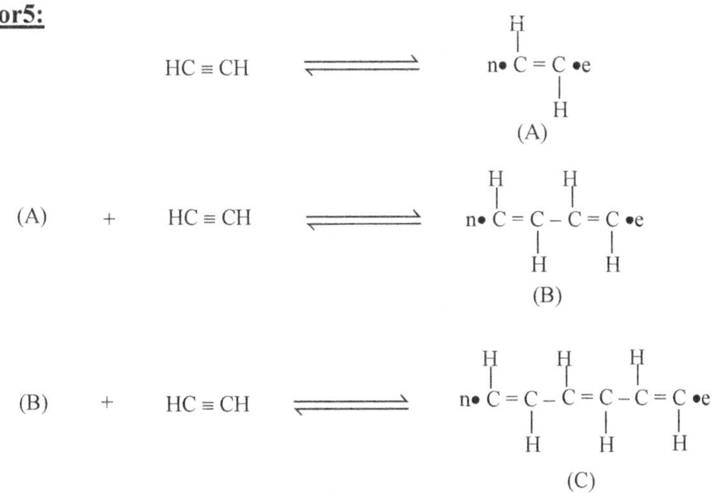

$$HC \equiv CH \quad \rightleftharpoons \quad n\bullet C = C \bullet e$$

(A)

$$(A) \quad + \quad HC \equiv CH \quad \rightleftharpoons \quad n\bullet C = C - C = C \bullet e$$

(B)

$$(B) \quad + \quad HC \equiv CH \quad \rightleftharpoons \quad n\bullet C = C - C = C - C = C \bullet e$$

(C)

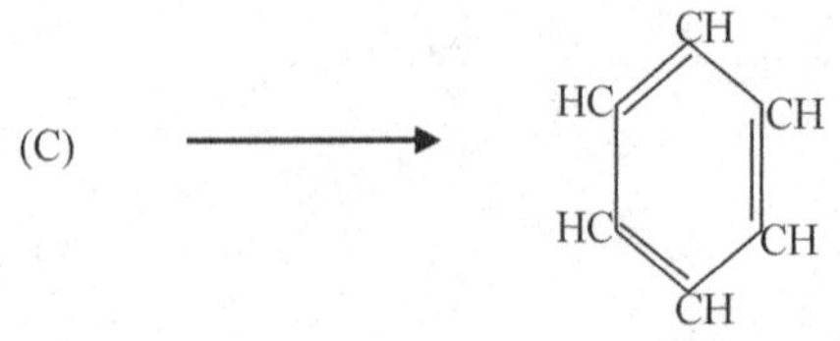

(C) ⟶ Benzene

11.144a

Overall equation: 3HC≡CH ⟶ Benzene 11.144b

Three moles of acetylenes in Activated state of existence based on the operating conditions (just a fraction since some are also in Equilibrium state of existence and probably none in Stable state of existence), combine together in one single stage to give benzene. Cyclobutadiene could not be formed because of the large SE in the ring. Reactions take place on molar basis. While one can observe that the ring has been expanded from five to six, as was observed during the reported ring expansion reactions in the thermal decomposition of tert-butyl-1,3-cyclopentadiene[22] and ring expansion in 1-, or 2- or 3-methylcyclopentadiene radicals; quantum chemical and kinetic calculations[23], *these have nothing to do with the mechanisms for ring expansion, wherein the original ring of five is taken as it is and worked on mechanically to expand it without opening the former ring. That is ring expansion.* The reactions in Stage 4 or 5 above like the cases just referred to above have nothing to do with expansion of rings. We did not expand the cyclopentadiene ring to benzene.

ii) When a fraction of the cyclopentadiene is in Stable state of existence leaving the remaining fraction in Equilibrium state of existence, then the followings take place.

EQUILIBRIUM MECHANISM

Stage 1:

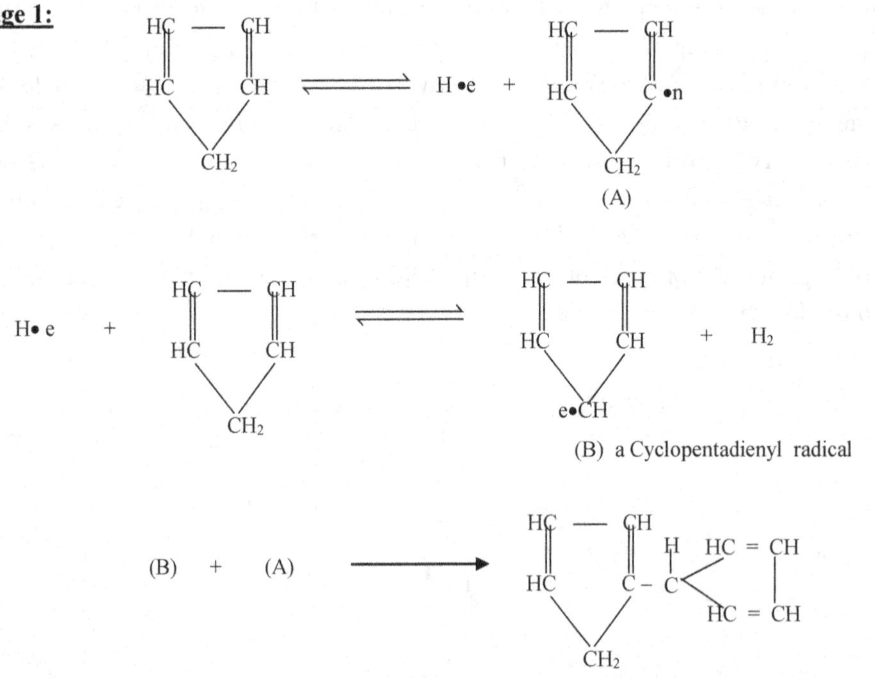

Overall equation: 2 Cyclopentadiene ⟶ H_2 + (C) 11.145b

Stage 2:

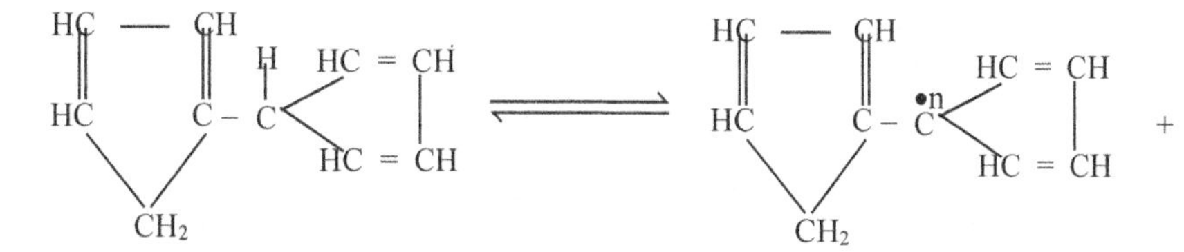

(C) 1-Cyclopentadienyl cyclopendiene   (D)

(D)   (E)

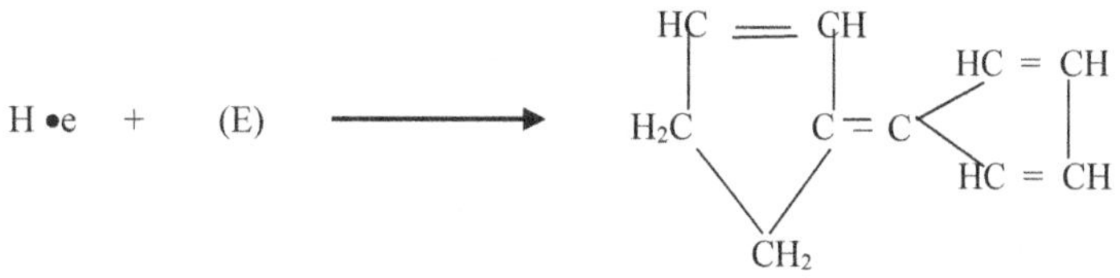

(F) Cyclopentadienylene
cyclopentene

11.146a

Overall equation: (C) $\longrightarrow$ (F)   11.146b

DECOMPOSITION MECHANISM

Stage 3:

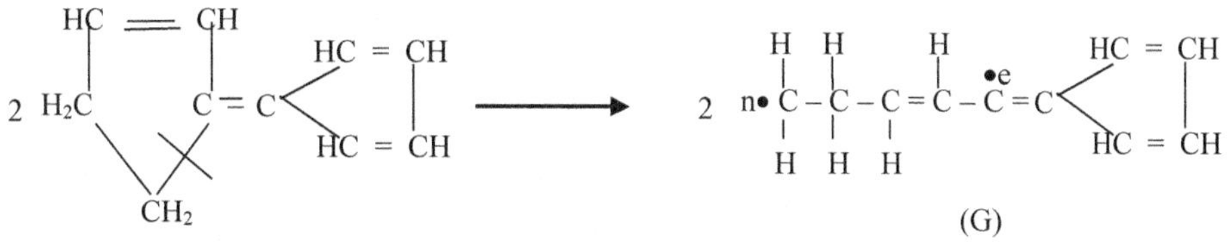

(F) Cyclopentadienylene
cyclopentene   (G)

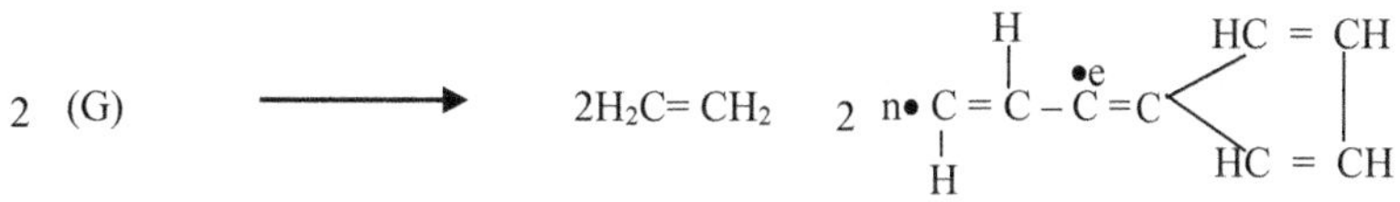

2 (G) $\longrightarrow$ $2H_2C = CH_2$ (H)

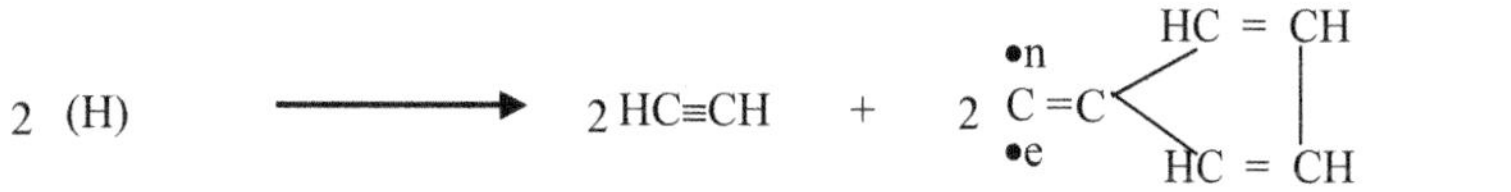

2 (H) $\longrightarrow$ $2 HC \equiv CH$ +   (I)

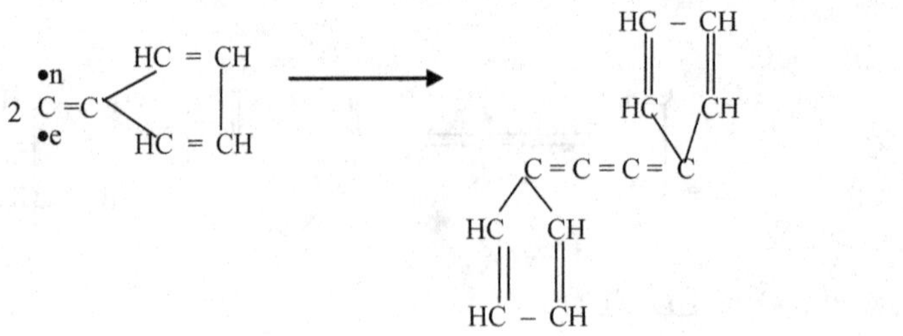

.(I*) 1,4-Di-cyclopentadienyl cumulene

11.147a

Overall equation: $\quad 2F \longrightarrow 2HC\equiv CH + 2H_2C=CH_2 + (I^*)$

11.147b

Overall overall equation: 12Cyclopentadiene $\longrightarrow 6H_2 + 6HC\equiv CH + 6H_2C=CH_2 + 3(I^*)$

11.148

EQUILIBRIUM MECHANISM

Stage 4:

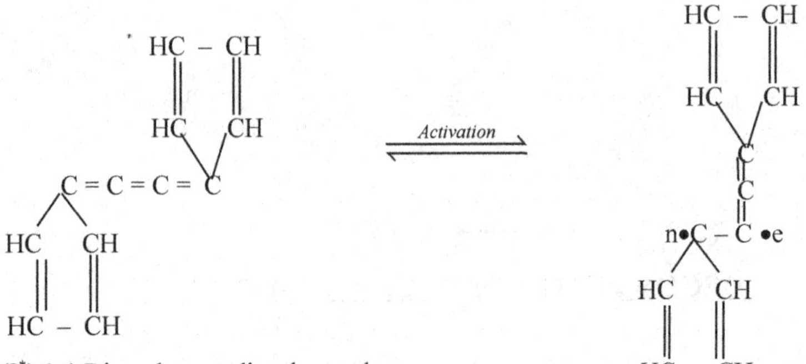

(I*) 1,4-Di-cyclopentadienyl cumulene

(J)

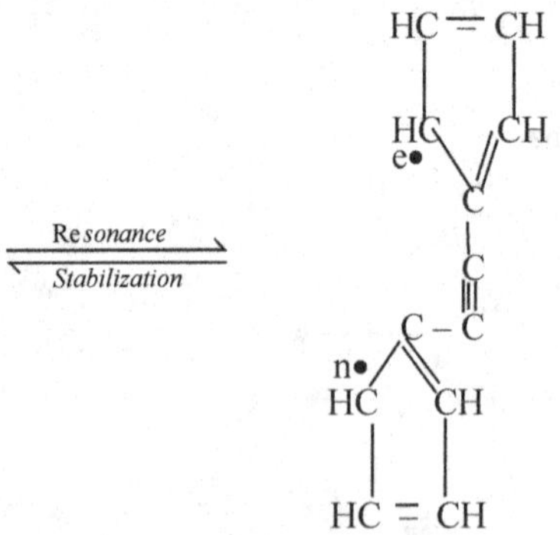

(K)

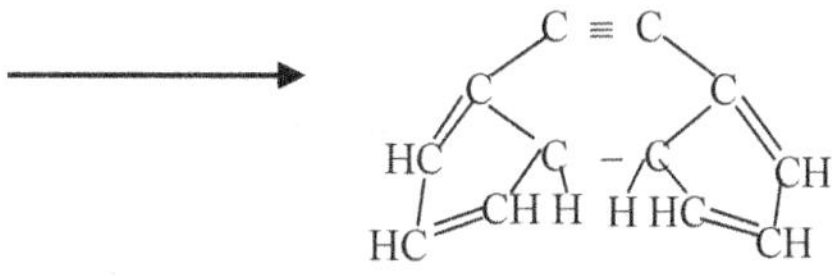

(L)

11.149a

Overall equation: (I) ⟶ (L)

11.149b

Stage 5

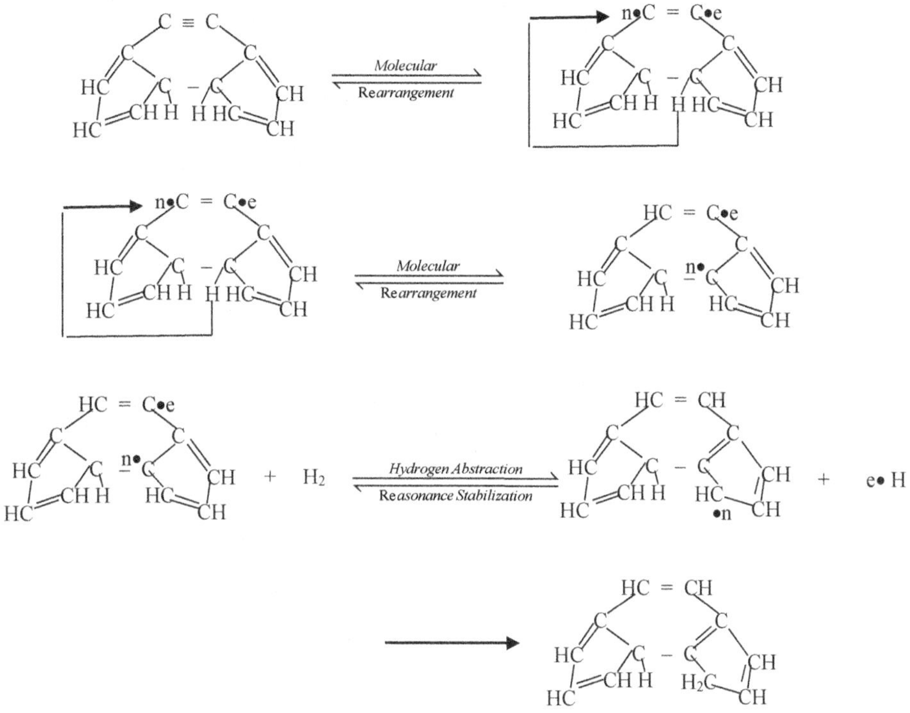

(M)

Overall overall equation: 12Cyclopentadiene ⟶ $6H_2C=CH_2$ + $6C_2H_2$ + 3(M)

11.150a

+ $3H_2$

11.150b

Stage 6:

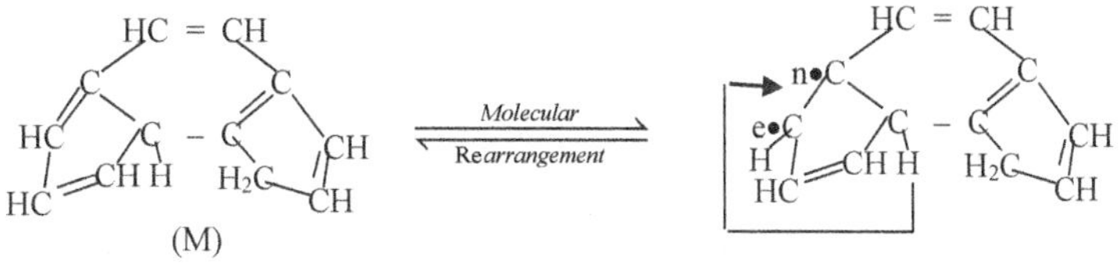

(M)

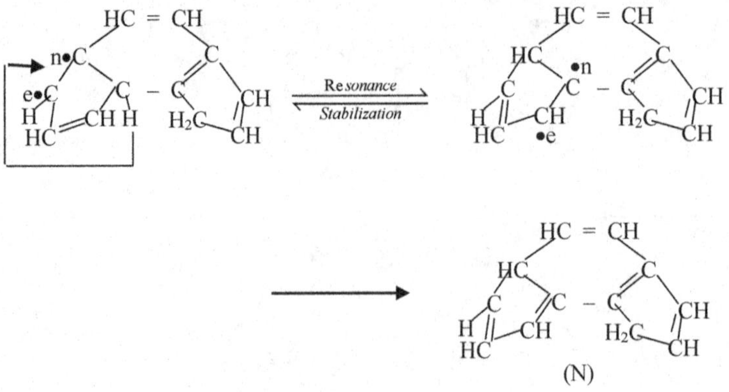

(N)   11.151

Stage 7:

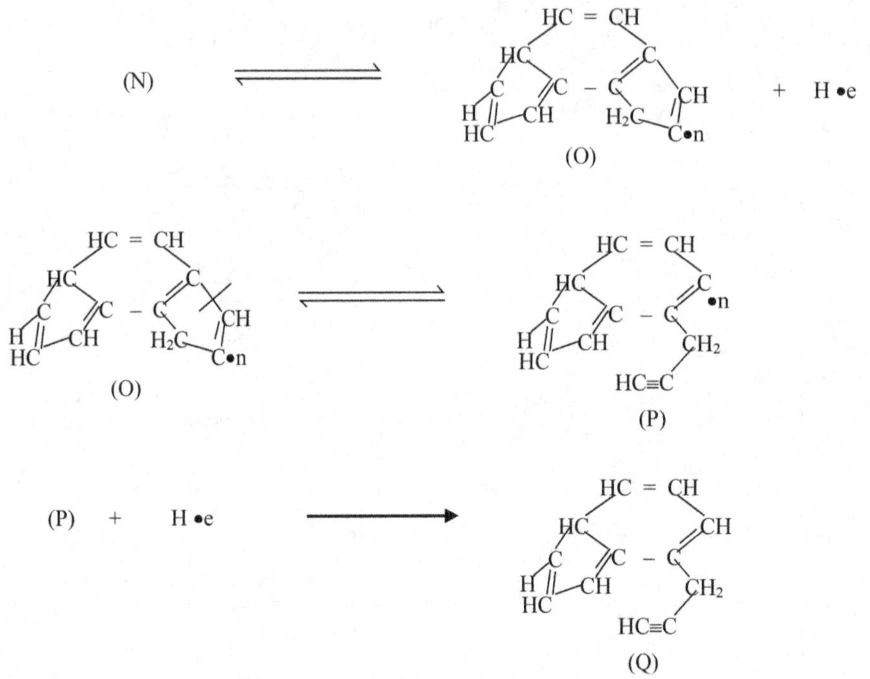

(N) ⇌ (O) + H •e

(O) ⇌ (P)

(P) + H •e → (Q)

11.152

Stage 8:

(R)

254

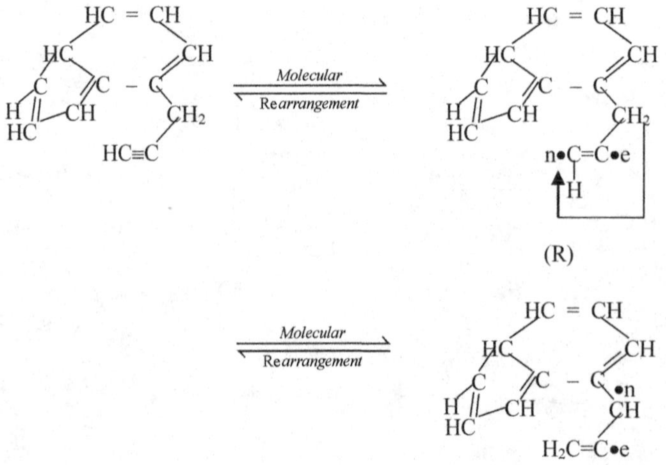

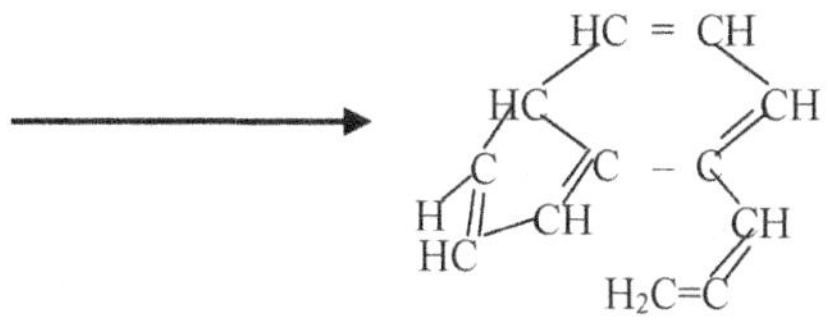

(S)

11.153

Stage 9:

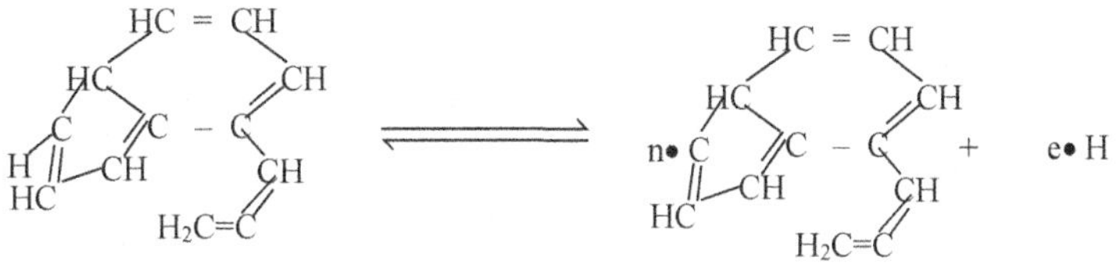

(T)

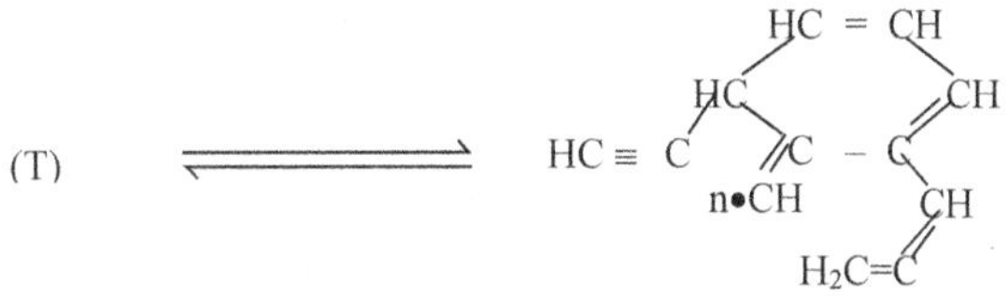

(T) ⇌

(U)

H •e + (U) ⇌

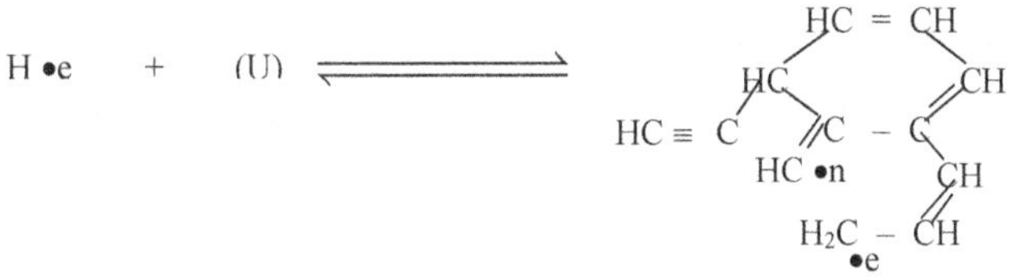

(V)

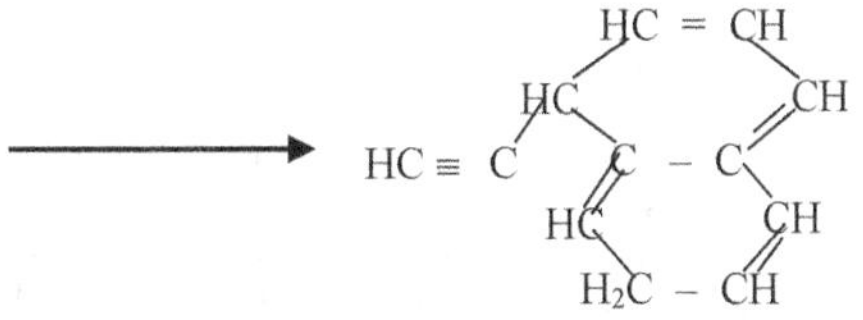

⟶

(W)

11.154

Stage 10:

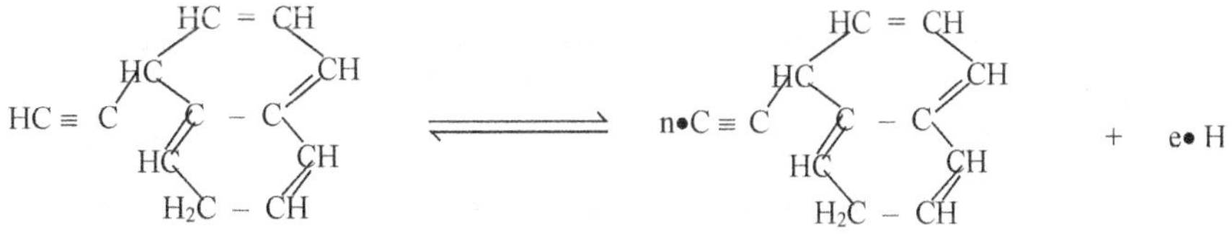

(X)

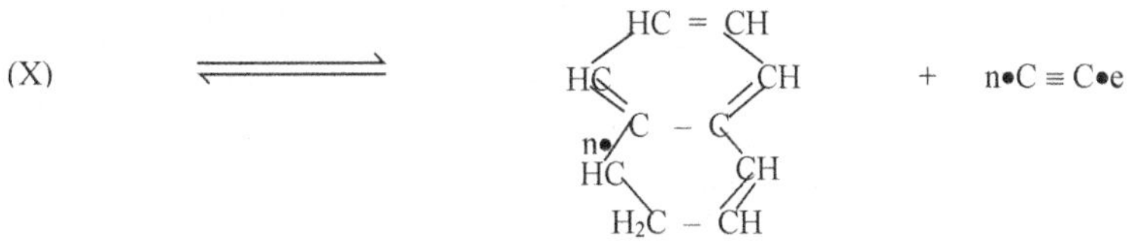

(X) ⇌

(Y)

255

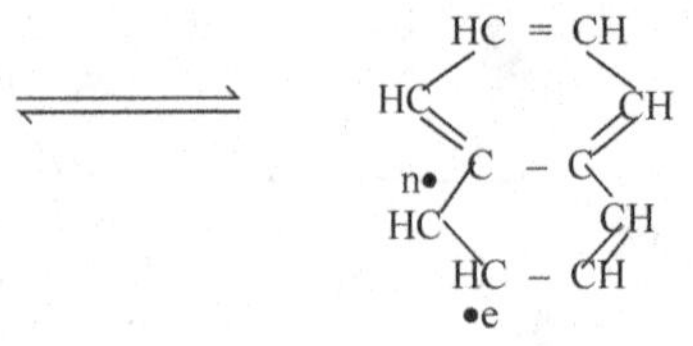

$$\rightleftharpoons \qquad \text{(Z)} \qquad + \quad n\bullet C \equiv CH$$

$$H\bullet e \quad + \quad n\bullet C \equiv CH \qquad \rightleftharpoons \qquad H C \equiv CH$$

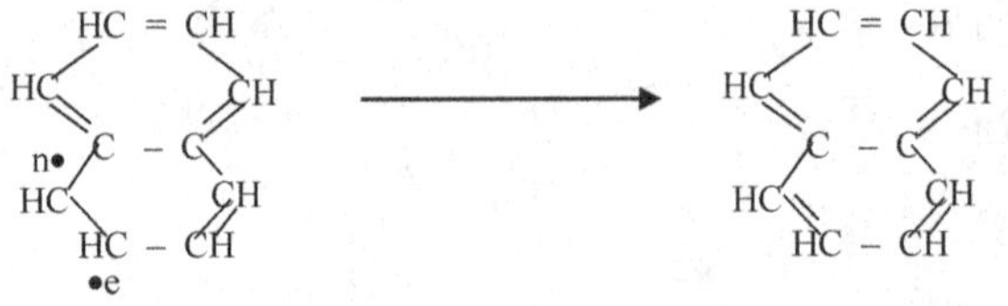

Z   Naphthalene   11.155a   **11.155a**

Overall overall equation: $12\text{Cyclopentadiene} \longrightarrow 6H_2C=CH_2 + 9C_2H_2 + 3H_2 +$

$$3C_{10}H_8 \qquad\qquad 11 \qquad\qquad \textbf{11.155b}$$

Stage 11: $H_2C = CH_2 \rightleftharpoons H\bullet e + n\bullet CH = CH_2$

$$n\bullet CH = CH_2 \rightleftharpoons n\bullet CH = HC\bullet e + H\bullet n$$

$$H\bullet e + H\bullet n \rightleftharpoons H_2$$

$$n\bullet CH = HC\bullet e \longrightarrow HC \equiv CH \qquad\qquad 11 \qquad\qquad \textbf{11.156a}$$

Overall overall equation: $12 \text{ Cyclopentadiene} \longrightarrow 15HC \equiv CH + 9H_2 + 3C_{10}H_8$

$$1 \qquad\qquad \textbf{11.156b}$$

Stage 12: Same as Stage 4 or 5 of Equation 11.144a.

Overall overall equation: $12\text{Cyclopentadiene} \longrightarrow 9H_2 + 5C_6H_6 + 3C_{10}H_8 \qquad 1 \qquad \textbf{11.156b}$

Twelve stages coincidentally were involved above for the synthesis of Naphthalene at such operating condition close to what obtains in the Petroleum industries with respect to Cracking of hydrocarbons. Note that throughout the stages, no ring was expanded in size. Though 9,10-dihydrofulvalene exists as shown below, it cannot be obtained during decomposition of cyclopentadiene, based on the new foundations being laid in the New Frontier. It is 1,10-dihydrofulvalene which was called 1-Cyclopentadienyl cyclopentadiene in Stage 1 that was first formed. In order to provide a point of scission for it, it was made to exist in Equilibrium state of existence to give Cyclopentadienylene cyclopentene in Stage 2, via resonance stabilization/Electroradicalization. Notice where the electro-free-radical came from to grab the nucleo-free-radical. It came from the second ring. In the third stage, decomposition began

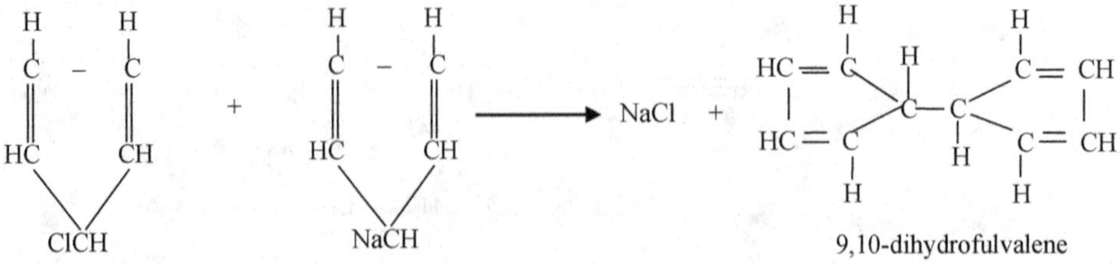

$\quad$ ClCH   NaCH   NaCl +   9,10-dihydrofulvalene   **11.157**

to start giving smaller products and most importantly (I'), herein called 1,4-Di-cyclopentadienylene cumulene. It from this naphthalene was formed. The first product was H_2 from Stage 1, followed by ethene and next acetylene and then the (I'). In Stage 4, with (I') activated via one of the externally located activation centers, rearranged the radicals via resonance stabilization to give well placed radicals which closed and formed a six-membered ring without yet opening the cyclopentadiene ring which are symmetrically placed on the ring (L). One can imagine how NATURE operates. None of the rings have been opened to form the six-membered ring carrying a triple bond well placed. At first one will wonder what the triple bond is doing there. In Stage 5, it was the triple bond the least nucleophilic of all the conjugatedly placed closed centers that was activated to begin molecular rearrangement and in order to provide a productive stage and much more, H_2 formed from Stage 1 was partly abstracted to form (M). In the process, one of the double bonds was removed from one of the five-membered rings and placed on the six-membered ring. Until this point, no point of scission has yet been provided. In Stage 6, (M) underwent both molecular rearrangement and resonance stabilization to give (N) which still has no point of scission. However, via Equilibrium state of existence, one of the five membered rings with only two double bonds was finally opened in Stage 7. In Stage 9, the second ring was opened with commencement of formation of a second six-membered ring. This continued to Stage 10 where the naphthalene was formed with release of acetylene. No soot was formed. In Stage 11, the ethene decomposed to acetylene and H_2. Finally, in Stage 12, all the acetylenes formed all along were now used to give benzene. The products obtained above will be the main products when the number of moles of cyclopentadiene is even. When odd, the situation is slightly different.

When the number of moles of cyclopentadiene involved in an "ideal reactor" is odd, then Indene and some other products begin to appear in the natural world. For the mechanism provided above, one used twelve moles of cyclopentadiene in a capillary reactor, the types that largely exist in NATURE. In the physical world of today, it does not matter whether it is odd or even, because most of the reactors are not close to what exist in Nature. For example, how many pumps are flexible? In fact, without the complex route above, once benzene and the smaller components have been formed, naphthalene, anthracene and more complex rings can readily be obtained. For example, from acetylene alone, the followings can be obtained-

(i)   Benzene, using three moles
(ii) Styrene, using four moles, from which Cyclooctatetraene is obtained, and this molecularly rearranges to styrene.
(iii) And many more complex products when combined with other components such as benzene.

It should be noted, that ethene is never a major product from methylene. The methylenes formed can be used with the benzene ring for example to give cycloheptatriene. This is where the real ring expansion takes place as shown below.

Stage 1:

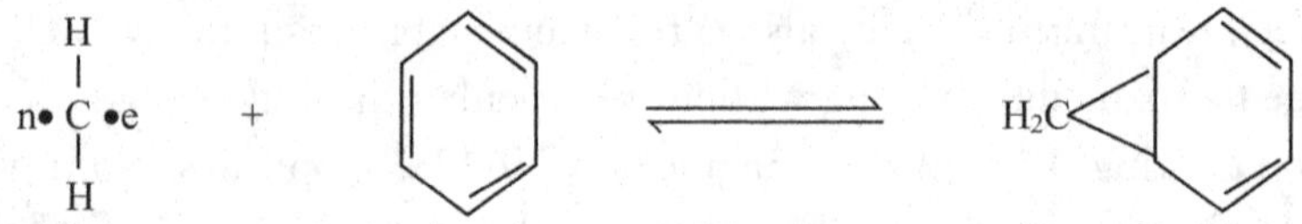

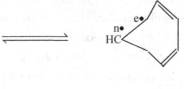

Stage 2:

11.158

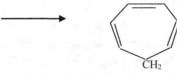

Cycloheptatriene

11.159a

Overall Equation: Methylene + Benzene ⟶ Cycloheptadiene

11.159b

The methylene can be used along with acetylene and benzene as follows.

Stage 1:

$$n\bullet \overset{\overset{\displaystyle H}{|}}{\underset{\underset{\displaystyle H}{|}}{C}} \bullet e \quad + \quad HC \equiv CH \quad \rightleftharpoons \quad n\bullet \overset{\overset{\displaystyle H}{|}}{\underset{\underset{\displaystyle H}{|}}{C}} - \overset{\overset{\displaystyle H}{|}}{\underset{\underset{\displaystyle H}{|}}{C}} = \overset{\overset{\displaystyle H}{|}}{C} \bullet e$$

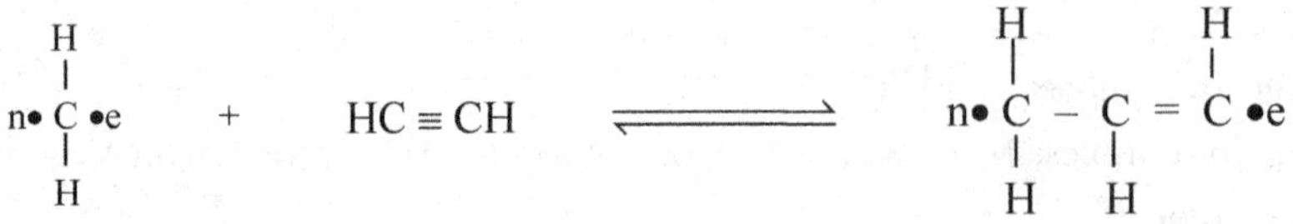

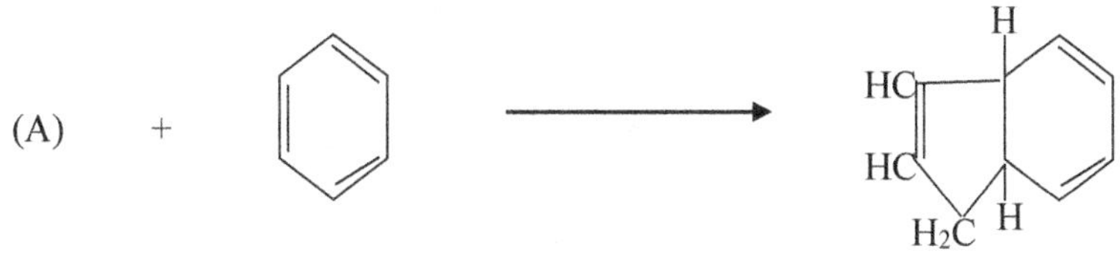

$$(A) \quad + \quad [\text{benzene}] \quad \longrightarrow \quad (B) \qquad 11.160$$

Stage 2:

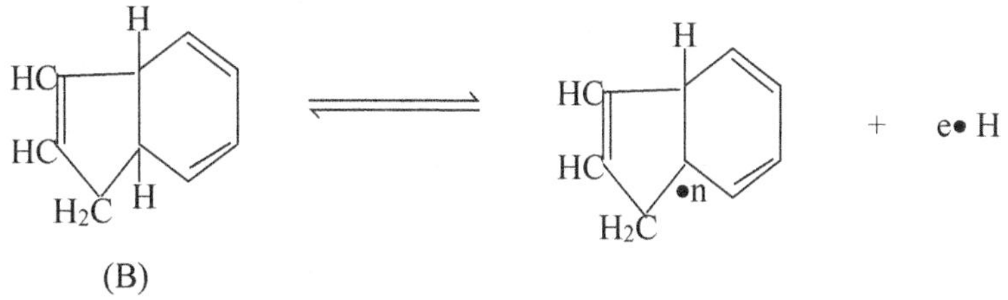

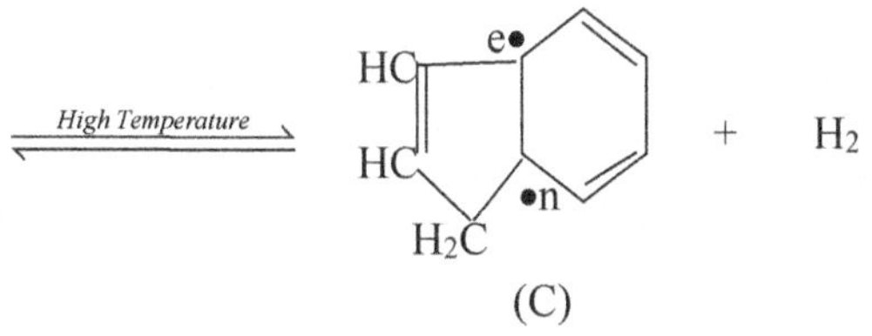

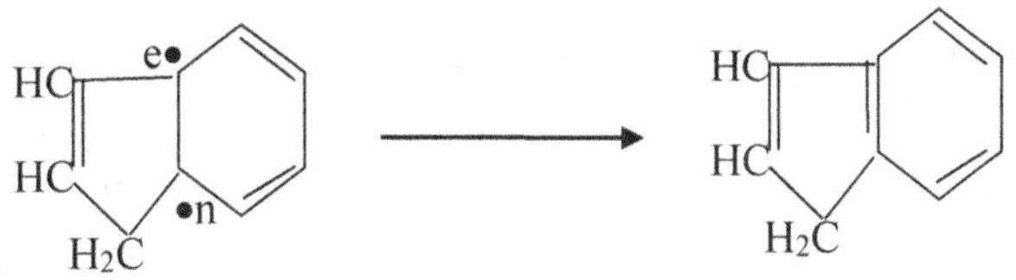

$$(B) \quad \rightleftharpoons \quad + \quad e\bullet H$$

$$\xrightarrow{\textit{High Temperature}} \quad (C) \quad + \quad H_2$$

$$(C) \quad \longrightarrow \quad \text{INDENE} \qquad 11.161a$$

Overall equation: Methylene + Acetylene + Benzene $\longrightarrow$ *Indene* + H_2 11.161b

When benzene and two mole of acetylene are available in the system, the followings are obtained.

Stage1:

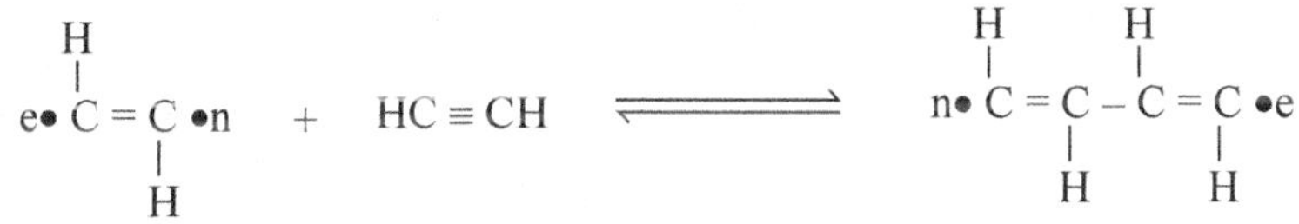

$$e\bullet C = C \bullet n \quad + \quad HC \equiv CH \quad \rightleftharpoons \quad n\bullet C = C - C = C \bullet e \qquad (A)$$

$$n\bullet C = C - C = C \bullet e \quad + \quad [\text{benzene}] \quad \longrightarrow \quad (B) \qquad 11.162$$

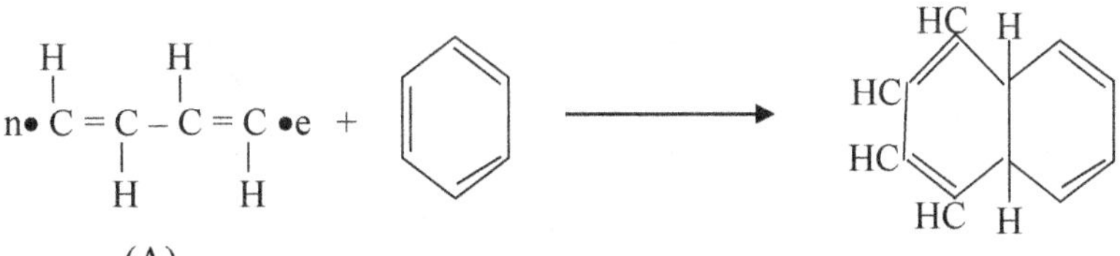

Stage 2:

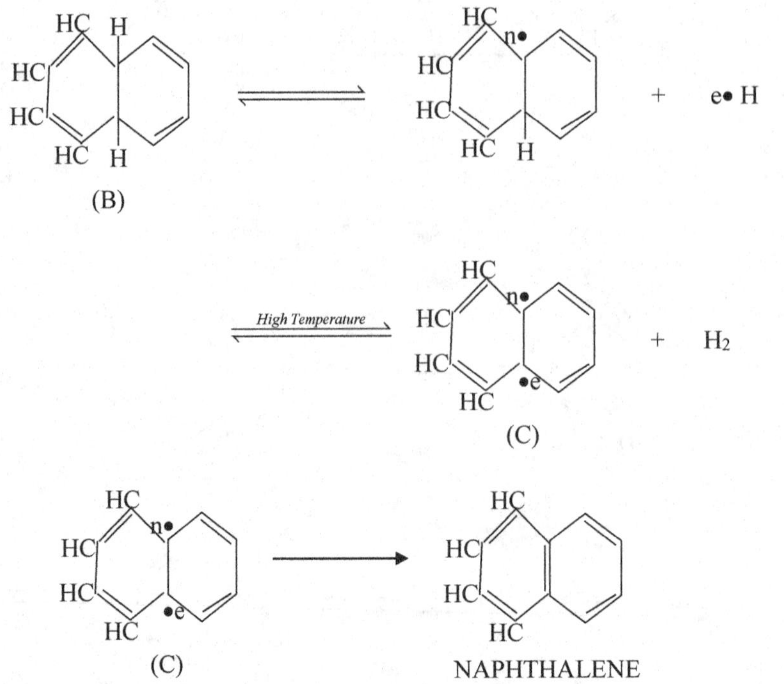

(B) ⇌ (C) + e• H

⇌ (High Temperature) → (C) + H_2

(C) ⟶ NAPHTHALENE   11.163a

Overall equation: 2Acetylene + Benzene ⟶ Naphthalene

11.163b

At the operating conditions, the monomers are more in self-activated states. It is the more nucleophilic that diffuses to the less nucleophilic monomer using the electro-free-radical ends.

In the so-called reported ring expansion reactions in the thermal decomposition of tert-butyl-1,3-cyclopentadiene[22] in single-pulse shock tube studies at shock pressures of 182-260kPa and temperature of 996-1127K, Isobutene (2-methylpropene), 1,3-cyclopentadiene and toluene were observed as the major products. The thermolysis of the dilute mixtures of the substrate was carried out in the presence of a free-radical scavenger. Based on current developments, the mechanisms of the reactions can be explained as follows.

EQUILIBRIUM MECHANISM

Stage 1:

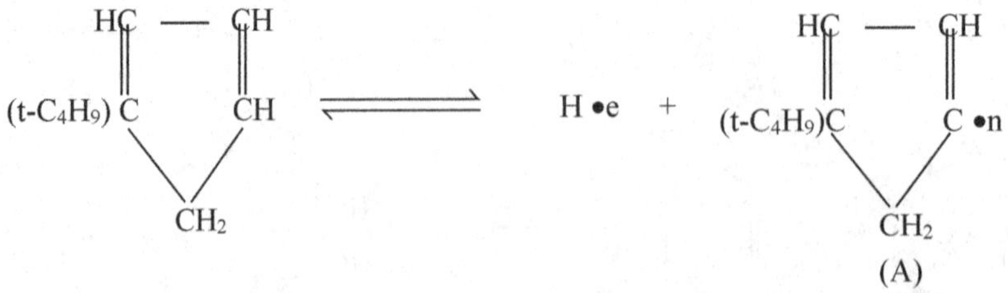

$H •e$ + (A)

260

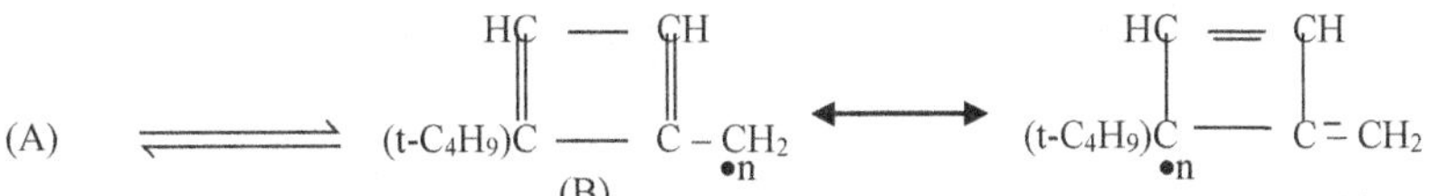

$$(A) \quad \rightleftharpoons \quad \underset{(B)}{(t\text{-}C_4H_9)C \overset{\displaystyle HC = CH}{\underset{\displaystyle \underset{\bullet n}{C - CH_2}}{\big|\qquad\big|}} } \quad \longleftrightarrow \quad \underset{(C)\ \text{More Stable}}{(t\text{-}C_4H_9)\underset{\bullet n}{C} \overset{\displaystyle HC = CH}{\underset{\displaystyle C = CH_2}{\big|\qquad\big|}}}$$

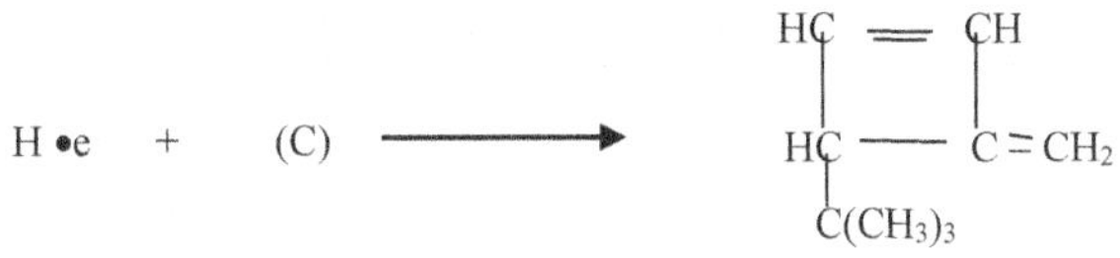

$$H \bullet e \quad + \quad (C) \quad \longrightarrow \quad \underset{(D)}{\underset{\displaystyle \overset{\displaystyle |}{C(CH_3)_3}}{HC \overset{\displaystyle HC = CH}{\underset{\displaystyle C = CH_2}{\big|\qquad\big|}}}}$$

11.164a

Overall equation: 2 t-Butyl Cyclopentadiene $\longrightarrow$ 2(D)

DECOMPOSITION MECHANISM

Stage 2:

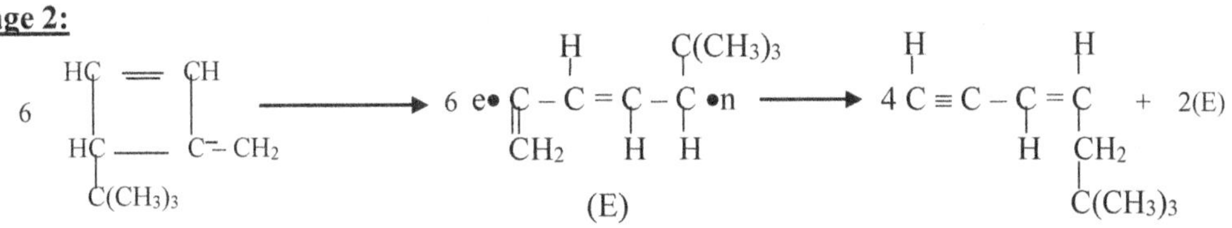

$$6\ \underset{\displaystyle C(CH_3)_3}{HC \overset{\displaystyle HC = CH}{\underset{\displaystyle C - CH_2}{\big|\qquad\big|}}} \longrightarrow 6\ e\bullet \underset{(E)}{\overset{\displaystyle H \qquad C(CH_3)_3}{\underset{\displaystyle CH_2 \quad H \quad H}{C - C = C - C \bullet n}}} \longrightarrow 4\ \underset{(F)}{\overset{\displaystyle H \qquad H}{\underset{\displaystyle H \quad CH_2}{C \equiv C - C = C}}} \underset{\displaystyle C(CH_3)_3}{} + 2(E)$$

(D)

11.165a

Overall equation: 6 t-Butyl Cyclopentadiene $\longrightarrow$ 4(F) + 2(E)   11.165b

Based on the types of products obtained, two thirds of (E) formed rearranged while the remainder was decomposed as shown below.

Stage 3a:

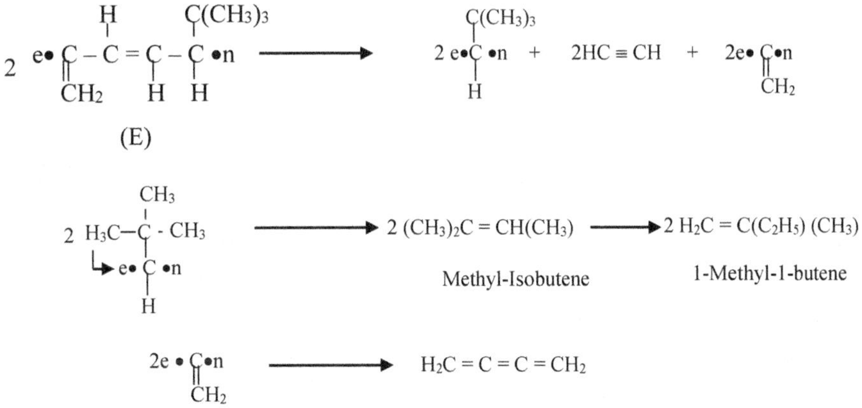

$$2\ e\bullet \underset{(E)}{\overset{\displaystyle H \qquad C(CH_3)_3}{\underset{\displaystyle CH_2 \quad H \quad H}{C - C = C - C \bullet n}}} \longrightarrow 2\ e\bullet\overset{\displaystyle C(CH_3)_3}{\underset{\displaystyle H}{C \bullet n}} + 2HC \equiv CH + 2e\bullet\overset{\displaystyle}{\underset{\displaystyle CH_2}{C \bullet n}}$$

$$2\ H_3C\text{-}\overset{\displaystyle CH_3}{\underset{\displaystyle \overset{\displaystyle |}{\underset{\bullet}{}}\ e\bullet \overset{}{\underset{\displaystyle H}{C} \bullet n}}{C}}\text{-}CH_3 \longrightarrow 2\ (CH_3)_2C = CH(CH_3) \longrightarrow 2\ H_2C = C(C_2H_5)(CH_3)$$

Methyl-Isobutene   1-Methyl-1-butene

$$2e \bullet \overset{}{\underset{\displaystyle CH_2}{C \bullet n}} \longrightarrow H_2C = C = C = CH_2$$

11.166a

Overall equation: 6 t-Butyl-Cyclopentadiene $\longrightarrow$ 2Acetylene + 2 [1-Methyl-1-butene] +

1,4-cumulene + 4(F)

11.166b

EQUILIBRIUM MECHANISM

Stage 3b:

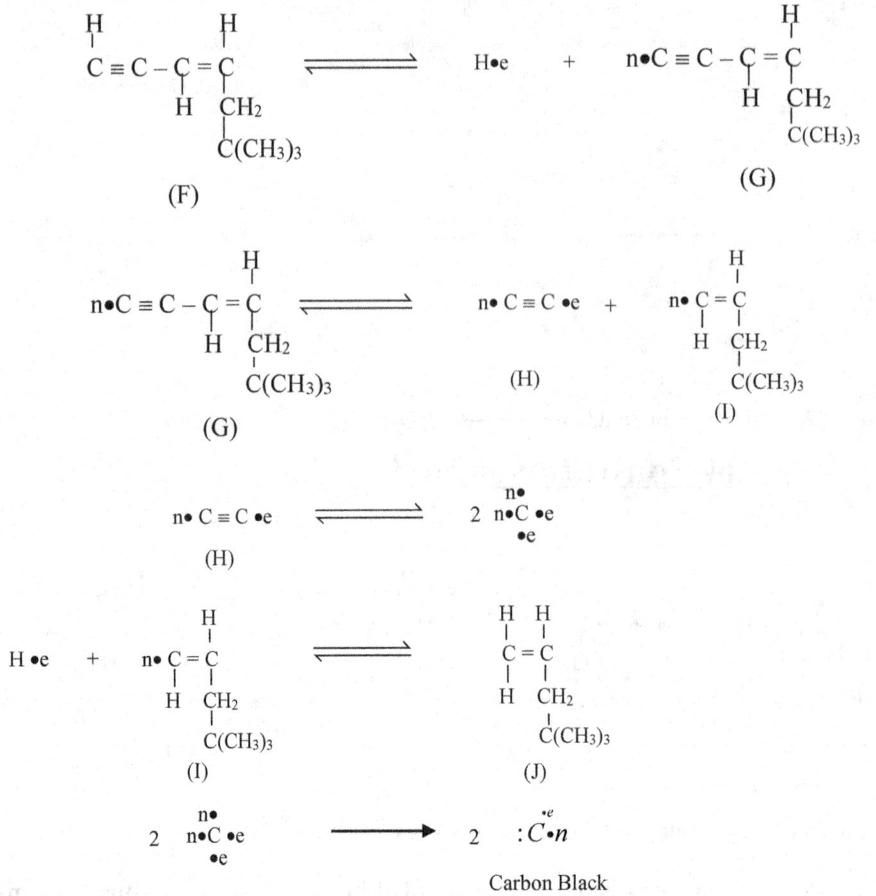

(F) ⇌ H•e + (G)

(G) ⇌ (H) n• C ≡ C •e + (I)

(H) n• C ≡ C •e ⇌ 2 (with n• C •e structure)

H •e + (I) ⇌ (J)

2 (n• C •e) ⟶ 2 : C•n   Carbon Black

11.167a

Overall equation: 6 t-butyl-Cyclopentadiene ⟶ 2 Acetylene + 2 [1-Methyl-1-butene] +

1,4-cumulene + 4 [t-butyl Propene] + 8 Carbon Black

11.167b

Stage 4:

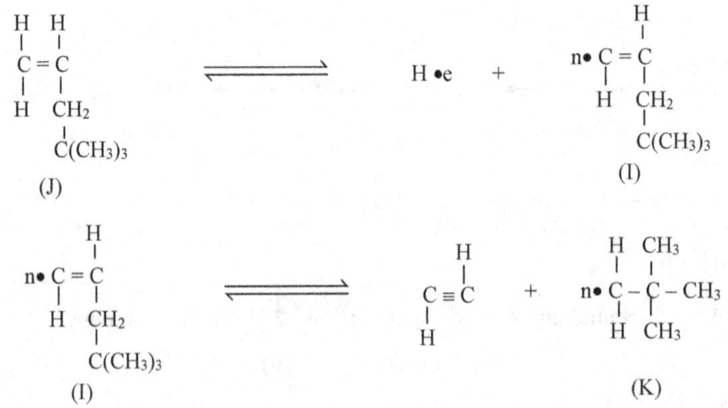

(J) ⇌ H •e + (I)

(I) ⇌ C ≡ C (H) + n• C – C – CH3 (K)

$$n\bullet \underset{\underset{\displaystyle CH_3}{|}}{\overset{\overset{\displaystyle CH_3}{|}}{C}} - \underset{\underset{\displaystyle CH_3}{|}}{\overset{\overset{\displaystyle CH_3}{|}}{C}} - CH_3 \rightleftharpoons n\bullet CH_3 \quad + \quad n\bullet \underset{\underset{\displaystyle CH_3}{|}}{\overset{\overset{\displaystyle CH_3}{|}}{C}} - \underset{\underset{\displaystyle CH_3}{|}}{\overset{\overset{\displaystyle CH_3}{|}}{C}} \bullet e$$

(K) (L)

$$H \bullet e \quad + \quad n\bullet CH_3 \rightleftharpoons CH_4$$

$$n\bullet \underset{\underset{\displaystyle H}{|}}{\overset{\overset{\displaystyle H}{|}}{C}} - \underset{\underset{\displaystyle CH_3}{|}}{\overset{\overset{\displaystyle CH_3}{|}}{C}} \bullet e \longrightarrow \underset{\underset{\displaystyle H}{|}}{\overset{\overset{\displaystyle H}{|}}{C}} = \underset{\underset{\displaystyle CH_3}{|}}{\overset{\overset{\displaystyle CH_3}{|}}{C}}$$

(L) Isobutene 11.168a

Overall equation: 6 t-Butyl-Cyclopendiene ⟶ 6Acetylene + 4Methane + 4Isobutene

+ 2[1-Methyl-1-Butene] + 1,4-cumulene + 8C

11.168b

Stage 5:

$$\underset{\underset{\displaystyle H}{|}}{\overset{\overset{\displaystyle H}{|}}{C}} = \underset{\underset{\displaystyle CH_3}{|}}{\overset{\overset{\displaystyle CH_3}{|}}{C}} \rightleftharpoons n\bullet \underset{\underset{\displaystyle H}{|}}{\overset{\overset{\displaystyle CH_3}{}}{C}} = \underset{\underset{\displaystyle CH_3}{|}}{C} \quad + \quad e\bullet H$$

Isobutene

$$n\bullet \underset{\underset{\displaystyle H}{|}}{\overset{\overset{\displaystyle CH_3}{}}{C}} = \underset{\underset{\displaystyle CH_3}{|}}{C} \rightleftharpoons n\bullet \underset{\underset{\displaystyle H}{|}}{\overset{\overset{\displaystyle CH_3}{}}{C}} = C \bullet e \quad + \quad n\bullet CH_3$$

$$H\bullet e \quad + \quad n\bullet CH_3 \rightleftharpoons CH_4$$

$$n\bullet \underset{\underset{\displaystyle H}{|}}{\overset{\overset{\displaystyle CH_3}{}}{C}} = C \bullet e \longrightarrow \underset{\underset{\displaystyle H}{|}}{\overset{\overset{\displaystyle CH_3}{}}{C}} \equiv C$$

11.169a

Overall equation: 2Isobutene ⟶ 2CH₄ + 2Propyne 11.169b

Note that only half of the isobutene was kept in Equilibrium state of existence, because of the large number of moles, leaving the remaining fraction in Stable state of existence.

Stage 6:

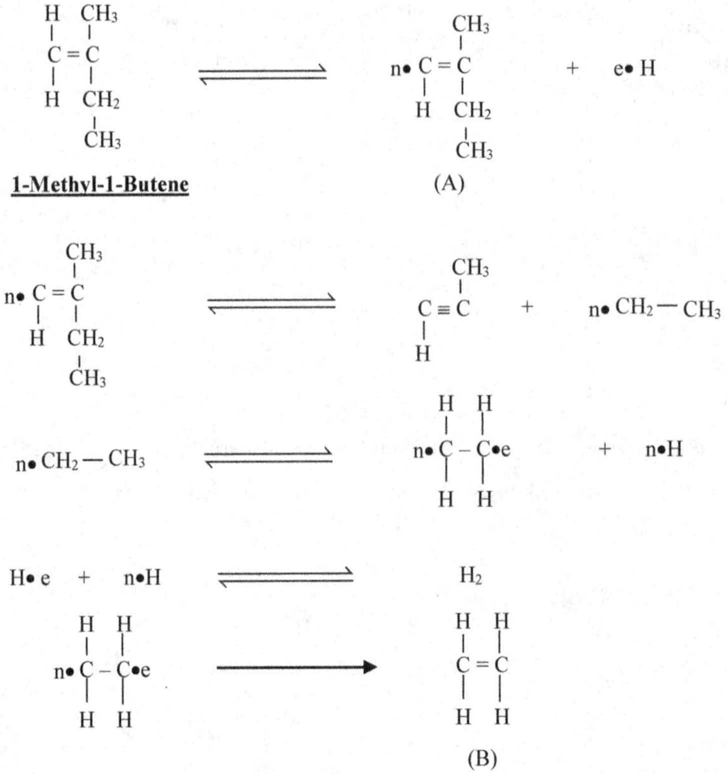

$$\underset{\textbf{1-Methyl-1-Butene}}{\underset{\overset{|}{CH_3}}{\underset{\overset{|}{CH_2}}{\overset{H\quad CH_3}{\underset{H}{\overset{|\quad\;\;|}{C=C}}}}}} \;\;\rightleftharpoons\;\; \underset{(A)}{\underset{\overset{|}{CH_3}}{\underset{\overset{|}{CH_2}}{\overset{CH_3}{n\bullet\underset{H}{\overset{|}{C}}=C}}}} \;\; + \;\; e\bullet H$$

$$\underset{\overset{|}{CH_3}}{\underset{\overset{|}{CH_2}}{\overset{CH_3}{n\bullet\underset{H}{\overset{|}{C}}=C}}} \;\;\rightleftharpoons\;\; \underset{H}{\overset{CH_3}{\underset{|}{C}\equiv C}} \;\; + \;\; n\bullet CH_2-CH_3$$

$$n\bullet CH_2-CH_3 \;\;\rightleftharpoons\;\; \underset{H\;\;H}{\overset{H\;\;H}{n\bullet\underset{|\;\;|}{\overset{|\;\;|}{C-C}}\bullet e}} \;\; + \;\; n\bullet H$$

$$H\bullet e \;\; + \;\; n\bullet H \;\;\rightleftharpoons\;\; H_2$$

$$\underset{H\;\;H}{\overset{H\;\;H}{n\bullet\underset{|\;\;|}{\overset{|\;\;|}{C-C}}\bullet e}} \;\;\longrightarrow\;\; \underset{(B)}{\underset{H\;\;H}{\overset{H\;\;H}{\underset{|\;\;|}{\overset{|\;\;|}{C=C}}}}}$$

11.169c

Overall equation: 2[1-Methyl-1-Butene] $\longrightarrow$ 2Propyne + 2Ethene + 2H$_2$

11.169d

Stage 7:

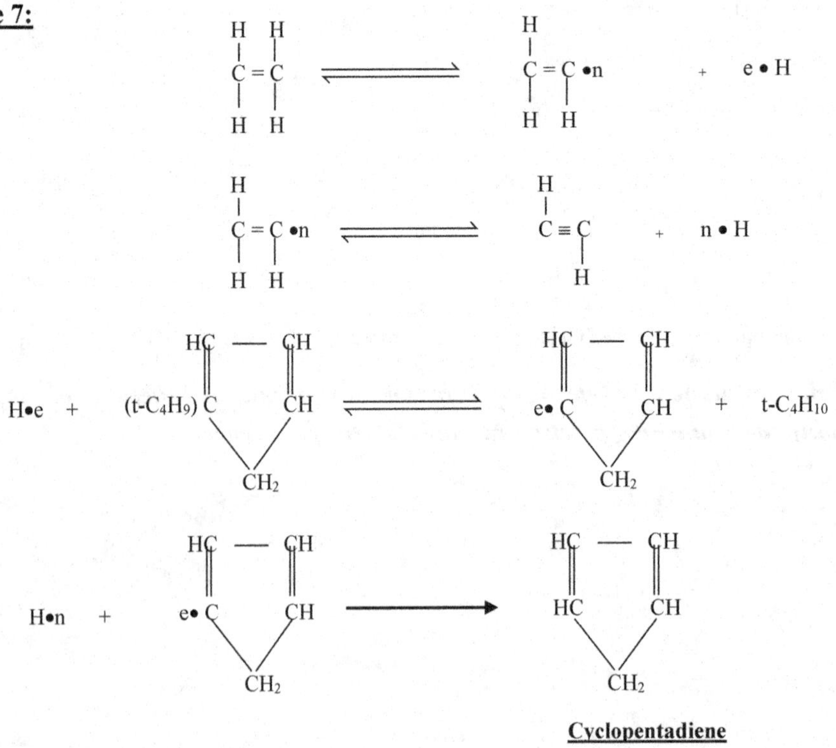

$$\underset{H\;\;H}{\overset{H\;\;H}{\underset{|\;\;|}{\overset{|\;\;|}{C=C}}}} \;\;\rightleftharpoons\;\; \underset{H\;\;H}{\overset{H}{\underset{|\;\;|}{\overset{|}{C=C}}\bullet n}} \;\; + \;\; e\bullet H$$

$$\underset{H\;\;H}{\overset{H}{\underset{|\;\;|}{\overset{|}{C=C}}\bullet n}} \;\;\rightleftharpoons\;\; \underset{H}{\overset{H}{\underset{|}{C}\equiv C}} \;\; + \;\; n\bullet H$$

$$H\bullet e \;\; + \;\; (t\text{-}C_4H_9)\,C \cdots \;\;\rightleftharpoons\;\; e\bullet C \cdots \;\; + \;\; t\text{-}C_4H_{10}$$

$$H\bullet n \;\; + \;\; e\bullet C \cdots \;\;\longrightarrow\;\; \underset{\textbf{Cyclopentadiene}}{\cdots}$$

11.170a

Overall equation: 2Ethene + 2t-Butyl-Cyclopentadiene $\longrightarrow$ 2Acetylene + 2t-Butane + 2Cyclopentadiene   11.170b

Overall overall equation: 8t-Butyl-Cyclopentadiene $\longrightarrow$ 8Acetylene + 6Methane + 4Propyne + 2H$_2$ + 8C + 2t-Butane + 2Cyclopentadiene + 1,4-Cumulene + 2 Isobutene   11.170c

Stage 8:

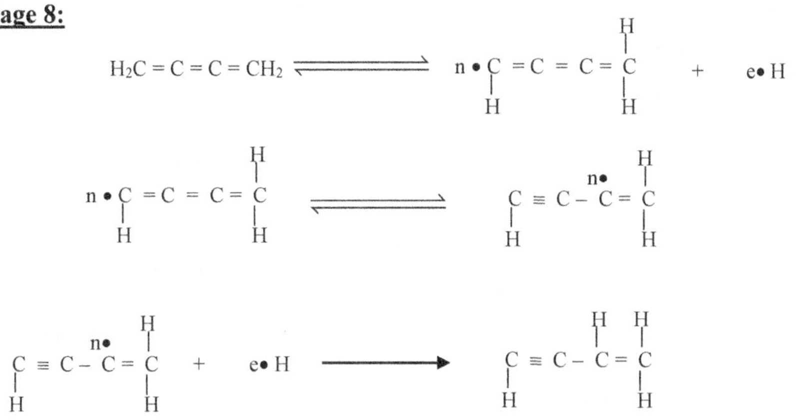

11.171a

Stage 9:

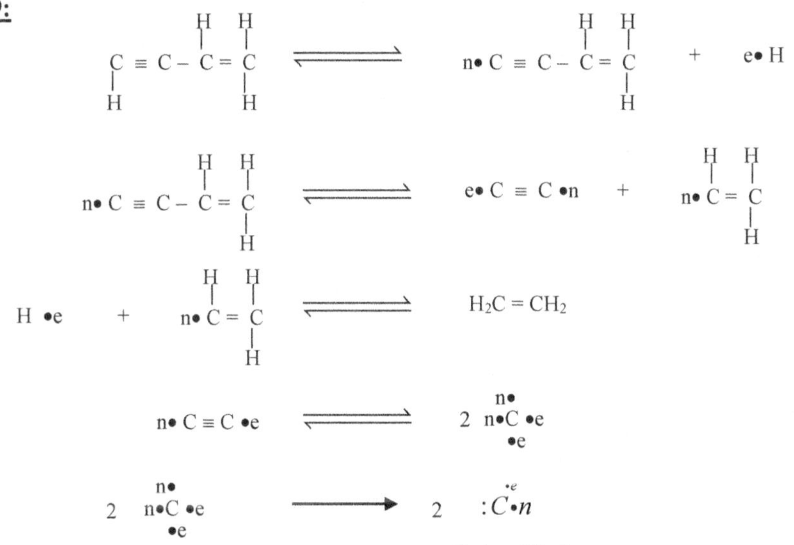

Carbon Black

11.171b

Overall equation: 1,4-Cumulene $\longrightarrow$ H$_2$C = CH$_2$ + 2C (Carbon black)   11.171c

Stage 10: Same as Stage 7 of Equation 11.170a.

Overall equation: Ethene + t-Butyl cyclopentadiene $\longrightarrow$ Acetylene + t-Butane + Cyclopentadiene   11.172

Stage 11:

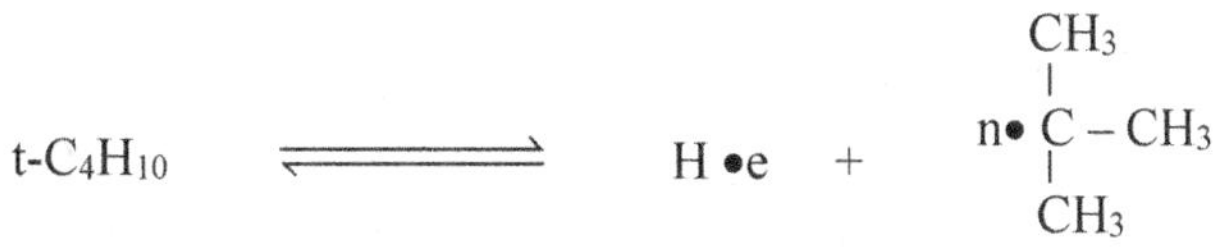

(C)

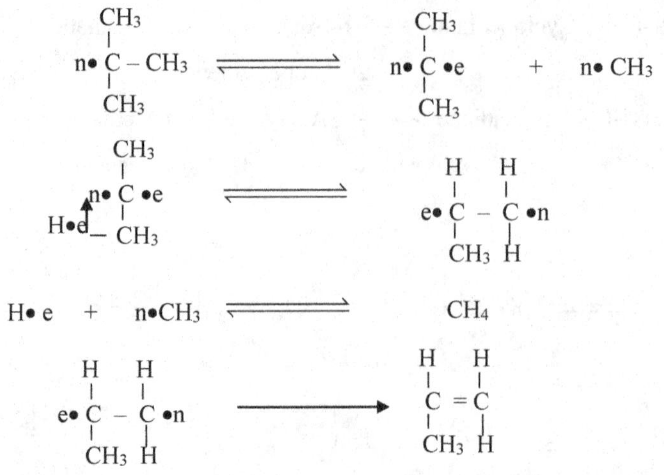

Propene   11.173a

Overall equation: $3t\text{-}C_4H_{10} \longrightarrow$ 3Propene + 3CH$_4$

Stage 12: The propene is broken down in a similar manner like the others.   11.173b

Overall equation: 3Propene $\longrightarrow$ 3Acetylene + 3CH$_4$ 11.173c

Overall overall equation: 9 t-Butyl-Cyclopentadiene $\longrightarrow$ 12Acetylene + 12Methane +

4Propyne + 2Isobutene + 2H$_2$ + 10C + 3Cyclopentadiene   11.173d

Stage 13:

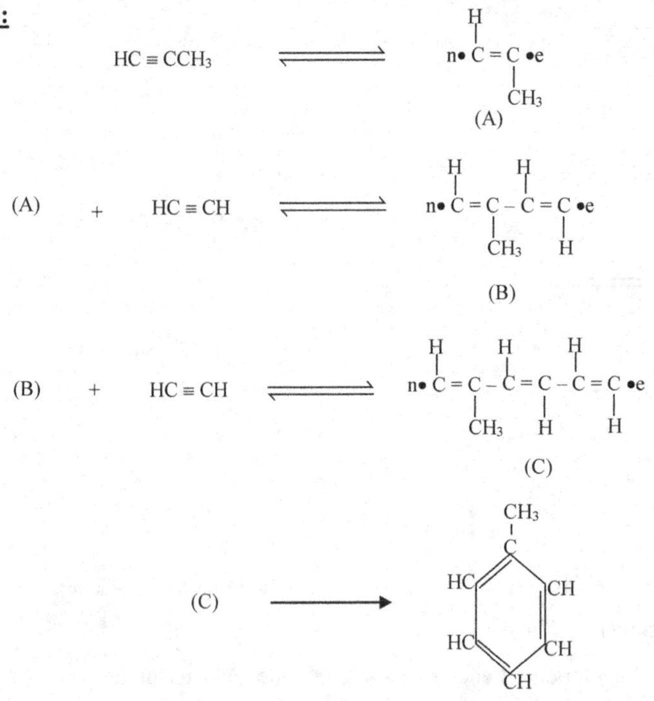

(A)

(B)

(C)

Toluene   11.174a

Overall equation: $8HC\equiv CH + 4Propyne \longrightarrow$ 4Toluene   11.174b

Overall overall equation: 9 t-butyl-Cyclopentadiene $\longrightarrow$ 4Acetylene + 12Methane +

2Isobutene + 4Toluene + 10C + 3H$_2$ + 3Cyclopentadiene   11.174c

Indeed, there are 14 stages so far, because, the molecular rearrangement of methyl isobutene to 1-methyl-1-butene in Equation 11.166a was not shown. How the major products were obtained, have been provided. Based on the operating conditions, below that used for cracking of hydrocarbons, the methane cannot be cracked. From the four moles of acetylene left behind, benzene is also one of the product. The fraction of isobutene left is that which could not exist in Equilibrium state of existence. If it had decomposed, then two additional moles of toluene would have been produced without the formation of isobutene and benzene as products. One does not need to describe the stages, because they speak for themselves and there are just too much to learn from based on types of rearrangements, the chemistry of Carbenes, Equilibrium states of Existence of compounds, and so much. ***What was interesting to note was the decomposition of t-butane in Stage 11. Its decomposition was thought not to be possible, because H.n could not be released from n.C(CH₃)₃. Instead n.CH₃ was released.*** When all the stages are viewed with the eyes of the needle, one will begin to see all disciplines coming into play. One cannot sit down to break a chemical compound indiscriminately or any how or via the easiest so-called visible means all against the laws of NATURE as we do in our world. One cannot move from Stage 1 to Stage 8 as is done in the Chemistry of today with respect to rate determining steps- all **absolute** nonsense.

Molecular rearrangement of the Second kind was applied in the provisions above in the second Step of Stage 3a of Equation 11.166a and third Step of Stage 11 of Equation 11.173a. Based on the ways by which they took place, the followings are valid.

$$e\bullet C_4H_9 \quad > \quad e\bullet C_3H_7 \quad > \quad e\bullet C_2H_5 \quad > \quad e\bullet CH_3 \quad > \quad H\bullet e$$

ORDER OF ELECTRO-FREE-RADICAL CAPACITY 11.175a

$$H\bullet n \quad > \quad n\bullet CH_3 \quad > \quad n\bullet C_2H_5 \quad > \quad n\bullet C_3H_7 \quad > \quad n\bullet C_4H_95$$

ORDER OF NUCLEO-FREE-RADICAL CAPACITY 11.175b

One can observe that the order is reversed nucleo-free-radically for these radical-pushing groups. While the former is a reflection of their radical-pushing capacities electro-free-radically, the latter is a reflection of their radical-pushing capacities nucleo-free-radically. This is a reflection of something else. For example H can never be transferred or rejected as a hydride. But it can be abstracted as a hydride under specific conditions.

Without going through the exercise of providing the mechanisms of reactions, one cannot see how NATURE operates. One can never assume thinking that it is so simple and obvious. From all the considerations so far, it is very important to note the FINGER PRINTS i.e. EQUILIBRIUM STATE OF EXISTENCE of all the compounds being encountered. However, downstream, an Encyclopedia will be provided for them, since they will be required for many applications.

There is need to search if this family of monomers can be made to undergo molecular rearrangement of the third kind or of another kind as this will shed more light about the character of this monomer and members of the family. Shown below is a cyclopendiene carrying just CH₃ group internally located. This is 2-methyl cyclopentadiene (Not 2-methylene cyclopentadiene[23], as this does not exist). Only 5-methylene cyclopentadiene exists.

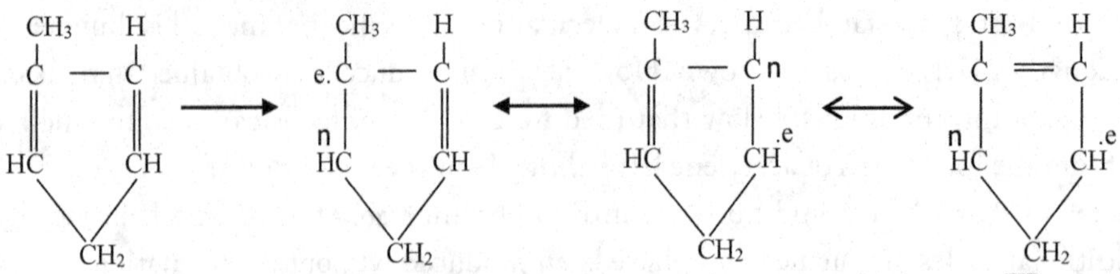

(I) Not a mono-form (II) (III)

Discrete closed- loop resonance stabilization

11.176

Note that (I) is not one of the mono-forms, since that center is more nucleophilic. Based on the radicals carried by the active centers and the conjugatedly placed character of the double bonds, the monomer is resonance stabilized and of the discrete closed loop type. All of them are resonance stabilized irrespective of the types of groups internally placed carried. At the beginning one was speculating that when some groups are carried, resonance stabilization ceases. So far, this has been found not be the case. The reason why this is so, is becoming obvious. If two directions exist for movement via a thin line- the sigma bond, one direction must always be available for use under certain conditions.

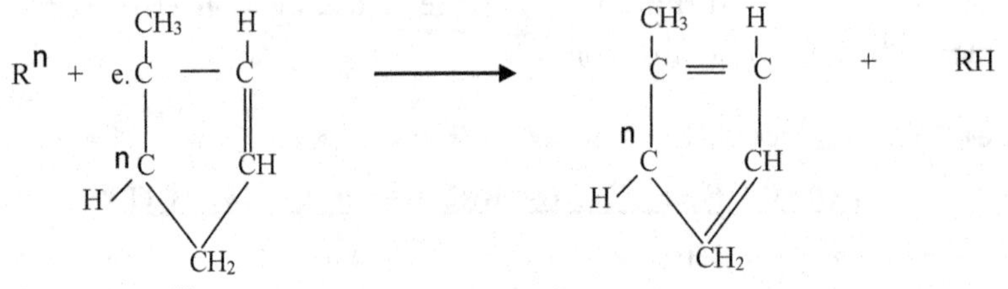

(I)

11.177a

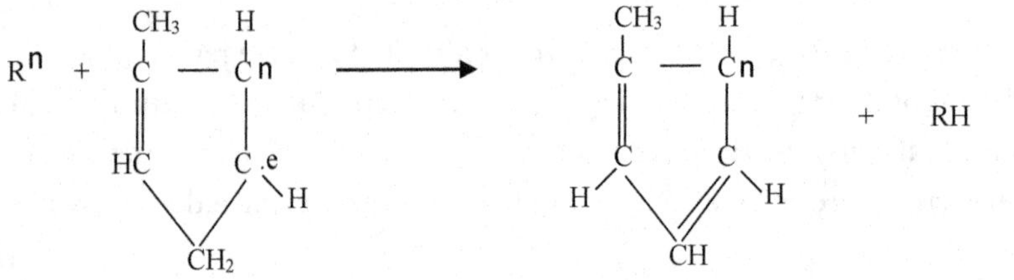

(II)

11.177b

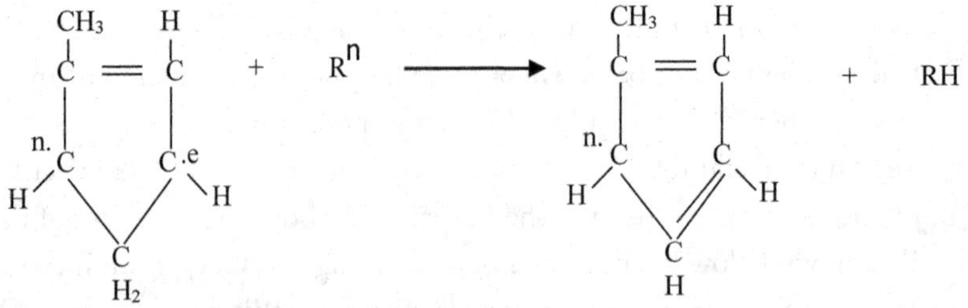

(III)

11.177c

From the above, one can see why the 1,2-mono-form is not one of the mono-forms. Like isoprene, the transfer species cannot be provided by the CH_3 group. CH_3 cannot therefore provide transfer species for molecular rearrangement of the first type, since the group is shielded and internally located. If externally located, it cannot also be used. Unlike isoprene, free-radically it can only be polymerized electro-free-radically and cannot be activated chargedly. Note that the CH_2 group is not resonance stabilized, since

268

it is externally located to the double bonds. Unlike in isoprene, the group is shared by two C centers externally located to the double bonds of a di-ene and not a mono-ene.

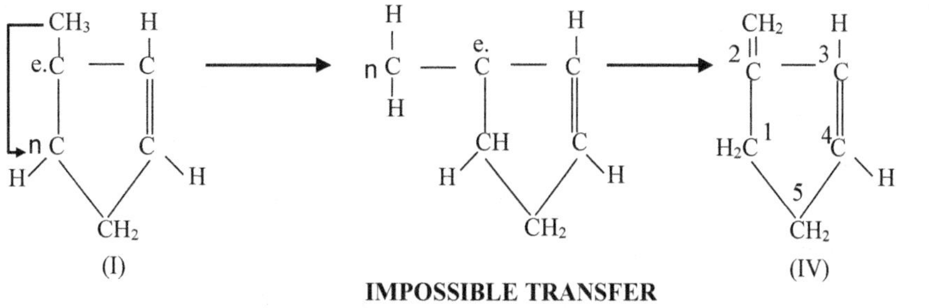

$$\text{(I)} \xrightarrow{\hspace{2cm}} \text{IMPOSSIBLE TRANSFER} \xrightarrow{\hspace{2cm}} \text{(IV)} \qquad 11.177d$$

Though (IV) has a point of scission, it should be noted that (I) cannot undergo molecular rearrangement of the first kind of the sixth type. If the right transfer species had been used, what is shown below (I) will be obtained. This looks more nucleophilic than the original monomer, (I) above.

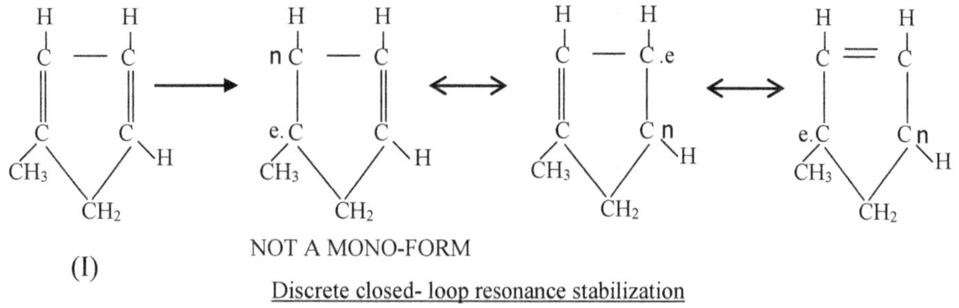

$$\text{(I)} \xrightarrow{\hspace{1cm}} \underset{\text{NOT A MONO-FORM}}{\hspace{1cm}} \longleftrightarrow \hspace{1cm} \longleftrightarrow \hspace{1cm}$$

Discrete closed- loop resonance stabilization   11.177e

Hence, one should expect that the 2-methyl-cyclopentadiene will rarely exist when activated, since it will molecularly rearrange to 1-methyl-cyclopentadiene.

Transfer species cannot be provided by the CH_3 group, but by the group internally located, because like cyclopropene, the CH_2 group connected to two double bonded centers is more radical pushing than any R group, more than that in cyclopropene which also is more radical-pushing than any R. While in cyclopropene CH_2 is connected to an alkene (Ethene), in cyclopentadience, the same CH_2 group is connected to an alkediene (Butadiene). Therefore, one should expect the following order in radical-pushing capacity, butadiene being more nucleophilic than ethene.

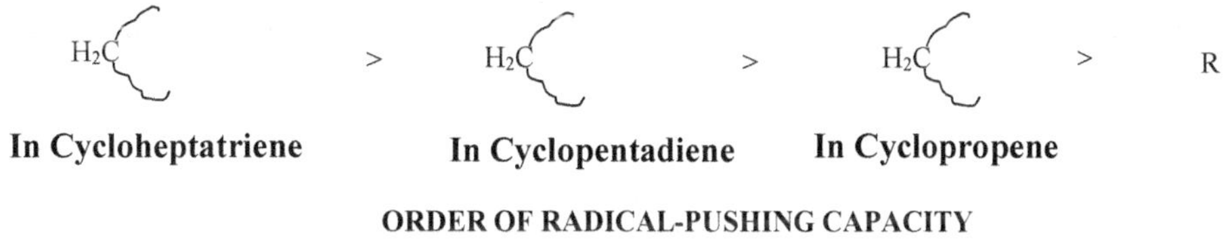

$$\underset{\text{In Cycloheptatriene}}{H_2C} \quad > \quad \underset{\text{In Cyclopentadiene}}{H_2C} \quad > \quad \underset{\text{In Cyclopropene}}{H_2C} \quad > \quad R$$

ORDER OF RADICAL-PUSHING CAPACITY   11.178

When considered for molecular rearrangement, the followings are possibilities.

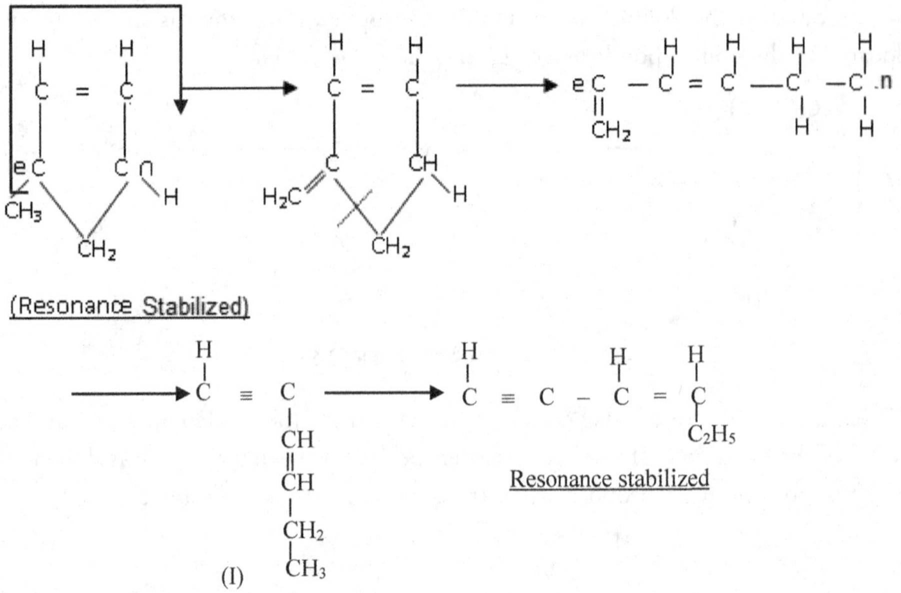

(Resonance Stabilized)

Resonance stabilized

(I)

IMPOSSIBLE TRANSFER   11.179a

Though exploratively, one of the isomers of the monomer can be identified, based on the fact that the group in the ring is more radical-pushing than any group externally and internally located on the double bonds, the movement of above is impossible. Therefore, the molecular rearrangement is not favored. Groups can only be moved from the CH_2 group. The same applies to the case below and to all other cases.

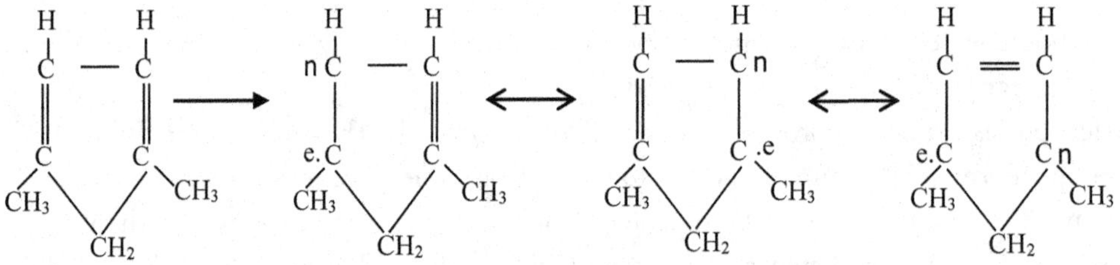

Discrete closed- loop resonance stabilization (symmetric)
                                                11.179b

It is important at this point to distinguish between opened/half-closed-loop resonance stabilization and discrete closed loop resonance stabilization. The closed type which cyclopenta-diene seems to favor is of the discrete type, since one CH_2 group breaks the continuity, unlike in cyclobutadiene.

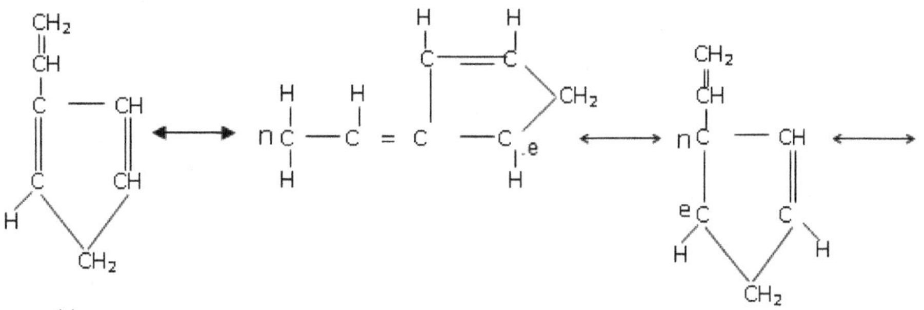

(I)

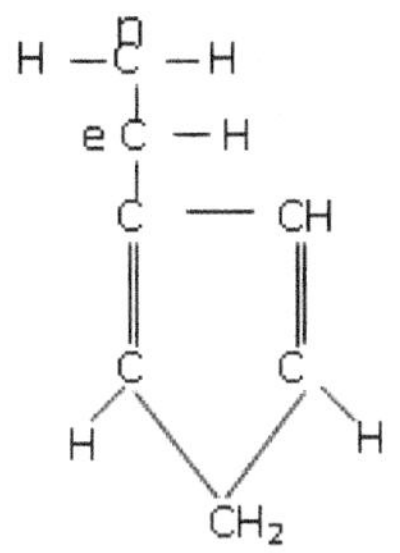

Opened /half closed- loop resonance Stabilization

11.179c

Though this is not new to us now, since it has been seen with vinyl cyclopentene and more, here it is like a case of two directions-one which is inside the ring and another half inside and half outside. The least nucleophilic center is the alkene, noting that the CH_2 group is not resonance stabilized.

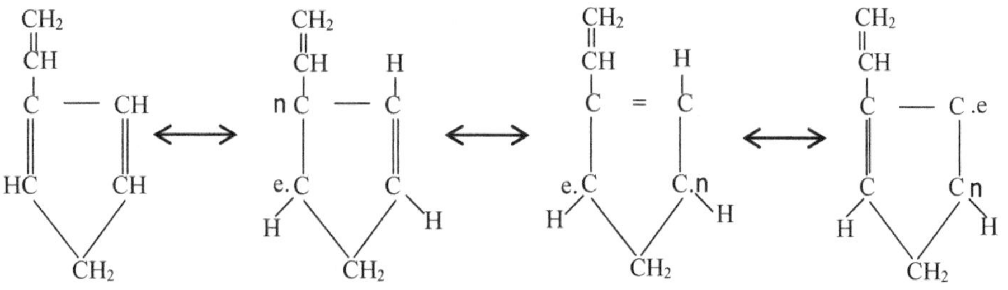

Impossible resonance stabilization

11.179d

With opened/half-closed loop resonance stabilization, the monomer a nucleophile has internal CH_2 group providing the transfer species. With closed loop resonance stabilization, the monomer also a nucleophile, free-radically can also have transfer species. As can be observed above, opened-loop and closed-loop resonance stabilization cannot take place at the same time. Of the two types, the one that predominates for the monomer above is that center which provides less or least nucleophilic character. That is, the one that will predominate will not depend on the types of groups carried by the different activation centers, but by the one externally located. As a monomer for polymerization purposes, it cannot easily be used, due to steric limitations.

Considering the molecular rearrangements of the resonance stabilized mono-forms of (I) of Equation 11.179c, the following is to be expected free-radically.

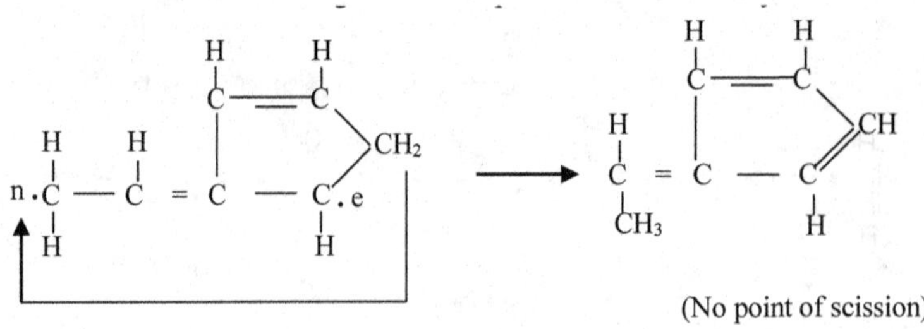

(No point of scission)   11.180a

When radical-pushing substituent groups are introduced externally, the followings are obtained beginning with CH_3.

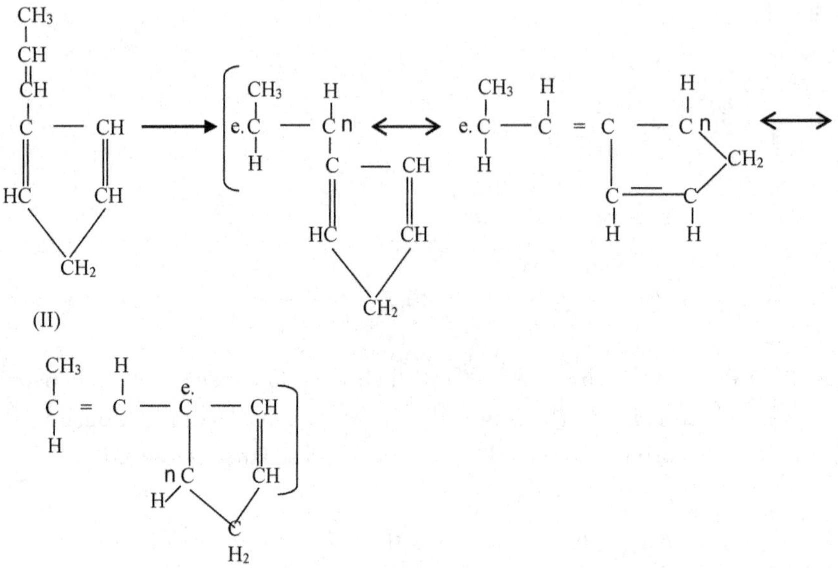

(II)

NOT FAVORED   11.180b

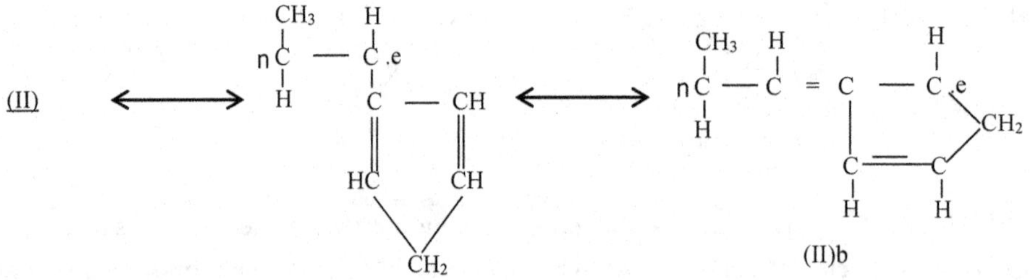

FAVORED   11.180c

As can be observed by the opened/half closed loop resonance stabilization systems indicated above, it seems that the active centers carry fixed charges, in view of the types of groups carried by them. It is the (a) and (b) that is favored above, in view of the capacity of the ring. Nevertheless, the ring can never be opened instantaneously after the rearrangement. Replacing CH_3 group with C_2H_5 or any higher, it still remains the same.

From the considerations so far, the following is valid when the resonance stabilization groups are used to provide resonance stabilization.

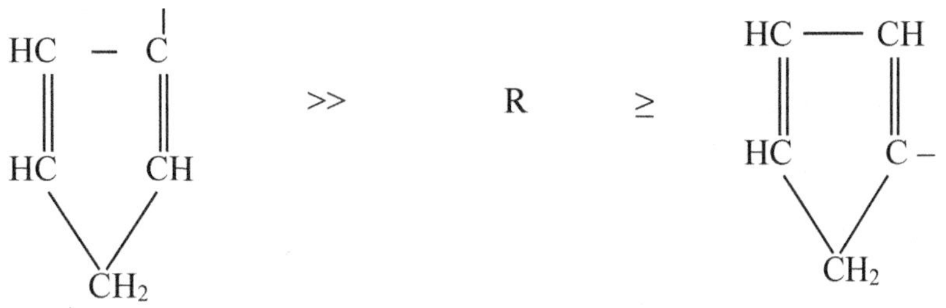

HC — C
‖   ‖
HC   CH
  \ /
  CH$_2$

$>>$ R $\geq$

HC —— CH
‖   ‖
HC   C —
  \   /
  CH$_2$

<u>Order of radical-pushing capacity of resonance stabilization groups</u>   11.181

With the monomer or compound represented below, the followings are obtained.

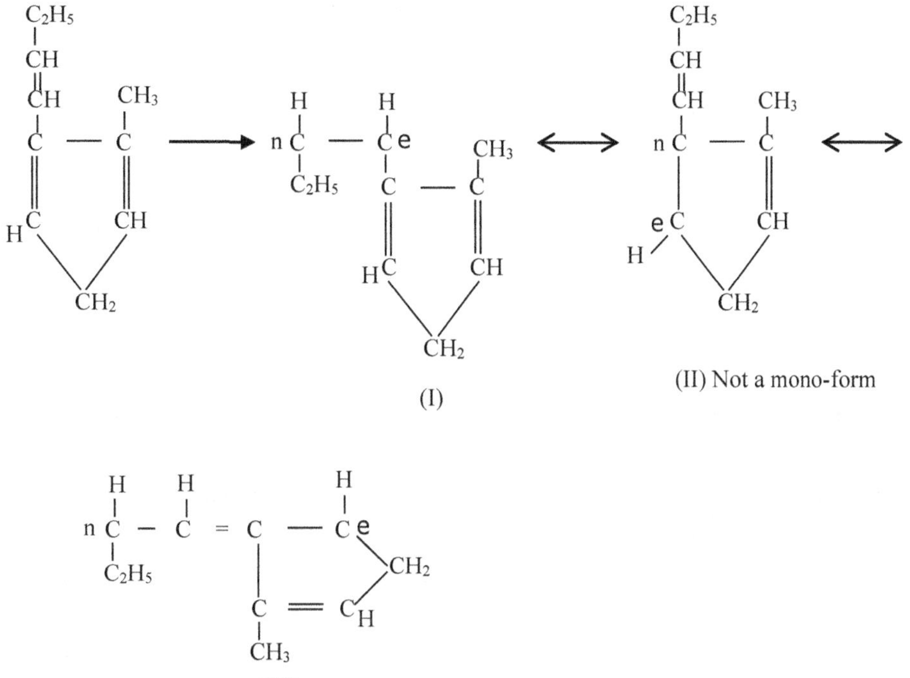

C$_2$H$_5$
|
CH
‖
CH   CH$_3$
| |
C ——— C
‖   ‖
H–C   CH
   \   /
   CH$_2$

$\longrightarrow$

n C — C·e   CH$_3$
| |
C$_2$H$_5$ C —— C
        ‖   ‖
      H–C   CH
         \   /
         CH$_2$

(I)

$\longleftrightarrow$

C$_2$H$_5$
|
CH
‖
CH   CH$_3$
| |
n C —— C
‖   ‖
·e C   CH
H/ \   /
    CH$_2$

(II) Not a mono-form

$\longleftrightarrow$

H   H   H
| | |
n C — C = C — C·e
| |
C$_2$H$_5$ CH$_2$
      | /
      C === C$_H$
      |
      CH$_3$

(III)

11.182

OPENED LOOP/HALF CLOSED RESONANCE STABILIZED

(I), (II) and (III) can favor rearrangement with no possibility of opening of the ring since no point of scission can yet be provided. When the alkene group is placed in 1-position in the ring, then it is not only externally located, but Opened/full closed-loop resonance stabilized as shown below. Worthy of note are the radicals carried by the centers. Only the most stable mono-form has been shown. Like cyclopropene, the CH$_2$ group in the ring is resonance stabilized. No molecular rearrangement of the third kind can take place, when CH$_3$ group internally located is greater than C$_2$H$_5$.

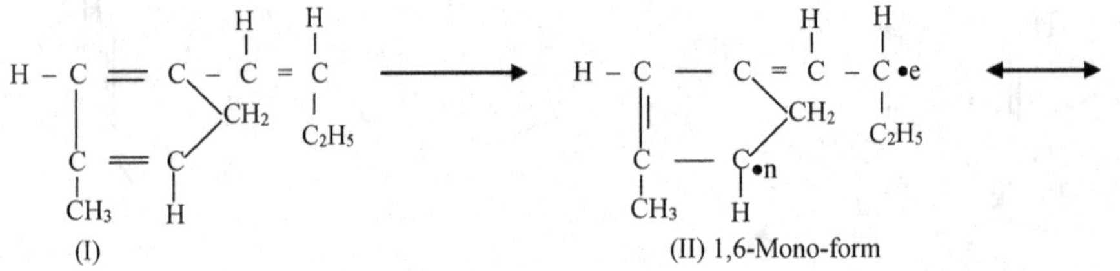

(I) → (II) 1,6-Mono-form

11.183

(II) 1,6-Mono-form → (III) →

(IV) → (V)

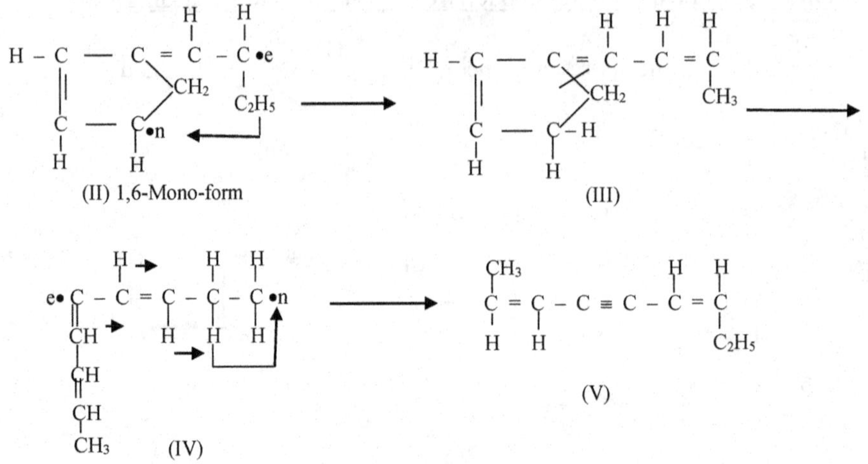

11.184

One can observe the conditions under which the ring has been opened. Note the type of product obtained, making it look as if the externally located alkene group was not tempered with. Compare the case above with the corresponding case of cyclopropene. Rearrangement stops when C_2H_5 is changed to C_3H_7.

With acetylenic resonance stabilization groups, one will begin with $H-C\equiv C-$ group internally placed.

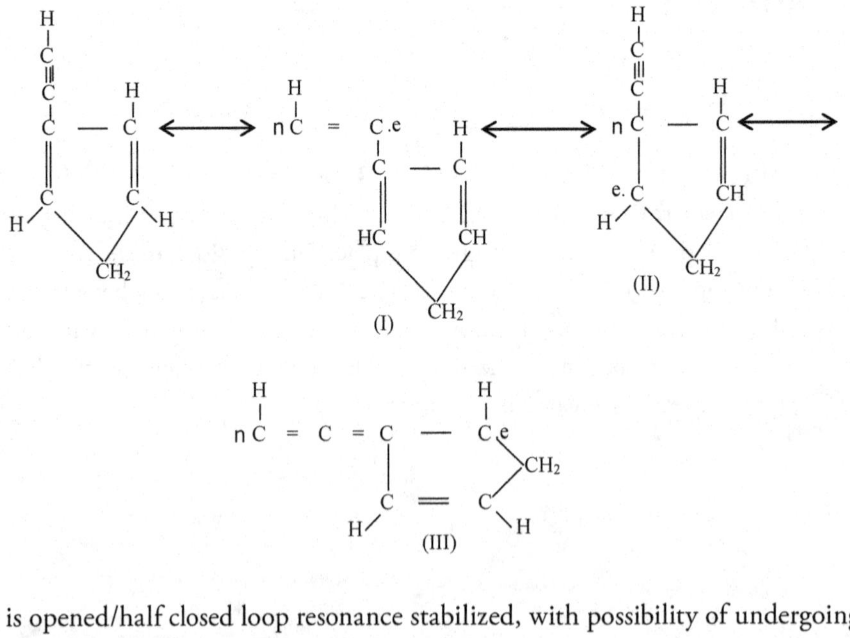

(I) ↔ (II)

(III)

11.185

The monomer is opened/half closed loop resonance stabilized, with possibility of undergoing molecular rearrangement of the first kind, but not of the third kind. Only (II) and (III) are the mono-forms. In fact, there may be need to replace the H atom on acetylene to keep the monomer stable. Now, one will replace the externally located H atom on acetylene with CH_3 group.

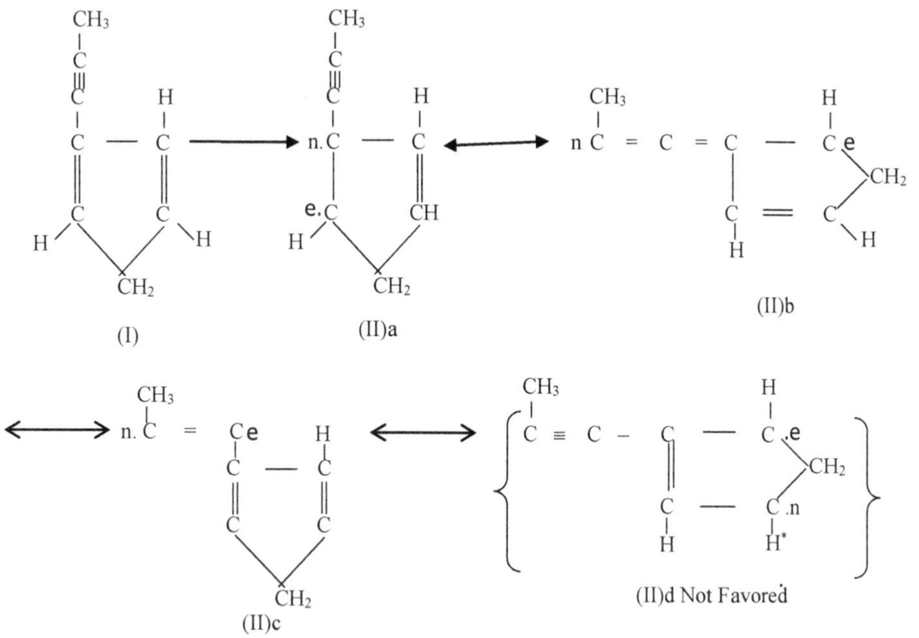

$$\text{(I)} \quad \text{(II)a} \quad \text{(II)b} \quad \text{(II)c} \quad \text{(II)d Not Favored} \qquad 11.186$$

(II)a, (II)b and (II)c are opened/half closed loop resonance stabilized while (II)d is discrete closed loop resonance stabilized and cannot exist unless when the H marked with asterisks is replaced with for example Cl. (II)c is not a mono-form. None of the rings can be opened after molecular rearrangements of the first kind. However, when the radicals are wrongly placed, molecular rearrangement of the third kind looks favored as shown below.

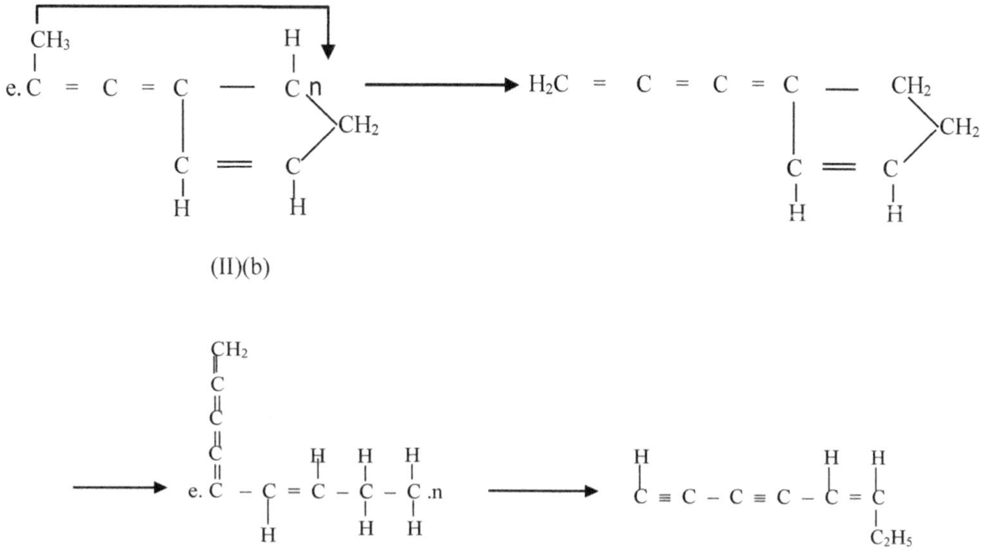

$$\text{(II)(b)}$$

$$\text{NOT FAVORED-RADICALs wrongly placed} \qquad 11.187$$

Via exploratory means, one can identify one of the isomers and the isomer looks real, since rearrangement ceases when C_3H_7 is put in place of CH_3. However, one had expected the presence of the non-resonance stabilized section of the ring placed alongside the resonance stabilized part. Thus, for the case above, the ring cannot be opened.

For the case shown below, the followings are to be expected, when suppressed.

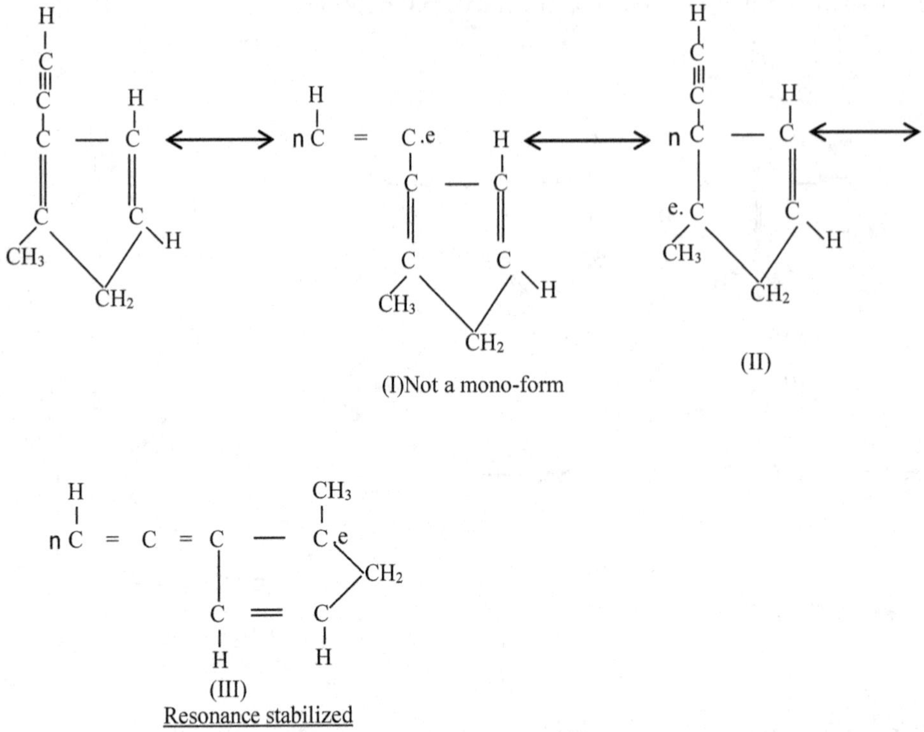

(I)Not a mono-form

(II)

(III)
Resonance stabilized

11.188

Depending on the types of groups carried by the centers, one can also have discrete closed loop resonance stabilization, that in which the acetylene is just a group internally located. This is only possible when the least nucleophilic center is in the ring, that is, the center not carrying acetylene. It can molecularly rearrange with no possibility of the ring opening as shown below, wherein CH_3 group has been taken to be the provider of transfer species. Since CH_2 group in the ring is externally

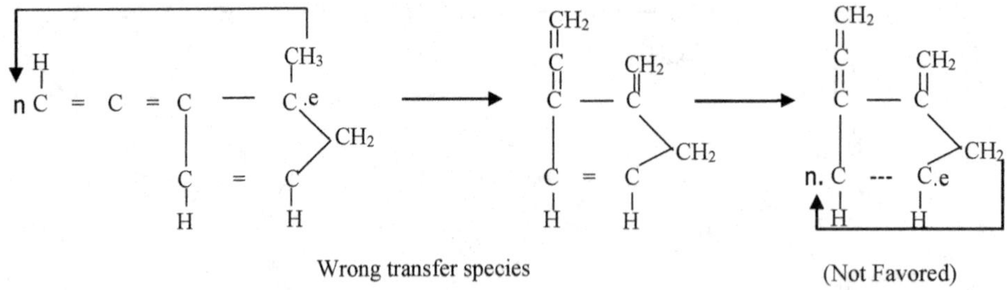

Wrong transfer species

(Not Favored)

11.189

located, and has great radical-pushing capacity than any R group, CH_3 group as used above is said to be wrong. However, when the right transfer species is used, the ring cannot be opened.

Now recalling the monomer of Equation 11.185, consider replacing the remaining externally adjacently located H atom outside the ring with CH_3 group. However, note that the acetylene is still internally located.

$$
\begin{array}{ccc}
\underset{\underset{\underset{\underset{\underset{CH_2}{|}}{C}}{\overset{|}{C}\diagdown_{H}}}{\overset{CH_3}{\overset{|}{\underset{\overset{|||}{C}}{C}}}}{C} - \underset{\underset{CH}{||}}{\overset{CH_3}{\overset{|}{C}}}
& \longleftrightarrow &
n.C = C\ e \quad CH_3 \\
\end{array}
$$

Not a mono-form

$$
\underset{\underset{CH_3}{|}}{\overset{CH_3}{\overset{|}{n.C}}} = C = \underset{\underset{CH_3}{|}}{\overset{}{C}} - \underset{\underset{\overset{C}{\diagdown} = \underset{H}{\overset{|}{C}} \diagup^{CH_2}}{}}{\overset{H}{\overset{|}{Ce}}}
$$

Opened/half closed-loop resonance stabilized   11.190

Like all the others, the ring cannot be opened after undergoing molecular rearrangement of the first kind. The provider of transfer species will always be the CH_2 group in the ring regardless the size of alkylane group carried by acetylene. In general, it can clearly be noted that cyclopentadiene rings cannot be instantaneously opened under any conditions when resonance stabilization groups are internally located.

When the groups are externally located, the situation is completely different. Based on what we saw in Stage 2 of Equation 11.141a, just as with cyclopropene, the CH_3 group is significant to them. While for cyclopropene, it is CH_3, for cyclopentadiene, it is $-CH=CH-CH_3$ the radical- pushing capacity of which has been shown to be greater than CH_3. The only difference is the presence of $-CH=CH-$ in cyclopentadiene, a group of great significance particularly when H on it is changed. With only H, the monomer is almost identical to cyclopropene in chemical behavior. Based on the above, it not surprising to see why cyclopentadiene is more nucleophilic than cyclopropene.

$$
\begin{array}{ccc}
\underset{\underset{\overset{|}{H}}{\overset{|}{C}} = \underset{\overset{|}{H}}{\overset{}{C}}}{H - \overset{}{C}} = \underset{\diagdown_{CH_2}}{C_{\diagdown}} - C \equiv C
& \longrightarrow &
\underset{\underset{\overset{|}{H}}{\overset{||}{C}} - \underset{\overset{\bullet e}{\diagdown}}{\overset{}{C}}}{H - \overset{}{C}} - \underset{\diagdown_{CH_2}}{C_{\diagdown}} = C = \overset{\overset{H}{|}}{C} \bullet n
& \longleftrightarrow
\end{array}
$$

(I)   (II) 1,6-Mono-form   11.191

This cannot undergo any rearrangement. In fact, it will highly be unstable, always in Equilibrium state of existence wherein the H held is that on the ring side, just like cyclopropene. Like cyclopropene, the CH_2 group is fully resonance stabilized. The opening of the rings when groups are carried only externally on the acetylene center will follow the same pattern as with cyclopropene. When groups are carried externally, their rings can be opened while still favoring rearrangement of the third kind, all with their different limitations.

Based on the analysis so far, there is no doubt that the following is valid.

$$Cyclodienes \quad > \quad Cycloalkenes \quad > \quad Cycloalkanes$$

<u>Order of nucleophilicity</u>

11.192a

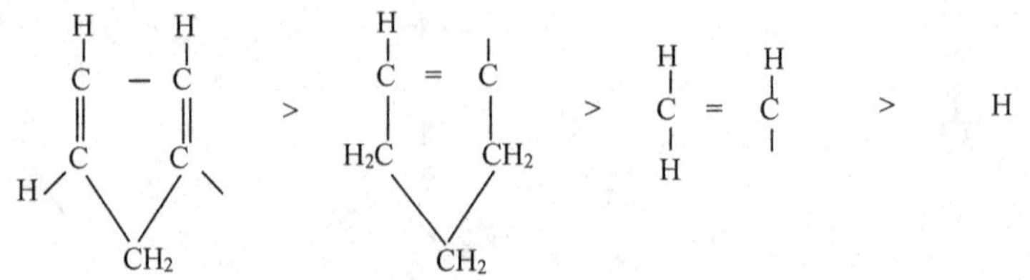

<u>Order of radical- pushing capacity</u>

11.192b

11.2.2. Cyclohexadiene

While points of scission exist for some of these cases and larger members, not all can however be opened. There seems to be three types of cyclohexadiene, depending on the location of the double bonds.

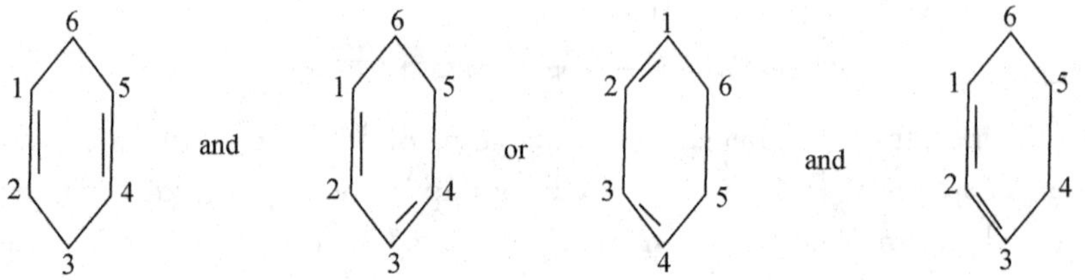

| (I) | (II)a | (II)b | (III) |

<u>Isolated type</u> (cyclohexa - 1,4- diene)   <u>Conjugated type</u> (cyclohexa - 1,3- diene)   Cumulated type (cyclohexa-1,3-Cumulene) 11.193

While (I) has no point of scission, the (II)s, have one point of scission- the 5, 6-bond. Though (III) which has one point of scission is not popularly known to exist, it has been recently identified as an intermediate in the reactions of 1-halocyclohexenes and methyl substituted 1-halocyclohexenes with potassium t-butoxide[24]. It readily molecularly rearranges to cyclohexyne. If the (II)s, can be provided with the minimum required strain energy, existence of 1,3,5 - hexatriene would be favored as shown below.

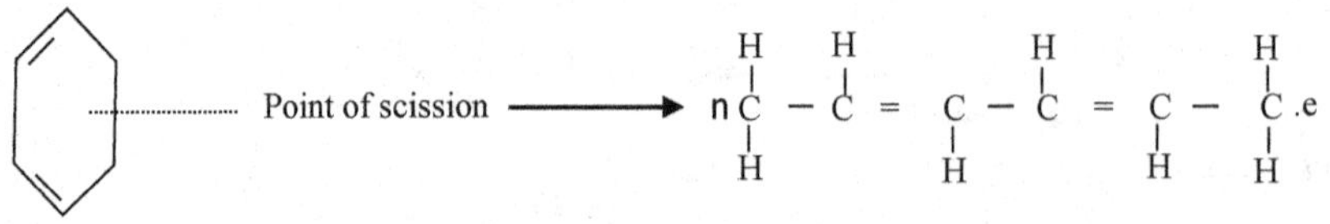

1, 3, 5- hexatriene

11.194

(I), in which the two double bonds are isolatedly placed is uniquely different from the (II)s, which are the same. Consider their activations.

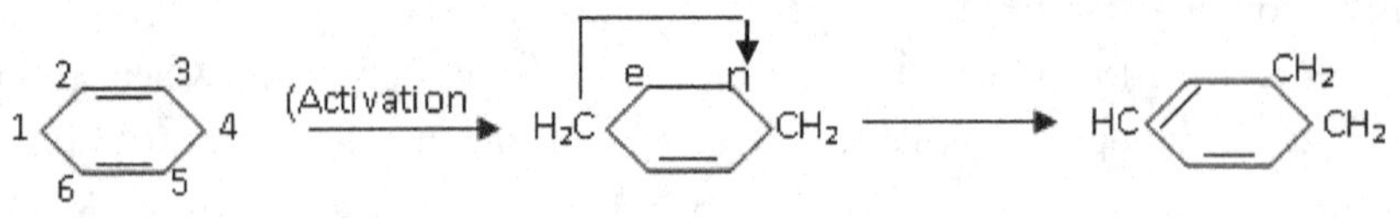

(I)   (III) (more stable)   (Favored)

11.195

Since the two activation centers are isolatedly placed, they are not resonance stabilized. Therefore, molecular rearrangement can take place to give the conjugated diene. With any of the (II)s, the followings are obtained when fully resonance stabilized.

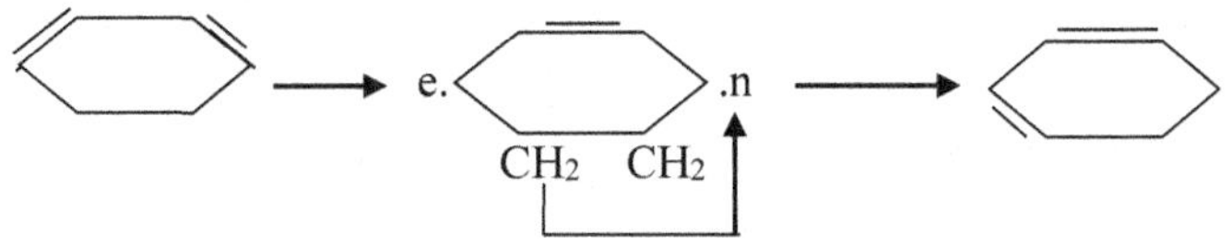

POSSIBLE TRANSFER STEPWISELY

11.196

Though the same conjugated diene is produced, the rearrangement will not take place as shown above, but stepwisely. Thus, (I) may not favor very stable existence, since what may be seen most of the time when heated is the conjugated cyclohexadiene[25].

Introducing a strong radical-pushing group externally, the following is obtained.

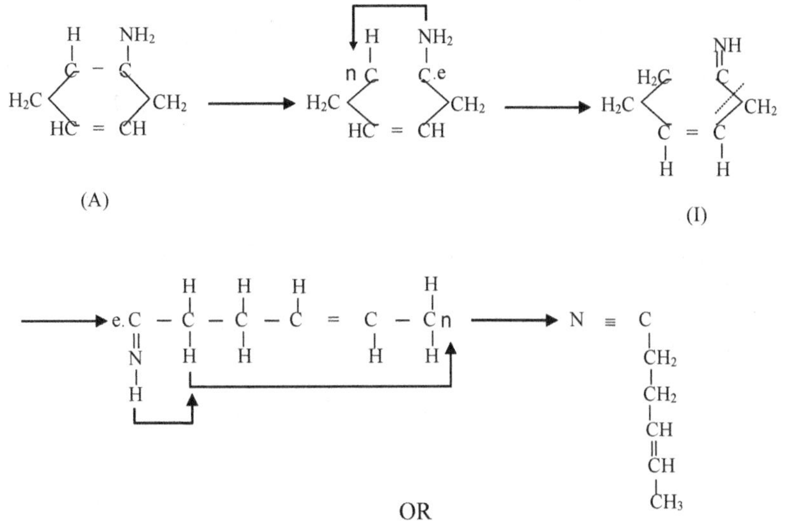

(A)

(I)

OR

(II) (Not Favored)   11.197

(A)

(I)

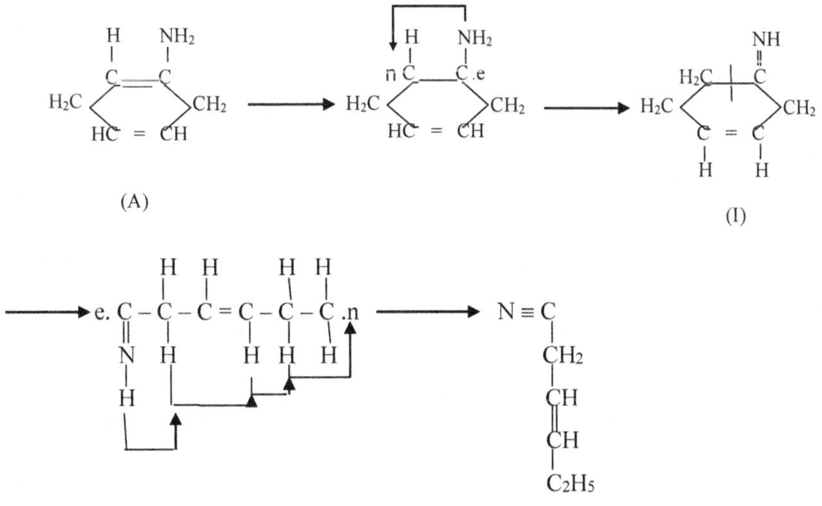

(III) Not Favored

11.198

Indeed, none of the above is favored, based on the applications of the Laws of Nature. However, it was done for exploratory purposes, in order to display how we operate in our world. On the other hand, from the exploration, one will notice why there were said not to be favored. In order words, we in our world have been on exploration since antiquity. Now we have begun to put all the pieces together. The center activated above is not the less nucleophilic center. It is the second center, which when used, no rearrangement is possible. When the 1,3- counterpart is used, the followings are obtained.

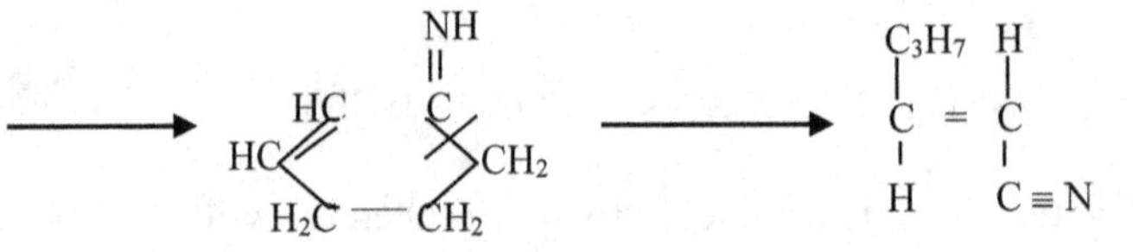

$$\text{(II)} \qquad \longrightarrow \qquad \text{An Electrophile} \qquad 11.199$$

[FAVORED]

With the use of the 1,3- conjugated amine, an electrophile which will not favor its natural route seems to be obtained. Originally, the monomer was a Nucleophile. For the 1,4- diene, it is not favored, because the first molecular rearrangement cannot take place as will be shown when the rule are stated. Therefore, with NH_2 group in 1- position for the 1,4- diene, the ring cannot be opened. Based on the experience acquired during exploration, the following is valid.

$$CH_3 - CH = CH - CH_2 - CH_2 - \qquad < \qquad C_2H_5 - CH = CH - CH_2 -$$

ORDER OF RADICAL- PUSHING CAPACITY $\qquad 11.200$

For the conjugated diene when NH_2 group is externally located, latently considered above in Equation 11.199, the followings are obtained.

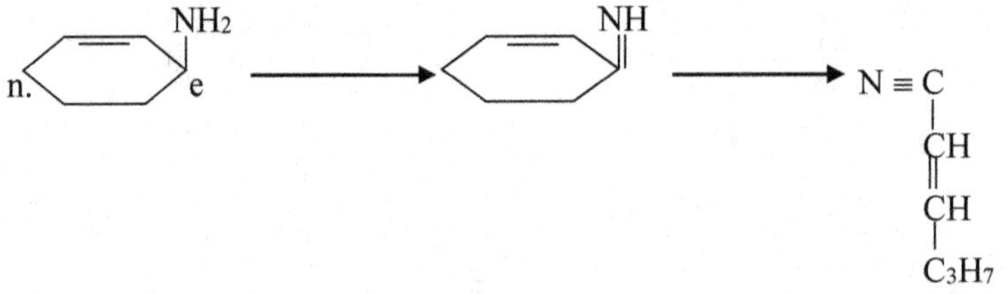

$$11.201$$

Notice that the product is the same as in Equation 11.199, that which could not be obtained from the isolated case.

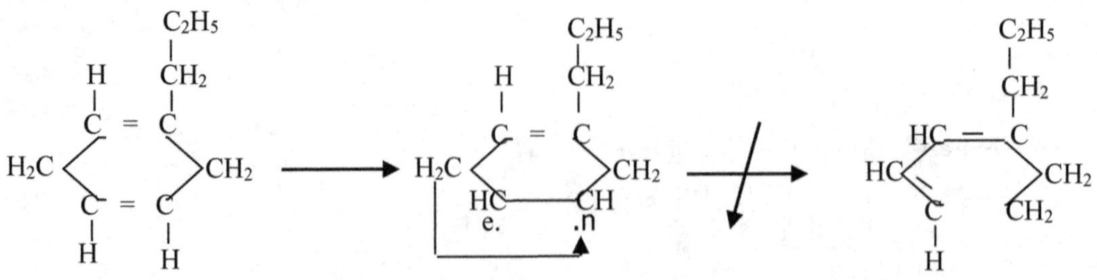

$$\text{(B)} \qquad \longrightarrow \qquad \text{(II)} \qquad \longrightarrow\!\!\!\!/ \qquad \text{(III)}$$

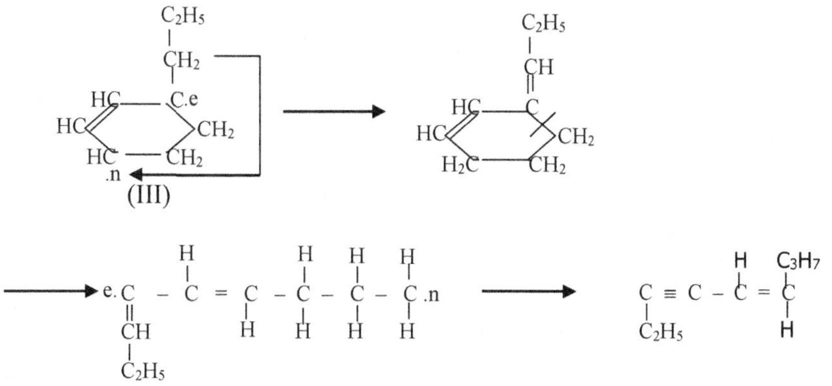

$$
\begin{array}{c}
\text{C}_2\text{H}_5 \\
|\\
\text{CH}_2 \\
\|\\
\text{HC} \quad \text{C.e} \\
\text{HC} \qquad \text{CH}_2 \\
\text{HC} - \text{CH}_2 \\
.n
\end{array}
\quad\longrightarrow\quad
\begin{array}{c}
\text{C}_2\text{H}_5 \\
|\\
\text{CH} \\
\|\\
\text{HC} \quad \text{C} \\
\text{HC} \qquad \text{CH}_2 \\
\text{H}_2\text{C} - \text{CH}_2
\end{array}
$$

(III)

$$
\longrightarrow
\begin{array}{c}
\text{H}\\|\\
\text{e.C} - \text{C} = \text{C} - \text{C} - \text{C} - \text{C}.n \\
\|\\
\text{CH}\\|\\
\text{C}_2\text{H}_5
\end{array}
\quad\longrightarrow\quad
\begin{array}{c}
\text{H} \quad \text{C}_3\text{H}_7 \\
\text{C} \equiv \text{C} - \text{C} = \text{C} \\
|\qquad\qquad|\\
\text{C}_2\text{H}_5 \qquad \text{H}
\end{array}
$$

11.202

As was the case in Equation 11.199, two molecular rearrangement steps cannot take place. The first step of rearrangement cannot take place to give (III), the 1,3- form. However, the 1,3- monomer will rearrange with opening of the ring made possible. After the opening of the ring, this was followed by molecular rearrangement of the third kind to give the resonance stabilized monomer above- a vinyl acetylene. Like the case of Equation 11.199 or 11.201, only one rearrangement is favored. One can observe two cases where the 1,4-cyclohexadiene is very stable, that is, cannot rearrange to the 1,3-monoform. From the observations so far, no isomer has been observed for it's 1,4-diene. The rearrangement above for the 1,3-diene stops when the C_3H_7 on the ringed monomer is changed to C_4H_9. This sends a message of the size of the CH_2 groups in the ring.

Like cyclobutene, the (II) of Equation 11.193 will favor instantaneous opening of its ring in the presence of strong coordination initiators, which also should not carry groups which will provide steric limitations. ***The activated state obtained in Equation 11.i94 radically only, is resonance stabilized. Chargedly, it is not resonance stabilized, since it cannot be deactivated to give 1,3,5-hexatriene. It will always remain in its activated state as shown below. The same applies also to cyclobutene.*** Therefore, the possibility of having 1,2-mono-form along the chain when the ring is opened instantaneously chargedly or radically, is impossible. If only one center is activated chargedly when ring has not been opened, it cannot resonance stabilize to the 1,4-monoform. This can only be done radically. Therefore chargedly, only one center can be activated and that center can be used in the absence of steric limitations as shown below.

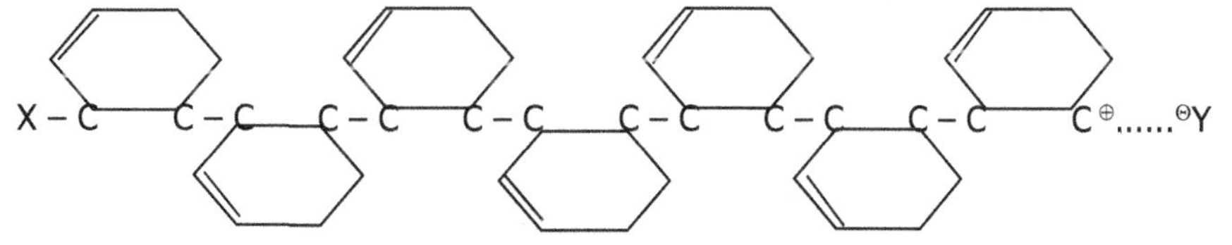

$$
X - C \overset{}{\underset{}{C}} - C \overset{}{\underset{}{C}} - C \overset{}{\underset{}{C}} - C \overset{}{\underset{}{C}} - C \overset{}{\underset{}{C}} - C \overset{}{\underset{}{C}} - C \overset{}{\underset{}{C}} - C \overset{\oplus}{\underset{}{}} \ldots\ldots {}^{\ominus}Y
$$

Syndiotactic-1,2-Poly (Cyclohexene)

11.203

When the ring is not opened instantaneously, chargedly, only the 1,2 –mono-form can exist along the chain as shown above. Resonance stabilization cannot take place chargedly. Therefore, 1,4- mono-form cannot exist along the chain, unless when opened instantaneously. Observe the type of initiator which has been used. It cannot be Lithium t-butyl alkyl, wherein the carrier of the chain is Li, since a cation cannot carry a chain. If it is Li, it can only be used radically and alone (i.e., not paired) Despite the absence of vacant orbitals or reservoirs for the monomers, the placement above is syndiotactic due to steric limitations. Presence of 1,2- and 1,4- mono-forms along the chain which was said to be obtained chargedly using alkyl lithium amine systems[25] can only be obtained free-radically as shown below.

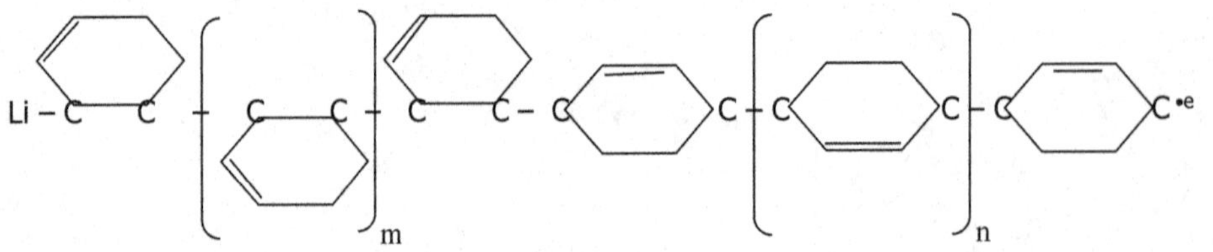

Poly(Cyclohexene)-1,2-/1,4- mono-forms 11.204

Notice again that the same LiC_4H_9 has been used above free-radically, but unpaired. Whether paired or not, the counter-center may not close the chain since C_8H_{18} may be formed. Presence of the 1,2-mono-form can only be favored at the beginning of growth when the initiator is weak. When the active growing center becomes stronger and stronger in strength, 1,4- mono-forms begin to appear along the chain continuously as shown above. The weaker the initiator, the more the presence of 1,2- mono-form. The stronger the initiator, the more the presence of 1,4-mono-form along the chain or indeed all will be 1,4-mono-forms. The chain can be killed either by starvation to form dead terminal double bonded ringed polymer or when optimum chain length has been reached or by use of foreign agent. The transfer species released when the chain is killed is the same transfer species that will prevent the initiation of the monomer nucleo-free-radically or "so-called anionically".

However, when the ring is opened instantaneously, both free-radical routes are favored. Chargedly the ring can be opened instantaneously, since the double bonds will not be the first to be activated, thus removing the influence of electrostatic forces of repulsion. The monomer being a Nucleophile will favor only the electro-free-radical route whether the nucleo-free-radical center is present or not. With the use of LiC_4H_9, as source of initiator, the route is electro-free-radical.

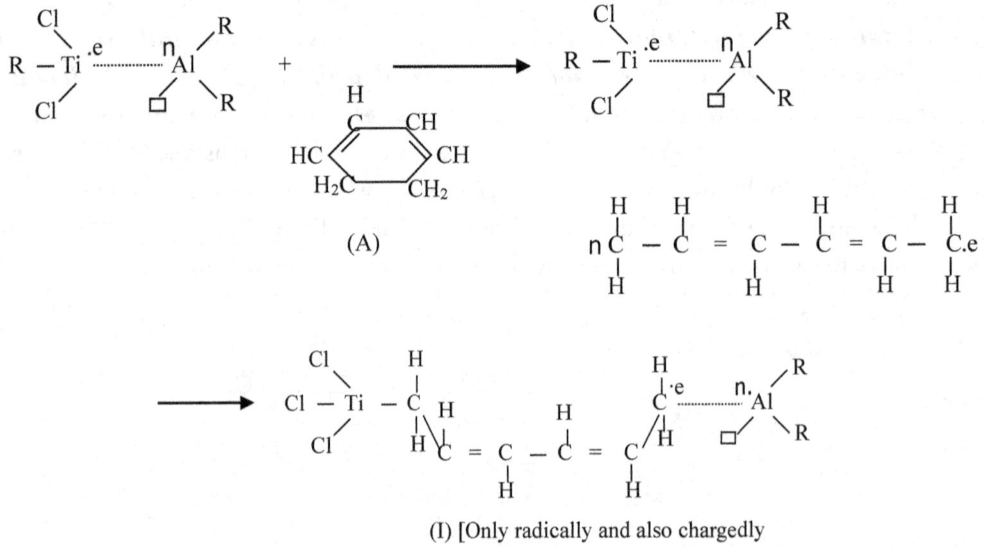

(I) [Only radically and also chargedly 11.205

The possibility of having 1,4- or 1,2- mono-form from 1,6-mono-form is impossible. Only the 1,6-mono-form will exist along the chain. ***With radical-pushing groups properly placed on the ring, instantaneous opening of the ring is far more readily favored than with cyclopentadiene, where the type of opening below or above is impossible.***

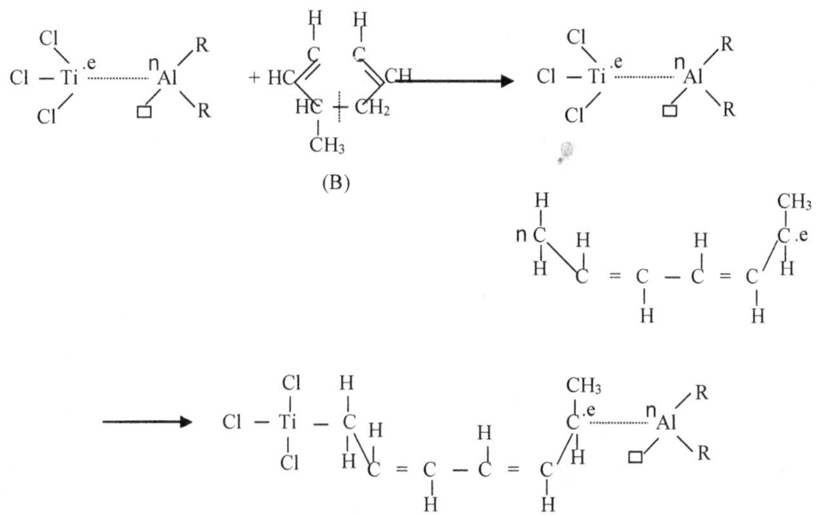

.(I) [Radically & Chargedly] 11.206

Existence of (I) will partly depend on the size of the R groups on the aluminum center. Nevertheless, the initiator must be strong to unzip the ring. In the presence of one vacant orbital as above, one should know the type of placement to expect- Iso-cis-tactic placement of the trans- monomer. While (A) will favor both electro-free and nucleo-free-radical routes, (B) will favor only the electro-free-radical route.

In the presence of weak initiators, the followings are obtained.

(II)a

Para -placement of radicals

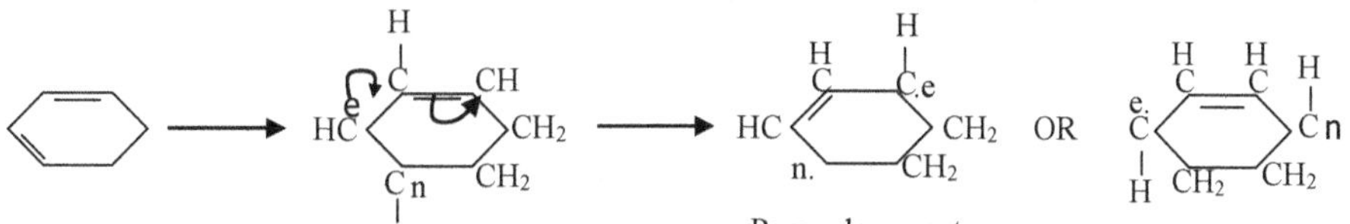

(1, 4 - activation) 11.207

(II)a

1,,2-Activation

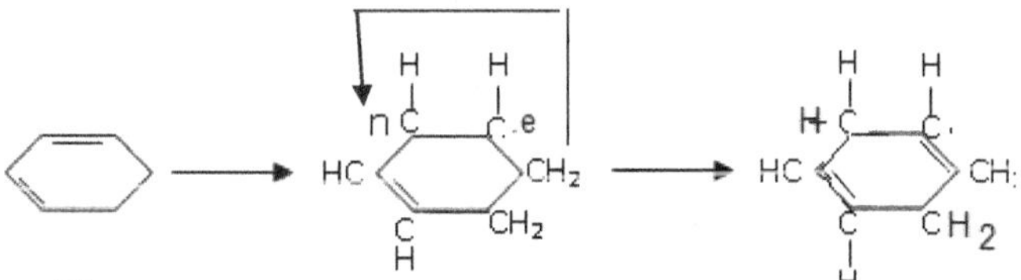

Isolated Cyclohexadiene 11.208

When molecular rearrangement of the first kind of the sixth type takes place in the presence of a weak initiator, isolated cyclohexadiene is obtained from the 1,2– activated mono-form that which is not to be expected. It will not be favored since the movement is symmetric and since the 1,4-mono-form based on Equation 11.196 cannot do it. Based on the same equation, ***the 1,4- mono-form is the most stable state of 1,3-cyclohexadiene,***

If the initiator is very strong, the cyclohexadiene unzips. When the ring is not opened, nucleo-free-radically and negatively as already said, no initiation is favored due to presence of transfer species of the tenth type of first kind. The same applies to the isolated case.

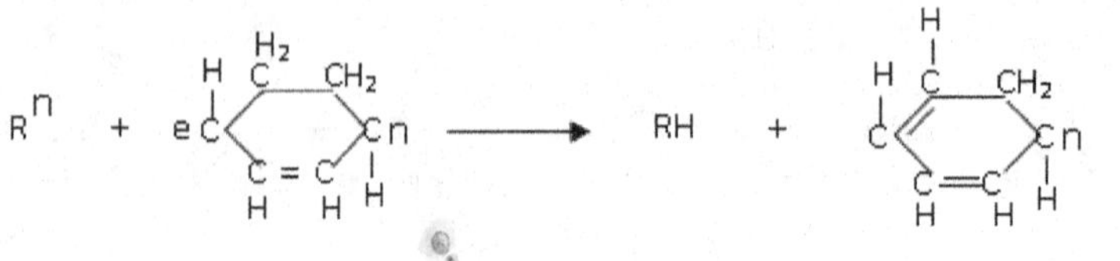

11.209

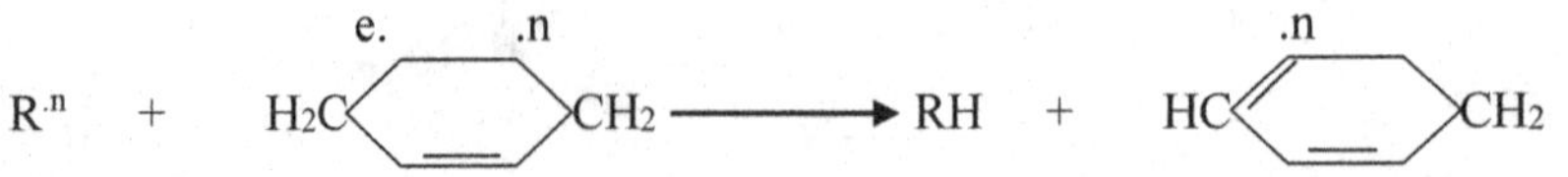

11.210

Polycyclohexene has been said to be obtained from cyclohexadiene using coordination initiators of the Ziegler-Natta type or cationic ion-paired types. [27,28] The routes favored by their polymerization is not even known. However, this can be explained as follows using $NaX/RhCl_3$ com-bination exploratively, since it cannot really be used radically. It was only used radically because of resonance stabilization and availability of reservoirs.

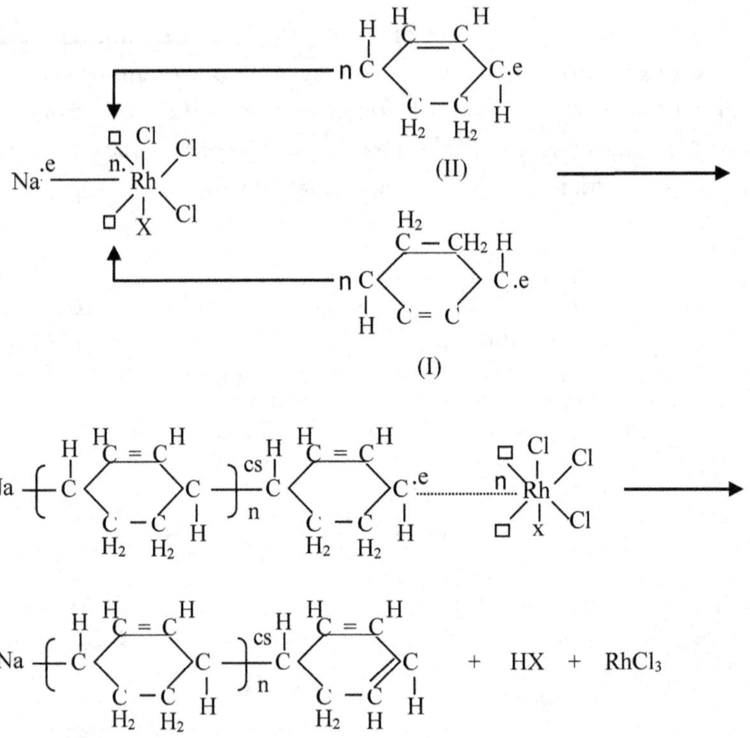

Transfer species of first kind of the tenth type

11.211

With the para-placement of the radicals and not charges for these six-membered resonance stabilized cases, polymerization is more readily favored than when not resonance stabilized. The polycyclo-hexene (which indeed is polycyclohexadiene) with a terminal cyclic double bond obtained above can be dehydrogenated also only radically to produce polyphenylene.

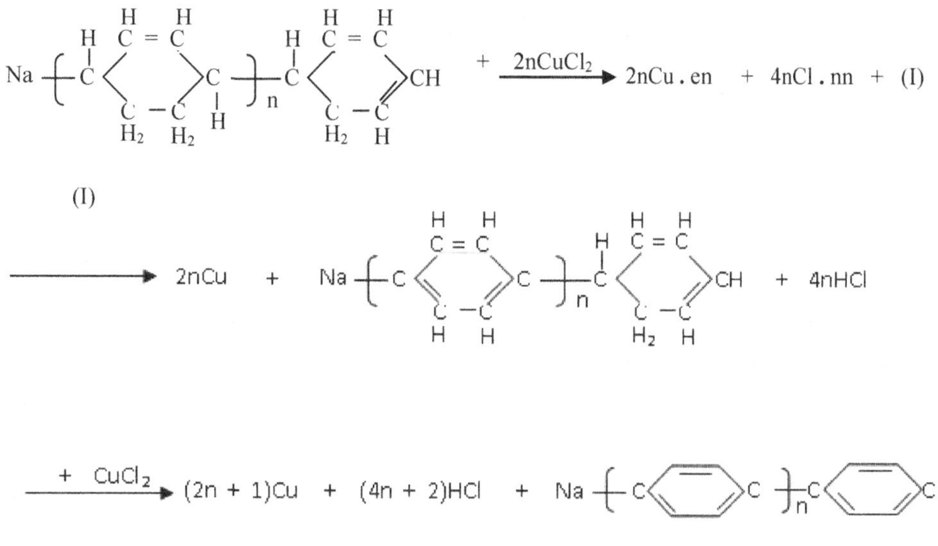

$$\text{Na}\left(...\right)_n ... \text{CH} + \underrightarrow{\;2n\text{CuCl}_2\;} 2n\text{Cu.en} + 4n\text{Cl.nn} + \text{(I)}$$

(I)

$$\longrightarrow 2n\text{Cu} + \text{Na}\left(...\right)_n ... \text{CH} + 4n\text{HCl}$$

$$\xrightarrow{\;+\;\text{CuCl}_2\;} (2n+1)\text{Cu} + (4n+2)\text{HCl} + \text{Na}\left(...\right)_n ... \qquad 11.212$$

This is an Equilibrium mechanism system with many stages. Each monomer unit is dehydrogenated sequentially to the last but one monomer unit in four stages as shown below.

Stage 1: $\quad \text{CuCl}_2 \;\underset{\textit{Existence}}{\overset{\textit{Equilibrium State of}}{\rightleftharpoons}}\; \text{ClCu} \bullet\text{en} + \text{nn}\bullet\text{Cl}$

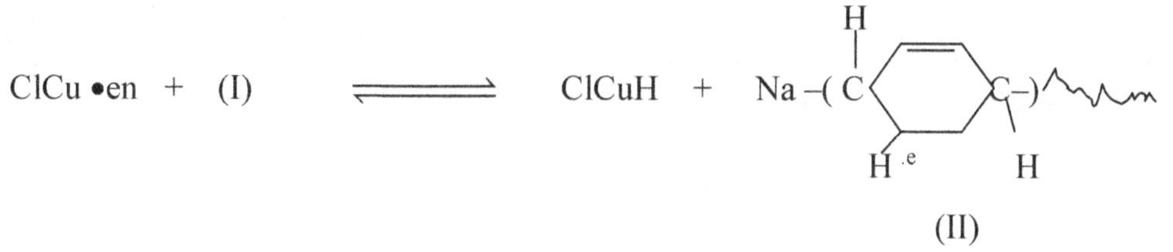

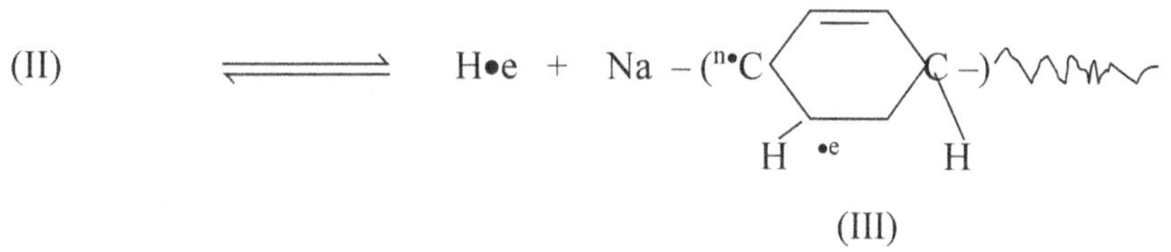

$\text{ClCu}\bullet\text{en} + \text{(I)} \;\rightleftharpoons\; \text{ClCuH} + \text{Na}-(\text{C} ... \text{C}-)\sim$

(II)

$\text{(II)} \;\rightleftharpoons\; \text{H}\bullet\text{e} + \text{Na} - (^{n\bullet}\text{C} ... \text{C}-)\sim$

(III)

$\text{H}\bullet\text{e} + \text{nn}\bullet\text{Cl} \;\rightleftharpoons\; \text{HCl}$

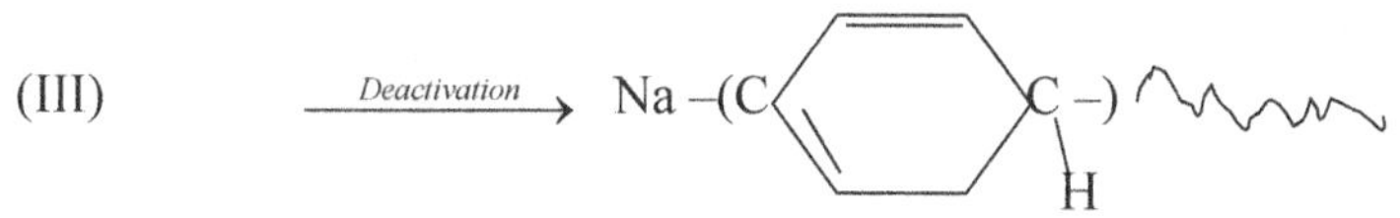

$\text{(III)} \;\xrightarrow{\;\textit{Deactivation}\;}\; \text{Na} - (\text{C} ... \text{C}-)\sim$

$$\qquad\qquad 11.213$$

285

Stage 2:   ClCuH $\rightleftharpoons$ (Equilibrium State of Existence)   Cl •nn   +   en• CuH

en• CuH $\rightleftharpoons$ en• Cu •nn   +   H •e

H •e   +   Cl •nn $\rightleftharpoons$ HCl

en• Cu •nn $\xrightarrow{Copper}$ Cu :   11.214a

11.214b

Overall Equation: ClCuH $\longrightarrow$ Cu   +   HCl   +   Heat

Stage 3:   Similar to Stage 1 for the third center to give

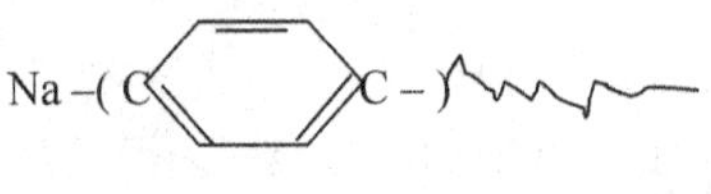

Na –(C ⬡ C – ⌇⌇⌇)

(IV)   11.215

Stage 4:   Same as Stage 2.

Overall Equation: (I) + 2CuCl$_2$ $\longrightarrow$ (IV)   +   4HCl   +   2Cu   11.216

The terminal ring with two double bonds will require two stages. Therefore, imagine if n equals 1000 units, then 4002 stages will be required. The CuCl$_2$ has been used here as a dehydrogenating agent. Note the unique Equilibrium state of Existence of ClCuH. It is an unstable compound. Chargedly, the reactions above are impossible, since copper cannot form an ionic bond. That is, it cannot carry ionic positive charges, but only radicals or electrostatic or polar or covalent positive charges.

Now consider replacing the H atom in 1- and 2- positions with substituent groups one at a time.
For 2-position:

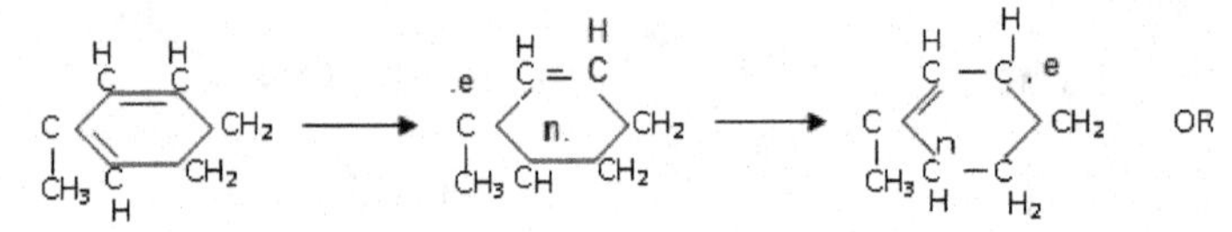

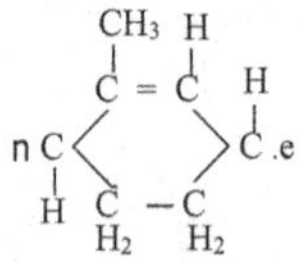

11.217

[Note: Wrong center activated]

For 1-position:

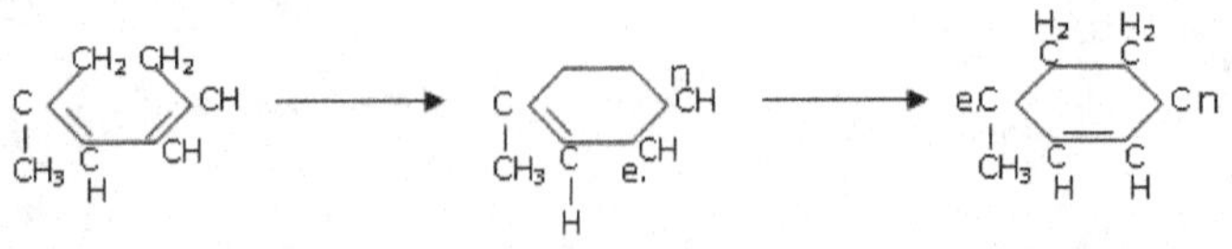

11.218a

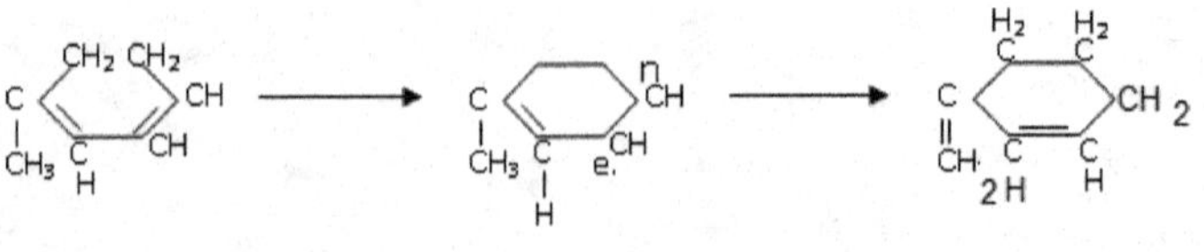

11.218b

NOT FAVORED

286

Though the CH_3 group is not resonance stabilized here, it cannot still provide transfer species because the radical-pushing capacity of the groups inside the ring is greater than that of CH_3. There-fore, the molecular rearrangement above cannot yet take place wherein the possibility of ring opening becomes possible to form acetylene with an alkene. Exploratively, assuming that the rearrangement above is favored, when the CH_3 group is replaced with higher groups rearrangement ceases when the group is C_4H_9. Indeed, rearrangement commences from C_2H_5.

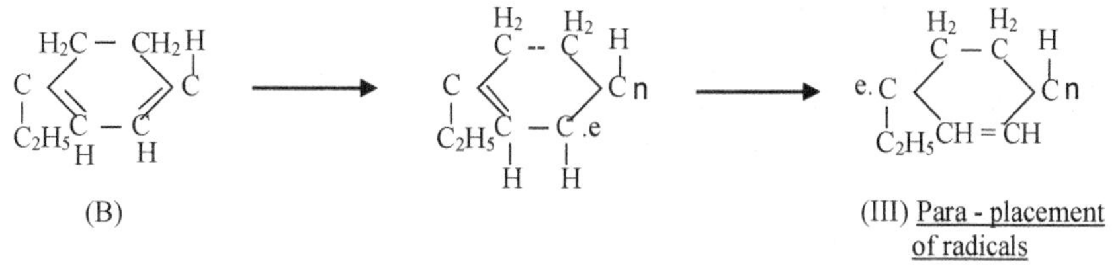

(B)   (III) <u>Para - placement
of radicals</u>   11.219

The C_2H_5 group may not yet be a provider of transfer species, based on the capacity of the closed group in the ring ($- CH_2 - CH_2 -$). The capacity of the substituent group carried by the ring can be observed to be important; so also is its location.

For the case shown below, the followings are equally obtained, noting that the wrong center has been activated.

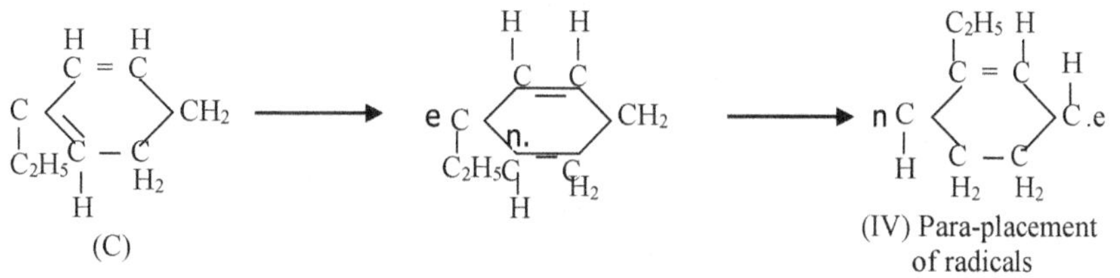

(C)   (IV) Para-placement
of radicals   11.220

While the C_2H_5 group in Equation 11.219 is not resonance stabilized, that in Equation 11.220 is and cannot readily be used. When the group is greater than CH_3 and it is located between the two conjugatedly placed double bonds, para-placement of the radicals is still possible. Nevertheless, these (i.e., when greater than C_2H_5) cannot molecularly rearrange to the more nucleophilic 1-position placement. It is important to note that it is the activation center carrying the substituent group externally located that is more nucleophilic and therefore never activated. When the group is internally located, the center carrying it is also more nucleophile and therefore cannot be the first to be activated. At this point in time, one will state additional laws guiding the activation of activation centers.

   i)   When a compound or monomer carries more than one activattion center of the same capacity and character, i.e., all X, then they can all be activated at the same time, provided they are isolatedly placed.

   ii)   When a compound or monomer carries more than one activation center cumulatively or conjugatedly placed, only one center can be activated, whether the centers are of the same nucleophilic capacity or not.

   iii) When a compound or monomer carries more than one activation center of different capacities but same character, i.e., all X, then they cannot be activated at the same time, but one at a time, and the first to be activated all the time is the least nucleophilic center, no matter how placed.

   iv)   When a compound or monomer carries more than one activation center of different capacities and character, i.e., X and Y conjugatedly placed, both center can be activated at the same time depending on the operating conditions or one at a time depending on the type of activator.

v) When a compound or monomer carries more than one activation center of different capacities and character, i.e., X and Y cumulatively placed, both can be activated at the same time separately on two monomers, but only one at a time on one monomer.

vi) When two or more compounds or monomers carrying one activation center which can only be X, are present together, they can all be activated at the same time based on the operating conditions or one at time beginning with the monomer carrying the least nucleophilic center depending on the operating conditions.

The para-placement of the monomer of Equation 11.220 cannot be reversed, when stronger radical-pushing groups are introduced into the ring as shown below.

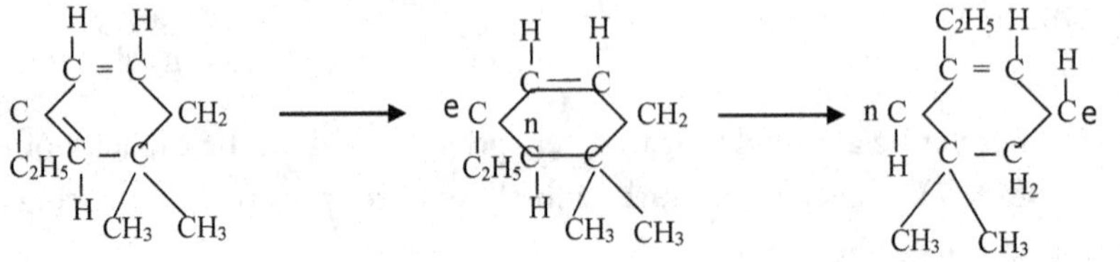

$$\underline{\text{Para - placement of radicals}} \qquad\qquad 11.221$$

When the conditions observed "exploratively" are applied, the followings are obtained.

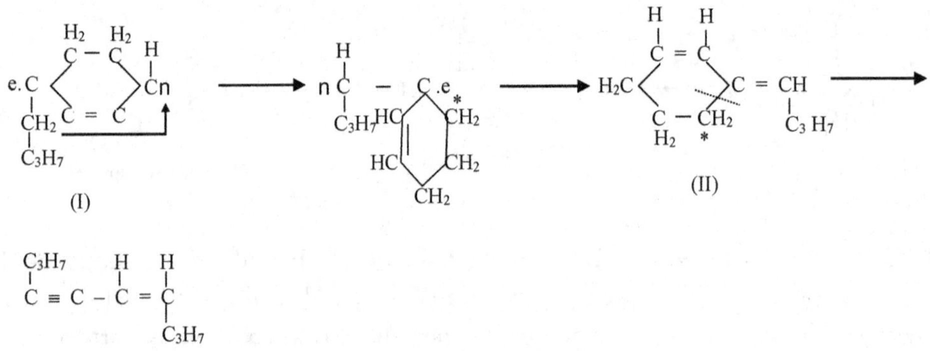

$$11.222$$

Molecular rearrangements of both the first and third kinds take place to give a resonance stabilized monomer obtained after scission. Molecular rearrangement of the third kind begins when CH_3 is C_2H_5 or C_3H_7 and stops at C_4H_9. For this reason, the following is valid.

Order of Radical-pushing capacity $\qquad\qquad 11.223$

The closed group above is that whose terminals are the terminals of a diene.

With radical-pulling groups, consider CF_3 type. The para-placement of the radicals is now reversed. The rings cannot be opened.

288

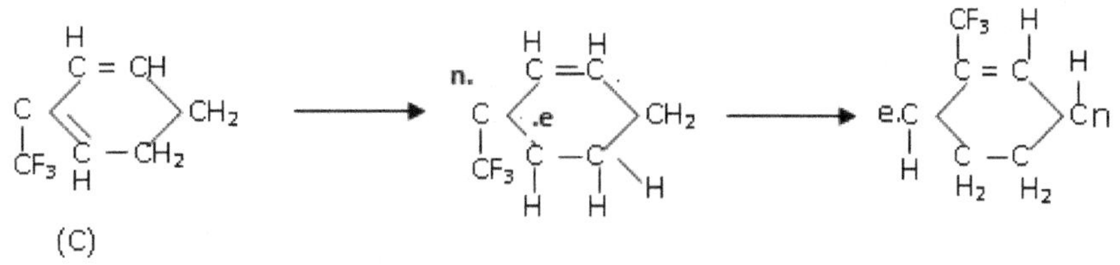

(C)

11.224

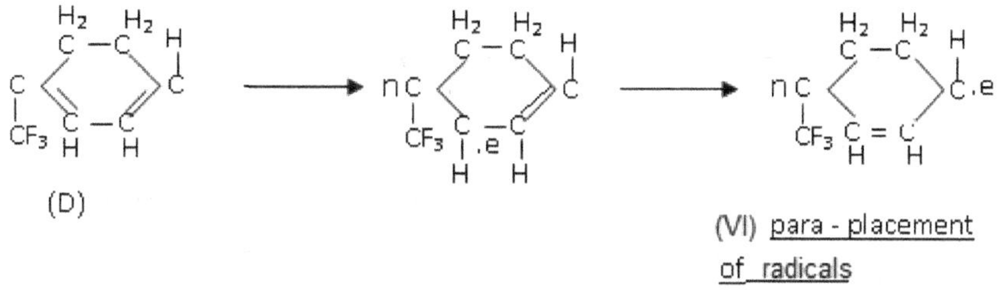

(D)

(VI) <u>para - placement
of radicals</u>

11.225

With free media initiators, no route is favored for the last case- the last equation. Only paired coordination initiators can be used to polymerize the monomer which is a nucleophile. Even when CF_3 is changed to C_2F_5 and higher, the ring cannot be opened via decomposition mechanism.

The movement of an electro-free-radical begins only after activation or when there is a nucleo-free-radical hanging around adjacently placed or when there is hole adjacently placed to it. Nevertheless, while CF_3 in (C) is resonance stabilized, that in, (D) is not resonance stabilized.

With external or internal resonance stabilization groups, the followings are obtained.

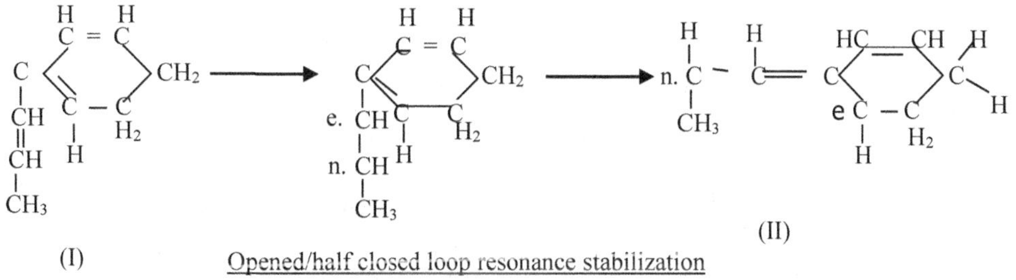

(I)   (II)

<u>Opened/half closed loop resonance stabilization</u>

11.226

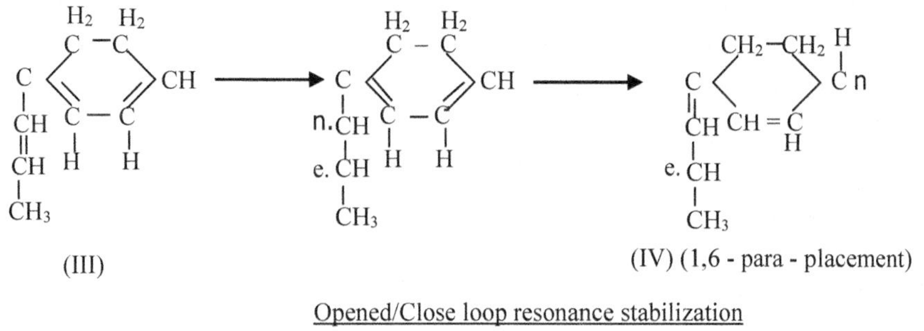

(III)   (IV) (1,6 - para - placement)

<u>Opened/Close loop resonance stabilization</u>

11.227

(V)   (VI) (1, 6 - para - placement)

NOT FAVORED   11.228

(V)   Opened/closed loop reonsance stabilization   (VI) (1, 6 - para - placement)

(VI)a Resonance stabilized   **FAVORED**   11.229

In the first case, the ring cannot be opened after molecular rearrangement of the first kind, since such transfer cannot take place, since the group is shielded. In the second case, no transfer species from inside the ring can be moved. In the third case, with the right source of transfer species, an acetylene carrying vinyl groups on both sides are obtained, clear indication of the larger order of nucleophilicity of the monomer. The movement was favored, because the radical-pushing capacity of half of what is inside the ring is less than C_2H_5 group, the other half of what is in the ring being resonance stabilized. The rearrangement above will stop when the C_2H_5 is changed to C_4H_9.

With acetylenic resonance stabilization group replacing alkenyl groups above, one should know what to expect based on the foundations laid so far. None of them can be opened instantaneously. The characteristic behavior of cyclobutene with two CH_2 groups internally located to an alkene or mono-ene is bound to be identical to that of cyclohexadiene also with two CH_2 groups internally located to a di-ene and this is also bound to be identical to that of cyclooctatriene with two CH_2 groups internally located to a tri-ene. The same applies to cyclopropene, cyclopentadiene, and cyclooctatriene. Yet all these belong to different families. One can observe a family of compounds coming from different families of compounds all coming from the Hydrocarbon family tree

11.2.3. Cycloheptadiene

Unlike cyclohexadiene, there is no para-placement of radicals, due to lack of symmetry. Like cyclohexadiene, there are three types -

(a)   One in which the double bonds are isolatedly placed
(b)   One in which the two double bonds are conjugatedly placed.
(c)   One in which the two double bonds are cumulatively placed (Very unstable)

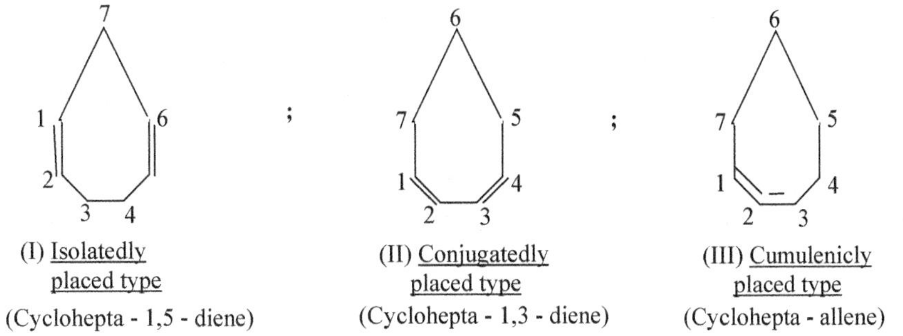

(I) Isolatedly placed type (Cyclohepta - 1,5 - diene); (II) Conjugatedly placed type (Cyclohepta - 1,3 - diene); (III) Cumulenicly placed type (Cyclohepta - allene)

11.230

There is no doubt that, minimum required strain energy can be provided for the rings. (I) is just one of the two isolated cases which are the same. While (I) has only one point of scission- the 3,4-bond, (II) has two identical points of scission- 5,6- or 6,7- bonds. (III) which is unstable and readily rearranges to cycloheptayne, has three points of scission. It was not easy to open 1,3-cyclohexadiene instantaneously for which in general for their polymerizations, special initiators are required to open their rings. Higher temperatures will obviously in addition be required to open their rings. The less difficult it will be for cyclohepta-diene, in view of the absence of symmetry. Nevertheless, despite the size, when conditions are adequately chosen, the (II) can be opened instantaneously to produce (IV) shown below under normal to moderate operating conditions. At higher operating conditions, methylene or ethylene, acetylenes and ethylene begin to appear as products. (III)a shown below without resonance stabilization to get (IV), cannot be deactivated. In the presence of an initiator, (III)a is used as shown electro-free-radically or nucleo-free-radically or after resonance stabilization before formation of (IV). (III)a or that obtained from (IV) will form the monomer units along the chain backbone under different operating conditions.

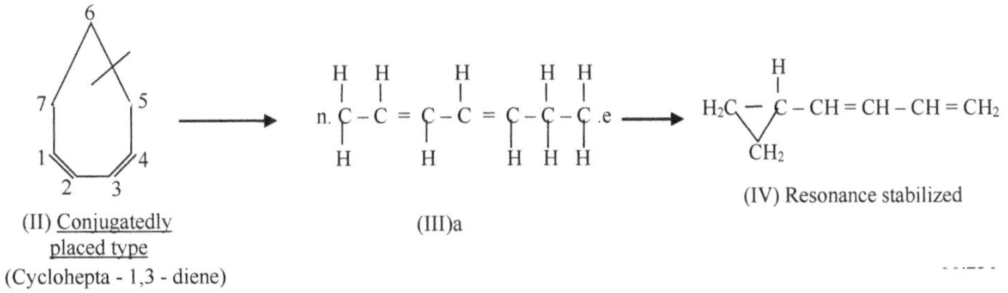

(II) Conjugatedly placed type (Cyclohepta - 1,3 - diene); (III)a; (IV) Resonance stabilized

11.231

The (IV) above is Cyclopropane-1,3-butadiene, that which is resonance stabilized. On the other hand, (III)a at close to cracking conditions, can further be broken down to give benzene, C and other products.

With (I), when the ring is instantaneously opened, the followings are obtained.

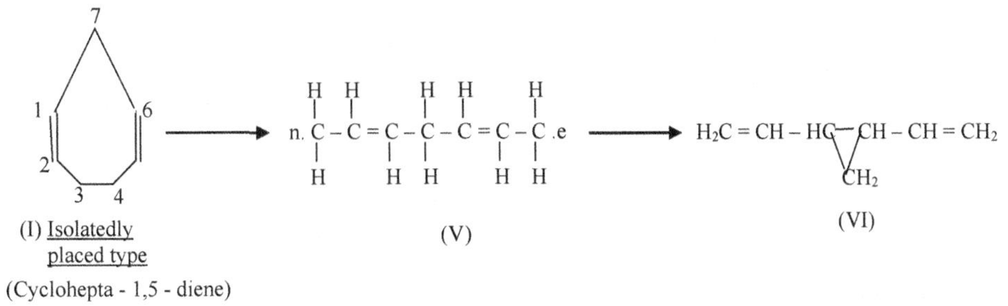

(I) Isolatedly placed type (Cyclohepta - 1,5 - diene); (V); (VI)

11.232

The (V) formed cannot readily be deactivated. However, via resonance stabilization, like the case of (II), (VI) a Divinyl-cyclopropane or a 4-vinyl cyclopentene is formed under normal to moderate operating conditions. In the presence of an initiator, (V) is either used as it is nucleo- and electro- free-radically or

291

used as (VI) after resonance stabilization, but before deactivation. When (V) is cracked, one knows what to expect, based on the experience acquired so far.

. Molecular rearrangement of the first kind of the tenth type involving transfer species of the first kind of the tenth types readily takes place when weak initiators are involved. While (I) will molecularly rearrange to itself, (II) cannot undergo molecular rearrangement when the 1,4- mono-form is not used. It can be observed in general that symmetric trans-placement of radicals on a monomer when activated, which odd-membered monomers do not have, is very important; so also is the trans-placement of monomers. Hence, in general odd-membered rings are not popularly known to be useful as monomers when the ring is not opened.

Very little or nothing new can be introduced in further investigation of this compound other than the application of the concepts which have been developed so far. Looking at these members of this family-Cyclodienes, one can observe a very logical network system. Cyclopentadienes, Cyclohexadienes and Cycloheptadienes, have different degrees of abilities to show their finger-prints, that is, exist in Equilibrium state of existence. It is easier for the five-membered ring than for the six-membered ring than for the seven-membered ring. We have seen that of cyclopentadiene and applied it to some extent only with respect to its decomposition, wherein one clearly identified a cyclopentadienyl group. Shown below are the Equilibrium states of existence of cyclohexadiene and cycloheptadiene.

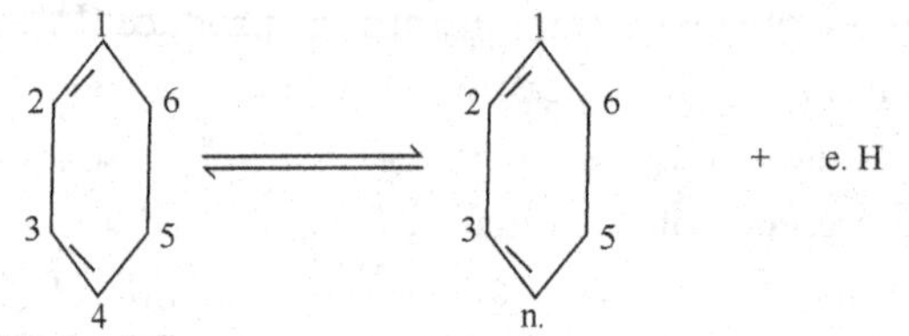

(I) Conjugatedly placed type
(Cyclohexa- 1,3 - diene)

+   e. H

11.233a

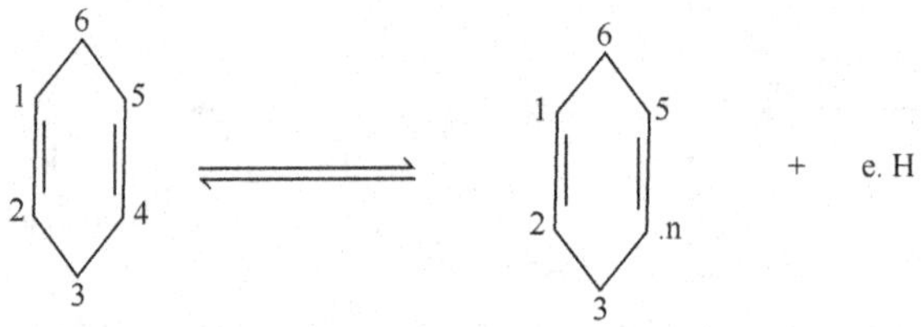

(II)
Isolated type (cyclohexa - 1,4- diene)

+   e. H

11.233b

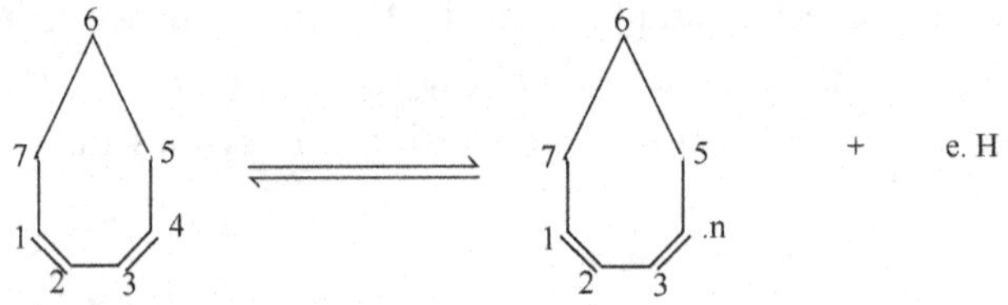

(III) Conjugatedly placed type
(Cyclohepta - 1,3 - diene)

+   e. H

11.234a

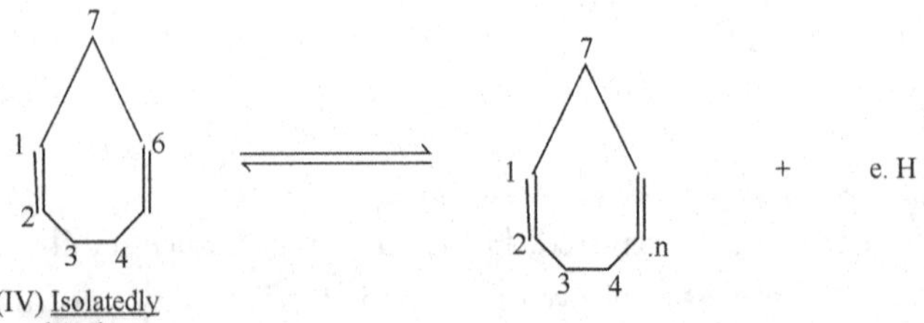

(IV) Isolatedly placed type
(Cyclohepta - 1,5 - diene)

+   e. H

11.234b

Note the H atoms held in Equilibrium state of existence. In the paper "Thermal aromatization of Methyl-1,3-cyclohexadienes- an important argument against commonly accepted sigmatropic 1,7-H-shift reactions", it was demonstrated that the methyl and C-atoms of methyl-cyclohexadienes interchange their positions intramolecularly during the thermal conversion to Toluene at tempera-tures above $600^{0}C$ in a quartz flow system[29]. When the methyl form of (I) is decomposed via Equilibrium mechanism, the followings are to be expected.

Stage 1:

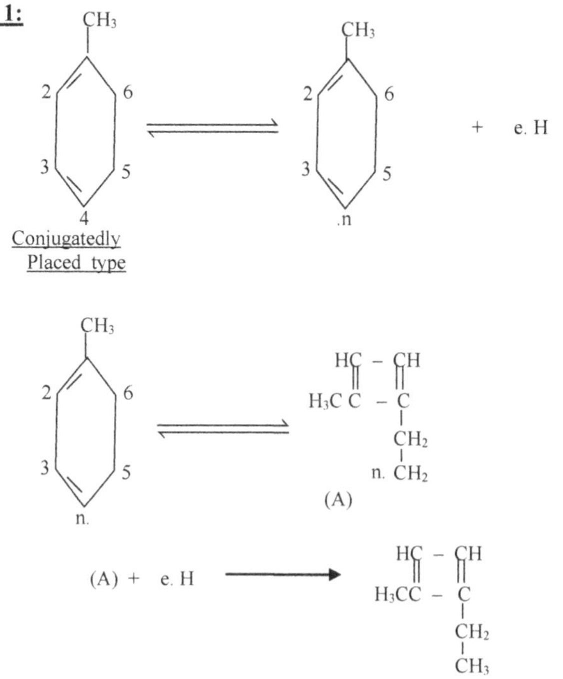

$$ 11.235a $$

Stage 2:

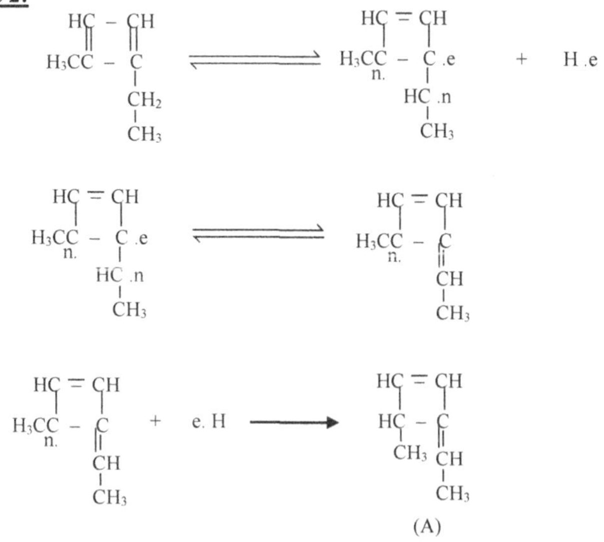

$$ 11.235b $$

Stage 3a:

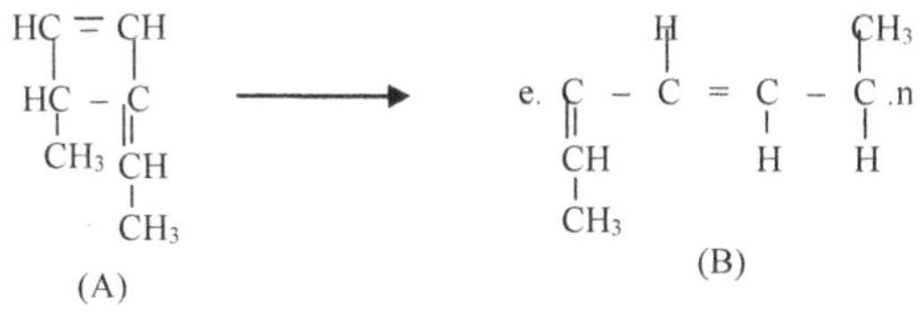

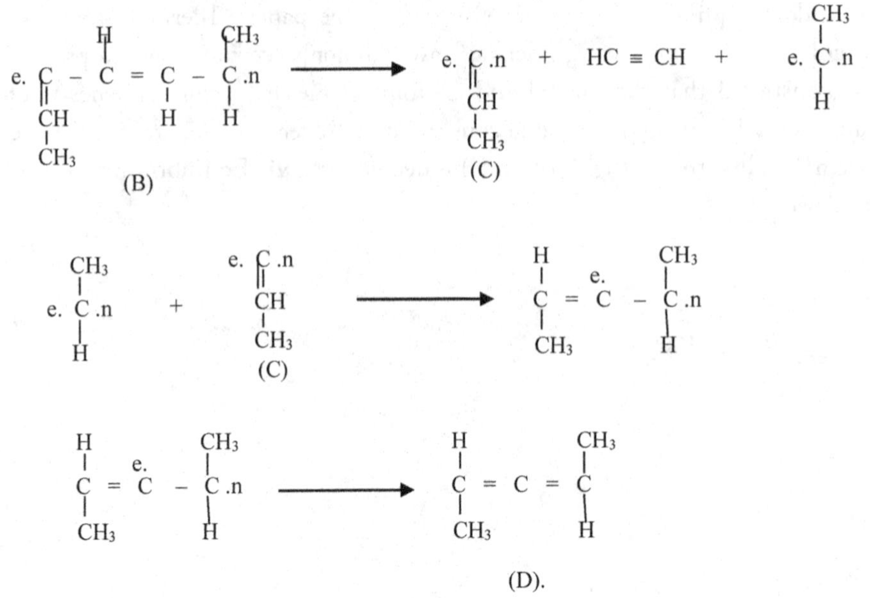

e.
$$C - C = C - C.n \longrightarrow e. C.n + HC \equiv CH + e. C.n$$

(B) → (C) + HC ≡ CH + (with CH₃/H substituent C)

Stage 3b: 11.236

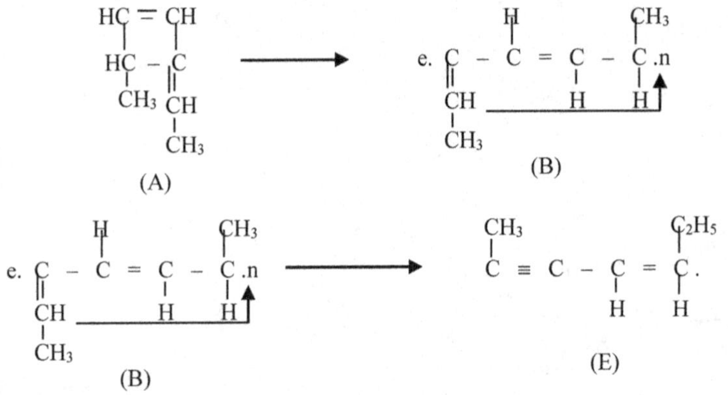

(A) → (B)

(B) → (E)

Overall equation: 2 (Methyl cyclohexadiene) $\longrightarrow$ **(D)** + **(E)** + **HC** $\equiv$ **CH** 11.237

11.238

Stage 4:

(E) $\rightleftharpoons$ (F)

(F) $\longrightarrow$ (G)

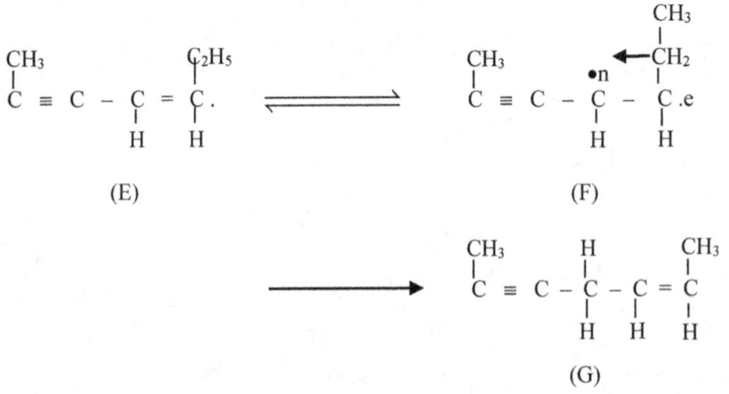

11.239

Stage 5:

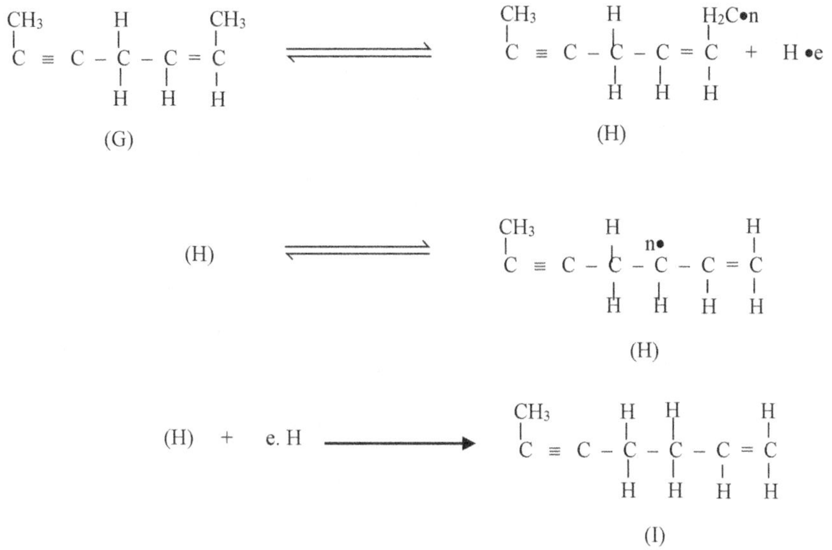

(G) ⇌ (H) + H •e

(H) ⇌ (H)

(H) + e. H ⟶ (I)   11.240

Stage 6:

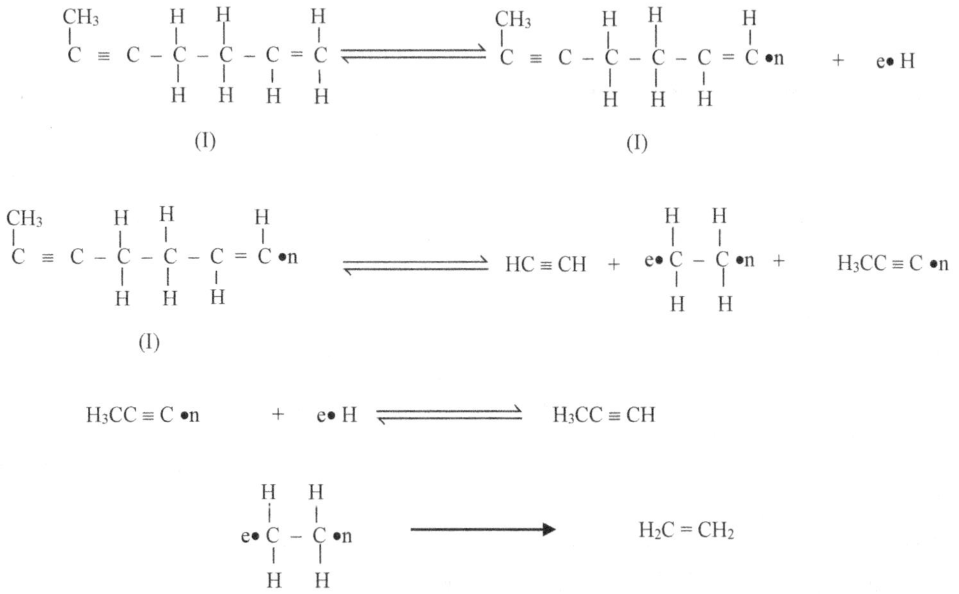

(I) ⇌ (I) + e• H

(I) ⇌ $HC \equiv CH$ + e• C – C •n + $H_3CC \equiv C$ •n

$H_3CC \equiv C$ •n + e• H ⇌ $H_3CC \equiv CH$

e• C – C •n ⟶ $H_2C = CH_2$ 11.241

Stage 7:   (D) molecularly rearranges to $H_3CC \equiv CC_2H_5$ via Equilibrium mechanism   11.242

Overall overall equation: 2 (Methyl Cyclohexadiene) ⟶ $2HC \equiv CH$ + $H_2C = CH_2$

   + $HC \equiv CCH_3$ + $H_5C_2C \equiv CCH_3$ 11.243

Stage 8:

$$HC \equiv CCH_3 \rightleftharpoons n.\ \overset{\overset{\displaystyle H}{|}}{C} = \underset{\underset{\displaystyle CH_3}{|}}{C}.e$$
(I)

$$(I)\ +\ HC \equiv CH \rightleftharpoons n.\ \overset{\overset{\displaystyle H}{|}}{C} = \underset{\underset{\displaystyle CH_3}{|}}{C} - \overset{\overset{\displaystyle H}{|}}{C} = \underset{\underset{\displaystyle H}{|}}{C}.e$$
(J)

$$n.\ \overset{\overset{\displaystyle H}{|}}{C} = \underset{\underset{\displaystyle CH_3}{|}}{C} - \overset{\overset{\displaystyle H}{|}}{C} = \underset{\underset{\displaystyle H}{|}}{C}.e\ +\ HC \equiv CH \rightleftharpoons n.\ \overset{\overset{\displaystyle H}{|}}{C} = \underset{\underset{\displaystyle CH_3}{|}}{C} - \overset{\overset{\displaystyle H}{|}}{C} = \underset{\underset{\displaystyle H}{|}}{C} - \overset{\overset{\displaystyle H}{|}}{C} = C.e$$
(J)　　　　　　　　　　　　　　　　　　　(K)

$$n.\ \overset{\overset{\displaystyle H}{|}}{C} = \underset{\underset{\displaystyle CH_3}{|}}{C} - \overset{\overset{\displaystyle H}{|}}{C} = \underset{\underset{\displaystyle H}{|}}{C} - \overset{\overset{\displaystyle H}{|}}{C} = C.e \longrightarrow$$ TOLUENE
(K)

11.244

Overall overall equation: 2 (Methyl Cyclohexadiene) $\longrightarrow$ Toluene + $H_3CC \equiv CC_2H_5$

+ $H_2C = CH_2$　　　11.245

Stage 9: $H_2C = CH_2$ decomposes further to give $HC \equiv CH$ + H_2　　　11.246

Overall overall equation: 2 (Methyl Cyclohexadiene) $\longrightarrow$ $HC \equiv CH$ + $H_5C_2C \equiv CCH_3$ +

Toluene + H_2　　　11.247

The order of nucleophilicity of some acetylenes when they are present is as follows.

$$H_3CC \equiv CC_2H_5\ >\ HC \equiv CC_2H_5\ >\ HC \equiv CCH_3\ >\ HC \equiv CH$$
Order of Nucleophilicity　　　11.249

In the final products obtained above when just two moles of methyl cyclohexadiene are involved, one can observe the presence of not only Toluene, but also a highly nucleophilic acetylene which at higher operating conditions can be broken down to give acetylene, H_2 and C. At the operating conditions, via instantaneous opening of its ring, toluene cannot be obtained. 2-methyl-1,3,5-hexatriene obtained cannot undergo further decomposition, but used as it is. When 5-methyl-1,3-cyclohexadiene is opened instantaneously, 1,3,5-heptatriene (i.e., 1-methyl-1,3,5-hexatriene) is obtained as will shortly be shown. However, one can observe that half of the product from stage 1, was allowed to undergo molecular rearrangement of the third kind. The remaining half decomposed via Decomposition mechanism. Like cyclopentadiene, after the first two stages, the next two stages take place in parallel- Stage 3a on one side and Stage 3b on the other side. From Stage 4, Equilibrium mechanism dominated in all the remaining stages. Worthy of note is that in Stages 4 and 5, the fully resonance stabilized monomer (E), started de-resonance stabilizing itself, in order to have the ability of existing in Equilibrium state of existence in Stage 6 and commence decomposition. Based on the manners by which activation of monomers take place, one was not surprised to find that the first ringed product was toluene. It is the more nucleophilic monomer, i.e., "the Female", that diffuses to the less nucleophilic monomer, i.e., "the Male" all the time in Nature. At the operating conditions, one expects that all the monomers will be in the activated state of existence. In NATURE, no matter the operating conditions, ORDERLINESS must be maintained. There are times when the monomer is stable; there are times when it is activated and times when it is kept in Equilibrium state of existence all taking place at the same operating conditions. Imagine how the Equilibrium mechanism is able to operate under such conditions (Very high temperatures). Nature is too much to comprehend.

While the 1-methyl cyclohexadiene considered above will favor the existence of toluene, 5-methyl cyclohexadiene will not via the same mechanism. In fact, the existence of Stage 3b of Equation 11.237 would be impossible. Only Stage 3a can take place. Indeed, in view of the location of the CH_3 group (i.e., in the 5-position), the ring can very readily be instantaneously opened to give 1,3,5- heptatriene as shown below under mild operating conditions.

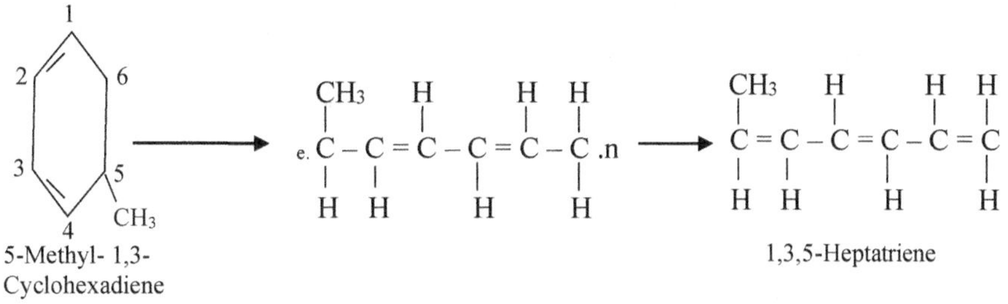

$$11.250$$

At higher operating conditions, the 1,3,5-heptatriene can readily be broken down. It is broken down via Decomposition mechanism in its activated state in a similar manner as shown above to give the following overall equation. At least, two stages all in series are involved. Note that Toluene cannot be obtained from it, unlike the case of 1-methyl-1,3-Cyclohexadiene.

Overall overall equation: 5-methyl-1,3-Cyclohexadiene $\longrightarrow$ Benzene + CH_4 11.251

Note that the Equilibrium state of existence of the 1,3,5-Heptadiene is as follows.

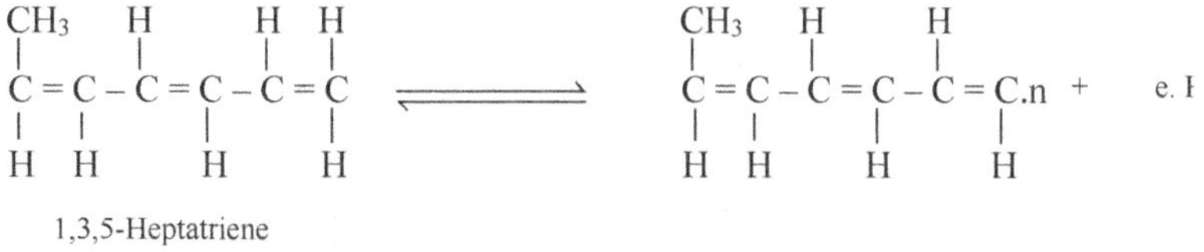

$$11.252$$

Thus, via instantaneous opening of the ring, 5-methyl-1,3-cyclohexadiene was decomposed. Unlike the case of 1-methyl-1,3-cyclohexadiene, its decomposition when the monomer itself is made to exist in Equilibrium state of existence to obtain toluene is impossible. The products obtained via Decomposition mechanism after opening of the ring is shown below.

Overall overall equation: 5-methyl-1,3-Cyclohexadiene $\longrightarrow H_5C_2C \equiv CCH_3$ + $HC \equiv CH$ 11.253

With 2- or 3- methyl cyclohexadiene, this can also be cracked to give toluene as one of the products.

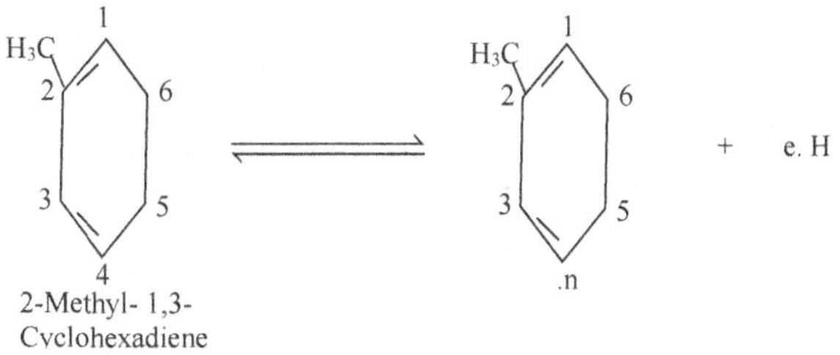

$$11.254$$

With the group internally located on the ring but unlike the case of 5-methyl-cyclohexadiene which has no pushing or pulling effect, when opened instantaneously, products similar to that of 1-methyl cyclohexadiene are obtained. Unlike 1-methyl cyclohexadiene, but like 5-methyl cyclohexadiene, Stage 3b of Equation 11.237 may not take place. However, based on instantaneous opening of the ring, via Equilibrium mechanism, the following is the overall equation.

Overall overall equation: 2(2-methyl-1,3-cyclohexadiene) $\longrightarrow$ Toluene + HC≡CCH$_3$ + $\qquad$ 11.255

$$H_2 \quad + \quad H_3CC{\equiv}CCH_3 \qquad\qquad 11.255$$

When 4-methyl-cyclohexadiene and 2-methyl-cyclohexadiene are instantaneously opened, the followings are obtained.

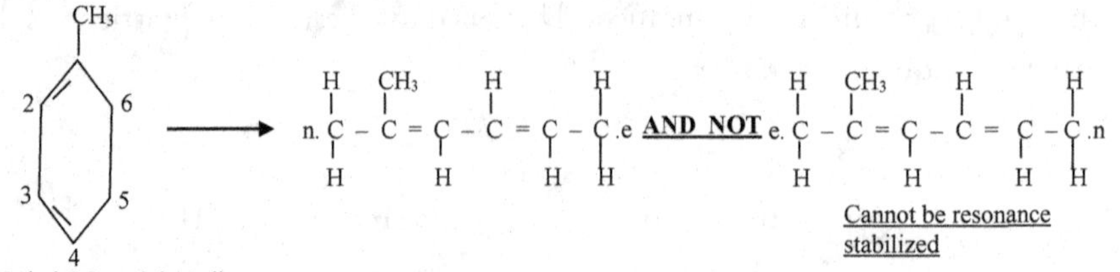

1-Methyl-1,3-cyclohexadiene

n.C–C=C–C=C–C.e **AND NOT** e.C–C=C–C=C–C.n

Cannot be resonance stabilized

11.256

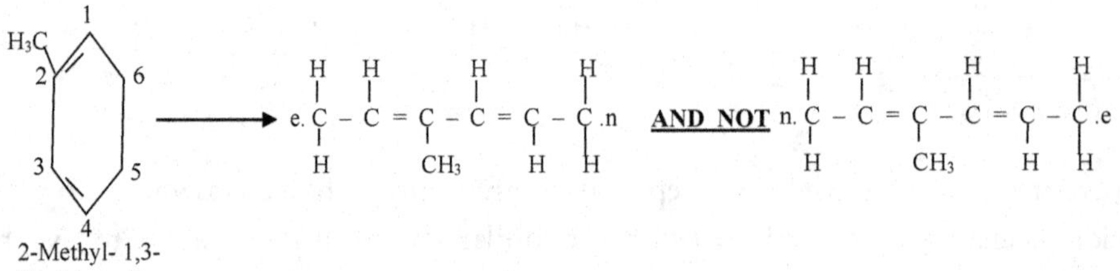

2-Methyl- 1,3-Cyclohexadiene

e.C–C=C–C=C–C.n **AND NOT** n.C–C=C–C=C–C.e

11.257

When a ring (with only one point of scission) that is resonance stabilized either from within or from outside, is opened instantaneously, the radical carried by the centers are such that, the resonance stabilized character of the opened monomer is still maintained. Also shown below are the Equi-librium states of existence of the hexatrienes above. They are all isomers of heptatriene the less nucleophilic of them, since they will favor both routes when instantaneously opened.

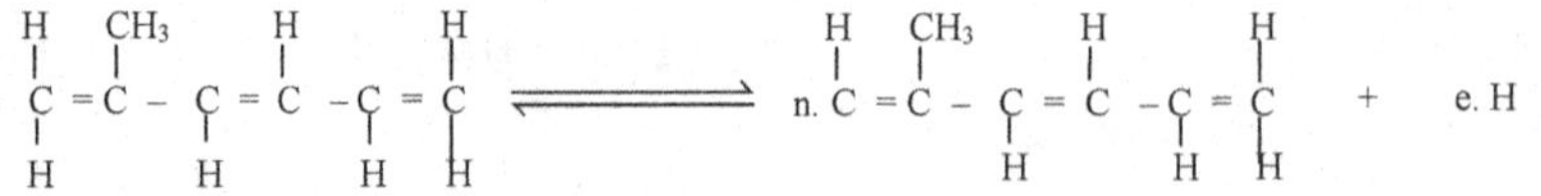

2-Methyl-hexatriene From 1-methyl-cyclohexadiene

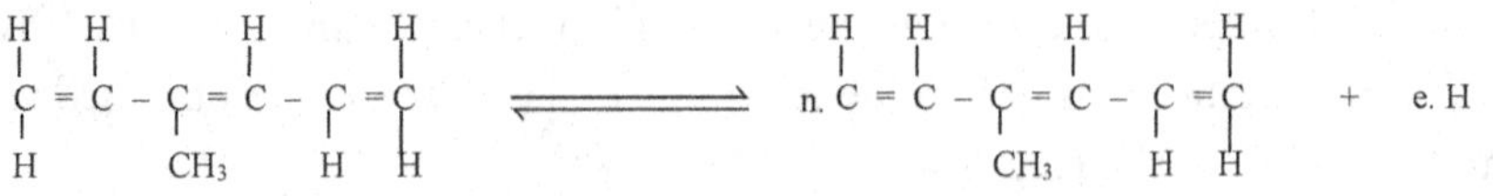

11.258

3-Methyl-hexatriene From 2-methyl-cyclohexadiene

11.259

Even 1-methyl-1,4-cyclohexadiene has been reported to undergo thermal dehydrogenation catalytically to give Toluene as one of the products[30].

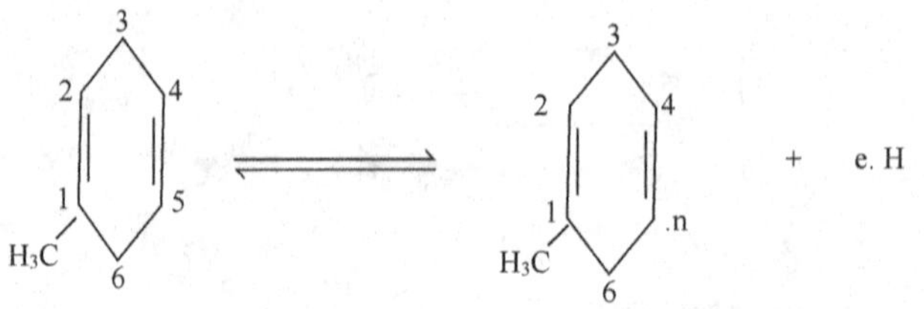

(I)
1-Methyl-Cyclohexa - 1,4- diene

11.260

Since it has no point of scission, it cannot be opened instantaneously in the absence of molecular rearrangement. In the absence of this rearrangement to give the 1,3-cyclohexadiene, the only way it can be opened is when it is in Equilibrium state of existence. However, when rearrangement is favored, catalyst can then be used for its dehydrogenation, (something yet to be reported). Since, one has provided its Equilibrium state of existence above, when decomposed, the followings are to be expected in the absence of catalysts, because of the operating conditions and all said above.

Stage 1:

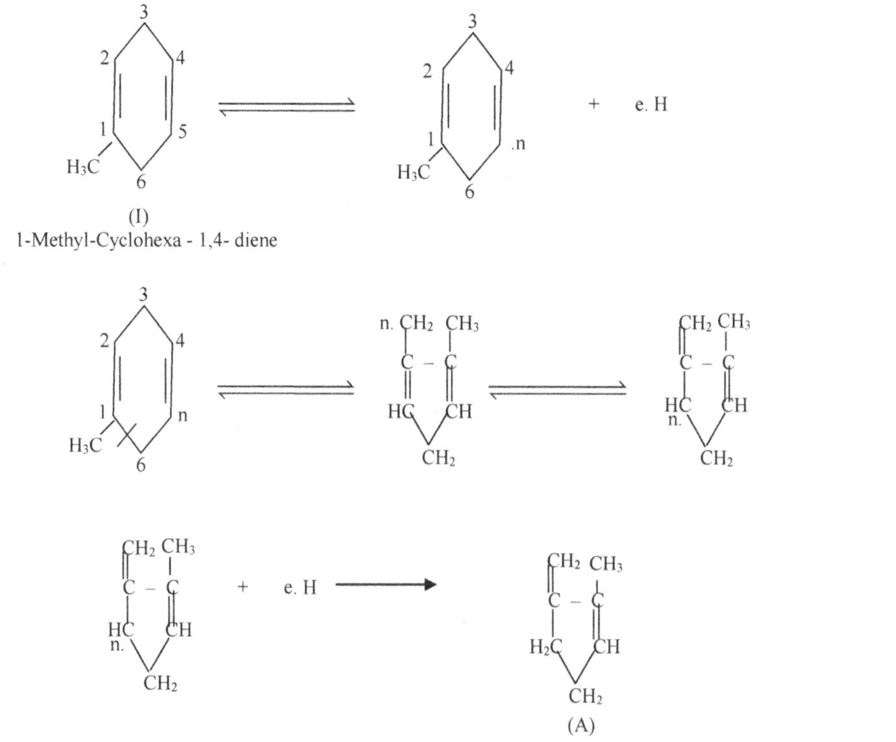

(I)
1-Methyl-Cyclohexa - 1,4- diene

(A)   11.261

Stage 2:

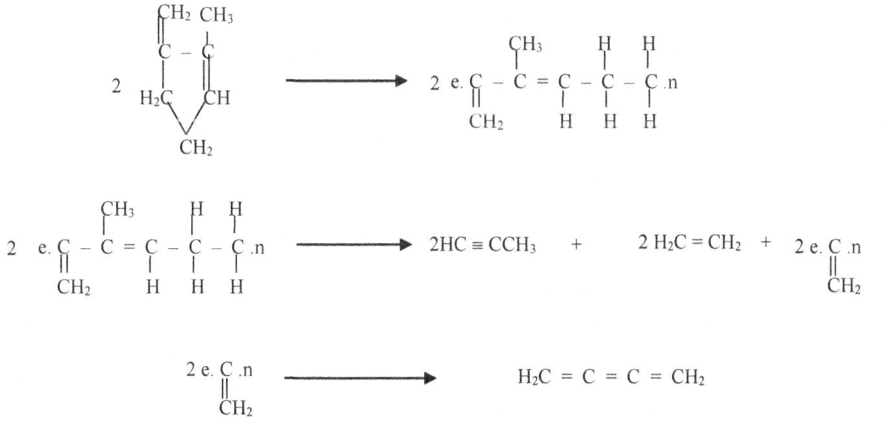

                                                                11.262

Stages 3 & 4: The Cumulene is broken down as already shown in Equations 11.171a and 11.171b

                                                                11.263

                                                                11.264

299

Overall equation: $3(1\text{-Methyl-cyclohexadiene}) \longrightarrow$ (A) $+$ $2HC \equiv CCH_3$ $+$

$$2C \ + \ 3H_2C = CH_2 \qquad\qquad 11.265$$

Stage 4: Ethenes break down to give $\quad 3HC{\equiv}CH \ + \ 3H_2$ $\qquad\qquad$ 11.266

Stage 5:

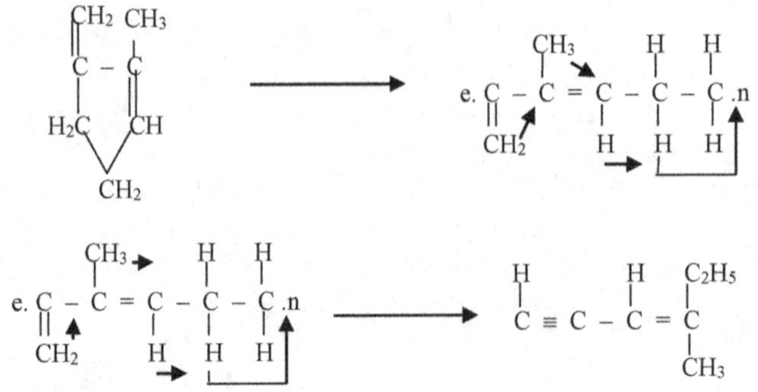

$\qquad\qquad\qquad\qquad\qquad\qquad\qquad\qquad\qquad\qquad\qquad\qquad$ 11.267

Stage 6:

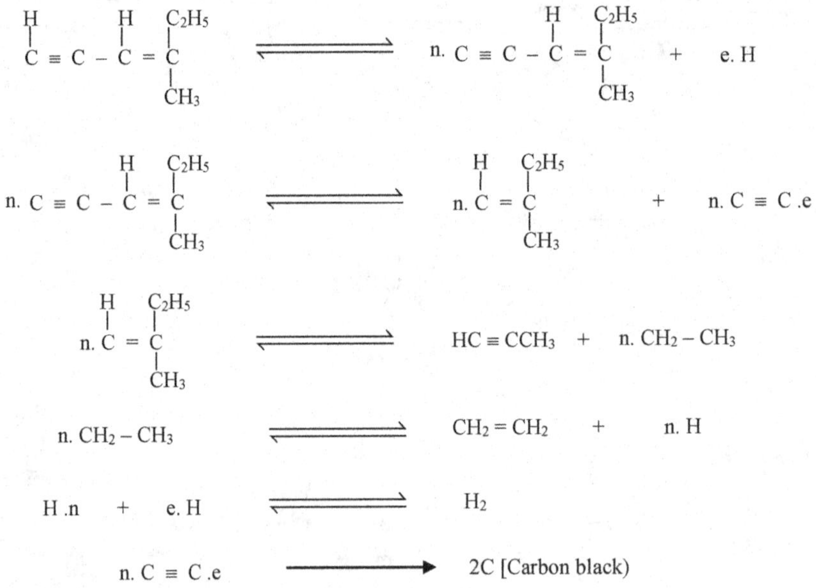

$\qquad\qquad\qquad\qquad\qquad\qquad\qquad\qquad\qquad\qquad\qquad\qquad$ 11.268

Stage 7: Ethene breaks down to $\ HC \equiv CH \ + \ H_2$ $\qquad\qquad$ 11.269

Overall equation: (A) $\longrightarrow HC \equiv CH \ + \ 2H_2 \ + \ 2C \ + \ HC{\equiv}CCH_3$ $\qquad$ 11.270

Stage 8: Toluene was obtained from the methyl acetylene and acetylene. $\qquad$ 11.271

Overall overall equation: $3(1\text{-Methyl-1,4-Cyclohexadiene}) \longrightarrow 2\text{Toluene} \ + \ 4C$

$$+ \ HC \equiv CCH_3 \ + \ 5H_2 \qquad\qquad 11.272$$

If the equilibrium state of existence had not been as shown, these products will not be obtained. Instead, one will be seeing xylene as one of the products. On the whole, one can see the number of stages involved. All the stages speak for themselves.

Because of the great importance of toluene, its chemistry has developed to the point where ringed isomers of toluene have been identified. These are non-aromatic in character. These are largely called ISOTOLUENES[31-36].

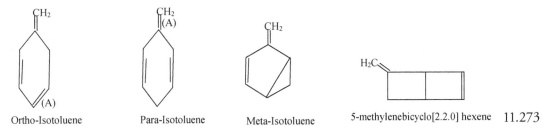

Ortho-Isotoluene   Para-Isotoluene   Meta-Isotoluene   5-methylenebicyclo[2.2.0] hexene   11.273

How the ortho- and para- isomerize to toluene are up till date been unknown. They do by undergoing molecular rearrangement of the first kind. The weakest nucleophilic center in the ortho- and weaker nucleophilic center in the para-, have been identified with (A). While the ortho- first rearranges to the para-, the para- rearranges straight to toluene. Their ability to rearrangement depends on the operating conditions. Only the first and the last have a point of scission

We will recall the case of cyclopentadiene in Equation 11.145a, where part of the cyclopentadiene was in Equilibrium state of existence, while the remaining fraction was in Stable state of existence. Why did it not happen with cyclohexadiene and cycloheptadiene? Beginning with cyclohexadiene, the followings are obtained.

Stage 1:

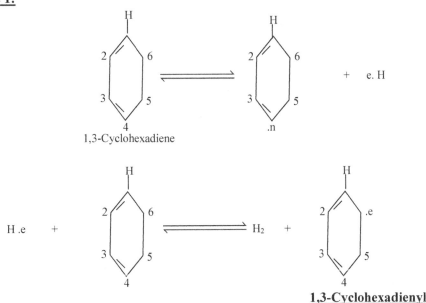

1,3-Cyclohexadienyl

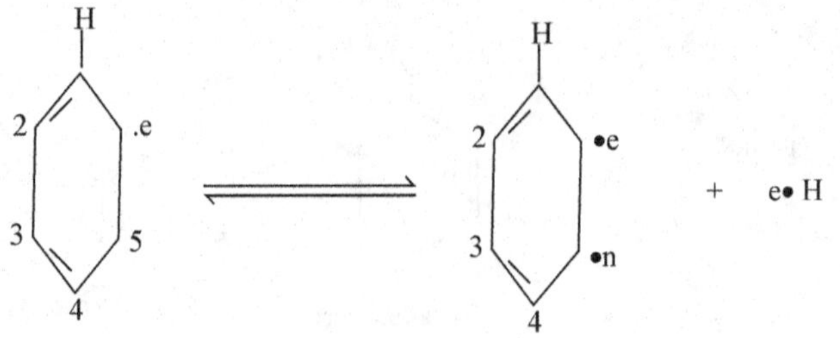

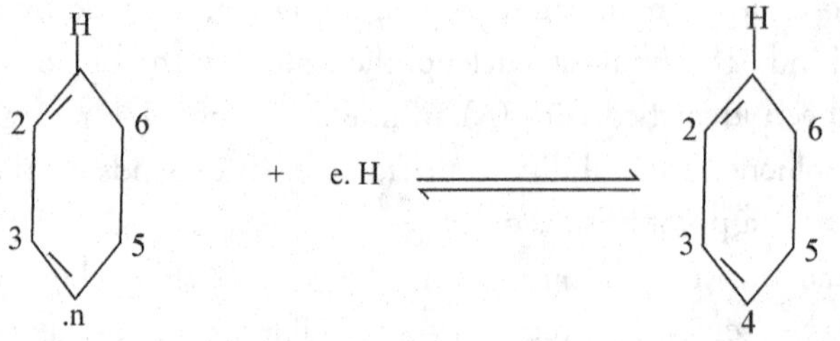

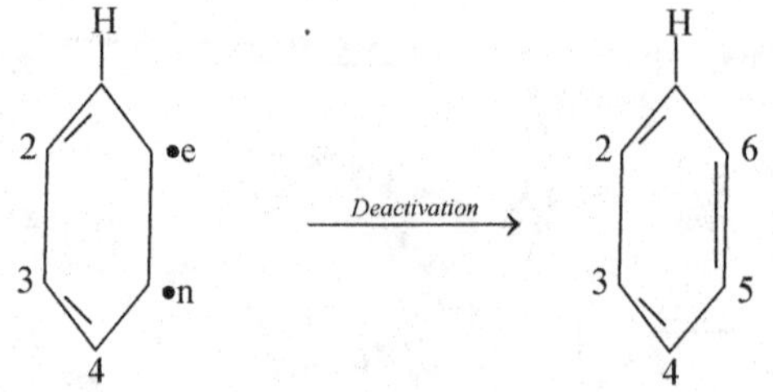

Benzene

Overall equation: 2(1,3-Cyclohexadiene) ⟶ 1,3-Cyclohexadiene + H₂ + Benzene

11.274

11.275

With cycloheptatriene, the followings are similarly obtained

Stage 1:

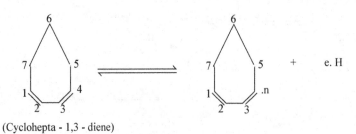

(Cyclohepta - 1,3 - diene)

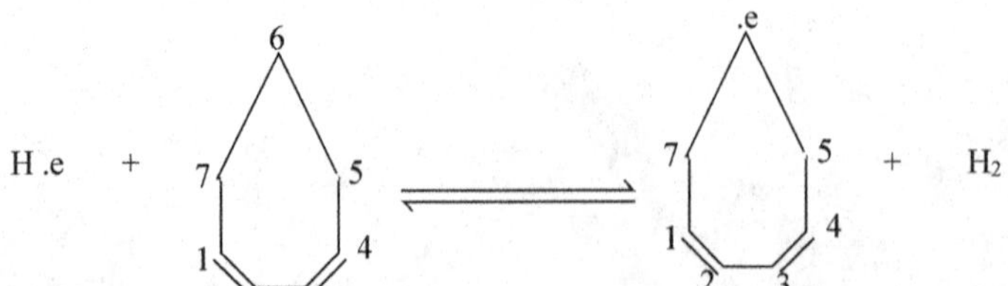

1,3-Cycloheptadienyl

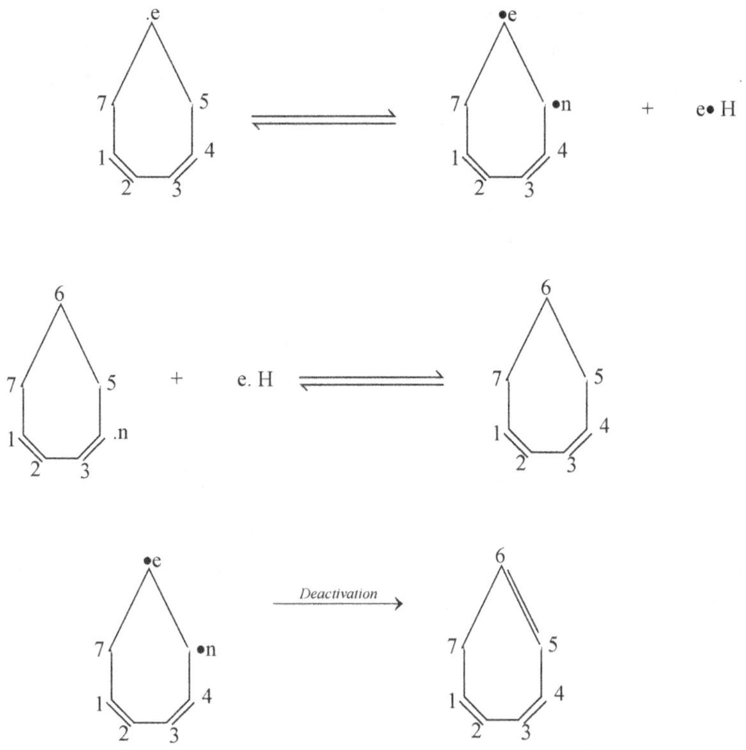

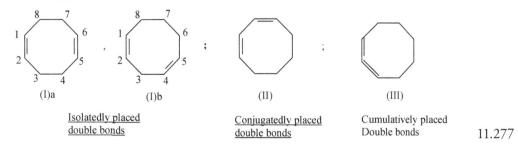

Cycloheptatriene   11.276a

Overall equation: 2(1,3-Cycloheptadiene) ⟶ 1,3-Cycloheptadiene + H_2 + Cycloheptriene   11.276b

One can imagine the wonderful world of Chemistry. **The monomers have dehydrogenated themselves by using itself to do the job.** The job was done in a single stage. If at the operating conditions reportedly used in the literature, a fraction of the monomer was still kept in stable of existence, then benzene will be largely present in the product for cyclohexadiene, and cycloheptatriene will also be largely present for the case of cycloheptadiene. *With operating conditions, one can change the state of existence of anything in life, living or non-living.* One can observe that the first member of this family, cyclopentadiene [Not cyclobutadiene] still remains very distinct from other members of the family. *Nothing in life is there for nothing, just as nothing goes for nothing-two different philosophical statements.* Also notice that for the first time, cyclohexa-dienyl and cycloheptadienyl radicals have been identified in the equations above.

11.2.4 Cyclooctadiene

There are three type of this family of monomers as shown below.

(I)a	(I)b	(II)	(III)
Isolatedly placed double bonds	Conjugatedly placed double bonds	Cumulatively placed Double bonds	11.277

There are two types of the isolatedly placed double bonds - symmetrically placed (I)a which can be called 1,5-Cyclooctadiene, and non-symmetrically placed (I)b which can be called 1,4-Cyclooctadiene. In both cases, for each one, their centers are of the same nucleophilicity. So also are the others. It has been in the past reported that 1,3-cyclooctadiene can be prepared by 1,5- cycloocta-diene[37]. This reaction is said to need the reagent H_2 and catalyst Polymer bound Ni^2 catalyst at temperature of 100^0C. The reaction time was said to be 15 hours with a yield of 48%[37]

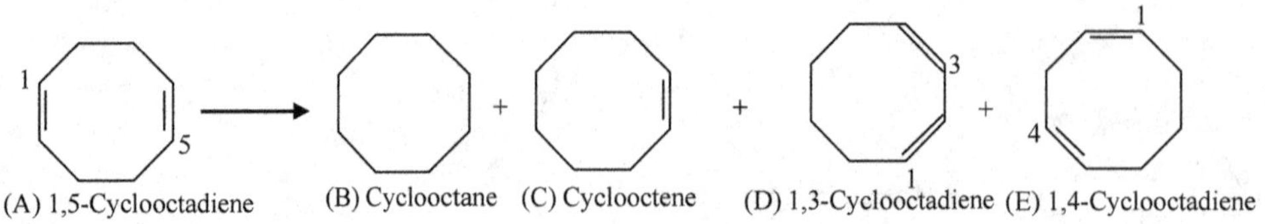

(A) 1,5-Cyclooctadiene   (B) Cyclooctane   (C) Cyclooctene   (D) 1,3-Cyclooctadiene   (E) 1,4-Cyclooctadiene   11.278

The reactions above can simply be explained as follows. With the presence of the Ni complex, hydrogenation of the 1,5-cycloocyadiene took place to produce (B) and (C) above. If there had been excess H_2 in the system, (D) and (E) will never be obtained. In fact, we don't need H_2 and the catalyst above to get (D) and (E). Without understanding the mechanisms of how NATURE operates, most of all what we are doing in our world today are absolute nonsense. (E) was first obtained as follows.

Stage 1:

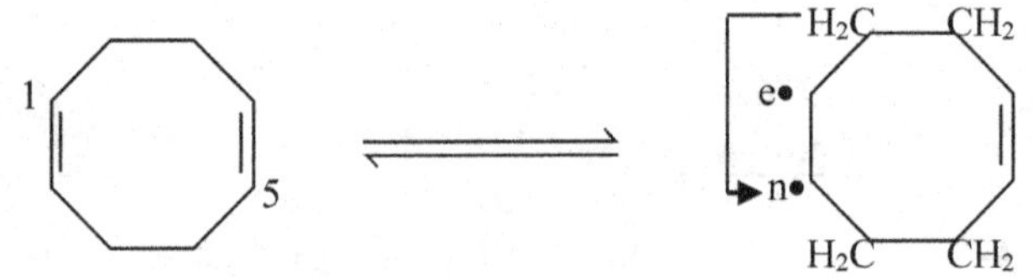

(A) 1,5-Cyclooctadiene

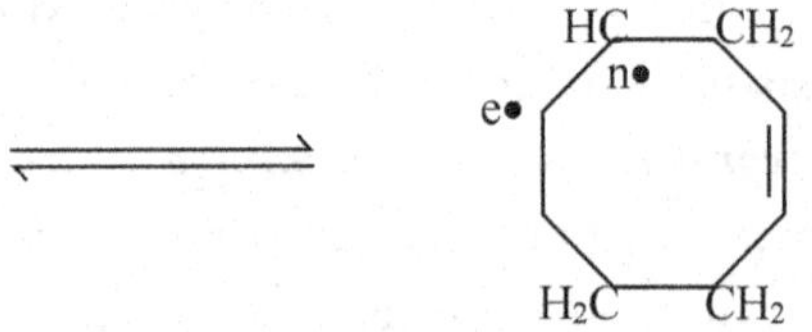

$$\xrightarrow{\textit{Deactivation}}$$

HC——CH$_2$

H$_2$C——CH$_2$

(E)

Overall equation: 1,5-Cyclooctadiene ——→ 1,4-Cyclooctadiene   11.279a

From (E), (D) was obtained as follows.   11.279b

Stage 2:

HC——CH$_2$ ⇌   HC——$\overset{n\bullet}{\text{CH}}$ +   H •e

H$_2$C——CH$_2$ H$_2$C——CH$_2$

(E)

HC——$\overset{n\bullet}{\text{CH}}$ ⇌   n •   HC——CH

H$_2$C——CH$_2$ H$_2$C——CH$_2$

(F)

(F)   +   H •e   ——→   H$_2$C   HC——CH

H$_2$C

H$_2$C——CH$_2$

(D)   11.280a

Overall equation: 1,4-Cyclooctadiene ——→ 1,3-Cyclooctadiene   11.280b

1,4-cyclooctadiene cannot molecularly rearrange to 1,3-cyclooctadiene or to 1,5-cyclooctadiene. 1,5-cyclooctadiene rearranged to 1,4-cyclooctadiene in Stage 1 above via Activation. Then in Stage 2, a fraction of the 1,4-cyclooctadiene obtained in Stage 1 was transformed to 1,3-cyclooctadiene via Equilibrium state of existence. It is irreversible. Note the location of the H held in Equilibrium state of existence in (E) of Stage 2. Thus, based on Equation 11.277, (I)a molecularly rearranged to (I)b and part of (I)b was converted to (II) via Equilibrium state of existence. (II) cannot molecularly rearrange to another form. Unlike cyclohexadiene of type (II), there is no way para-placement of the radicals or charges can be made possible. Hence, cases such the (II), will not be useful as monomers for polymerization purposes, in the absence of para-placement of the charges or radicals; unless the ring is opened instantaneously. The existence of (III) is not well known. Since six-membered case has been transiently isolated, there is no doubt that (III) will exist, most likely as Cyclooctayne via molecular rearrangement.

Like (I) of Equation 11.193 for cyclohexadiene, when (I)a above is activated the following is obtained.

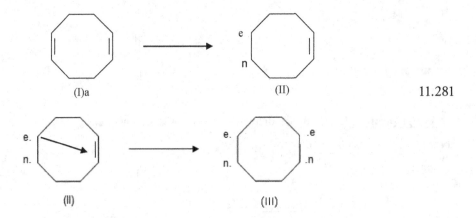

(I)a   (II)   11.281

(II)   (III)

Diffusion/Activation   11.282

Because of the distantly isolated placement of the double bonds, the electro-free-radical end of one center diffuses to the next activation center of equal nucleophilicity to give (III). This takes place with a weak initiator. In fact, with a stronger initiator, both centers can be activated at the same time since they are of the same nucleophilicity and they are not conjugatedly or cumulatively placed.

(I)a has been known in the past to favor some polymerizations, depending on the particular initiator system, with trans-annular "migration" taking place during addition.[1] This can readily be explained as follows, noting that it can be radically and chargedly.

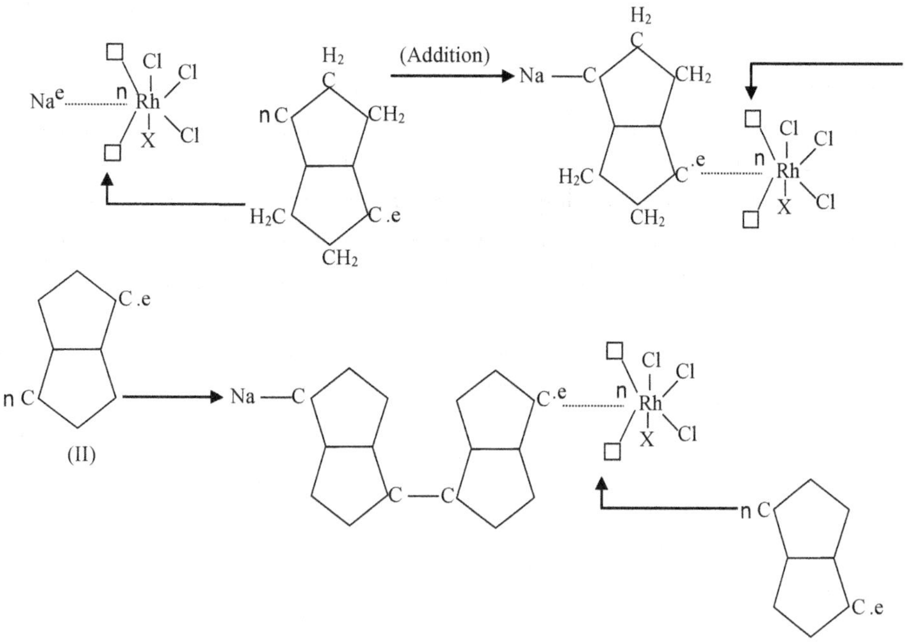

11.283

Note that the initiator above has been used radically exploratively. It cannot be used radically and in the structure shown above, Na is not the carrier of the chain as will be shown downstream when new concepts are introduced. But it still remains as an Electrostatically charged-paired initiator.

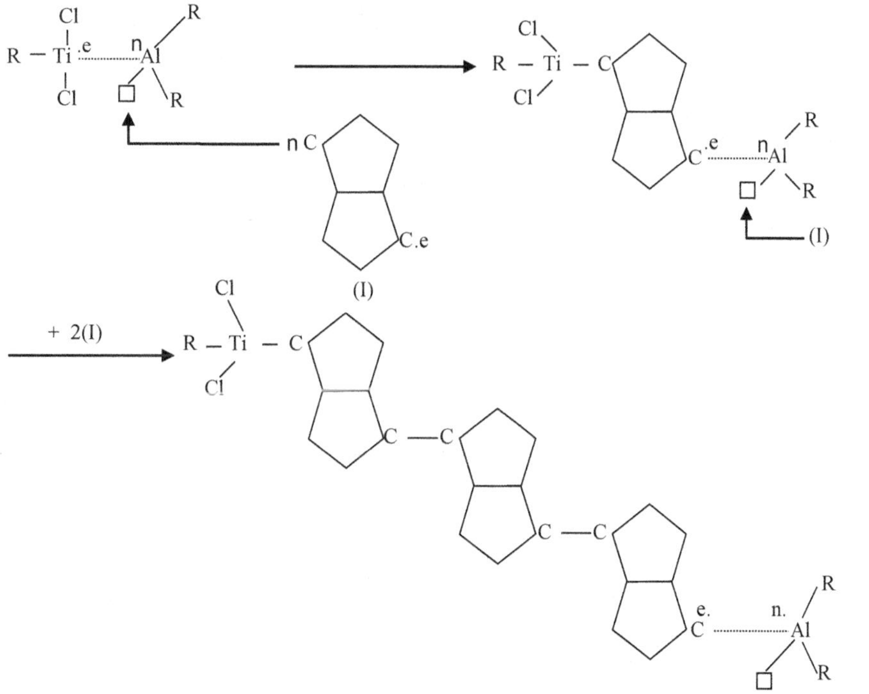

11.284

While the reaction can readily take place chargedly here, they have been displayed radically, subject to finding the conditions under which pairing of radicals can readily take place radically. The first initiator used above is of the Electrostatic type, while the second initiator is of the Covalent type. One can observe the different types of polymeric products obtained when one or two vacant orbitals are present. In view of the somehow non-linear character of addition, coordination initiators may not be easy to use. The best initiators to use are the electro-free-radical paired initiators of transition metal types with two vacant orbitals.

The existence of the polymeric products is favored, since the five membered fused rings cannot easily be provided with the minimum required strain energy (MRSE). That is, the rings cannot be instantaneously opened. It is also for this reason the six-membered counterpart, that is, (I) of Equation 11.193 will favor trans-annular addition, despite the fact that MRSE can readily be provided for the four-membered fused rings with points of scission. If the fused four-membered ring with a double bond in one of them of Equation 11.273 can exist, then why should this not exist?

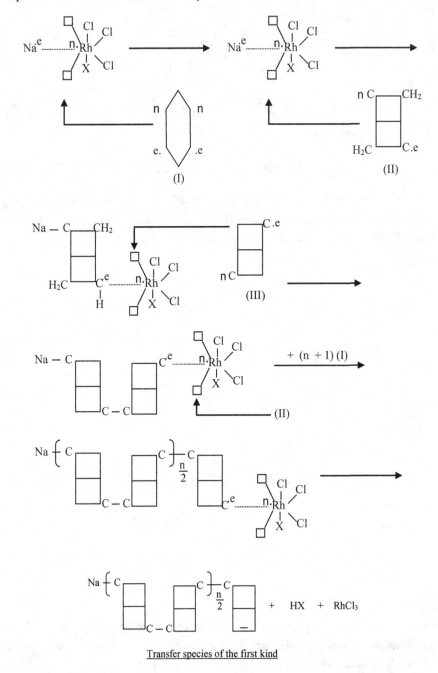

Transfer species of the first kind

11.285

Next in the series to favor the similar type of initiation reactions of cylcooctadiene are ten-membered, twelve-membered rings and above (even-membered rings).

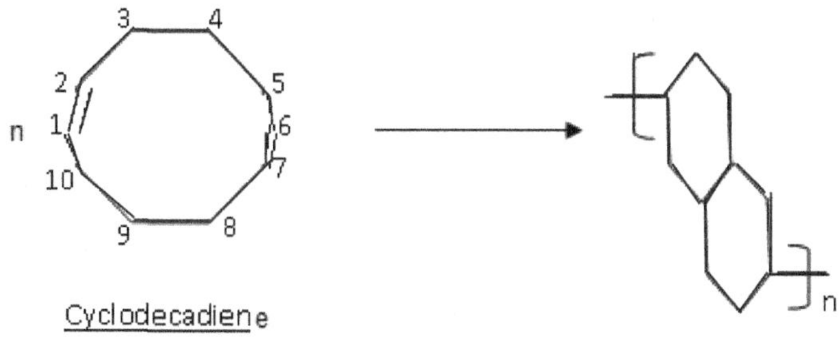

Cyclodecadiene

11.286

The initiator has to be strong in strength for trans-annular addition to take place if there is to be any further polymerization, just like for trans-placement of ringed monomers along the chain. Electro-free-radically and with positively charged initiators, these monomers can readily be polymerized. Nevertheless, in view of the route being natural to the monomer, one can observe why high molecular weight polymers can readily be obtained electro-free-radically and positively, since the activating centers will keep increasing in strength for every addition of monomer.

For the seventh and eight membered rings to favor molecular rearrangement of the first type, one of the active centers must carry radical-pushing groups externally located with capacity greater than C_3H_7 and C_4H_9 respectively.

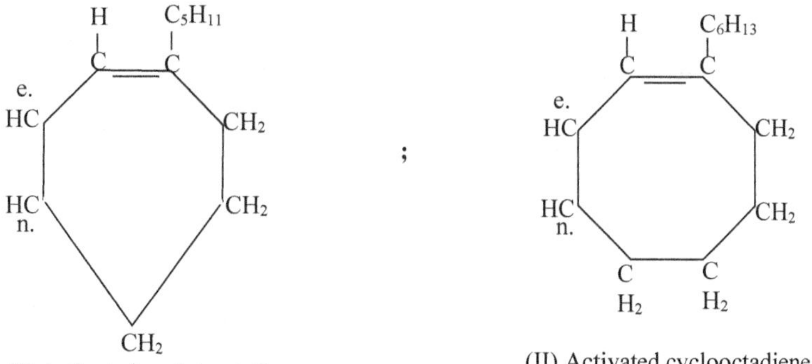

(I) Activated cycloheptadiene   (II) Activated cyclooctadiene   11.287

After the molecular rearrangement, the rings can be opened, as will shortly be shown. When the double bonds in the dienes are conjugatedly placed, molecular rearrangement of the third kind is readily favored.

In the presence of weak initiators, a resonance-stabilized monomer can be obtained when molecular rearrangements are allowed.

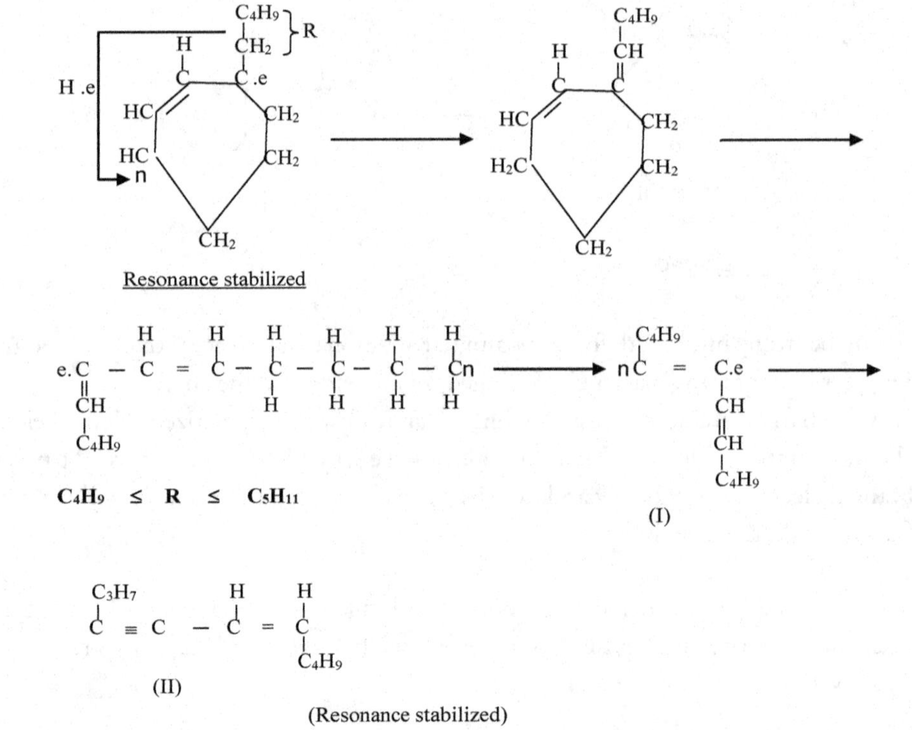

Resonance stabilized

$C_4H_9 \leq R \leq C_5H_{11}$

(I)

(II)

(Resonance stabilized)

11.288

For eight-membered rings, in the presence of strong initiators, the following is obtained only free-radically. The instantaneous opening of the ring of such size has been shown to be readily possible for 1,3-cyclooctadiene[38], 1,4-[39] and 1,5-[40-43]cyclooctadienes.

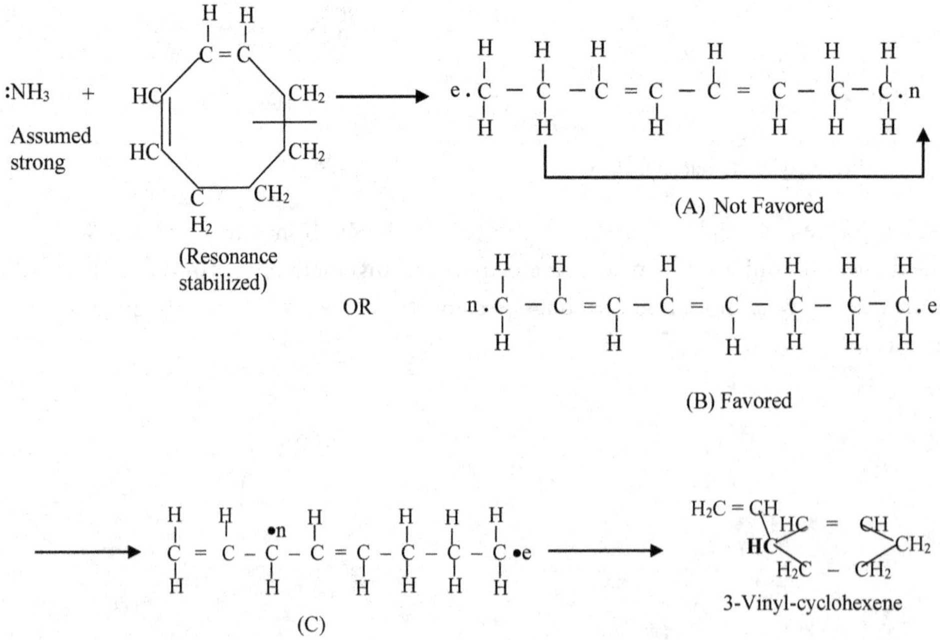

(A) Not Favored

(B) Favored

(C)

3-Vinyl-cyclohexene

11.289

Based on the current developments, (A) is not favored, since the radical potential difference between the two carbon centers where scission took place is zero. The favored one is (B) which cannot undergo molecular rearrangement of the third kind to give a triene or deactivate. After resonance stabilizing, 3-Vinyl-cyclohexene or Butadienyl cyclobutane may be obtained. In the presence of an initiator, either

(B) or (C) can be used depending on the strength of the initiator. In general however, (B) with conjugated double bond will largely form the monomer units along the growing chain free-radically[38]. *It is believed that during resonance stabilization, electro-free-radicals do not diffuse from a π-bond when an electro-free-radical is visibly present and functional.* For the case above, it is not functional. Though the monomer is a nucleophile, it can be polymerized nucleo- and electro-free-radically.

In the presence of weak initiators, the following is to be expected for this type of ring when a radical-pushing group is placed on the 1-position.

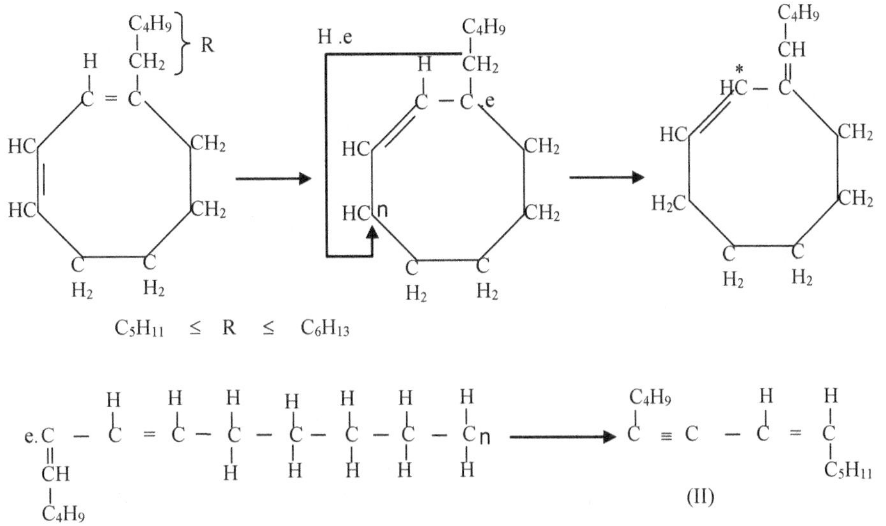

$$C_5H_{11} \leq R \leq C_6H_{13}$$

(II)

(Resonance stabilized)

11.290

Like the seven-membered ring, the rearrangements will take place, since a resonance stabilized monomer is obtained for the range of alkylane group specified.

For the other non-resonance stabilized cyclooctadienes, consider the instantaneous opening of 1-,5-cyclooctadiene which is a product of dimerization of butadiene in the presence of Ni catalyst, a coproduct being vinylcyclohexene[44]. While 1,5-cyclooctadiene is obtained by 1,4- to 1,4- addition of butadiene, the vinylcyclohexene is obtained by 1,4- to 1,2- addition of the butadiene.

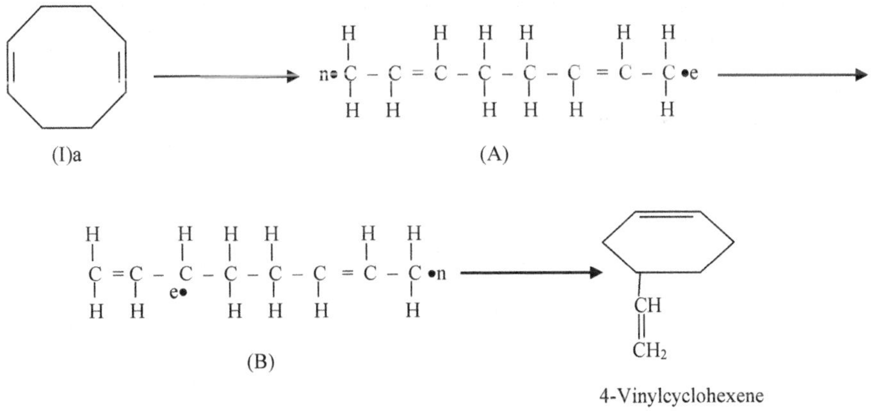

(I)a

(A)

(B)

4-Vinylcyclohexene

11.291

Chargedly, (A) can exist, but cannot resonance stabilize to give (B). In the monomer unit there are two isolatedly placed double bonds. (A) radically cannot be deactivated such that there are three double bonds. At best, in the absence of a strong initiator, it can be deactivated to form a 4-Vinylcyclohexene, the same as was obtained in the production of the cyclooctadiene from butadiene. With a strong initiator radically,

311

(A) will be the monomer unit along the chain. In most of our reactors universally, wherein it is difficult to maintain 100% **CONTACT** between components, one can imagine the types of products obtained from such systems. **CONTACT** real and imaginary is one of the most important variables in humanity. *In Chemical and Electrical Engineering disciplines for example, it is an important variable of great concern.*

Finally, for 1,4-cyclooctadiene, the followings are to expected when the ring is opened instantaneously.

1,4-Cyclooctadiene   (A)

(B)

4-Vinylcyclohexene   11.292

Chargedly, (A) the monomer unit along the chain can exist, since resonance stabilization cannot take place chargedly. Radically, in the monomer unit there are two isolatedly placed double bonds. (A) radically cannot be deactivated such that there are three double bonds. At best, in the absence of a strong initiator, it may be deactivated to form a 4-Vinylcyclohexene, the same as was obtained from 1,5-cyclooctadiene, that which is not to be expected. It is believed that, the (B)s shown so far can exist, since only the visible electro-radicals diffuse when present and when there is a nucleo-radical to grab. *Electro-radicals from π-bonds diffuse only when a visible one cannot function.* All their (A)s are different. With a strong initiator radically, (A) will be the monomer unit along the chain.

In general, it can be observed that cyclodienes are unique, in view of the types of initiations involved, methods of deactivation, para-placement of the active centers and trans-placement of the active centers. This is unique for only the even-membered rings. For example 1,5-cyclooctadiene can favor both para-placement of active centers during polymerization to form fused five-membered rings along the chain and also trans-placement of a group. *1,5-Cyclooctadiene (COD)* adds SCl_2 (or similar reagents) to give 2,6-dichloro-9-thiabicyclo*[3,3,1]* nonane[43].

(I)a   TRANS-PLACEMENT OF SCl_2 11.293

It is a two-stage Equilibrium mechanism system. COD also reacts with borane to give 9-borabicyclo*[3,3,1]* nonane[44].

(I)a

TRANS-PLACEMENT OF BH₃   11.294

It is a three-stage Equilibrium mechanism system. It is important to note the structure of the 9-borabicyclo nonane, wherein there is an electrostatic bond between the boron centers. This reagent commonly known as **9-BBN,** is used in Organic Chemistry for hydroboronations.

Some cyclodienes are discrete, close-loop resonance stabilized, depending on the locations of the two double bonds and substituents groups. Para-placement is favored in the presence of the resonance stabilization only for six-membered rings, while trans-annular addition (not migration as will be explained downstream) is favored for six, eight, ten and larger even-membered rings.

Like cycloalkanes, cycloalkenes, the order of nucleophilicity of the members of cyclodienes follows the same pattern as the others.

Cyclooctadiene > Cycloheptadiene > Cyclohexadiene > Cyclopentadiene > Cyclobutadiene

Order of nucleophilicity   11.295

This is the reverse of the order of strain energy in their rings. Between the resonance stabilized and non-resonance stabilized members of cyclohexadiene members and above, the non-resonance stabilized members should be of lesser nucleophilicity than their resonance stabilized counterparts. However, as is now obvious the following order is clearly valid.

> Tetraenes > Trienes > Dienes > Ethenes

Order of nucleophilicity   11.296

One should know the order to expect of triynes, diynes and ethynes (acetylenes). While there could be changed orders within family members based on the groups carried by them, there is a fixed order from family to family.

11.3. Cyclotrienes

These are monomers with three double bonds in their rings. Due to influence of strong presence of maximum required strain energy (Max.RSE), existence of 3-, 4, and 5- membered rings are impossible.

Very highly strained – CANNOT EXIST   11.297

11.3.1 Cyclohexatriene (Benzene)

Hence one can only begin with benzene for which even if the minimum required strain energy exists, the ring cannot be opened instantaneously, since there is no point of scission. However, in view of the fact

313

that it is resonance stabilized as shown below, only one activation center can be activated one at a time radically and indeed also chargedly. Chargedly, it is meaningless.

11.3.1.1 Resonance Stabilization and Electro-radicalization phenomena in Benzene rings.

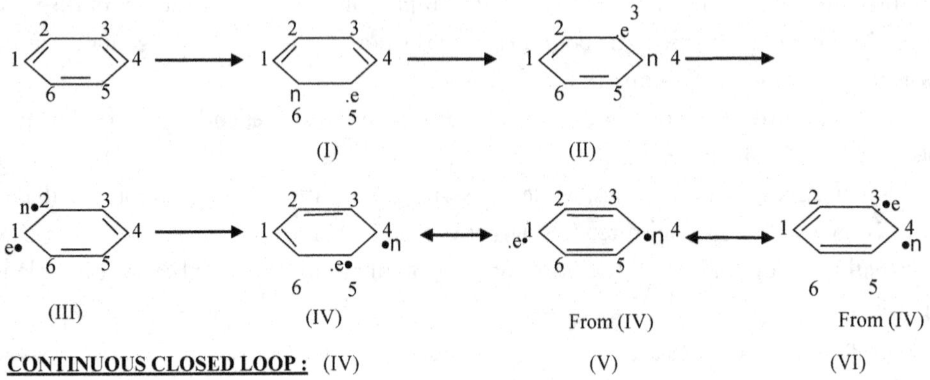

CONTINUOUS CLOSED LOOP : (IV)

RESONANCE STABILIZED FORMS

11.298

It is important to note that the resonance stabilization here is different, because it is a two-step process.

For (I)

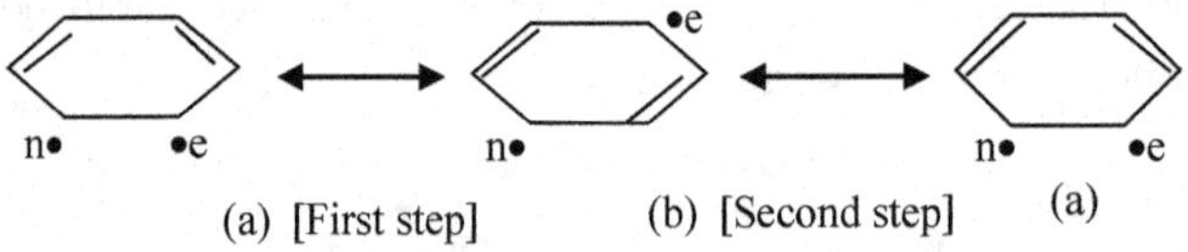

(a) [First step]   (b) [Second step]   (a)

11.299

For (II) and (III), it is the same. The isolated case shown above, (b), is only transient in character. ***Trans-placement of the radicals was impossible because of the presence of an adjacently located double bond in A CLOSED-LOOP. If, it was an OPENED-LOOP system it will stop after activating the center if the initiator is weak, just as in 1,2- and 1,4-mono-forms in butadienes.*** As is already obvious, the reactions above are not possible chargedly. However, the type of resonance stabilization is of the closed-loop type and continuous in character. In view of its continuity and other factors, it is difficult to have the following activated states when resonance stabilization is present in such systems.

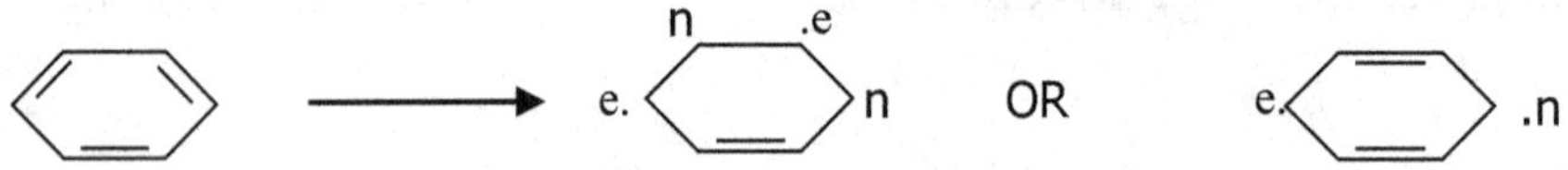

Impossible activated states

11.300

In order words para-placement of the radicals on active centers is virtually impossible, unlike in cyclohexadiene.

Now, one will introduce one substituent group into the ring. using radical-pushing and pulling groups.

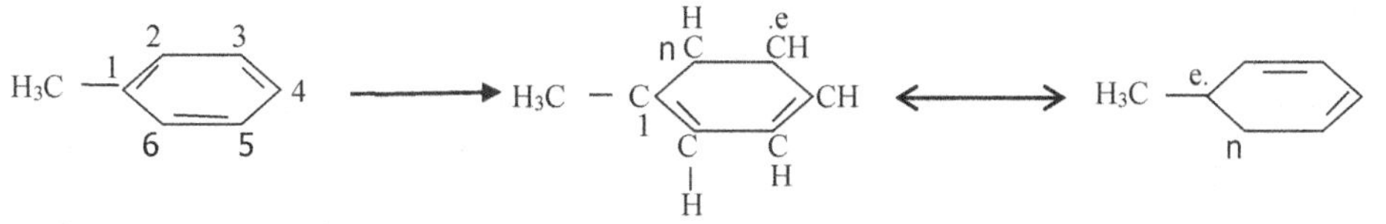

H$_3$C — ... CH$_3$ — C ... CH ⟷ H$_3$C — ...

11.301

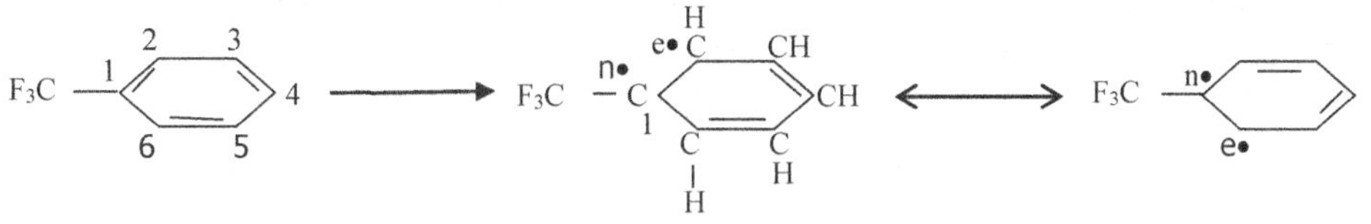

F$_3$C — ... F$_3$C — C ... CH ⟷ F$_3$C — ...

Nucleophile

11.302

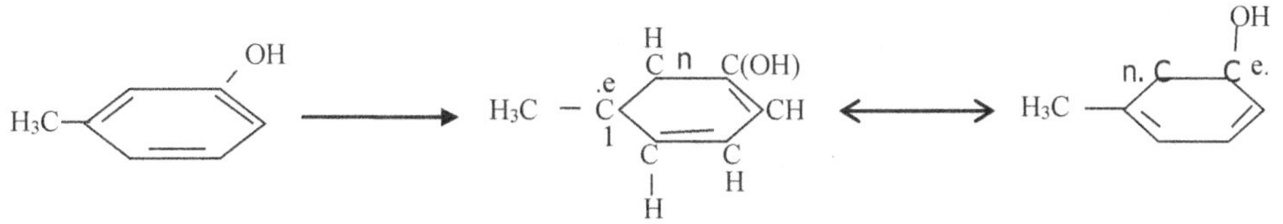

H$_3$C — OH ⟶ H$_3$C — C ... C(OH) ... CH ⟷ H$_3$C — ... OH

11.303

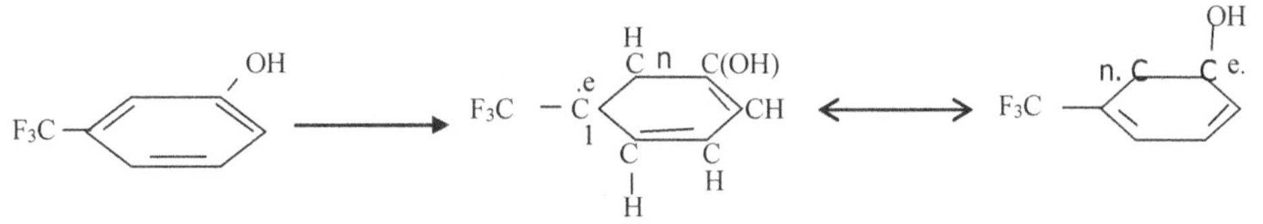

F$_3$C — OH ⟶ F$_3$C — C ... C(OH) ... CH ⟷ F$_3$C — ... OH

11.304

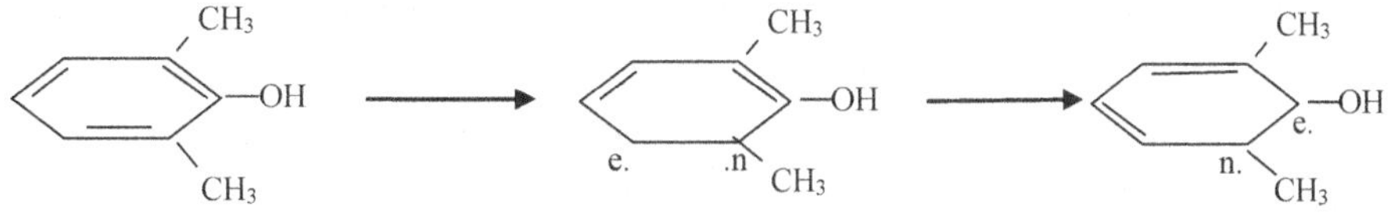

CH$_3$ ... OH ⟶ CH$_3$ ... OH ⟶ CH$_3$ ... OH

CH$_3$ ... CH$_3$ ... CH$_3$

11.305

CH$_3$ ⟶ CH$_3$ ⟶ CH$_3$

Cl ... Cl ... Cl

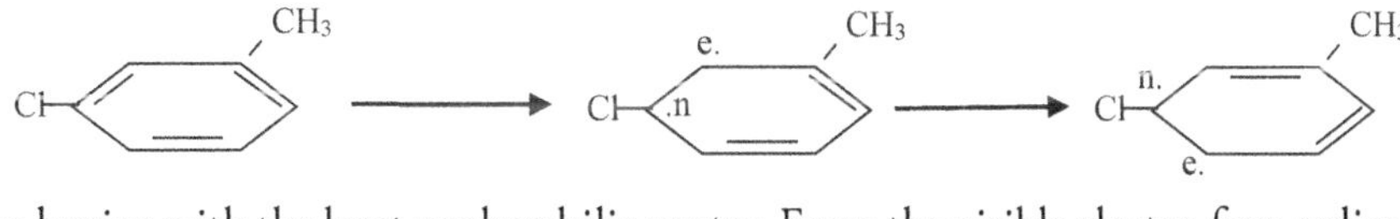

11.306

Activation begins with the least nucleophilic center. From the visible electro-free-radical, movement begins and eventually ends on the center carrying the largest radical-pushing group. When a radical-pulling group is present, activation begins from the center carrying it, followed by movement of electro-free-radical to end at the center next to the carbon center carrying the radical-pulling group on the other side. While the center carrying radical-pulling group if it is the only one present will continue to carry only nucleo-free-radicals, centers carrying radical-pushing groups can carry any radical, depending on where it is placed. Unlike cyclohexa-1,3-dienes, where when a group is placed externally to and after the two double bonds, the group (and not the monomer) ceases to be resonance stabilized. With benzene, the situation is different because, while cyclohexadiene has discrete type of resonance stabilization, benzene has the continuous type. This is one major reason why benzene is very unique. It is a closed loop system with three double bonds conjugatedly placed without any gap. ***Nevertheless, with or without radical-pushing or pulling substituents group, when resonance stabilized, there is indeed only one***

mono-form, obtained from transfer of visible electro-free-radicals to grab an adjacently located nucleo-free-radical in a π-bond or transfer of an electro-free-radical from a π-bond to grab a visible nucleo- free- or non-free-radical (Resonance stabilization). In styrene, electro-free-radical begins movement from one spot and come back to the same spot. Here, it begins from one side and ends on the next side. While the substituent groups carried by them cannot be tempered with as source of transfer species, they can be the source of Equilibrium/Combination/Decomposition States of Existence, depending on what is carried by them.

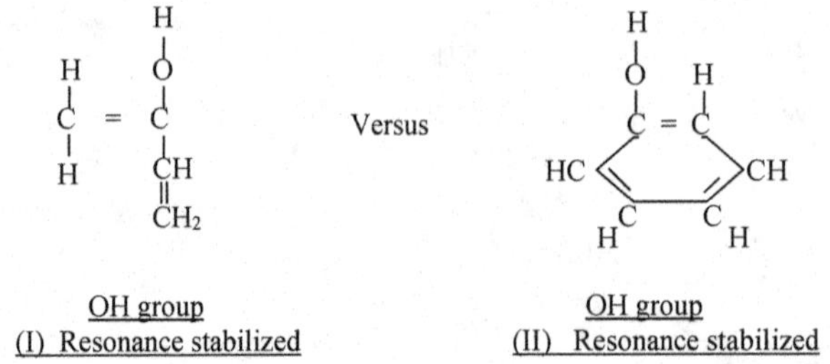

OH group
(I) Resonance stabilized

Versus

OH group
(II) Resonance stabilized

11.307

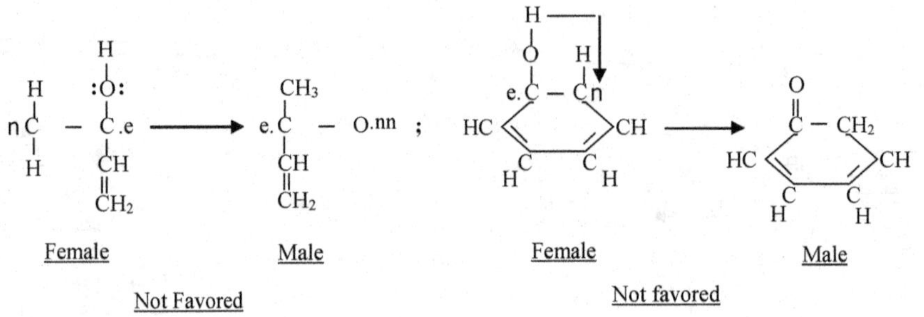

Female Male

Female Male

Not Favored

Not favored

11.308

Just as the OH group in (II) is resonance stabilized, so also is the OH group in (I), despite the continuous closed loop and (discrete) opened-loop nature of both systems. On the other hand, in both cases above the wrong center has been activated. Also when the monomer is not activated, the OH group is not resonance stabilized as follows ionically and radically.

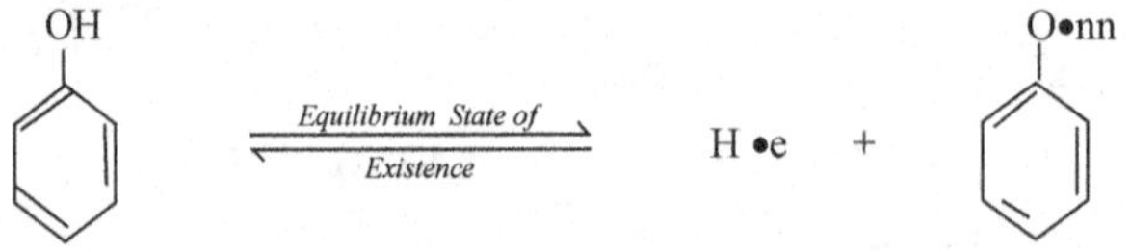

11.309

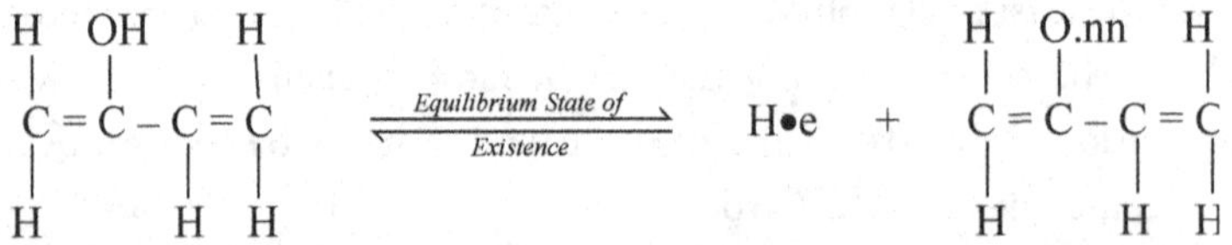

11.310

Hence in the unactivated states, the H atom on OH is loosely bonded to the oxygen center ionically and radically. Only free-radicals can stay on the carbon centers in the ring. Free or Non-free "anions", non-free "cations" and non-free-radicals cannot stay on the carbon centers in the ring. However, the nucleo-non-free-radical on the O center in Equation 11.309 is resonance stabilized since it will come back to where it started.

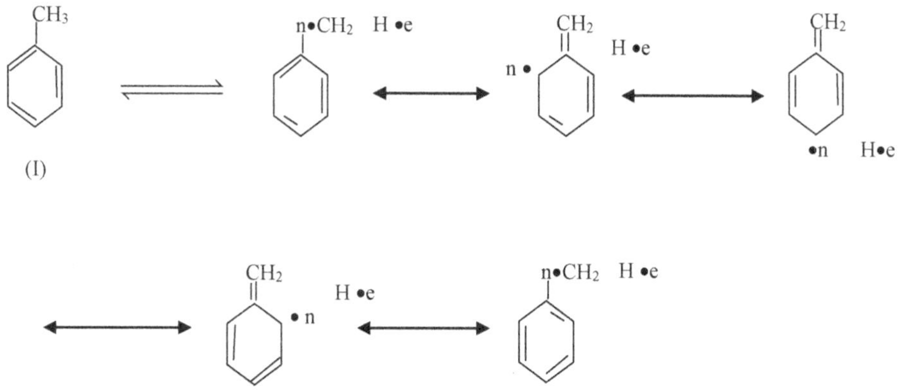

(I)

11.311

Electroradicalization/Equilibrium State of Existence

None of these can be opened instantaneously, because while it will apply to the ortho-, it will not apply to the para- positions when H •e is added. When in Equilibrium state of existence in *the Energised state of existence,* the H held is that from the ring beginning from the ortho- positions and when filled, the next to be held is that in the para- position as will be shown downstream in the volumes when the mechanisms of Chemical reactions including bio-chemical ones will be provided. *For the first time to Humanity, the structures of Enzymes will be provided. All the ringed structural backbones of Living systems will also be provided.*

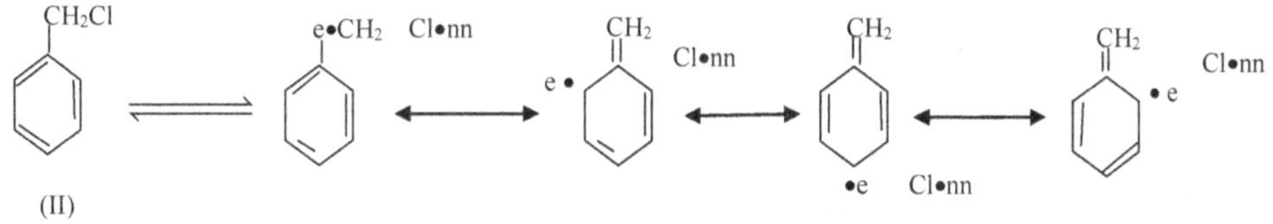

(II)

Electroradicalization/Equilibrium State of Existence

11.312

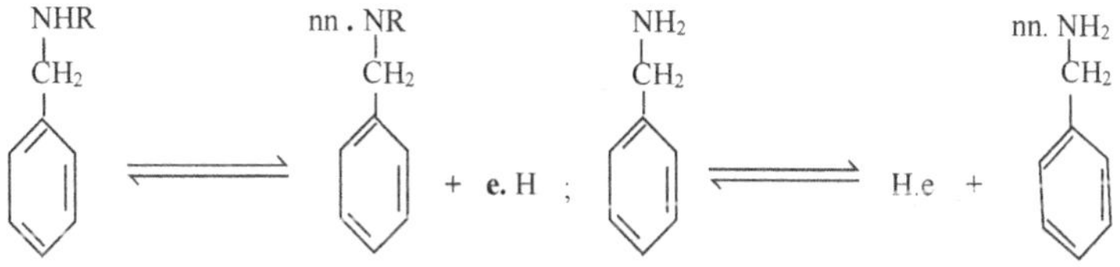

(III)   **Cannot be resonance stabilized**

Equilibrium State of Existence

11.313

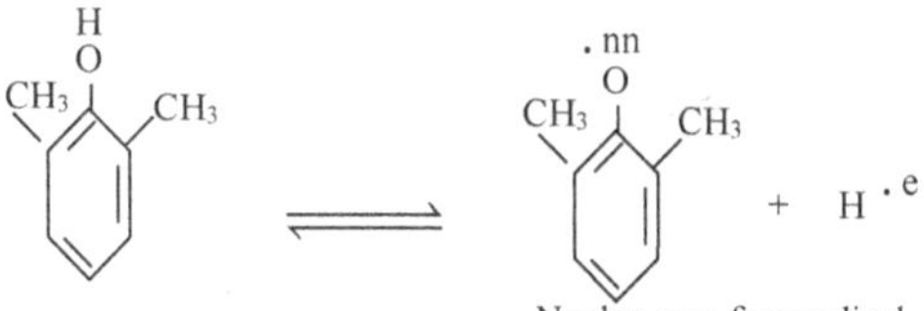

(IV)

Nucleo-non-free-radical

(Can be resonance stabilized)

Equilibrium State of Existence

11.314

317

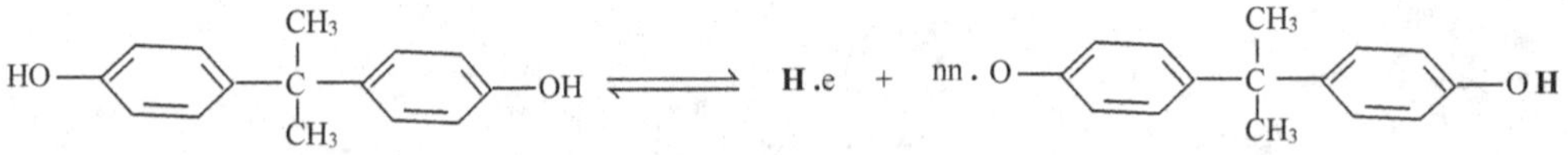

$$HO-C_6H_4-C(CH_3)_2-C_6H_4-OH \rightleftharpoons H.e + nn.O-C_6H_4-C(CH_3)_2-C_6H_4-OH$$

(V)

Nucleo-non-free-radical
(Can be resonance stabilized)

Equilibrium State of Existence　　　　　11.315

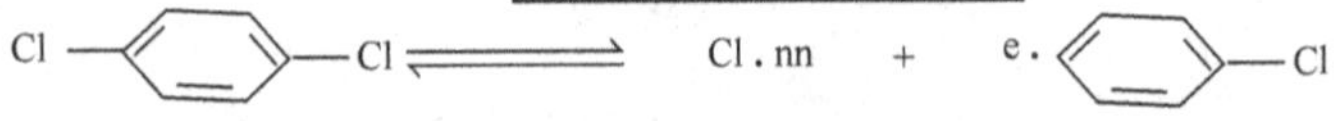

$$Cl-C_6H_4-Cl \rightleftharpoons Cl.nn + e.C_6H_4-Cl$$

(VI) Non-resonance stabilized monomer

Equilibrium State of Existence　　　　　11.316

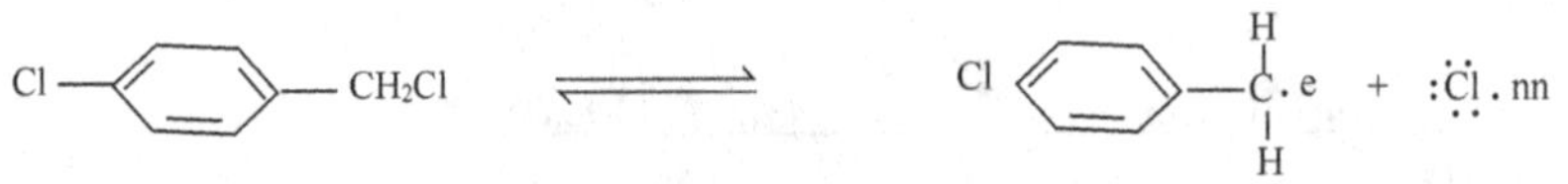

$$Cl-C_6H_4-CH_2Cl \rightleftharpoons Cl-C_6H_4-\overset{H}{\underset{H}{C}}.e + :\ddot{C}l.nn$$

(VII) Resonance stabilized monomer

Equilibrium State of Existence　　　　　11.317

All the states of existence above are considered in the absence of activation of the ring. ***This is what has been called the De-Energized state of existence of the compounds.*** There are some cases above that can exist in Equilibrium/Activated State of Existence. For all of them with the exception of (I), (II) and (III) (i.e. Equations 11.311 to 11.313), none of their rings can be opened instantaneously for polymerization purposes as will shortly be explained.

Depending on the type, number and positional placement of substituent groups in a benzene ring, it can be resonance stabilized or not resonance stabilized. For example, (IV) and (V) above are said to be resonance stabilized free-radically, because the nucleo-non-free-radical grabbed will still come back to its original position. The case of (IV) and (V) is like the case of benzoquinone. The case of (III) is obvious – the nucleo-non-free-radical not adjacently placed to the ring. The case of (VI) is limitations from the concept of Strain Energy. All these apply to monomers shown below, when not activated.

$$HO-C_6H_4-OH \quad ; \quad H_2N-C_6H_4-O-C_6H_4-NH_2 \quad ;$$

(VIII) 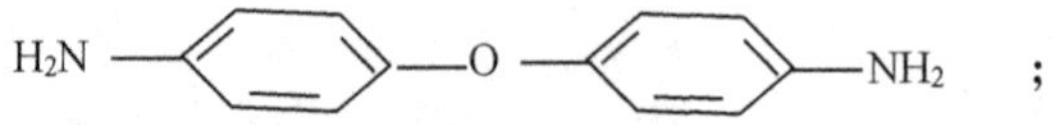　　　　　　　　　　　　　　　　(IX)

$$KO-C_6H_4-C(CH_3)_2-C_6H_4-OK$$

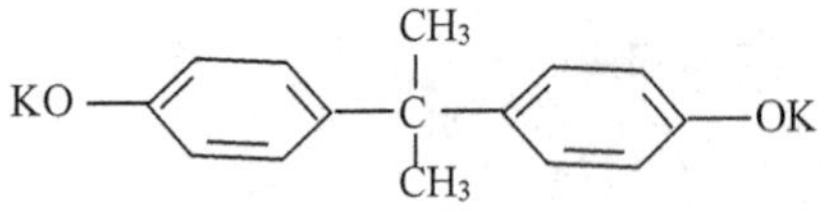

(X)

Resonance stabilized monomers　　　　　11.318

One can begin to identify cases above which can be used as Step monomers.

Monomers such as shown below are resonance stabilized when they are in any particular state of existence. For one particular state, the groups cannot be tempered with.

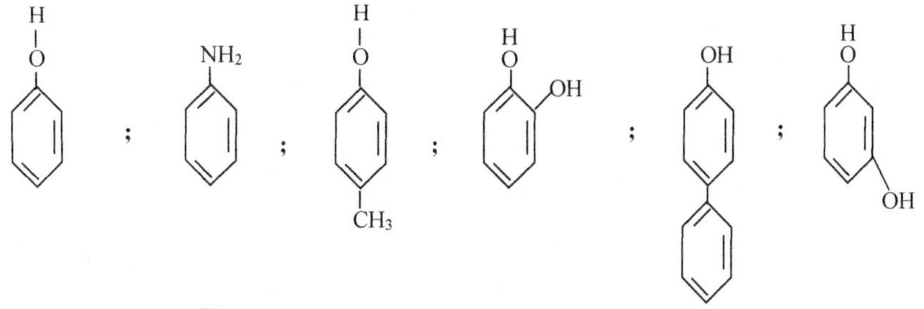

(I)   (II)   (III)   (IV)   (V)   (VI)

<u>Resonance stabilized monomers (When activated)</u>

11.319

Based on the mechanism of resonance stabilization, the character of the monomer, and the location or type of substituent group, (IV) of Equation 11.314 can be resonance stabilized when activated as already shown in Equation 11.305. .

For (V) of Equation 11.315, the followings are obtained when activated.

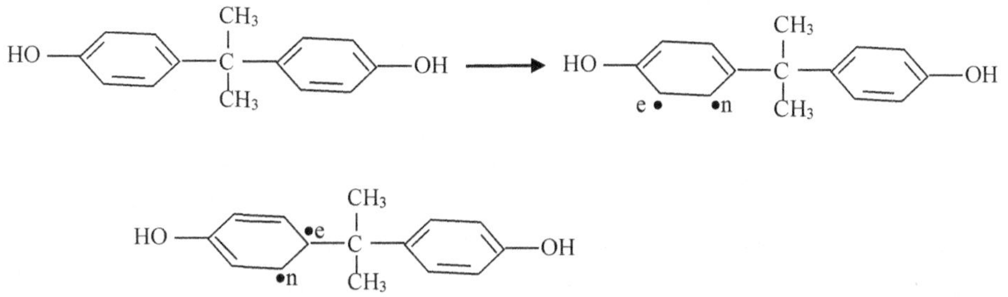

[ONE-SIDED RESONANCE STABILIZATION]

11.320

Chargedly, the monomer can be activated, but not resonance stabilized. With (IX) of Equation 11.318, none of the carbon centers in the para-positions carrying NH_2 group, can carry negative charges. The same applies to other cases in the Equations. In (IV) and (VI) of Equation 11.319, the followings are obtained.

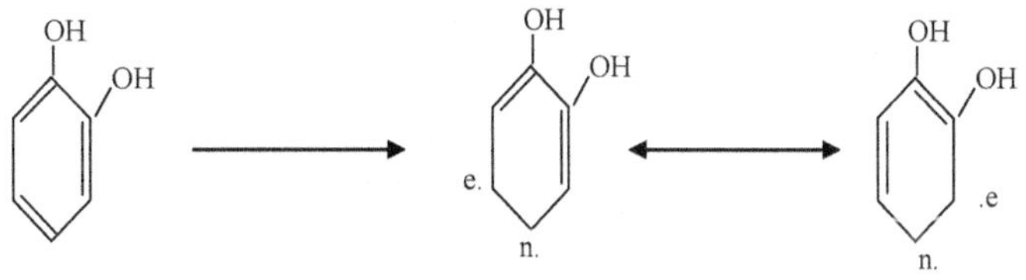

Resonance Stabilized

11.321

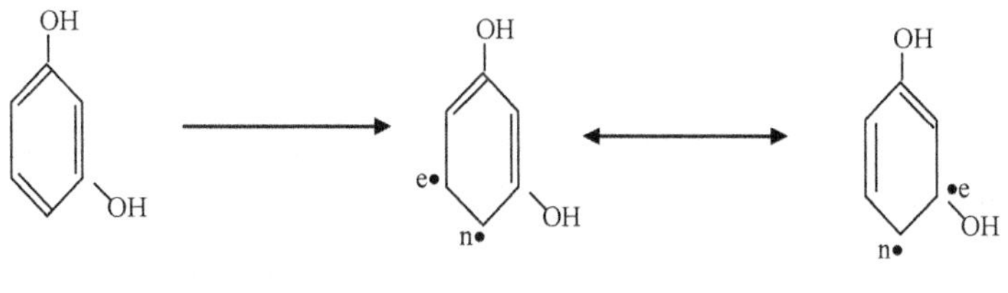

(Resonance Stabilized)

11.322

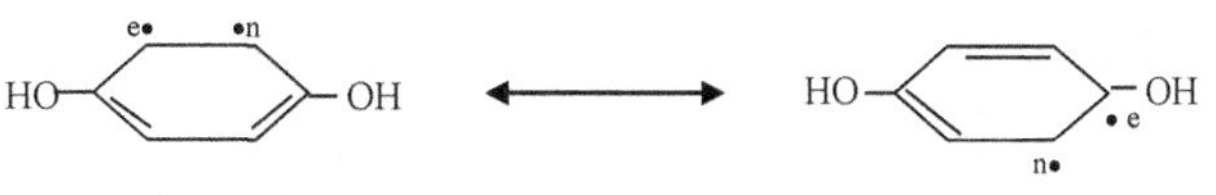

(VIII) Eqn. 11.318

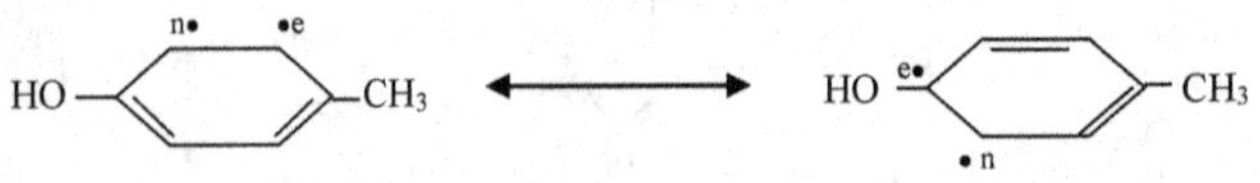

(III) Eqn. 11.319   (Resonance Stabilized)

11.323

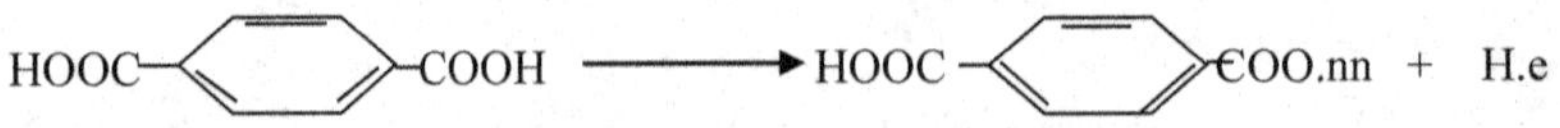

[Equilibrium State of Existence- A Step monomer]   11.324

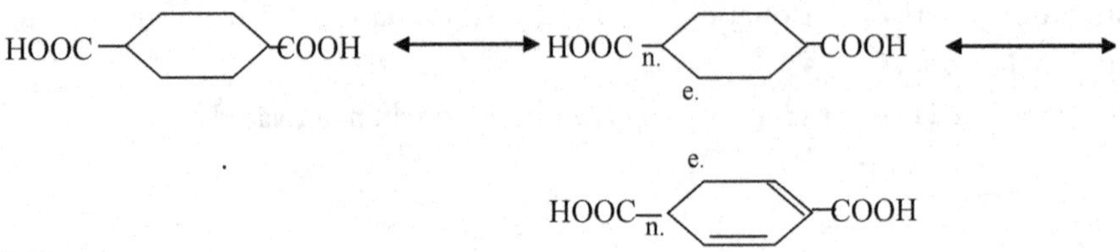

11.325

Worthy of note is the last example above, embracing what one has been doing so far. However, it can thus be observed that all benzene rings are resonance stabilized when activated. One can observe why the benzene monomers behave as have been observed over the years, particularly with the concept of meta-, ortho- and para- placements on the rings. All these will completely and unquestionably be explained as we move down the Series and Volumes.

11.3.1.2 Degree of Unsaturation in Benzene Rings.

Also unique with benzene rings is the degree of unsaturation in the ring, compared to that of acetylene, as measured by the H to C ratio in the first members as well as strain factors.

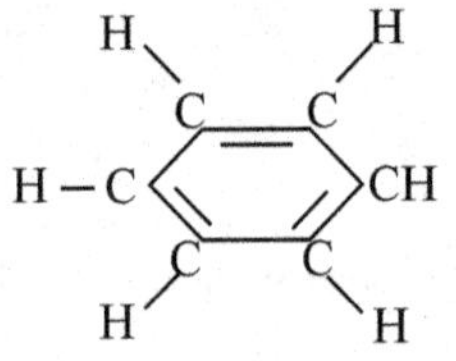

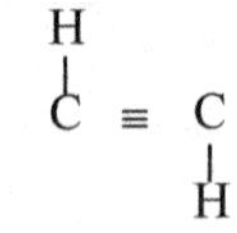

H : C ratio: 1 : 1 (strained)   1 : 1 (Unstrained)

11.326

Despite the 1 to 1 ratio, benzene is far more unsaturated than acetylene. One should therefore expect at least three hydrogen atoms in benzene to be loosely bonded to the carbon centers free-radically one at a time. But, if the benzene is resonance stabilized, only one hydrogen atom is loosely bonded to the ring as shown below free-radically only, in its unactivated or Energized state of existence, one at a time.

True state of Benzene ring

11.327

The free-radical existence of the benzene rings is not only limited to unsubstituted benzene ring, but also to substituted benzene rings when it is in the Energized state of existence.

For example, for (IV) of Equation 11.314 which is resonance stabilized when in Equilibrium state of existence, the H atom para-placed to the OH group is loosely bonded free-radically to the ring, because of the strong unsaturation of the ring when in the energized state.

Equilibrium state of existence in the Energized state of benzene ring

11.328

All compounds which have been encountered so far, have only one Equilibrium state of existence, i.e., one finger print when they exist as such. For the first time, we are experiencing a unique case which has two finger prints. Benzene alone has only one finger print. But, when substituent groups are carried by them, Equilibrium state of existence can be favored by them either from the group as shown for (IV) of Equation 11.314, when the benzene ring is in the De-energized state, i.e., unactivated state and at normal to moderate operating conditions, or from the benzene ring in view of its highest level of unsaturation and this is said herein to be the Energized state, i.e., unactivated state but at higher operating conditions not sufficient enough to activate the ring; noting that once the ring is activated all the groups carried are resonance stabilized.

With (VII) of Equation 11.317, the situation is slightly different, since the substituent group is an allylic group. The resonance stabilization of an electro-free-radical in the ring has already been shown in Equation 11.312 wherein the electro-free-radical comes back to where it started its movement.

With divinyl benzene a unique monomer, the followings are also obtained.

(A) Resonance stabilized

11.329

Worthy of note as will be fully shown downstream in this volume, is that one vinyl group is not involved in the resonance stabilization of the monomer. Like the case of styrene, the visible electro-free-radical comes and grabs the nucleo-free-radical from the π-bond and this continues back to where it was. If the vinyl groups were both involved, then how could divinyl benzene function as a unique cross-linking agent? One can observe that divinyl benzene is not symmetric in character. This clearly indicates the continuous closed loop character of benzene.

We saw the uniqueness of allylic groups in (II) of Equation 11.312 where its Equilibrium state of existence in the De-energized state was shown. Shown below is that in the Energized state, wherein resonance stabilization takes place to identify which H is next to be held. The H first held in Equilibrium state of existence is that in one of the ortho- positions, when a radical pushing group is carried, depending on the operating conditions. This is followed next by that in the second ortho-position. This is followed next by that in the para-position. There are countless numbers of ways by which groups can be placed on benzene rings. All operate without breaking any of the Laws of NATURE.

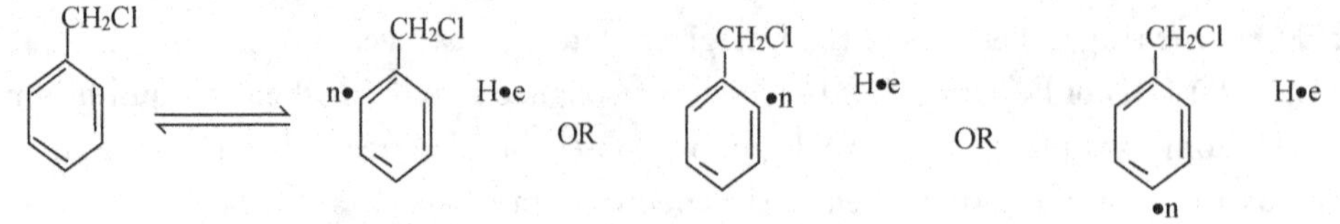

Equilibrium State of Existence in the Energized state 11.330

It is believed therefore that with allylic groups, the benzyl group is very reactive, because the electro-free-radical obtained can readily be resonance stabilized but still to come back to where it started. This was shown in Equation 11.312. Shown below is an exemplary case[45], the mechanisms of which are unknown, just like for all chemical reactions.

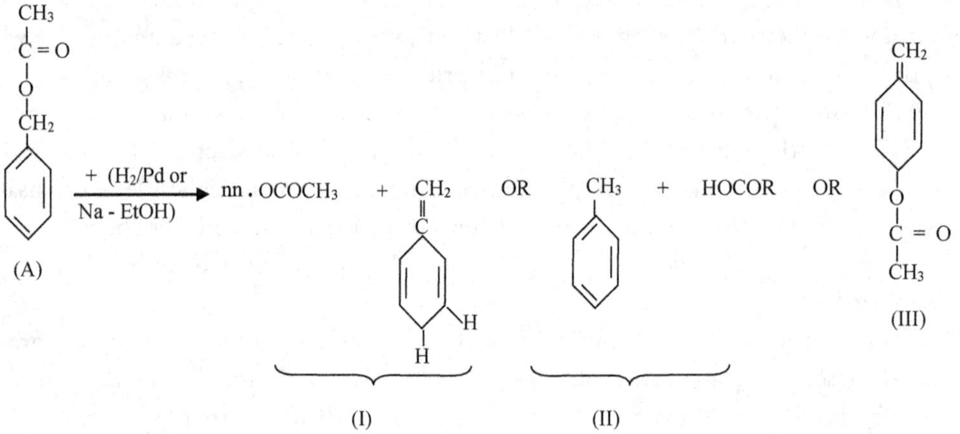

11.331

All these are Equilibrium mechanism reactions of one, two or more stages. The EtOH is ethanol which when used with Na is a hydrogenation catalyst as probably already shown and recalled below.

Stage 1: $4Na \xrightleftharpoons{Excited} 4Na \bullet e$

$4Na.e + 4C_2H_5OH \xrightleftharpoons{Abstraction} 4C_2H_5ONa + 4H \bullet e$

$4H \bullet e \longrightarrow 2H_2$ 11.332a

Stage 2: $2Na \xrightleftharpoons{Excited} 2Na \bullet e$

$2Na \bullet e + 2H_2 \rightleftharpoons 2NaH + 2H \bullet e$

$2H \bullet e \longrightarrow H_2$

11.332b

Stage 3: $NaH \rightleftharpoons Na \bullet e + H.n$

$Na \bullet e + H_2 \rightleftharpoons NaH + H \bullet e$

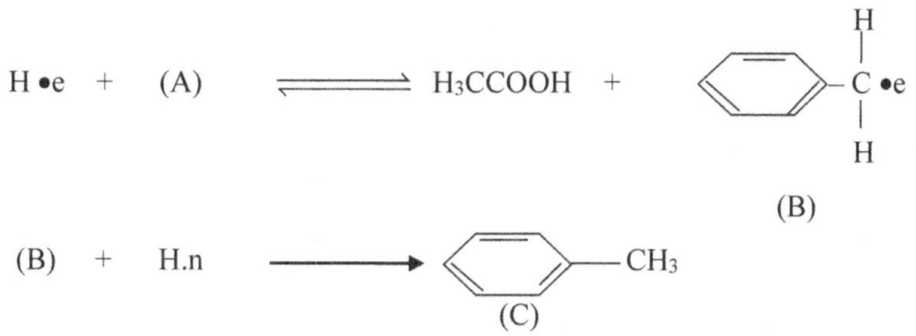

$H \bullet e + (A) \rightleftharpoons H_3CCOOH +$

(B)

$(B) + H.n \longrightarrow$

(C)   11.332c

Overall Equations: $6Na + 4C_2H_5OH + (A) \longrightarrow 4C_2H_5ONa + 2NaH + (C)$

$+ H_3COOH$ 11.332d

In the absence of resonance stabilization in the unactivated state, it is (II) of Equation 2.331 that is favored. (I) cannot be obtained, since the electro-free-radical must always come back to where it started. Otherwise, para-placement of radicals on benzene ring would have been possible. Hence, (III) cannot exist as a product. Hence also, allylic groups on benzene rings cannot be resonance stabilized in the unactivated state radically. Candidly speaking, chargedly, they cannot be activated at all. With the use of hydrogen catalysts such as Pd, the numbers of stages are less than above, noting that the Pd is just a passive catalyst. On the other hand, imagine the numbers of moles of Na and ethanol required for the hydrogenation. As a matter of fact, the reaction above is not hydrogenation wherein H_2 is added to a single compound via a double bond, but an abstraction reaction to form an acid followed by the formation of an alipharomatic compound.

With (VI) of equation 11.316, the ring can be resonance stabilized as shown below. In Equation 11.316, the Equilibrium state of existence of (VI) in the De-energized state was shown and it is not its Equilibrium state of existence in the Energized state, because the Cl group which is an atom is directly connected to the benzene ring just like H.

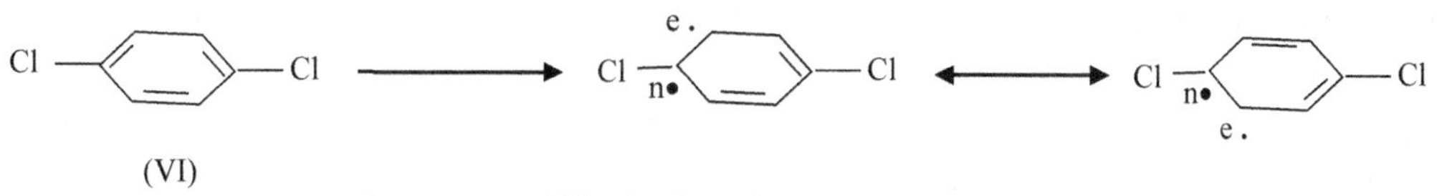

(VI)   Resonance stabilization in Activated state   (VI)b   11.333

The activation center first to be activated is what has been used. Based on the movement of the electro-free-radical, only the meta- position with respect to one Cl center can be used.

It is in view of the free-radical existence of benzene ring in its unactivated state as shown in Equation 11.327, that Addition reactions so called "Electrophilic substitution reactions" - via nitration, halogenation, acylation, alkylation and sulfonation reactions are made possible. For example, for the alkylation reaction, the followings are obtained.

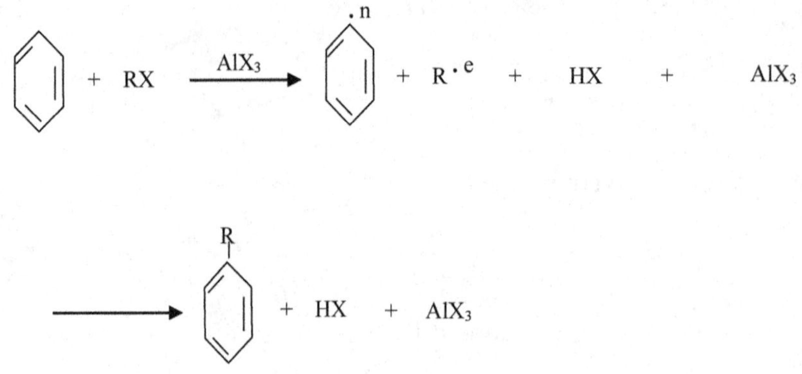

(Where Xs are halogen atoms)   11.334

For the sulfonation reaction, the followings are also obtained.

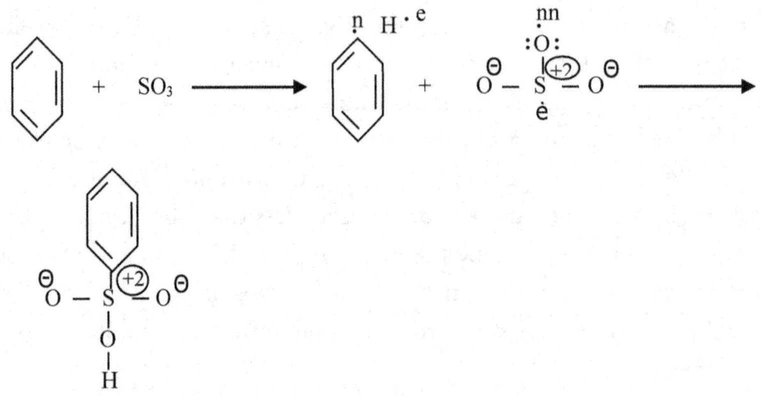

Benzenesulfonic acid   11.335

It is important to note the activated states of SO_3. The charges on the oxygen and sulfur centers are not ionic, but polar in character. The two reactions shown so far are just one stage Equilibrium mechanism systems.

Like in acetylene, after replacement of one hydrogen atom in benzene with a radical-pushing group, a second hydrogen atom is still slightly loosely bonded to a carbon center. But unlike acetylene, this depends more on the capacity of the group already carried by the ring, and not that of the activating species. Unlike acetylene, benzene has more C and H atoms of the same ratio with that of acetylene, used to form a highly strained ring. Unlike in acetylene, benzene is resonance stabilized. Unlike in acetylene, when they are activated, while in benzene the H or groups carried are shielded, this is not the case with acetylene. What acetylene has, benzene has far more than three-fold of it. The type of group in terms of pulling and pushing abilities and capacities initially carried by benzene, determine how, ortho-, meta - and para- placements are obtained. With radical-pushing groups, the followings are to be expected. While benzene can be activated chargedly, acetylene cannot be activated chargedly.

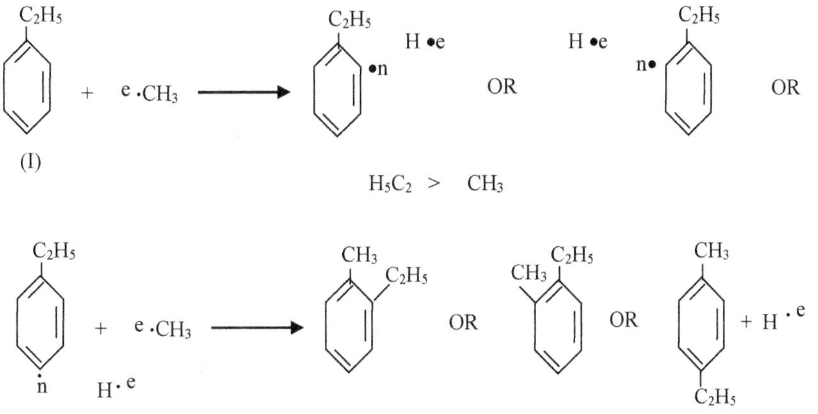

(II)a (II)b 11.336

Note that to obtain (II)b, the two ortho-positions if not occupied must be shielded as will be shown downstream.

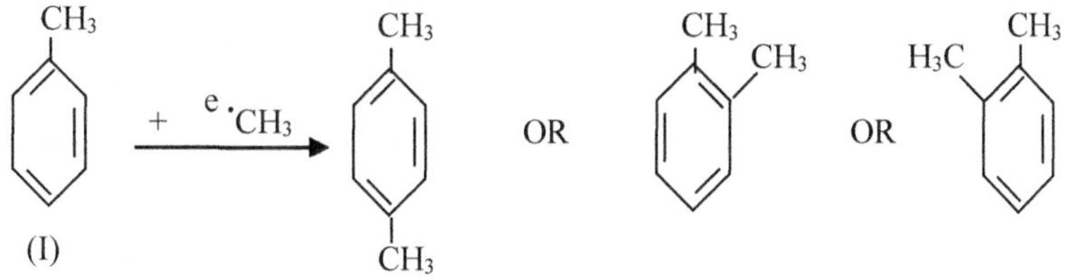

(III)a (III)b

(CH_3 = CH_3) 11.337

The same as above also applies to (III)a. All the above are resonance stabilized monomers. So also is the one shown below.

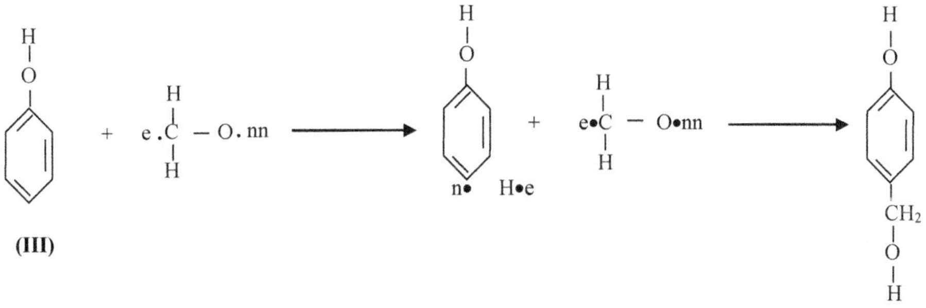

(IV) 11.338a

The above is favored only when the two ortho-positions have been blocked with for example the use of $AlBr_3$; after which the two positions are deblocked after filling the para-position.

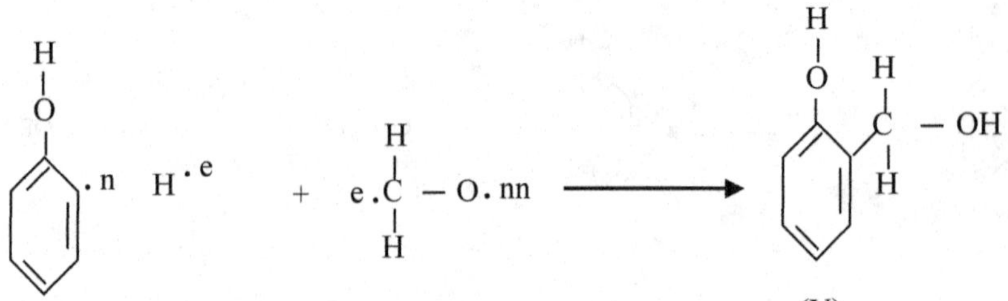

(V)   11.338b

The above reactions involving displacement reactions are favored, only when the CH_2OH group from the formaldehyde for example is of weaker radical-pushing capacity than that of OH group, which indeed is the case. It is important to note that the meta-positions are not involved in the displacement reactions above, in view of the type of radical carried by that center when resonance stabilized. All the reactions above are Equilibrium mechanisms systems with just one stage.

Unlike in acetylene, after the first replacement, there is a second one - the second ortho-position. After this replacement, this is finally followed by the third replacement - the one and only para-position. For (II) b of Equation 11.336 and (IV) of Equation 11.338a, the two hydrogen atoms in the two different ortho-positions with respect to C_2H_5 after deblocking are still very loosely held one at a time. For (V) of Equation 11.338b, the hydrogen atoms in one ortho-position and para-position are also still very loosely held one at a time with no deblocking necessary free-radically, when the monomers are placed in the Energized state.

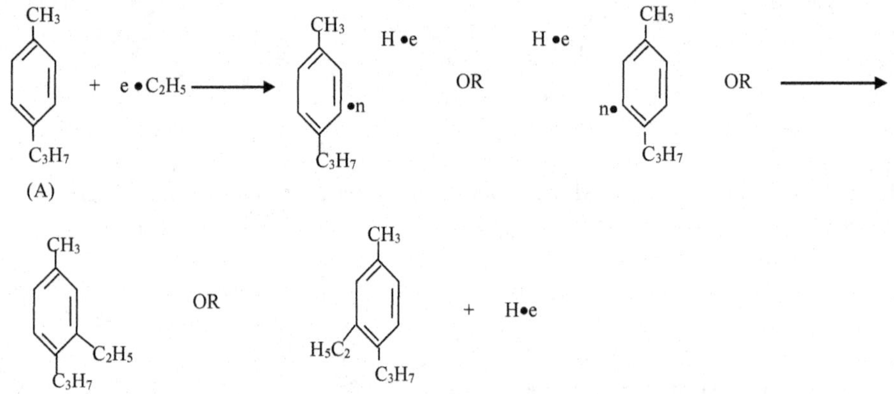

(A)

(B)   11.339

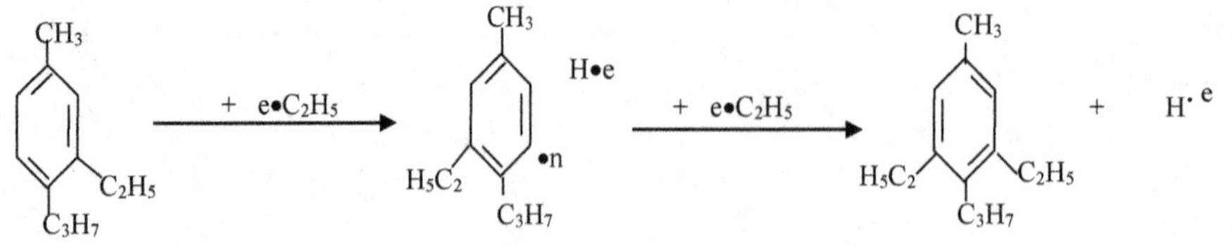

(C)   (D)   11.340

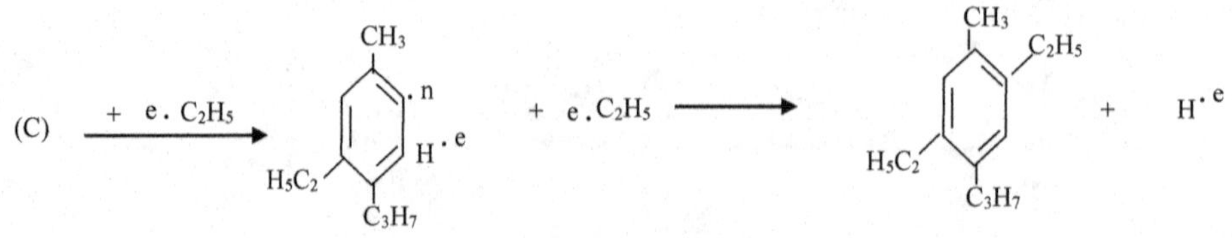

IMPOSSIBLE PLACEMENT   (E)   11.341

It is said to be impossible, because C_3H_7 group, the largest radical-pushing group is the group in control of the system. It is not C_2H_5 which is next to it. If C_4H_9 was used in place of C_2H_5, no replacement will ever take place, because the group is more radical-pushing than the controlling group in the ring. This is how NATURE operates.

For (IV) of Equation 11.338a, after deblocking the two ortho-positions, the followings are next obtained.

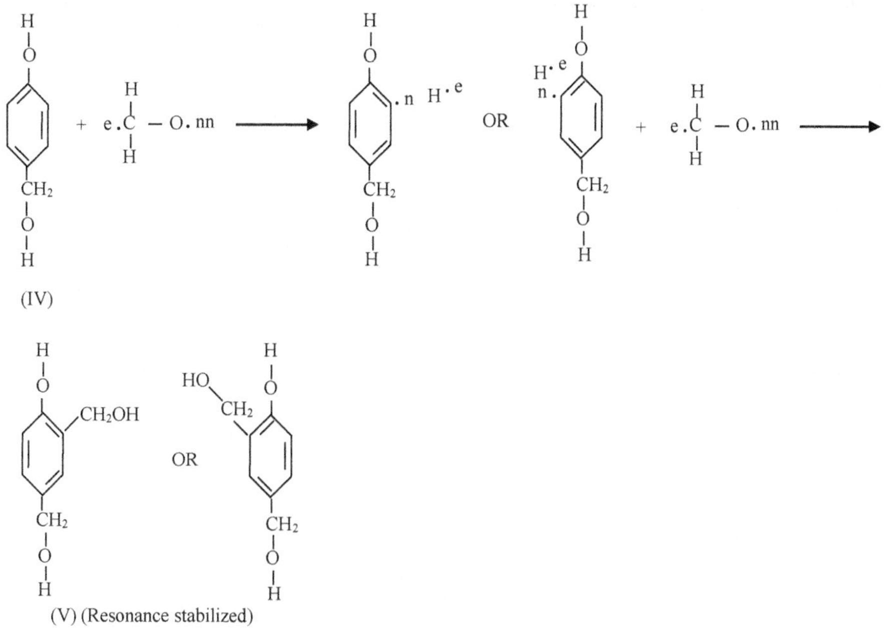

(IV)

(V) (Resonance stabilized)

11.342

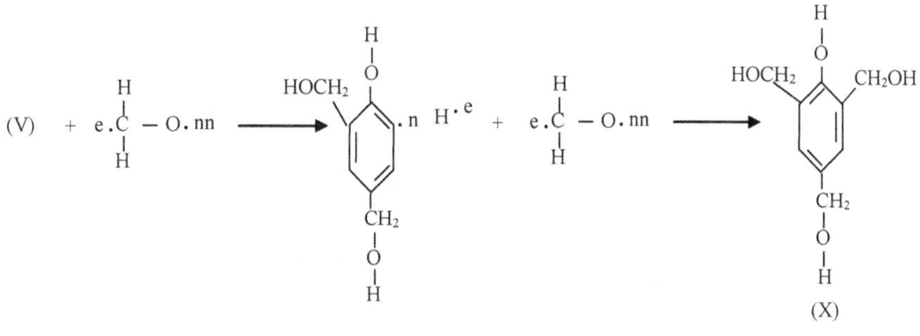

(X)

11.343

(II)a Eqn. 11.336

(III)a

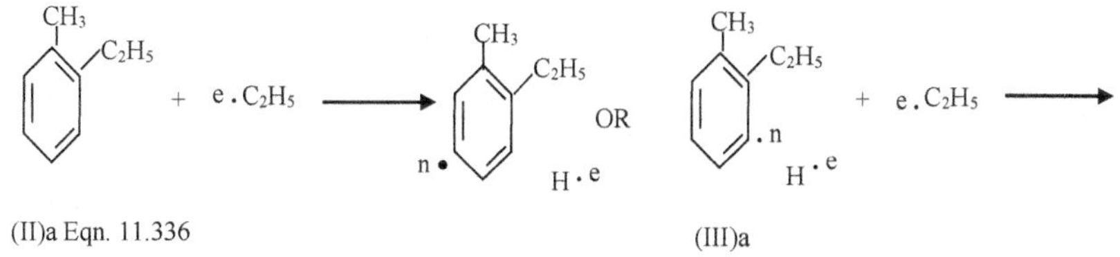

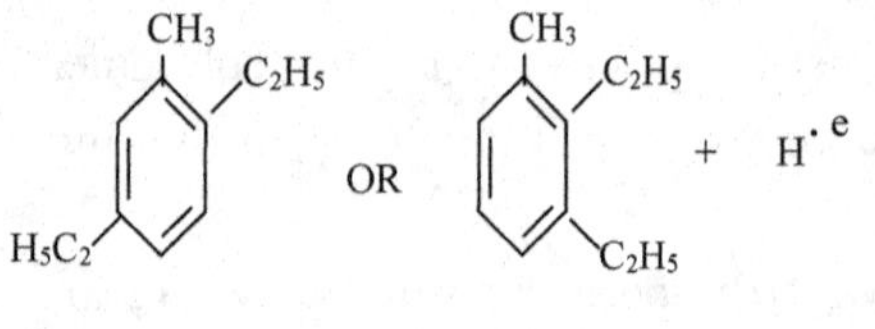

The structures (III)b with CH₃, C₂H₅, and H₅C₂, C₂H₅ substituents OR structure with CH₃, C₂H₅ and C₂H₅ substituents, plus $H \cdot^e$ are shown.

$$11.344$$

The favored one is (III)b and not that which is para-placed.

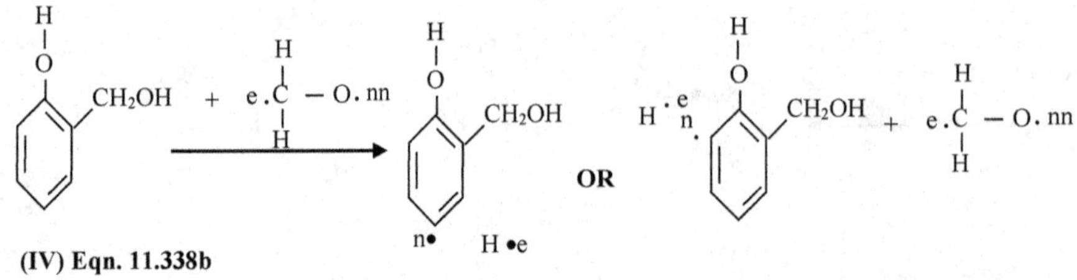

(IV) Eqn. 11.338b

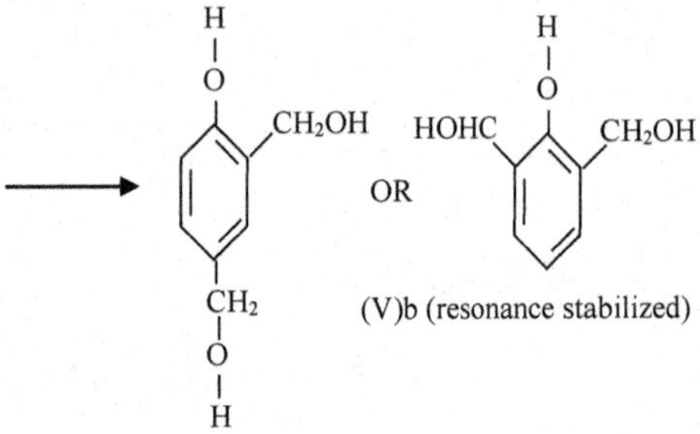

$$11.345$$

The favored one is (V)b and not the one which is para-placed.

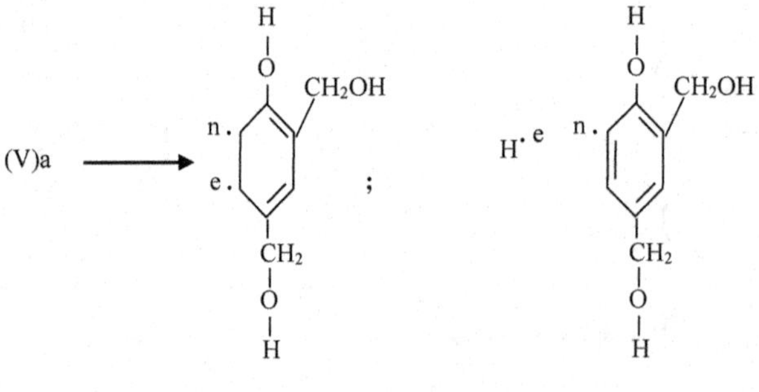

Activated State   Energized Equilibrium States of existence

$$11.346$$

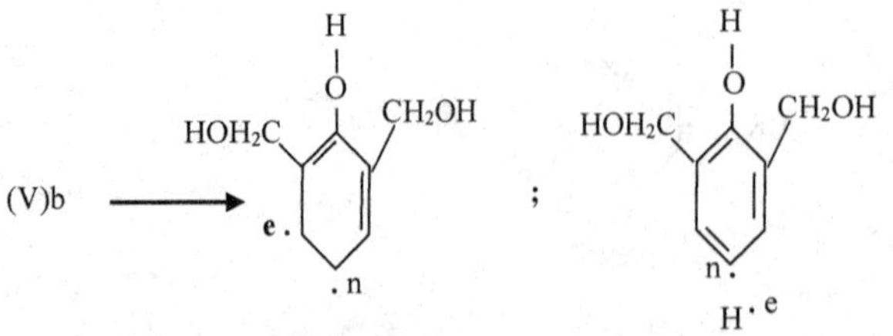

Activated State   Energized Equilibrium State of Existence

$$11.347$$

It is important to note the activation centers involved. It is also important to note that all the reactions are radical in character. (II)b or (IV) of Equation 11.336 and 11.338b will shortly be visited. The types of free-radicals carried by the active centers when activated, are worthy of note. They are fixed. As can be observed and from experience, there are marked differences between OH, NH_2 and their allylic groups such as CH_2OH group. With all groups of greater capacity than H, the ortho- and para- positions can be observed to be uniquely different in terms of orderliness.

Based on the groups already present in the ring, hence Equation 11.340 is favored as opposed to Equation 11.341. It is the largest group in capacity that decides the new para- and ortho- positions. Based on the types of reactions favored above, the following relationships are valid free-radically.

$$OH \; > \; C(OH)_3 \; > \; CH(OH)_2 \; > \; CH_2(OH) \; > \; CH_3$$

<u>Order of radical-pushing capacity</u>   11.348

$$CH_3 \; > \; CH_2F \; > \; CHF_2 \; > \; CF_3$$

<u>Order of radical-pushing capacity</u>   11.349

The validity of the last equation above has already been established. For radical-pulling and pushing groups, one should also expect the followings.

$$CF_3 \; > \; CF_2H \; > \; CFH_2 \; > \; F$$

<u>Order of charge-pulling capacity</u>   11.350

$$CHF_2 \; > \; CH_2F \; > \; CHF_2 \; > \; H \; > \; CF_3$$

<u>Order of radical-pushing capacity</u>   11.351

The same applies when OH is replaced with NH_2 group and etc. Some ways by which substituent groups are placed on benzene ring, have begun to be shown.

While in (I) of Equation 11.336, two ortho- and one para-positions are available for use, for the cases shown below, only the two-ortho positions are available for use, because the para-position to OH is already occupied.

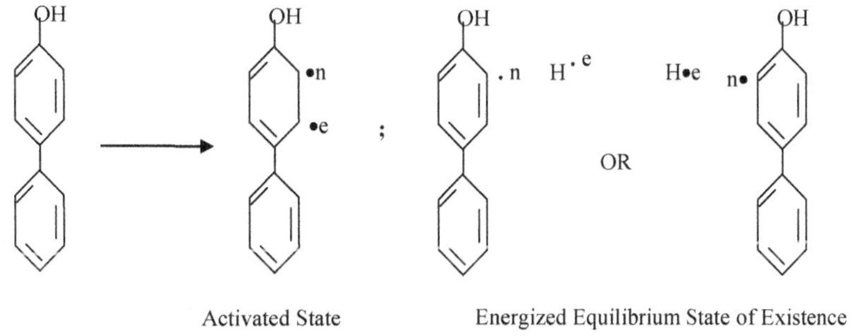

Activated State   Energized Equilibrium State of Existence   11.352

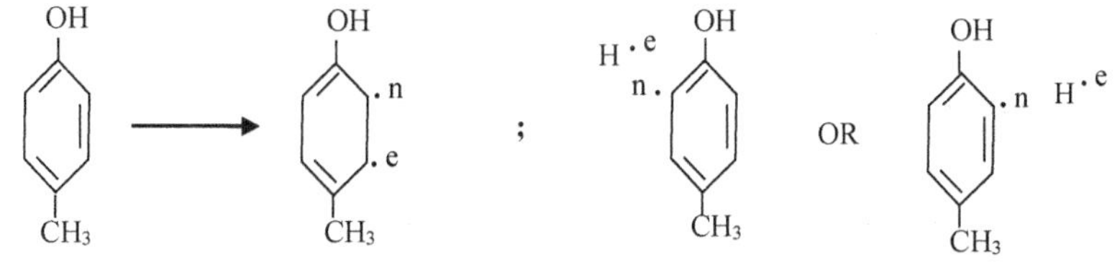

11.353

For the first reaction above, it should be noted that the bond activated is that not carrying substituent group(s) .

With a monomer such as resorcinol, it will be important to note that four hydrogen atoms are loosely bonded free-radically, one at a time to the carbon centers, when energized with the use of reagent, heat or other means, in view of its symmetric character.

11.354

11.355

Due to the symmetry, the two active centers in (A) and (B) above can carry any free-radical. Hence, the four positions which include meta-, ortho- and para- positions are available for use. This has nothing to do with disproportionation or isomerism of ortho-xylene to meta-xylene and to para-xylene. The same applies to (III)b of Equation 11.337 or any symmetric benzene. One can begin to observe how some of these compounds are used as STEP monomers. There have been great needs to show these unique features of benzene ring, in order to fully validate the mechanisms of activation (when more than one center is present) and the mechanisms of resonance stabilization phenomena and etc. Without establishing these orders, the new definitions for ions, radicals, atoms, molecules, bonds and so on would not be valid. Reactions do not just take place like that. There must be an order based on natural laws and not material laws, or rule of thumbs or empirical rules and so on, which are all still material laws.

It will be recalled in Chapter five of Volume (II) that groups such as NO_2, SO_2R etc. were identified as strong charge-pulling groups and weak radical-pushing groups with respect to F and H respectively. These groups are less radical-pushing in capacity than hydrogen atom. It is for this reason that nitro-groups enter a nitrated ring via the meta position, as opposed to the ortho-and para-positions for more radical-pushing groups. It is important to note that just as a radical-pushing group when present alone on the ring, makes the ortho- and para- positions available, so also a charge-pulling group or radical-pushing group of lower radical-pushing capacity than H when present alone on the ring, makes only the meta-position available.

11.356

When (I) is activated and placed in the energized state, the followings are obtained.

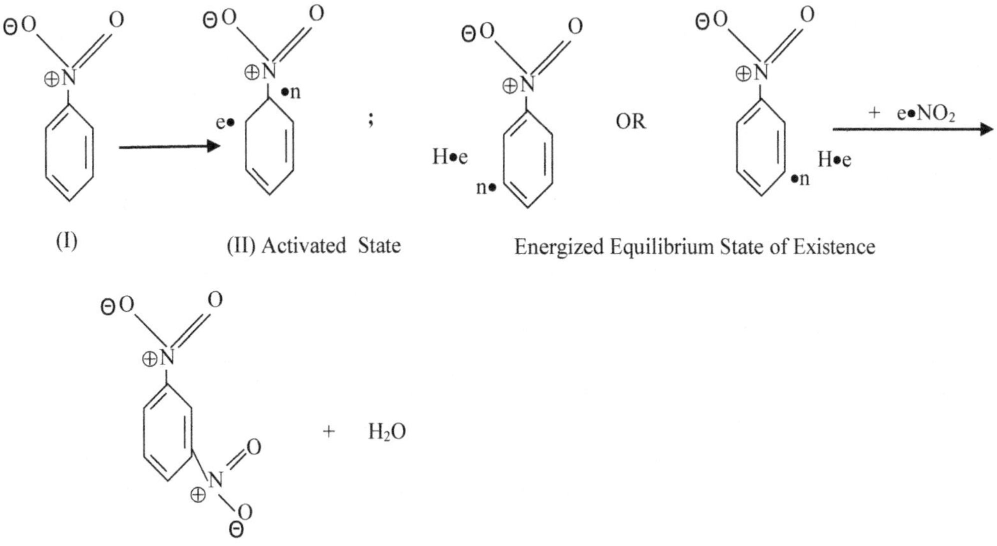

(I)   (II) Activated State   Energized Equilibrium State of Existence

(III)   11.357

This is an Equilibrium mechanism system with just one stage. The activated state of (I), that is, (II) is worthy of note since NO_2 is less radical-pushing than H. Hence, it is the hydrogen atom in the meta-position that is still loosely bonded free radically. The reactions above cannot take place chargedly, since the invisible π-bond in the ring is far stronger than that in cyclopropene. It is the two meta-positions that are available for use, the second one under higher operating conditions. After this, no hydrogen atom becomes loosely bonded to the ring, except when initially there was a strong radical-pushing group in the ring. It is important to note that (I) above is a resonance stabilized compound free-radically only when activated.

When (III) is activated and placed in the energized state, the followings are obtained.

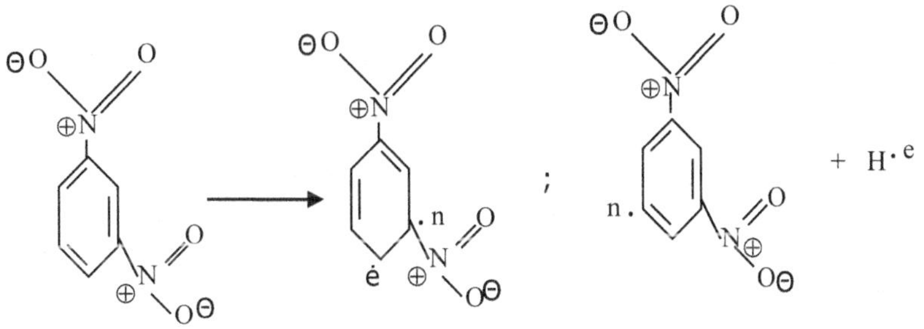

(III)   (III)a Activated State   (III)b Energized Equilibrium State   11.358

By the continued direct nitration above, it may be difficult to introduce the third nitro group into the ring, either because the capacities of the nitro groups in the ring are too small to keep the benzene ring in (III) in an Energized state, or that the nitrating agent cannot keep (III) in Equilibrium state existence as shown above i.e., (III)b. Instead, (III) is kept in Stable state of existence by these strong nitrating agents. A special catalyst as will shortly be shown, will be required to keep (III) in Equili-brium state of existence. However, it can be introduced into the ring by other means such as shown below.

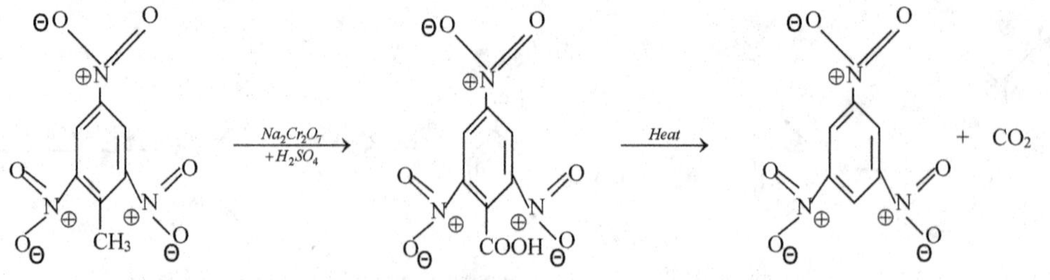

(A)   (B)   (C)   11.359

Above, through the CH_3 group initially placed on the ring, three NO_2 groups were first easily placed on the ring to give the 2,4,6-Trinitrotoluene (TNT) an explosive (A). Using oxidizing agents $Na_2Cr_2O_7$, the CH_3 group was changed to COOH to give (B). With heat and the presence of COOH, the benzene ring was put into the De-energized state to release CO_2 to give (C)- 1,3,5-Trinitro-benzene a more powerful explosive than TNT. There are many other ways the same product can be obtained without contravening any of the Laws of Nature.

Considering p-dinitrobenzene and o-dinitrobenzene, the followings are obtained.

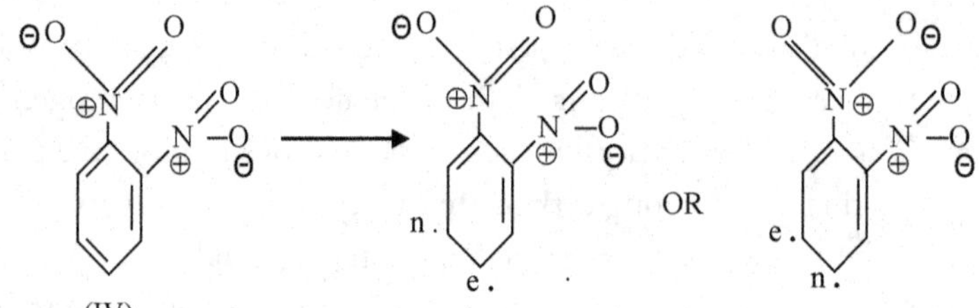

(IV)   (V)   11.360

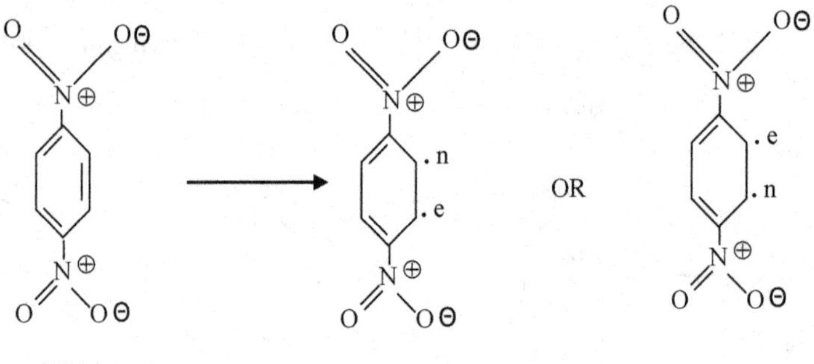

(VI)   (VII)   11.361

Since H.e > e. NO_2 in capacity and n.H < n.NO_2 in capacity as is already obvious, it may be possible that instead of H being held, it is NO_2 that is held when in Equilibrium state of existence, noting that the NO_2 groups are meta- placed. Based on literature data, this is not the case, since the ring is always in Stable state of existence when involved in Chemical reactions. Just like the meta- case, with the presence of two NO_2 groups on the ring, the ring cannot be put in the Energized state. Unlike the meta- case however, the second NO_2 group can be abstracted, because the para- and ortho- positions are not the natural position in the ring when a nitro- group was originally present on the ring. Note also that the nitrogen center in the NO_2 group cannot carry free negative charge due to electrostatic forces of repulsion.

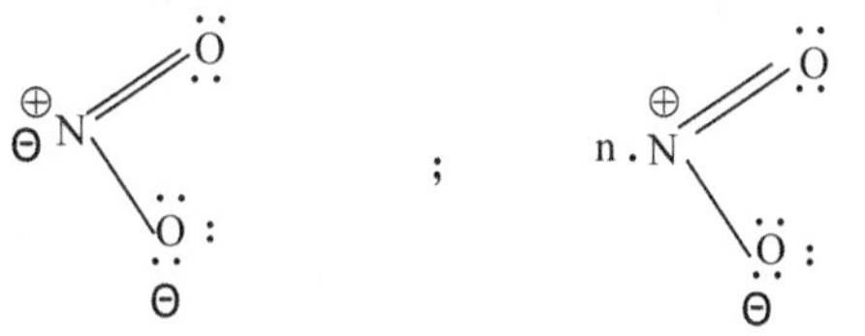

(Not favored)   ;   (Favored)   11.362

Thus, while the nitro- groups in (III) of Equation 11.358 cannot be displaced, the nitro- group in (IV) and (VI) can be displaced as shown below.

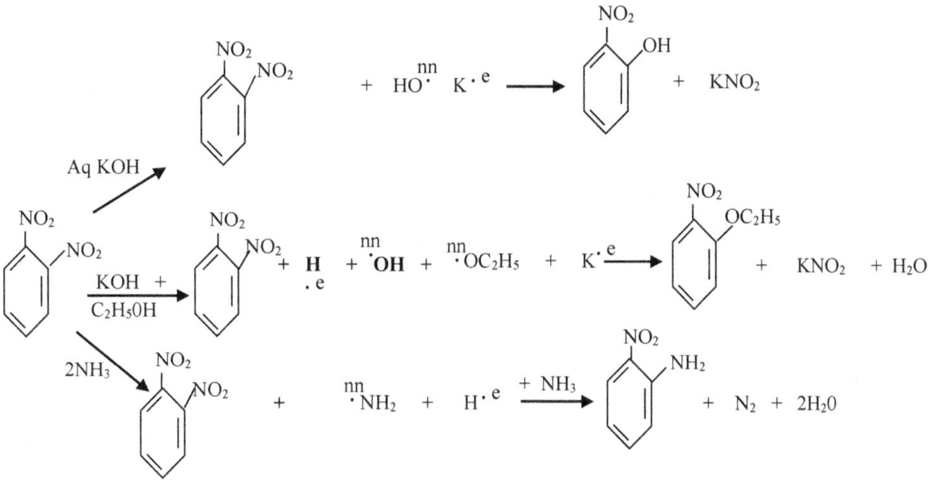

11.363

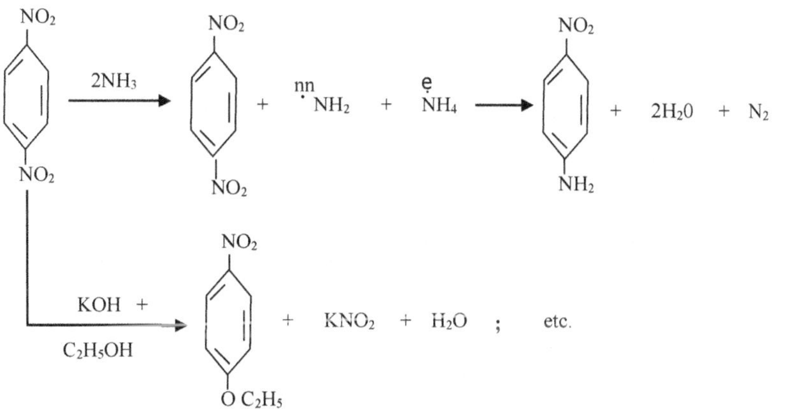

11.364

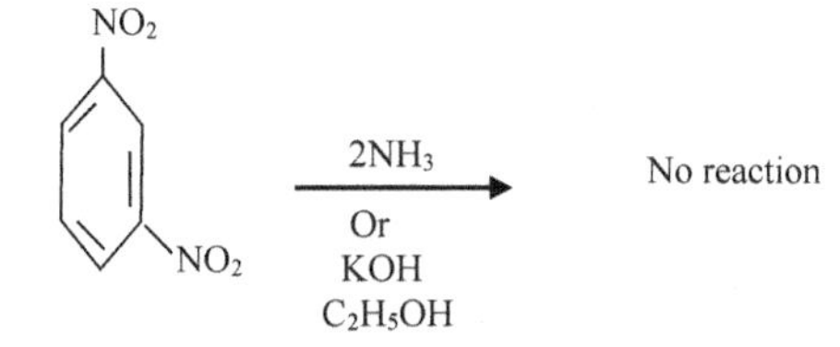

11.365

It is very important to note that all the reactions are radical in characters, different from what has been known to be the case for many years. They are all Equilibrium mechanism systems of one, two or more stages. While the ortho- and para- nitro groups can be abstracted, the one meta- placed cannot be abstracted, because the ortho- and para- placed nitro- groups are not in their natural positions.

An example of a nucleophilic substitution reaction, which is believed to be ionic in character, is the formation of o-nitrophenol when nitrobenzene is heated with potassium hydroxide in the presence of air.

The reaction cannot take place chargedly, since charged hydrides do not exist under normal or abnormal conditions.

$$+ \quad KOH \quad + \quad 1/2O_2 \qquad\qquad 11.366$$

Stage 1:

$$KOH \;\rightleftharpoons\; K\bullet e \;+\; nn\bullet OH$$

$$K\bullet e \;+\; O_2 \;\rightleftharpoons\; K - O - O\bullet en$$

$$K - O - O\bullet en \;+\; \text{(ring)} \;\rightleftharpoons\; \text{(ring)}\bullet e \;+\; KOOH$$

$$HO\bullet nn \;+\; \text{(ring)}\bullet e \;\longrightarrow\; \text{(B)}$$

$$11.367a$$

Overall equation: $\quad$ (I) $\quad+\quad$ KOH $\quad+\quad$ O$_2$ $\quad\longrightarrow\quad$ (B) $\quad+\quad$ KOOH $\qquad$ 11.367b

Stage 2a: $\quad^*$KOOH $\;\rightleftharpoons\;$ K$\bullet$e $\;+\;$ nn$\bullet$ OOH

$$H - O - O\bullet nn \;\rightleftharpoons\; en\bullet O - O\bullet nn \;+\; n\bullet H$$

$$K\bullet e \;+\; n\bullet H \;\rightleftharpoons\; KH$$

$$en\bullet O - O\bullet nn \;\longrightarrow\; O_2 \;(AIR) \qquad\qquad 11.368a$$

Overall equation: (I) $\;+\;$ KOH $\;+\;$ O$_2$ $\;\longrightarrow\;$ (B) $\;+\;$ KH $\;+\;$ O$_2$ $\qquad$ 11.368b

Stage 2b: $\quad^*$KOOH $\;\rightleftharpoons\;$ KO$\bullet$nn $\;+\;$ en$\bullet$ OH

$$H - O\bullet en \;\rightleftharpoons\; en\bullet O\bullet nn \;+\; e\bullet H \;+\; Heat$$

$$KO\bullet nn \;+\; e\bullet H \;\longrightarrow\; KOH \qquad\qquad 11.369a$$

Overall equation: 2(I) $\;+\;$ 2KOH $\;+\;$ 2O$_2$ $\;\longrightarrow\;$ 2(B) $\;+\;$ 2KOH $\;+\;$ ***O$_2$***

$$\text{(AIR)} \qquad\qquad\qquad \text{(Oxidizing Oxygen)} \qquad 11.369b$$

The two stage Equilibrium mechanism system above speaks for itself. In the first stage the oxidizing oxygen carried by the potassium was used to grab H from the ring to form a peroxide. Worthy of note is that the H removed is that located on the ortho- position, because the meta- position is the natural position for NO$_2$

group. In the second stage (Stage 2a or 2b), it is Stage 2b that is favored, based on the real Equilibrium state of existence of KOOH. It is Oxidizing oxygen molecule highlighted above that is released and not molecular oxygen. In Stage 2a, molecular oxygen was recovered and KH was formed. With this, KOH is not the real catalyst, but O_2 from the air; while with Stage 2b, KOH is a catalyst. If water had been present in the system, based on Stage 2a, the presence of KOH as a final product would have been possible as shown below.

Stage 1:

$$KH \rightleftharpoons K\bullet e + n\bullet H$$

$$K\bullet e + H_2O \rightleftharpoons KH + en\bullet OH + Heat$$

$$HO\bullet en \rightleftharpoons H\bullet e + en\bullet O\bullet nn + Heat$$

$$nn\bullet O\bullet en + KH \rightleftharpoons HO\bullet nn + K\bullet e$$

$$H\bullet e + n\bullet H \rightleftharpoons H_2$$

$$\bullet K\bullet e + n\bullet OH \rightleftharpoons KOH \,^*$$

"Stable, Reactive and Soluble"

11.369c

Overall equation: $\quad KH + H_2O \rightleftharpoons KOH + H_2$ 11.369d

It is indeed Stage 2b that is favored, that wherein concentrated moles are involved since oxidizing oxygen molecule is one of the products. ***The potassium hydride which is non-polar/non-ionic is soluble in water, but does not dissolve in it, since water is polar/ionic.*** Hence, the Equilibrium state of existence of KOH is suppressed in the sixth step above, since before a compound can be soluble in a solvent, it must first dissolve or miscibilize in the solvent.

One can observe why there is strong need to look aggressively at the past. Even the reactions for reduction of nitro groups, which were thought to be ionic in character, is very impossible, since it is only in our world we make impossible things possible as against the Laws of Nature.

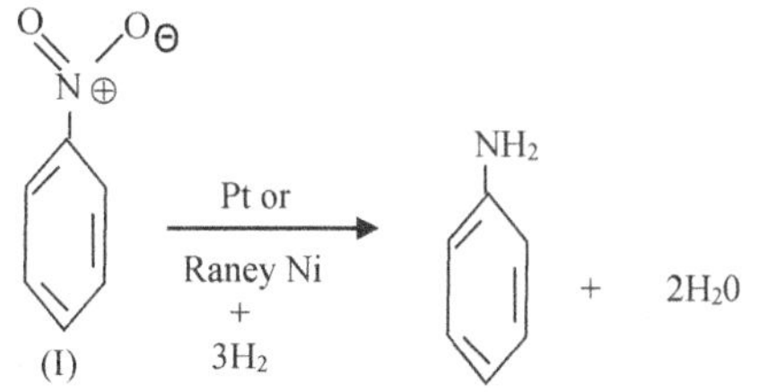

11.370

Stage 1:

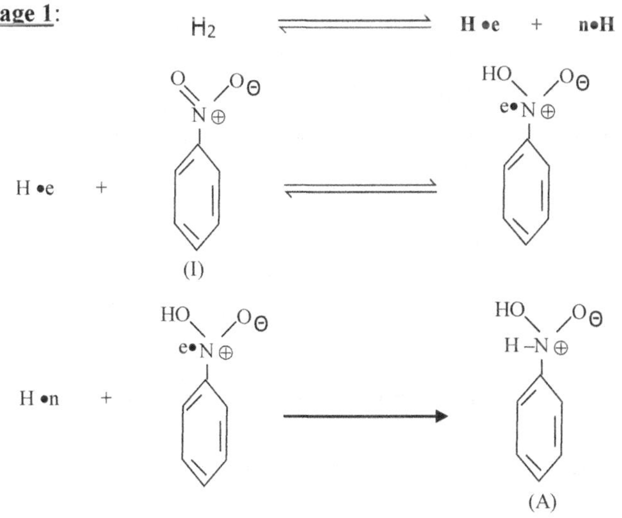

$$H_2 \rightleftharpoons H\bullet e + n\bullet H$$

11.371a

Overall equation: (I) + H₂ ⟶ (A) 11.371b

Stage 2:

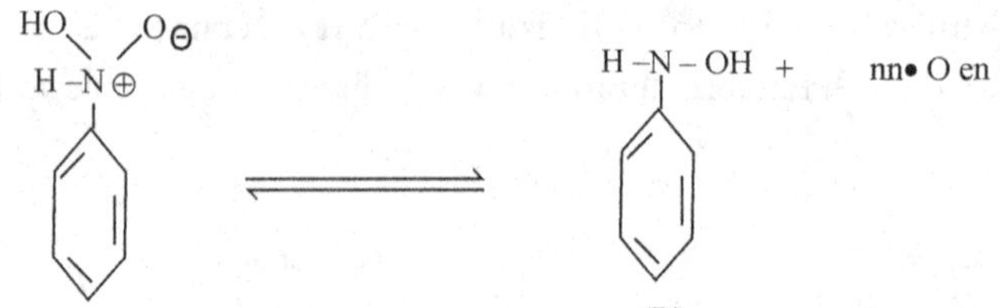

11.372a

Overall equation: (I) + 2H₂ ⟶ (B) + H₂O

11.372b

Stage 3:

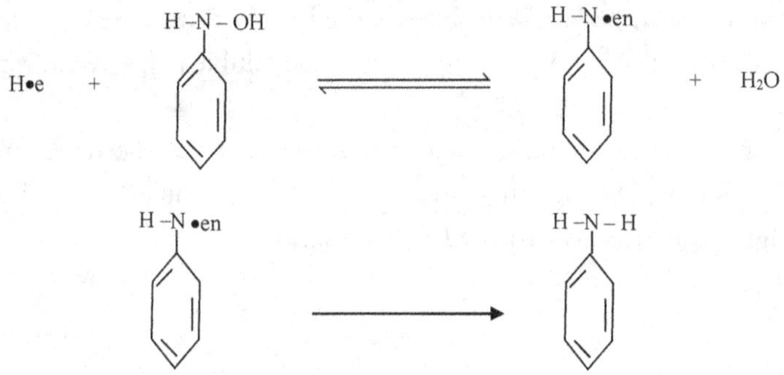

11.373a

Overall equation: (I) + 3H₂ ⟶ Aniline + 2H₂O 11.373b

It is a three stage Equilibrium mechanism system wherein the benzene ring was kept in a De-energized state of existence throughout the three stages. The N = O bond a strong nucleophilic center can only be activated radically and not chargedly, due to the presence of the polar bond in the NO_2 group which has been confirmed to be a weak radical-pushing group, free radically and strong charge-pulling group chargedly. Though strong in nucleophilicity, it is weaker than what the ring is carrying. The polar bond cannot be activated, due to the laws placed by the boundaries In the first stage, the N = O bond was first activated by H •e which was provided by H_2 kept in Equilibrium state of existence by the pores inside the transition metals. At the end, (A) was formed. In Stage 2, (A) being unstable, releases oxidizing oxygen atom which then grabs hydrogen from H_2 to form water, in order to complete a stage when (B) was formed. In Stage 3, with (B) kept in Stable state of existence, H came again to grab the OH group to form water and an electro-non-free-radical carrying compound from which aniline was obtained. Our can see the wonders of NATURE.

If two nitro groups are present in the ring, it is the meta-dinitrobenzene type that is therefore readily involved, in which only one of the nitro groups can be reduced. It is only the meta-position type, because the ortho- and para-positions are not natural to the ring when NO_2 group is carried. Note that this is not like the case where the NO_2 group is abstracted.

$$NO_2\text{-benzene-}NO_2 \quad + \quad 3(NH_4)_2S \quad \longrightarrow \quad NO_2\text{-benzene-}NH_2 \quad + \quad 6NH_3 \quad + \quad 3S \quad + \quad 2H_2O$$

$$(H_4N^{\oplus}\cdots\ ^{\ominus}S^{\ominus}\cdots\ ^{\oplus}NH_4)$$

$$11.374$$

The reason why the second nitro group cannot be hydrogenated is because, once the NH_2 is put in place on the ring, it takes control over the actions of the ring, being far more radical-pushing than the nitro group. In order words if the ring is kept in the Energized state of existence, the positions available for use will then be the ortho- and para- positions with respect to NH_2. Since the structure of so many compounds such as the ammonium sulfide above are unknown in present day Science, the Equilibrium mechanism of the reaction above is herein provided below. All the others where not fully provided are left as exercises for the readers using the foundations which have been laid so far. The main bond in the ammonium sulfide above is the electrostatic bond and not covalent or ionic bonds; yet they are still important in the compound in different ways. For example, it is via its ionic character that Equilibrium state of Equilibrium state of existence is favored.

Stage 1: $\quad H_4N^{\oplus}\cdots^{\ominus}S^{\ominus}\cdots^{\oplus}NH_4 \quad \underset{Existence}{\overset{Equilibrium\ State\ of}{\rightleftharpoons}} \quad H\bullet e \ + \ nn.NH_3^{\oplus}\cdots^{\ominus}S^{\ominus}\cdots^{\oplus}NH_4$

(A)

$$(A) \quad \rightleftharpoons \quad NH_3 \ + \ nn.\overset{\bullet\bullet\ \ominus}{\underset{\bullet\bullet}{S}}\cdots^{\oplus}NH_4$$

(B)

$$H\bullet e \ + \ \text{(NO}_2\text{, }N^{\oplus}=O\text{, }O^{\ominus}\text{ ring)} \quad \rightleftharpoons \quad \text{(NO}_2\text{, }N^{\oplus}-OH\text{, }O^{\ominus}\text{ ring)}$$

(C)

$$(B) \ + \ (C) \quad \longrightarrow \quad \text{(NO}_2\text{, }N^{\oplus}-OH\text{, }N^{\oplus}\cdots^{\ominus}S^{\ominus}\cdots^{\oplus}NH_4\text{, }^{\ominus}O\text{ ring)}$$

(D)

$$11.375$$

Stage 2: $\quad H_4N^{\oplus}\cdots^{\ominus}S^{\ominus}\cdots^{\oplus}NH_4 \quad \underset{Existence}{\overset{Equilibrium\ State\ of}{\rightleftharpoons}} \quad H\bullet e \ + \ nn.NH_3^{\oplus}\cdots^{\ominus}S^{\ominus}\cdots^{\oplus}NH_4$

(A)

$$(A) \quad \rightleftharpoons \quad NH_3 \ + \ nn.\overset{\bullet\bullet\ \ominus}{\underset{\bullet\bullet}{S}}\cdots^{\oplus}NH_4$$

(B)

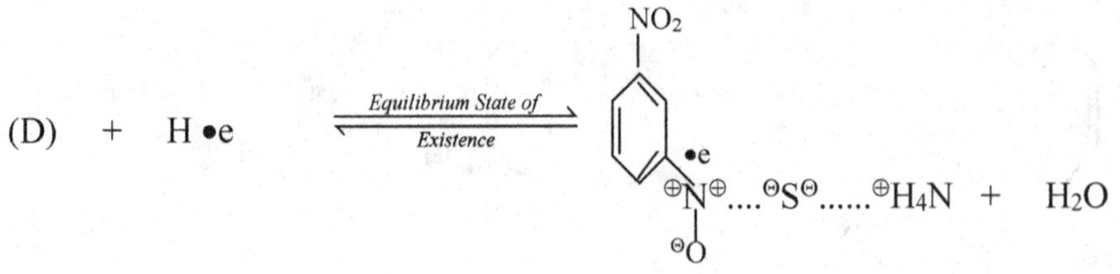

(D) + H •e ⇌ (*Equilibrium State of Existence*)

(E)

(E) + (B) →

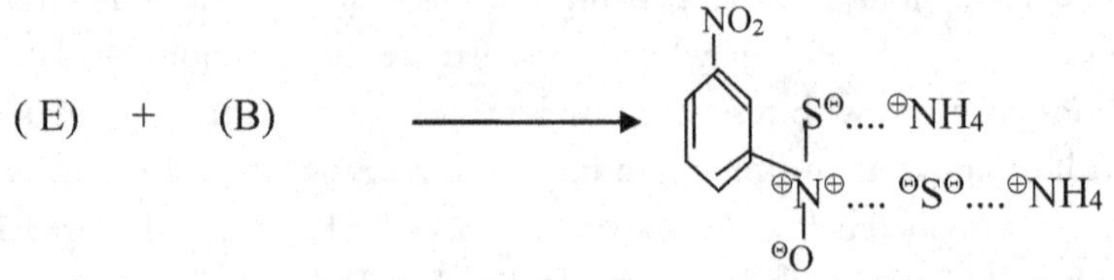

(F)

Stage 3: $H_4N^{\oplus}...^{\ominus}S^{\ominus}....^{\oplus}NH_4$ ⇌ (*Equilibrium State of Existence*) $H \bullet e + nn.NH_3^{\oplus}......^{\ominus} S^{\ominus}....^{\oplus}NH_4$

(A)

11.376

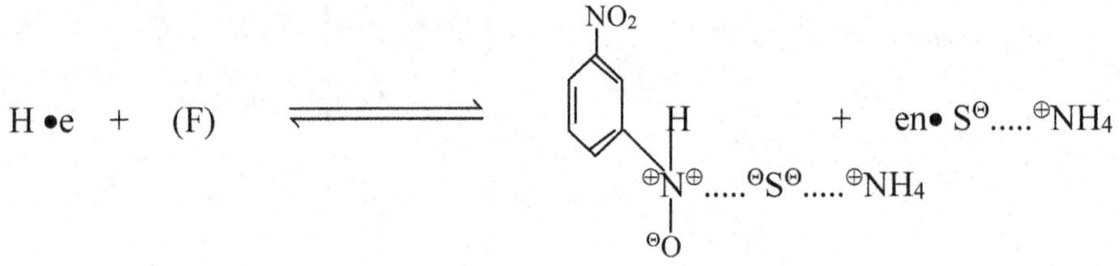

$H \bullet e$ + (F) ⇌

(G) + $en\bullet S^{\ominus}.....^{\oplus}NH_4$

(A) ⇌ NH_3 + $nn.\overset{..}{\underset{..}{S}}{}^{\ominus} .....^{\oplus}NH_4$

(B)

$en\bullet S^{\ominus}.......^{\oplus}NH_4$ ⇌ S + NH_3 + $H \bullet e$

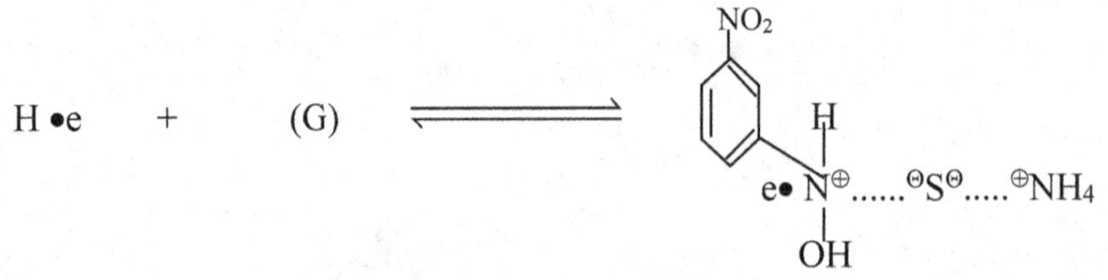

$H \bullet e$ + (G) ⇌

(H)

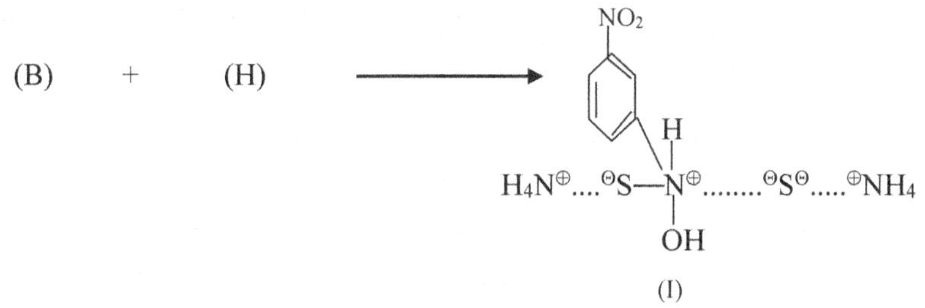

$$(B) \quad + \quad (H) \quad \longrightarrow$$

(I)

11.377

Stage 4:

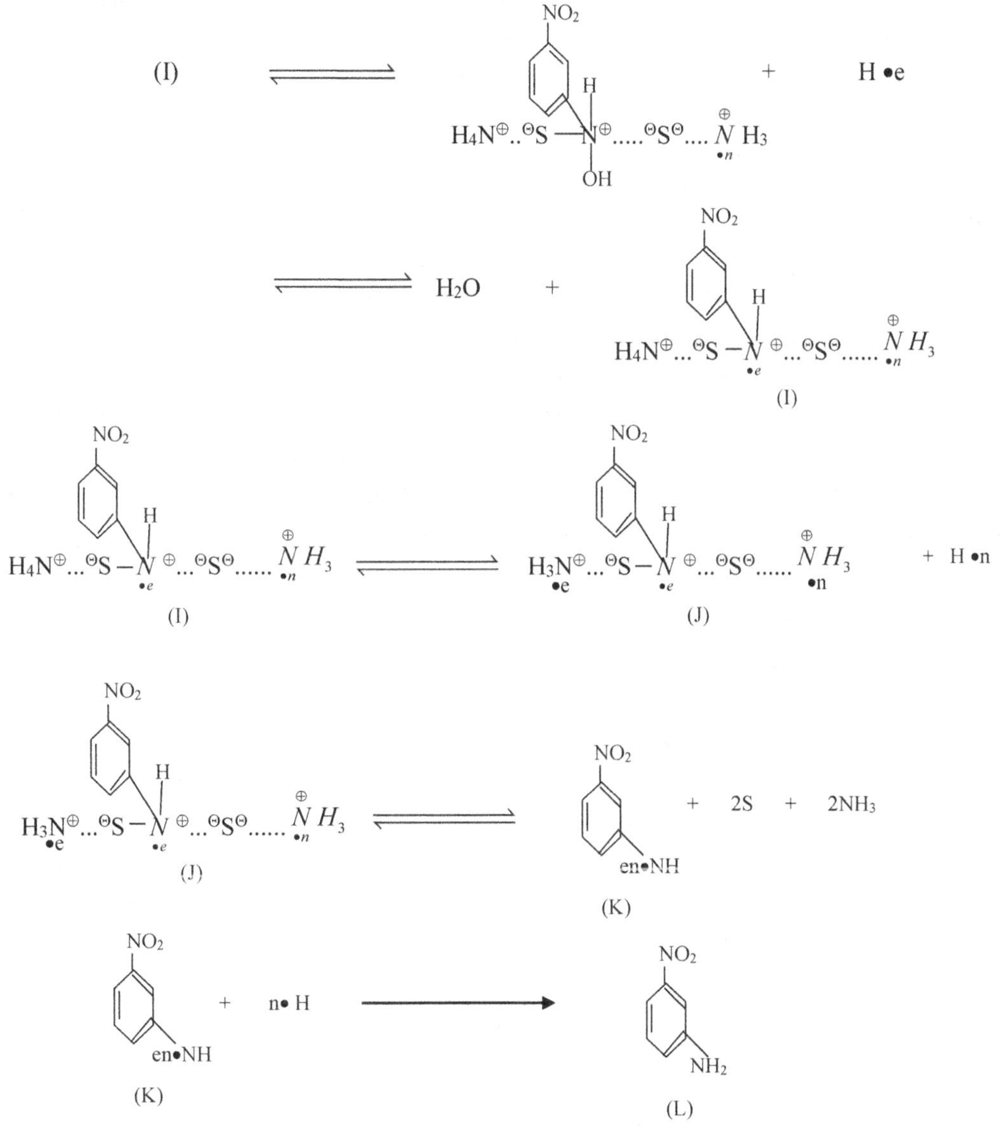

(I) $\rightleftharpoons$ ... $+ \quad H \bullet e$

$\rightleftharpoons H_2O \quad +$

(I)

(I) $\rightleftharpoons$ (J) $+ \ H \bullet n$

(J) $\rightleftharpoons$ ... $+ \ 2S \ + \ 2NH_3$

(K)

(K) $+ \ n \bullet H \quad \longrightarrow$ (L)

11.378

Overall Equation: Same as Equation 11.374.

Worthy of note are the followings-

i)   The structure and Equilibrium State of Existence of ammonium sulfide.

ii)   The hydrogenation catalytic character of the ammonium sulfide. It is not a catalyst, but a reactant.

iii) The existence of electrostatic and polar bonds and showing how the polar bond was activated.

339

iv) The ammonium sulfide as a source of S and ammonia. Hydrogen molecule were not externally provided, but by the ammonium carried by the sulfide.

On the whole, there are four and only four stages, which began with activation of the N = O bond, using H provided by the sulfide. In the process, one molecule of ammonia was released and (D) was formed. In the second stage, another sulfide came again to replace the OH group with a more electropositive group with respect to the trios- N, O, S, and group came from itself. In the process, another ammonia molecule and water were released and (E) was formed. In the third stage, the third sulfide came again to begin the hydrogenation and release two molecules of ammonia, one atom of sulfur and (H). Worthy of note in that stage is how the polar bond was activated without breaking the Boundary laws. The most unique stage is how (D) was decomposed to release two molecules of ammonia, two atoms of sulfur, one molecule of water and finally form (L), the desired product. Always note the last step of all stages, the only step which carries a single right double headed arrow, if the stage is productive. The state of the product obtained can never be that of the Equilibrium state of existence of the compound, unless when suppressed by another component in the system with stronger Equilibrium state of existence.

Thus, in general, one can observe the strong free-radical character of benzene ring. When a radical-pushing group whose capacity is less than that of H is in the ring (such as NO_2), it becomes impossible to introduce any other radical-pushing group of greater capacity than NO_2 into the ring directly via the meta- positions. This can only be done indirectly as we have seen above.

With benzene ring, it can be observed that no molecular rearrangement of third kind can take place, in view of the fact that all the groups are resonance stabilized (i.e., well shielded) when activated, all as a result of the continuous closed-loop features of the resonance stabilization. Molecular rearrangement can only take place after partial hydrogenation of the ring as shown below.

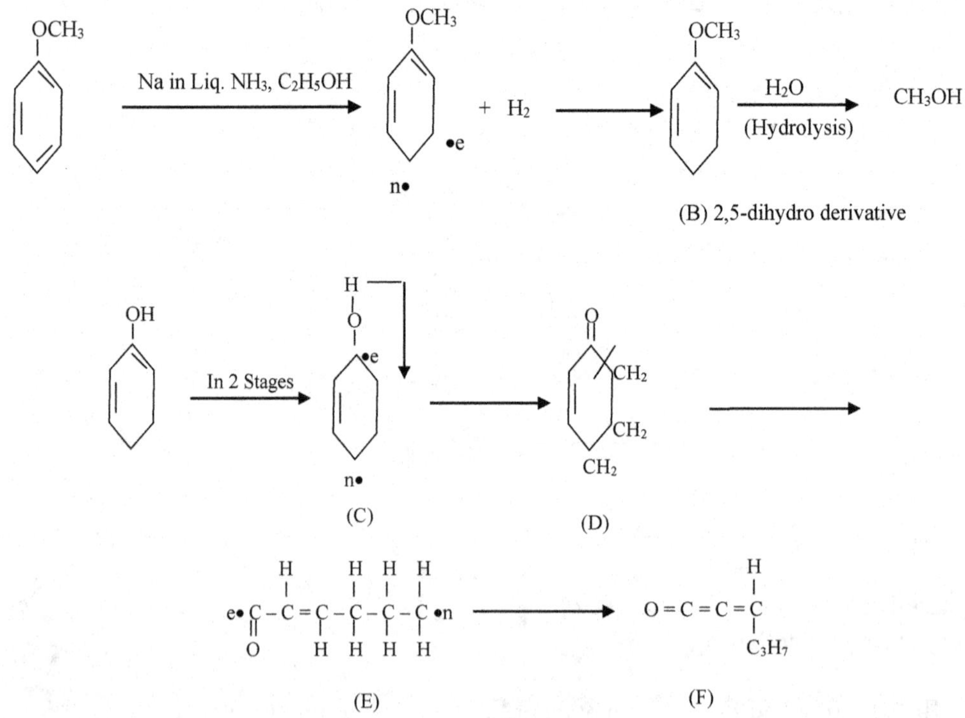

11.379

The hydrogenation reaction to produce (B) commonly referred to as a Birch reduction[46], does not exist as indicated by the mechanism above. On the other hand, (B) above which is said to be 2,5-di-hydro derivative[46] is not. The (B) above is 3,4- di-hydro derivative. (C) was obtained in two stages, in which what was indeed called 2,5-dihydro derivative is 3,6-dihydro derivative, while (C) is 5,6-di-hydro derivative. After hydrogenation similar to that shown in Equations 11.332a to 11.332d, this was then followed by the third stage, where the OCH_3 group was next replaced with OH group via hydrolysis, after it has been externally located. In the next stage, the alcohol obtained molecularly rearranged to give α,β-unsaturated ketone (D) an Electrophile, noting that there is no β,γ-unsaturated ketone as an intermediate. The (D) formed can readily be opened using higher operating conditions, since the ring is well strained with a point of scission to give (F) also an Electrophile – a 1,4-Ketene. One can thus observe the unique features of benzene ring never before known.

11.3.1.3 Ring Expansion of Benzene ring.

Though benzene cannot be opened instantaneously, it is known to favor the existence of cycloheptatriene using cuprous-chloride as catalyst with diazomethane. This obviously is via a radical mechanism as already shown and recalled below.

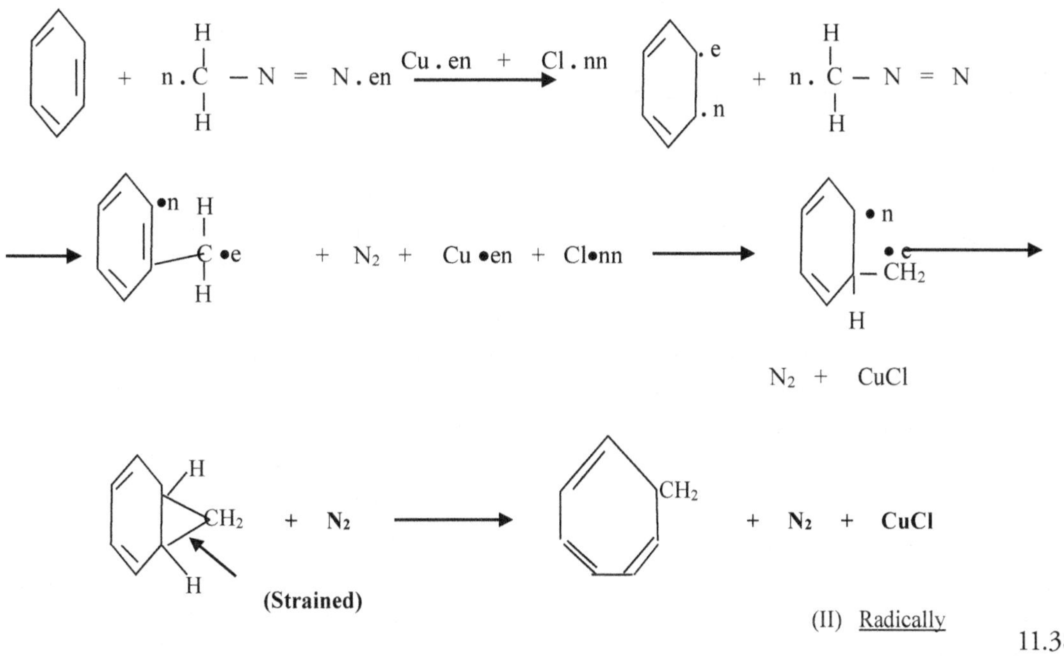

$$11.380$$

The reactions are not favored chargedly since the C center in methylene cannot carry charges and, in view of the type of catalyst involved -a non-ionic metallic catalyst. Without the release of strain energy inside the three-membered ring the presence of the seven-membered ring would have been impossible. It is opened inside at the common boundary, the bond with the maximum radical potential difference, followed by transfer of transfer species of the sixth type. In (II), in the last step where the ring is opened, this is done radically via Cu.en center which has paired unbonded radicals,

11.381

(I) Cycloheptatriene

one of the sources of electrostatic forces to unzip the three-membered ring.

This brings one to the next member of the family cycloheptatriene, noting that the reaction above is a ring expansion phenomenon of Benzene in the de-energized state, activated state in this case. The point of scission is a common boundary to the two rings. Benzene rings can also be reduced in size only when it is non-resonance stabilized. The four-membered condensed ring obtained are strained, but with no point of scission.

11.3.1.4 Polymerization of Benzene

The polymerization of benzene for so many years has never been of major concern, until in recent years[47-56]. It has always been believed not to undergo Addition polymerization reactions, either via opening of its ring or without opening of the ring. It has also been believed that olefins such as cyclohexenes and benzene have little or no ring strain and therefore said not to undergo polymerization because there is no thermodynamic preference for polymer versus monomer[57]. With increasing number of π-bonds in a ring, the more strained the ring becomes. It is therefore very impossible to say that benzene ring is stainless! On the other hand, why is it that cyclobutadiene C_4H_4 is very unstable, cyclohexatriene or benzene (C_6H_6) looks stable and cyclooctatetraene (C_8H_8) is mildly unstable? Since they have the same C to H ratio of 1 to 1 and the number of π-bonds present in their rings looks almost proportional to their size, one should expect them to have the same strain energy. But, this is not the case based on the levels of their stabilities.

The polymerization of cyclobutadiene has never been as issue of concern, because it favors only transient existence under our operating conditions (i.e., at STP). While it can carry radical-pulling groups for it to exist, it cannot carry radical-pushing groups. Cyclohexatriene can carry any group without its existence being made impossible. The polymerization of cyclooctatetraene has been reported to be possible via opening of its ring[58-61], using special so-called alkilidene catalyst carrying a metallic center. Based on the mechanism provided for the opening of the ring, it is impossible to open a four-membered ring fused to the activated cyclooctatetraene at two points of scission at the same time, trying to make something which is impossible possible. When the ring is opened, polyacetylene is the product. Same should almost apply to benzene which is not the first member of this peculiar family; the first member being cyclobutadiene. If the ring of cyclooctatetraene can be opened as if done instantaneously, then what should prevent the opening of benzene in the same manner? Imagine the difference it will be when the ring is made to be more strained by putting just a radical-pushing group on the ring. ***Groups such as methyl, isopropyl, n-butyl, t-butyl, neopentyl, 2-ethylhexyl, n-octadecyl, cyclopentyl, phenyl, methoxy, t-butoxy and CH₃)₃Si, have been reported to be used in the past[60].*** With such type of groups on a benzene ring, there is no doubt that, when the right operating conditions are used, the ring will more readily be opened far more than the case with the use of cyclooctatetraene alone.

Under vibratory milling conditions, benzene and pyridine were said to lead to solid products consisting of a mixture of fractions soluble in methanol and DMF and an insoluble fraction. Chromatographic and spectroscopic analyses indicated that the soluble fraction is a mixture of compounds with linear structures obtained by opening of the aromatic ring. Interaction of these compounds with O_2, light, or high temperatures, leads to cross-linked structures[47]. This was said to be a mechanochemical polymerization. It is mechano- in the sense that a mechanical means has been used in place of an initiator to open the ring. How can such rings with no point of scission in its stable state, be opened? No matter what visible force is used, such rings can never be opened. To open such rings, one must create a point of scission. This can be done either by saturating the ring, that which we don't want to do here, or keep the ring in Equilibrium state of existence or activate the ring and do something immediately. With the case above, in the absence of any initiator (Chemical means), the benzene under such conditions was kept in Equilibrium state of existence and opened as shown below.

Stage 1:

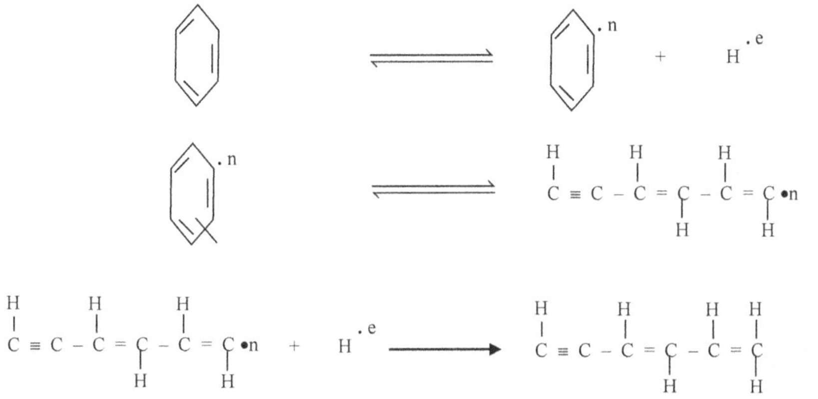

$$11.382a$$

(A) Butadienyl acetylene 11.382

$$11.382b$$

Overall equation: 20 Benzene ⟶ 20 Butadienyl acetylene   11.382b

Stage 2:

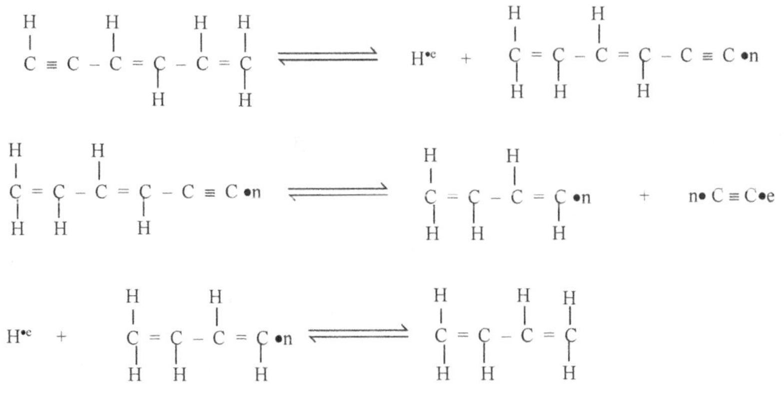

(B) Butadiene

$$11.383a$$

n• C ≡ C•e ⟶ 2C (Carbon black)   11.383

$$11.383b$$

Overall equation: 10 Butadienyl acetylene ⟶ 20C + 10 Butadiene   11.383l

Stage 3:

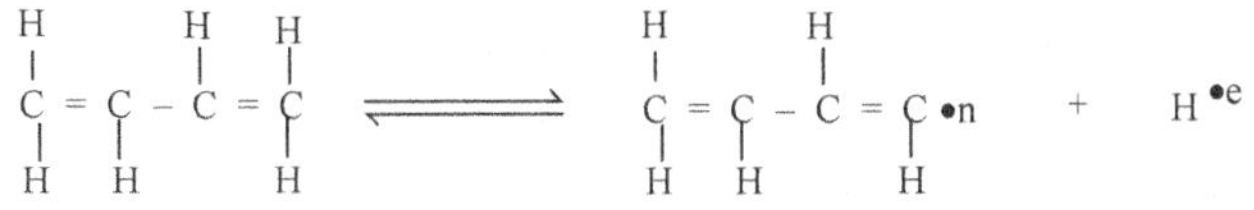

(C)

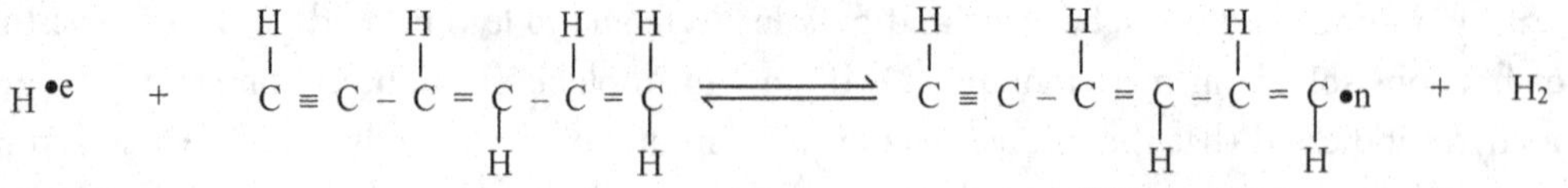

$$H^{\bullet e} \quad + \quad C \equiv C - C = C - C = C \rightleftharpoons C \equiv C - C = C - C = C \bullet n \quad + \quad H_2$$

(D)

$$(C) \quad + \quad (D) \longrightarrow C \equiv C - C = C - C = C - C = C - C = C$$

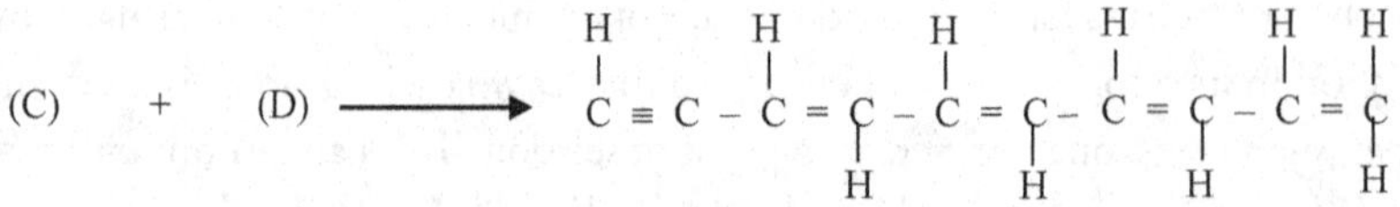

(E) Octatetraenyl acetylene 11.384a

Overall equation: **10 Butadiene + 10 Butadiene acetylene $\longrightarrow$ 10 Octatetraenyl acetylene**

+ 10 H$_2$ 11.384b

Stage 4:

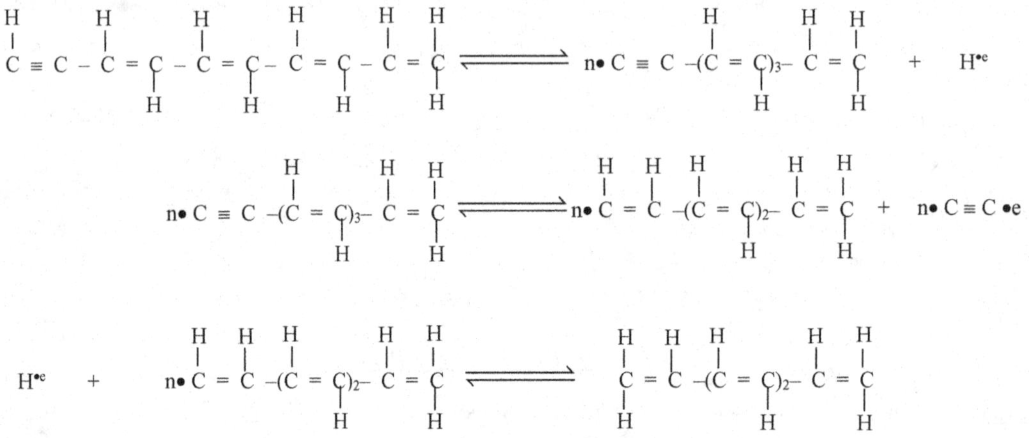

$$C \equiv C - C = C - C = C - C = C - C = C \rightleftharpoons n\bullet C \equiv C - (C = C)_3 - C = C \quad + \quad H^{\bullet e}$$

$$n\bullet C \equiv C - (C = C)_3 - C = C \rightleftharpoons n\bullet C = C - (C = C)_2 - C = C \quad + \quad n\bullet C \equiv C \bullet e$$

$$H^{\bullet e} \quad + \quad n\bullet C = C - (C = C)_2 - C = C \rightleftharpoons C = C - (C = C)_2 - C = C$$

(F)

$$n\bullet C \equiv C \bullet e \longrightarrow 2C \text{ (Carbon black)} \qquad 11.385a$$

Overall equation: 5 Octatetraenyl acetylene $\longrightarrow$ 10 C + 5 Octatetraene 11.385b

Stage 5:

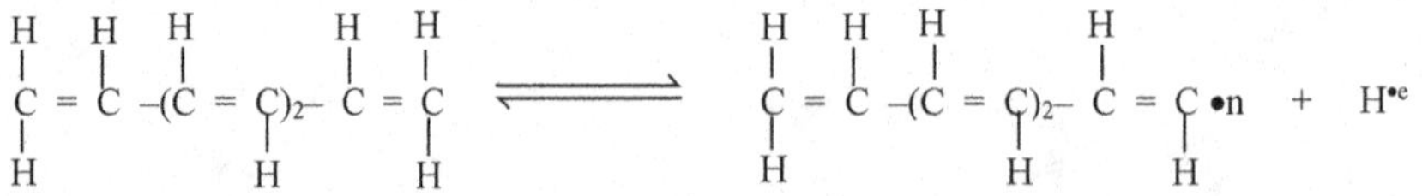

$$C = C - (C = C)_2 - C = C \rightleftharpoons C = C - (C = C)_2 - C = C \bullet n \quad + \quad H^{\bullet e}$$

(G)

$$H^{\bullet e} + C \equiv C - C = C - C = C - C = C - C = C \rightleftharpoons C \equiv C - (C = C)_3 - C = C \bullet e \quad + \quad H_2$$

(H)

344

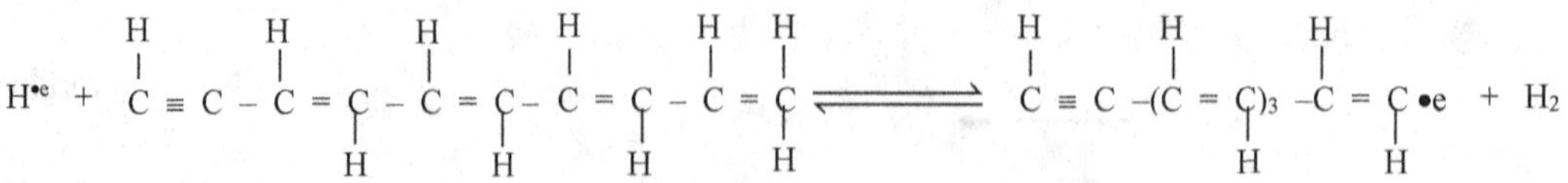

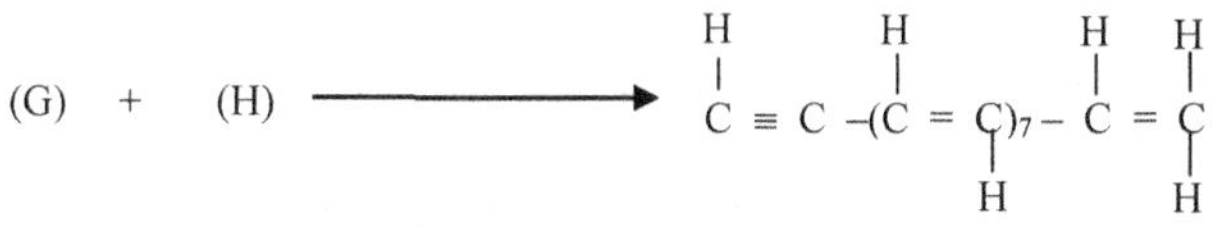

$$(G) \ + \ (H) \longrightarrow \underset{(I)}{H-C \equiv C-(C=C)_7-C=C-H}$$

$$\text{11.386a}$$

Overall equation: 5 Octatetraene + 5 Octatetraenyl acetylene $\longrightarrow$ 5H$_2$ + 5Polyacetylene

$$\text{11.386b}$$

Overall overall equation: 20 Benzene $\longrightarrow$ 30 C + 15H$_2$ + 5Polyacetylene $\qquad$ 11.387

One can see the number of stages involved in producing the Living polyacetylene from 20 molecules of benzene. The Living polyacetylene is said to be living, because the carrier of the chain is acetylene with H at its end. With this type of carrier, at least a good fraction of the polymer will always exist in Equilibrium state of existence, depending on the operating conditions, the fraction decreasing as the chain continues to grow. From the five stages above with the use of 20 moles of benzene, one molecule of the polymer produced has eight acetylenic monomer units with acetylene being the carrier and H the terminating agent. With the 20 molecules of benzene used, if polymerization was allowed to continue, only one molecule of the type of (I) above will be left behind along with more C and H$_2$, bringing the total number of stages to nine. That type of (I) left alone will subsequently break down to a longer diene and C. Above, one has used an even number of moles of benzene. If the number of moles is odd, one knows what to expect.

A look at the five stages above, clearly indicates that this is not an Addition polymerization system as has been mistakenly thought to be the case in the past. It is not a Combination mechanism system as can be seen above. ***This is a Step polymerization system wherein all the stages take place via Equilibrium mechanism.*** In Stage 1, benzene was put in Equilibrium state of existence- the Energized state, and with force resulting from vibrations, the ring was opened to give (A)- a butadienyl acetylene. In Stage 2, a fraction of this was kept in Equilibrium state of existence. Above, one used fifty percent, in order to minimize the numbers of stages one will go through and for simplicity. That fraction broke down to give (B) butadiene and Carbon black (not Coke). In Stage 3, the remaining fraction in Stable state of existence, was next attacked by the butadiene which was put in Equilibrium state of existence. In the process, H$_2$ was formed along with (E)- octatetraenyl acetylene. Just like (A), a fraction of (E) was kept in Equilibrium state of existence, and this decomposed to give an extended butadiene and carbon black in Stage 4. In Stage 5, just like Stage 3, (I) was produced. This continued until the process could not be repeated anymore. The polymer formed has very many cross-linking sites. It will not be surprising therefore that when the polymer is heated, cross-linked products will be obtained. But when it interacts with O$_2$ and light, the H on the acetylene will create some other problems, leading to formation of cross-links.

Some studies on plasma polymerization of benzene and cyclohexane in the gas phase, obtained film powder and oily polymeric products from benzene and only films from cyclohexane[48]. Therein, it was found that with the exception of the oily polymeric products, all the other products were highly cross-linked and contained free-radicals and that contrary to the plasma polymerization oil of benzene, all the others showed no aromatic character[48]. Under the conditions of the use of glow discharge in a plug flow reactor and where cyclohexane whose ring is difficult to open was polymerized, then, there is no doubt that both the ring of benzene and cyclohexane were opened. While to the case of cyclohexane, the opening was instantaneous, for the case of benzene, it was not instantaneous, but as shown above. If any aromatic character was still present in the system, then it is coming from the benzenes which could not be energized and that is the unreacted benzenes. Conversion was not hundred percent.

Benzene accommodated within interlayer space of graphite-alkali metal (Potassium and rubidium) intercalation compounds was found to be polymerized not only to biphenyl, but also to terphenyl and quarterphenyl, while only biphenyl was formed by the action of potassium or rubidium metal alone under the same conditions[55]. The formation of higher oligomers of benzene in the interlayer space in graphite intercalation compounds was ascribed to the amphoteric nature of the compounds which were said to be capable of both accepting and donating electrons[55]. In the presence of ionic metals such as Na, K, Rb, benzene ring can easily be kept in a de-energized state of existence. Hence the followings take place when benzene is with K alone.

Stage 1:

(A) Biphenyl   11..388a

Overall equation: 2 K + 2 Benzene $\longrightarrow$ 2KH + Biphenyl   11.388b

It is only just a one stage Equilibrium mechanism system in which KH is a second product not identified. Usually all components in many experiments are never completely identified. If complete, so many doors will be opened. On the other hand, analytical equipment wherein oxidizing oxygen can be clearly distinguished from molecular oxygen or wherein radicals can clearly be identified and so many other cases have not yet been built. Most of these equipments already seem to exist. The only problem is that we yet do not know how to interpret the data we see, something which one did in Size Exclusion Chromatography many years ago, where the universal equation for separation according to size in a porous media was developed[61]. In the interlayer space of graphite intercalation compounds, a fraction of the benzene was forced to be in the energized state. After the first stage above, the followings were next obtained.

Stage 2:

(B)

(C)

(B) + (C) $\longrightarrow$

(D)   11..389a

Overall equation: Benzene + Biphenyl $\longrightarrow$ Terphenyl + H_2 11..389b

Stage 3:

(B)

346

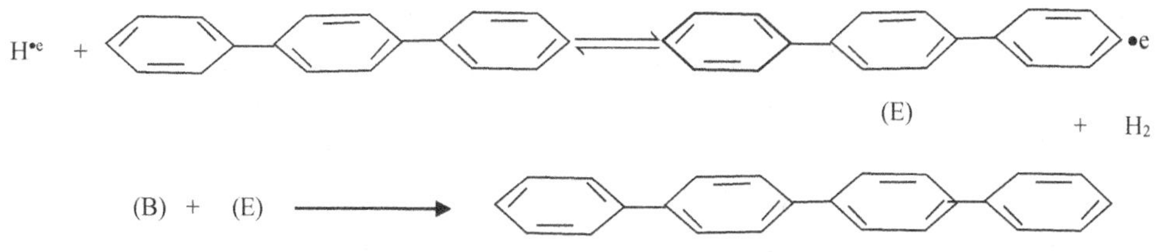

$$H^{\bullet e} + \text{(hexaphenyl structure)} \rightleftharpoons \text{(triphenyl structure)} \bullet e$$

(E)   + H$_2$

(B) + (E) $\longrightarrow$ (quaterphenyl structure)

11..390a

Overall equation: Benzene + Terphenyl $\longrightarrow$ Quaterphenyl + H$_2$ 11..390b

Overall overall equation: 4 Benzene + K $\longrightarrow$ KH + 2H$_2$ + Quaterphenyl   11..391

With presence of more benzene in the polymerization system, a long chain of p-polyphenyl will eventually be obtained along the chain. Like the case above, this is a STEP polymerization system, The carrier of the chain is H and the terminating agent is H. The statement given for the difference in the reactions above viz- "amphoteric nature of the graphite intercalation compounds, which are capable of both donating and accepting electrons" is a clear reflection of the development of present day Science, in which none of the statements makes sense. Electrons are inside the Nucleus of an atom. Only paired unbonded radicals can be donated to fill a vacant orbital. In fact, they are not donated but shared to form a dative bond making sure that the Boundary laws are not broken. What the graphite is doing can unquestionably be seen above. All these compounds just like humans have different "personalities". These we have always known without knowing that we know them. The only difference is that we do not ask questions. Otherwise, what is the meaning of an Oxidizing Agent when there is plenty of oxygen around us, Hydrogen catalysts when we generate large quantities of H$_2$ every day from the refineries alone, a nitrating agent when there is plenty of N$_2$ and O$_2$ around us and so on?

In the electrochemical polymerization of benzene in the presence of aluminum chloride (AlCl$_3$)[49], AlCl$_3$/CuCl$_2$/H$_2$O[50], ferric chloride (FeCl$_3$)[51-53], Molybdenum pentachloride (MoCl$_5$)[54,56], p-polyphenyl was the polymeric product obtained. However unlike the case above, the polymer was said to contain a small amount of Chlorine, with the exception of the use of AlCl$_3$. In addition with respect to the others, small amounts of low molecular weight organic products containing 4,4$^{/}$-dichlorobiphenyl, dichlorobenzene, and chlorobenzene, were also identified as products. Facile dehydrogenation of I,4-cyclohexadiene to benzene using these "catalysts" was said to lend support to the proposal that the reaction proceeds by OXIDATIVE CATIONIC POLYMERIZATION. Like the case above also, the benzene ring could not be opened. In the first case, it is not oxidative, because there is no oxygen involved in the reactions. Secondly, no charges are involved in the system. All the reactions take place radically. The overall reactions provided for their polymerization reactions using some of these catalysts were as follows[51,54].

$$n\,C_6H_6 + n\,MoCl_5 \longrightarrow \text{(polyphenyl structure)}_n + n\,MoCl_3 + 2n\,HCl$$

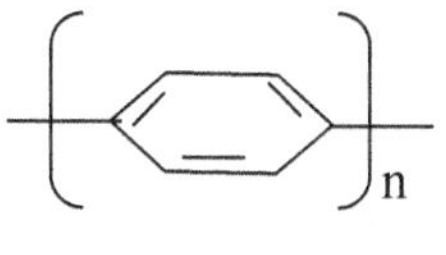

11..392a

$$n\,C_6H_6 + 2n\,FeCl_3 \longrightarrow \text{(polyphenyl structure)}_n + 2n\,FeCl_2 + 2n\,HCl$$

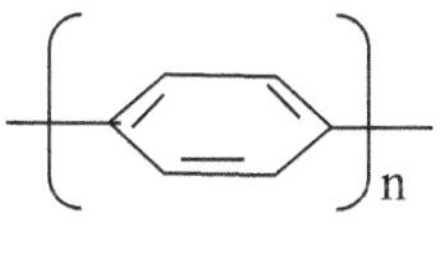

11..392b

Without any doubt, this again is a STEP POLYMERIZATION SYSTEM, since we have a situation where there are small molecular by-products. For the cases above, while Al can form hydrides with H, Mo and Fe cannot form hydrides with H. While Mo is a Transition metal, Fe is a Transition transition metal. On the other hand, nothing was said about the influence of the water added with the non-ionic metallic chlorides as was highlighted in the water cocatalysis in the polymerization of benzene by ferric chloride[53]. As will be shown downstream in this Volume, when water is present with salts such as $ZnCl_2$, no initiator can be obtained, but with ZnR_2, Electrostatically or Covalently charged-paired initiators are formed. With $FeCl_3$, apart from hydration resulting from the vacant orbitals in $FeCl_3$, there is no reaction at the operating conditions of below 100^0C. The equations above do not show the presence of Cl in the polymeric chain, if a small fraction was detected. The pyrolysis of the polymer in vacuo at 750^0-800^0C were said to give a sublimate in addition to residual material resembling carbon black. The sublimed product were said to contain biphenyl, low molecular weight p-polyphenyls including terphenyl, quarterphenyl, and quinquephenyl, in addition to uncharacterized higher molecular weight substances[51]. These are the products which have already been identified above (Equations 11.388 to 11.390a) via the use of K/Graphite. If Chlorine was identified with the polymers, this can only be found at the terminals of the chain or along the ortho-positions of some of the benzene rings along the chain. If carbon black was present during pyrolysis, then the benzene ring must have been opened as already reflected from Equations 11.382 to 11.387.

With the use of $FeCl_3$ wherein its molar ratio with benzene is 2 to 1 as shown in Equation 11.392b above[51], the followings are obtained.

Stage 1:

(A)

$$H^{\bullet e} + FeCl_3 \rightleftharpoons HCl + en\bullet FeCl_2$$

$$en\bullet FeCl_2 + \text{(benzene)} \longrightarrow \text{(B)} \quad FeCl_2$$

(B)

11..393a

Overall equation: Benzene + $FeCl_3 \longrightarrow HCl$ + (B)

11..393b

Stage 2:

$$HCl \rightleftharpoons H\bullet e + nn\bullet Cl$$

$$H\bullet e + \text{(B)} \rightleftharpoons e\bullet \text{(C)} FeCl_2 + H_2$$

(C)

$$Cl\bullet nn + \text{(C)} \longrightarrow Cl- \text{(D)} -FeCl_2$$

(D)

11..394a

Overall equation: HCl   +   (B) $\longrightarrow$ H$_2$ +   (D)   11..394b

Overall overall equation: 100 Benzene + 200 FeCl$_3$ $\longrightarrow$ 100H$_2$ + 100(D) + 100 FeCl$_3$

11..394c

With no benzene any more in the system, the polymerization route proceeds differently as shown below.

Stage 3:

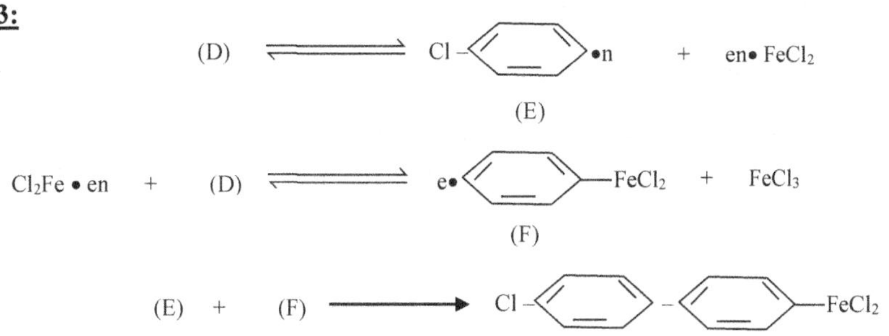

(E)

(F)

(F)   11.395

Stage 4:

(E)

(G)

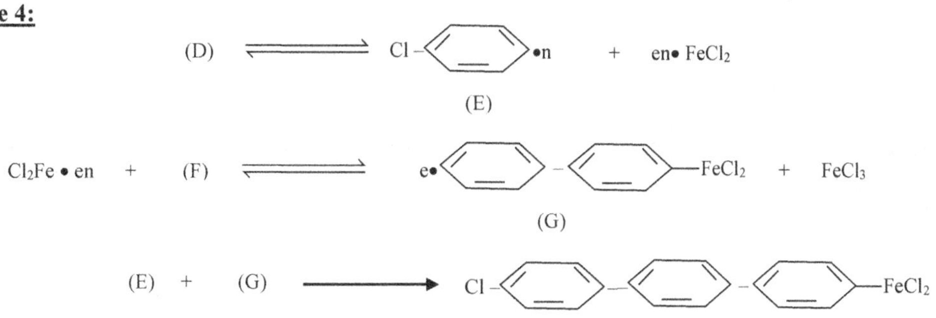

(H)   11.396

Addition continues in a stepwise fashion until all (D)s are consumed. This will involve 102 stages to give the following polymeric chain and overall equation.

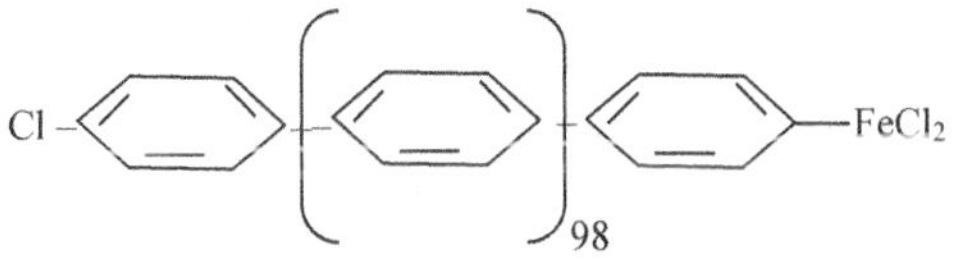

(I)

Overall overall equation: 100 Benzene + 200 FeCl$_3$ $\longrightarrow$ 100 H$_2$ + (I) + 199 FeCl$_3$ 11.397

Since FeCl$_3$ is a hydrogenation catalyst, H$_2$ formed can readily be kept in Equilibrium state of existence. For this reason, the reactions continue as follows

Stage 103:

$$2H_2 \rightleftharpoons 2\,H\bullet e \;+\; 2\,n\bullet H$$

$$2H\bullet e \;+\; 2FeCl_3 \rightleftharpoons 2HCl \;+\; 2\,en\bullet FeCl_2$$

$$2en\bullet FeCl_2 \rightleftharpoons 2\,FeCl_2$$

$$2n\bullet H \longrightarrow H_2$$

Overall equation: $100\,H_2 + 100\,FeCl_3 \longrightarrow 100\,HCl + 100\,FeCl_2 + 50H_2$ 11.398a

Stage 104: This is the same as Stage 103 above to give the overall equation below.   11.398b

Overall equation: $50\,H_2 + 50\,FeCl_3 \longrightarrow 50\,HCl + 50\,FeCl_2 + 25\,H_2$

Stage 105: This is also the same as Stage 103 to give the overall equation below.   11.399

Overall equation: $24\,H_2 + 24\,FeCl_3 \longrightarrow 24\,HCl + 24\,FeCl_2 + 12H_2$

Stage 106: This is also the same as Stage 103 above to give the overall equation below.   11.400

Overall equation: $12H_2 + 12\,FeCl_3 \longrightarrow 12\,HCl + 12\,FeCl_2 + 6\,H_2$

Stage 107: This again is the same as Stage 103 above with the following overall equation.   11.401

Overall equation: $6H_2 + 6\,FeCl_3 \longrightarrow 6\,HCl + 6\,FeCl_2 + 3H_2$ 11.402

Recalling that we left one H_2 molecule from Stage 105, we now have $4H_2$ molecules left.

Stage 108: This is the same as the last stage.

Overall equation: $4H_2 + 4FeCl_3 \longrightarrow 4HCl + 4FeCl_2 + 2H_2$ 11.403

Stage 109: This is the same as Stage 103.

Overall equation: $2H_2 + 2FeCl_3 \longrightarrow 2HCl + 2FeCl_2 + H_2$ 11.404a

Overall overall equation: $100\,H_2 + 199\,FeCl_3 \longrightarrow 198\,HCl + 198\,FeCl_2 + H_2 + FeCl_3$

   11.404b

Overall overall equation: $100\,C_6H_6 + 200\,FeCl_3 \longrightarrow (I) + 198HCl + 198\,FeCl_2 + H_2 + FeCl_3$

Stage 110:

$$H_2 \rightleftharpoons H\bullet e \;+\; n\bullet H$$

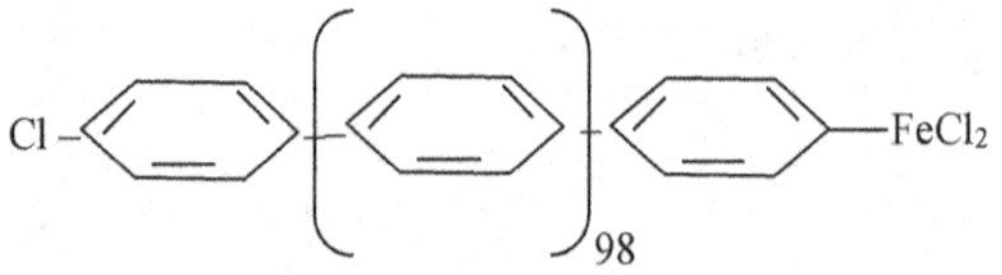

(I)

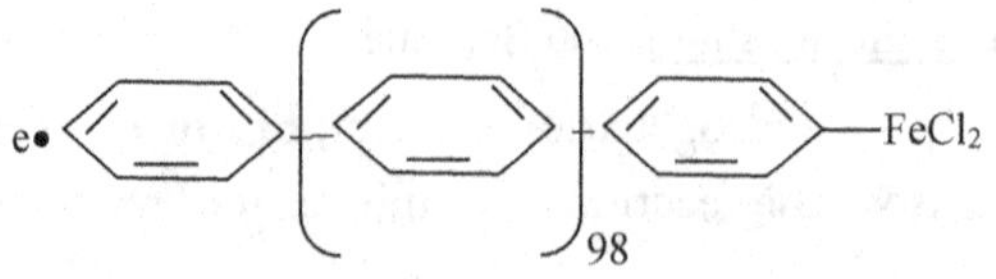

$$(J) \quad + \quad n \bullet H \longrightarrow \quad \text{H-} \underset{(K)}{\underbrace{\hexagon \left[\hexagon \right]_{98} \hexagon}} \text{-FeCl}_2 \qquad 11.405$$

Stage 111:

$$\text{H-} \underset{(K)}{\underbrace{\hexagon \left[\hexagon \right]_{98} \hexagon}} \text{-FeCl}_2 \quad \rightleftharpoons$$

$$n \bullet \underset{(L)}{\underbrace{\hexagon \left[\hexagon \right]_{98} \hexagon}} \text{-FeCl}_2 \quad + \quad \text{H} \bullet e$$

$$\text{H} \bullet e \quad + \quad \text{FeCl}_3 \quad \rightleftharpoons \quad \text{HCl} \quad + \quad en \bullet \text{FeCl}_2$$

$$\text{Cl}_2\text{Fe} \bullet en \quad + \quad (L) \longrightarrow \quad \text{Cl}_2\text{Fe} - \underset{(M)}{\underbrace{\hexagon \left[\hexagon \right]_{98} \hexagon}} \text{-FeCl}_2$$

$$11.406a$$

Overall equation: $\text{H}_2 + (I) + \text{FeCl}_3 \longrightarrow 2\text{HCl} + (M)$ \qquad 11.406b

Final overall equation: $100\ \text{C}_6\text{H}_6 + 200\ \text{FeCl}_3 \longrightarrow 200\text{HCl} + (M) + 198\text{FeCl}_2$ \qquad 11.407

Indeed the products identified in Equation 11.392b are the product which have been obtained above under the following operating condition-

i)   Temperature at 50^0C - 80^0C and I atmospheric pressure. If the temperature is raised to the hundreds, the situation will change drastically as was reportedly observed during pyrolysis of the products.

ii)   Perfect mixing, whereby there is 100% contact between components, not as exists in many largely used reactors where there is a mixer or stirrer. For example the bottom layer of a stomachs of living systems is a mixing tank with no stirrers.

iii) Use of exact molar ratios. If there was more benzene than the FeCl_3, the situation will change drastically.

Equation 11.392b is almost well balanced because there is no chlorine in the polymer. The poly-meric products obtained which have in the past been identified as "rust colored"[51] and "black solid"[52], is very well reflected by what the chain (M) is carrying at the terminals; the carrier of the chain being FeCl_2 and the terminating agent being FeCl_2. For the chain above the C/(H + Cl) atomic ratio is 1.485 for 100 monomer units. For 200 monomer units it is 1.49 and for 500 monomer units, it is 1.497. The limiting value of 1.5 is that wherein there is no Cl. Where the polymer is the only product, the limit 1.5 cannot be attained if there must be Cl in the chain.

On the whole there are one hundred and eleven stages. In Stage 1, the FeCl_3 was suppressed by the energized state of existence of benzene to form (B) and hydrochloric acid, two products which can still react with themselves to form (D) and H_2 in Stage 2. If there had been no transition metal in the system, the H_2 formed would have remained in Stable state of existence throughout the course of the reaction. However, it can readily be kept in Equilibrium state of existence with the presence of the transition metal. The same will apply if $MoCl_5$ is used in place of $FeCl_3$. Though the H_2 is in equilibrium state of existence, it is not strong enough to suppress that of (D) which is now to be used as a STEP MONOMER. Polymerization begins from Stage 3 and continued until all the (D)s are consumed. The (D) is di-bi-functional in character. One of the functional groups being metallic in character (-$FeCl_2$), makes it impossible for the dimer, tetramers and so on to exist in Equilibrium state of existence. Hence, the (D)s which are the only ones that can exist in Equilibrium state of existence, were consumed one at a tme in stages as shown above. They were consumed in series and not in series and parallel. If not all are consumed, because of the type of reactor, other small products such chlorobenzene begin to appear. For the number of moles of benzene and $FeCl_3$ used, polymerization stopped in Stage 102. The same will apply with the case of $MoCl_5$ with its (D) given as shown below.

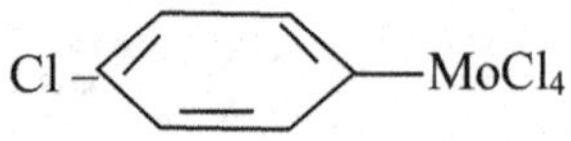

$$Cl - \langle \text{\underline{}} \rangle - MoCl_4$$

(D) – For $MoCl_5$ 11.408

With the full consumption of (D) and formation of the dead gigantic Step monomer (I), the H_2 molecules which have been sitting, now begins to work not yet with the polymer but with the excess $FeCl_3$. Because Fe cannot form hydrides with H, in Stage 103, Ferrous chloride ($FeCl_2$) and H_2 were formed from H_2 and Ferric chloride ($FeCl_3$). One can see herein the wonders of NATURE, just as downstream, from some chemical reactions one was able to see that infinity does not exist as a number for any variable, but only for time and variables that are functions of time. The same Stage 103 was repeated until only one mole of H_2 and $FeCl_3$ were left behind, because both cannot react together and be productive. These were next used in the last two stages to stabilized the (I) formed from Stage 3 in Stage 102. It is from Stage 103, that the use of $FeCl_3$ begins to differ from the use of $MoCl_5$ as shown below.

Overall equation: 100 Benzene + 100$MoCl_5$ ⟶ 100 H_2 + (I) + 99$MoCl_5$

11.409

Stage 103:

$$2H_2 \rightleftharpoons 2\,H\bullet e + 2\,n\bullet H$$

$$2H\bullet e + 2MoCl_5 \rightleftharpoons 2HCl + 2\,en\bullet MoCl_4$$

$$2en\bullet MoCl_4 \rightleftharpoons MoCl_3 + MoCl_5$$

$$2n\bullet H \longrightarrow H_2 \qquad\qquad 11.410a$$

Overall equation: 100 Benzene + 100 $MoCl_5$ ⟶ 51H_2 + 50 $MoCl_5$ + 49$MoCl_3$ +

(I) + 98HCl 11.410b

Stage 104: Same as above.

Overall equation: 100 Benzene + 100 $MoCl_5$ ⟶ 26 H_2 + 25 $MoCl_5$ + 74 $MoCl_3$

+ (I) + 148 HCl 11.411

Stage 105: Same as above.

Overall equation: 100 Benzene + 100 $MoCl_5$ ⟶ 14 H_2 + 13 $MoCl_5$ + 86 $MoCl_3$

+ (I) + 172 HCl 11.412

Stage 106: Same as above.

Overall equation: 100 Benzene + 100 $MoCl_5 \longrightarrow 8H_2 + 7MoCl_5 + 92\ MoCl_3$ + (I) + 184 HCl 11.413

Stage 107: Same as above.

Overall equation: 100 Benzene + 100 $MoCl_5 \longrightarrow 5H_2 + 4MoCl_5 + 95\ MoCl_3$ + (I) + 190 HCl 11.414

Stage 108: Same as above.

Overall equation: 100 Benzene + 100 $MoCl_5 \longrightarrow 3\ H_2 + 2MoCl_5 + 97\ MoCl_3$ + (I) + 194 HCl 11.415

Stage 109: Same as above.

Overall equation: 100 Benzene + 100 $MoCl_5 \longrightarrow 2H_2 + MoCl_5 + 98\ MoCl_3$ + (I) + 196HCl 11.416

Like the case of $FeCl_3$, the followings are obtained.

Stage 110:

$H_2 \rightleftharpoons H\bullet e + n\bullet H$

H•e +

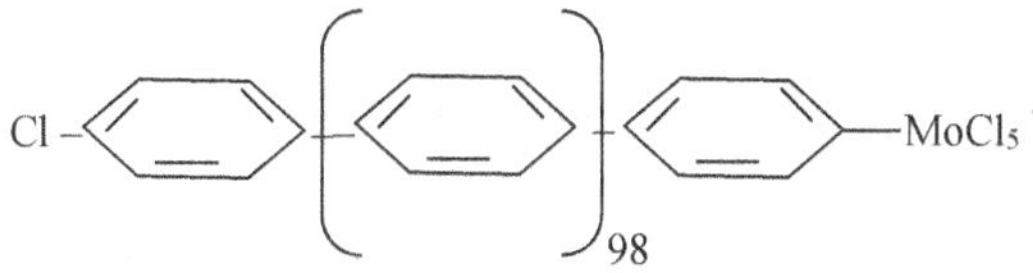

(I)

$\rightleftharpoons$ HCl +

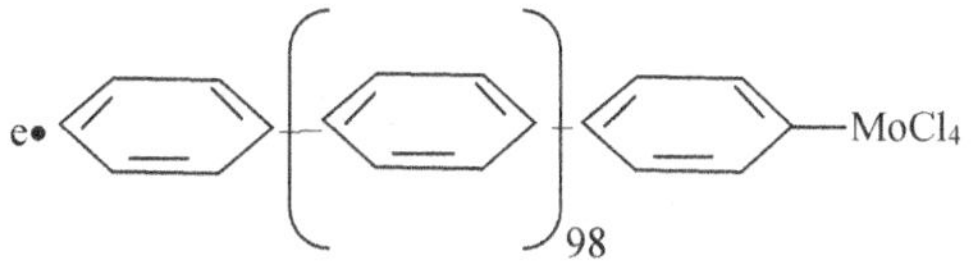

(J)

(J) + n• H $\longrightarrow$

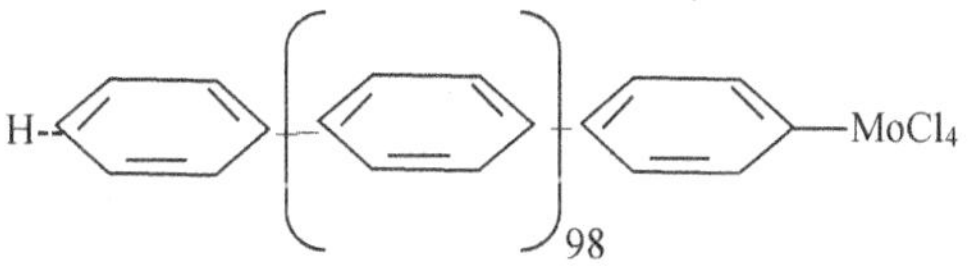

(K) 11.417

Stage 111:

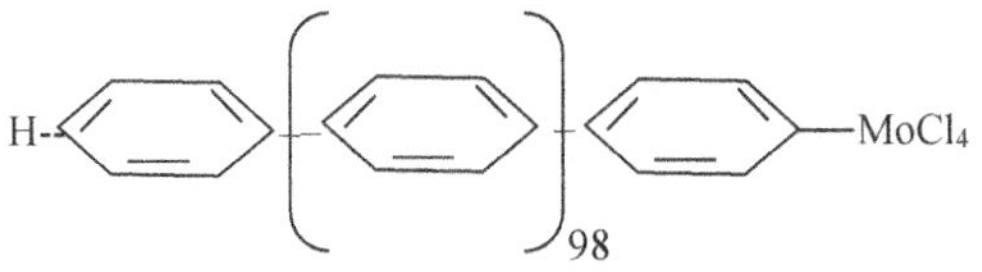

(K)

$\rightleftharpoons$

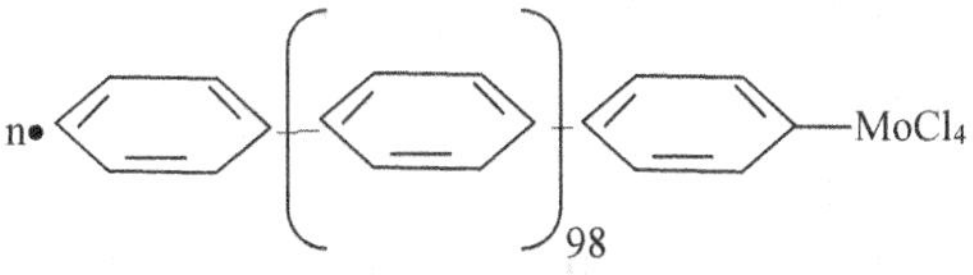

(L)

+ H •e

$$H \bullet e \quad + \quad MoCl_5 \quad \rightleftharpoons \quad HCl \quad + \quad en\bullet MoCl_4$$

$$Cl_4Mo\bullet en \quad + \quad (L) \longrightarrow Cl_4Mo - \langle \rangle - \left(\langle \rangle \right)_{98} \langle \rangle - MoCl_4$$

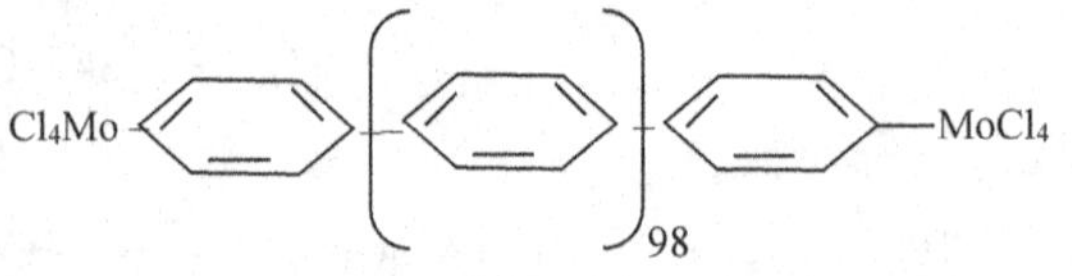

$$(M) \hspace{5cm} 11.418a$$

Overall equation: $H_2 + MoCl_5 + (I) \longrightarrow 2HCl + (M)$ 11.418b

Overall overall equation: $100\ C_6H_6 + 100\ MoCl_5 \longrightarrow H_2 + 198\ HCl + 98\ MoCl_3$

$$+ \quad (M) \hspace{5cm} 11.419$$

Because Fe and Mo are transition metals, it is not surprising to see that the same number of stages are involved; yet $FeCl_3$ and $MoCl_5$ are uniquely different as reflected in Stage 103 of Equations 11.398a and 11.410a respectively. At the operating conditions of around 60^0C, $MoCl_4$ should not disproportionate, unless under pyrolytic conditions. But it did here because of the presence of an even numbered of moles in that stage and under equilibrium conditions inside the stage. It cannot disproportionate when there is only one mole. That was why half number of moles of what was used for $FeCl_3$ was used with $MoCl_3$. The mechanism provided for all these systems are worthy of note, because they open new doors to greater understanding of how Nature operates.

When $AlCl_3$ is used[51,52], the situation is completely different, because Al can form hydrides with H and therefore cannot be used as an hydrogen catalyst, i.e., cannot keep H_2 in Equilibrium state of existence. Hence Stages such as Stages 103 to 109 will not take place. $AlCl_3$ cannot react with H_2 in its Stable state of existence and become productive as shown below, if $AlCl_3$ is made to exist in Equilibrium state of existence.

Stage 1: $AlCl_3 \rightleftharpoons Cl_2Al \bullet e + nn\bullet Cl$

$$Cl_2Al \bullet e + H_2 \rightleftharpoons Cl_2AlH + H \bullet e$$

$$H \bullet e + nn\bullet Cl \rightleftharpoons HCl$$

[Reactive, stable and soluble] 11.420a

Overall equation: $AlCl_3 + H_2 \rightleftharpoons HCl + AlHCl_2$ 11.420b

With $AlCl_3$, after the production of the same type of (D) for Al in the first two Stages, Step polymerization commences in Stage 3 in the same fashion until all (D)s are consumed. This will involve 102 stages to give the following polymeric chain and overall equation.

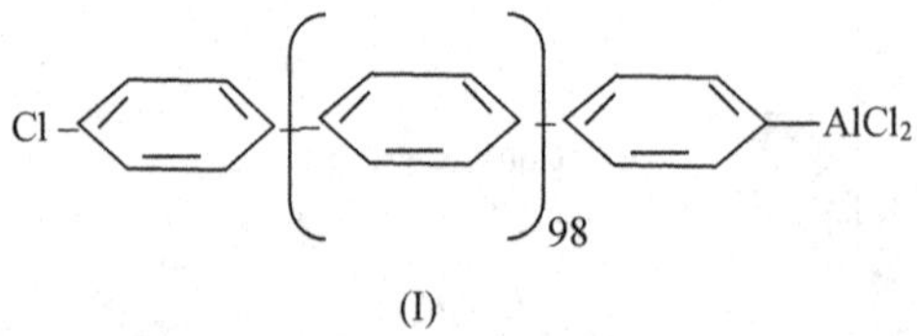

$$Cl - \langle \rangle - \left(\langle \rangle \right)_{98} \langle \rangle - AlCl_2$$

$$(I)$$

Overall overall equation: $100\ Benzene + 100\ AlCl_3 \longrightarrow 100\ H_2 + (I) + 99\ AlCl_3$ 11.421

The chain cannot react with H_2 or $AlCl_3$ or both and be productive. No HCl can be formed in the real world. In the absence of a passive hydrogen catalyst, nothing can be done at the operating conditions. Here, the chain can be observed to be terminally chlorinated. Until now, universally, these reactions were believed to proceed by "Oxidative Cationic Polymerization of the aromatic nuclei". Only THE ALMIGTHY INFINITE GOD and HIS Messengers know what this means. And for THEM, it is meaningless and a display of IGNORANCE in our world which do not still know what AN ATOM is. Yet, we in our world think we have advanced when we have not started. Even from what we have seen so far, the ATOM has not yet been fully defined, because just as we have RADICAL configuration for the outer shell of ATOM, so also we have ELECTRONIC configuration for the Nucleus of the ATOM, because electrons only reside inside the Nucleus of the ATOM. It is from the ELECTRONIC configuration, we begin to see what an ISOTOPE, MATTER, ANTI-MATTER AND MORE are.

Just as the ring of cyclooctatetraene has been reported to be opened[60] when the ring is made to carry a radical-pushing group, so also it can be done for benzene using special initiators. Though how it is opened has been unknown till date, this has been explained above. The answer like all cases which humanity cannot provide answers to, is always so simple- something which we classify as "COMMON SENSE". Things we ignore, is the origin of IGNORANCE. When radical pushing groups are placed on the ring, the ring becomes so strained to the point of attaining Max.RSE inside the ring. When that energy is made to exist in the ring, then the ring can no longer exist as a ring. It will open explosively.

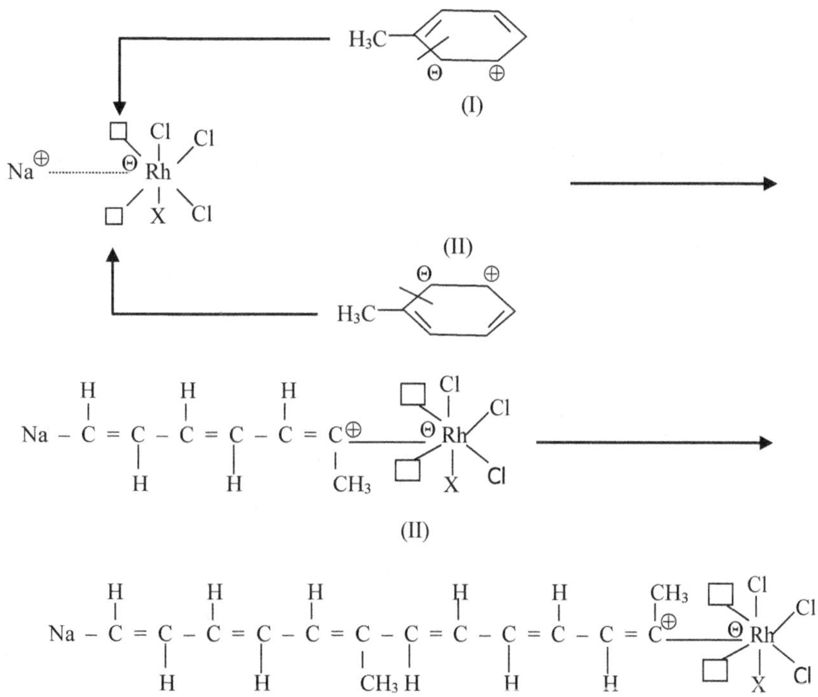

Syndiotactic poly(Methyl-tri-acetylene)   11.422

The initiator used above (which has been said not to be the real one from the Na/Rh combination) would not have been ideal for the system, because of the strong affinity for H by Na to form NaH radically. It is better to use that center which cannot form hydrides with H, such as Ti, V, etc. ***However, it has been used above to illustrate some basic fundamental principles, that is, that the activation can still be done chargedly since resonsnce stabilization cannot take place with the ring, but cannot be opened due to electrostatic forces of repulsion.*** What has been shown above can best be done radically. Note the point of scission on the ring and the center first activated. The point of scission is next to the point of attack. After opening of the ring, note the center carrying the positive charge. It is the center carrying the radical-pushing group which made it possible to open the ring in this manner. Because of the presence of two reservoirs on the counter-center, hence syndiotactic placement of the group can be observed. The larger the radical-pushing capacity of the group placed on the ring, the easier it is for the ring to be opened radically. The above looks like cationic polymerization via Combination mechanism, that which has been said to be impossible. Cations cannot carry a chain. Indeed, in the real initiator from the combination (Na/Rh), it is Rh a non-ionic metal that is in place of the Na as will be shown downstream. The only induction period for the case above is the time for the initiator to be prepared in-situ.

For the first time, one has shown how benzene ring can be used as a monomer to produce unique polymers. Step by step, one is beginning to show how NATURE operates in the real and imaginary domains. Without going through each one of them, one will just be scratching the surface. All atoms, molecules, compounds, species both known and unknown, all have their different personalities, far greater than the issues of Chemical and Physical properties. The issues of Chemical and Physical properties all largely based on laboratory or field data, though very important, are still but a grain of sand in our abilities to understand what these things are.

11.3.2. Cycloheptatriene

Like benzene, there is only one type of cycloheptatriene. Unlike benzene it is discretely closed-loop resonance stabilized. This has already been identified in Equation 11.381 the case in which the three double bonds are conjugatedly placed. There is no point of scission in the ring, though it has the MRSE.

$$11.423$$

(III) is the stable mono-form, which however cannot readily be used, due to steric limitations. Special coordination initiators will be required to polymerize (III). Cyclopentadiene is almost similar to the case here. If the center placed in the middle is the least nucleophilic center, then the monomer becomes a conjugated diene via resonance stabilization, that which cannot be the case. Hence (I) is the least nucleophilic center.

With the introduction of substituted groups into the ring, the followings are to be expected.

$$11.424$$

The CH_3 group is resonance stabilized wherein the transfer species cannot be tempered with. As a strong Nucleophile, there is only transfer species inside the ring- the CH_2 group whose capacity is greater than any R carried externally located to the double bonds in the ring (See Equation 11.178).

The next case is also where the R group (CH_3) is also internally located in the ring.

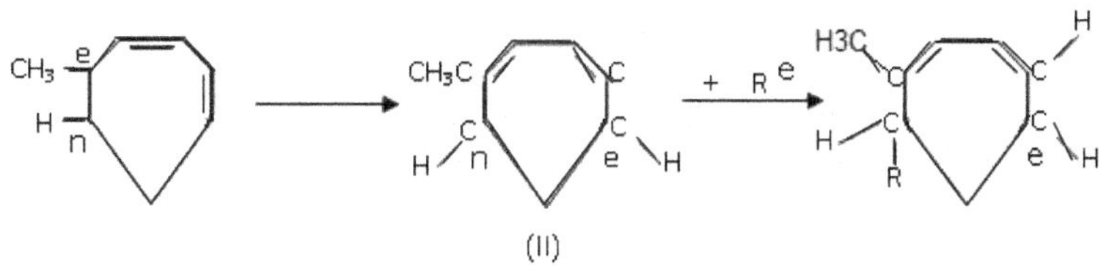

(II)

11.425

Thus, when the CH_3 group is internally located in the ring no matter where it is placed, it is resonance stabilized. However, when the CH_3 group is externally located, the group is not resonance stabilized. ***The center activated above is the wrong center, being the most nucleophilic.***

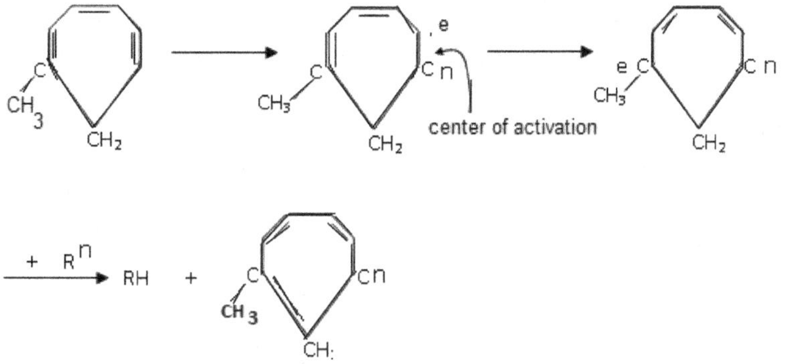

11.426

The same applies to higher radical-pushing group and radical-pulling groups.

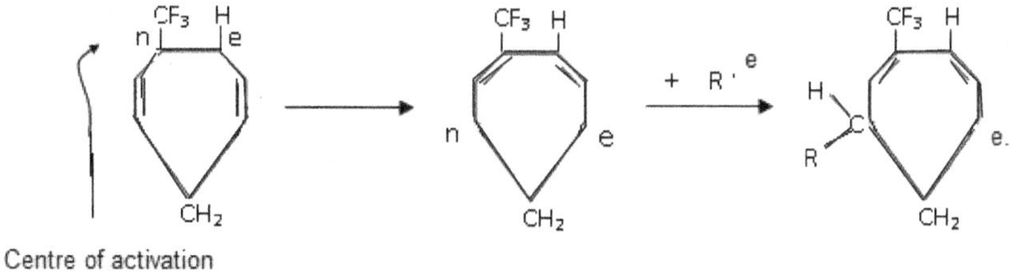

11.427

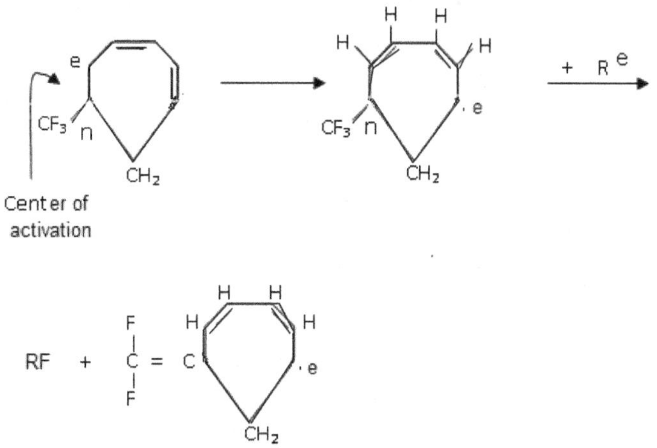

11.428

357

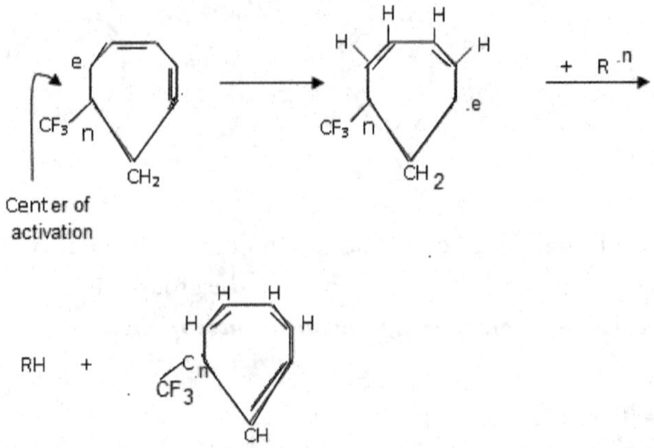

11.429

Like what has been observed so far, this monomer is neither a Nucleophile nor an Electrophile. If polymerization ever takes place, it has to be with the use of paired electro-free-radical initiator.

Nevertheless, it is important to note that, some nucleophiles will readily favor molecular rearrangement of the first kind, with the following possibilities existing exploratively.

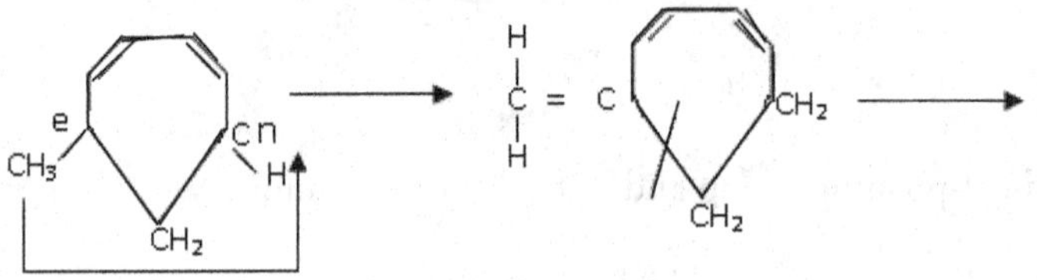

(I)

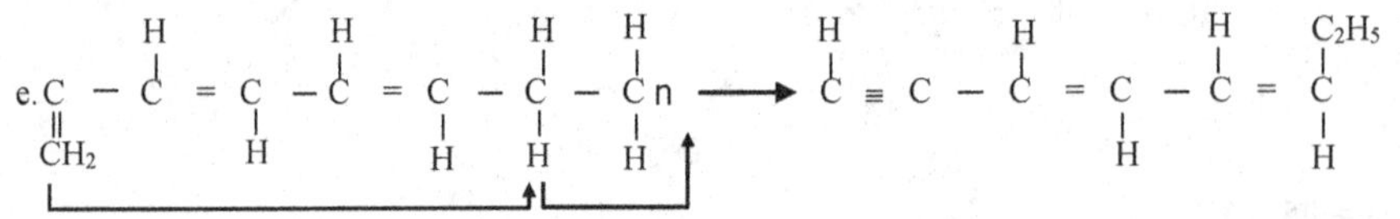

(II) Resonance stabilized

NOT FAVORED

11.430

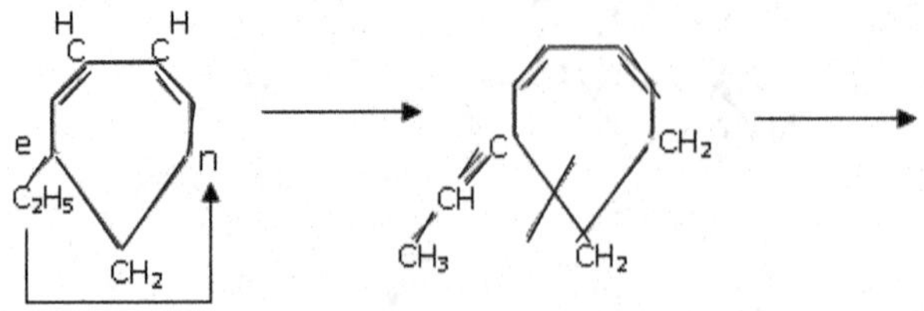

Resonance stabilized

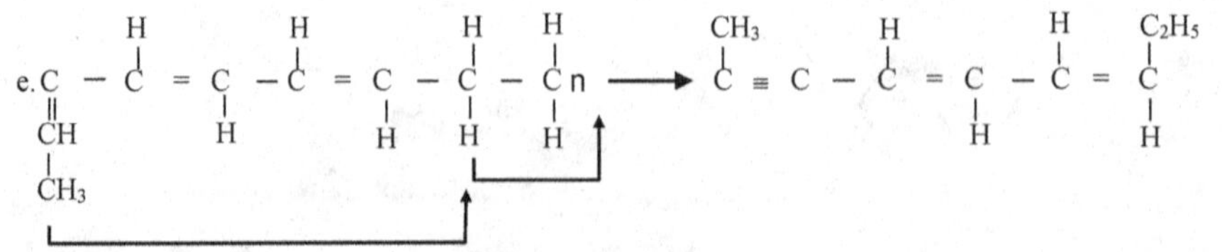

Resonance stabilized

NOT FAVORED

11.431

358

It can be observed that molecular rearrangements look favored only for CH_3 to C_3H_7 groups since no molecular rearrangement can take place for larger groups. However, based on the experiences gathered so far, molecular rearrangement of the first type can never take place from outside as shown, because the capacity of the ringed CH_2 group is larger than that of any alkylane group. Since the source of transfer species is from the CH_2 group, it cannot take here for these cases, since the monomers obtained will become less nucleophilic. Nevertheless, the exploratory cases above have offered us an opportunity of knowing some of their isomers.

By the character and type of resonance stabilized product obtained in Equations 11.425 to 11.428 above, there is no doubt that the following is valid.

$$Cyclotrienes \quad > \quad Cyclodienes \quad > \quad Cycloalkenes \quad > \quad etc.$$

<u>Order of nucleophilicity</u>   11.432

Cyclooctatriene > Cycloheptatriene > Cyclohexatriene > etc.

<u>Order of nucleophilicity</u>

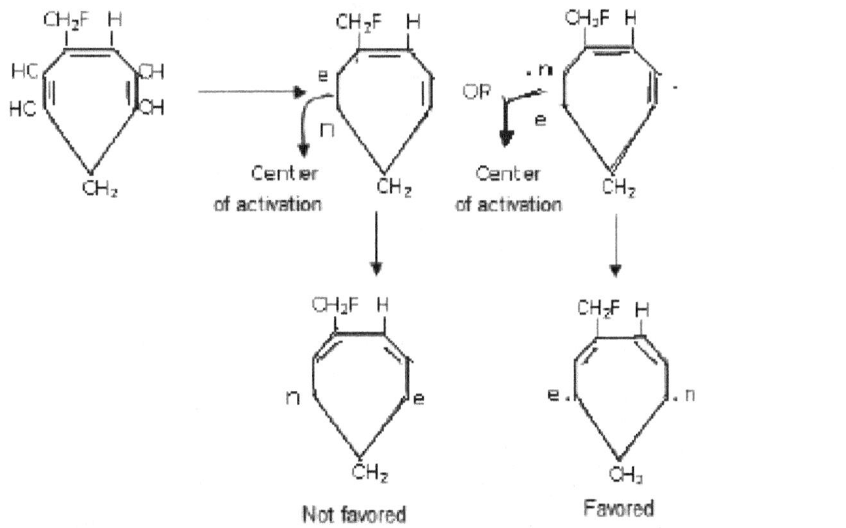

11.433

Not favored   Favored

11.434

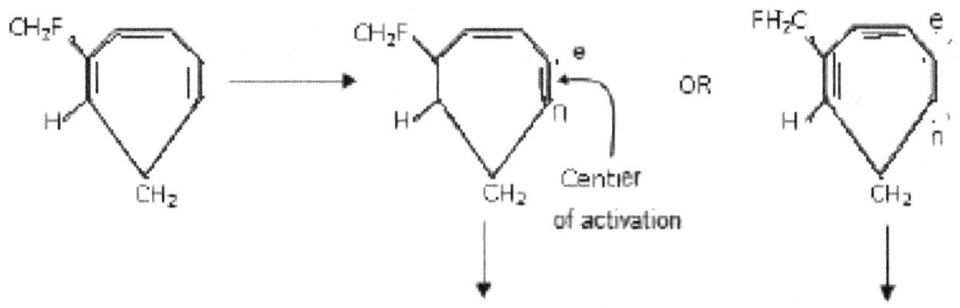

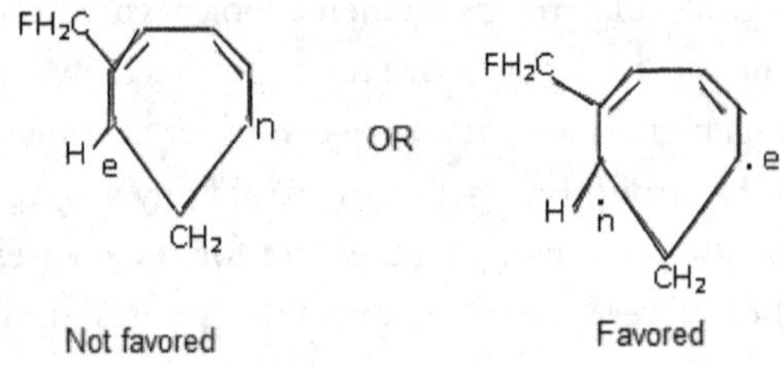

Not favored   OR   Favored

11.435

Not favored   Favored

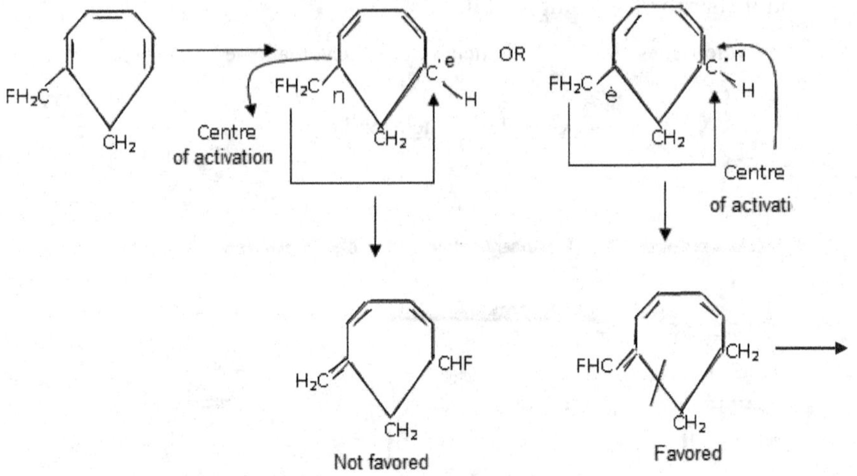

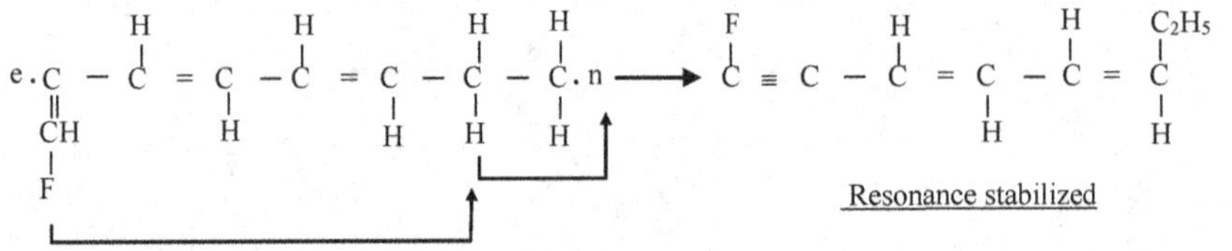

NOT FAVORED   11.436

As already established, the rearrangement above is not favored. But one can see one of its isomers. The same identical characters exhibited by cyclopentadiene are almost to be expected here. When in Equilibrium state of existence, one knows what to expect. The ability of cyclopropene to favor Equilibrium state of existence is far stronger than that of cyclopentadiene (as we have already encountered during its decomposition at higher operating conditions) and is far stronger than that of cycloheptatriene. This is because of the order of the radical-pushing capacities of their CH_2 groups as already shown in Equation 11.178. Thus, the possibility of cycloheptatriene cracking in the same way cyclopentadiene did, may be possible to give what is shown below.

OR/AND

(i)   (II)

11.437

Their favored existence is limited by the fact that cyclooctatetraene itself is unstable, that which rearranges to give styrene as will shortly be shown. One expects (I) or (II) above to be also unstable. These look like bigger Naphthalenes. As an exercise, show if (I) or (II) can be obtained from cycloheptatriene just as naphthalene was obtained from cyclopentadiene.

11.3.3 Cyclooctatriene

There is one point of scission in one of the types. Therefore, opening of the ring should be possible for that one, since the SE in the ring is close to the minimum required strain energy. There are two types of cyclooctadiene -

(i)   One in which the three double bonds are conjugatedly placed, (I).

(ii)   One in which two are conjugatedly placed and one is isolatedly placed, (II).

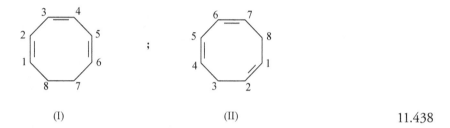

$$(I) \qquad\qquad ; \qquad\qquad (II) \qquad\qquad\qquad 11.438$$

In view of the size of the ring, the three bonds cannot be isolatedly placed. With easy provision of minimum required strain energy for the following should be obtained when opened instantaneously for (I) using strong initiators.

$$\text{(I)} \longrightarrow n\,\underset{H}{\overset{H}{C}} - \underset{H}{\overset{H}{C}} = C - \underset{H}{\overset{H}{C}} = C - \underset{H}{\overset{H}{C}} = C - \underset{H}{\overset{H}{C}}.e$$

(I)   1,3,5,7 - Octatetraene   11.439

Thus while cyclobutene favors the existence of 1,3 - butadiene and cyclohexadiene favors the existence of 1,3,5- hexatriene, cyclooctatriene favors the existence of 1,3,5,7- octatetraene or 1,3,5- substituted cyclohexadienes, all of them only radically and not chargedly.

In the presence of weak initiators, the following are obtained for (I).

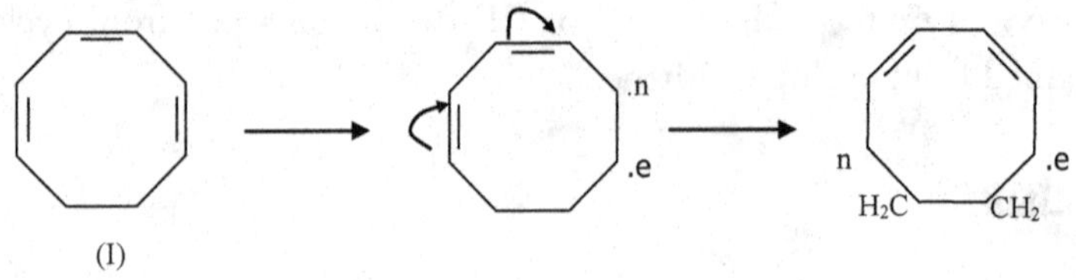

$$11.440$$

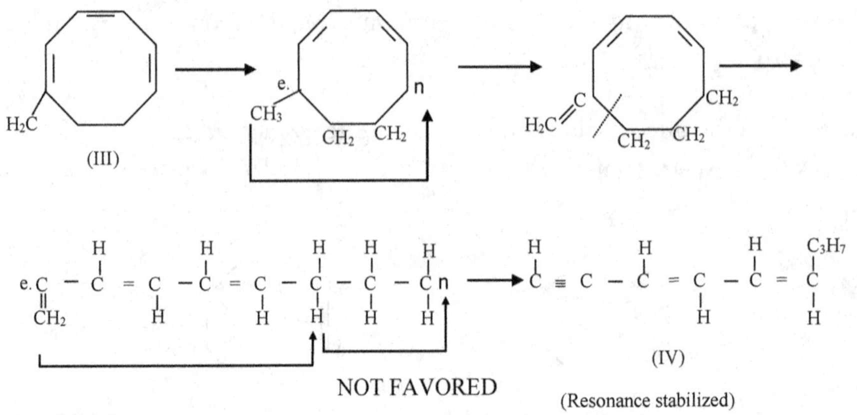

NOT FAVORED (Resonance stabilized)

$$11.441$$

The case above will not take place. It was done for explorative purpose. The reason is because of the capacity of the closed - $CH_2 - CH_2 -$ group, that which is greater than CH_3. When the CH_3 group is replaced with C_3H_7, the molecular rearrangement of the third kind would have been favored (up to C_4H_9 group). With nine-membered ring, the rearrangement of the first kind above will also not be favored. Molecular rearrangement of the third kind in which the CH_3 group will become internally located, will also not be favored.

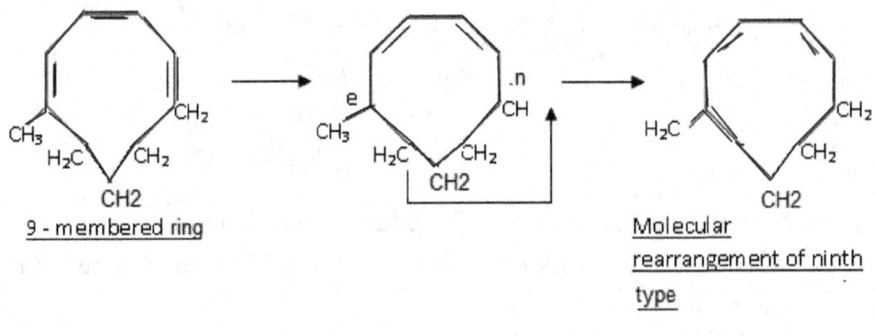

$$11.442$$

If molecular rearrangement of the first kind of the first type is however possible (CH_3 group is less in capacity than those in the ring) and there is MRSE in the ring, a more nucleophilic monomer than that of Equation 11.441 is obtained.

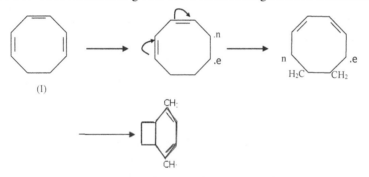

NOT FAVORED   11.443

With groups higher than C_5H_{11}, the rearrangement ceases. The case above is not favored, but exploratory. (I) may favor trans-annular addition to give four-membered ring fused to a six-membered cyclohexadiene.

11.444a

For (II) of Equation 11.438, the followings are obtained when activated.

(A) NOT FAVORED   11.444b

Internal trans-annular addition of the radicals above, will not take place, because the least nucleophilic center is the one isolatedly placed, wherein any of the centers can carry any radical. It can therefore molecularly rearrange to give 1,3,5- cyclooctatriene. However, when radical-pushing substituent group is introduced into the isolated part of the ring, the trans placement above becomes favored. The same applies when it is internally located in the conjugatedly placed double bond as shown below.

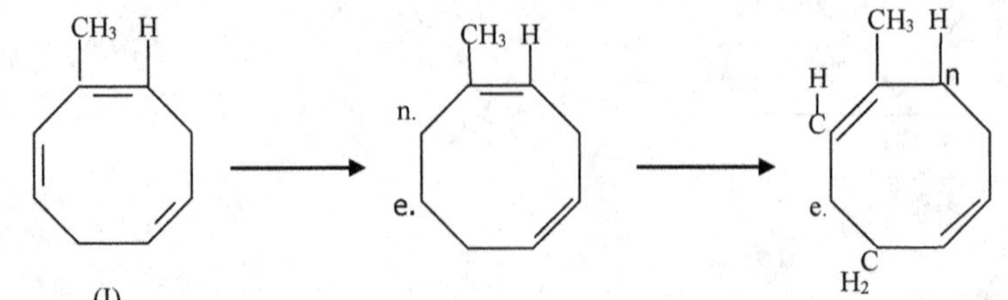

(I)   (I)a   11.445

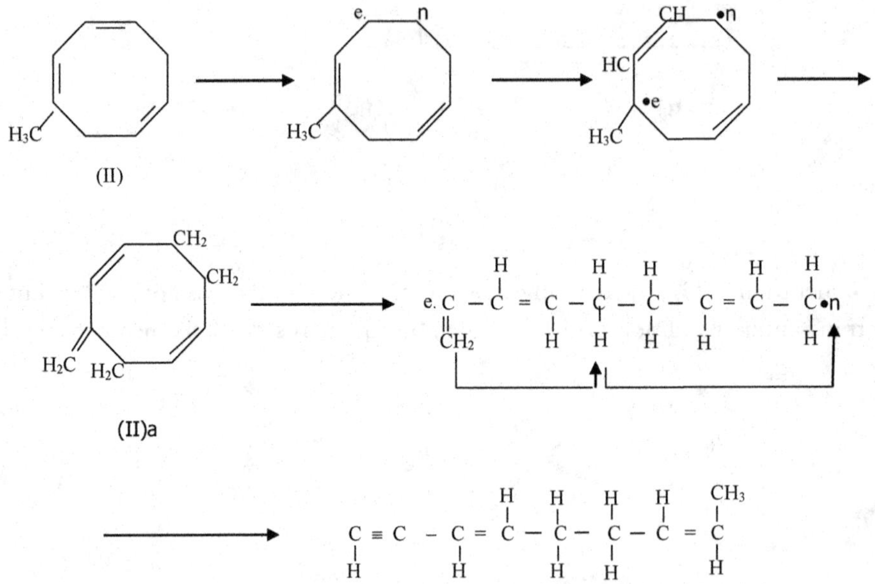

(II)

(II)a

$$C \equiv C - C = C - C - C - C = C$$

(II)b <u>Resonance stabilized</u>   **NOT FAVORED**   11.446

With radical-pushing group externally located to the conjugatedly placed double bond as shown above, molecular rearrangement of the third kind is favored for R group beginning from C_2H_5 to C_4H_9. Unfortunately however, the center used above is not the least nucleophilic center. The least nucleophilic center is the isolatedly placed double bond

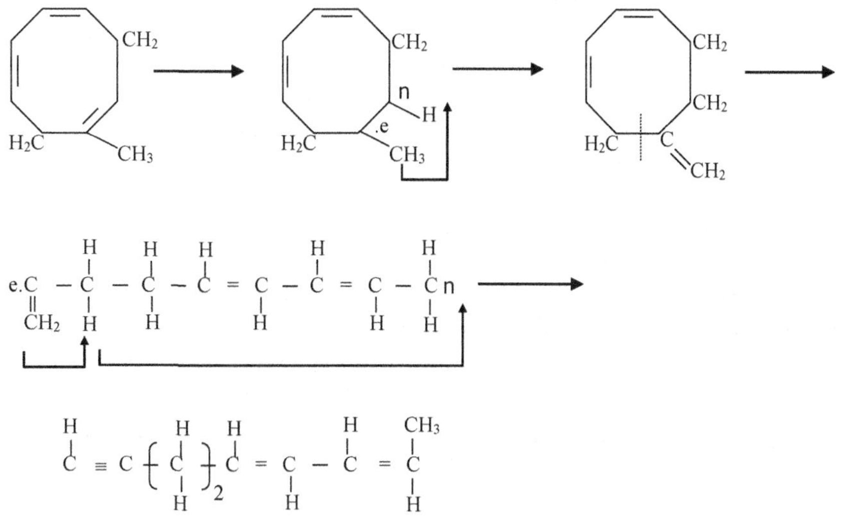

(III)b Resonance stabilized

FAVORED 11.447

If the center activated above is the least nucleophilic center, then the ring opening above is favored. At least, one should note the point of scission in the ring..

While there will be no need to continue with higher cyclotrienes, based on the foundation which has been laid so far, it is important to note that from cyclononatrienes upwards, the existence of the third kind of molecular rearrangement is very well favored when the right groups are well placed.

11.4 Cyclotetraenes

Like cyclotrienes where the most important member is benzene a six-membered ring, with the present family under consideration, one of the most important members is cycloocta-tetraene - an eight-membered ring. Smaller sized-rings do not exist, because they possess more than the Max.RSE.

11.4.1 Cylooctatraene

Cylooctatraene was originally prepared to determine whether its chemical properties would resemble those of benzene[62]. From the estimated resonance stabilization energy, it is far less stable than benzene. This was confirmed by the ready and complete rearrangement of the cyclooctatetrane to styrene. Comparing benzene and cyclooctatraene, by artificial activation of all the activation centers, which according to one of the rules is impossible in the presence of resonance stabilization, the followings are obtained.

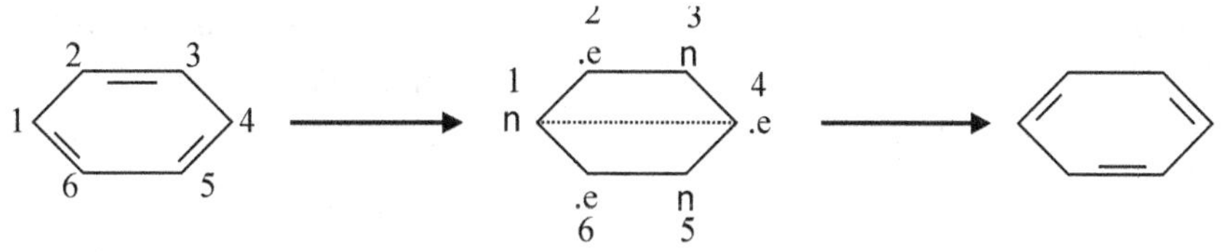

11.448

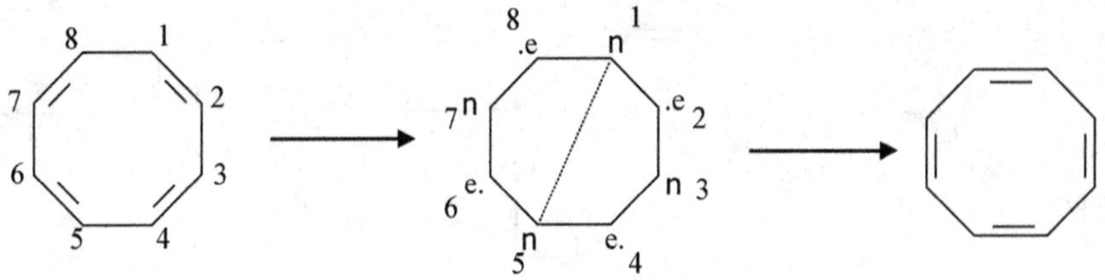

<u>Artificial Activations</u>   11.449

While in benzene the linearly externally located active centers (para-placement) carry opposite radicals, in cyclooctatetraene the linearly externally located active centers (1,5 - centers) cannot carry opposite radicals. Unlike cyclobutadiene (which favors transient or no existence), cyclo-octatetraene is weakly resonance stabilized in view of its size.

Therefore, when cyclooctatetraene is weakly or strongly activated, the followings are obtained.

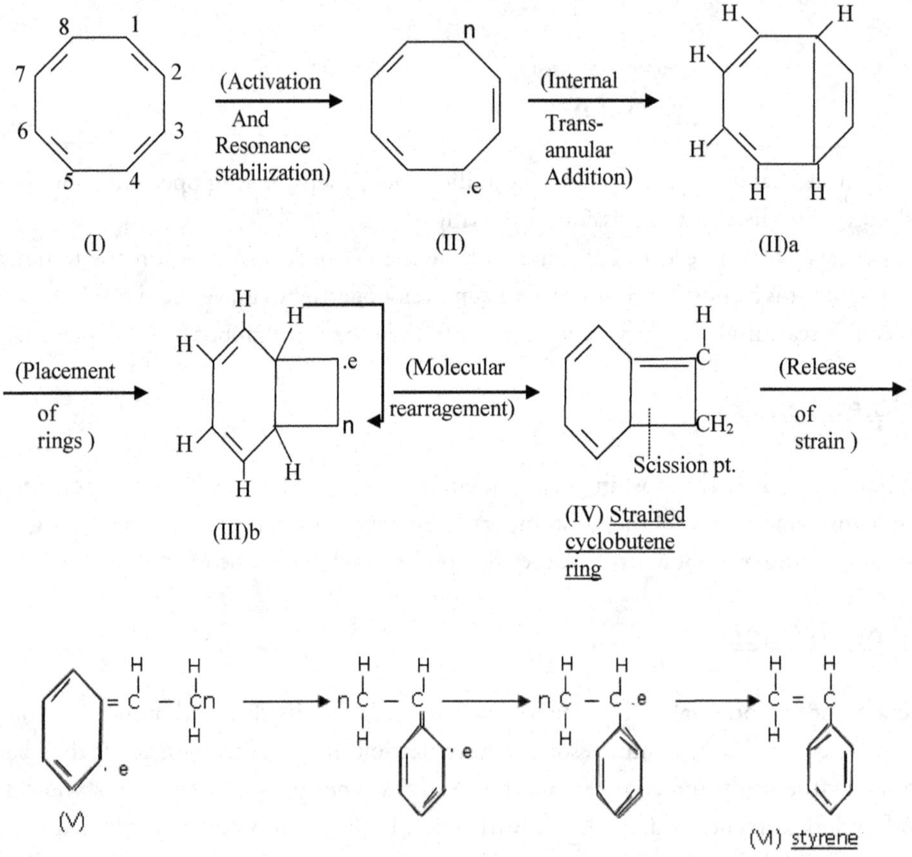

11.450

If (I) had favored full or continuous close-loop resonance stabilization, the existence of (II) would have been impossible. In order words, one can confidently say that complete continuous closed-loop resonance stabilization is impossible regardless the strength of initiators. Thus, one can clearly see a clear distinction between benzene (Continuous) and Cyclootatetraene (Discrete).

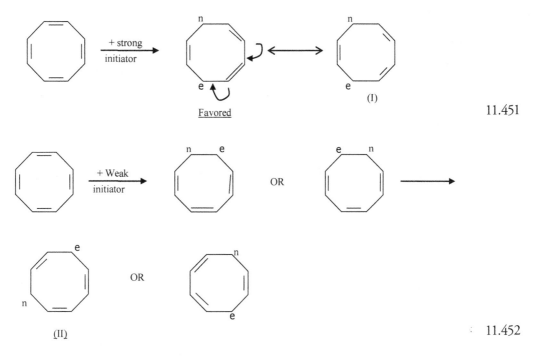

11.451

11.452

But, looking critically at the case of benzene, the following is obtained.

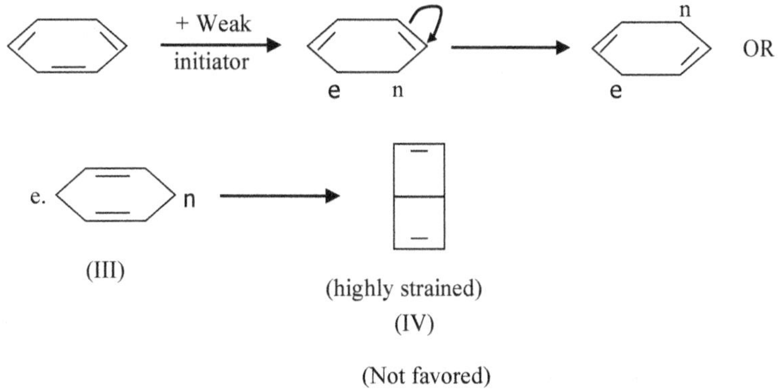

11.453

Note that (IV) cannot exist, since it is highly strained for its size. It is almost like cyclobutadiene, since a single bond is serving as a common boundary between two four membered cyclobutene rings. Since benzene if resonance stabilized may not favor the existence of (III) where there is para-placement of radicals on the active centers, whether the initiator is weak or strong, then the favored existence of (II) and non-favored existence of full movement has little to do with just the strength of the initiators, under the present circumstances, but with size. ***No matter how large the fully unsaturated ring may be and no matter how strong the initiator is, for very large fully unsaturated rings, there must be a limit to the transfer of electro-free-radicals during resonance stabilization phenomena. Hence the limit seems to end with six-membered rings above wherein trans-annular addition cannot take place for substituted and non-substituted fully unsaturated rings, regardless the conditions.*** The limit does not apply to linear tetraenes, pentaenes and higher, all opened-loop systems.

For example, consider a cyclodecapentaene shown below. This is a fully unsaturated ten-membered -ene ring.

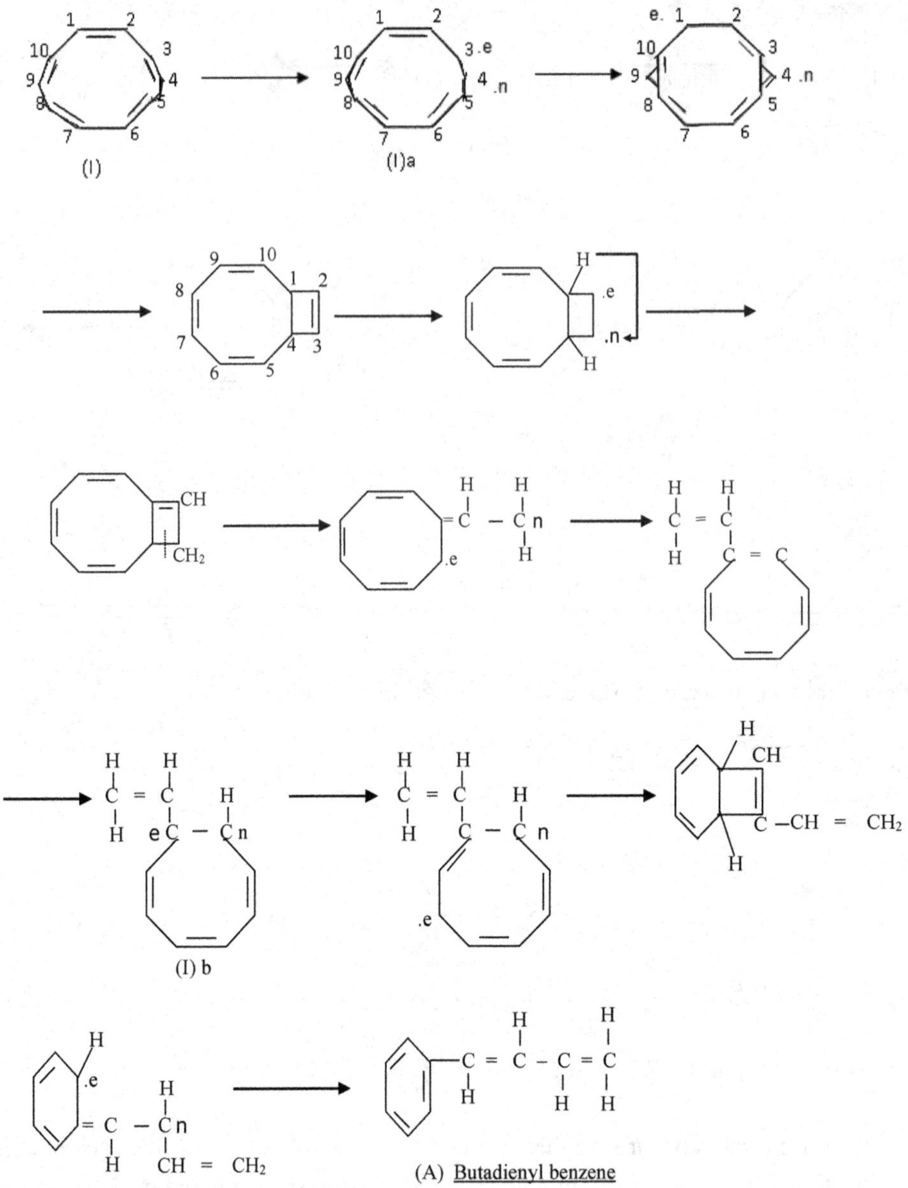

$$11.454$$

The octo- gave us styrene which is vinyl benzene and the deca- is giving us something new which is butadienyl benzene. Downstream, the name "Vinyl" will be changed to "Ethenyl", since vinyl, like acryl are specifically for something else, such as in vinyl chloride, vinyl alcohol methyl acrylate, acrolein, acrylic acid and so on. These are the LOGICS in Chemistry too much to comprehend. Existence of (A) is favored, but not via the route above, since the externally located activated center in (I)b will be the first to be activated, just like in styrene. If electro-free-radical had moved two steps instead of one, then (C) shown below would have been obtained. This is an important exercise.

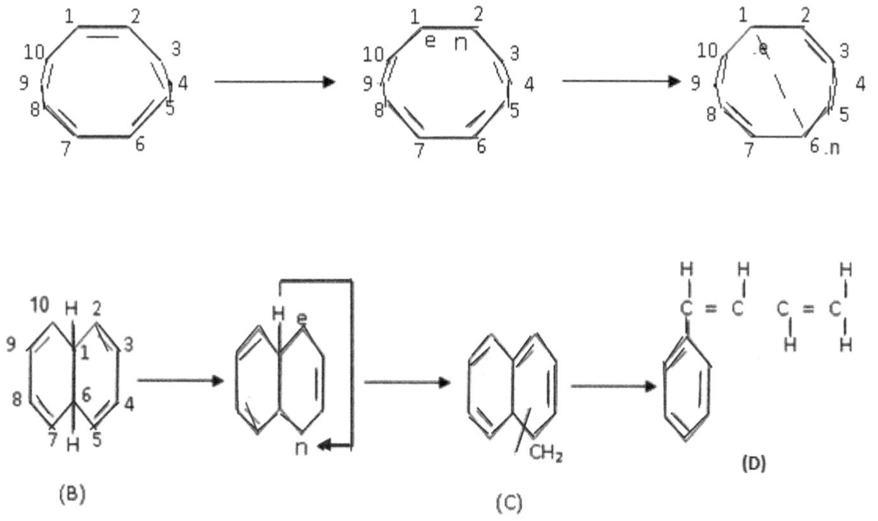

(B)   (C)

11.455

The same (A) [i.e., (D)] as above is obtained. Here, the electro-free-radical moved two steps instead one step as above. Hence the last mechanism is the favored one, ***since resonance stabilization in cyclopentaene is less than that in cyclotetraene, which in turn is less than that in cyclotriene and which in turn is less than that in cyclodiene that which has transient existence. While the electro-free-radical moved two steps in full cyclotriene, it moved the same two steps in full cyclotetraene and same two steps in full cyclopentaene.*** Thus, one can observe that the movement of visible-electro-free- or non-free-radicals is not limited by the size of the ring when resonance stabilized, but by how far it can go in cyclic system – only two steps – The Trinity. For everything in life, there are limitations, but not for the linear case which is a function of time and space. The limitation in the linear case has to do not only with space, i.e., the size of the linear chain or system for transmission of electro-radicals, but also with time. In cyclic systems, as the rings become very large, the influence of resonance stabilization is diminished when the double bonds are conjugatedly placed and is full. (D) being favored from (C) depends on the strength of the activating initiator and the temperature. Thus, not only is cyclooctatetraene unstable; so also is cyclodecapentaene and higher members, due to the limit placed by six-membered rings. When (B) is dehydrogenated free-radically, naphthalene is obtained.

Now, consider introducing one radical-pushing group into the ring beginning with CH_3 type.

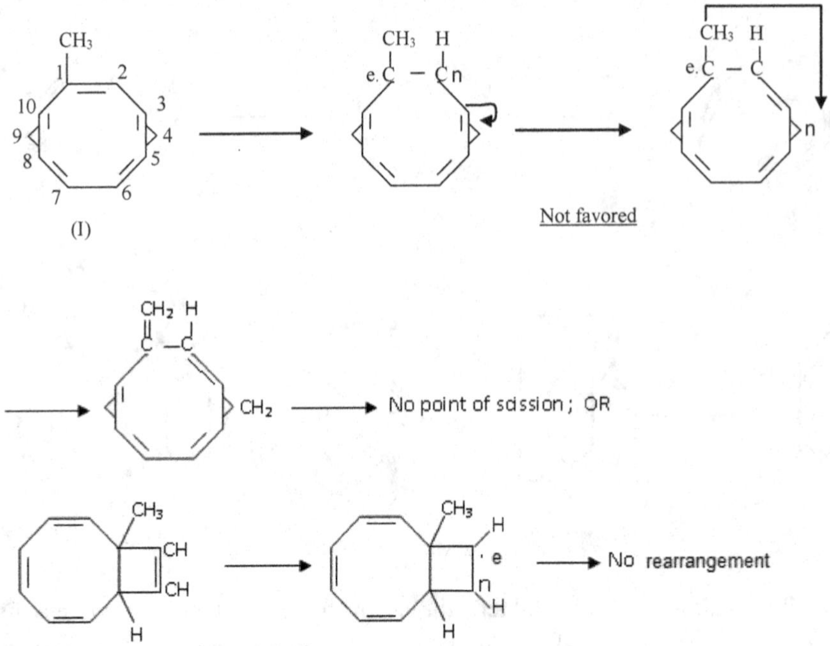

11.456

In the first case, the wrong center has been first activated above. Secondly, no transfer of transfer species can take place as shown above, since group of any size is resonance stabilized. Thirdly, the movement of the electro-free-radical when that center is activated in one step, cannot reduce the ring as shown above. As it seems, the least nucleophilic center is that behind it – the 9,10—center. Even when activated, no transfer species exist either from inside or outside. The cases above have as usual been shown, in order to show some basic fundamental principles, which have been stated into rules or laws. When that center is activated and resonance stabilized in two steps, the followings are obtained.

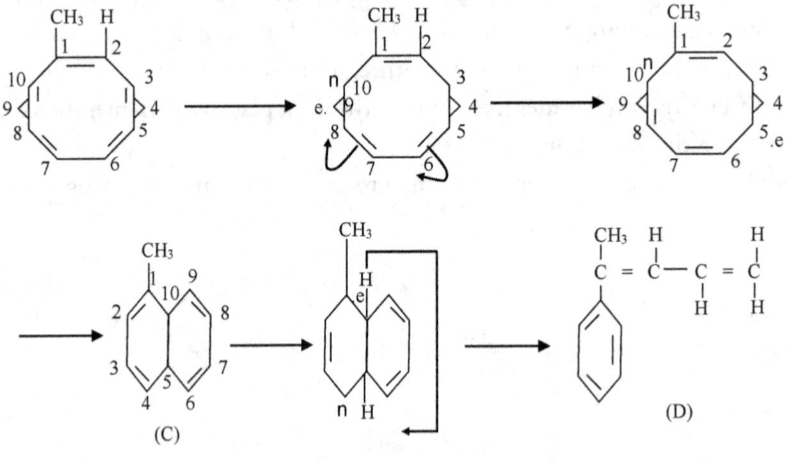

11.457

This is more favored than the case of Equation 11.455, because of the unique location of CH_3 group which remains shielded in (D). The presence of the CH_3 group where placed, has made the second six-membered ring to be more strained to demand its instantaneous opening, The larger the CH_3 type of group carried, the more it is to keep the second ring more strained and therefore be opened instantaneously.

Worthy of note as already said, is the activation center activated and the fact that in (D), the CH_3 group is shielded. Secondly note that the movement of electro-free-radicals cannot go beyond two steps after activation. Unlike the case in Equation 11.455, (C) above may not need additional higher operating conditions to open its second ring. If CH_3 is replaced with OR NR_2 types of groups, no additional driving force will be required to open the ring instantaneously, because of the strong radical-pushing capacities of these groups. From the already established traditional order, the follow-ing is valid.

Cyclopentaenes > Cyclotetraenes > Cyclotrienes > Cyclodienes > etc.

Order of Nucleophilicity

11.458

Based on this order and present analysis, it is important to note that styrene and related monomers are not truly olefinic in family classification. They are far more nucleophilic than olefinic monomers. That is, the following order is obvious.

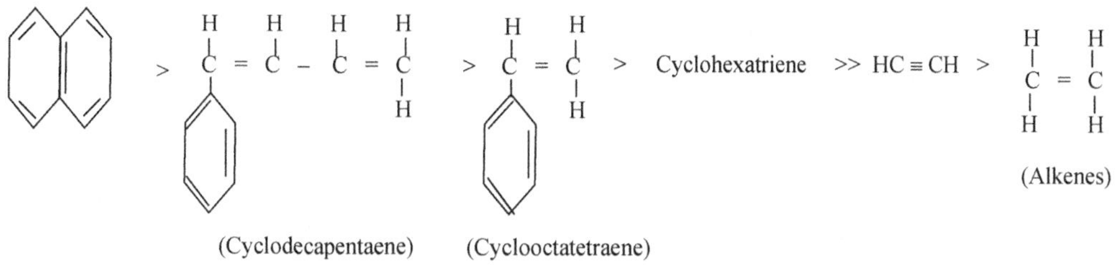

(Cyclodecapentaene)   (Cyclooctatetraene)   (Alkenes)

Order of Nucleophilicity

11.459

While the orders above seem to be the real order for their monomers, the situation is different for their corresponding substituent or substituted groups, when the resonance stabilization is of the continuous closed-loop type.

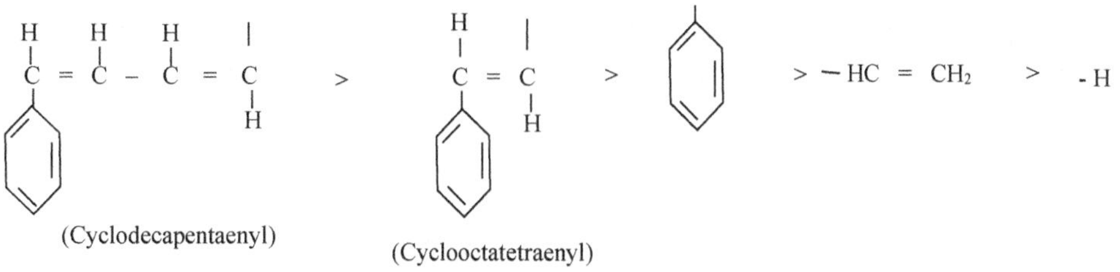

(Cyclodecapentaenyl)   (Cyclooctatetraenyl)

Order of radical-pushng capacity of resonance stabilization groups

11.460

The order remains the same as have been established so far with non-continuous closed loop resonance stabilization groups.

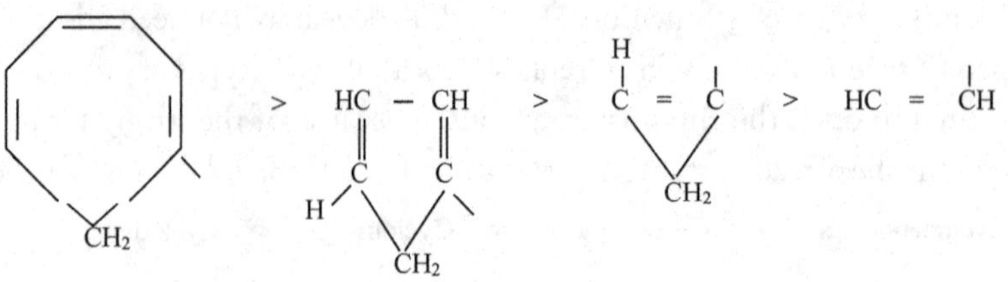

Order of radical-pushing capacity of substituent groups   11.461

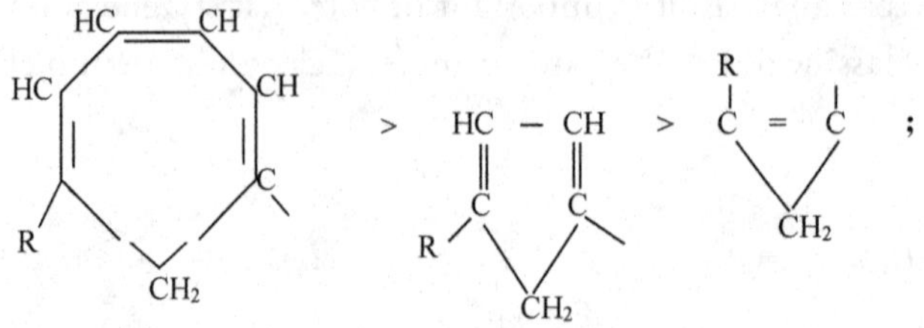

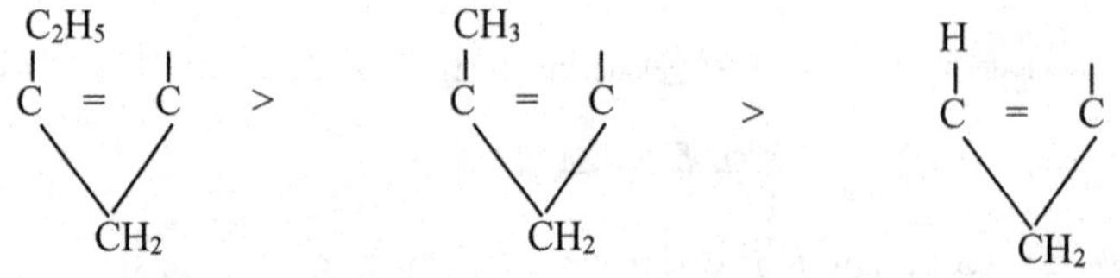

Order of radical-pushing capacity of substituent groups   11.462

The orders above are obvious from all the considerations so far.

Now, consider the use of radical-pushing groups of lower capacity than H, on the eight-membered rings.

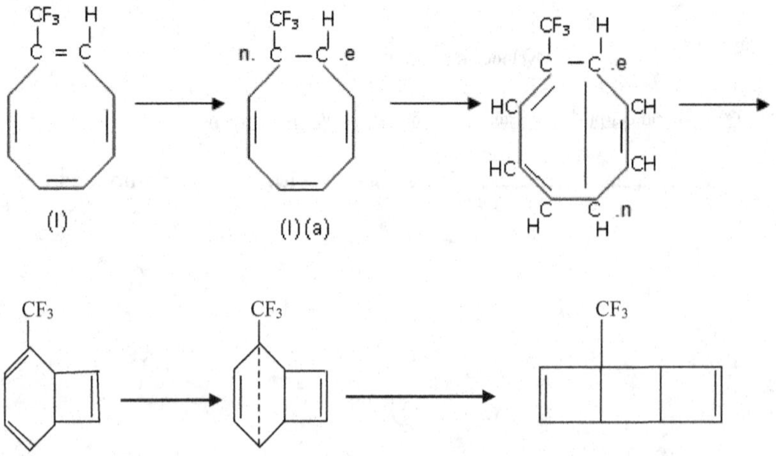

NOT FAVORED   11.463

OR

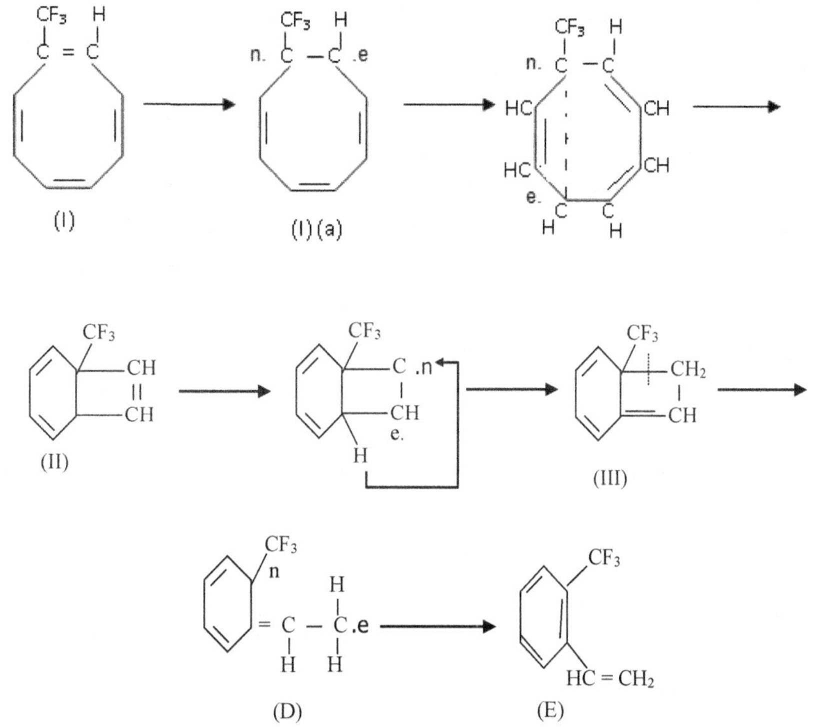

(I)   (I)(a)

(II)   (III)

(D)   (E)   11.464

In resonance stabilization, it is the visible electro-free- or non-free-radical that first moves when there is a π-bond adjacently placed. When there is no π-bond, then the electro-free- or non-free-radical from a π-bond migrates to grab a visible nucleo-radical adjacently placed. Hence, it is the second case above that is favored to produce (E). These are important to note, because in some cases, whichever way you move, the same product is obtained, while in other cases, the same product cannot be obtained. All these movements under different operating conditions, are made possible, because NATURE abhors a Vacuum, Non-linearity, differentiation and many others.

Imagine if the wrong activation center had been activated, the following would have been obtained.

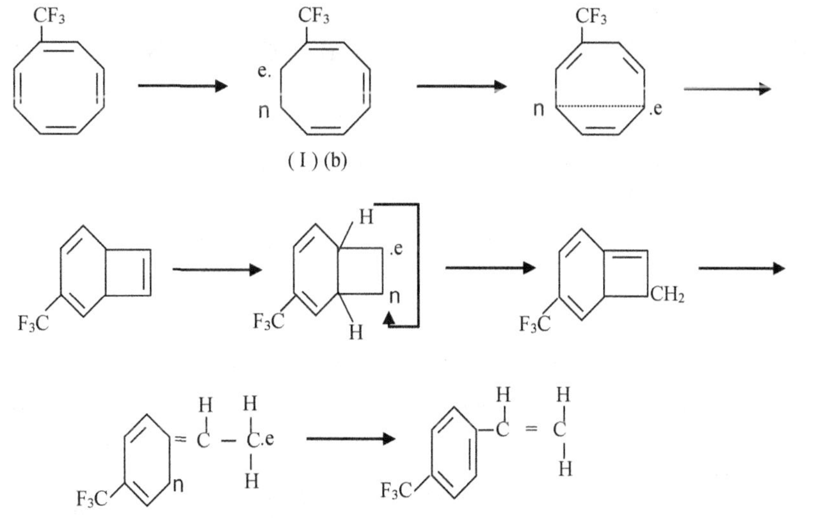

(I)(b)

NOT FAVORED   11.465

Knowing the activation center first activated is very important. Is it that in (I)a or (I)(b) or any other? It can never be any center other than (I)a, being the least nucleophilic center. There are remarkable differences between opened and closed-loop continuous or discrete systems in terms of characters radically and not chargedly. Countless world data or research works are required in order to establish the natural order. When one radical-pushing group replaces one H atom, the followings are obtained.

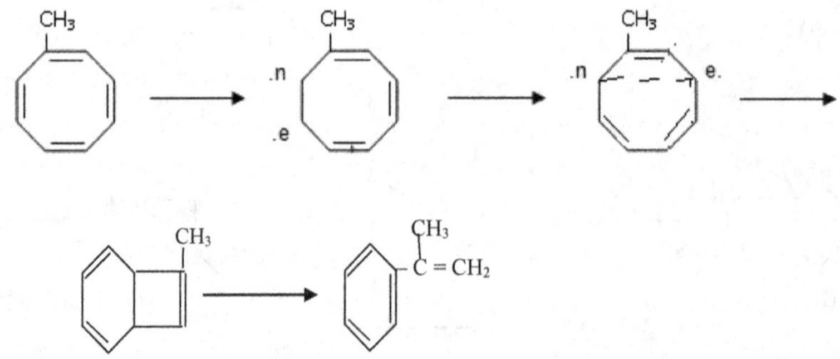

11.466

Above with or without CH_3 group, only two steps can be taken . If a stronger radical-pushing group such as NR_2 is put in place of CH_3, the same two steps will be taken. With the two steps taken above, 1-methyl styrene is the product. If a wrong center had been activated, the followings would have been obtained if only one step is taken.

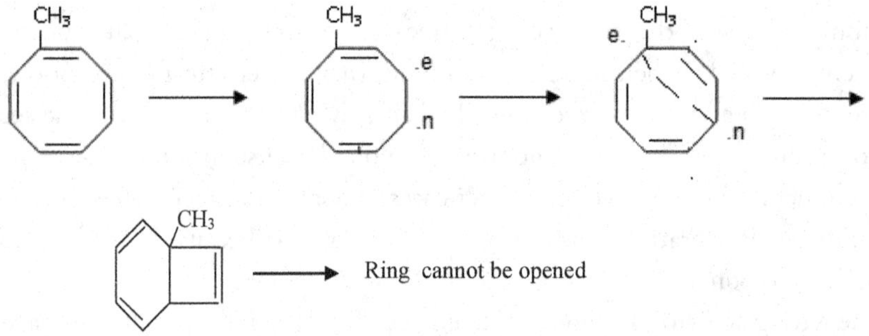

Ring cannot be opened

11.467

If two steps were taken, the ring would have been opened to give a different styrene type of product. With CFH_2 which is more radical-pushing than H but less than CH_3, the following is obtained.

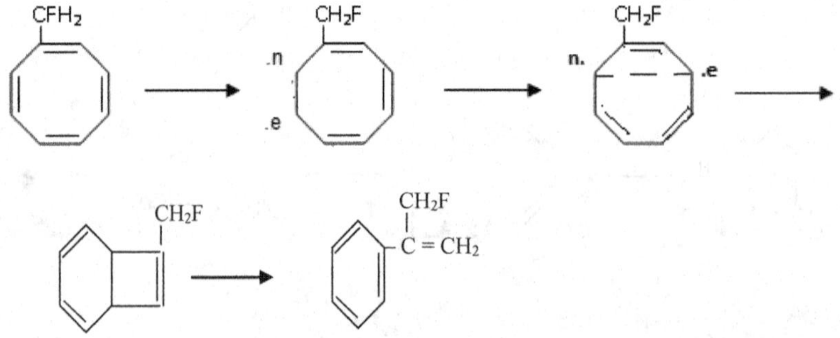

11.468

Thus, all these reactions can only be free-radical in character. Free-radically, internal trans-annular addition and rearrangement are possible with or without any possibility of the four-membered ring being opened. While CH_2X is allylic in character, CH_3 is not, where X is for example Cl, but not CHF_2 or

374

CH_3. Without application of already proposed rules in a surgical and ordered manner, all these unique observations and revelations would not have been made possible.

Because of the size of cyclooctatetraene, cyclodecapentaene, compared to the existing first member cyclohexatriene (Benzene) [Not cyclobutadiene which favors transient existence at STP], hence the order of their ability to favor Equilibrium state of existence is as follows-

[Cyclobuta-di-ene] > Cyclohexa-tri-ene > Cycloocta-tetra-ene > Cyclodeca-penta-ene >

Cycloduodeca-hexa-ene >

ORDER OF ABILITY TO EXIST IN EQUILIBRIUM STATE OF EXISTENCE   11.469

It is only benzene that favors continuous closed-loop resonance stabilization in this family. This seems to be one of the origins of its aromaticity. Though cyclooctatetraene is known not to be aromatic, **its so-called dianion $C_8H_8^{-2}$ called cyclooctatetraenide is aromatic.** It is said to be so-called, because the carbon center cannot carry an anion. It can carry a negative covalent charge, but not under this condition. What it is carrying are electro-free-radicals as shown below.

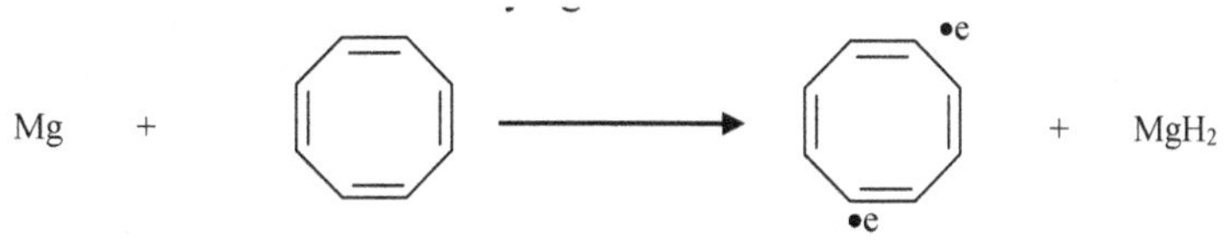

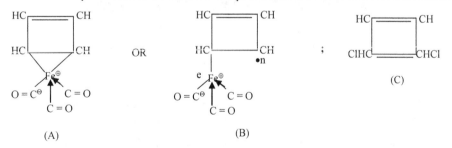

Cyclooctatetra-dienide   11.470

With two electro-free-radicals placed on the ring, resonance stabilization in the closed system becomes continuous and no longer discrete. The two radicals do the movement at the same time to go round once and change their positions. Hence, as it seems, it shows the aromatic character. One should therefore not expect cyclobutadiene to show some aromatic character if it can be fully isolated. In fact, cyclobutadiene-iron tricarbonyl, has been identified to be "aromatic" in character in the sense that it undergoes facile attack by electrophilic reagents to produce monosubstituted cyclobutadiene-iron tricarbonyl complexes[63]. Functional groups in the substituents are said to display many of their normal chemical reactions which can be used to prepare further types of substituted cyclobutadiene-iron tricarbonyl complexes. Even up to date the structure of the cyclobutadiene-iron tricarbonyl is unknown. Shown below is a tentative structure.

Cyclobutadiene-iron tricarbonyl   11.471

Although there are two holes on the iron center, (A) is not resonance stabilized. It is (B) that is more readily favored, because of the SE in the three-membered ring of (A). With the presence of the nucleo-free-radical on the C center, (B) is not still discretely or continuously resonance stabilized. Nevertheless, since Fe is a part of the three-membered ring, it is less strained than when C had been there. Hence as it seems, it is (A) that is close to the structure if $Fe(CO)_5$ was used. Cyclobutadiene-iron tricarbonyl is used in Organic Chemistry as a precursor for cyclobutadiene. It was first synthesized by Rowland Pettit starting from cyclooctatetraene[64-66]. The method of synthesis is important to attend to. The cyclooctatetraene is first chlorinated to the [4.2.0]-bicyclic compound which reacts further with the alkyne dimethyl acetylenedicarboxylate in a so-called Diels-Alder (DA) reaction followed by a reverse –DA reaction by pyrolysis at 200^0C releasing cis-dichlorocyclobutene (C) shown above and below. The (C) reacts with di-iron nonacarbonyl (obtained from photolysis of iron penta- carbonyl) to **Cyclobutadiene iron tricarbonyl.**

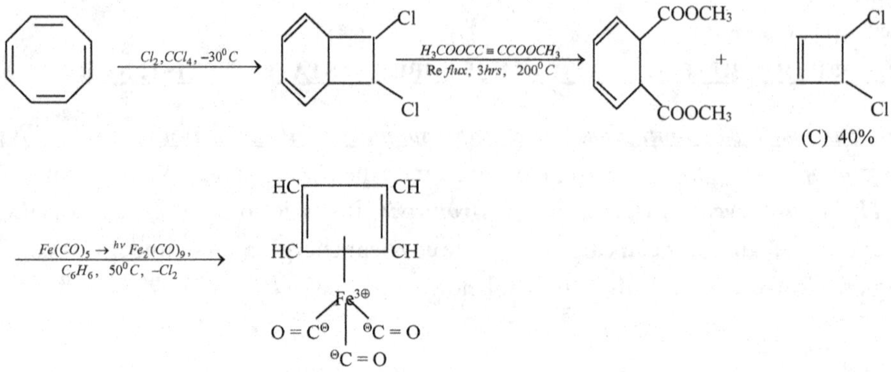

(C) 40%

(A) 45%

11.472

Looking at the equation above, though yet not fully correct as an ART, without it, one cannot understand how NATURE operates. It has been an enormous task providing such details. However, the product said to be obtained (A), whose structure is incorrect, is not the product. The real product is that shown below.

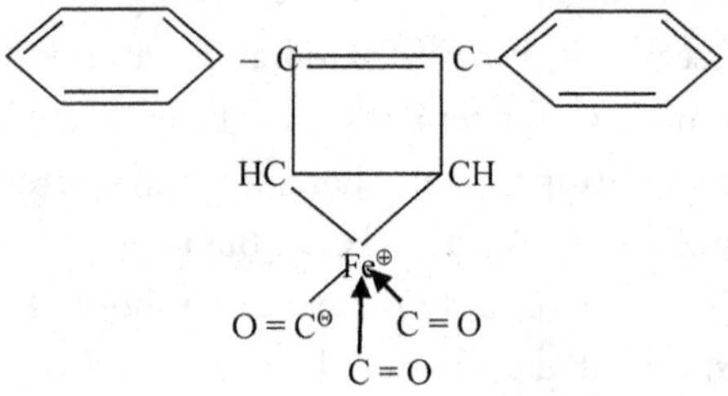

(B) Cyclo-Di-phenyl-butene-iron tricarbonyl

11.473

Looking at Equation 11.472 which is an Equilibrium mechanism system involving many stages with the eye of the needle, without Benzene, the Cl atoms in (C) cannot be easily removed. When they were removed, HCl was formed with phenyl group placed on the ring in two or more stages. Cl_2 was not the product said to be removed as shown in the equation. Note how the reaction began with cyclooctatetraene in the presence of Cl_2 and carbon tetrachloride at such low operating conditions well identified with temperature (-30^0C). We should ask ourselves, why the presence of CCl_4? Is it there as a "Policeman" or a "Security Guard" to guide the security or actions of Cl_2? It does not partake in the reactions. It is therefore a passive catalyst to do what? It helps to keep Cl_2 very readily in Equilibrium state of existence, because the policeman (CCl_4) is carrying four Cl atoms. In fact, it forces the cyclooctatetraene to rearrange to the point where styrene was not allowed to exist. Instead of undergoing molecular rearrangement of the first kind in order to create a point of scission, the policeman prevented it and let Cl add to the four-membered ring after its activation on both C centers. With the presence of Cl in the four–membered ring which has been de-unsaturated (saturated), the SE in the four-membered ring was greatly reduced. With a deep explanation of what was provided above by asking questions, this was followed by other stages bringing into play all disciplines as was just slightly done above. The first things to note are the operating conditions particularly with respect to temperature and concentrations of the reacting components. Imagine what was added to the fused six- and chlorinated four- membered rings to form (C).

It is because of the presence of the phenyl groups connected to the four-membered ring, hence it was observed to be aromatic, the aromatic character being provided by any of the phenyl groups in (B) of Equation 11.473. Its display of aromaticity as said to be evidenced by some of its reactions classified as electrophilic aromatic substitution[67], are indeed correct (aromatic) taking place not on the four-membered ring as was thought to be the case, but on the benzene ring of (B) of Equation 11.473. When cyclobutadienes are said to exist, it is not in the form of cyclobutadiene, but in the form of cyclobutene fused to a three- (with $Fe(CO)_5$) or four-(with $Fe_2(CO)_9$) membered ring. The presence of some specific metals in place of C as central atoms in ringed compounds, greatly helps to reduce the SE in their rings. This is what we see in the building blocks of living systems which essentially contain ringed compounds which cannot be opened. ***Thus, one of the most important variables for aromaticity, is presence of continuous closed–loop resonance stabilization system, that wherein when activated, in the first movement the electro-free-radical takes two steps and form a bond to become dead with the three π-bonds now placed differently; followed by next activation and second movement of the electro-free-radical in two steps to form a bond to become dead with the three π-bonds now placed to where they were at the beginning. The cycle of the two by two movement and transformation continues until the operating conditions are changed negatively, for example reducing the temperature.***

The structures of the iron carbonyls like many other compounds are either unknown or wrongly represented. Shown below are the real structures of the iron carbonyls in the reactions of Equation 11.473.

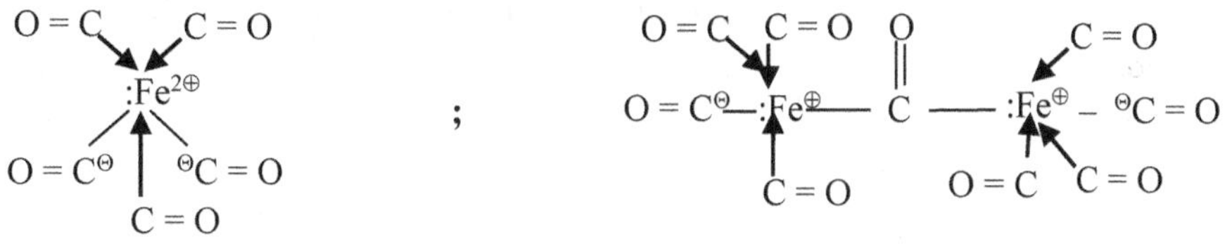

Iron pentacarbonyl   Di-iron nonacarbonyl   11.474

When structures are not properly shown or are unknown, then the situation is worse than gambling, because the situation is a blind one leading to a great fall into an endless pit. Even the structure of carbon monoxide when shown as shown universally is incomplete and meaningless. First show the real structure; from which after that, one can now abbreviate it showing its full identity. How can one show the structure of CO without the paired unbonded radicals and vacant orbital (the killer source of living systems) carried by the carbon center? The ALMIGTHY INFINITE GOD did not put for example paired unbonded radicals and vacant orbitals on the atoms that carry them there for nothing. ***Nothing exists here for nothing positively and negatively.***

When cyclooctatetraene is made to exist in Equilibrium state of existence, it can be made to undergo Step polymerization like the case of Benzene in Equations 11.382a to 11.387. However, never is it possible to polymerize it without opening the ring unlike the case of Benzene, because apart from steric limitations, there is nothing like meta-, ortho- and para- positions concepts with eight- membered rings and higher. Steric limitations should not be a limitation, since horizontal, diagonal, vertical or indeed zigzag placements all exist. Also another probable reason is as a result of what has been shown in Equations 11.448 and 11.449. ***When the rings are activated from one activation center, the movement of electro-free-radical is such that trans- placement of electro-free- and nucleo-free- radicals is only possible with benzene and not possible for cycloocta-tetraene. When not trans-placed, then***

the role of steric limitations becomes meaningful for a ring with more than one activation center.
All reported cases of the polymerization of cyclooctatetraene involve ring openings[60,68-70] of their rings.
Where only cyclooctatetraene is involved[68,69], the method of ring opening is different from when the ring
is carrying a radical-pushing group[60,70]. One is using at this point in time the eye of the needle to classify
them without going through all the works. The reason is so clear- the foundations laid so far. The need
for the necessity to bring orderliness is because all of the different methods in terms of Addition or Step
polymerization are said to involve so-called abbreviated standard method – ROMP that is Ring Opening
Metathesis Polymerization. The word Metathesis in which charges are involved –Diels-Alder reaction is
there to create a state of confusion, insanity and disorderliness. They are all meaningless.

In one method[68], a tungsten carbene catalyst was used to open the ring, a situation which cannot
be instantaneous because there is no point of scission in the ring. Did the carbene when activated add
to the ring to expand it before opening? No because when the ring is expanded, no point of scission is
created in the ring in the process. The product obtained is said to be a highly lustrous silver film and the
mechanism as usual-Oxidative..... Invariably, everybody follows the line which is not thin but wide. The
catalyst is nothing else than a passive one still doing something. Instead of the cyclooctatetraene remaining
in its Stable state of existence, it keeps it in the Equilibrium state of existence, since in the Activated
state of existence, nothing can be done. The tungsten carbene catalyst has the ability of keeping the
cyclooctatetraene in Equilibrium or Activated state of existence, because of what it is carrying. Hence, its
polymerization is similar to that of benzene of Equations 11.382a to 11.387 which is not only free-radical
in character, but also Step addition of monomers along the chain, since small molecular byproducts are
obtained.

When two undergraduate experiments in Organic Polymers were reportedly carried out on such
ringed compounds[69], it was more of a shock, because both the teacher and the students know very
little of what was being done. They plate polymerized the monomer to give polyacetylene. This again is
similar to the case above. For them, the monomer cyclooctatetraene is forced to be kept in Equilibrium
state of existence, something which could not be done when radical-pushing groups start appearing on
the ring. When a radical-pushing group is now placed on the ring[60,70], the ring not only becomes more
strained, but also more placed in Stable state of existence, the only state in which Addition polymerization
via Combination mechanism becomes possible. This was how the polyacetylenes substituted, on the
average, every eight carbon atoms were obtained. It is not by ROMP method, but by the use of a very
strong initiator which will not allow for movement of the electro-free-radical after activation. In fact
cyclooctatetraene will more readily favor being polymerized via Addition polymerization than Benzene,
because of the continuous closed-loop resonance stabilization favored by it, unlike with cyclooctatetraene.
However, it has been used in Equation 11.422 to show exactly how cyclooctatetraene can be polymerized
in Addition polymerization which takes place via Combination mechanism.

11.4.2 Cyclononatetraene

Only one type exists- that in which the four double bonds are conjugatedly placed. The ring has more
than the MRSE, but less than the MaxRSE, with no point of scission on the ring. Unlike cyclooctatetraene,
it is stable when activated, since it cannot produce styrene derivatives as shown below, in the presence of
a weak coordination initiator.

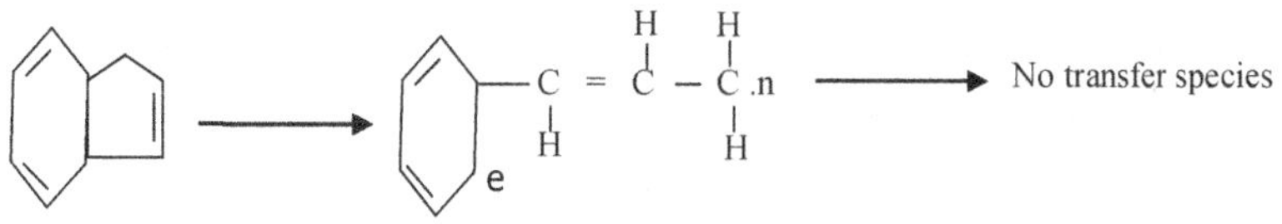

11.475

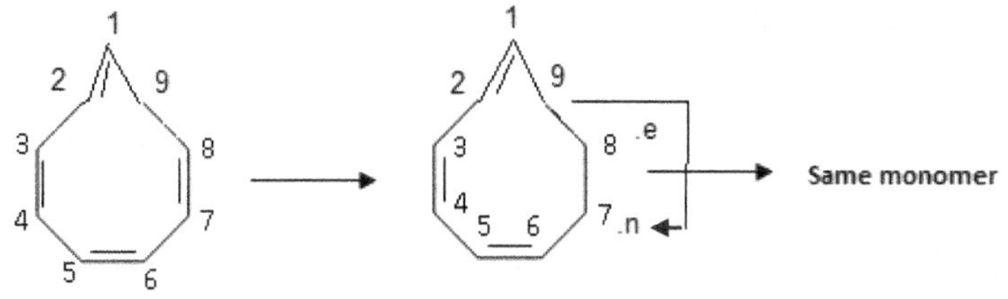

11.476

Very shockingly, the ring cannot be opened unless when the compound is made to exist in Equilibrium State of Existence. Like cyclopropene, cyclopentadiene, cycloheptatriene, the CH_2 group in the ring is still of greater capacity than any R group but more than those in the smaller ringed members of this family (See Equation 11.178).

Introducing radical-pushing groups, the followings are obtained when resonance stabilized with rearrangement.

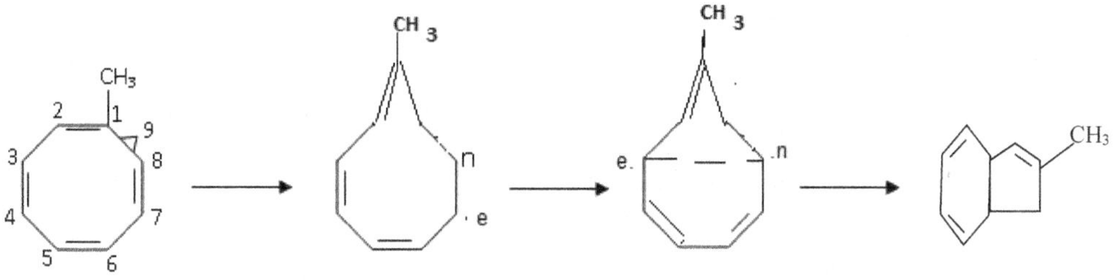

11.477

Shown below is another possibility when the wrong activation center is involved. This is the center which is the most nucleophilic, unlike the case above.

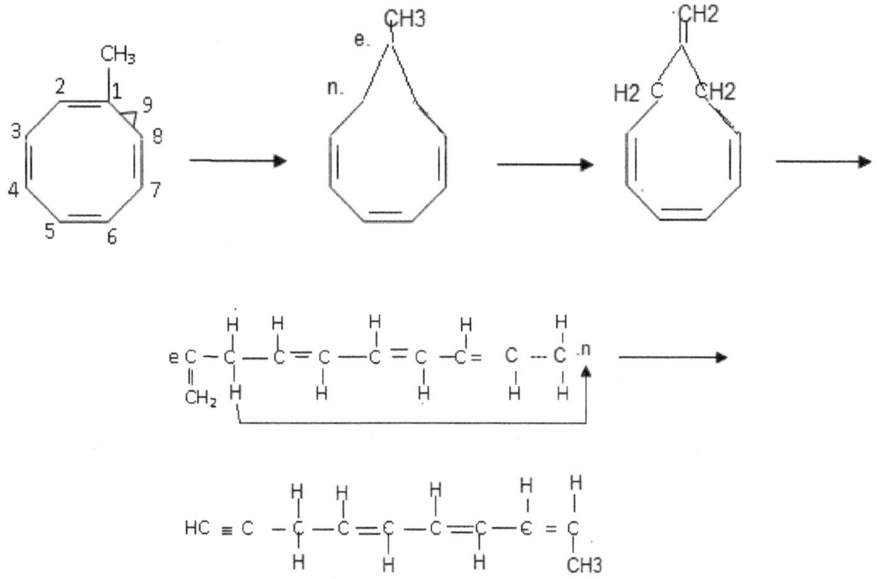

NOT FAVORED

11.478

Apart from activating the wrong center, the wrong transfer species was involved. However, exploratively, one can identify one of its isomers existing or not.

Imagine the case where the CH$_3$ group is internally located, the followings are obtained.

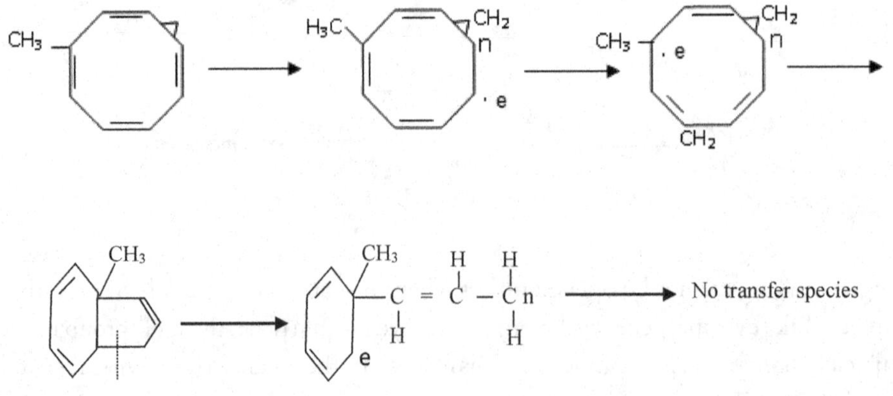

$$11.479$$

When the CH$_3$ group is internally located, no rearrangements are favored. Based on the product obtained in the analysis above, the following order is obvious for the members of the family.

Cyclooctatetraene   <   Cyclononatetraene   <   Cyclodecatetraene   <

Order of nucleophilicity

$$11.480$$

In general, one can observe the unique qualities of cycloalkanes, cycloalkenes, cyclodienes, cyclotrienes and cyclotetratraenes, as far as resonance stabilization and molecular rearrangements phenomena are concerned. These have been very complete departures from what have been known for so many years, even to the beginning of development of science or beginning of our time.

11.5 Multiple Condensed Cyclic Monomers

Finally considering specific monomers with several rings, two cases will be used for study -norbornene and a - pinene.

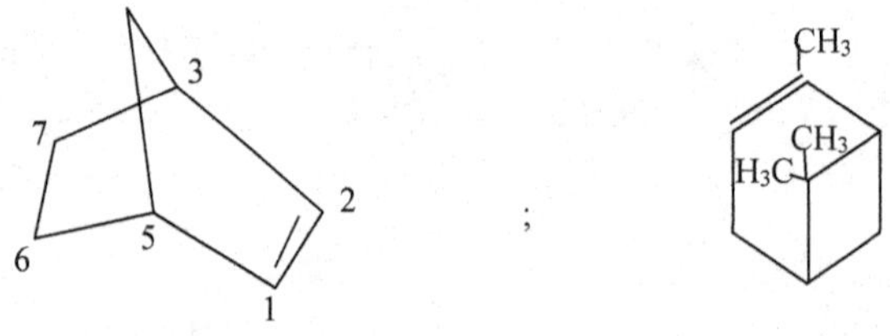

Norbornene   ;   α - Pinene

$$11.481$$

Norbornene is said to favor the following reaction[71]. This is just one of the types favored by it, based on the operating conditions.

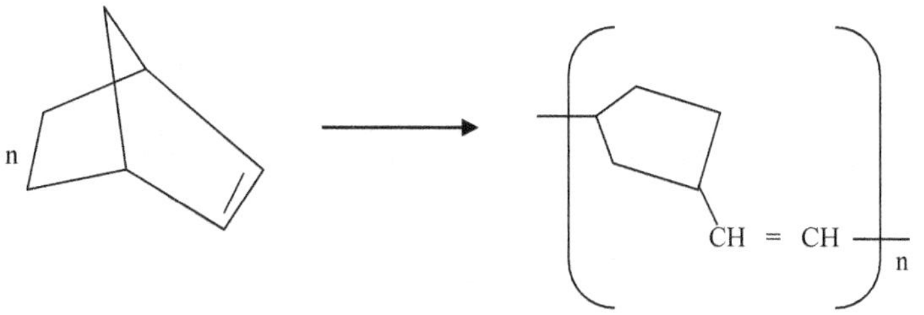

11.482

By the nature of the polymeric product, this looks like scission of 1,5- or 2,3- carbon - carbon single bond, points which are no scission points for instantaneous opening of the ring when strained. In the monomer, one will think that there are two five-membered rings in which one is more strained, in view of presence of a double-bond and two sides commonly shared by them. No, there is a cyclohexene ring fused to cyclopentane which externally "saturates" it with three common boundaries. The "saturation" is done on a cyclcohexene ring by methylene, just as acetylene did on cyclohexadiene to give Norbornadiene from which norbornene was obtained by hydrogenation. Scission cannot take place from any of the common boundaries 3,4- or 4,5- carbon-carbon single bond, or from the common boundaries 3,7- or 5,6- carbon-carbon single bond. These points of scission do not favor the existence of the observed monomer unit along the chain. The reason for the impossible use of any of these points, is because with the saturation, the original cyclohexene ring has been made in such a way that the overall SE has been greatly reduced, making it more difficult to provide the MRSE for the system to open the ring instantaneously when there still exist a better possibility. If none had existed, then NATURE will use one of the identical points of scission- 3,7- or 5,6- carbon-carbon single bond. Points of scission with maximum radical potential difference can be provided as shown below under specific operating conditions.

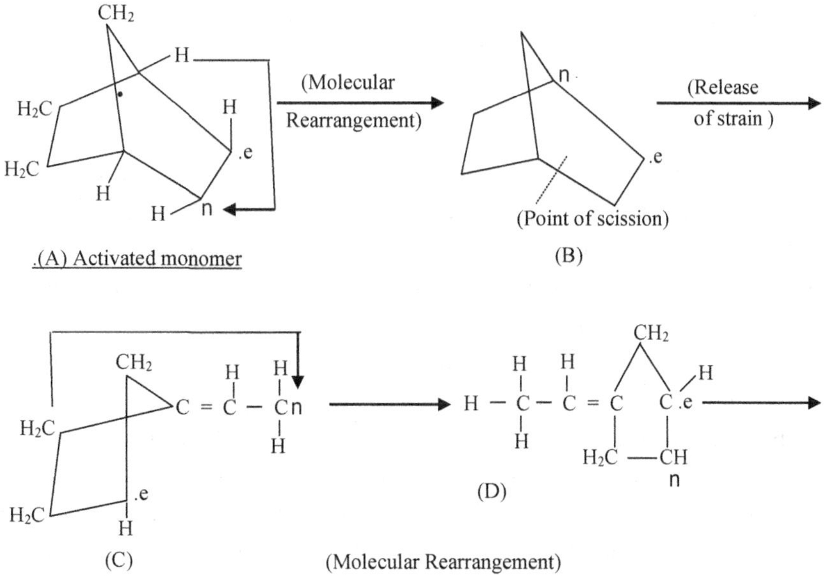

.(A) Activated monomer

(B)

(C)

(D)

(Molecular Rearrangement)

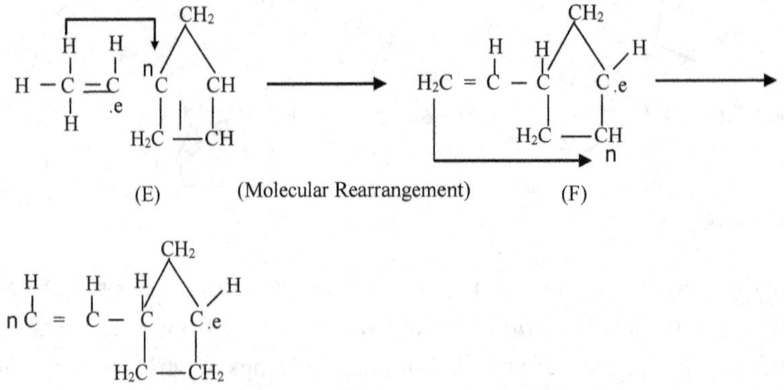

(E)   (Molecular Rearrangement)   (F)

(G)   11.483

It is after molecular rearrangement of different types that an adequate point of scission was provided. Only weak positively charged (because of absence of resonance stabilization) and in particular electro-free- radicals coordination initiators can be used for the polymerization of the monomer, for which for their growing polymer chain, the followings are obtained.

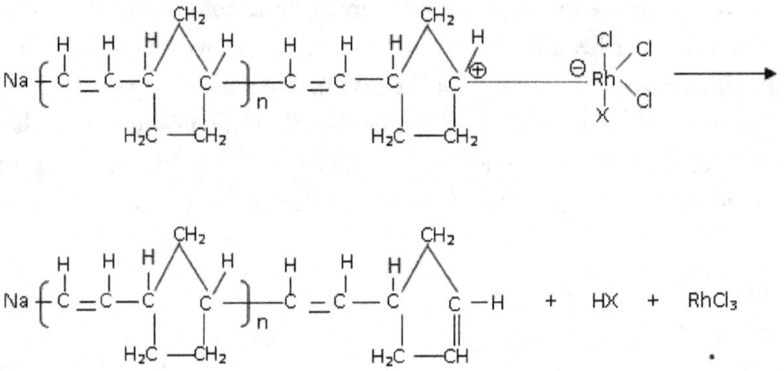

11.484

It is important to note that the growing polymer chain above is the same as that of Equation 11.482 reported in the literature[71]. It is (G) of Equation 11.483 that is favored, which can be observed not to be resonance stabilized. This has in the past been said to be one of the methods of polymerization called ROMP, Ring Opening Metathesis Polymerization. The second method called the Vinyl type of polymerization[72-75], is that in which molecular rearrangement was not allowed to take place at all, clear indication of the use of a very strong initiator. The monomer unit is (A) of Equation 11.483. Alternating placement of the monomer (a Female) with radical-pushing group via the use of that center (the so-called vinyl center) with maleic anhydride (a Male) has in the past been reported[76]. It is the female that diffuses to the male to form a couple which forms the alternating monomer unit along the chain and the initiator used is that which is unreactive to each one of them or at least one of them and these are nucleo-free-radicals. or free negatively charged centers. The third method of polymerization is that in which there was time to allow only one molecular rearrangement step to take place as shown in Equation 11.483 to give (B). No scission was allowed to take place. That is the monomer unit along the growing polymer chain at the beginning, next to be followed by (A). That monomer unit (B), has been wrongly represented in the literature[73,77] just like many others. How can there be a monomer unit carrying the only single π-bond when none of the rings has been opened? Such methods were classified as Cationic polymerization method[73].

Looking critically at Equation 11.483, one should indeed have expected (D) which looks strained to instantaneously open if temperature had been higher as shown below to give another type of monomer unit.

382

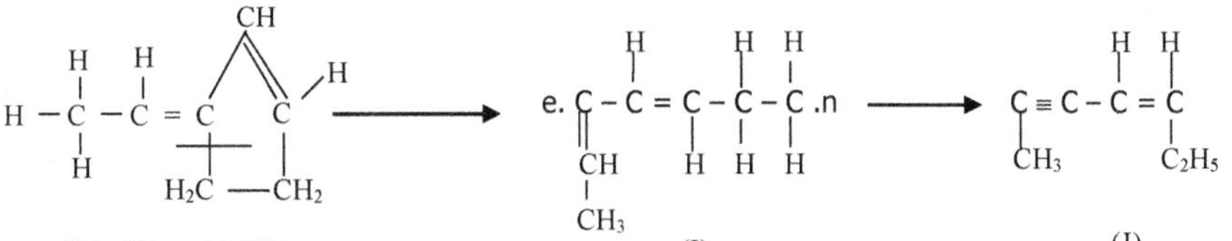

(D) of Eqn. 11.483   (I)   (J)   11.485

When radical-pushing substituent groups are present on the ring, (I) above will be the monomer unit along the chain if molecular rearrangement of the third kind is not alowed take place to produce (J). When it does take place based on the operating conditions, then (J) when activated can also be a monomer unit along the chain. A situation where double bonds are placed at the terminals of a monomer unit is least to be expected as shown below and impossible[78].

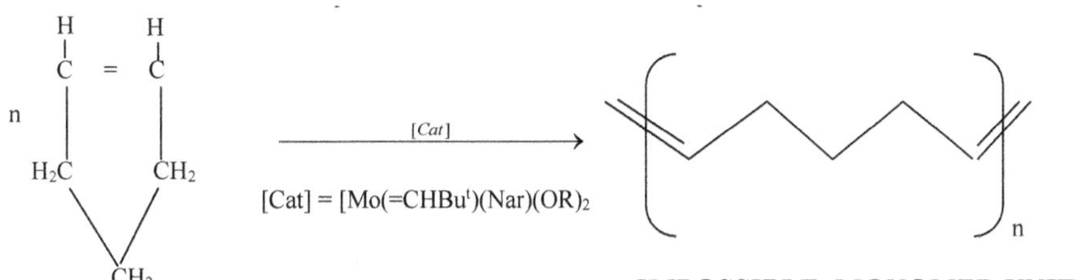

$[Cat] = [Mo(=CHBu^t)(Nar)(OR)_2$

IMPOSSIBLE MONOMER UNIT   11.486

Polyterpenes are also produced by low temperature polymerization of dipentene (i.e., α-pinene of Equation 11.481) in the presence of aluminum chloride[79--81]. This can be explained as follows.

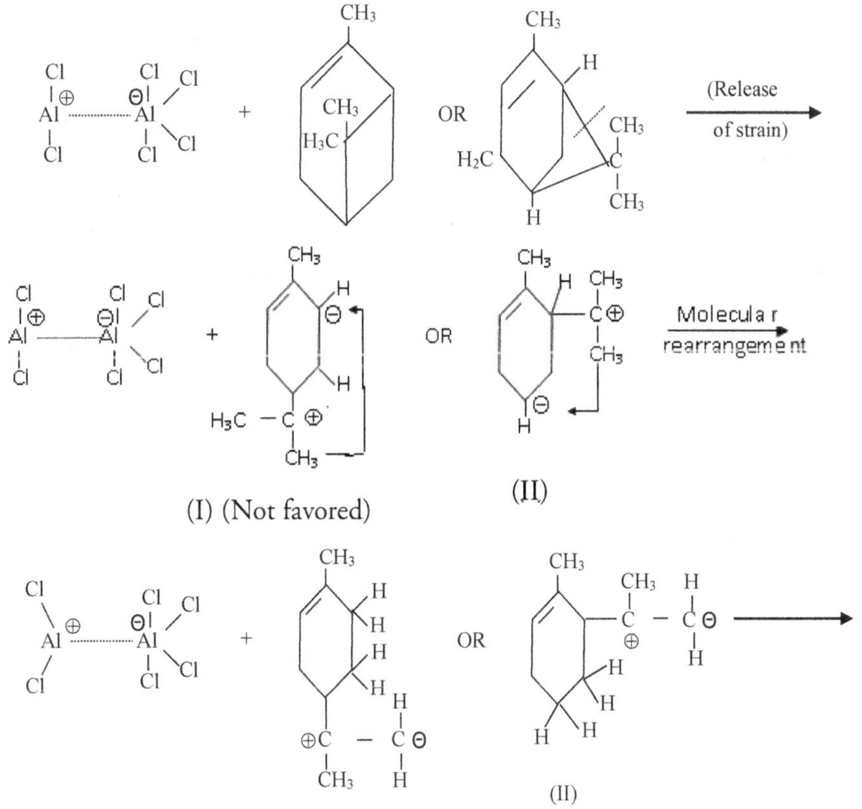

(I) (Not favored)   (II)

(I)   (II)

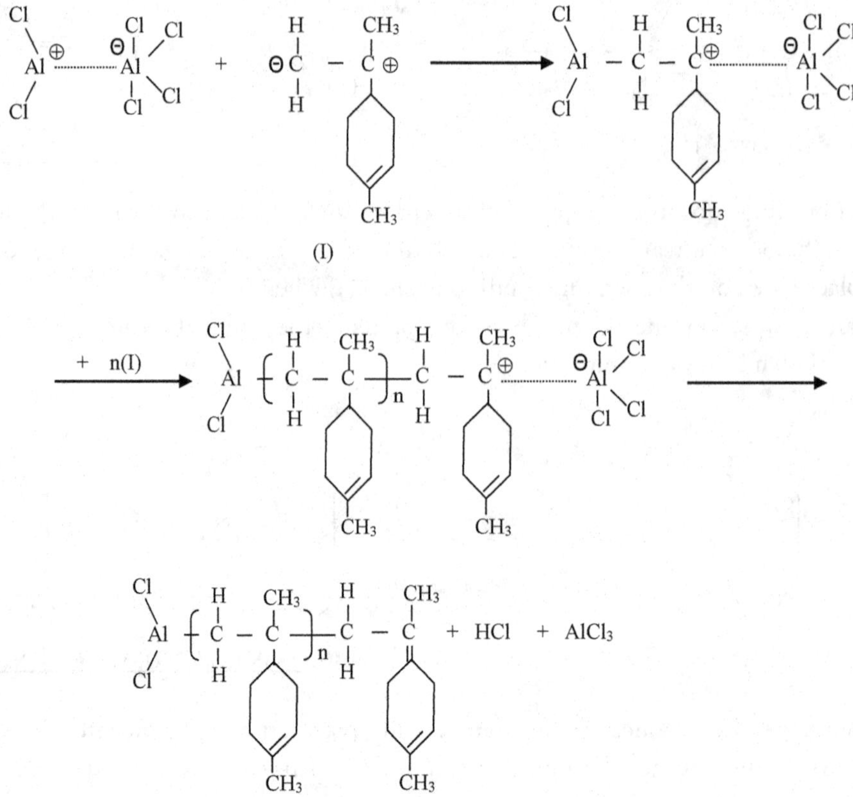

(I)

11.487

The four-membered cyclo-butane ring is strained because it has two common boundaries with the six-membered ring. However, it still requires more energy to attain the MRSE to open the ring instantaneously. That energy cannot be obtained at low polymerization temperature or in the presence of so many undesired components added to the system. Otherwise very low molecular weight products will be obtained. It should be noted like the case above, scission cannot take place at the common boundaries except during ring expansion when maximum radical potential difference exists. Between (I) and (II) above, it is (I) that is favored in view of the fact that the single bond has the larger radical-pushing potential difference. After release of strain, molecular rearrangement immediately takes place to produce a more stable molecule, which becomes the monomer. It is important to note that the transfer species involved in killing the growing chain is the transfer species of the first kind of the ninth type and not of the first type. Note the presence of cyclohexene and endo-olefinic group[80,81] in the monomer units.

β-Pinene the only isomer to a-pinene used above is very more readily polymerizable than a-pinene[82-84], but more expensive.

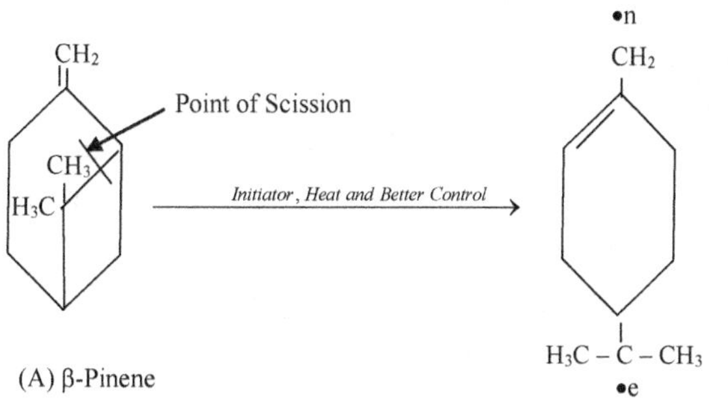

(A) β-Pinene

(B) Monomer Unit

11.488

$AlCl_3$, AlR_2Cl[82], Gamma-radiation[83] and Z/N types of initiators[84] have in the past been used as initiators. As a strong Nucleophile, its natural routes are electro-free-radical and positively charged routes. ***It is strong, because it cannot be deactivated when activated.*** Just as one expected that β-pinene will molecularly rearrange to α-pinene, so also one expected that (B) the monomer unit above will rearrange to the monomer unit (I) of Equation 11.487. β-pinene like the case of methylene cyclopropane, cannot rearrange to α-pinene, because the radical-pushing capacity of what is carried inside the ring, is greater than the radical-pushing capacity of CH_3 externally located and when done, it is found not to be possible. All these rules which look knitty gritty in character are very important to note. They are so because, Nature is nothing else than orderliness in .... With respect to the monomer units, rearrangement from transfer of transfer species from one of the CH_3 groups in (B) to the center carrying the nucleo-free-radical is against the laws of molecular rearrangement, noting that the distance between them should not be a limitation. Shown below is the growing poly chain from β-pinene compared with that from α-pinene.

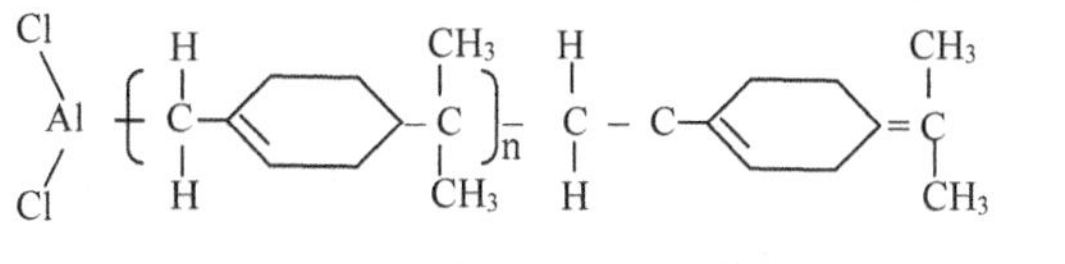

Dead polymers from β-Pinene

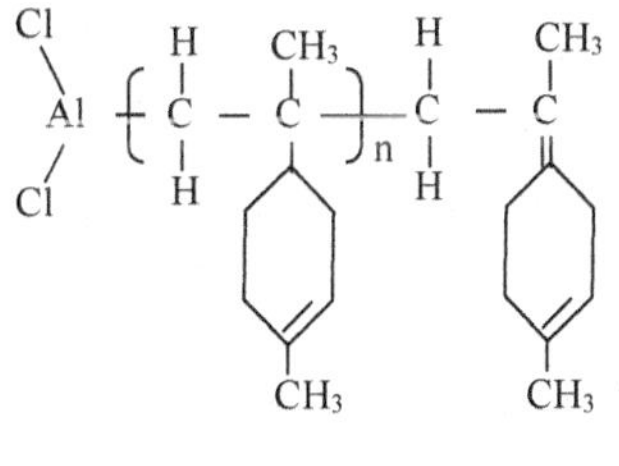

Dead polymers from α-Pinene

11.489

The transfer species in the first case rejected to kill the chain is coming from the ring, while that rejected in the second case above to kill the chain is coming from the ring, the same via molecular rearrangement of the third kind as that abstracted in the route not natural to it. It is therefore not easy to see why very high molecular polymers are obtained more readily from β-pinene than from α-pinene. In fact, only dimers and oligomers can readily be obtained from α-pinene.

From the New Frontiers so far shown, most of the important controversies which have been raised over the years with respect to what has been covered so far have been laid to rest. Even then this is just but the beginning. It is very important to say that the data which have been provided over the years about these compounds have been very useful in throwing more light to these new concepts which are natural in totality. In fact without the data, these developments would have been impossible to confirm or give birth to. Their confirmations make the current developments very unquestionable, because whatever questions asked, the solutions are right there in front of your nose. From all these developments, **infinite** new numbers of research works are bound to arise. For example, there is need to measure the strain energies in all ringed monomers or compounds (that which has begun), identify what are the minimum required strain energies for different families of monomers, measure the capacities of radical-pushing and pulling groups, measure the Nucleophilic and Electrophilic characters of compounds that carry activation centers, take a look at new classifications for different families of compounds and so on too countless to list. This indeed one has begun to do mathematically with respect to determination of SE in ringed compounds. Now we can start synthesizing new monomers which exist in Nature, but unknown to us, new polymers which exist in Nature but unknown to us. Yet animals like the spiders, bees, dogs, and plants and etc. know them! Looking at the ways NATURE operates, there are countless numbers of unknown compounds and indeed of any type of chemicals which can be synthesized all under specific operating conditions. All things are possible only and only when the Laws of Nature are obeyed.

11.6 Proposition of Rules of Chemistry and Concluding Remarks.

Carbon-carbon ringed monomers of alkanes, alkenes, dialkenes, trialkenes and tetraalkenes etc. have been considered for study from a very different point of view. For the first time, the phenomena of free-radical resonance stabilization have begun to be fully defined with respect to their mechanisms and types, because we are yet to see more about them downstream. The phenomenon when the compound is activated begins with transfer of visible electro-free-radicals to grab nucleo-free-radicals from π-bonds to form a new bond or when no visible electro-free-radical exists but a visible nucleo-free-radical exists, then transfer of electro-free-radical from a π-bond to grab the adjacently located nucleo-free-radical to form a new bond. For the first time, one has come to realize that the benzene ring is the unique hydrocarbon ring of all rings. It is the only one that has dual characters- the Energized and De-energized states. In the Energized state all the groups carried by the ring are resonance stabilized and therefore cannot be tempered with. In the De-energized state, the groups when they carry substituted group can have the ability to control the affairs of the ring. The reason why these abilities exhibited by benzene ring exists is because only benzene ring has been found to have Continuous Closed-loop type of resonance stabilization character. It is this character that makes it aromatic. The same applies to all fully unsaturated rings where benzene is present, e.g., Naphthalene, Anthracene. The aromatism manifests itself only when the ring is in the ENERGIZED Activated state of Existence or De-energized state with a bond conjugatedly placed to it, such as in styrene. When in Energized Equilibrium state of existence, the aromaticity cannot readily be exhibited such as the case with concentrated styrene solution-unpleasant odor. For the first time, how ortho, meta- and para-placements are obtained, have been shown. In fully unsaturated rings, transfer of electro-radicals do not go beyond two carbon activation centers, which is the size of a benzene ring. With larger sized fully unsaturated rings, the movement is even reduced to one or two steps. It is not continuous as in benzene. If opened the movement is not limited by step movement.

When some unsaturated rings are opened instantaneously via Equilibrium state of existence or use of an initiator, the number of double bonds remains the same, as opposed to reduction by one exploratively. There are opened-loop, discrete closed-loop and continuous closed-loop radical resonance stabilization types of phenomena.

For the first time, the phenomena of ring expansion and ring reduction are being explained. Why some rings do not exist have been explained through the existence of maximum required strain energy (MaxRSE) and influence of molecular rearrangements. For the first time, the nucleophilic capacity of styrene and related monomers, carbon-carbon ringed monomers or compounds have been provided. The radical-pushing capacities of newly identified substituted or substituents groups have also been provided. All these capacities are qualitative orders. There is therefore the need to know their quantitative measures.

The manners by which activation centers in a resonance stabilized monomer or a non-resonance stabilized monomer or compound (where more than one exist), are activated have fully been identified. All these orders are important, particularly when co-polymerization systems will be considered in totality. In all the considerations so far, no new driving forces favoring the instantaneous opening of a ring has been identified. However, all factors determining points of scission in a ring have been identified. Most of the rules which have been proposed in the last chapter and last Volume, have indeed been applied to open new doors particularly with respect to resonance stabilization phenomena of different types.

Rule 744: This rule of Chemistry for **Ringed-monomers with fused rings,** states that, *the sixth factor determining the ability of a ring to possess the MRSE,* is the presence of three or four membered ring in the system; for which when externally located to larger sized rings, the more strained are the smaller sized rings. [See Rules 594 to 598]
(Laws of Physics in Chemistry)

Rule 745: This rule of Chemistry for **Unsaturated C/H ring compounds,** states that, *with increasing presence of double bonds in the rings, the less the presence of small membered rings,* for example, while for cyclodienes, there is no three-membered ring, with cyclotrienes, there are no three-, four- or even five-membered rings, with cyclodienes there are four (transient), five, six and higher membered rings, with cyclotrienes, there are six, seven, and higher membered rings; all due to limitations placed by SE.
(Laws of Physics in Chemistry)

Rule 746: This rule of Chemistry for **Ringed cyclo-enes,** states that, *if the nucleophilicty of one visible π-bond in a ring with one or more points of scission had been greater than that of the invisible π-bond inside the ring, then the ring will not exist as a ring.*
(Laws of Physics in Chemistry)

Rule 747: This rule of Chemistry for **Cyclobutenes,** states that, when the ring is opened instantaneously, this can be done radically and chargedly; and when the double bond, that is, the visible π- bond is activated, this can also be done chargedly and radically; all under different operating conditions, noting that when opened instantaneously chargedly the 1,4- mono-form cannot be deactivated, while radically it can be deactivated.
(Laws of Creations for Cyclobutenes)

Rule 748: This rule of Chemistry for **Cyclobutene,** states that, when mildly heated in the absence of initiators, it readily decomposes to give 1,3-Butadiene as shown below only radically-

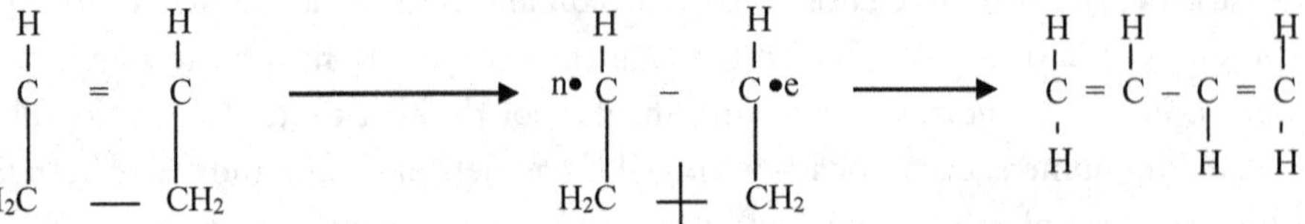

Cyclobutene   **1,3- Butadiene**

for which it can be observed that cyclobutene belongs to a family which is a hybrid of Alkenes and Alkynes ($HC \equiv CC_2H_5$).

(Laws of Creations for Cyclobutene)

Rule 749: This rule of chemistry for **The first member of Cyclobutenes,** states that, when weak positively-charged or weak electro-free-radical-paired initiators are involved, only trans- placement is favored during polymerization; for which the followings are obtained noting that the monomer is a complete Nucleophile.

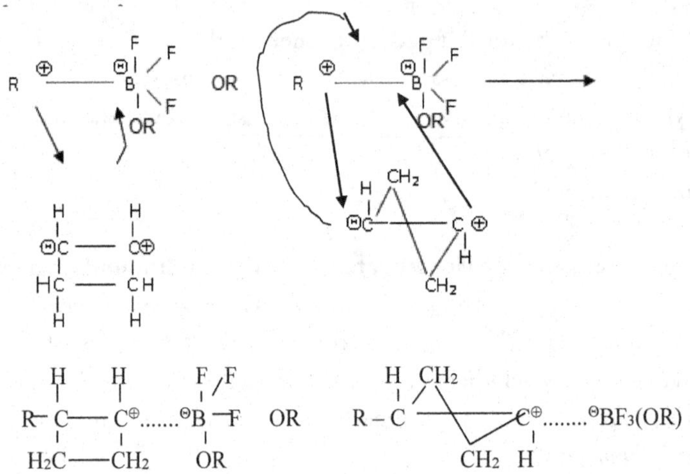

(I) [Least favored] –Cis-   (II) [More favored] – Trans-

(Laws of Creation for Cyclobutene)

Rule 750: This rule of Chemistry for **The first member of Cyclobutenes,** states that, when weak positively charged coordination initiators with reservoirs for the monomers are involved, the followings are to be expected,

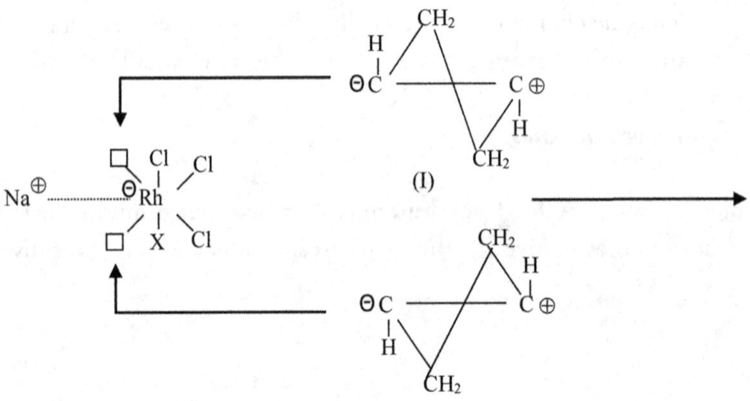

(I)

(II)

388

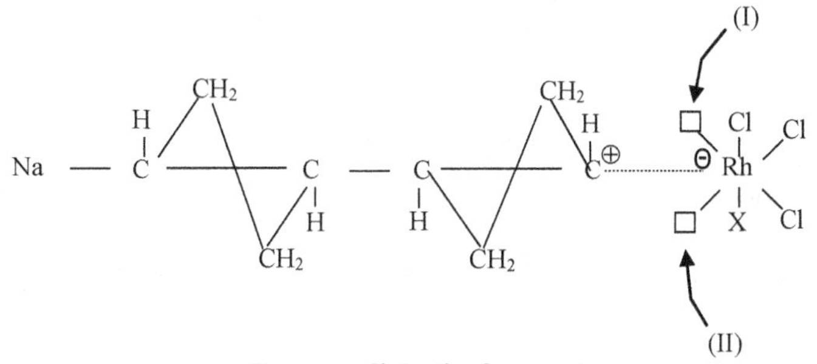

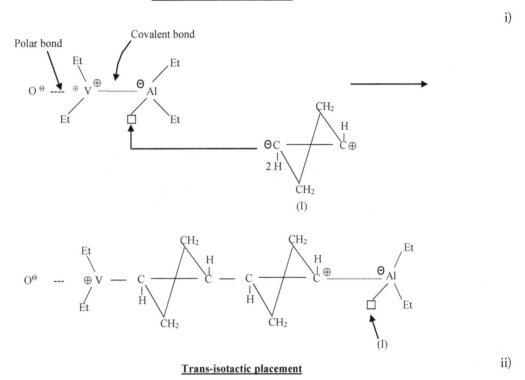

Trans-syndiotactic placement.   i)

Trans-isotactic placement   ii)

for which the first initiator is obtained from Na/Rh combination noting that the carrier of the chain is indeed Rh (not Na as shown above) as will be shown down stream while NaX is used for forming dative bonds on the Rh center, while the second is from V/Al combination to give two different structures as indicated.

(Laws of Structural configurations for polymers)

<u>Rule 751:</u> This rule of Chemistry for **Cyclobutenes,** states that, since the only single visible double bond is not always the first to be activated when activated, and since there is only one point of scission on the ring, not all the rings carrying radical-pushing groups such as the types shown below can be opened instantaneously chargedly-

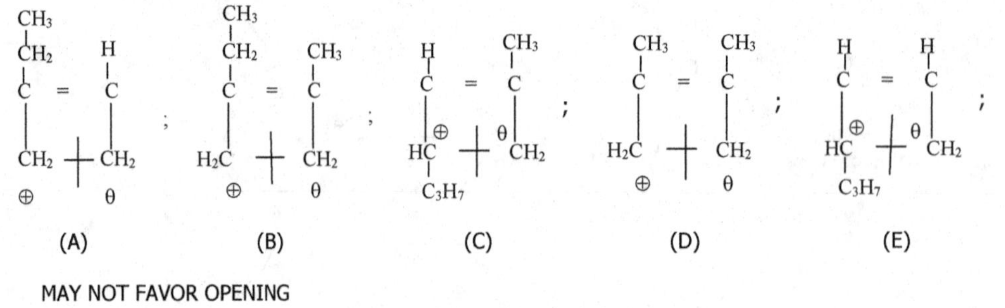

The structures (A) through (F) with their chemical diagrams:

(A)   (B)   (C)   (D)   (E)

MAY NOT FAVOR OPENING

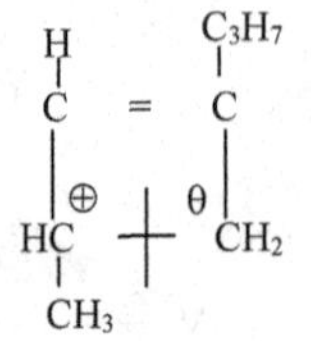

(F)

CAN BE INSTANTANEOUSLY OPENED CHARGEDLY

noting that the charges carried by the active centers when opened instantaneously have been shown above based on what the centers are carrying, for which only (C), (D), (E), and (F) are the only ones which can be opened instantaneously chargedly and radically, while (A) and (B) may not favor opening chargedly, but only radically with the $\oplus$ replaced with .n and θ replaced with .e in view of resonance stabilization required for deactivation when desired.
(Laws of Creations for Cyclobutenes)

Rule 752: This rule of Chemistry for **The first member of Cyclobutenes,** states that, when catalysts based on metals such as tungsten, titanium and molybdenum are used as strong electro-free-radical-paired coordination initiator, the followings are obtained, for which in view of the strength of the coordination center and other operating conditions, the minimum required strain energy for the ring was attained to unzip instantaneously at the point of scission indicated below with adequate

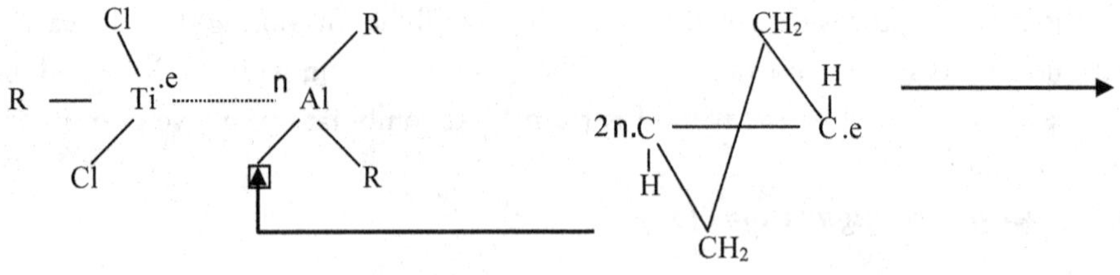

(I)

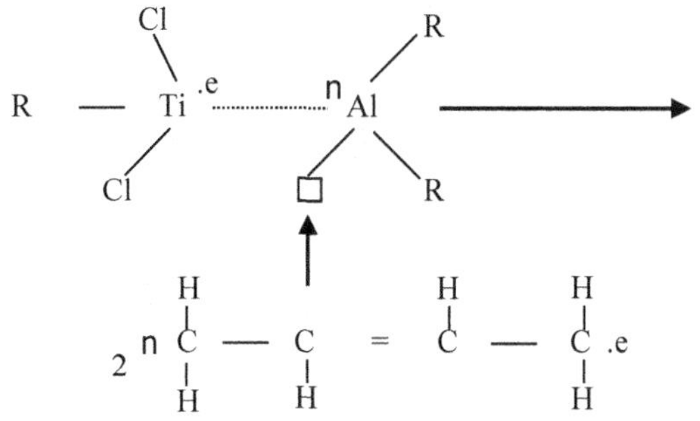

(I)

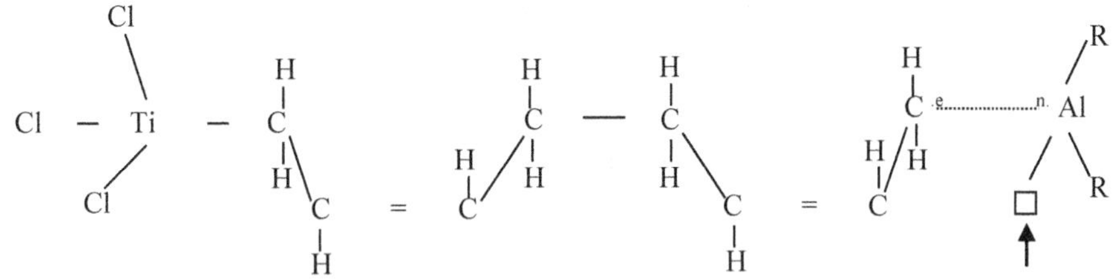

(Cis - placement)

(I)

orientation before addition, to give cis-polybutadiene when only one vacant orbital is present on the counter-radical center or trans-polybutadiene when two vacant orbitals are present.
(Laws of Structural configuration for polymers)

Rule 753: This rule of Chemistry for **First member of Cyclobutenes,** states that, free-radically in the absence of electrostatic forces, activation can only take place via the double bond as follows-

i)

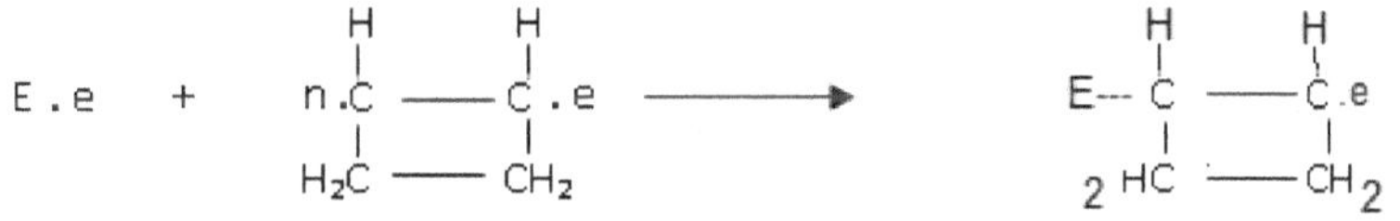

ii)

in which due to the presence of transfer species of the first kind of the tenth type, no initiation is favored nucleo-free-radically, but electro-free-radically the route natural to the monomer, with the possibility of having trans-placement of the ring or syndiotactic placement due to steric limitations.
(Laws of Creations for Cyclobutene)

Rule 754: This rule of Chemistry for **Cyclobutenes carrying at least one radical-pushing alkylane group internally located on the ring and not on the double bond such as 3-methyl-cyclo-butene,** states that, these are strong Nucleophiles which can either be opened instantaneously or used via the double bond depending on the operating conditions; for which the route favored by them are electro-free-radical and positively charged routes as shown below in the route not natural to them.

391

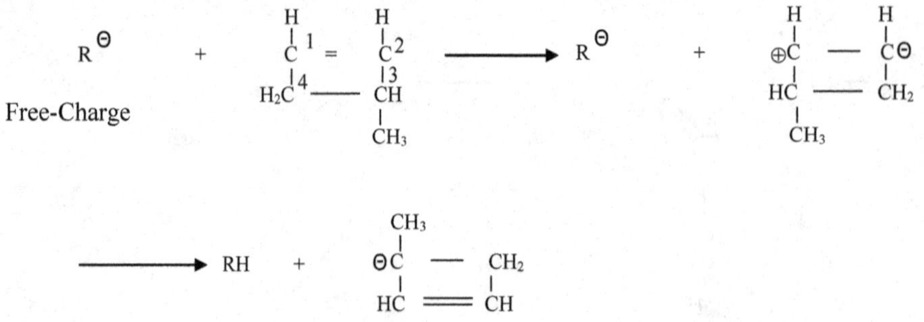

Transfer species of first kind of tenth type

i)

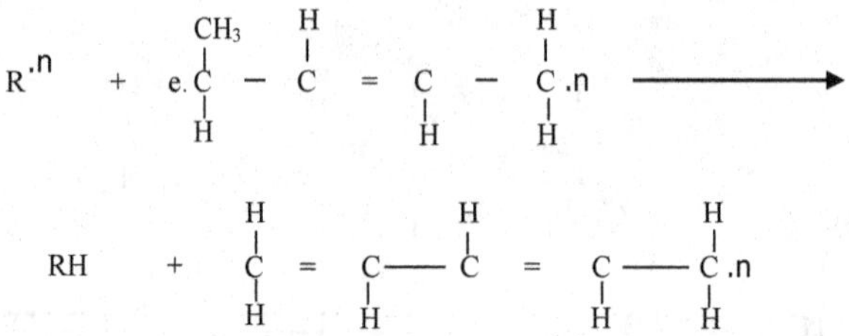

Transfer species of first kind of first type

ii)

(Laws of Creation for Cyclobutenes)

Rule 755: This rule of Chemistry for **Cyclobutenes with R radical-pushing groups internally located but not on the double bond,** states that, when weak positively charged or weak electro-free-radical initiators are involved, the followings are to be expected; and this takes place only when the nucleophilicity of the invisible π-bond remains stronger than that of the visible π- bond; from which one can see the distinctions between the structural configurations- tds and tdi versus ts and ti. *(See Rule 750).*

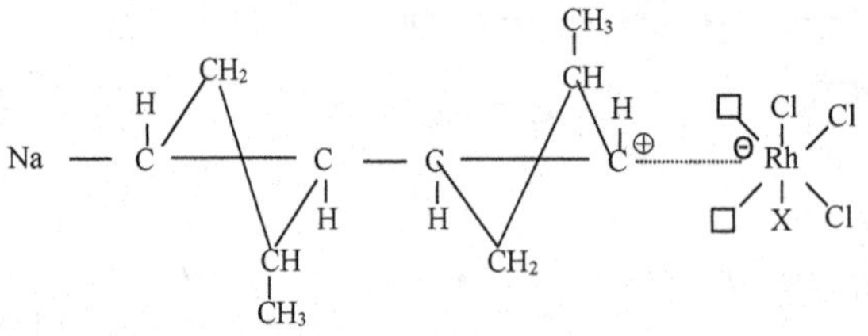

Trans-di-syndiotactic placement. (tds)

i)

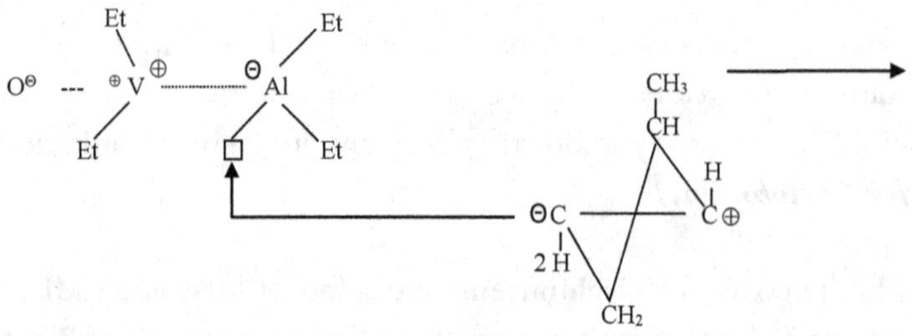

(I)

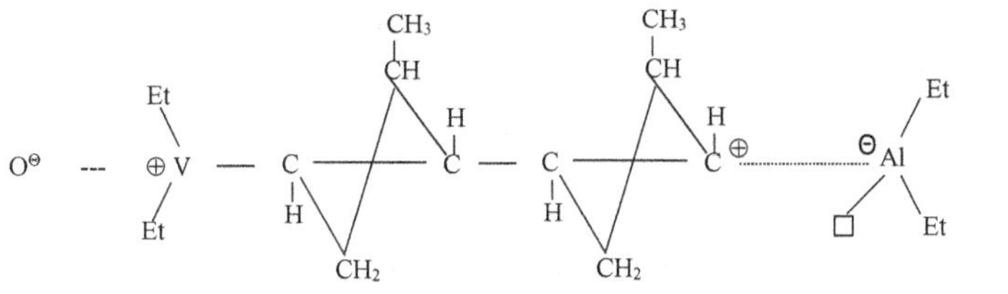

Trans-di-isotactic placement (tdi)

ii)

(Laws of Creations for Cyclobutenes)

Rule 756: This rule of chemistry for **Cyclobutenes with R radical-pushing groups inside the ring but not on the double bond,** states that, when catalysts based on metals such as tungsten, titanium and molybdenum are used as strong electro-free-radical-paired coordination initiator, the followings are obtained, for which in view of the strength of the coordination center and other operating conditions, cis- and trans- placements are obtained, depending on the number of vacant orbitals present as shown below for the cis, and as can be observed, only poly(1-methyl butadiene) or polypentadiene can be obtained when the ring is opened instantaneously.

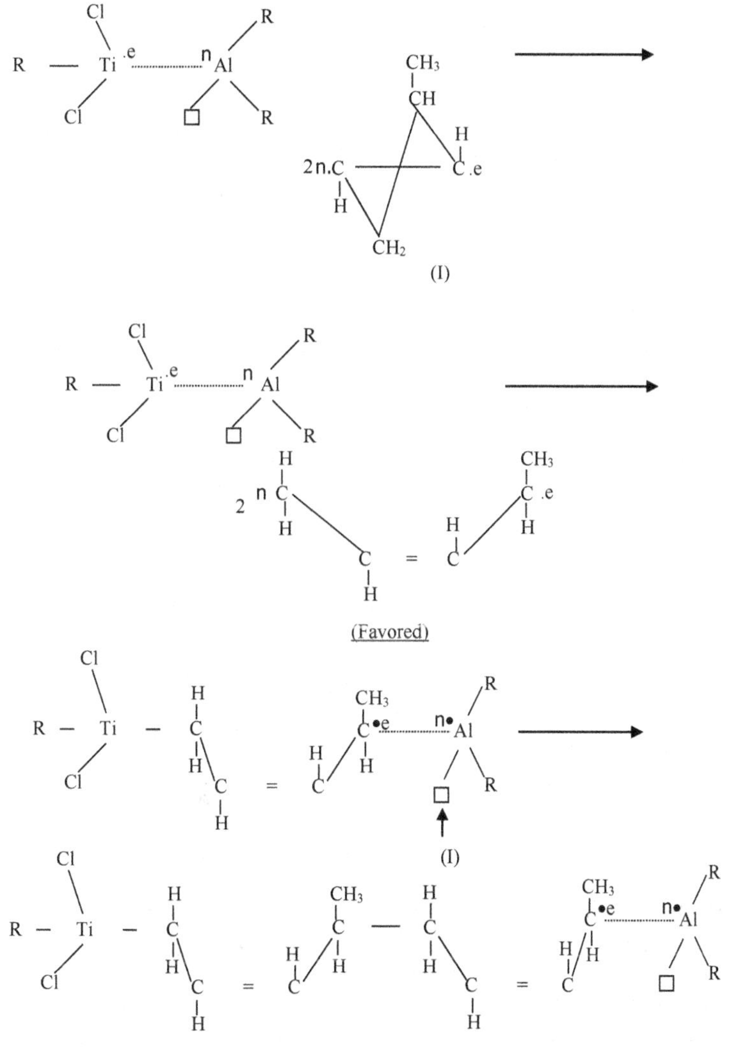

(I)

(Favored)

(I)

Cis - 1, - methyl butadiene polymer

(Laws of Creations for Cyclobutenes)

Rule 757: This rule of Chemistry for **Cyclobutenes with radical-pushing group on the double bond,** (such as 1-methyl cyclobutene), states that, for their growing polymer chains the followings are to be expected when weak positively charged or weak electro-free radical paired initiators are used-

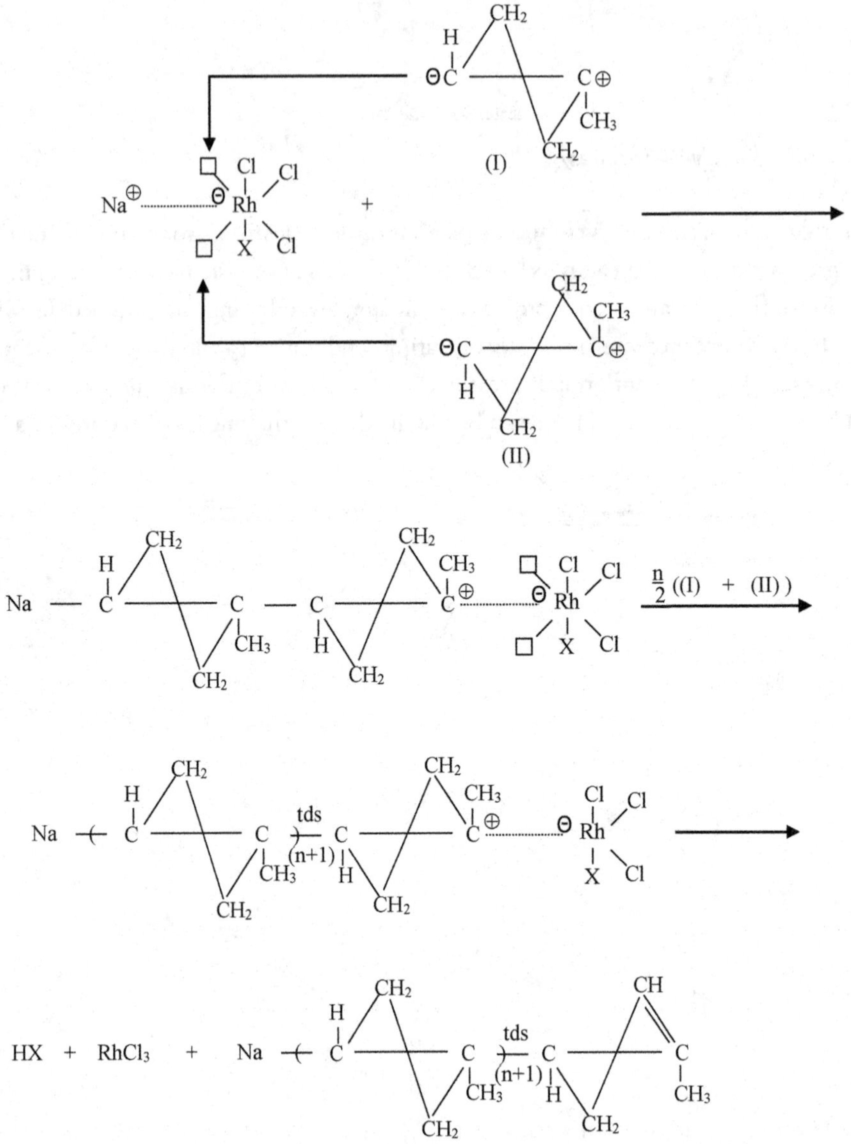

Transfer species of the first kind of the tenth type

i)

for which the reactions above will be more favored the larger R becomes, while with strong free-radical coordination initiators, the followings are obtained; for which polyisoprene can be observed to be produced, cis-placed in view of the presence of one vacant orbital on the counter- center, noting that unlike the case where the ring is not opened, here both natural and unnatural routes are favored for other initiators, but not for the Z/N types of initiators which can identify which monomer is Male or Female.

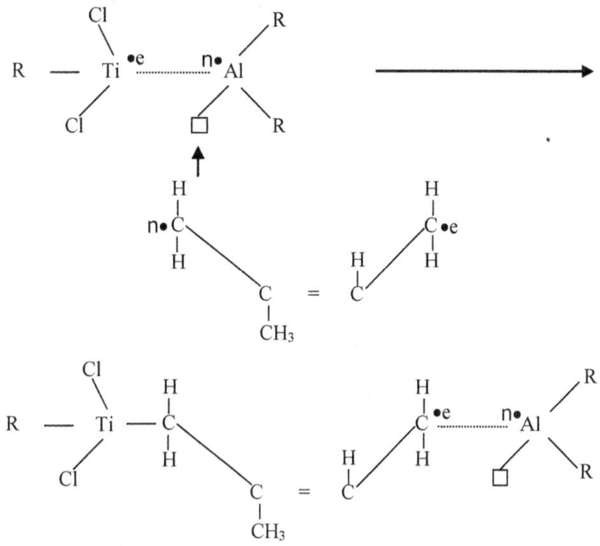

(Laws of Creations for Cyclobutenes)

Rule 758: This rule of Chemistry for **2-Methyl Cyclobutene and 1-Methyl Cyclobutene,** states that, when they are made to undergo molecular rearrangements, while the latter will rearrange back to itself, the former cannot rearrange at all as shown below for them.

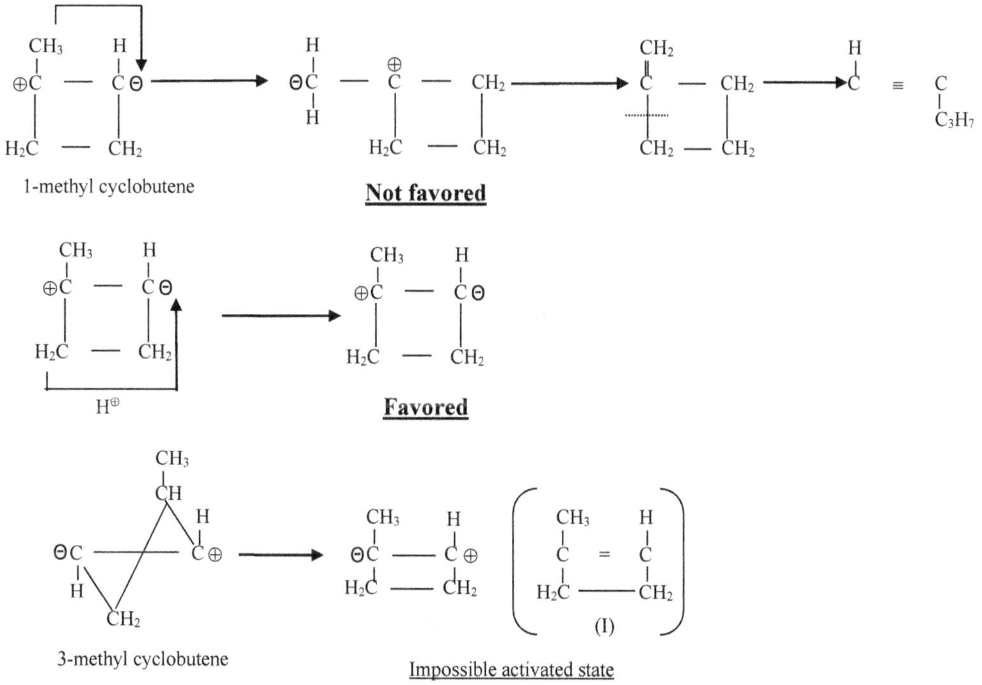

for which as shown by the first equation, methyl cyclobutene cannot molecularly rearrange to give methylenecyclobutane, since the radical-pushing capacity of the group in the ring is greater than that of CH_3.

(Laws of Creations for Cyclobutenes)

Rule 759: This rule of Chemistry for **1-Methyl Cyclobutene and Methylenecyclobutane,** states that, these are isomers wherein while 1-methyl cyclobutene cannot rearrange to methylene-cyclobutane, the reverse is the case as shown below; for which it can be observed that unlike in

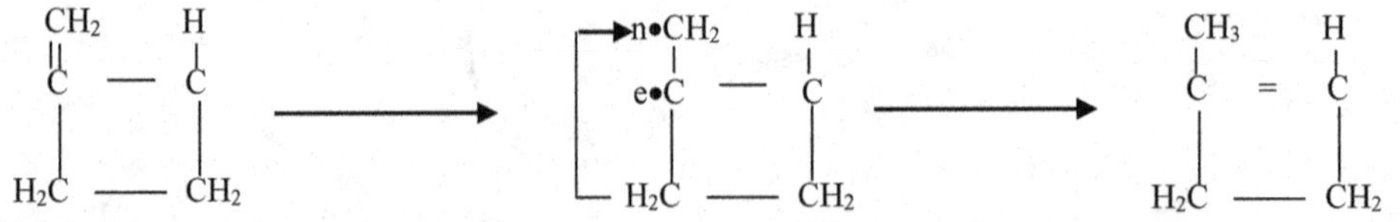

Methylenecyclobutane   Methyl cyclobutene

Cyclopropene where the CH_2 is connected to two active centers, here it is connected to one active center, making the capacity of the H inside the ring equal to that externally located.
(Laws of Creations for 1-methyl cyclobutene and methylenecyclobutane)

Rule 760: This rule of Chemistry for **Cyclobutenes with radical-pushing groups greater than CH_3 and up to C_4H_9,** states that, these can readily undergo molecular rearrangements of the first and third kinds to give acetylenes as shown below-

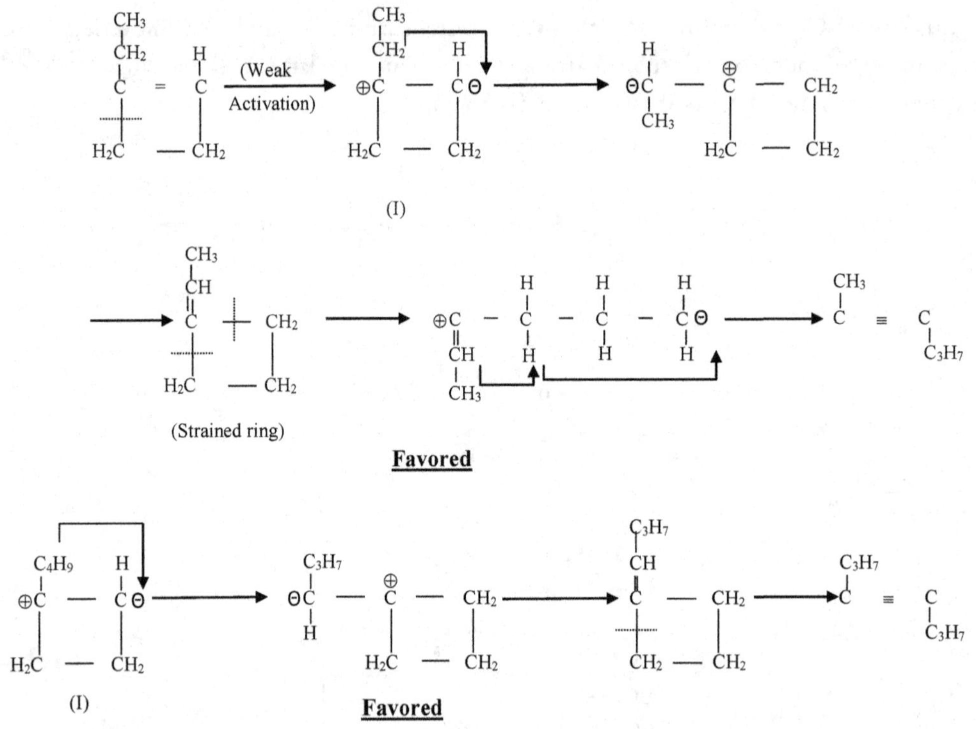

(I)

(Strained ring)

Favored

(I)   **Favored**

noting the increasing order of nucleophilicity as R increases, reaching a limit wherein the rearrangement can no longer take place from C_5H_{11} and above, despite the increasing order of nucleophilicity.
(Laws of Creations for Limitations-Logics of Chemistry)

Rule 761: This rule of Chemistry for **Cyclobutenes with R radical-pushing groups of different capacities on both C centers of the double bond,** states that, rearrangements are favored up to a particular R size group as shown below-

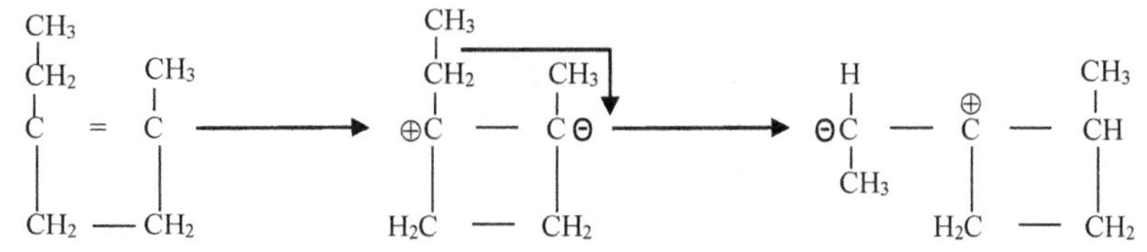

$$CH_3 \quad CH_3$$

(I)

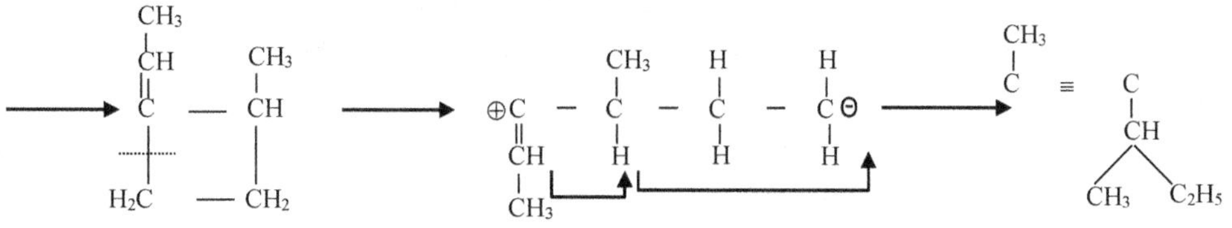

(I)

noting that if they are of equal radical-pushing capacities, rearrangement will still take place with limitations, as long as R is greater than CH_3.

(Laws of Creations in Humanity-Logics of Chemistry)

Rule 762: This rule of Chemistry for **1,2,3-Trimethylcyclobutene, that with three-radical-pushing groups of equal capacity,** states that, when it rearranges, it rearranges back to itself and when mildly heated, it can readily be instantaneously opened as shown below to give 2,3-Dimethyl pentadiene.

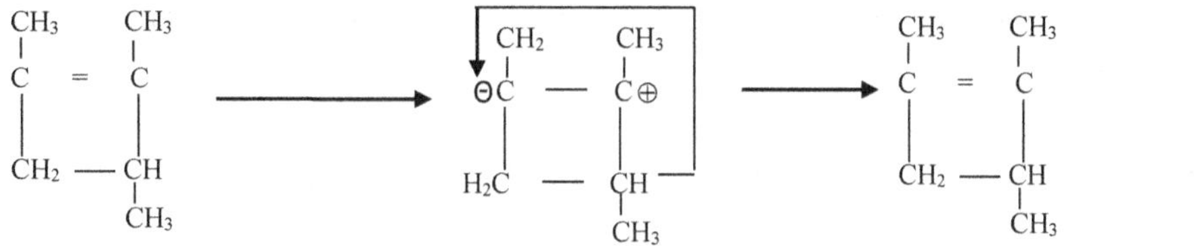

1,2,3-Trimethylcyclobutene

1,2,3-Trimethylcyclobutene

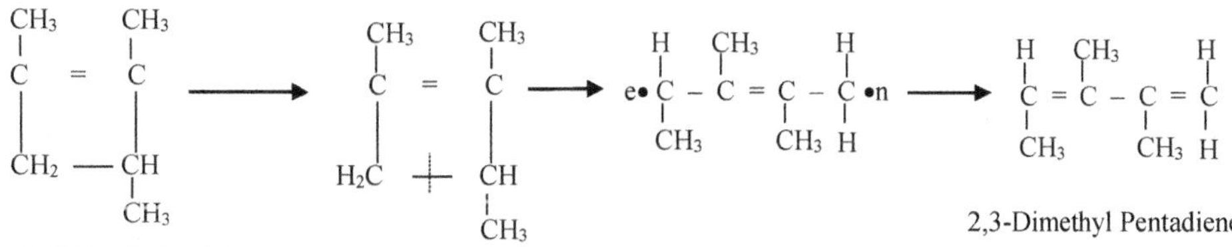

1,2,3-Trimethylcyclobutene

2,3-Dimethyl Pentadiene

(Laws of Creations for Cyclobutenes)

Rule 763: This rule of Chemistry for **Cyclobutene**, states that, the radical-pushing capacity of the groups internally located, which is a measure of nucleophilicity inside the ring, is given as follows-

$$H_2C \quad - \quad CH_2 \qquad \geq \qquad C_2H_5$$

Closed-group

ORDER OF RADICAL-PUSHING CAPACITY

for which when compared to that of cyclopropene, it is very small, because of **the unique character of a Triangle** where the three C centers are too close for comfort, worsened further by the presence of a double bond inside the ring.
(Laws of Creations for Cyclobutene)

Rule 764: This rule of Chemistry for **Cyclopropenes and Cyclobutenes with R radical-pushing groups,** states that, the followings are valid-

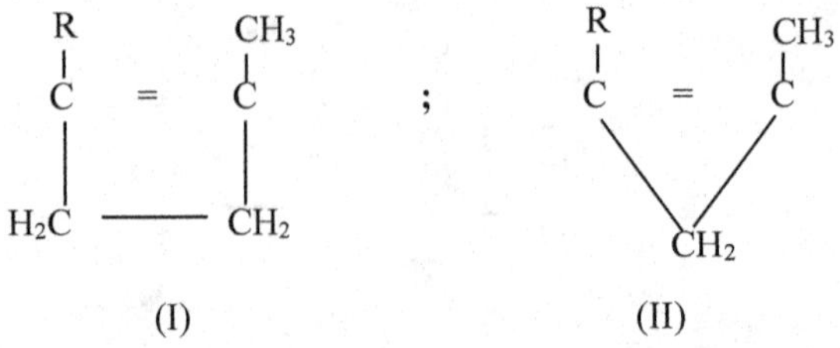

(I)   (II)

while (I) will favor existence of acetylenic monomers for R up to C_5H_{11} beginning from C_2H_5, (II) will not favor their existence, *yet it is closer to acetylene than cyclobutene is and* when R is changed to H, both will decompose via Equilibrium mechanism to highly nucleophilic acetylenes with (I) being more nucleophilic than (II).
(Laws of Creations in Humanity-Logics of Chemistry)

Rule 765: This rule of Chemistry for **Cyclobutenes carrying radical-pulling halogens such as Cl on the double bond,** states that, though this looks like chloroprene when the ring is instantaneously opened, the followings are worthy of note-

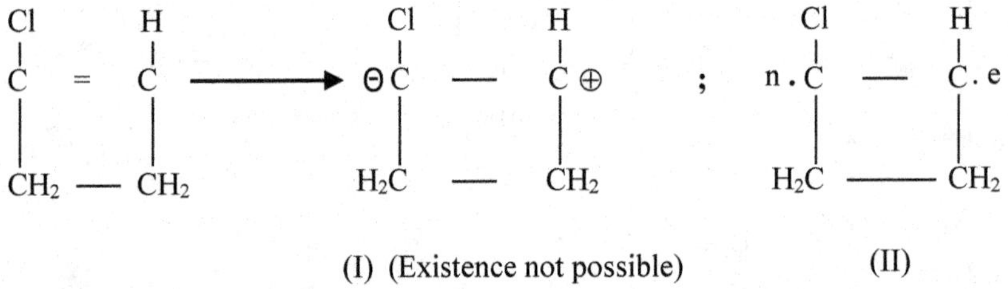

(I) (Existence not possible)   (II)

radically, polymerization is favored only electro-free-radically, while chargedly the monomer cannot be activated due to electrostatic forces of repulsion; it can be opened free- radically as shown below under harsh operating conditions and when opened chloroprene is produced and this can be used as a

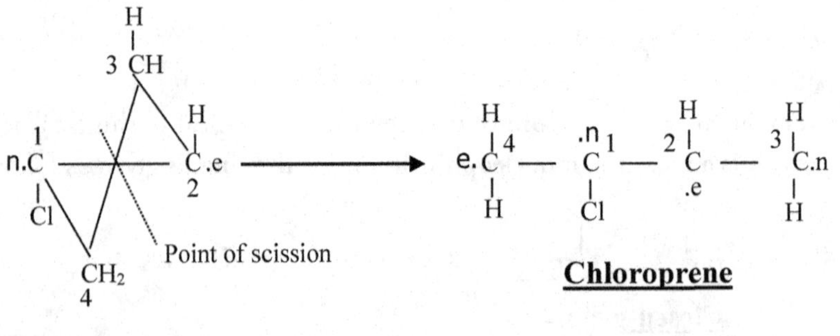

Chloroprene

monomer favoring both electro- and nucleo-free-radical routes, *unlike when the ring is not opened;* noting that it can also be opened instantaneously chargedly and *that the double activation shown above is obviously impossible (for it was only shown above to show some fundamental principles).*
(Laws of Creations for Cyclobutenes)

Rule 766: This rule of Chemistry for **Cyclobutenes carrying radical-pushing groups of lower capacity than H such as halogenated groups,** states that, these monomers/compounds when activ-

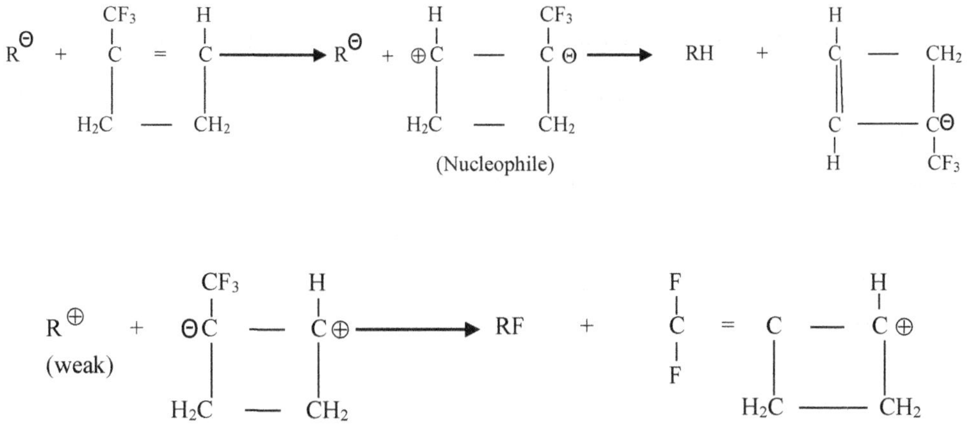

(Nucleophile)

(weak)

ated cannot be polymerized, and when opened instantaneously free-radically, it can be made to favor both free-radical routes; and the same may also almost apply chargedly, favoring the positively and negatively charged routes.

(Laws of Creations for Cyclobutenes)

Rule 767: This rule of Chemistry for **Cyclobutenes with radical-pulling groups with transfer species (e.g. COOR, CONH$_2$,) and those without transfer species (e.g. COR, CN),** states that, these are Electrophiles which favor no route in particular for those with transfer species, and cannot readily be instantaneously opened and when those with transfer species rearrange, cannot undergo molecular rearrangement of the third kind when the ring is opened

(Laws of Creations for Cyclobutenes)

Rule 768: This rule of Chemistry for **Vinylcyclobutene,** states that, this is a resonance stabilized monomer only radically as shown below-

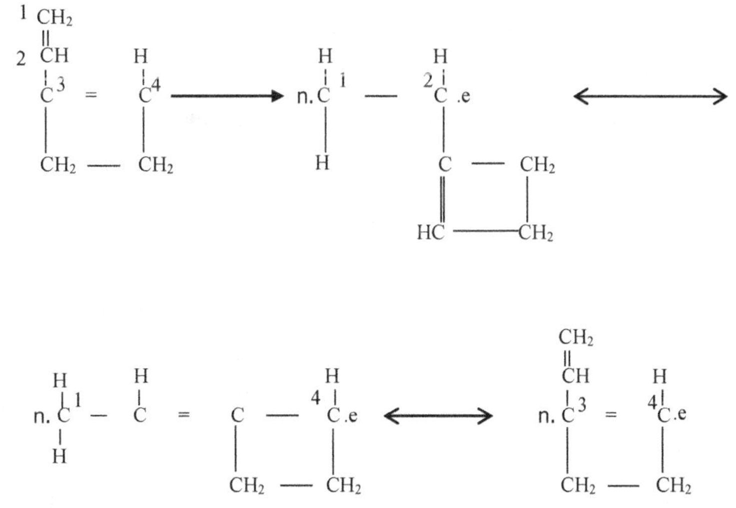

Not a monoform   i)

from which the following is valid.

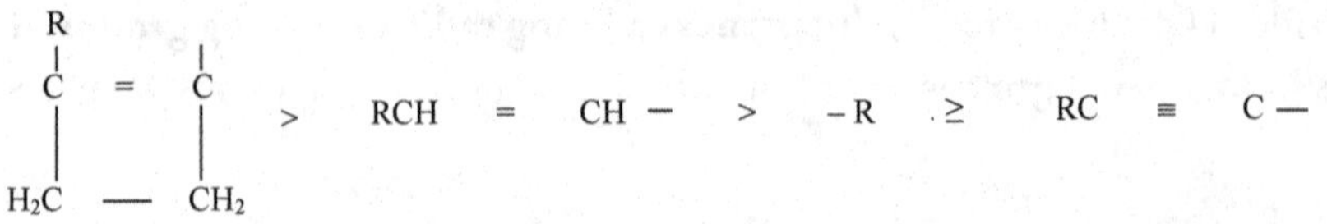

$$\begin{array}{c} R \\ | \\ C = C \\ | \quad\quad | \\ H_2C \;-\; CH_2 \end{array} \quad > \quad RCH \;=\; CH - \quad > \quad -R \quad .\geq \quad RC \equiv C - \qquad\qquad \text{ii)}$$

Order of radical-pushing capacity of resonance stabilization groups

(Laws of Creations for Vinylcyclobutene)

Rule 769: This rule of Chemistry for **Vinylcyclobutenes wherein the H on the C = C activation center on the ring is made to carry CH_3 and higher groups,** states that, no molecular rearrangement of the third kind can take place, like the corresponding case of cyclopropene and they can readily be instantaneously opened to favor the presence of a diene where the ring still remains the resonance stabilization provider as shown below-.

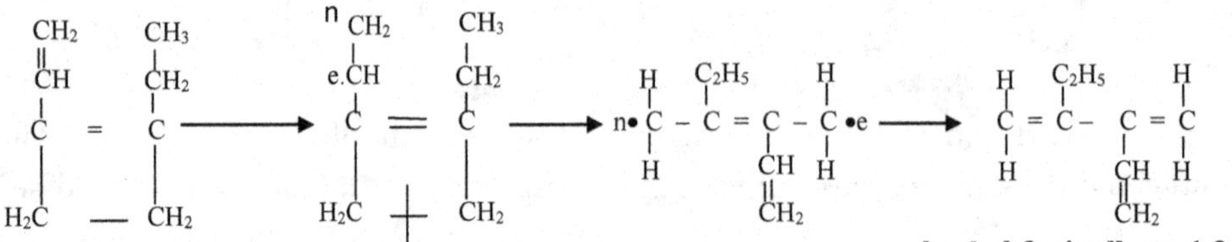

2-ethyl-3-vinylbuta-1,3-diene

wherein the resonance stabilization provider before the ring was opened is the ring and after the ring was opened is the invisible π-bond in the ring, with the vinyl and C_2H_5 groups now shielded or resonance stabilized.

(Laws of Creations for Vinylcyclobutenes)

Rule 770: This rule of Chemistry for **Vinylcyclobutene wherein only the alkene group is allowed to carry radical-pushing groups,** states that, for this resonance stabilized monomer, the followings are to be expected with respect to molecular rearrangement-

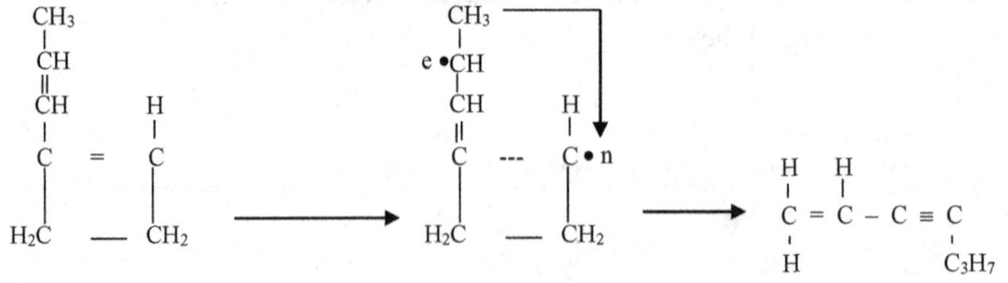

i)

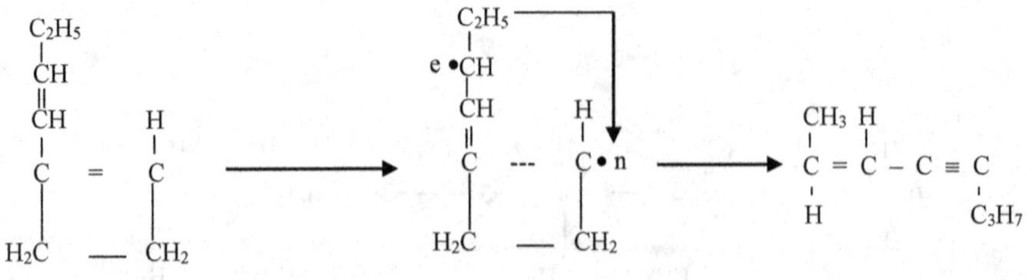

ii)

for which molecular rearrangement of the third kind ceases when R group on the alkene side becomes greater than C_4H_9 and the products obtained are vinyl acetylenes which are strongly Nucleophilic. [See Rule 758]

(Laws of Creations for Vinylcyclobutenes)

Rule 771: This rule of Chemistry for **Vinylcyclobutene wherein only the alkene group is allowed to carry radical-pushing group of the type OH,** states that, for this resonance stabilized monomer, the followings are to be expected with respect to molecular rearrangement of the third kind-

$$
\begin{array}{ccc}
\begin{array}{c}
OH \\ | \\ CH \\ \| \\ CH \\ | \\ C \;=\; C \\ | \qquad | \\ H_2C \;—\; CH_2
\end{array}
&
\longrightarrow
\begin{array}{c}
OH \\ | \\ e\,\bullet CH \\ | \\ CH \\ \| \\ C \;---\; C\bullet n \\ | \qquad | \\ H_2C \;—\; CH_2
\end{array}
&
\longrightarrow
\begin{array}{c}
H \\ | \\ O = C \\ | \\ C \equiv C \\ | \\ C_3H_7 \\ \text{Electrophile}
\end{array}
\end{array}
$$

for which an electrophilic acetylene is produced.

(Laws of Creations for Vinylcyclobutene)

Rule 772: This rule of Chemistry for **Vinylcyclobutene wherein only the alkene group is allowed to carry radical-pushing groups,** states that, for this resonance stabilized monomer, the followings are to be expected with respect to instantaneous opening of its ring-

$$
\begin{array}{ccc}
\begin{array}{c}
CH_3 \\ | \\ CH \\ \| \\ CH \\ | \\ C \;=\; C \\ | \qquad | \\ H_2C \;—\; CH_2
\end{array}
&
\longrightarrow
\begin{array}{c}
CH_3 \\ | \\ e\,\bullet CH \\ | \\ n\bullet\; CH \\ | \\ C \;=\; C \\ | \qquad | \\ H_2C \;+\; CH_2
\end{array}
&
\longrightarrow
\begin{array}{c}
CH_2 \\ \| \\ H\;\; CH \quad CH_3 \\ | \quad | \qquad | \\ C = C - C = C \\ | \qquad | \quad | \\ H \qquad H \quad H
\end{array}
\end{array}
$$

IMPOSSIBLE INSTANTANEOUS OPENING OF THE RING

for which though the vinyl group still remains not involved in the decomposition, it is now the resonance stabilization provider in the butadiene and no longer the ring, with the invisible π-bond in the ring now becoming the least nucleophilic center via the group which was originally resonance stabilized, while the group which was not resonance stabilized when the ring was not opened is now resonance stabilized after opening of the ring as a vinyl group.

(Laws of Creations for Vinylcyclobutenes)

Rule 773: This rule of Chemistry for **Vinylcyclobutenes wherein R is carried by the vinyl center and R^1 is carried by the ring,** states that, when R is greater than or equal to R^1, molecular rearrangement of the third kind to produce vinyl acetylene is favored with no instantaneous opening of the ring made possible, while when R is less than R^1, instantaneous opening of the ring is favored to produce double headed butadienes with no ability to undergo molecular rearrangement of the third kind-

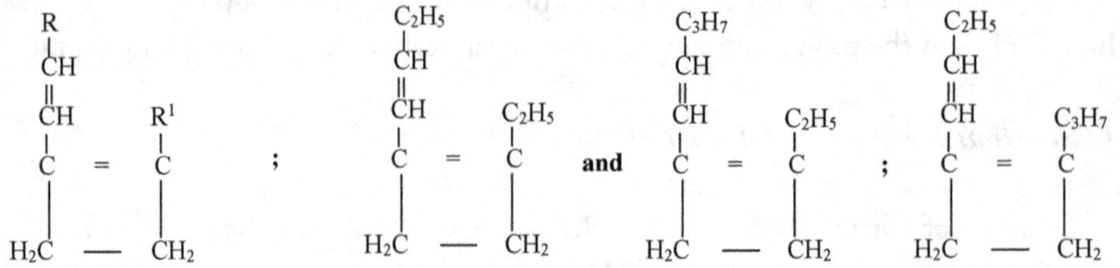

$$R-CH=CH->R$$

Order of Radical-pushing capacity

(Laws of Creations for Vinylcyclobutenes)

for which the following can be found to be valid-

Rule 774: This rule of Chemistry of **Cyclobutenes with acetylenic resonance stabilization groups replacing alkenic groups,** states that, the followings are to be expected for the first member, for which the monomer can be seen to be resonance stabilized favoring only the electro-free-radical

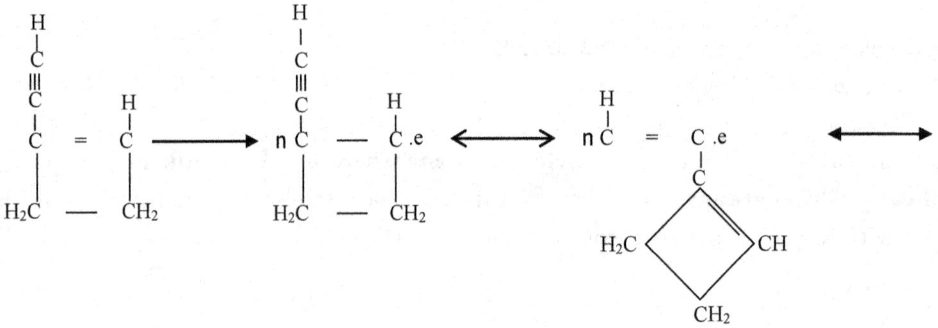

Not a mono-form

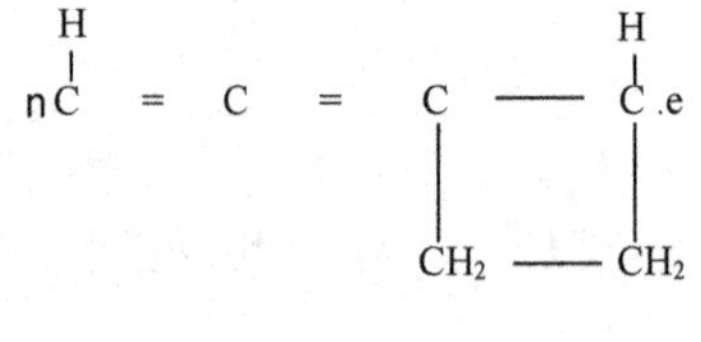

route, with no ability to undergo molecular rearrangement of the third kind to give resonance stabilized monomer or favor instantaneous opening of its ring, from which the following relation-ships are valid.

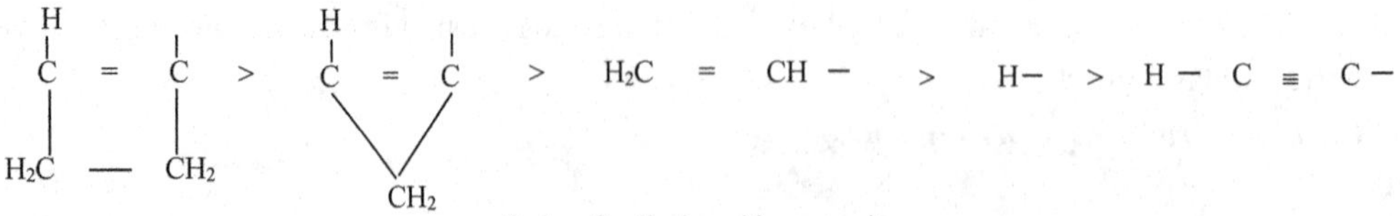

Order of radical-pushing capacity

$$HC(C_4H_9)=CH- \quad > \quad \overset{H}{\underset{H_2C-CH_2}{C}}=C \quad \geq \quad (C_2H_5)HC=CH-$$

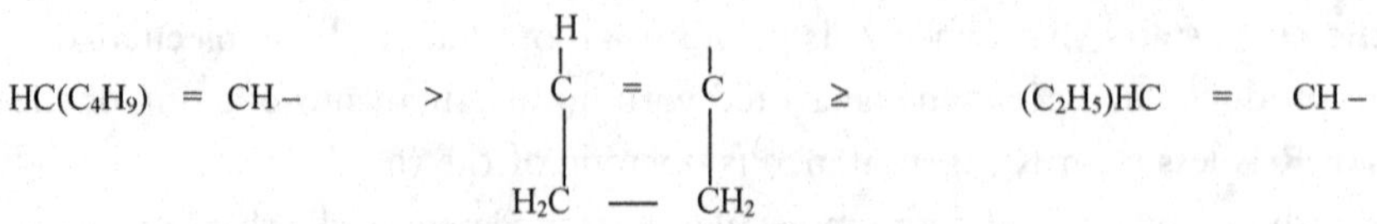

Transition point for Molecular rearrangement of the 3rd kind

(Laws of Creations for Acetylenyl cyclobutene)

402

Rule 775: This rule of Chemistry for **Acetylenic Cyclobutenes where the externally located H atoms are replaced one at a time with alkylane groups,** states that, the followings are obtained when the group is on acetylene, for which molecular rearrangement can be favored by it followed by opening of the ring to give a Diyne with limitations (C_4H_9)-

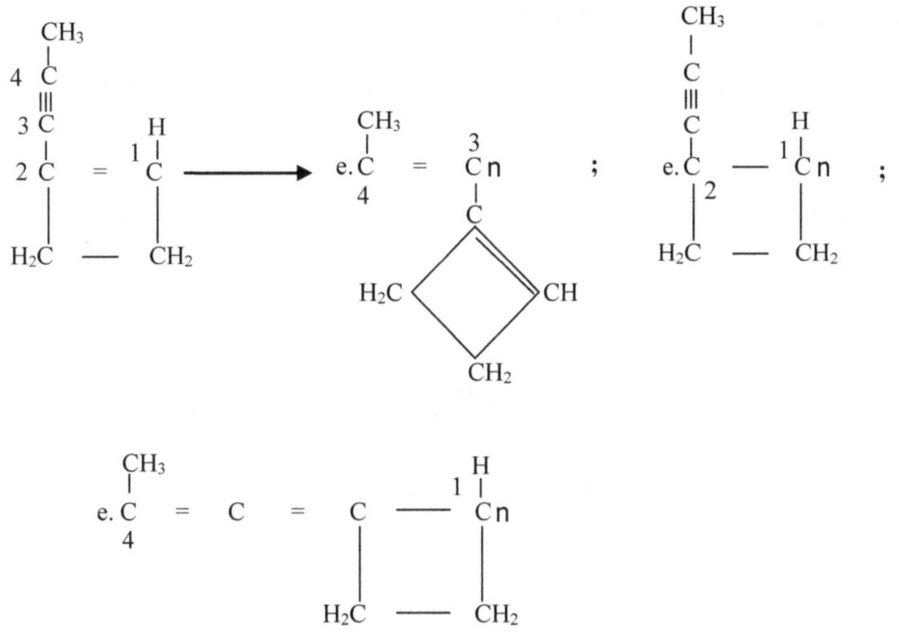

Resonance stabilized

noting that the first case above is not a mono-form.

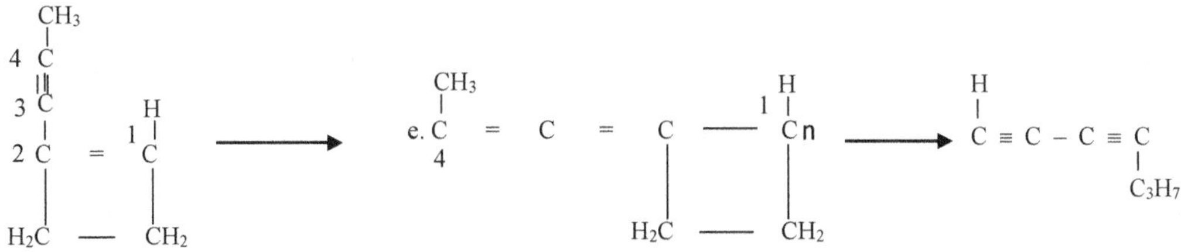

Resonance stabilized

Resonance stabilized

(Laws of Creations for Methyl acetylenyl cyclobutene)

Rule 776: This rule of Chemistry for **Acetylenic Cyclobutenes where the externally located H atoms are replaced one at a time with alkylane groups,** states that, the followings are obtained when the group is on the ring, for which molecular rearrangement of the third kind cannot take place to produce a non-resonance stabilized cumulenic acetylene as shown below-noting that the first case above is not a mono-form, for which the rearrangement shown below is not possible, since the monomer is not fully resonance stabilized;

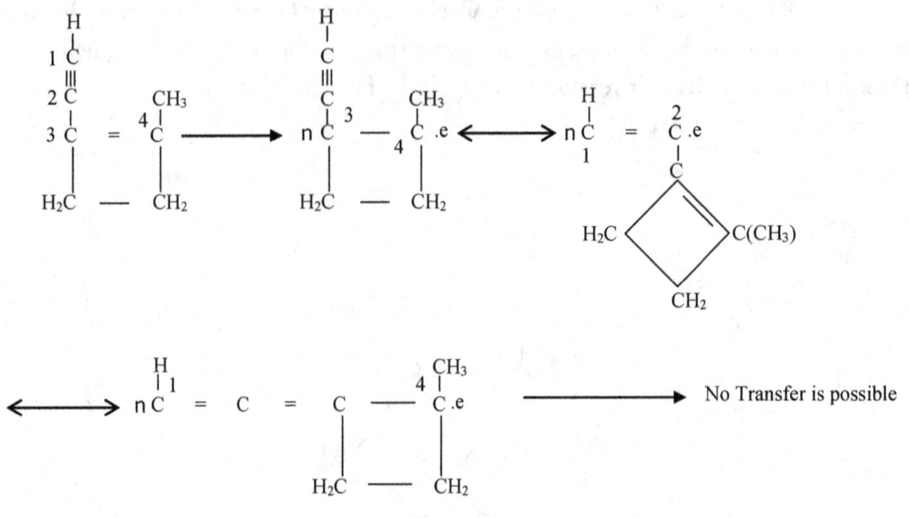

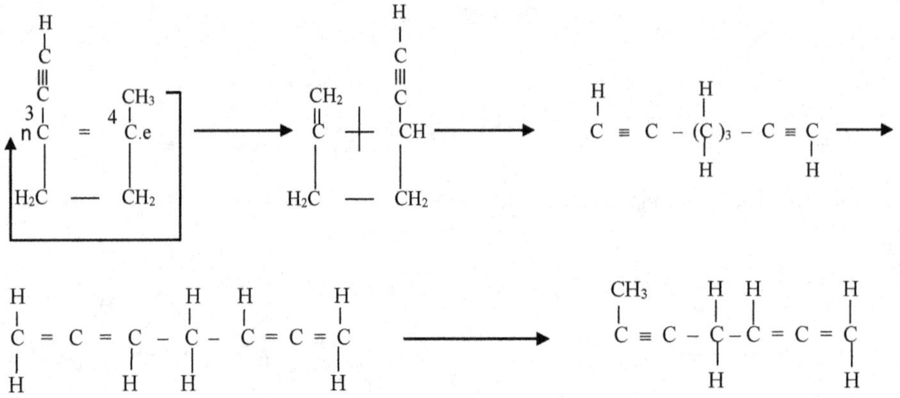

NOT FAVORED

Non-Resonance stabilized

noting that the wrong center has been used above, and when the right center is uased, no original or ringed monomer can be obtained.

(Laws of Creations for Acetylenyl methyl cyclobutene)

Rule 777: This rule of Chemistry for **Acetylenic Cyclobutenes wherein R is carried by the acetylene center and R¹ is carried by the ring,** states that, when R is greater than R¹, molecular rearrangement of the third kind to produce vinyl acetylene is favored with no instantaneous opening of the ring made possible, while when R is less than or equal to R¹, neither is instantaneous opening of the ring favored nor the ability to undergo molecular rearrangement of the third kind made possible-

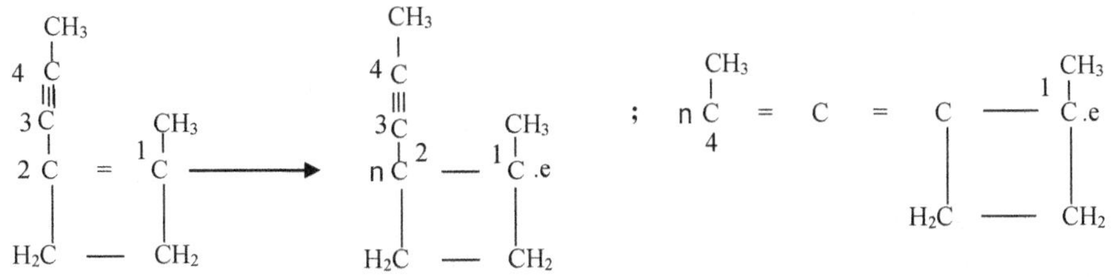

Resonance stabilized

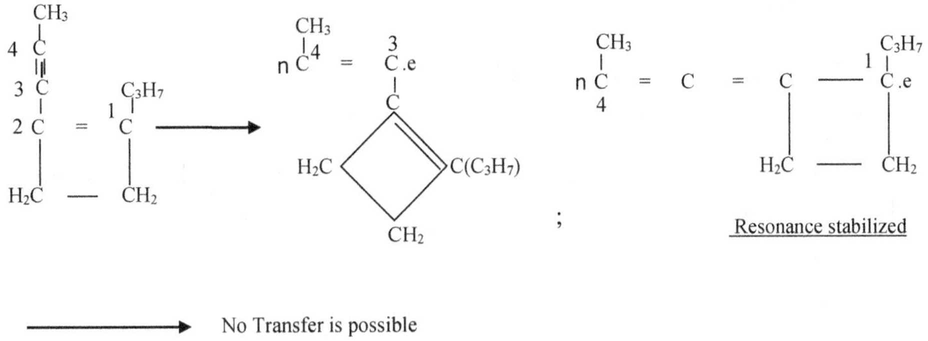

Resonance stabilized

No Transfer is possible

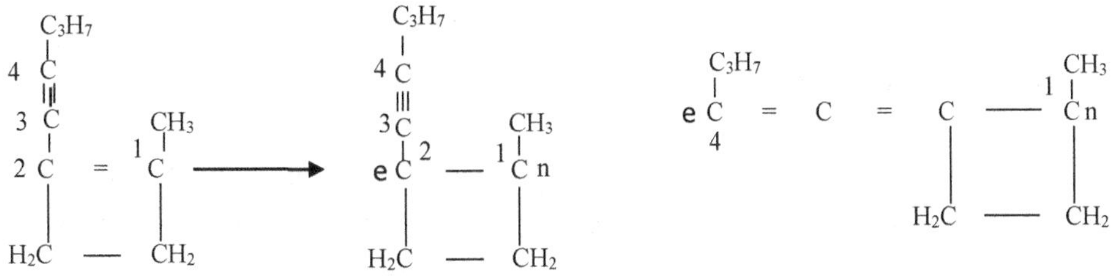

Resonance stabilized

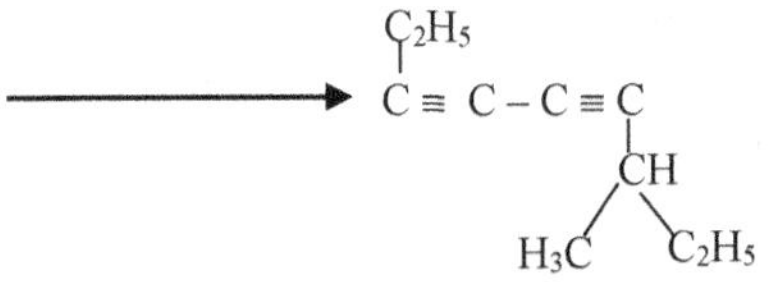

Resonance stabilized

for which the following can be seen to be valid-

$$R - CH = CH - \quad > \quad R \quad \geq \quad R - C \equiv C -$$

(Laws of Creations for Acetylenic cyclobutenes)

<u>Rule 778:</u> This rule of Chemistry for **Acetylenic Cyclobutenes where the externally located H atoms are both replaced with alkylane groups of the same or different capacities,** states that, it is only when the H atom in the CH_2 specially located inside the ring is replaced with a resonance stabilization group, that it can be made possible for the ring to be opened instantaneously as shown below-

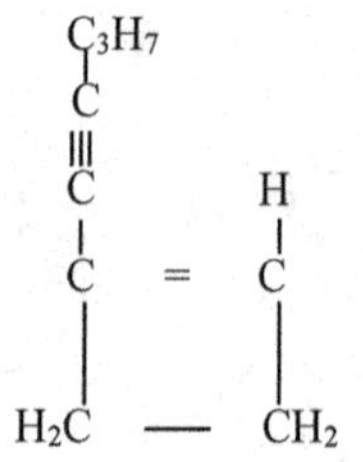

$$C_3H_7 - C \equiv C - C(=CH-H)(-H_2C-CH_2-)$$

;

Acetylene- Resonance stabilization provider

CANNOT BE OPENED

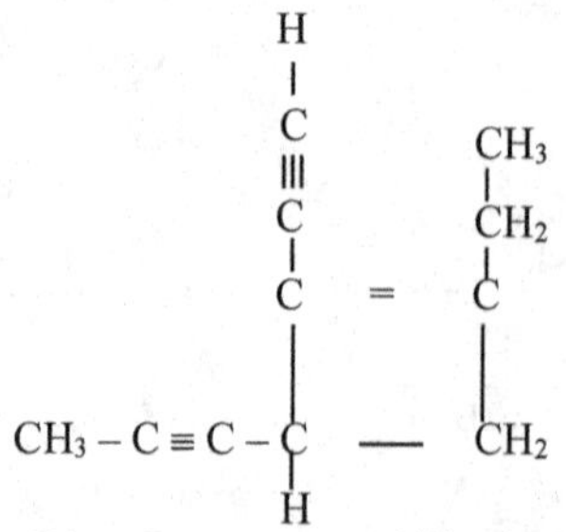

Ring- Resonance stabilization provider

CAN BE OPENED

(Laws of Creations for Acetylenic cyclobutenes)

Rule 779: This rule of Chemistry for **Cyclobutene,** states that, from the order of nucleophilicity of its two isomers shown below, the order of nucleophilicity between members of the three families can readily be obtained, noting that these are isomers not obtained by molecular rearrangement, but by other means.

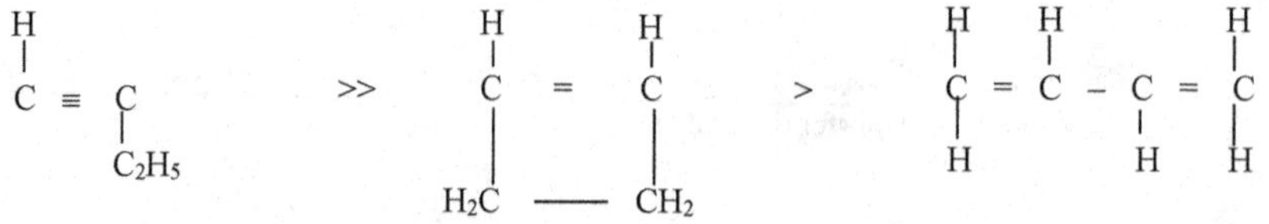

$$HC \equiv C - C_2H_5 \quad \gg \quad [\text{cyclobutene}] \quad > \quad H_2C=CH-CH=CH_2$$

ORDER OF NUCLEOPHILICITY

(Laws of Creations for Cyclobutenes)

Rule 780 This rule of Chemistry for **Cyclobutenes with externally located OH group,** states that, the followings are obtained for them, for which worthy of note is the fact that the two Nucleophilic

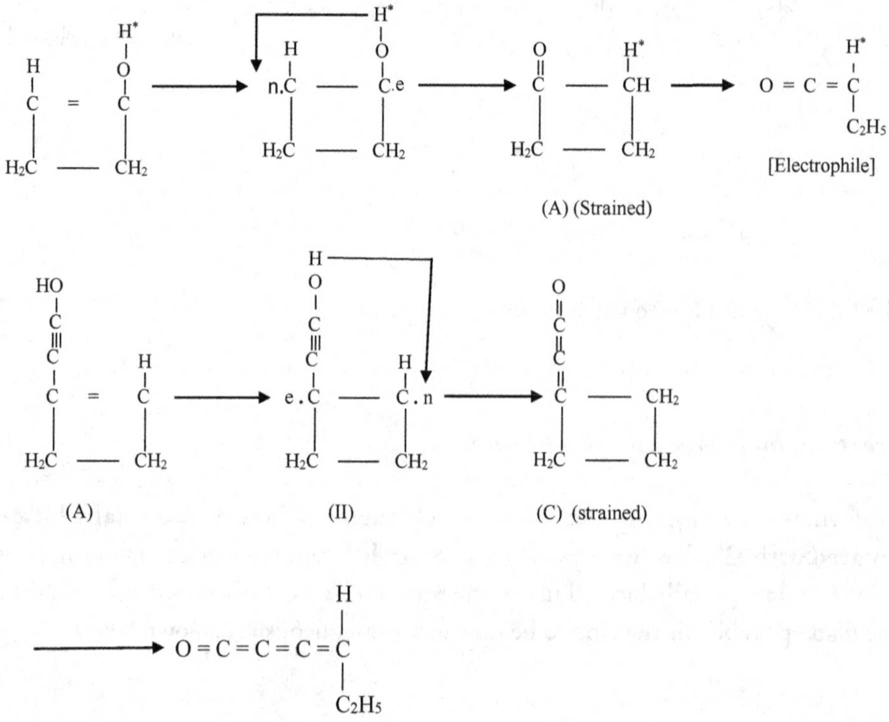

(A) (Strained)

[Electrophile]

(A)   (II)   (C) (strained)

$$\longrightarrow \quad O=C=C=C=C(-H)(-C_2H_5)$$

[Electrophile]

monomers above were converted or rearranged to give Electrophiles. *[See Rule 771]*
(Laws of Creations for Cyclobutenes)

Rule 781: This rule of Chemistry for **Cyclobutenes with NH_2 type of group placed on the acetylene center,** states that, this is a Nucleophile wherein their two mono-forms molecularly rearrange to produce non-resonance-stabilized acetylenic monomers of same characters (Electrophiles), with no limitations placed on R (CH_3) carried by the ring.

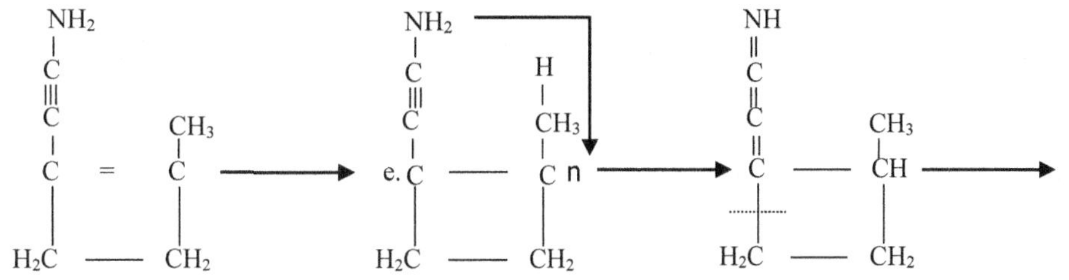

(A) (Resonance stabilized) - <u>Nucleophile</u>

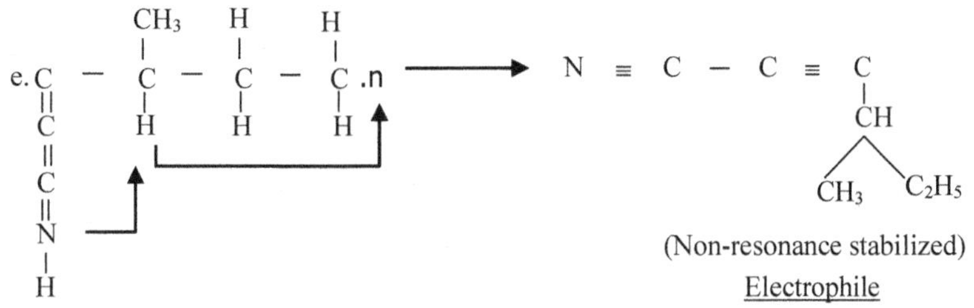

(Non-resonance stabilized)
<u>Electrophile</u>

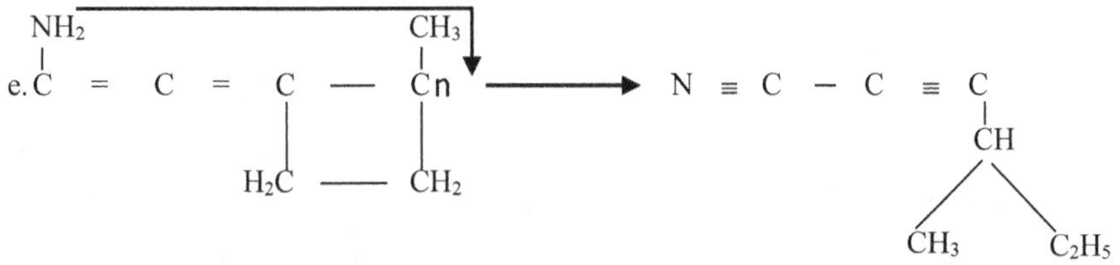

(Laws of Creations for Cyclobutenes)

Rule 782: This rule of Chemistry for **Cyclobutene NHR types of groups placed on the acetylene center,** states that, these are Nucleophiles wherein their two mono-forms molecularly rearrange to produce non-resonance stabilized Electrophiles

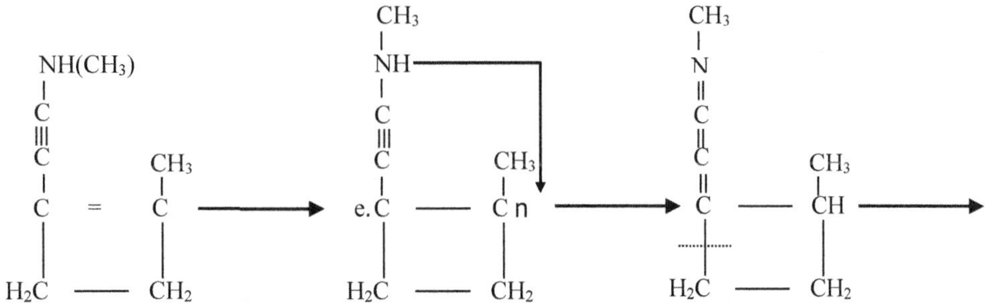

<u>Nucleophile</u>

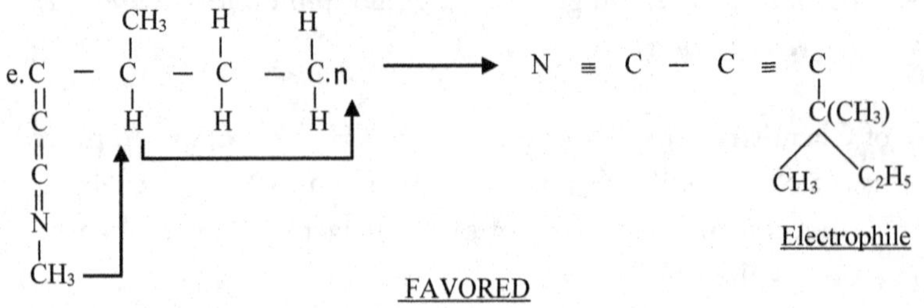

$$e.C - \overset{\overset{\displaystyle CH_3}{|}}{\underset{\underset{\underset{\underset{CH_3}{|}}{N}}{\overset{\|}{\underset{\|}{C}}}}{\overset{\|}{C}}} - \overset{\overset{\displaystyle H}{|}}{\underset{\underset{H}{|}}{C}} - \overset{\overset{\displaystyle H}{|}}{\underset{\underset{H}{|}}{C}}.n \longrightarrow N \equiv C - C \equiv \overset{\overset{\displaystyle C}{}}{\underset{\underset{\underset{\overset{CH_3 \quad C_2H_5}{\diagdown}}{C(CH_3)}}{|}}{}}$$

Electrophile

FAVORED

for which all the cases shown below will favor this rearrangement to give Electrophiles.

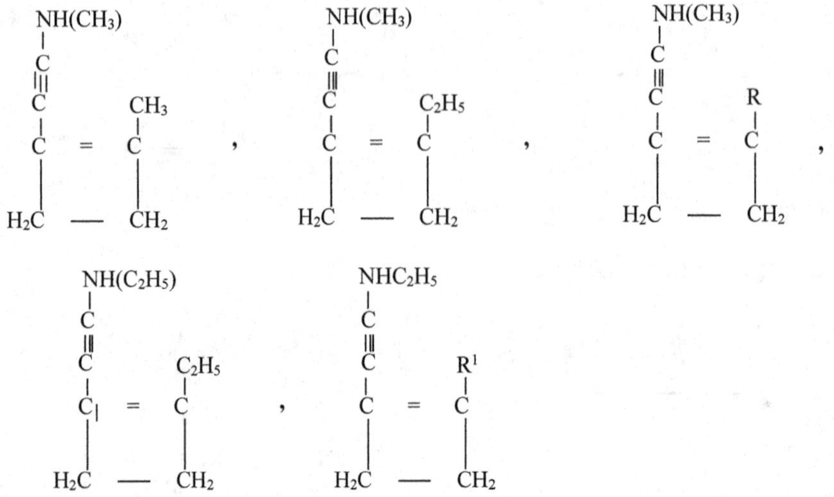

$$\underset{H_2C \;-\!\!\!-\!\!\!- \; CH_2}{\overset{NH(CH_3)}{\underset{\underset{|}{\overset{|}{C}}}{\overset{|}{\underset{C}{\overset{|||}{C}}}}}} \overset{CH_3}{\underset{|}{=}} C \quad , \qquad \underset{H_2C \;-\!\!\!-\!\!\!- \; CH_2}{\overset{NH(CH_3)}{\underset{\underset{|}{\overset{|}{C}}}{\overset{|}{\underset{C}{\overset{\|}{C}}}}}} \overset{C_2H_5}{\underset{|}{=}} C \quad , \qquad \underset{H_2C \;-\!\!\!-\!\!\!- \; CH_2}{\overset{NH(CH_3)}{\underset{\underset{|}{\overset{|}{C}}}{\overset{|}{\underset{C}{\overset{\|}{C}}}}}} \overset{R}{\underset{|}{=}} C \quad ,$$

$$\underset{H_2C \;-\!\!\!-\!\!\!- \; CH_2}{\overset{NH(C_2H_5)}{\underset{\underset{|}{\overset{|}{C}}}{\overset{|}{\underset{C}{\overset{\|}{C}}}}}} \overset{C_2H_5}{\underset{|}{=}} C \quad , \qquad \underset{H_2C \;-\!\!\!-\!\!\!- \; CH_2}{\overset{NHC_2H_5}{\underset{\underset{|}{\overset{|}{C}}}{\overset{|}{\underset{C}{\overset{\|}{C}}}}}} \overset{R^1}{\underset{|}{=}} C$$

Where   R $\leq$ CH$_3$,   R^1 $\leq$ C$_2$H$_5$

(Laws of Creations for Cyclobutenes)

Rule 783: This rule of Chemistry for **Cyclobutenes with NH$_2$ group placed on the ring side,** states that, the followings are to be expected; for which when the ring is opened instantaneously via molecular rearrangement of the third kind, non-resonance stabilized Nitrile may be favored.

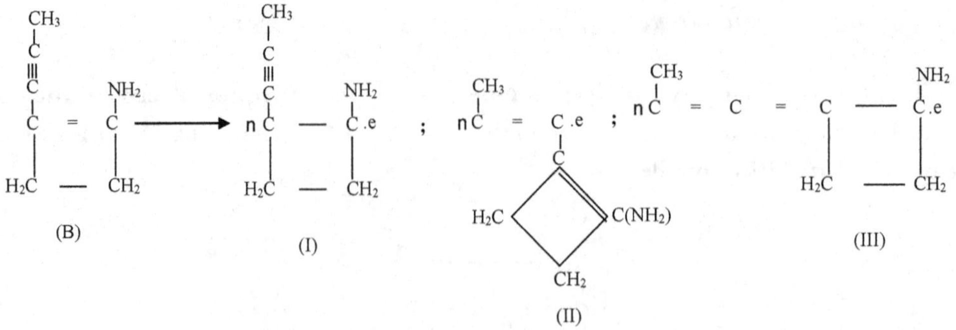

$$\underset{H_2C \;-\!\!\!-\!\!\!- \; CH_2}{\overset{CH_3}{\underset{\underset{|}{\overset{|}{C}}}{\overset{|}{\underset{C}{\overset{|||}{C}}}}}} \overset{NH_2}{\underset{|}{=}} C \longrightarrow \underset{H_2C \;-\!\!\!-\!\!\!- \; CH_2}{\overset{CH_3}{\underset{\underset{|}{\overset{|}{nC}}}{\overset{|}{\underset{C}{\overset{|||}{C}}}}}} - \overset{NH_2}{\underset{|}{C}}.e \quad ;$$

(B)   (I)

$$nC \overset{CH_3}{=} C.e \qquad ; \qquad$$

(II)

$$nC \overset{CH_3}{=} C = \underset{H_2C \;-\!\!\!-\!\!\!- \; CH_2}{C} - \overset{NH_2}{\underset{|}{C}}.e$$

(III)

(For structure (II)):

$$\underset{\underset{CH_2}{|}}{\overset{CH_3}{\underset{\underset{H_2C}{}}{\underset{|}{\overset{C}{\diamondsuit}}}C(NH_2)}}$$

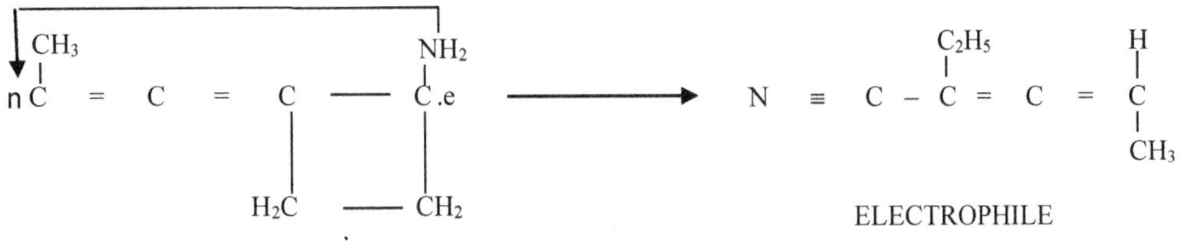
CH3
NH2
n C = C = C — C .e
H2C — CH2
N ≡ C – C = C = C
C2H5
H
CH3
ELECTROPHILE

(III)

(Laws of Creations for Cyclobutenes)

<u>Rule 784:</u> This rule of Chemistry for **Cyclobutenes with NR_2 placed on the acetylenic side,** states that, for them to be useful, H on CH_2 groups in the ring must be changed as shown below for specific cases.

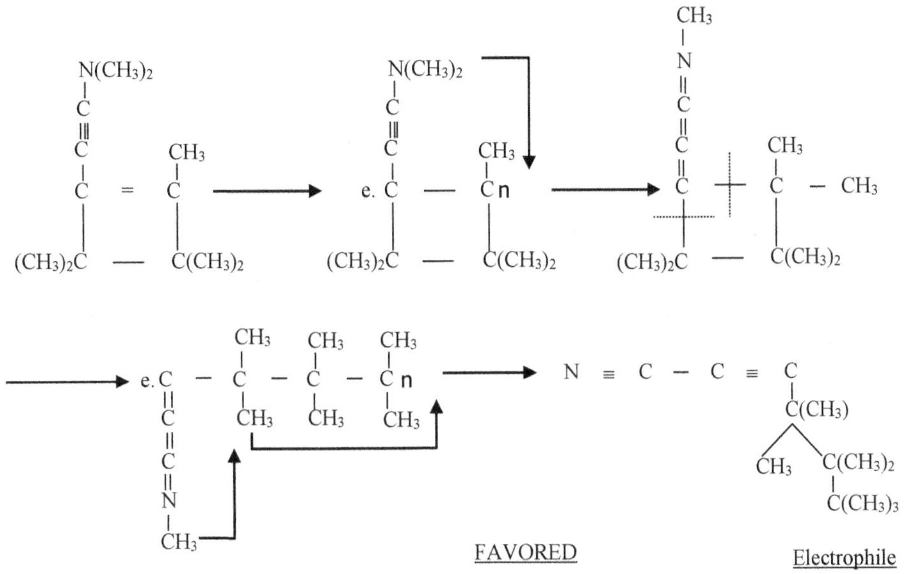
N(CH3)2
(CH3)2C — C(CH3)2
FAVORED
Electrophile

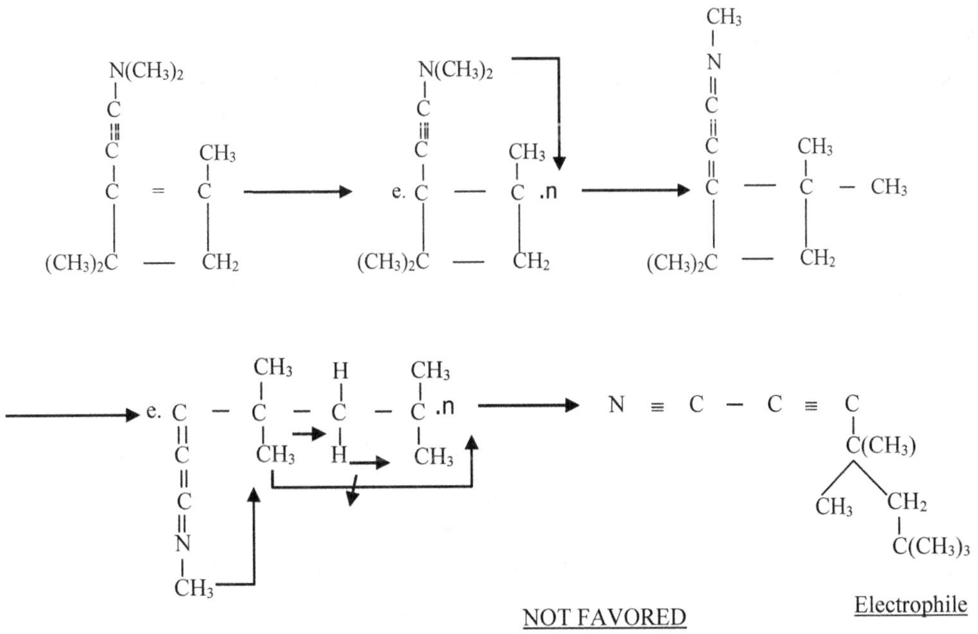
N(CH3)2
(CH3)2C — CH2
NOT FAVORED
Electrophile

in which the last case is said not to be favored because, CH_3 group cannot move to a center carrying H after the first movement.
(Laws of Creations for Cyclobutenes)

Rule 785: This rule of Chemistry for **first members of Cyclopropenes, Cyclobutenes, Cyclo-pentenes and higher,** states that, the point of scissions in their rings are as shown below,

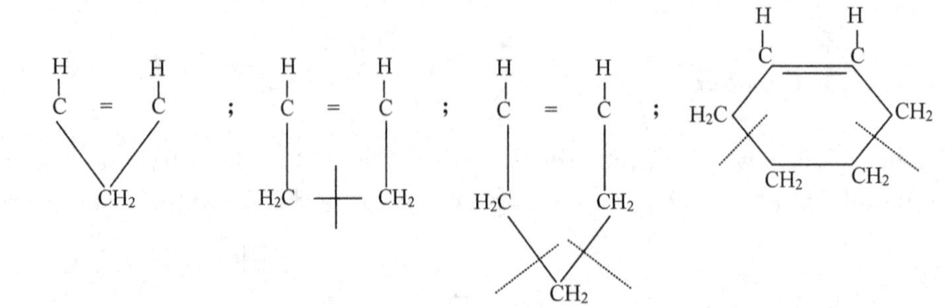

(I) <u>None</u>	(II) <u>1 - Point</u>	(III) <u>2- Points</u>	(IV) <u>2- Points</u>
<u>Cyclopropene</u>	<u>Cyclobutene</u>	<u>Cyclopentene</u>	<u>Cyclohexene</u>

<u>Points o f scission</u>

noting that for these first members, the most difficult to open is the four-membered ring since the radical-pushing potential difference between the only point of scission is zero, while larger sized rings have only two points of scissions, the points adjacently located to the double bond.
(Laws of Creations for Cycloalkenes)

Rule 786: This rule of Chemistry for **Cycloalkenes with or without radical-pushing alkylane groups,** states that, their rings can only be opened instantaneously or via molecular rearrangement of the third kind, only if the product obtained can either be resonance stabilized after opening with movement of electro-free-radicals or electro-free-radical transfer species or be an Electrophile.
(Laws of Creations for Cycloalkenes)

Rule 787: This rule of Chemistry for **Cyclopentenes with radical-pushing groups,** states that, the number of points of scission can be reduced by the presence and location of the substituents groups as shown below; for which in the absence of symmetry, the point of scission should be determined

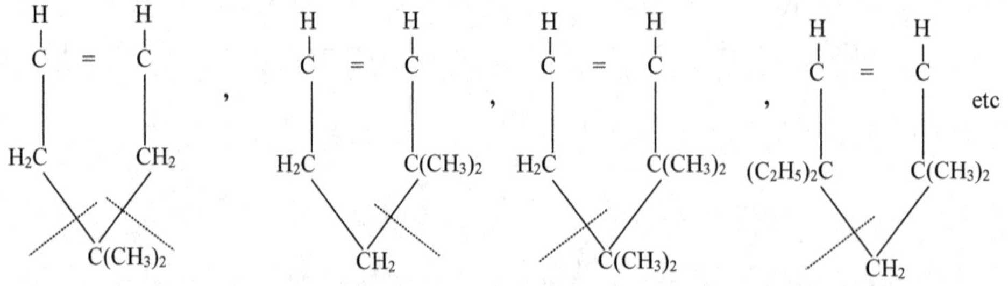

(I) <u>2- Points</u>	(II) <u>1- Point</u>	(III) <u>1- Point</u>	(IV) <u>1- Point</u>

<u>Points of scission</u>

by the single bond with *the largest radical-pushing potential difference,* noting that, even where many of them are opened instantaneously, the best that can be obtained are butadienes and acetylenes.
(Laws of Creations for Cyclopentenes)

<u>Rule 788:</u> This rule of Chemistry for **Family members of Cyclopentenes,** states that, when the ring is activated using a weak initiator, the followings are to be expected.

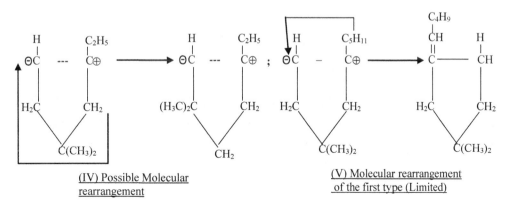

(I) Molecular rearrangement of the tenth type

(II) Impossible Molecular rearrangement

(III) Impossible Molecular rearrangement

(IV) Possible Molecular rearrangement

(V) Molecular rearrangement of the first type (Limited)

noting when transfer species of the sixth and tenth types are involved.

(Laws of Creations for Cyclopentenes)

Rule 789: This rule of Chemistry for **Cyclopentene,** states that, the possibility of having trans-or cis-isotactic placement of the monomer units along a chain is quite impossible; only trans-syndiotactic placement may be possible at very low polymerization temperatures,

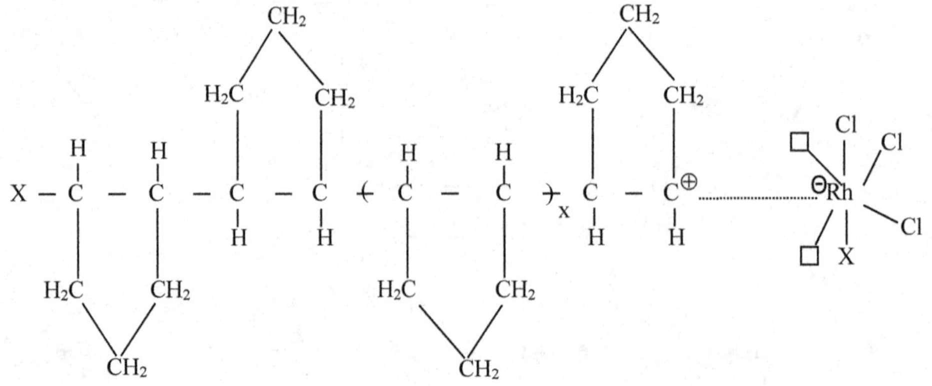

Trans- syndiotactic polycyclopentene growing chain

for which for the growing polymer chain above, transfer species of the first kind of the tenth type can be released to kill the chain electro-free-radically or positively as shown above, the same transfer species that will prevent the initiation of the monomer with negative charges and nucleo-free-radicals.
(Laws of Creations for Cyclopentene)

Rule 790: This rule of Chemistry for **The first member of cyclopentenes,** states that, if the ring is to be opened instantaneously, harsh operating conditions will be required, not the type for its decomposition-

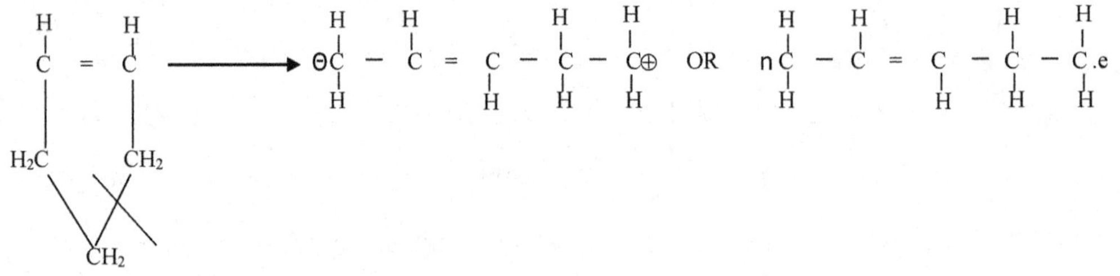

for which when possible without further decomposition, *vinyl cyclopropane* may be obtained or it can be used as a monomer via both routes to give polyenes.
(Laws of creations for Cyclopentene)

Rule 791: This rule of Chemistry for **Cyclopentenes carrying free-radical-pushing groups specially internally located as shown below,** states that, when their rings are opened instantaneously, substituted butadienes or cyclobutenes or vinyl cyclopropanes can be obtained from them depending on the operating conditions-

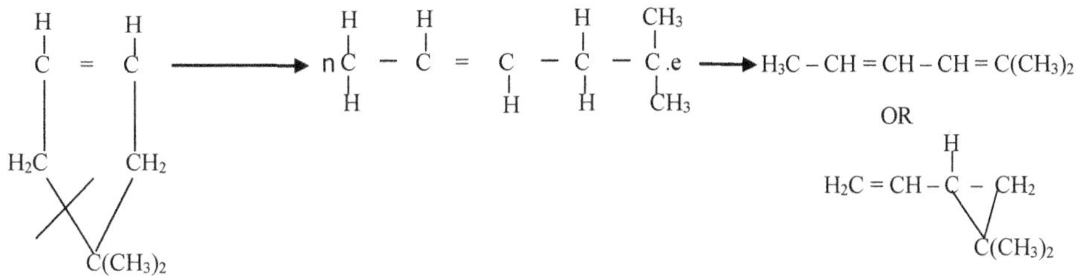

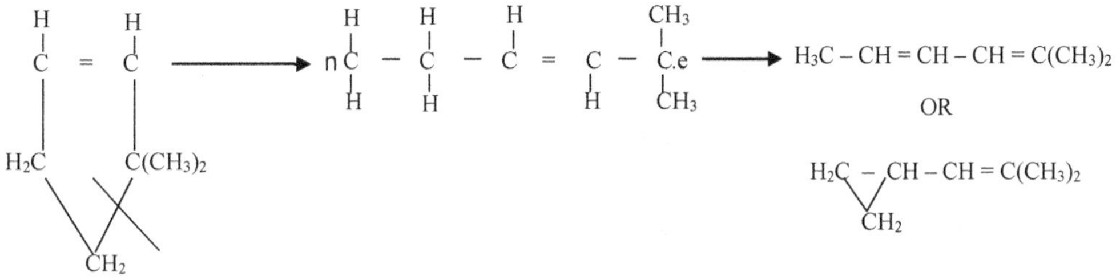

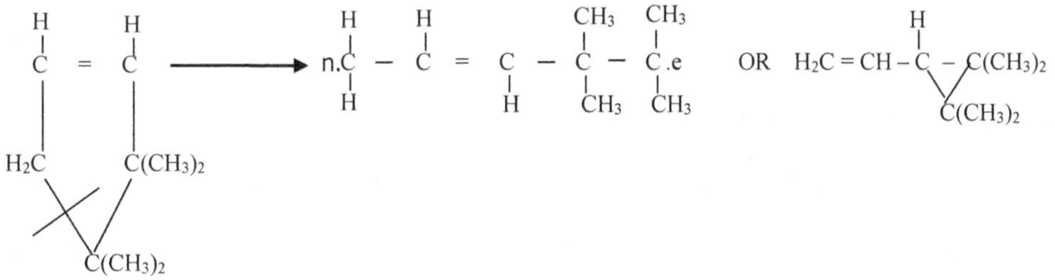

$$\begin{array}{ccc}
\overset{H}{\underset{|}{C}} = \overset{H}{\underset{|}{C}} & \longrightarrow & n\overset{H}{\underset{|}{\underset{H}{C}}} - \overset{H}{\underset{|}{C}} = \overset{}{C} - \overset{H}{\underset{|}{\underset{H}{C}}} - \overset{CH_3}{\underset{|}{\underset{CH_3}{C}}}.e \longrightarrow H_3C - CH = CH - CH = C(CH_3)_2
\end{array}$$

OR

$$H_2C = CH - \overset{H}{\underset{|}{C}} - CH_2 \quad \overset{|}{C(CH_3)_2}$$

$$\begin{array}{ccc}
\overset{H}{\underset{|}{C}} = \overset{H}{\underset{|}{C}} & \longrightarrow & n\overset{H}{\underset{|}{\underset{H}{C}}} - \overset{H}{\underset{|}{\underset{H}{C}}} - \overset{H}{\underset{|}{C}} = \overset{}{C} - \overset{CH_3}{\underset{|}{\underset{CH_3}{C}}}.e \longrightarrow H_3C - CH = CH - CH = C(CH_3)_2
\end{array}$$

OR

$$H_2C - CH - CH = C(CH_3)_2 \quad \underset{CH_2}{}$$

$$\begin{array}{ccc}
\overset{H}{\underset{|}{C}} = \overset{H}{\underset{|}{C}} & \longrightarrow & n.\overset{H}{\underset{|}{\underset{H}{C}}} - \overset{H}{\underset{|}{C}} = \overset{}{C} - \overset{CH_3}{\underset{|}{\underset{CH_3}{C}}} - \overset{CH_3}{\underset{|}{\underset{CH_3}{C}}}.e \quad OR \quad H_2C = CH - \overset{H}{\underset{|}{\underset{C(CH_3)_2}{C}}} - C(CH_3)_2
\end{array}$$

noting how and when transfer species and electro-free-radicals have been moved in all of them.
(Laws of Creations for Cyclopentene)

Rule 792: This rule of Chemistry for **Cyclopentenes,** states that, the second means by which the rings can be opened instantaneously is via the use of a ring obtained via molecular rearrangement of the first kind followed by molecular rearrangement of the third kind as shown below-

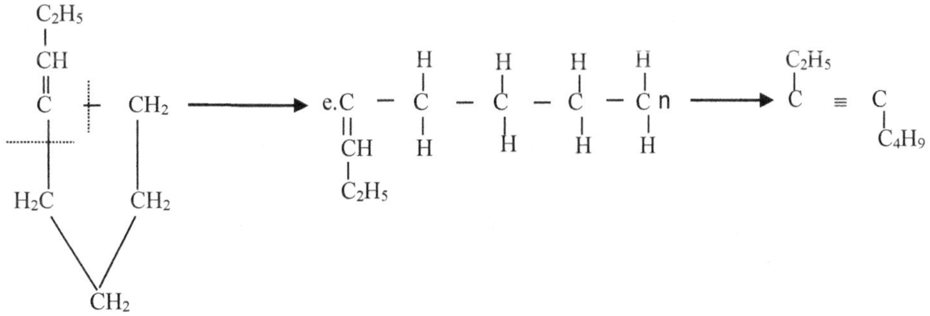

$$\begin{array}{ccc}
\overset{C_2H_5}{\underset{|}{\underset{CH}{\overset{||}{C}}}} + CH_2 & \longrightarrow & e.\overset{}{\underset{\underset{C_2H_5}{\overset{|}{CH}}}{\overset{||}{C}}} - \overset{H}{\underset{|}{\underset{H}{C}}} - \overset{H}{\underset{|}{\underset{H}{C}}} - \overset{H}{\underset{|}{\underset{H}{C}}} - \overset{\bar{H}}{\underset{|}{\underset{H}{C}}} n & \longrightarrow & \overset{C_2H_5}{\underset{|}{C}} \equiv \underset{C_4H_9}{\overset{}{C}}
\end{array}$$

$$H_2C \qquad CH_2$$
$$CH_2$$

for which the size of R group cannot exceed C_5H_{11} for the case above to give Acetylenes.
(Laws of Creations for Cyclopentene)

Rule 793: This rule of Chemistry for **Cyclopentenes shown below,** states that, in general the followings are to be expected of them-

413

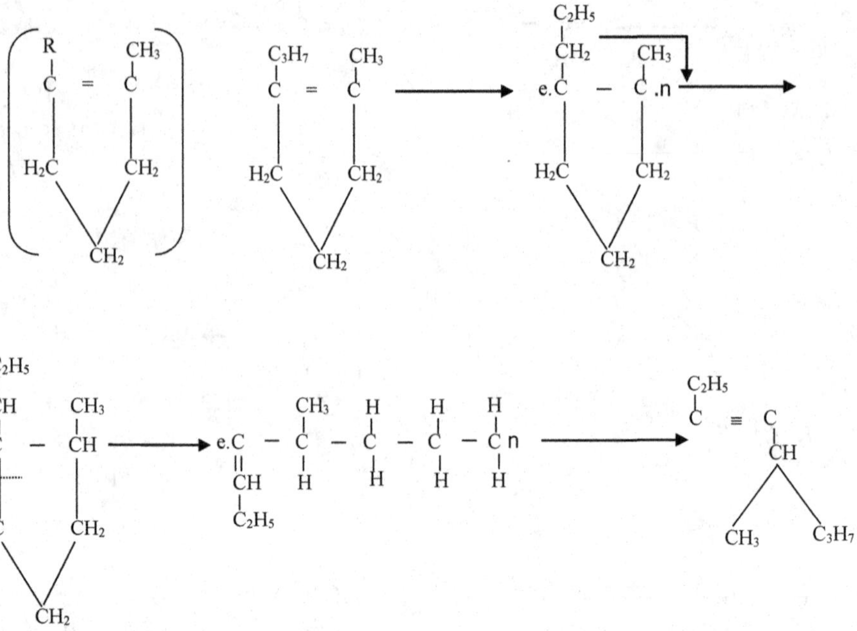

and the rearrangement stops here when the R group is C_6H_{13}.
(Laws of Creations for Cyclopropenes)

Rule 794: This rule of Chemistry for **Cyclopentene shown below**, states that, the followings are also to be expected of them-

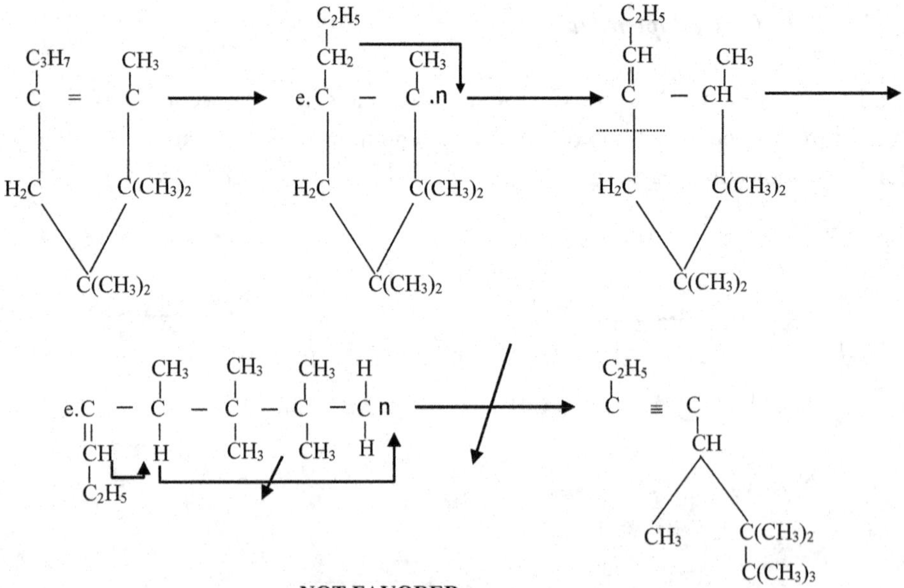

NOT FAVORED

for which molecular rearrangement of the third kind is not favored, since CH_3 which should be the first to move cannot be moved, its capacity being greater than H.
(Laws of Creations for Cyclopropenes)

Rule 795: This rule of Chemistry for **Cyclopentenes,** states that, following the same pattern of analysis of cyclopropenes and cyclobutenes, the following is obvious.

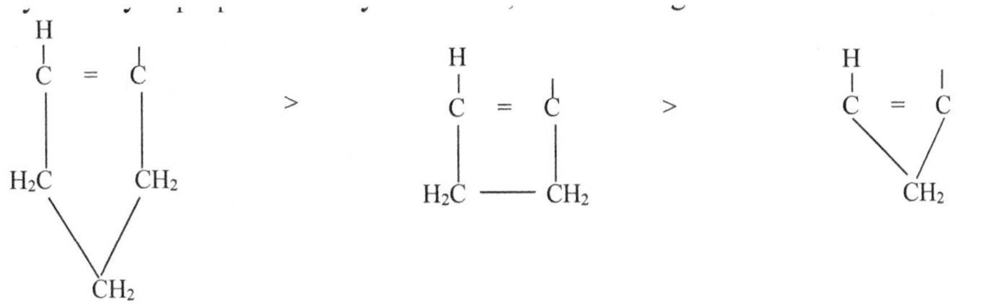

Order of radical-pushing capacity of resonance stabilization groups

(Laws of Creations for Cycloalkenes)

Rule 796: This rule of Chemistry for **Vinyl cyclopentenes carrying alkene resonance stabilization group,** states that, the provider of resonance stabilization for all of them is the ring which when the R

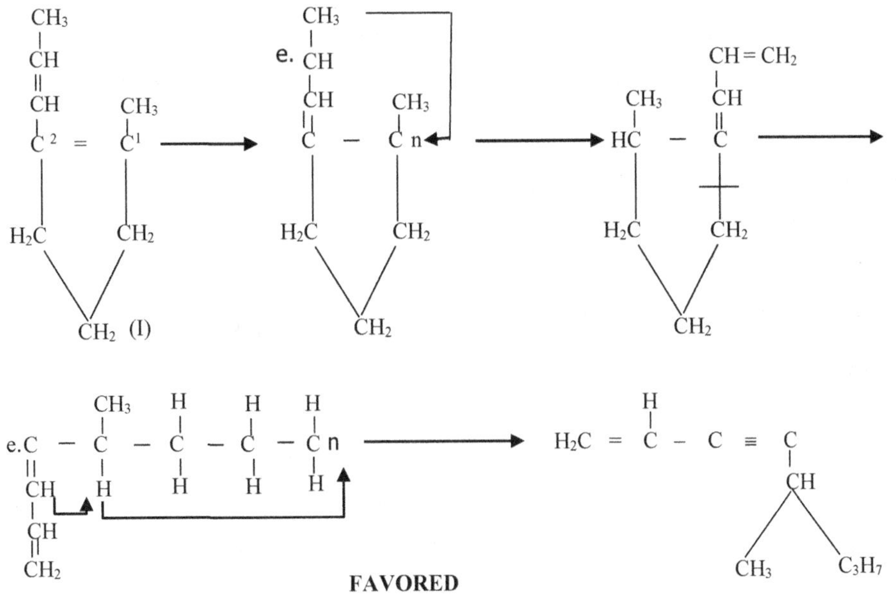

groups are equal, molecular rearrangement of the third kind can take place as shown and the ring cannot be opened instantaneously on its own when mildly decomposed without rearrangement.

(Laws of Creations for Vinyl cyclopentenes)

Rule 797: This rule of Chemistry for **Vinyl Cyclopentenes carrying alkene resonance stabiliza-tion groups with R group on the alkene greater in radical-pushing capacity than the R¹ group on the ring,** states that, while the ring cannot be opened instantaneously, it can however be made to

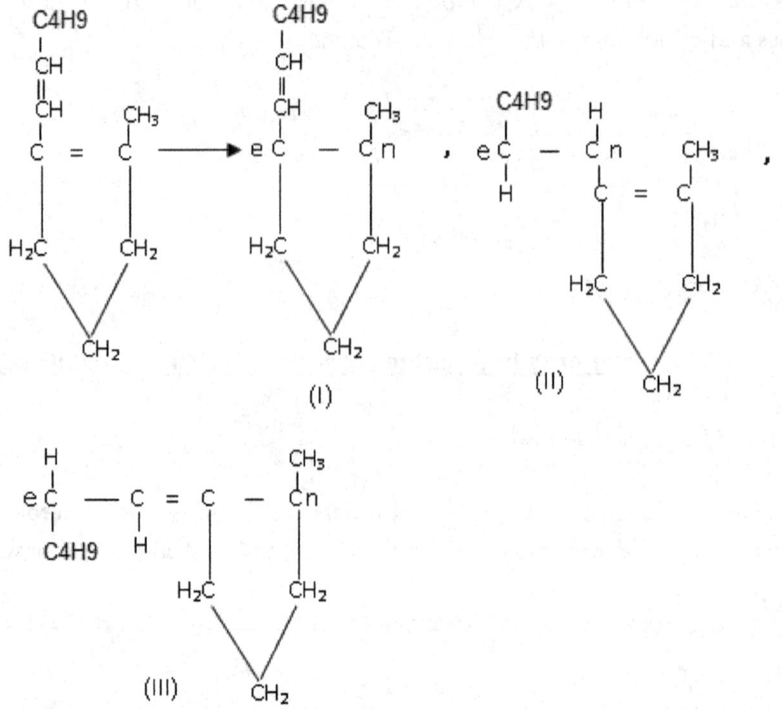

$$
\begin{array}{ccc}
\underset{\overset{|}{\underset{\overset{|}{C}}{CH}}}{\overset{C_4H_9}{\underset{\parallel}{\overset{}{}}}}{\underset{\overset{}{}}{\overset{}{}}} & & \\
\end{array}
$$

(Structure showing: C4H9–CH=CH–C(=C with CH3)– ring with H2C, CH2, CH2 cyclopentene) →
$e\overset{}{C} - \overset{}{C}n$ (I), $e\overset{}{C} - \overset{}{C}n$ with CH3, C=C (II)

(III) $e\overset{H}{\underset{C_4H_9}{C}} - C = C \overset{CH_3}{\underset{H}{-}} \overset{}{C}n$ with H2C, CH2, CH2

undergo molecular rearrangements to give resonance stabilized monomer or compound (vinyl acetylene) which can only be polymerized electro-free-radically.

(Laws of Creations for Vinyl cyclopentenes)

Rule 798: This rule of Chemistry for **Cyclopentenes**, states that, the followings are valid for groups of cyclopentenes when H is the group held in Equilibrium state of existence-

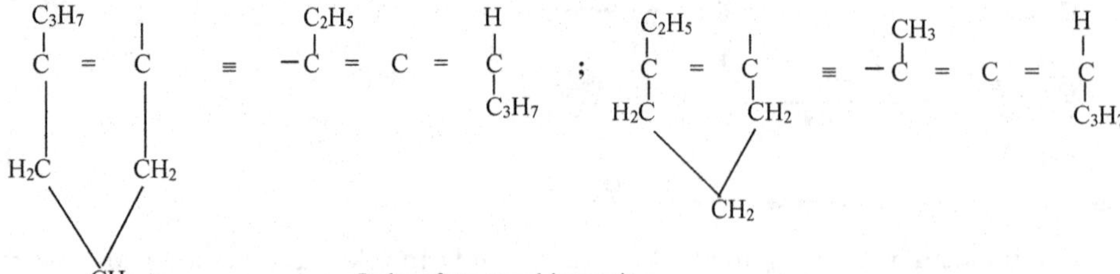

$$
\underset{\overset{|}{H_2C}\quad\overset{|}{CH_2}}{\overset{C_3H_7}{\underset{|}{C}} = \overset{|}{C}} \equiv -\overset{|}{C} = C = \overset{H}{\underset{C_3H_7}{\overset{|}{C}}} \quad ; \quad \underset{\overset{|}{H_2C}\quad\overset{|}{CH_2}}{\overset{C_2H_5}{\underset{|}{C}} = \overset{|}{C}} \equiv -\overset{CH_3}{\underset{|}{C}} = C = \overset{H}{\underset{C_3H_7}{\overset{|}{C}}}
$$

with CH2 ring

Order of structural isomerism.

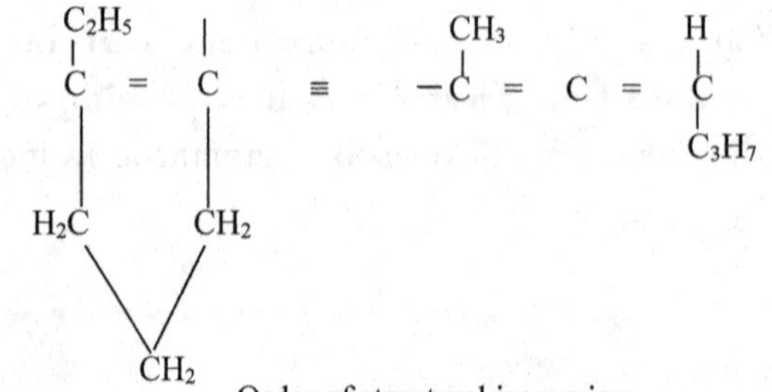

$$
\underset{\overset{|}{H_2C}\quad\overset{|}{CH_2}}{\overset{C_2H_5}{\underset{|}{C}} = \overset{|}{C}} \equiv -\overset{CH_3}{\underset{|}{C}} = C = \overset{H}{\underset{C_3H_7}{\overset{|}{C}}}
$$

with CH2 ring

Order of structural isomerism.

(Laws of Creations for Cyclopropenes)

416

Rule 799: This of Chemistry for **Cyclopentene,** states that, based on the character of the compound, the followings are valid-

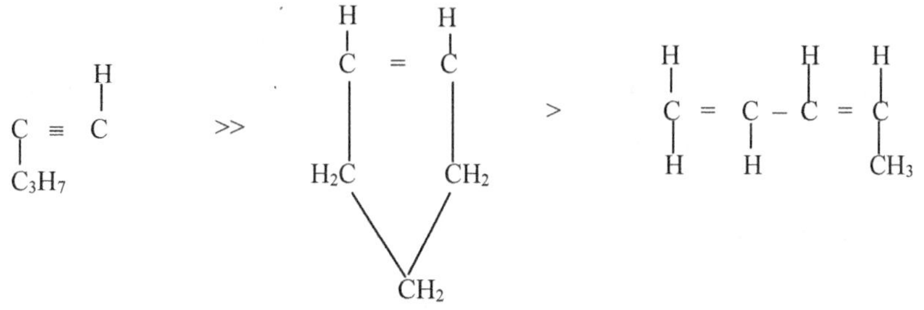

ORDER OF NUCLEOPHILICITY

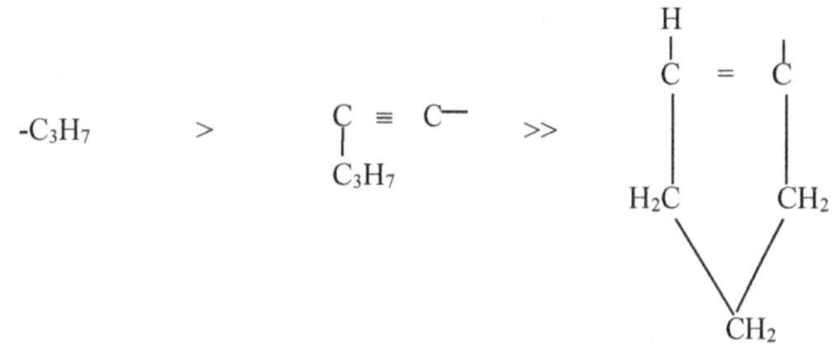

ORDER OF RADICAL-PUSHING CAPACITY

for which one can observe that the larger the size of the ring, the more nucleophilic is the monomer and the second π-bond in the triple bond is indeed the invisible π- bond in the ring.
(Laws of Creations for Cyclopentene)

Rule 800: This rule Chemistry for **Vinyl Cyclopentenes in which the R carried by the alkene is less in radical-pushing capacity than R^1 carried by the ring,** states that, neither instantaneous opening of the ring nor molecular rearrangement of the third kind favored if resonance stabilized monomers are to be produced.

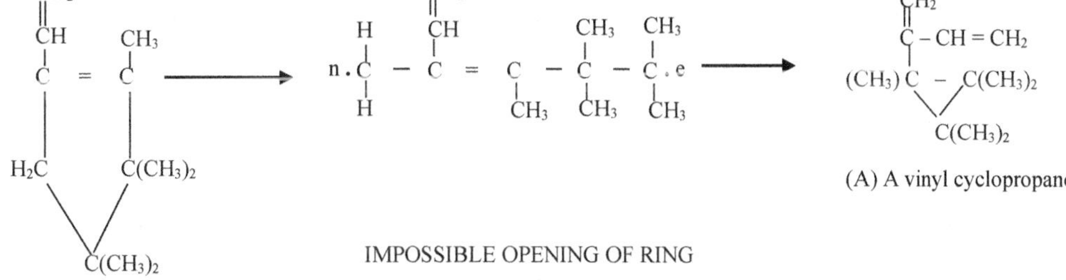

IMPOSSIBLE OPENING OF RING

(A) A vinyl cyclopropane

for which the alkenyl group can be observed to be the resonance stabilization provider as opposed to the ring, making the opening above to be impossible.
(Laws of Creations for Vinyl Cyclopentenes)

Rule 801: This rule of Chemistry for **Cyclopentene with acetylenic resonance stabilization group in which both the ring and group are carrying R^1 and R radical-pushing groups respectively of equal capacity,** states that, these cannot be made to undergo molecular rearrangement of the third kind as shown below and their rings cannot also be opened instantaneously-

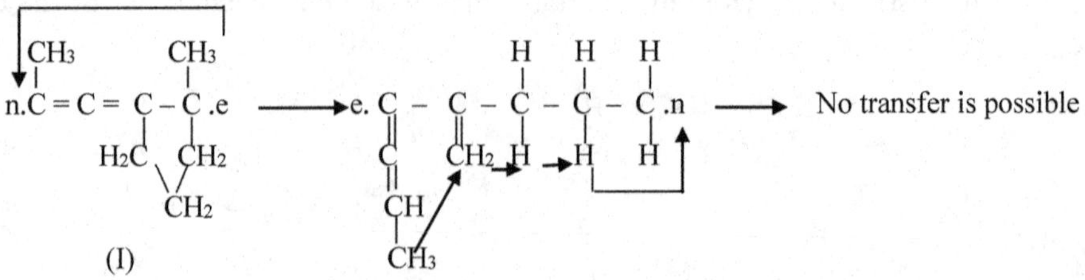

$$n.C = C = C - C.e \longrightarrow e.C - C - C - C - C.n \longrightarrow \text{No transfer is possible}$$

(I)

for which this can be observed to be unlike the case of alkenic group when R is equal to R', taking note of the steps involved during opening of the rings and transfer of transfer species.
(Laws of Creations for Acetylenic Cyclopropenes)

Rule 802: This rule of Chemistry for **Cyclopentenes with acetylenic resonance stabilization group of the type shown below that in which R is H, and R^1 on the ring is an alkylane group**, states that, no molecular rearrangement is favored to give a diyne as shown below –

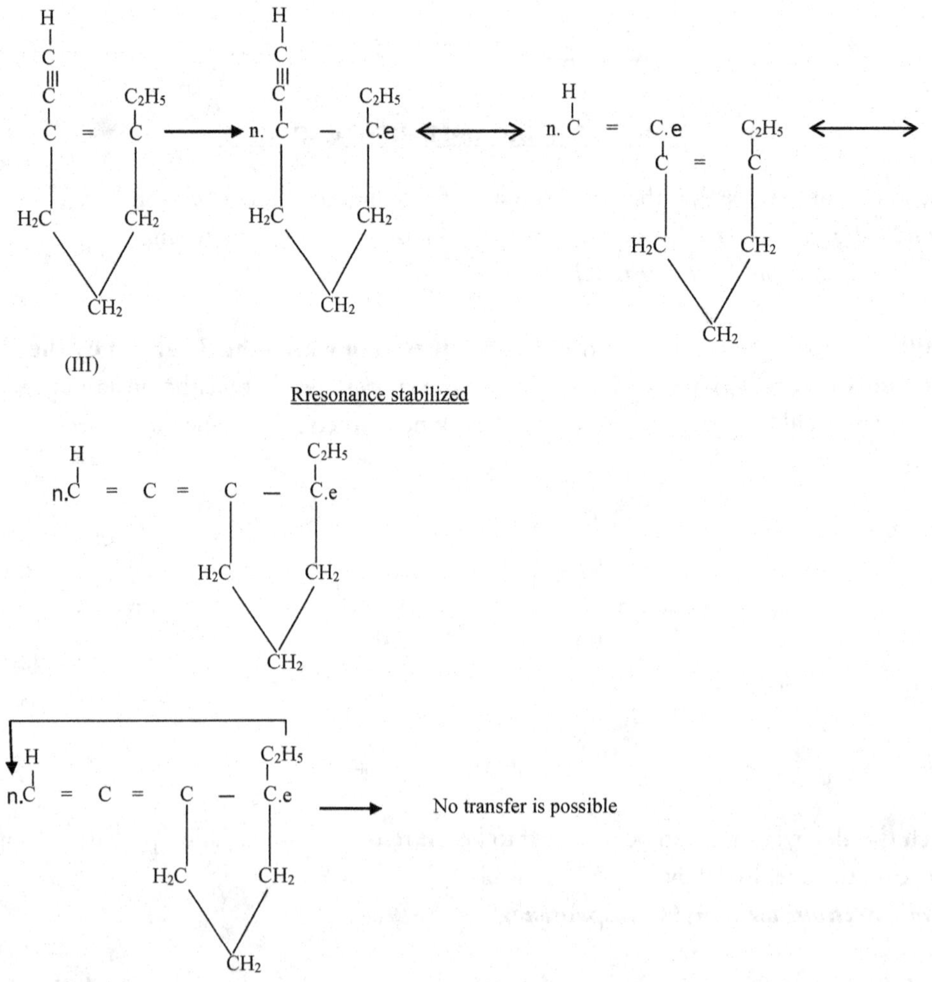

(III)

Rresonance stabilized

(Laws of Creations for Acetylenic Cyclopropenes)

418

Rule 803: This rule of Chemistry for **Cyclopropenes with acetylenic resonance stabilization groups in which R on acetylene is greater than R^1 on the ring,** states that, the followings are obtained for them using H for R^1 to produce diynes-

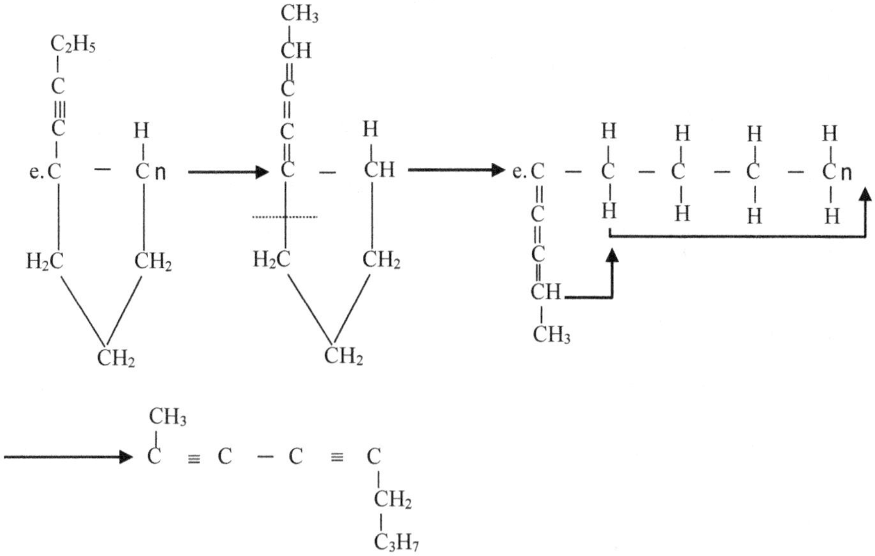

[MOLECULAR REARRANGEMENT OF THIRD KIND]

only radically, noting that for the case above rearrangement ceases when R is greater than C_5H_{11}.

(Laws of Creations for Acetylenic Cyclopropenes)

Rule 804: This rule of Chemistry for **Cycloalkenes,** states that, the order of the nucleophilicity of the members increases, with increasing size of the rings when same groups are carried.

| Cyclohexenes | > | Cyclopentenes | > | Cyclobutenes | > | Cyclopropenes |

Order of nucleophilicity

(Laws of Creations for Cycloalkenes)

Rule 805: This rule of Chemistry for **Electrophilic Cyclopentenes,** states that, only the types shown below will favor polymerization via their natural routes via the double bond, for which (III) will be more difficult to polymerize than (II) which in turn will be more than (I), due to steric limita-

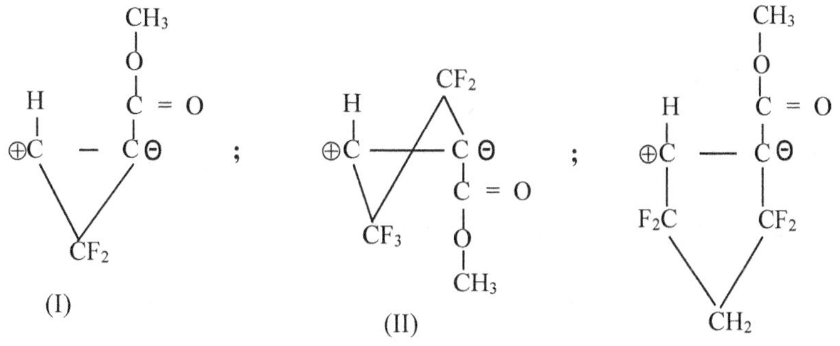

Possibke Polymerizable Electrophiles via double bond

tions and countless numbers of conformations that can be possible, increasing with increasing size of the ring, noting that, it is largely for this reason, cycloheptene (seven-membered ring) cannot be polymerized via the double bond.
(Laws of Creations for Electrophilic Cycloalkenes)

Rule 806: This rule Chemistry for **Cyclohexene,** states that, when decomposed in a single-pulse shock tube, ethene and 1,3-butadiene are the products as shown below-

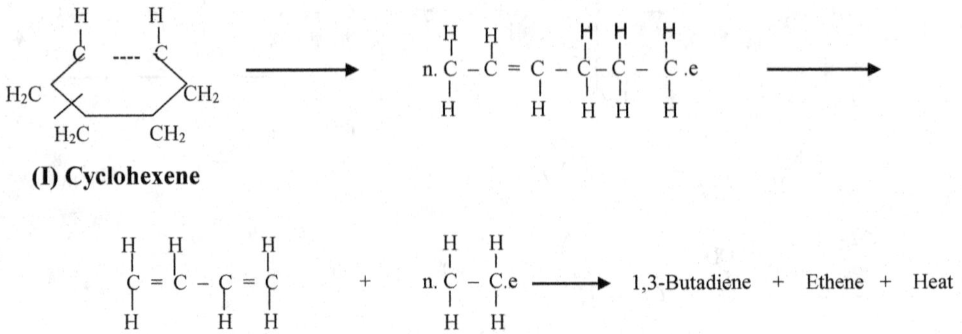

(I) Cyclohexene

noting that, the ring was instantaneously opened, followed by loss of ethene and formation of butadiene a resonance stabilized product.
(Laws of Creations for Cyclohexene)

Rule 807: This rule of Chemistry for **1-methylcyclohexene,** states that, when decomposed in a single-pulse shock tube, 1,3-butadiene and propene are the products as shown below-

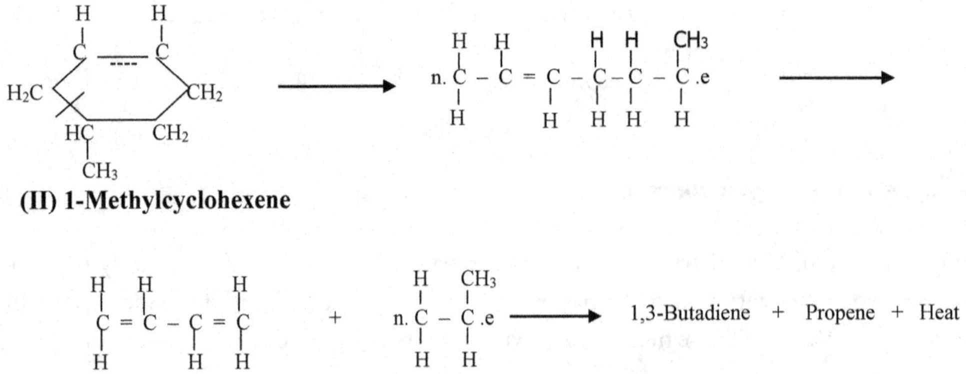

(II) 1-Methylcyclohexene

noting that the ring was opened instantaneously, followed by release of propene and then the formation of 1,3-butadiene a resonance stabilized product.
(Laws of Creations for Cyclohexenes)

Rule 808: This rule of Chemistry for **1-Vinyl cyclohexene,** states that, when decomposed in a single –pulse shock tube, 1,3-butadiene is the only product as shown below-

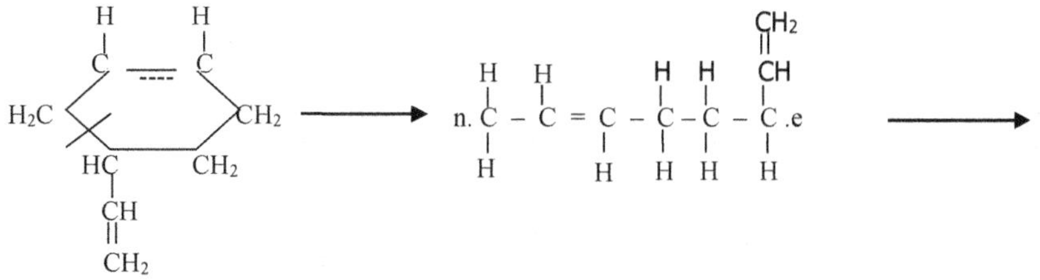

$$\text{(III) 4-vinylcyclohexene}$$

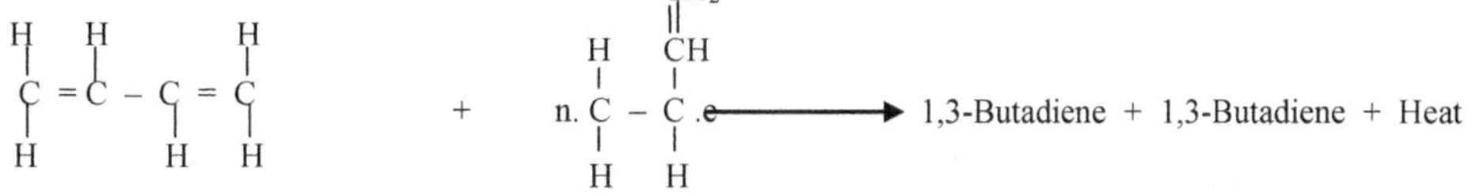

noting that the ring was opened instantaneously, followed by release of 1,3-butadiene and then the formation of 1,3-butadiene both resonance stabilized products all these being Decomposition mechanisms. in a single stage.

(Laws of Creations for Cyclohexenes)

Rule 809: This rule of Chemistry for **Cyclohexene,** states that, if operating conditions can be found to keep the cyclohexene in Equilibrium state of existence, then the followings will be obtained.

Stage 1:

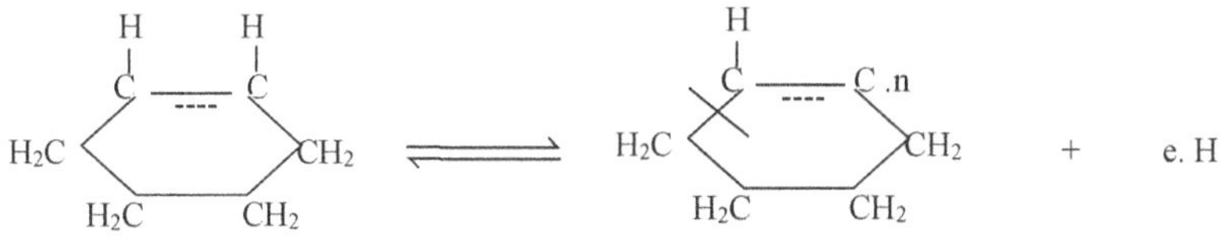

(I) Cyclohexene　　　　　　　　　　(A)

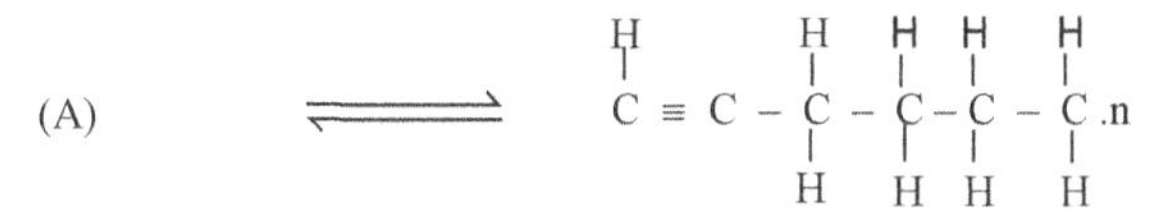

(A) ⇌ (B)

(B)　+　H.e　⟶　$HC \equiv C(C_4H_9)$　　　　　i)

Overall equation:　Cyclopentene　⟶　Butyl acetylene　　　　　ii)

noting that, just as cyclopentene will decompose via Equilibrium mechanism to propyl acetylene, so also cyclohexene will decompose to give butyl acetylene.

(Laws of Creations of Cyclohexene)

Rule 810: This rule of Chemistry for **Cyclohexene,** states that, when the ring cannot be opened instantaneously, when positively-charged coordination initiators without vacant orbitals are used, the followings are obtained-

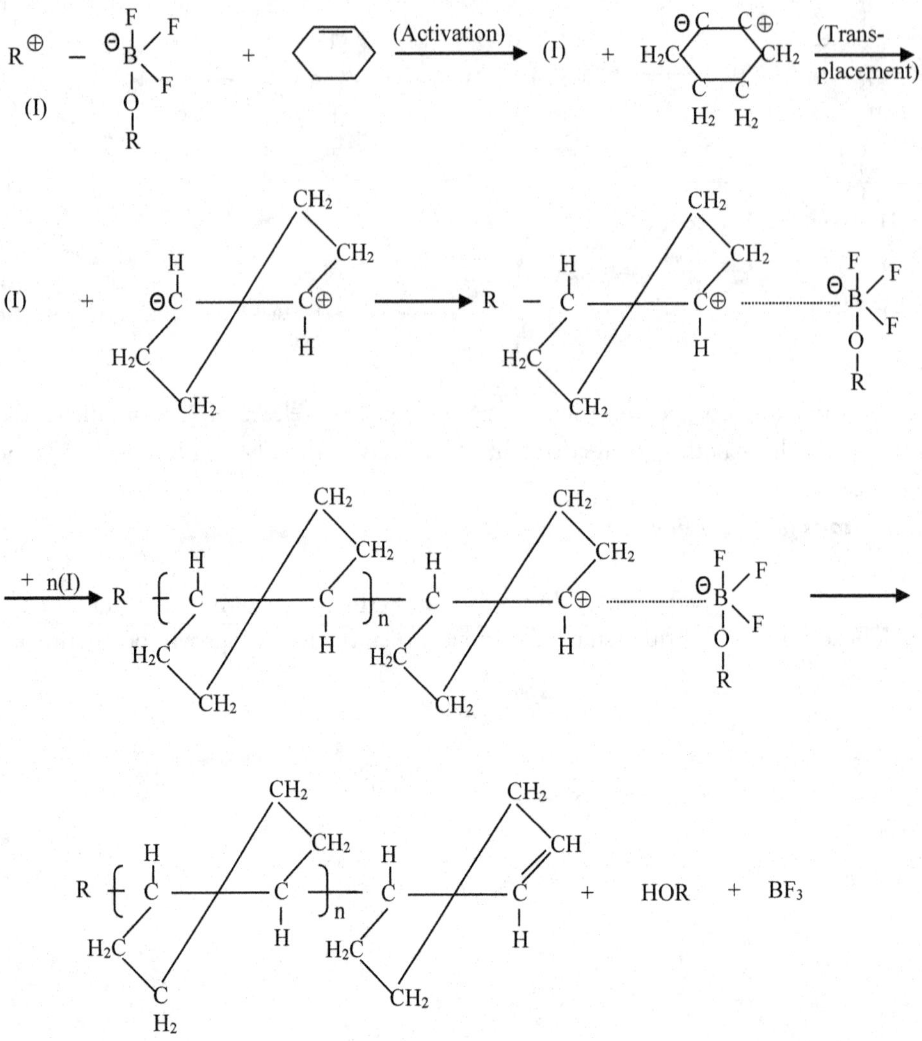

Transfer species of 1ˢᵗ kind of tenth type

for which mixed trans-di-isotactic and trans-di-syndiotactic placements are obtained in the absence of reservoir(s).

(Laws of Creations for Cyclohexene)

Rule 811: This rule of Chemistry for **The first member of cyclohexenes,** states that, molecular rearrangement of the tenth type readily takes place to produce same activated monomer, taking note of the manner in which H atoms on the positive end have been moved in two steps beginning

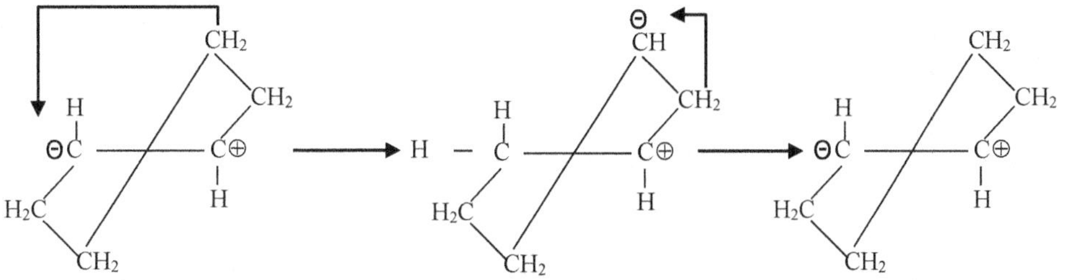

Molecular rearrangement of the tenth type

with the closest to the negative center; for which with cyclopropene and cyclobutene, one step is needed, while with cyclopentene and cyclohexene, two steps are needed and so on.
(Laws of Creations for cycloalkenes)

Rule 812: This rule of Chemistry for **Cyclohexene where one H atom located on the double bond is replaced with alkylane groups,** states that, the followings are obtained with CH_3 and with the use of an initiator with two vacant orbitals for which ***trans-di-syndiotactic*** placement is obtained and

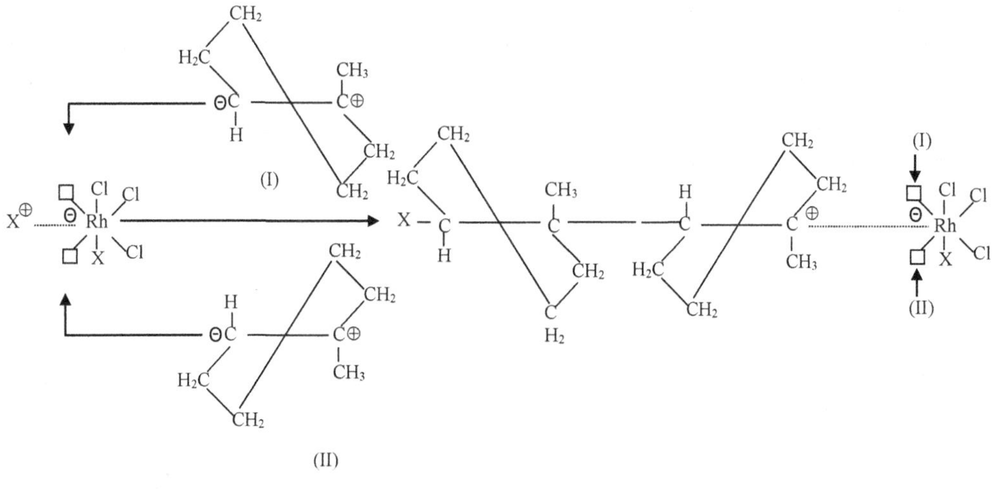

with one vacant orbital, ***trans –di-isotactic*** placement is obtained.
(Laws of Creations for Cyclohexenes)

Rule 813: This rule of Chemistry for **Cyclohexenes,** states that, it will need a group such as C_5H_{11} to alter the orientation as shown below; for which transfer species of the sixth type will be involved instead of the tenth type; for which in view of the size of C_5H_{11} group, full propagation may not be

favored.
(Laws of Creations for Cyclohexenes)

Rule 814: This rule of Chemistry for **Cyclohexenes with free-radical-pushing group of greater capacity greater than those carried in the ring,** states that, if the ring is strained enough after molecular rearrangement of the first kind, then it can be opened as follows-

("Asssumed" strained)

noting that the rearrangement is favored up C_6H_{13}
(Laws of Creations for Cyclohexenes)

Rule 815: This rule of Chemistry for **Cyclohexene,** states that, when made to carry a radical-pushing group of greater capacity than alkylane groups, such as NH_2, OH, the followings are to be expected for type of cyclohexene used below-for which rearrangement is not favored, because of the type of cyclohexene used; otherwise if it was just cyclohexene, it would have been favored to give a Nitrile.

(Laws of Creations for Cyclohexenes)

Rule 816: This rule of Chemistry for **Cyclohexenes,** states that, in the presence of resonance stabili-zation groups, the followings are obtained for the first and other members -

Vinyl cyclohexene ; ; ($R^1 > R$) ; etc

Cyclohexene diyne ; ; Cyclohexene Triyne ; etc

Examples of some resonance stabilized cyclohexenes.

for which like cyclobutene, one knows what to expect for all of them, noting that though they look similar, they can never be the same for all operating conditions.
(Laws of Creations for Vinyl and Acetylenic Cyclohexenes)

Rule 817: This rule of Chemistry for **Cyclohexenes,** states that, when radical-pulling groups such as COOR, CONR$_2$, CF$_3$ etc. types are involved, no polymerization route will be favored by the monomers, due to presence of transfer species via both routes and steric limitations, noting that in the presence of strong pulling forces provided by these unique groups on the activation centers (visible and invisible), their rings can never be opened instantaneously.
(Laws of Creations for Cyclohexenes)

Rule 818: This rule of Chemistry for **Cyclodienes,** states that, these are cases with two double bonds in the ring conjugatedly placed, for which this is not possible with three-membered ring (i.e. cyclopropadiene does not exist), but only for four-membered rings and above.
(Laws of Creations for Cyclodienes)

Rule 819: This rule of Chemistry for **Cyclodienes,** states that, these are also cases with two double bonds in the ring cumulatively placed, for which this is not possible for three- and four- membered rings, in view of the very large strain energy that exists in their rings and therefore the existence of cycloallenes is impossible for them.
(Laws of Creations for Cyclodienes)

Rule 820: This rule of Chemistry for **Cyclodienes,** states that, these are also cases with two double bonds isolatedly placed, for which this is not possible for three-, four- and five- membered rings, but only for six-membered rings and above.
(Laws of Creations for Cyclodienes)

Rule 821: This rule of Chemistry for **Cyclodienes,** states that, using three-membered ring, the followings are valid for them-

Cycloallene - Non existent

Order of strain energy in three-membered rings

(Laws of Creations for Cyclodienes)

Rule 822: This rule of Chemistry for **Cyclopropene and Cyclobutadiene,** states that, based on the closed groups carried by the C = C bond centers, the followings are valid for them-

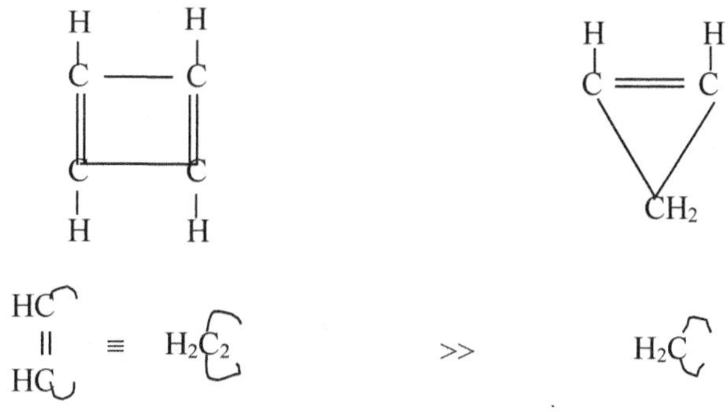

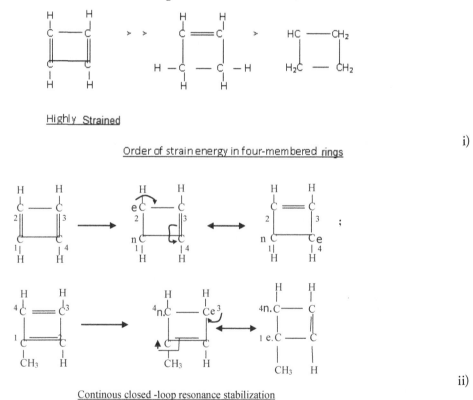

ORDER OF RADICAL-PUSHING CAPACITY

for which due to the large SE provided by the group in cyclobutadiene, the favored full existence of cyclobutadiene at STP is not possible.

(Laws of Creations for Cyclobutadiene)

Rule 823: This rule of Chemistry for **Cyclobutadiene,** states that, this compound which favors only transient existence favors ***continuous closed-loop resonance stabilization*** as shown below.

Highly Strained

Order of strain energy in four-membered rings

i)

Continous closed -loop resonance stabilization

ii)

(Laws of Creations for Cyclobutadiene)

Rule 824: This rule of Chemistry for **Cyclobutadiene,** states that, the transient existence favored by this compound is as a result of the fact that at STP, the compound is always in Activated/Equilibrium states of existence all the time for which the followings take place-

Stage 1:

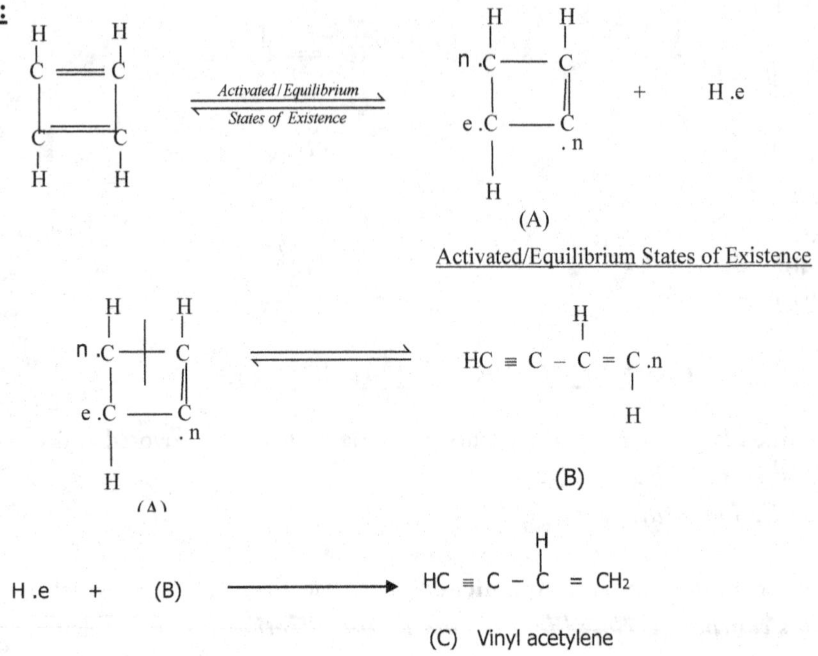

(A)

Activated/Equilibrium States of Existence

(B)

$$HC \equiv C - \overset{H}{\underset{H}{C}} = CH_2$$

(C)   Vinyl acetylene   i)

Stage 2:

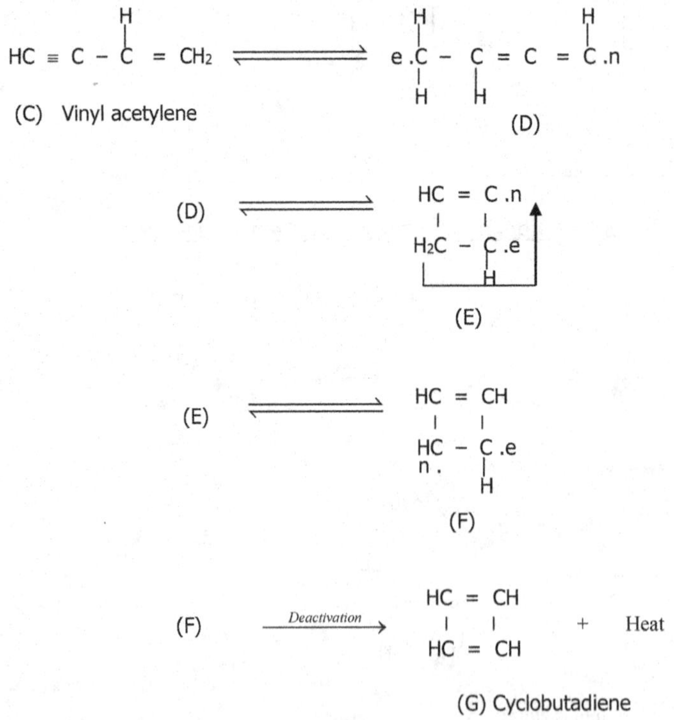

$$HC \equiv C - \overset{H}{C} = CH_2 \rightleftharpoons e.\overset{H}{\underset{H}{C}} - C = C = \overset{H}{C}.n$$

(C)   Vinyl acetylene   (D)

$$(D) \rightleftharpoons \begin{array}{c} HC = C.n \\ | \quad | \\ H_2C - C.e \\ H \end{array}$$

(E)

$$(E) \rightleftharpoons \begin{array}{c} HC = CH \\ | \quad | \\ n.HC - C.e \\ H \end{array}$$

(F)

$$(F) \xrightarrow{Deactivation} \begin{array}{c} HC = CH \\ | \quad | \\ HC = CH \end{array} + \text{ Heat}$$

(G) Cyclobutadiene   ii)

Overall equation: Cyclobutadiene $\xrightarrow[\text{Disappearance}]{\text{Time for}}$ Cyclobutadiene   iii)

for which, we can see why its presence can be identified- being so stained not to the point of having MaxRSE, but to always exist in Activated/Equilibrium state of existence all the time, for which it disappears in Stage 1 to form vinyl acetylene which based on the operating conditions, is kept in the activated state of existence in Stage 2 to form an activated four-membered cycloallene because it can immediately molecularly rearrange back to cyclobutadiene, which again begins another cycle.

(Laws of Creations for Cyclobutadiene's Transient existence)

428

Rule 825: This rule of Chemistry for **Cyclodienes**, states that, where Cycloallenes cannot exist for higher membered rings, the reason is because they molecularly rearrange to conjugatedly placed Cyclodienes as already shown for cumulenic four-membered ring [See Rule 824].
(Laws of Creations for Cyclodienes)

Rule 826: This rule of Chemistry for **Cyclopentadienes and higher family membered rings,** states that, these are the ones that favor stable unactivated existence in view of the absence of point of scission in the five membered ring and absence of MRSE in the rings of higher members when they were formed.
(Laws of Creations for Cyclodienes)

Rule 827: This rule of Chemistry for **Cyclopropadiene,** states that, this is a cyclic conjugated diene, whose ring cannot be opened, since like cyclopropene, there is no point of scission and there is no doubt that the monomer does not possess the MaxRSE as shown below, noting that the monomer favors discrete closed-loop resonance stabilization wherein for this case,

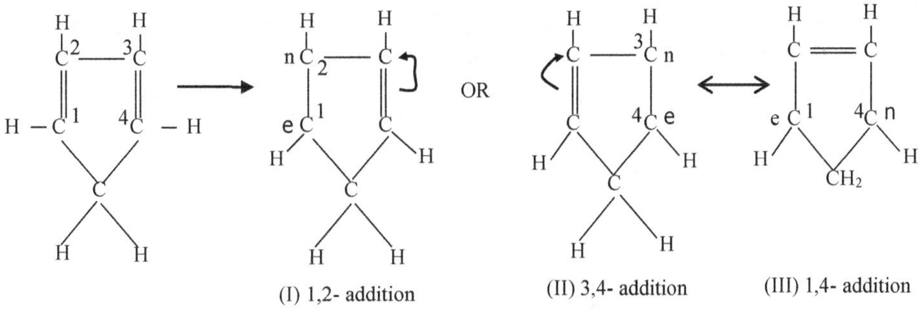

| (I) 1,2- addition | (II) 3,4- addition | (III) 1,4- addition |

Discrete closed loop resonance stabilized

1,2 - and 3,4 - addition mono-forms are the same and it is the electro-free-radical in the 1,2- or 3,4-π-bond inside the ring that initiates resonance stabilization.
(Laws of Creations for Cyclopentadiene)

Rule 828: This rule of Chemistry for **Resonance Stabilization phenomena,** states that, *when a visible electro-free-radical and an adjacently located π-bond are present*, resonance stabilization is more favored than *when a visible nucleo-free-radical and an adjacently located π-bond are present,* because it is more difficult for an electro-free-radical to leave the π- bond than when it is visibly placed; for which when the driving force for activation is weak, only 1,2 – or 3,4 – mono-form, the less nucleophilic center can be used as an addition monomer with limitations as oppsed to the use of 1,4 – mono-form.
(Laws of Creations for Resonance Stabilization phenomena)

Rule 829: This rule of Chemistry for **Cyclopentadiene,** states that, though it is 1,2- addition that will be more readily favored when the force providing activation is weak because of the lower favored state for provision of resonance stabilization, it can readily readily be made to undergo 1,4 - addition to double bonds with for example maleic anhydride at higher operating conditions, for which the followings are obtained, noting that the reaction cannot be ionic but only free-radically in a one stage Equilibrium mechanism system.

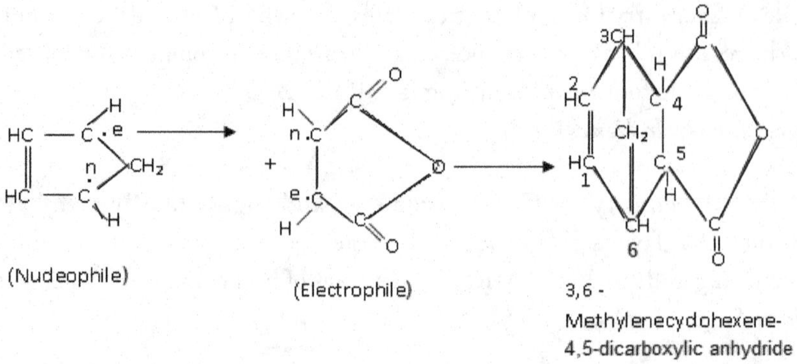

(Nucleophile)　　　　(Electrophile)　　　3,6 -
Methylenecyclohexene-
4,5-dicarboxylic anhydride

(Laws of Creations for Cyclopentadienes)

Rule 830: This rule of Chemistry for **Cyclopentadiene,** states that, though this monomer is resonance stabilized like 1-methyl-vinyl acetylene an isomer, unlike 1-methyl-vinyl acetylene which has transfer species of the first kind of first type, this has transfer species of the first kind of the tenth type on substituent group equally shared by two active centers in the 1,4 -mono-form as shown below.

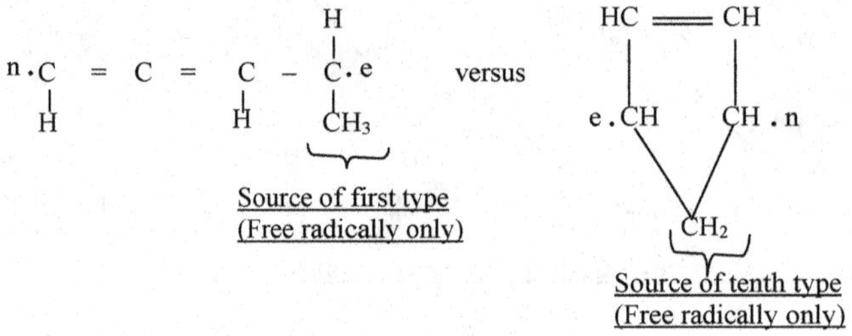

(Laws of Creations for Cyclopentadiene)

Rule 831: This rule of Chemistry for **Cyclopentadiene,** states that, this resonance stabilized monomer which can favor only electro-free-radical homopolymerization route readily can undergo molecular rearrangement to give the same monomer as shown below for the two mono-forms.

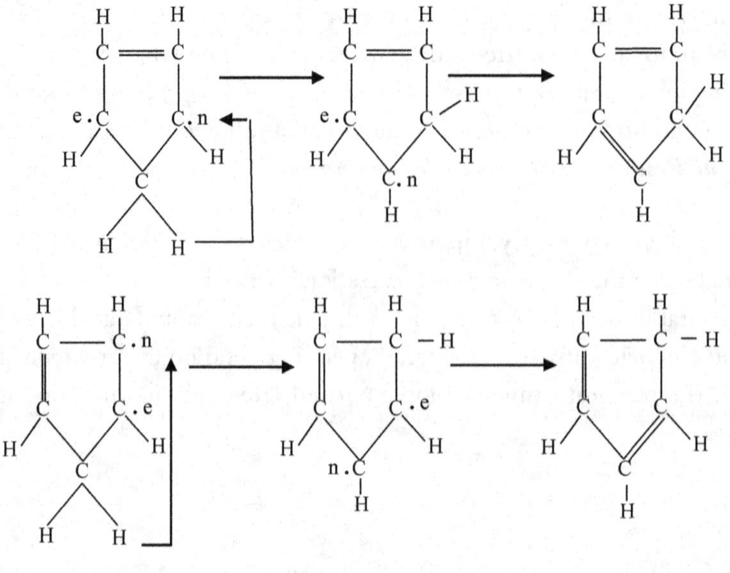

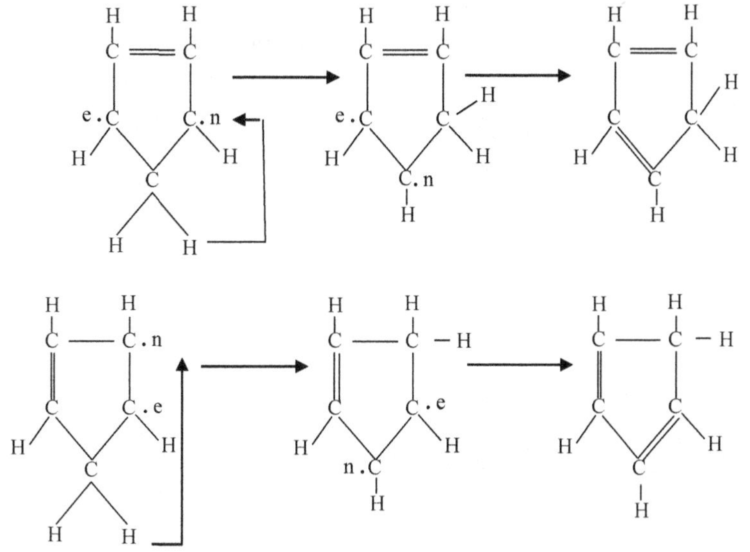

noting that they are all the same, since there are two monoforms.
(Laws of Creations for Cyclopentadiene)

Rule 832: This rule of Chemistry for **Cyclopentadiene,** states that, this monomer or compound is known *to dimerize slowly on standing to dicyclopentadiene* which can dissociate at its boiling point to the monomer, noting that the reaction can only be radical in character and that there is no transfer species; for which it is a fraction of the 1,4 – mono-form being **more nucleophilic** that diffuses to the stable ones mildly (i.e., 1,2-mono-form) and add to it to favor the existence of (A), as shown below.

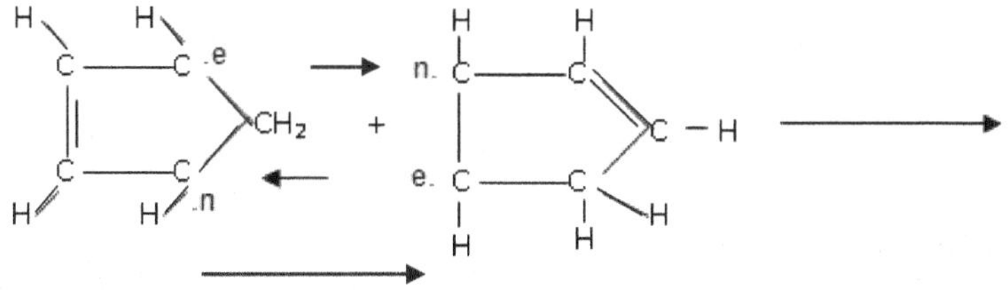

(Nudeophile)   (Nudeophile)

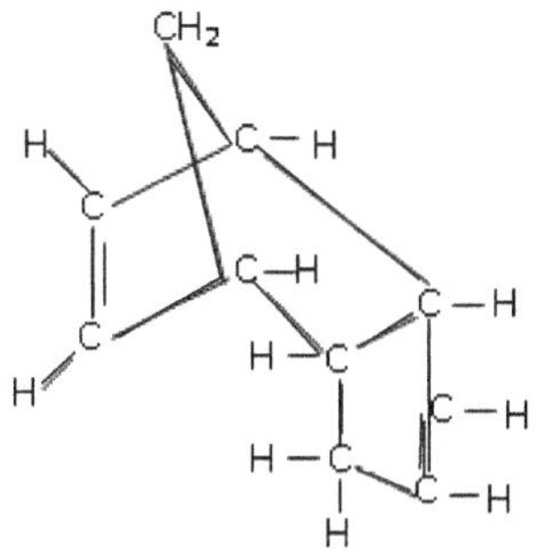

(A)

all these taking place via Equilibrium mechanism in one single stage.
(Laws of Creations for Dimerization of Cyclopentadiene)

Rule 833: This rule of Chemistry for **Cyclopentadiene,** states that, with electro-free-radical initiator, only the use of special coordination initiators will favor the polymerization of the monomer via only 1,4 - addition mono-form as indicated below, due to steric limitations, being the most stable of the two mono-forms,

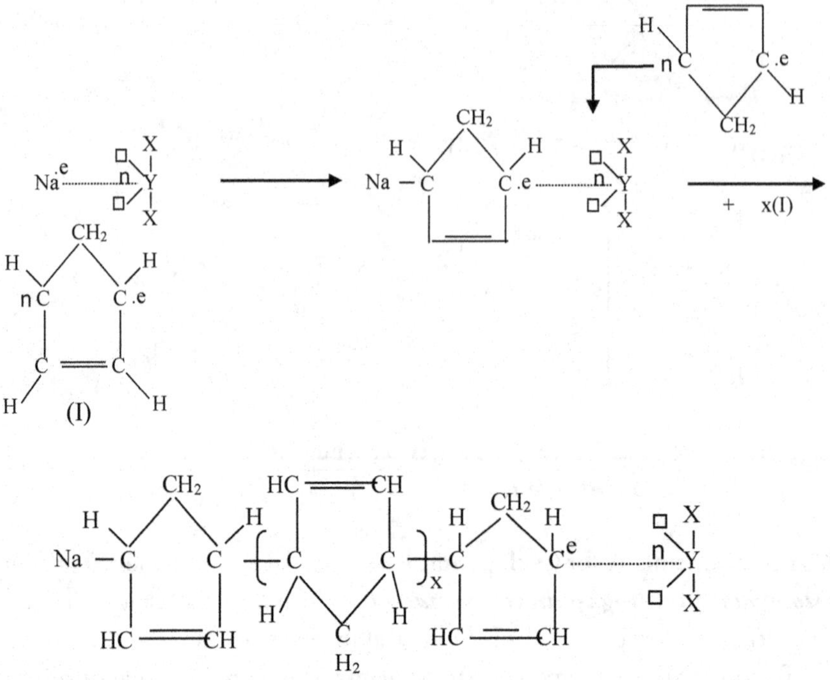

1,4 - addition - polymerization

noting that the placement above is syndiotactic placement cis in character and low polymerization temperatures may not be required to achieve the 1,4- addition polymerization and presence of reservoirs may not be required for the placement, such as the use of sodium naphthalene, unlike the use of the case above tentatively.

(Laws of Creations for Cyclopentadiene)

Rule 834: This rule of Chemistry for **Cyclopentadiene,** states that, only the groups carried inside the double bond (internally located groups) are resonance stabilized; for which when the CH_2 group is changed to CHF group, no route will be favored.

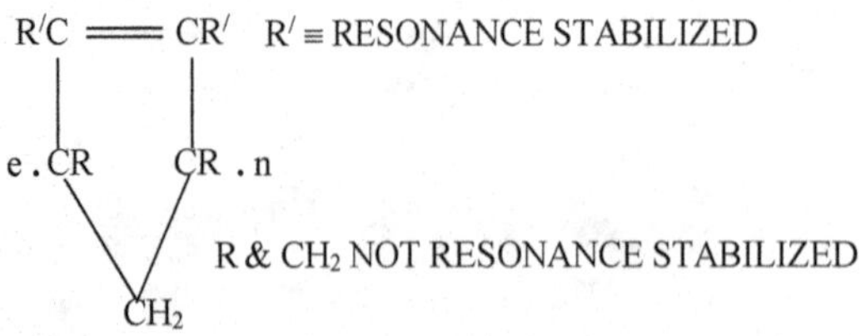

FAVORS ONLY ELECTRO-
FREE-RADICAL ROUTE.

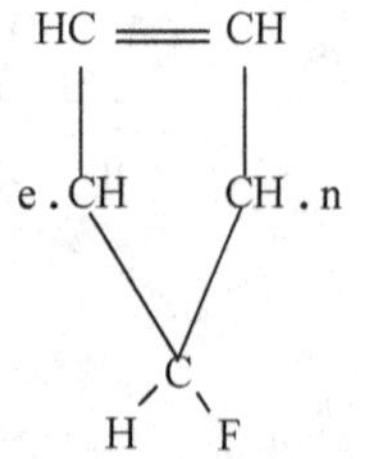

FAVORS NO ROUTE

(Laws of Creations for Cyclopentadiene)

432

Rule 835: This rule of Chemistry for **Initiators,** states that, when initiators such as HCl adduct of a vinyl ether, $SnCl_4$ (so-called Lewis acid catalyst) are used to polymerize cyclopentadiene, this can only be done electro-free- radically by the initiators as shown below for them-

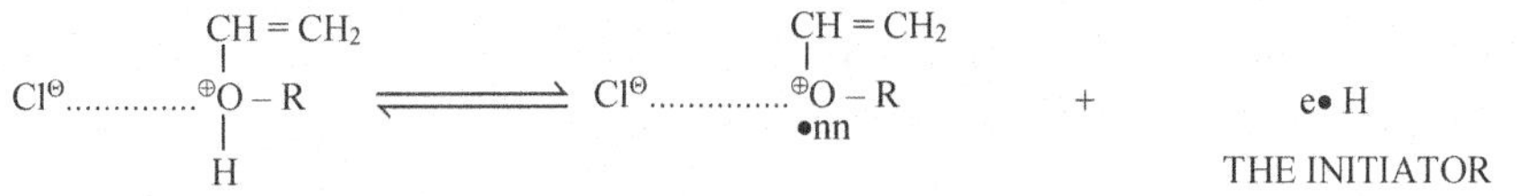

$$Cl^\ominus............\overset{CH=CH_2}{\underset{H}{\overset{|}{O}-R}} \rightleftharpoons Cl^\ominus............\overset{CH=CH_2}{\underset{\bullet nn}{\overset{|}{O}-R}} + \quad e\bullet H$$
THE INITIATOR   i)

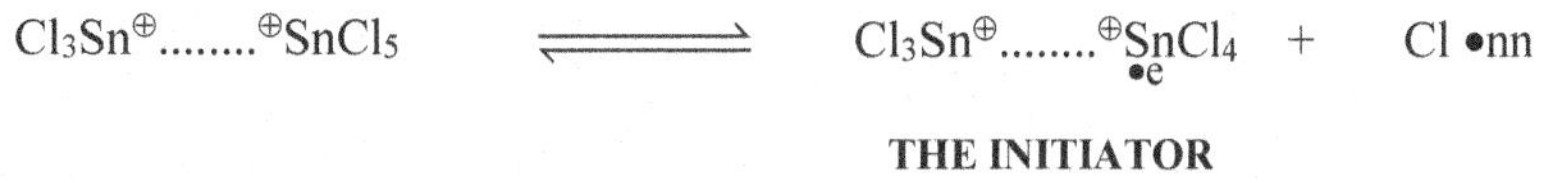

$$Cl_3Sn^\oplus........^\oplus SnCl_5 \quad \rightleftharpoons \quad Cl_3Sn^\oplus........\underset{\bullet e}{^\oplus SnCl_4} + \quad Cl \bullet nn$$
THE INITIATOR
ii)

noting that the nucleo-free- or non-free-radicals cannot diffuse or move to kill the growing chain and that the paired unbonded radicals on the Sn centers have been blocked datively using the same $SnCl_4$; it is only the chain that can diffuse to the nucleo-radicals and kill itself, noting also that the propagation is via Combination mechanism.
(Laws of Creations for Initiators)

Rule 836: This rule of Chemistry for **Initiators,** states that, when an initiator cyclopentadienyl sodium is used to polymerize a monomer such as acrylonitrile (A Male) and cyclopentadiene (A Female), this can only be done nucleo-free-radically or negatively for the former and electro-free-radically alone for the latter by this same type of initiator shown below-

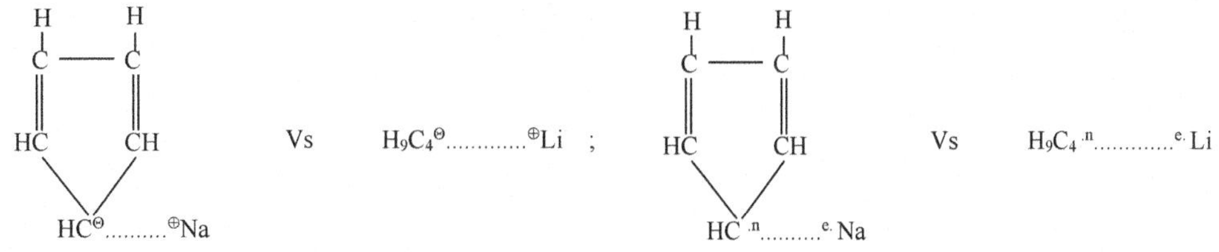

$HC^\ominus..........^\oplus Na$ Vs $H_9C_4^\ominus............^\oplus Li$; $HC^{\cdot n}..........^e Na$ Vs $H_9C_4^{\cdot n}............^e Li$

COVALENTLY CHARGED PAIRED INITIATORS   **COVALENTLY RADICAL PAIRED INITIATORS**

in which the carrier of the chain for the former is cyclopentadienyl or C_4H_9 and the carrier for the latter is Na or Li; noting that like the Z/N types, these are all Covalently free-radical- or charged-paired initiators with dual characters only radically, i.e. two active centers- one for a Male (Electrophiles) and the other for a Female (Nucleophiles), since $Na^\oplus$ cannot be used polymerically.
(Laws of Creations for Initiators)

Rule 837: This rule of Chemistry for **Initiators,** states that, when an initiator such as n-Bu_4NCl (said to be used as an additive to other initiators!) in dichloromethane (CH_2Cl_2) at -78°C, is used to polymerize cyclopentadiene, this can only be done electro-free-radically as shown below-

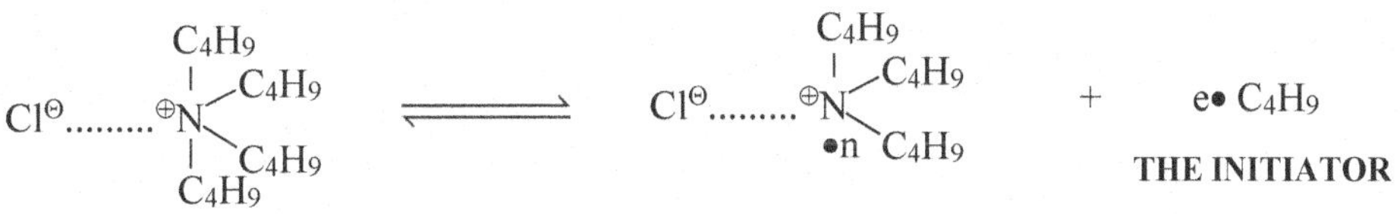

$$Cl^\ominus.........\overset{C_4H_9}{\underset{C_4H_9}{\overset{|}{N}}}{\overset{C_4H_9}{\diagup}}_{\diagdown C_4H_9} \rightleftharpoons Cl^\ominus.........\overset{C_4H_9}{\underset{\bullet n \, C_4H_9}{\overset{|}{N}}}{\overset{\diagup C_4H_9}{\diagdown C_4H_9}} + \quad e\bullet C_4H_9$$
THE INITIATOR

in which the initiator here can be observed to be nBu carrying the electro-free-radical for the polymerization of the monomer, noting that the counter-charged center carrying the nucleo-free-radical on the N center cannot close the chain since it cannot diffuse and the mechanism is not Equilibrium mechanism, but

433

Combination mechanism, the CH_2Cl_2 acting as the real additive and solvent which made it possible to operate at such a low temperature.
(Laws of Creations for Initiators)

Rule 838: This rule of Chemistry for **Cyclopentadiene,** states that, when made to exist in Equilibrium state of existence, the followings are obtained for its finger print-

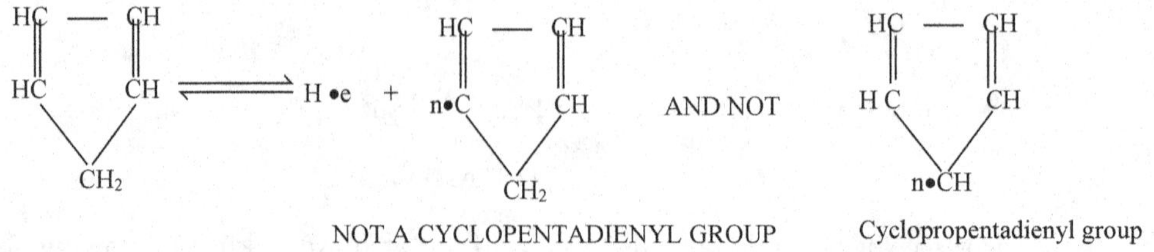

NOT A CYCLOPENTADIENYL GROUP   Cyclopropentadienyl group

for when I_2 like some others are made to react with the cyclopentadiene when kept in Stable state of existence, the H removed is not that held in Equilibrium state of existence, but that in CH_2 and *that is what makes it cyclopentadienyl in character* as shown above.
(Laws of Creations for Cycloclopentadiene)

Rule 839: This rule of Chemistry for **Cyclopentadiene,** states that, when all fractions of Cyclopentadiene are held in Equilibrium state of existence at specific operating conditions, and decomposed, the followings are obtained-

EQUILIBRIUM MECHANISM

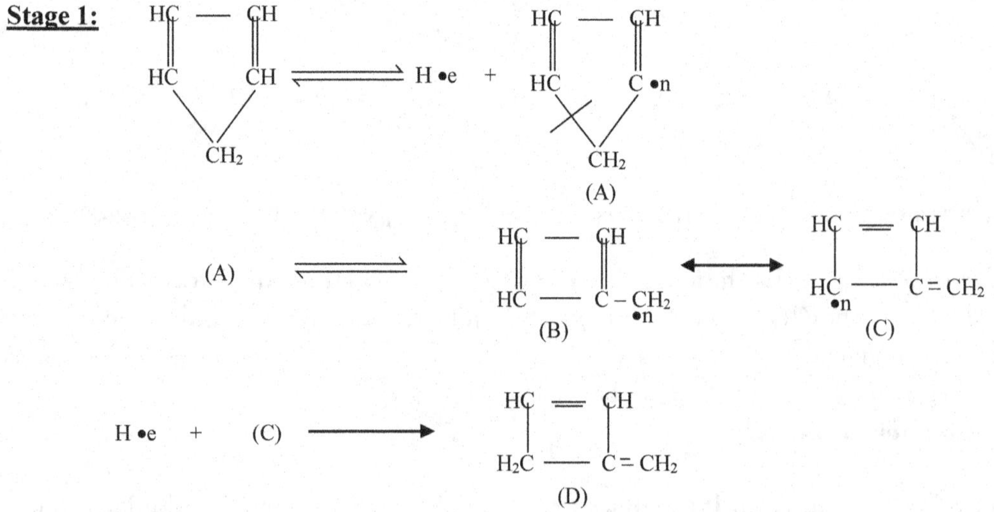

Stage 1:

1a

Overall equation: 2Cyclopentadiene $\longrightarrow$ 2(D)   1b

DECOMPOSITION MECHANISM

Stage 2:

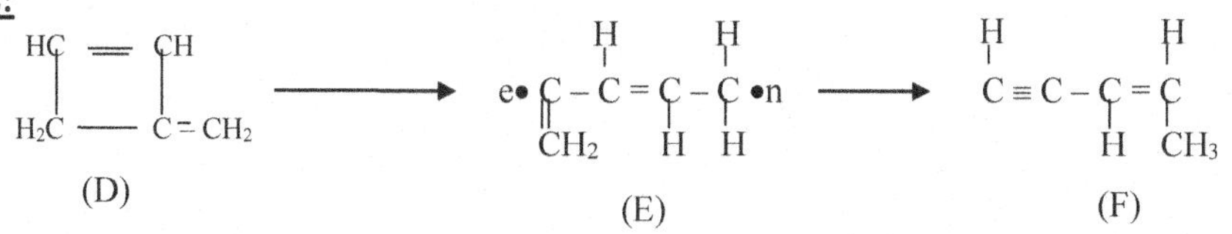

$$HC = CH \quad / \quad H_2C - C = CH_2 \quad (D) \longrightarrow e\bullet C - C = C - C \bullet n \quad (E) \longrightarrow C \equiv C - C = C \quad (F)$$

2a

Overall equation: 2Cyclopentadiene ⟶ (E) + (F)

2b

Stage 3a:

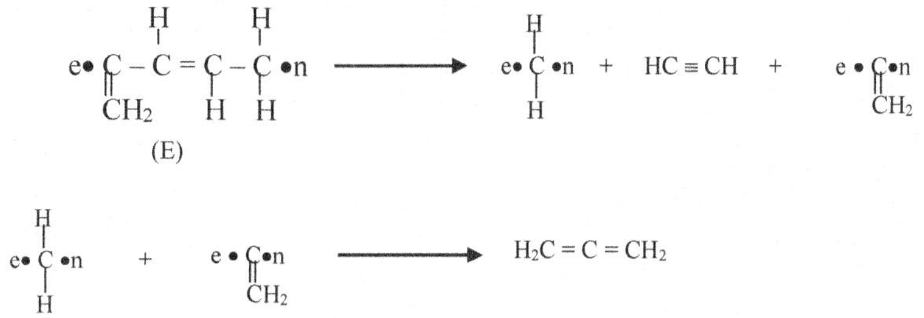

$$e\bullet C - C = C - C \bullet n \quad (E) \longrightarrow e\bullet C \bullet n + HC \equiv CH + e\bullet C \bullet n$$

$$e\bullet C \bullet n + e\bullet C \bullet n \longrightarrow H_2C = C = CH_2$$

3a

Overall equation: 2Cyclopentadiene ⟶ Acetylene + Propyne + (F)

3b

The rearrangement of 1,3-cumulene to propyne is a stage which has not been shown.

EQUILIBRIUM MECHANISM

Stage 3b:

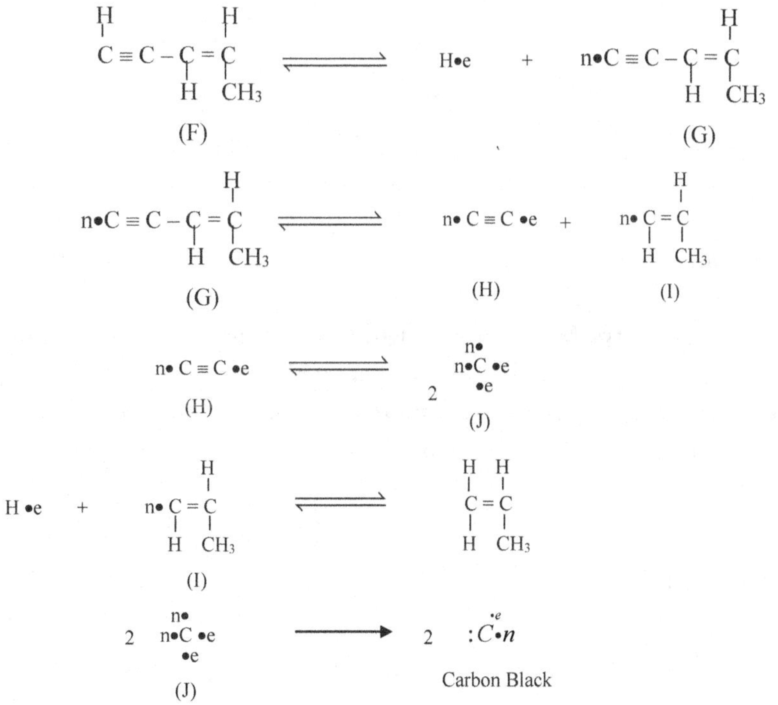

$$C \equiv C - C = C \quad (F) \rightleftharpoons H\bullet e + n\bullet C \equiv C - C = C \quad (G)$$

$$n\bullet C \equiv C - C = C \quad (G) \rightleftharpoons n\bullet C \equiv C \bullet e \quad (H) + n\bullet C = C \quad (I)$$

$$n\bullet C \equiv C \bullet e \quad (H) \rightleftharpoons 2 \quad n\bullet C \bullet e \quad (J)$$

$$H\bullet e + n\bullet C = C \quad (I) \rightleftharpoons C = C$$

$$2 \quad n\bullet C \bullet e \quad (J) \longrightarrow 2 \quad :\ddot{C}\bullet n \quad \text{Carbon Black}$$

4a

Overall equation: 6Cyclopentadiene ⟶ 3Acetylene + 3Propyne + 3Propene +

6Carbon Black   4b

Stage 4 or5:

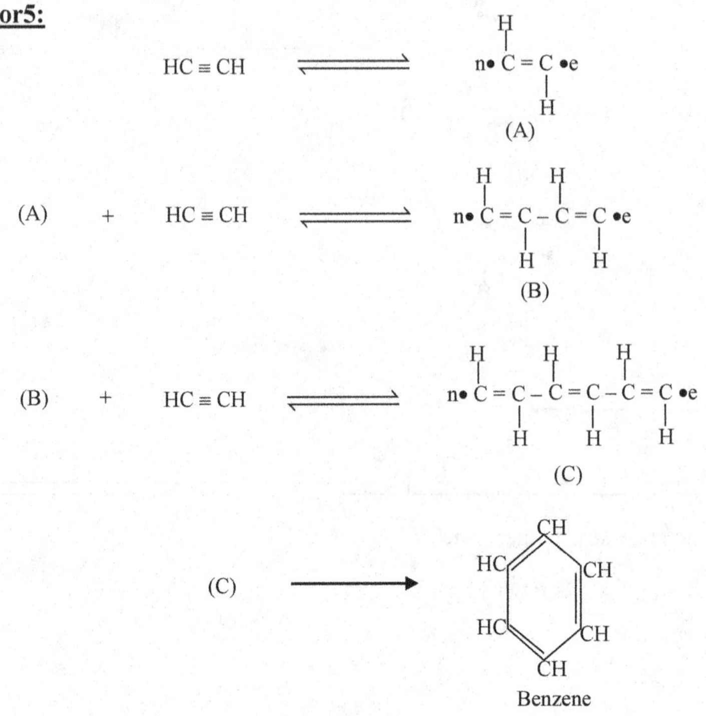

$$HC \equiv CH \rightleftharpoons n\bullet \overset{\overset{\displaystyle H}{|}}{C} = \overset{\underset{\displaystyle H}{|}}{C} \bullet e$$
(A)

$$(A) \quad + \quad HC \equiv CH \rightleftharpoons n\bullet \overset{\overset{\displaystyle H}{|}}{C} = \overset{\underset{\displaystyle H}{|}}{C} - \overset{\overset{\displaystyle H}{|}}{C} = \overset{\underset{\displaystyle H}{|}}{C} \bullet e$$
(B)

$$(B) \quad + \quad HC \equiv CH \rightleftharpoons n\bullet \overset{\overset{\displaystyle H}{|}}{C} = \overset{\underset{\displaystyle H}{|}}{C} - \overset{\overset{\displaystyle H}{|}}{C} = \overset{\underset{\displaystyle H}{|}}{C} - \overset{\overset{\displaystyle H}{|}}{C} = \overset{\underset{\displaystyle H}{|}}{C} \bullet e$$
(C)

$$(C) \longrightarrow \text{Benzene}$$

Benzene   5a

Overall equation: 3HC≡CH ⟶ Benzene   5b

noting the number of stages in series and parallel involved, the Equilibrium states of existence of some components, the types and manners of mechanisms involved and the fact that there is no ring expansion from five to six here.

(Laws of Creations for Cyclopentadiene)

Rule 840: This rule of Chemistry for **The Mechanism of Ring Expansion,** states that, *ring expansion is that wherein the original ring of say five is taken as it is and worked on <u>mechanically</u> to expand it to six without opening the former ring and letting it decompose;* and the ring must be such that is carrying at least one externally or internally located double bond as shown below for expansion of Benzene-

Stage 1:

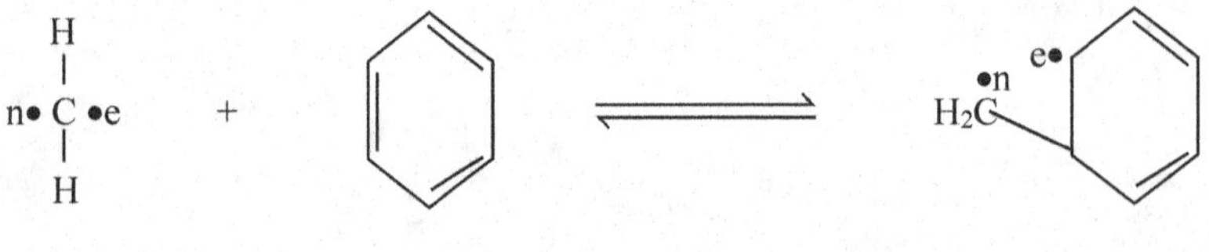

$$n\bullet \overset{\overset{\displaystyle H}{|}}{\underset{\underset{\displaystyle H}{|}}{C}} \bullet e \quad + \quad \bigcirc \quad \rightleftharpoons \quad H_2\overset{\bullet n \quad e\bullet}{C}\bigcirc$$

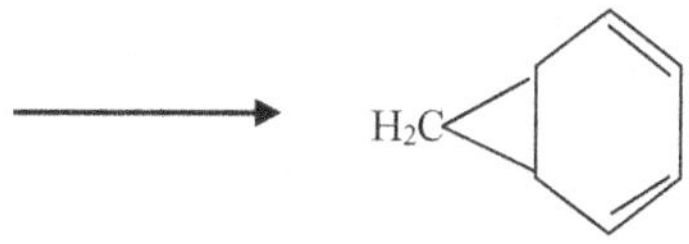

1

Stage 2:

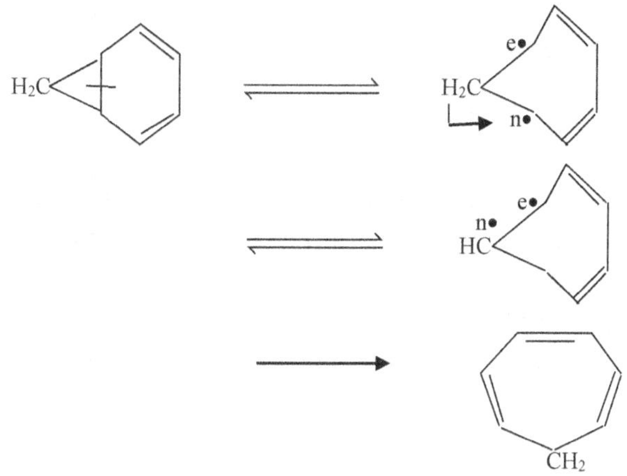

Cycloheptatriene

2a

Overall Equation: Methylene + Benzene ⟶ Cycloheptadiene

2b

(Laws of Creations for Ring Expansion)

Rule 841: This rule of Chemistry for **The Mechanism of Ring Reduction,** states that, *ring reduction in size is that wherein the original ring of say five is taken as it is and worked on mechanically to reduce it to four with instantaneous opening and rearrangement of the former ring without allowing decomposition to take place* to form smaller compounds, as shown below for t-Butyl Cyclopentadiene whose Equilibrium State of Existence is stronger than that of Cyclopentadiene-

EQUILIBRIUM MECHANISM

Stage 1:

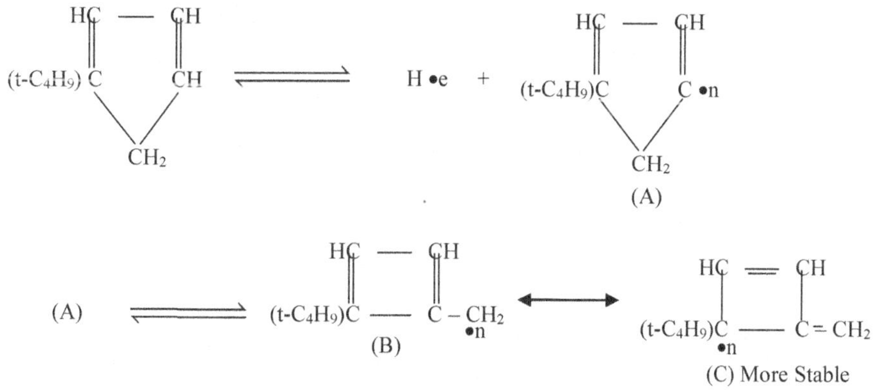

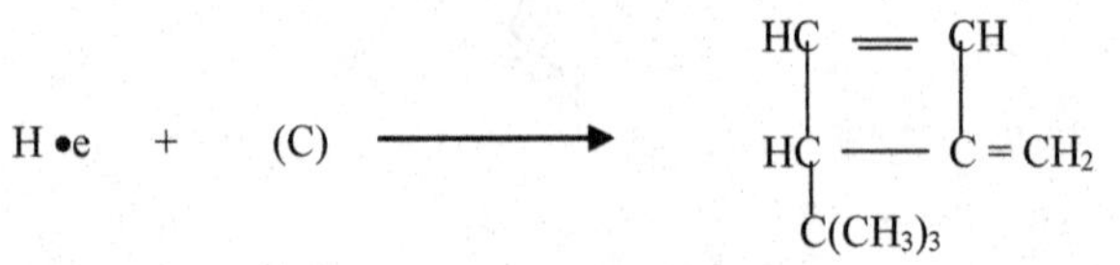

$$HC = CH$$
$$H \bullet e \;+\; (C) \longrightarrow \; HC - C = CH_2$$
$$C(CH_3)_3$$

(D)   1a

Overall equation: t-Butyl Cyclopentadiene $\longrightarrow$ 4-Methylene 3-t-butyl Cyclobutene (D)

(Laws of Creations for Ring Reduction)   1b

Rule 842: This rule of Chemistry for **Cyclopentadiene,** states that, when half fraction of the cyclopentadiene is in Stable state of existence leaving the remaining half in Equilibrium state of existence at specific operating conditions, then the followings take place.

EQUILIBRIUM MECHANISM

Stage 1:

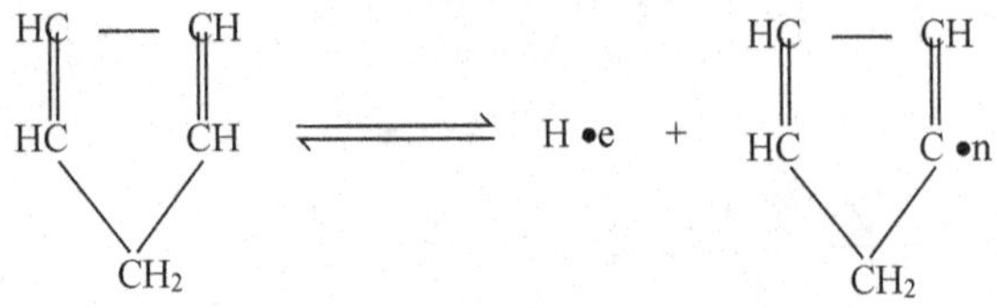

(A)

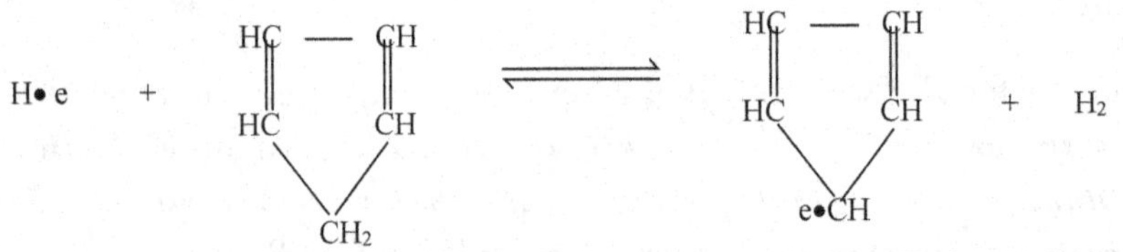

(B) a Cyclopentadienyl radical

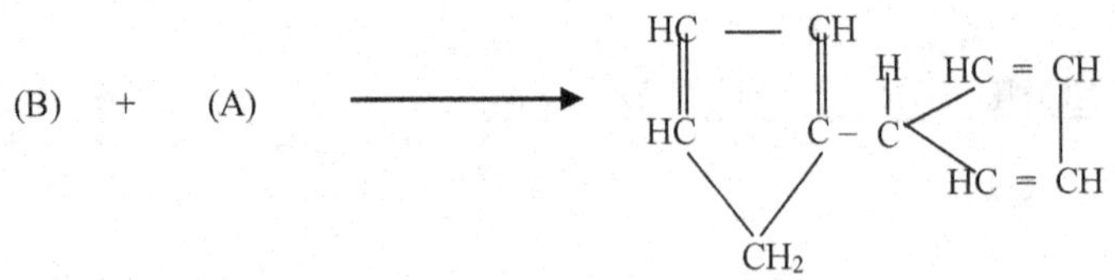

(B)   +   (A) $\longrightarrow$

(C) 1-Cyclopentadienyl cyclopentadiene   1a

Overall equation: 2 Cyclopentadiene $\longrightarrow$ H$_2$ + (C)   1b

Stage 2:

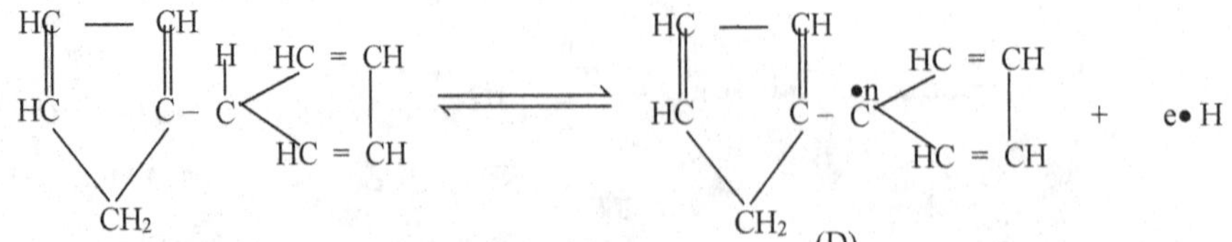

(C) 1-Cyclopentadienyl cyclopendiene

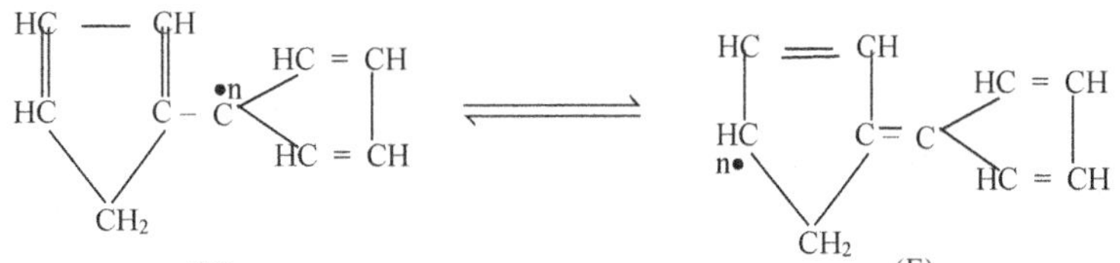

(D)   (E)

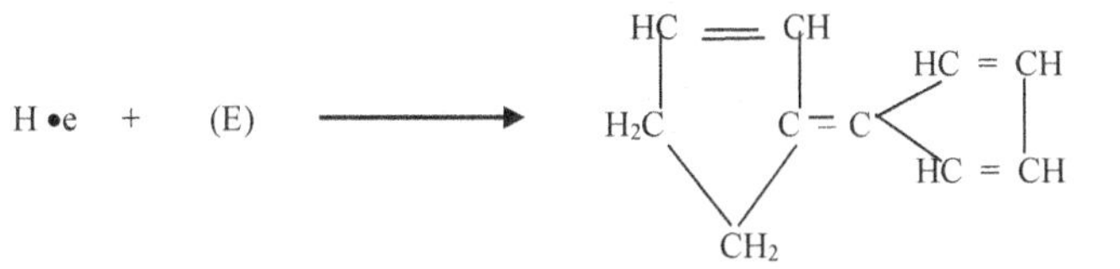

H •e   +   (E) ⟶

(F) Cyclopentadienylene
cyclopentene

2a

2b

Overall equation: (C) ⟶ (F)

DECOMPOSITION MECHANISM

Stage 3:

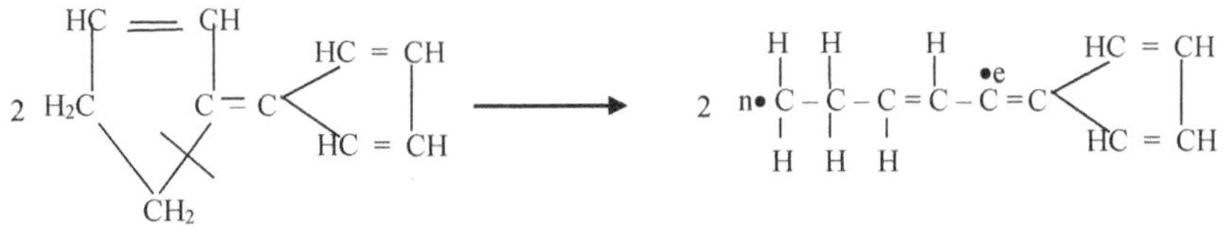

(F) Cyclopentadienylene
cyclopentene   (G)

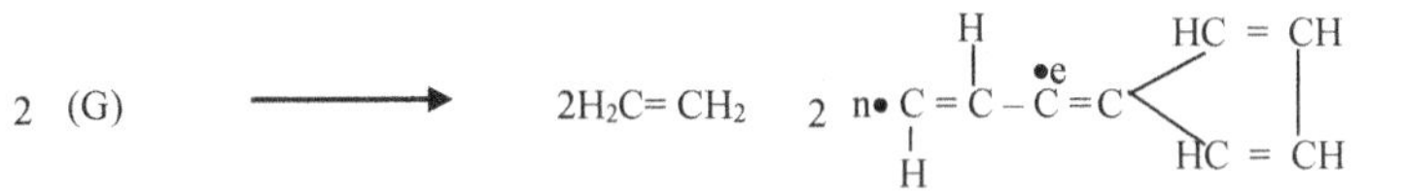

2 (G) ⟶ $2H_2C = CH_2$ (H)

(H)

2 (H) ⟶ 2 HC≡CH   +   2 (I)

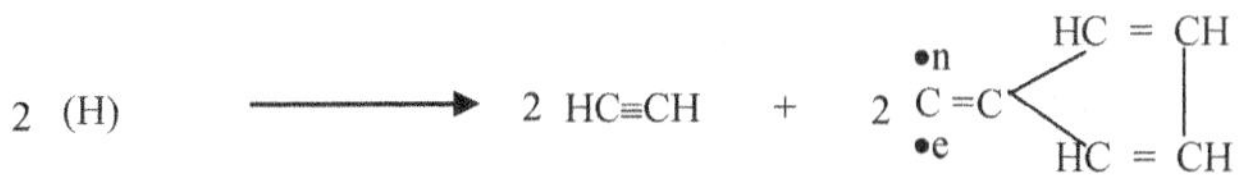

(I)

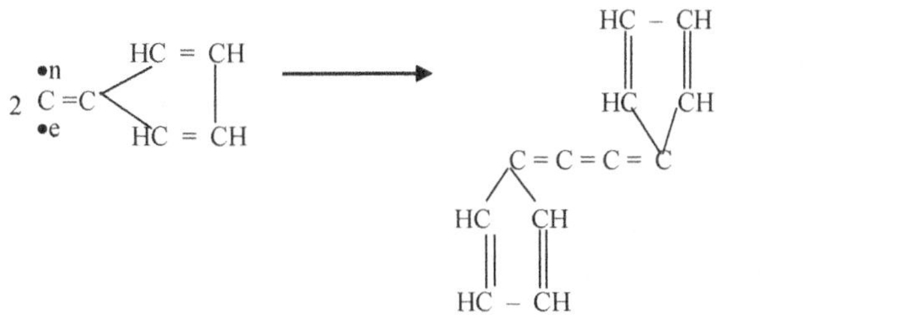

(I*) 1,4-Di-cyclopentadienyl cumulene

3a

439

Overall equation: $2F \longrightarrow 2HC\equiv CH + 2H_2C=CH_2 + (I^*)$ 3b

Overall overall equation: $12\text{Cyclopentadiene} \longrightarrow 6H_2 + 6HC\equiv CH + 6H_2C=CH_2 + 3(I^*)$ 3c

EQUILIBRIUM MECHANISM

Stage 4:

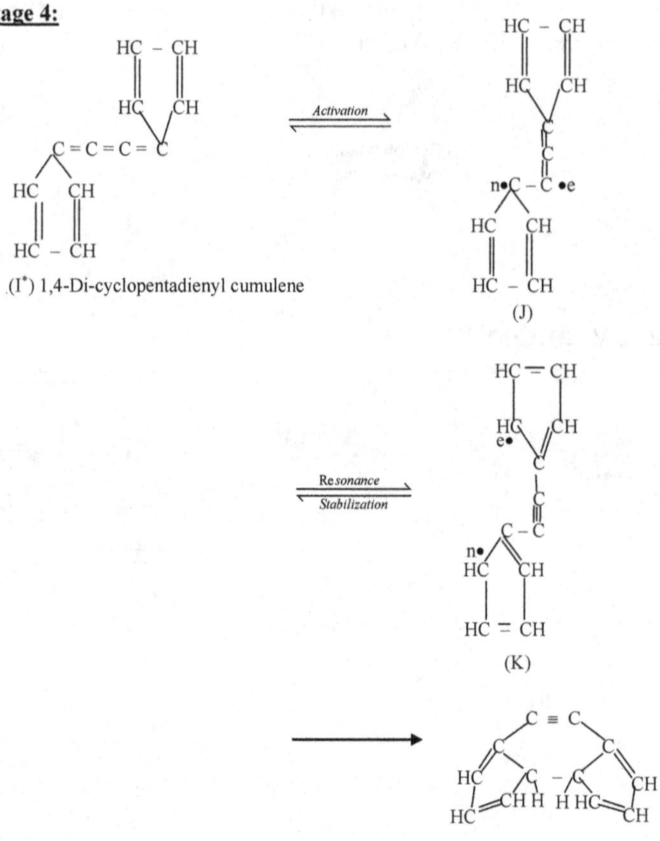

(I^*) 1,4-Di-cyclopentadienyl cumulene

(J)

(K)

(L)   4a

Overall equation: (I) $\longrightarrow$ (L)   4b

Stage 5

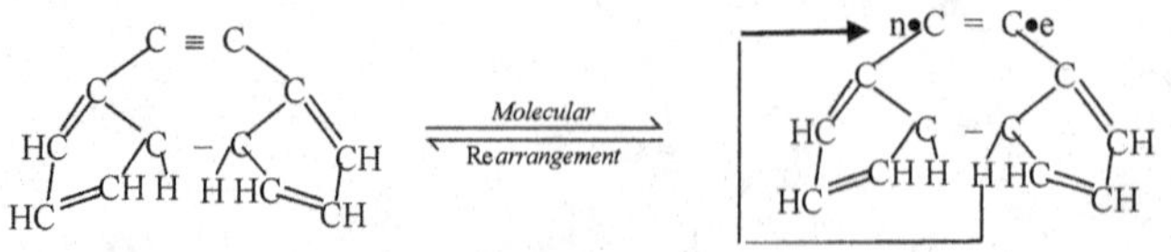

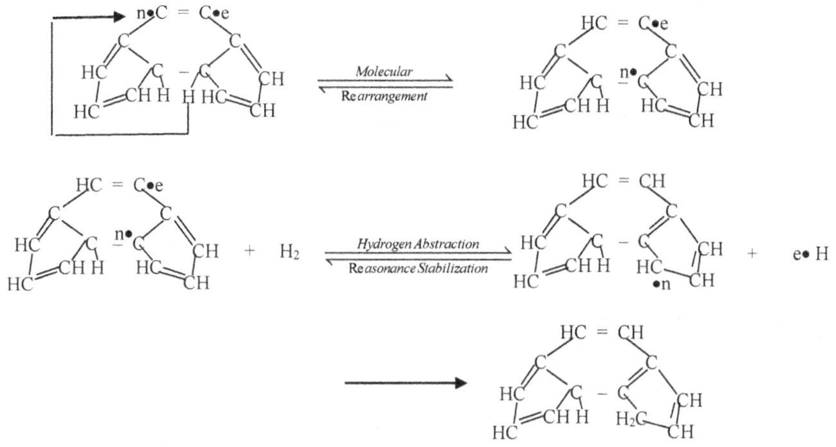

Overall overall equation: 12Cyclopentadiene $\longrightarrow$ 6H$_2$C=CH$_2$ + 6C$_2$H$_2$ + 3(M)

$$+ \quad 3H_2$$

5a

5b

Stage 6:

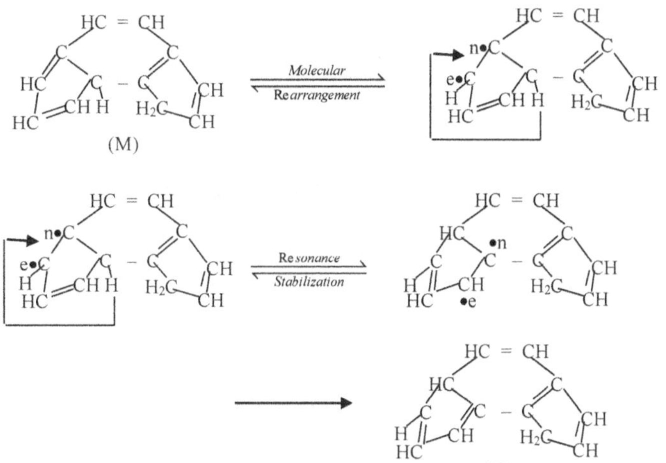

6

Stage 7:

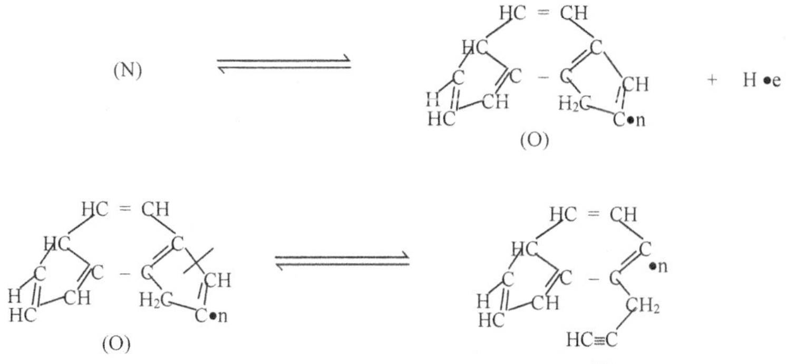

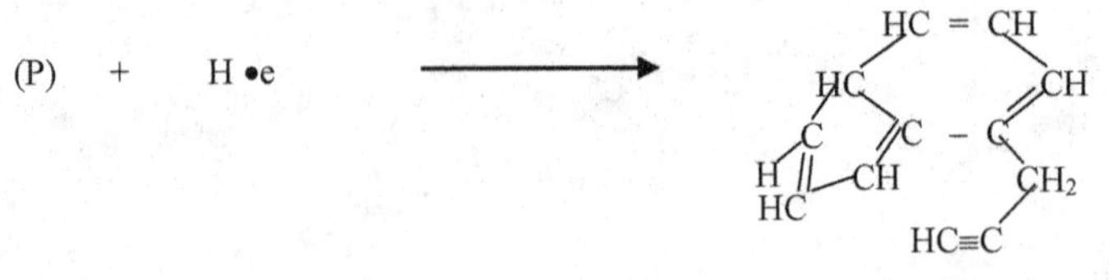

(P)　　+　　H •e　　⟶

$$
\begin{array}{c}
HC = CH \\
HC \qquad CH \\
C \quad C - C \\
H \quad CH \qquad CH_2 \\
HC \\
HC \equiv C
\end{array}
$$

(Q)

7

Stage 8:

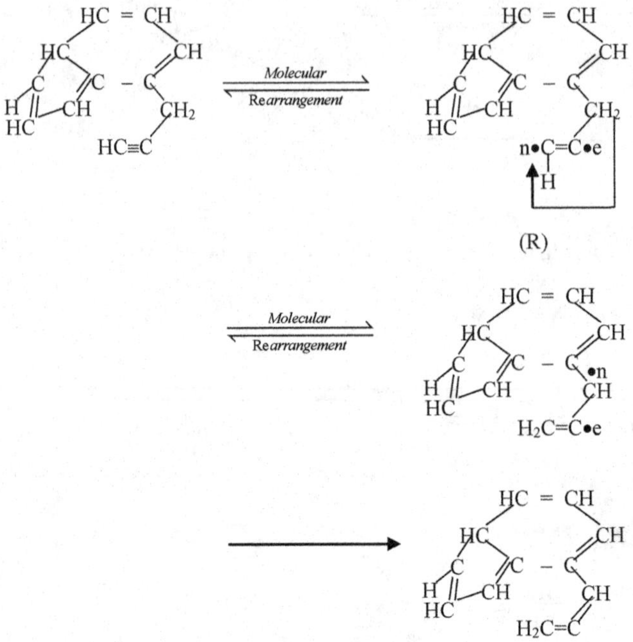

$$
\begin{array}{c}
HC = CH \\
HC \qquad CH \\
C \quad C - C \\
H \quad CH \qquad CH_2 \\
HC \\
HC \equiv C
\end{array}
\quad
\xrightleftharpoons[\textit{Rearrangement}]{\textit{Molecular}}
\quad
\begin{array}{c}
HC = CH \\
HC \qquad CH \\
C \quad C - C \\
H \quad CH \qquad CH_2 \\
HC \\
n•C = C•e \\
H
\end{array}
$$

(R)

$$
\xrightleftharpoons[\textit{Rearrangement}]{\textit{Molecular}}
\quad
\begin{array}{c}
HC = CH \\
HC \qquad CH \\
C \quad C - C \quad •n \\
H \quad CH \qquad CH \\
HC \\
H_2C = C•e
\end{array}
$$

$$
\xrightarrow{\hspace{2cm}}
\quad
\begin{array}{c}
HC = CH \\
HC \qquad CH \\
C \quad C - C \\
H \quad CH \qquad CH \\
HC \\
H_2C = C
\end{array}
$$

(S)

8

Stage 9:

$$
\begin{array}{c}
HC = CH \\
HC \qquad CH \\
C \quad C - C \\
H \quad CH \qquad CH \\
HC \\
H_2C = C
\end{array}
\quad
\rightleftharpoons
\quad
\begin{array}{c}
HC = CH \\
HC \qquad CH \\
n• C \quad C - C \\
CH \qquad CH \\
HC \\
H_2C = C
\end{array}
\quad + \quad e• H
$$

(T)

$$
\rightleftharpoons
\quad
HC \equiv C
\begin{array}{c}
HC = CH \\
HC \qquad CH \\
C - C \\
n•CH \qquad CH \\
H_2C = C
\end{array}
$$

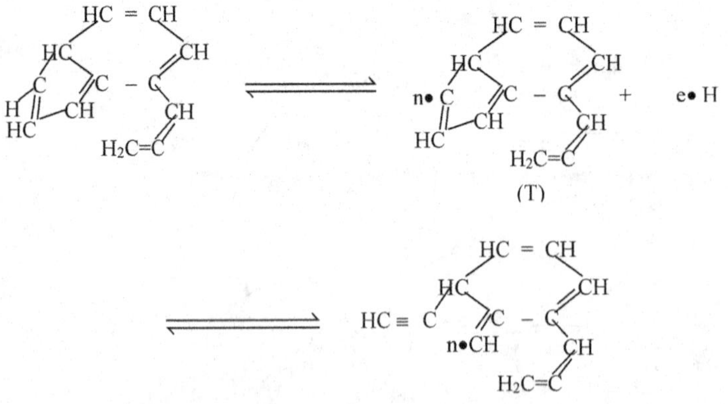

(U)

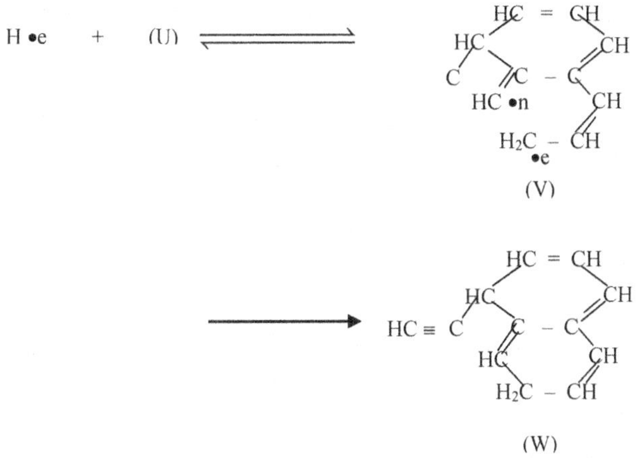

$$H \bullet e \quad + \quad (U) \rightleftharpoons$$

(V)

$$\longrightarrow$$

(W)

9

Stage 10:

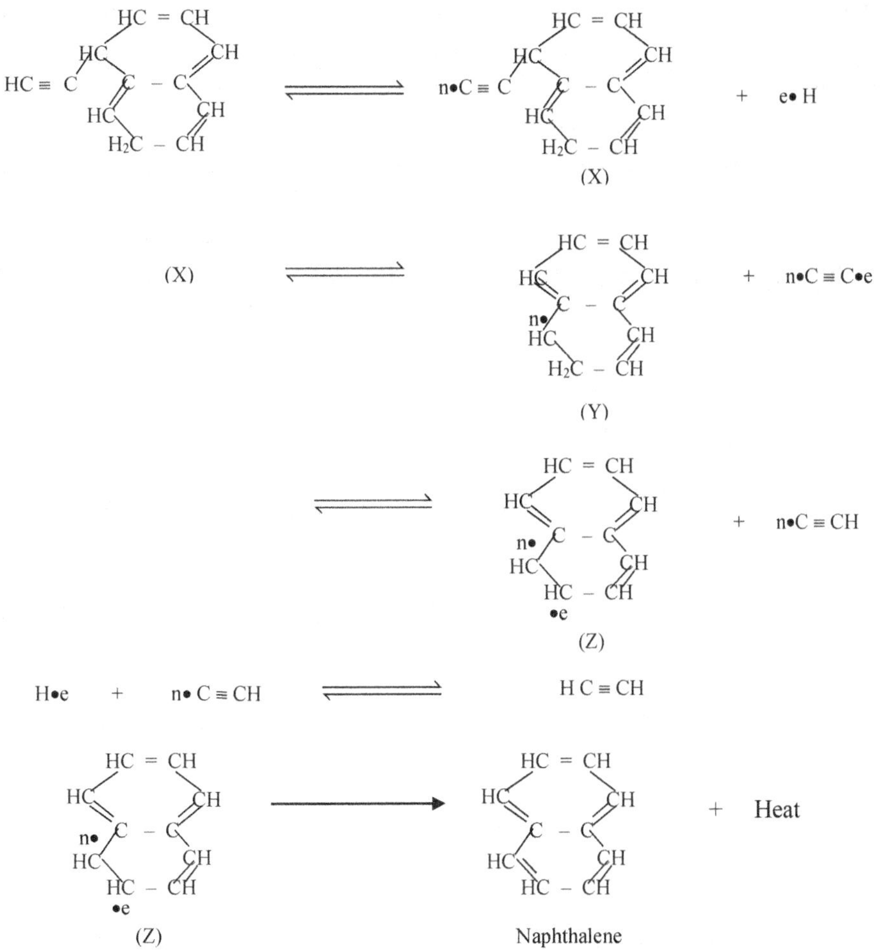

$$\rightleftharpoons \quad (X) \quad + \quad e \bullet H$$

(X)

$$(X) \quad \rightleftharpoons \quad + \quad n \bullet C \equiv C \bullet e$$

(Y)

$$\rightleftharpoons \quad + \quad n \bullet C \equiv CH$$

(Z)

$$H \bullet e \quad + \quad n \bullet C \equiv CH \quad \rightleftharpoons \quad HC \equiv CH$$

$$\longrightarrow \quad + \quad Heat$$

(Z)   Naphthalene

10a

Overall overall equation: 12Cyclopentadiene $\longrightarrow$ 6H$_2$C=CH$_2$ + 9C$_2$H$_2$ + 3H$_2$ +

$$3C_{10}H_8 \qquad\qquad 10b$$

Stage 11: $\quad$ H$_2$C = CH$_2$ $\rightleftharpoons$ H$\bullet$e + n$\bullet$CH = CH$_2$

$\qquad\qquad$ n$\bullet$CH = CH$_2$ $\rightleftharpoons$ n$\bullet$CH = HC$\bullet$e + H$\bullet$n

$\qquad$ H$\bullet$e + H$\bullet$n $\rightleftharpoons$ H$_2$

$\qquad\quad$ n$\bullet$CH = HC$\bullet$e $\longrightarrow$ HC $\equiv$ CH $\qquad\qquad$ 11a

Overall overall equation: 12 Cyclopentadiene $\longrightarrow$ 15HC $\equiv$ CH + 9H$_2$ + 3C$_{10}$H$_8$

$$11b$$

Stage 12: Same as Stage 4 or 5 of Equation 5a.of Rule 839. $\qquad\qquad$ 12a

Overall overall equation: 12Cyclopentadiene $\longrightarrow$ 9H$_2$ + 5C$_6$H$_6$ + 3C$_{10}$H$_8$ $\qquad$ 12b

in which on the whole, twelve stages are involved for the synthesis of Naphthalene at such operating conditions below what obtains in the Petroleum industries with respect to Cracking of hydrocarbons, noting that throughout the stages, no ring was expanded in size; noting that (M) should indeed be symmetrically placed, for if so, no naphthalene would have been obtained as shown above.
(Laws of Creations for Cyclopentadiene)

Rule 843: This rule of Chemistry for **The thermal decomposition of tert-butyl-1,3-Cyclopen-tadiene in single-pulse shock tube studies at shock pressures of 182-260kPa and temperature of 996-1127K,** states that, the mechanism of the reactions wherein Isobutene (2-methylpropene), 1,3-cyclopentadiene and toluene are the major products can be explained as follows, beginning with a large part kept in Equilibrium state of existence.

EQUILIBRIUM MECHANISM

Stage 1:

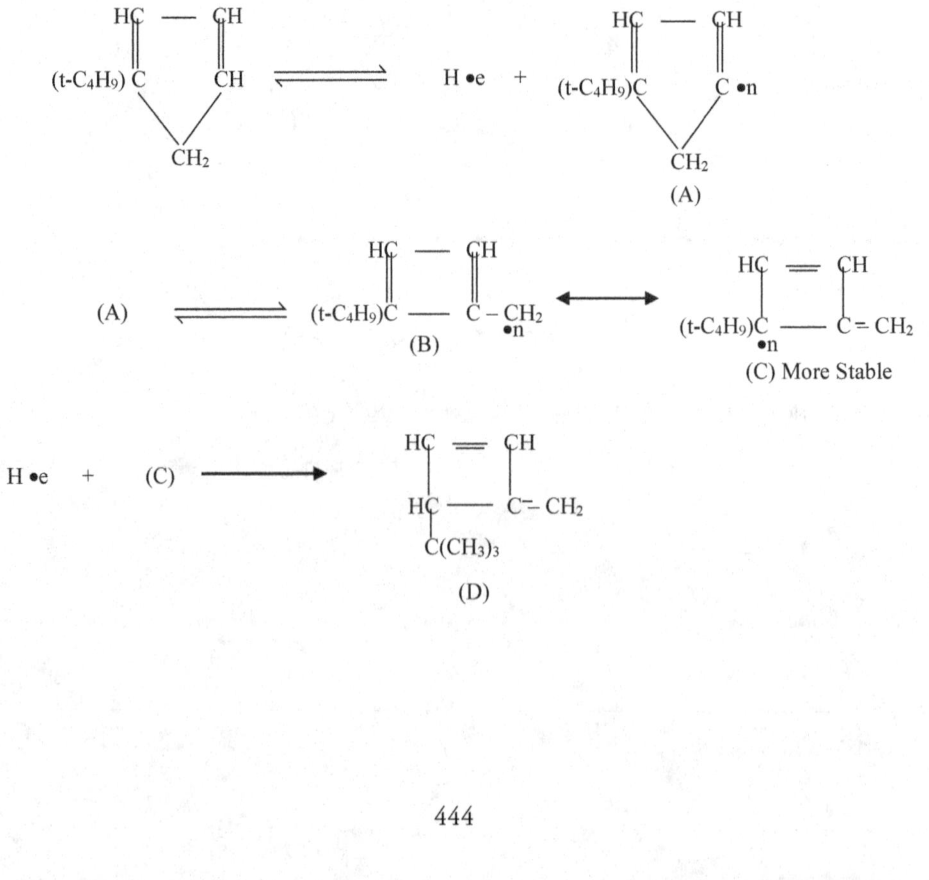

1a

Overall equation: 2 t-Butyl Cyclopentadiene $\longrightarrow$ 2(D)　　　　　　1b

DECOMPOSITION MECHANISM

Stage 2:

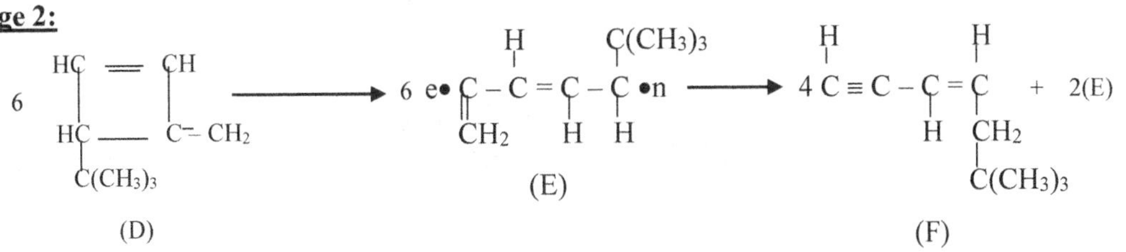

$$6\ \begin{array}{c} HC = CH \\ | \quad\quad | \\ HC \underline{\quad} C - CH_2 \\ | \\ C(CH_3)_3 \end{array} \longrightarrow 6\ e\bullet \begin{array}{c} H \quad\quad C(CH_3)_3 \\ | \quad\quad\quad | \\ C - C = C - C \bullet n \\ | \quad\quad\quad | \quad | \\ CH_2 \quad H \quad H \end{array} \longrightarrow 4\ \begin{array}{c} H \quad\quad H \\ | \quad\quad | \\ C \equiv C - C = C \\ \quad\quad | \quad | \\ \quad\quad H \quad CH_2 \\ \quad\quad\quad\quad | \\ \quad\quad\quad\quad C(CH_3)_3 \end{array} + 2(E)$$

(D)　　　　　　　　　　(E)　　　　　　　　　(F)　　　　　　2a

Overall equation: 6 t-Butyl Cyclopentadiene $\longrightarrow$ 4(F) + 2(E)　　　　　　2b

and based on the types of products obtained, two thirds of (E) formed rearranged while the remainder was decomposed as shown below-

Stage 3a:

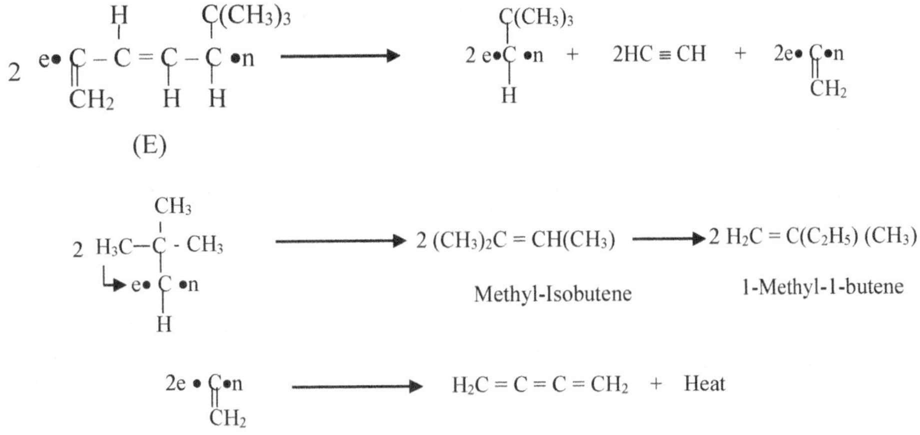

$$2\ e\bullet \begin{array}{c} H \quad\quad C(CH_3)_3 \\ | \quad\quad\quad | \\ C - C = C - C \bullet n \\ | \quad\quad\quad | \quad | \\ CH_2 \quad H \quad H \end{array} \longrightarrow 2\ e\bullet \begin{array}{c} C(CH_3)_3 \\ | \\ C \bullet n \\ | \\ H \end{array} + 2HC \equiv CH + 2e\bullet \begin{array}{c} C \bullet n \\ || \\ CH_2 \end{array}$$

(E)

$$2\ \begin{array}{c} CH_3 \\ | \\ H_3C - C - CH_3 \\ \quad\quad\downarrow e\bullet\ C \bullet n \\ \quad\quad\quad | \\ \quad\quad\quad H \end{array} \longrightarrow 2\ (CH_3)_2C = CH(CH_3) \longrightarrow 2\ H_2C = C(C_2H_5)(CH_3)$$

　　　　　　　　　　　　　Methyl-Isobutene　　　　　1-Methyl-1-butene

$$2e \bullet \begin{array}{c} C \bullet n \\ || \\ CH_2 \end{array} \longrightarrow H_2C = C = C = CH_2 + Heat \quad\quad 3a$$

Overall equation: 6 t-Butyl-Cyclopentadiene $\longrightarrow$ 2Acetylene + 2 [1-Methyl-1-butene] +

　　　　　　　　　　　　1,4-cumulene + 4(F)　　　　　　3b

EQUILIBRIUM MECHANISM

Stage 3b:

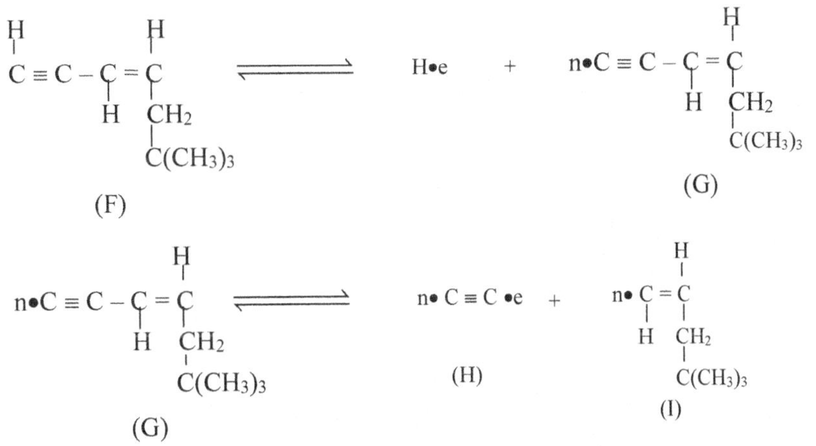

$$\begin{array}{c} H \quad\quad H \\ | \quad\quad | \\ C \equiv C - C = C \\ \quad\quad | \quad | \\ \quad\quad H \quad CH_2 \\ \quad\quad\quad\quad | \\ \quad\quad\quad\quad C(CH_3)_3 \end{array} \rightleftharpoons H\bullet e + n\bullet \begin{array}{c} H \\ | \\ C \equiv C - C = C \\ \quad\quad | \quad | \\ \quad\quad H \quad CH_2 \\ \quad\quad\quad\quad | \\ \quad\quad\quad\quad C(CH_3)_3 \end{array}$$

(F)　　　　　　　　　　　　　　　(G)

$$n\bullet \begin{array}{c} H \\ | \\ C \equiv C - C = C \\ \quad\quad | \quad | \\ \quad\quad H \quad CH_2 \\ \quad\quad\quad\quad | \\ \quad\quad\quad\quad C(CH_3)_3 \end{array} \rightleftharpoons n\bullet C \equiv C \bullet e + n\bullet \begin{array}{c} H \\ | \\ C = C \\ | \quad | \\ H \quad CH_2 \\ \quad\quad | \\ \quad\quad C(CH_3)_3 \end{array}$$

(G)　　　　　　　　　(H)　　　　　　　(I)

$$n \bullet C \equiv C \bullet e \;\rightleftharpoons\; 2 \; n \bullet \overset{n \bullet}{\underset{\bullet e}{C}} \bullet e$$

(H)

$$H \bullet e \;+\; n \bullet \overset{H}{\underset{\underset{C(CH_3)_3}{\overset{|}{CH_2}}}{\overset{|}{C}}} = \overset{}{C} \;\rightleftharpoons\; \overset{H}{\underset{H}{\overset{|}{C}}} = \overset{H}{\underset{\underset{C(CH_3)_3}{\overset{|}{CH_2}}}{\overset{|}{C}}}$$

(I)   (J)

$$2 \; n \bullet \overset{n \bullet}{\underset{\bullet e}{C}} \bullet e \;\longrightarrow\; 2 \; :\overset{\bullet e}{C} \bullet n \;+\; \text{Heat}$$

Carbon Black

3c

Overall equation: 6 t-butyl-Cyclopentadiene $\longrightarrow$ 2 Acetylene + 2 [1-Methyl-1-butene] +

1,4-cumulene + 4 [t-butyl Propene] + 8 Carbon Black   3d

Stage 4:

$$\overset{H}{\underset{H}{\overset{|}{C}}} = \overset{H}{\underset{\underset{C(CH_3)_3}{\overset{|}{CH_2}}}{\overset{|}{C}}} \;\rightleftharpoons\; H \bullet e \;+\; n \bullet \overset{H}{\underset{\underset{C(CH_3)_3}{\overset{|}{CH_2}}}{\overset{|}{C}}} = \overset{H}{C}$$

(J)   (I)

$$n \bullet \overset{H}{C} = \overset{H}{\underset{\underset{C(CH_3)_3}{\overset{|}{CH_2}}}{\overset{|}{C}}} \;\rightleftharpoons\; \overset{H}{\underset{H}{\overset{|}{C}}} \equiv C \;+\; n \bullet \overset{H}{\underset{\underset{CH_3}{\overset{|}{C}}}{\overset{|}{C}}} - CH_3$$

(I)   (K)

$$n \bullet \overset{H}{\underset{\underset{H}{\overset{|}{C}}}{\overset{|}{C}}} - \overset{CH_3}{\underset{\underset{CH_3}{}}{\overset{|}{C}}} - CH_3 \;\rightleftharpoons\; n \bullet CH_3 \;+\; n \bullet \overset{H}{\underset{\underset{H}{\overset{|}{C}}}{\overset{|}{C}}} - \overset{CH_3}{\underset{\underset{CH_3}{}}{\overset{|}{C}}} \bullet e$$

(K)   (L)

$$H \bullet e \;+\; n \bullet CH_3 \;\rightleftharpoons\; CH_4$$

$$n \bullet \overset{H}{\underset{\underset{H}{\overset{|}{C}}}{\overset{|}{C}}} - \overset{CH_3}{\underset{\underset{CH_3}{}}{\overset{|}{C}}} \bullet e \;\longrightarrow\; \overset{H}{\underset{H}{\overset{|}{C}}} = \overset{CH_3}{\underset{\underset{CH_3}{}}{\overset{|}{C}}} \;+\; \text{Heat}$$

(L)   Isobutene   4a

Overall equation: 6 t-Butyl-Cyclopendiene ⟶ 6Acetylene + 4Methane + 4Isobutene

+ 2[1-Methyl-1-Butene] + 1,4-cumulene + 8C   **4b**

Stage 5:

$$
\begin{array}{l}
\underset{\text{Isobutene}}{
\begin{array}{c}
\text{H} \quad \text{CH}_3 \\
| \qquad | \\
\text{C} = \text{C} \\
| \qquad | \\
\text{H} \quad \text{CH}_3
\end{array}}
\quad \rightleftharpoons \quad
n\bullet \begin{array}{c}
\text{CH}_3 \\
| \\
\text{C} = \text{C} \\
| \qquad | \\
\text{H} \quad \text{CH}_3
\end{array}
\quad + \quad e\bullet\text{H}
\end{array}
$$

$$
n\bullet \begin{array}{c}
\text{CH}_3 \\
| \\
\text{C} = \text{C} \\
| \qquad | \\
\text{H} \quad \text{CH}_3
\end{array}
\quad \rightleftharpoons \quad
n\bullet \begin{array}{c}
\text{CH}_3 \\
| \\
\text{C} = \text{C} \bullet e \\
| \\
\text{H}
\end{array}
\quad + \quad n\bullet\text{CH}_3
$$

$$
\text{H}\bullet e \quad + \quad n\bullet\text{CH}_3 \quad \rightleftharpoons \quad \text{CH}_4
$$

$$
n\bullet \begin{array}{c}
\text{CH}_3 \\
| \\
\text{C} = \text{C} \bullet e \\
| \\
\text{H}
\end{array}
\quad \longrightarrow \quad
\begin{array}{c}
\text{CH}_3 \\
| \\
\text{C} \equiv \text{C} \\
| \\
\text{H}
\end{array}
\quad + \quad \text{Heat}
$$

     5a

Overall equation: 2Isobutene ⟶ 2CH$_4$ + 2Propyne   **5b**

Note that only half of the isobutene was kept in Equilibrium state of existence, because of the large number of moles, leaving the remaining fraction in Stable state of existence.

Stage 6:

$$
\underset{\textbf{1-Methyl-1-Butene}}{
\begin{array}{c}
\text{H} \quad \text{CH}_3 \\
| \qquad | \\
\text{C} = \text{C} \\
| \qquad | \\
\text{H} \quad \text{CH}_2 \\
\qquad | \\
\qquad \text{CH}_3
\end{array}}
\quad \rightleftharpoons \quad
\underset{(A)}{
n\bullet \begin{array}{c}
\text{CH}_3 \\
| \\
\text{C} = \text{C} \\
| \qquad | \\
\text{H} \quad \text{CH}_2 \\
\qquad | \\
\qquad \text{CH}_3
\end{array}}
\quad + \quad e\bullet\text{H}
$$

$$
n\bullet \begin{array}{c}
\text{CH}_3 \\
| \\
\text{C} = \text{C} \\
| \qquad | \\
\text{H} \quad \text{CH}_2 \\
\qquad | \\
\qquad \text{CH}_3
\end{array}
\quad \rightleftharpoons \quad
\begin{array}{c}
\text{CH}_3 \\
| \\
\text{C} \equiv \text{C} \\
| \\
\text{H}
\end{array}
\quad + \quad n\bullet\text{CH}_2 - \text{CH}_3
$$

$$
n\bullet\text{CH}_2 - \text{CH}_3 \quad \rightleftharpoons \quad
n\bullet \begin{array}{c}
\text{H} \quad \text{H} \\
| \qquad | \\
\text{C} - \text{C} \bullet e \\
| \qquad | \\
\text{H} \quad \text{H}
\end{array}
\quad + \quad n\bullet\text{H}
$$

$$
\text{H}\bullet e \quad + \quad n\bullet\text{H} \quad \rightleftharpoons \quad \text{H}_2
$$

$$
n\bullet \begin{array}{c}
\text{H} \quad \text{H} \\
| \qquad | \\
\text{C} - \text{C} \bullet e \\
| \qquad | \\
\text{H} \quad \text{H}
\end{array}
\quad \longrightarrow \quad
\underset{(B)}{
\begin{array}{c}
\text{H} \quad \text{H} \\
| \qquad | \\
\text{C} = \text{C} \\
| \qquad | \\
\text{H} \quad \text{H}
\end{array}}
\quad + \quad \text{Heat}
$$

     6a

Overall equation: 2[1-Methyl-1-Butene] ⟶ 2Propyne + 2Ethene + 2H$_2$

Stage 7:

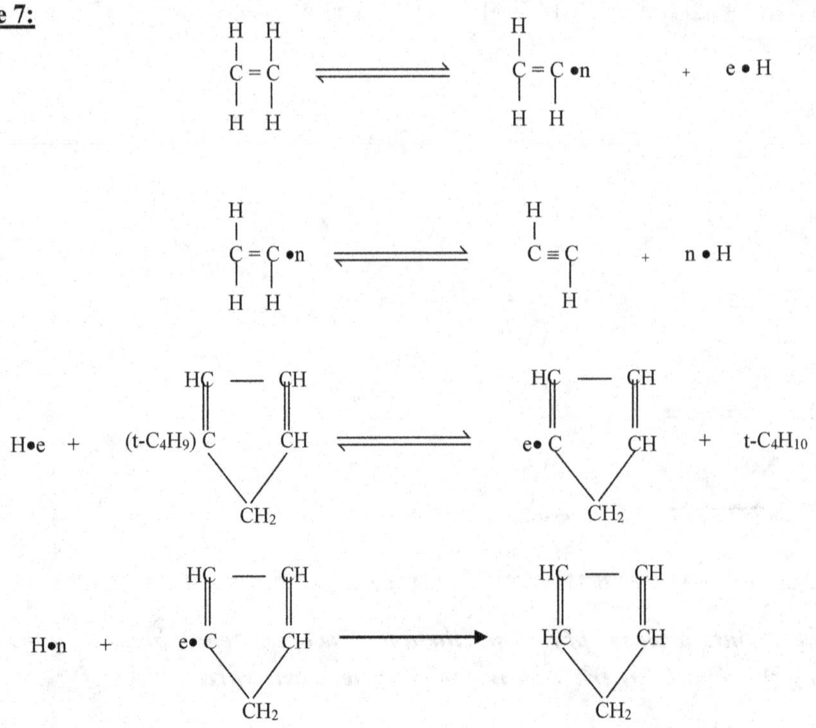

6b

Cyclopentadiene

7a

Overall equation: 2Ethene + 2t-Butyl-Cyclopentadiene ⟶ 2Acetylene + 2t-Butane + 2Cyclopentadiene

7b

Overall overall equation: 8t-Butyl-Cyclopentadiene ⟶ 8Acetylene + 6Methane + 4Propyne + 2H$_2$ + 8C + 2t-Butane + 2Cyclopentadiene + 1,4-Cumulene + 2 Isobutene

7c

Stage 8:

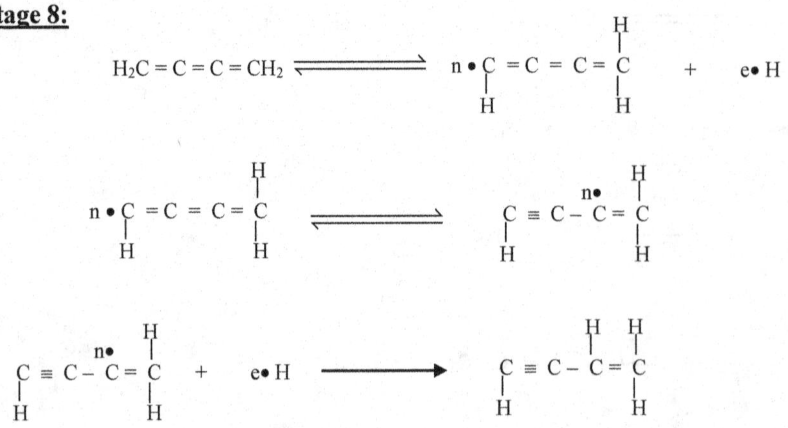

8

Stage 9:

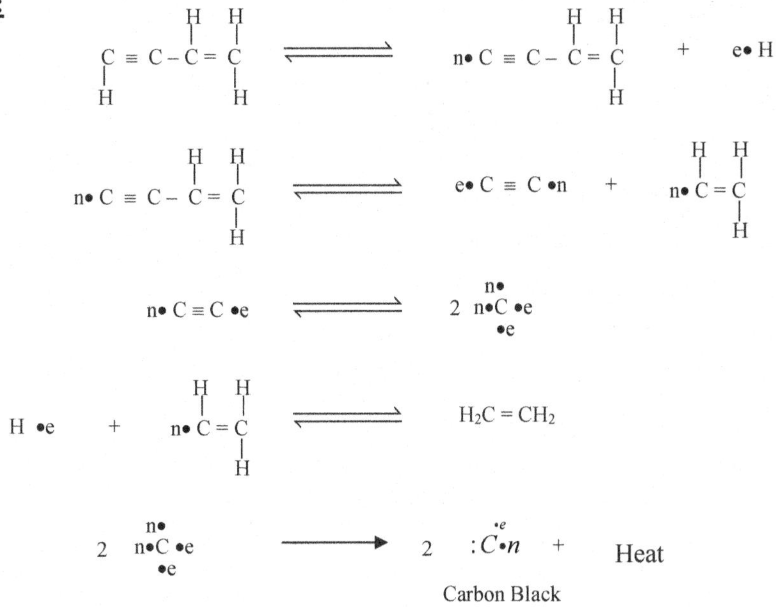

$$C \equiv C - C = C \rightleftharpoons n \bullet C \equiv C - C = C \quad + \quad e \bullet H$$
(with H atoms on the carbons)

$$n \bullet C \equiv C - C = C \rightleftharpoons e \bullet C \equiv C \bullet n \quad + \quad n \bullet C = C$$
(with H atoms on the carbons)

$$n \bullet C \equiv C \bullet e \rightleftharpoons 2 \; n \bullet C \bullet e$$
(with n• and •e)

$$H \bullet e \quad + \quad n \bullet C = C \rightleftharpoons H_2C = CH_2$$
(with H atoms on the carbons)

$$2 \; n \bullet C \bullet e \longrightarrow 2 \; :C \bullet n \quad + \quad \text{Heat}$$

Carbon Black   9a

Overall equation: 1,4-Cumulene $\longrightarrow$ $H_2C = CH_2$ $\quad + \quad$ 2C (Carbon black)   9b

Stage 10:   Same as Stage 7 of Equation 7a.

Overall equation: Ethene $+$ t-Butyl cyclopentadiene $\longrightarrow$ Acetylene $+$ t-Butane $+$
Cyclopentadiene   10

Stage 11:

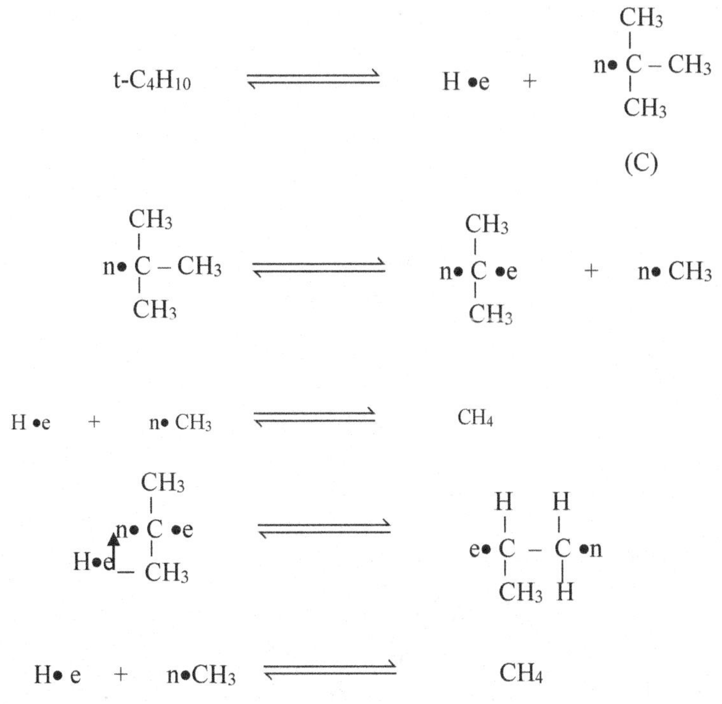

$$t\text{-}C_4H_{10} \rightleftharpoons H \bullet e \quad + \quad n \bullet \underset{CH_3}{\overset{CH_3}{C}} - CH_3$$

(C)

$$n \bullet \underset{CH_3}{\overset{CH_3}{C}} - CH_3 \rightleftharpoons n \bullet \underset{CH_3}{\overset{CH_3}{C}} \bullet e \quad + \quad n \bullet CH_3$$

$$H \bullet e \quad + \quad n \bullet CH_3 \rightleftharpoons CH_4$$

$$\underset{H \bullet e - CH_3}{n \bullet \overset{CH_3}{C} \bullet e} \rightleftharpoons e \bullet \underset{CH_3}{\overset{H}{C}} - \underset{H}{\overset{H}{C}} \bullet n$$

$$H \bullet e \quad + \quad n \bullet CH_3 \rightleftharpoons CH_4$$

449

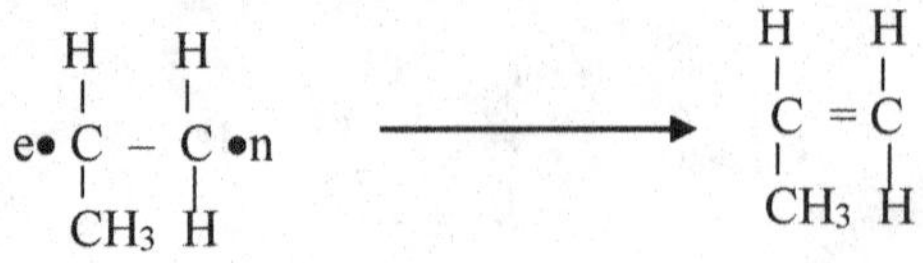

Propene

$$H_2C = CH(CH_3) + Heat \qquad \text{11a}$$

Overall equation: $3t\text{-}C_4H_{10} \longrightarrow 3Propene + 3CH_4$ \qquad 11b

Stage 12: The propene is broken down in a similar manner like the others.

Overall equation: $3Propene \longrightarrow 3Acetylene + 3CH_4$ \qquad 12a

Overall overall equation: 9 t-Butyl-Cyclopentadiene $\longrightarrow$ 12Acetylene + 12Methane +

4Propyne + 2Isobutene + $2H_2$ + 10C + 3Cyclopentadiene \qquad 12b

Stage 13:

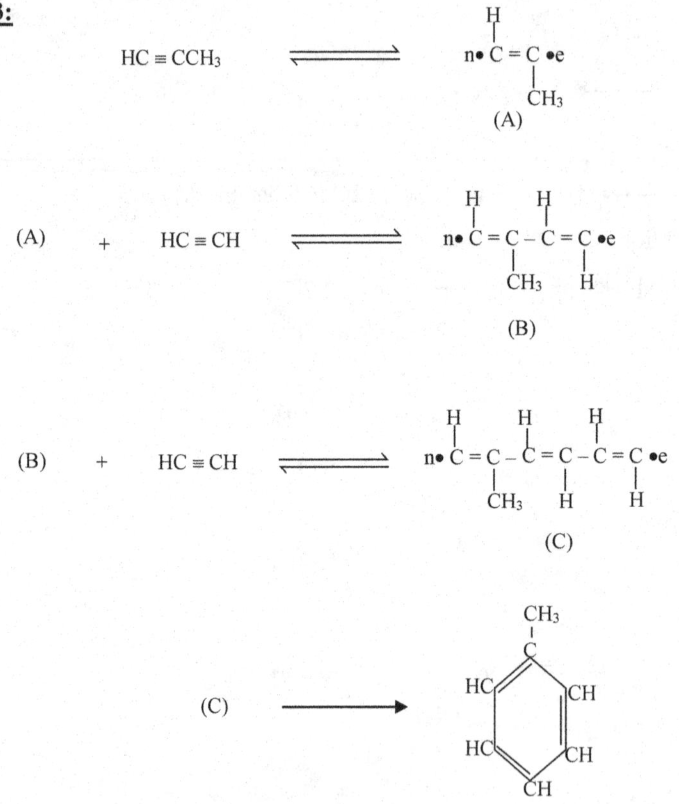

(A)

(B)

(C)

Toluene \qquad 13a

Overall equation: $8HC\equiv CH + 4Propyne \longrightarrow 4Toluene$ \qquad 13b

Overall overall equation: 9 t-butyl-Cyclopentadiene $\longrightarrow$ 4Acetylene + 12Methane +

2Isobutene + **4Toluene** + 10C + $3H_2$ + **3Cyclopentadiene** \qquad 13c

noting that there are indeed 14 stages so far based on the number of moles used, because the molecular rearrangement of methyl isobutene to 1-methyl-1-butene in Equation 3a was not shown, and if the number of moles had been larger than as shown above, then benzene would have been one of the main products with one more stage; and *finally noting how t-butane was decomposed in Stage 11 to give propene and methane and finally to acetylene and methane, based on the operating conditions,* in addition to all new concepts in all the stages.

(Laws of Creations for t-butyl 1,3-Cyclopentadiene)

Rule 844: This rule of Chemistry for **a monomer such as t-Butyl-1,3-Cycopentadiene,** states that, if an operating condition can be found wherein very small fraction is kept in Equilibrium state of existence leaving a very large fraction in Stable state of Existence, this monomer can be made to undergo **Step polymerization** to produce a polymer as shown below-

Stage 1:

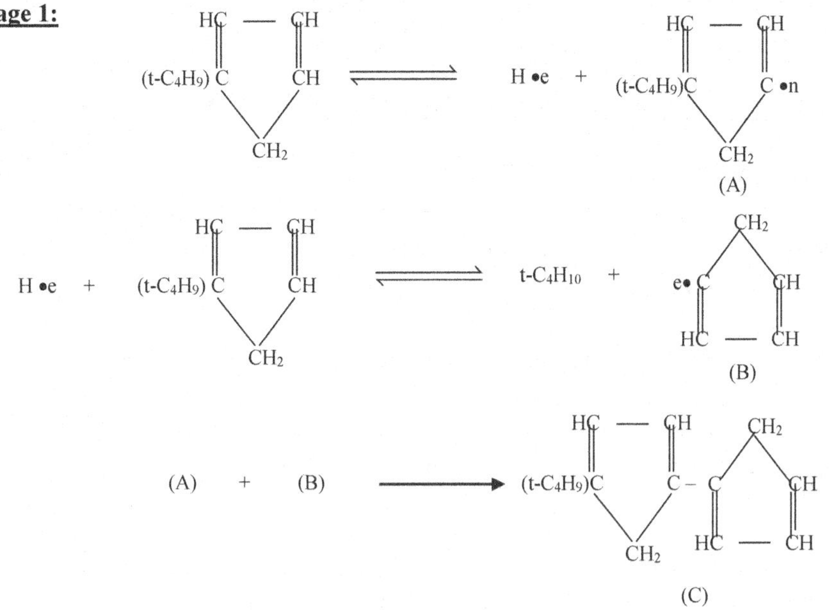

$$(t\text{-}C_4H_9)C \rightleftharpoons H \bullet e + (t\text{-}C_4H_9)C \quad (A)$$

$$H \bullet e + (t\text{-}C_4H_9)C \rightleftharpoons t\text{-}C_4H_{10} + e\bullet C \quad (B)$$

$$(A) + (B) \longrightarrow (t\text{-}C_4H_9)C \quad (C)$$

1

Stage 2:

$$(C) \rightleftharpoons (D) \bullet n + e \bullet H$$

$$H \bullet e + (t\text{-}C_4H_9)C \rightleftharpoons t\text{-}C_4H_{10} + e\bullet C \quad (B)$$

$$(B) + (D) \longrightarrow (E)$$

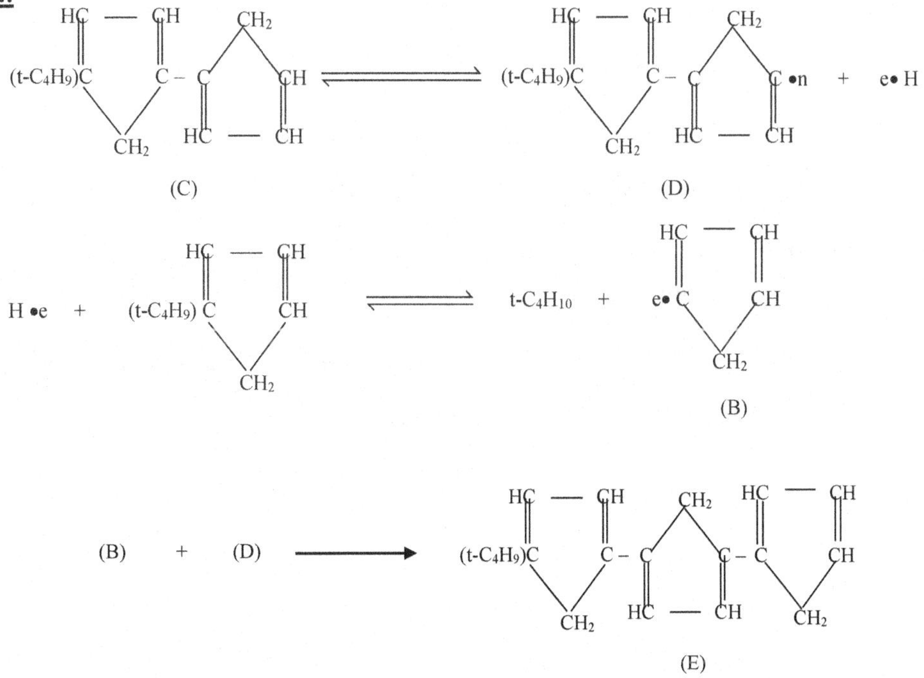

2

3

Overall equation: 3t-Butyl Cyclopentadiene $\longrightarrow$ (E) + 2 t-Butane

for which the small molecular byproducts is t-butane along with the syndiotactically placed polymeric chain, i.e., the monomer is a di-bi-functional STEP monomer

(Laws of Creations for Cyclopentadienes)

Rule 845: This rule of Chemistry for **Radical-pushing alkylane groups,** states that, the order of their capacities when they carry electro-free-radicals and nucleo-free-radicals are as shown below-

$$e\bullet C_4H_9 \quad > \quad e\bullet C_3H_7 \quad > \quad e\bullet C_2H_5 \quad > \quad e\bullet CH_3 \quad > \quad H\bullet e$$

ORDER OF ELECTRO-FREE-RADICAL CAPACITY

$$H\bullet n \quad > \quad n\bullet CH_3 \quad > \quad n\bullet C_2H_5 \quad > \quad n\bullet C_3H_7 \quad > \quad n\bullet C_4H_9$$

ORDER OF NUCLEO-FREE-RADICAL CAPACITY

for which their movements during rearrangement depend on how the active centers are located and what they are carrying (Carbenes, Alkenes and Alkynes), their States of Existence (Equilibrium, Decomposition and Combination) and whether they are being rejected or abstracted as transfer species.
(Laws of Creations for Radical-pushing groups)

Rule 846: This rule of Chemistry for **Cyclopentadiene of the type shown below,** states that, this monomer cannot be made to undergo molecular rearrangement of the third kind or of another kind, based on the radicals carried by the active centers and the fact that there is no transfer species from the CH_3 group or any other higher group which is resonance stabilized since it is internally located.

Discrete closed- loop resonance stabilization

for which the two mono-forms are (II) the least nucleophilic center and (III).
(Laws of Creations for Cyclopentadienes)

Rule 847: This rule of Chemistry for **Cycloalkenes such as Cyclopentadiene as shown below,** states that, when a substituent group is internally located, just as in isoprene, no transfer species can

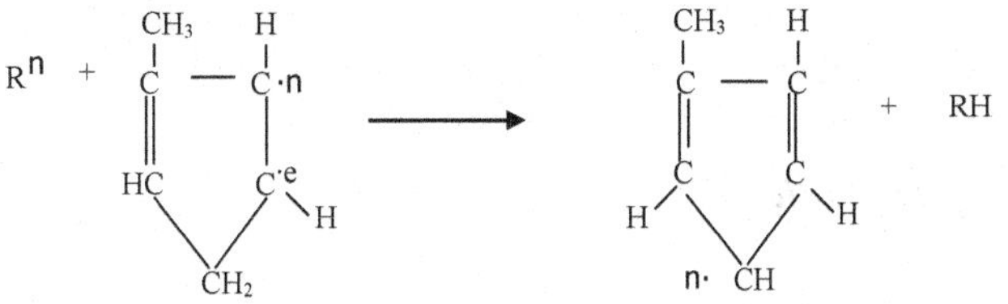

i)

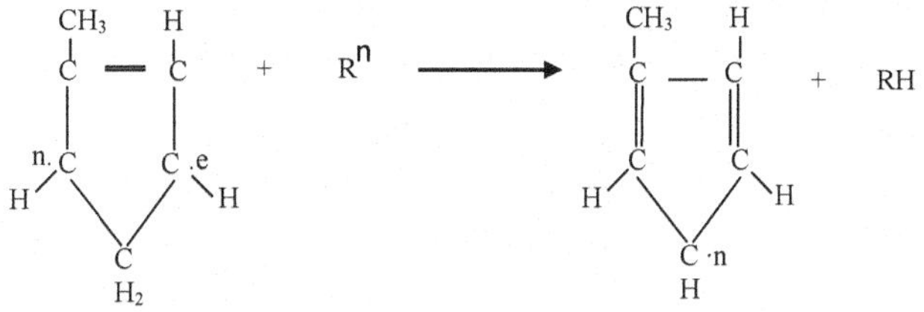

ii)

be abstracted from the group and no rearrangement can take via the group as shown below.

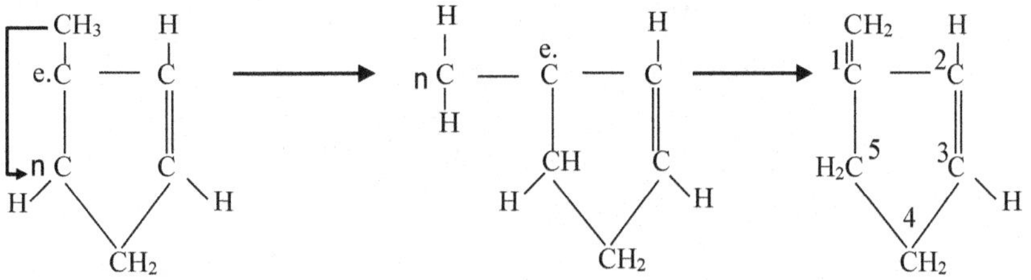

IMPOSSIBLE TRANSFER

(Laws of Creations for Cyclopentadienes)

Rule 848: This rule of Chemistry for **2-methyl-cyclopentadiene,** states that, when made to undergo molecular rearrangement of the first kind, the followings are obtained-

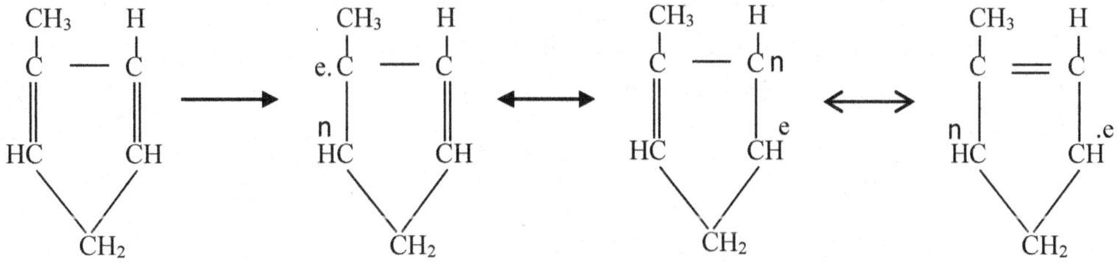

(I) Not a mono-form   (II)   (III)

Discrete closed- loop resonance stabilization

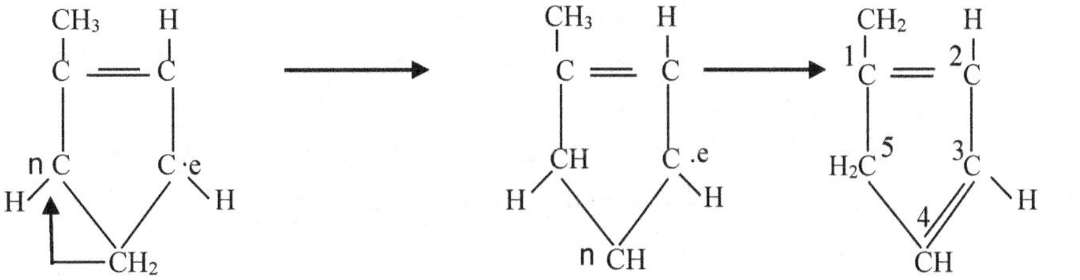

2-Methyl-Cyclopentadiene   1-Methyl-Cyclopentadiene

453

for which all two mono-forms including the non-mono-form (I), molecularly rearrange to give a more Nucleophilic 1-methyl-cyclopentadiene; making the existence of 2-substituted-cyclopenta-diene impossible when weakly activated.
(Laws of Creations for Cyclopentadienes)

Rule 849: This rule of Chemistry for **the types of cyclopentadiene shown below,** states that, the following is obtained, that wherein transfer species externally located from the R group (e.g. CH$_3$) *like in Cyclopentene does not still exist and cannot be used for rearrangement;* noting that for all

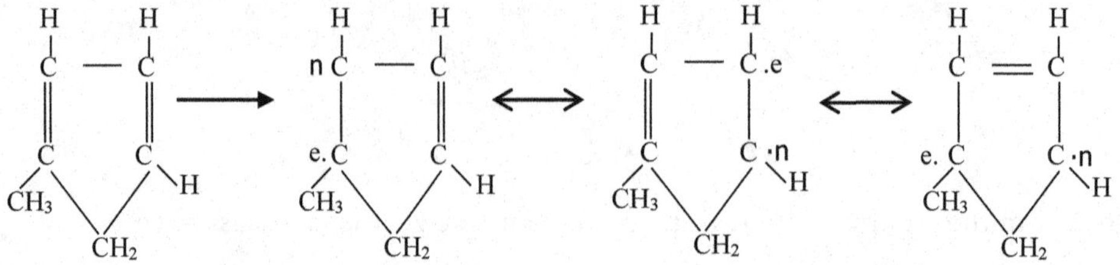

NOT A MONO-FORM

Discrete closed- loop resonance stabilization

of them, only the electro-free-radical route is favored by them.[Influence of H$_2$C group]
(Laws of Creations for Cyclopentadienes)

Rule 850: This rule of Chemistry for **1-Methyl-Cyclopentadiene,** states that, this cannot undergo molecular rearrangement of the first kind, since while all the two mono-forms will rearrange to less Nucleophilic 2-methyl-cyclopentadiene, the non-mono-form will rearrange back to itself; clear indication that the two mono-forms will not rearrange.
(Laws of Creations for Cyclopentadienes)

Rule 851: This rule of Chemistry for **Cyclopropene, Cyclopentadiene, and Cycloheptatriene,** states that, while in cyclopropene CH$_2$ is connected to an alkene (Ethene), in cyclopentadience, the same CH$_2$ group is connected to an alkediene (Butadiene) and in cycloheptatriene, the same CH$_2$ group is connected to an alketriene (Hexatriene), one should expect the following order in radical-pushing capacity, hexatriene being more nucleophilic than butadiene which in turn is more nucleophilic than ethene-

In Cycloheptatriene　　**In Cyclopentadiene**　　**In Cyclopropene**

ORDER OF RADICAL-PUSHING CAPACITY

for which these will always remain as the source of transfer species.
(Laws of Creations for Cyclopropene, Cyclopentadiene and Cycloheptadiene)

Rule 852: This rule of Chemistry for **Vinyl cyclopentadiene,** states that, for the type shown below, the resonance stabilization favored by it is called *Opened/half closed-loop resonance stabilization,* since the least nucleophilic center is that externally located-

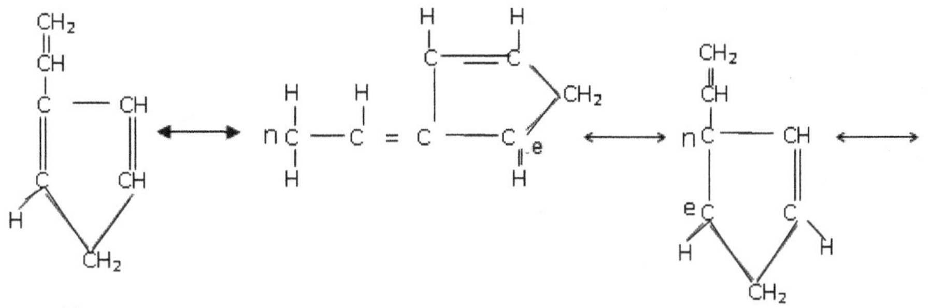

(I)

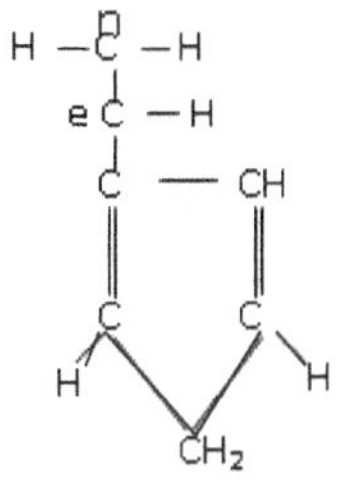

Opened /half closed- loop resonance Stabilization

the first and the last being the two mono-forms; noting that if it was not, then the vinyl group would then be a group resonance stabilized as shown below-

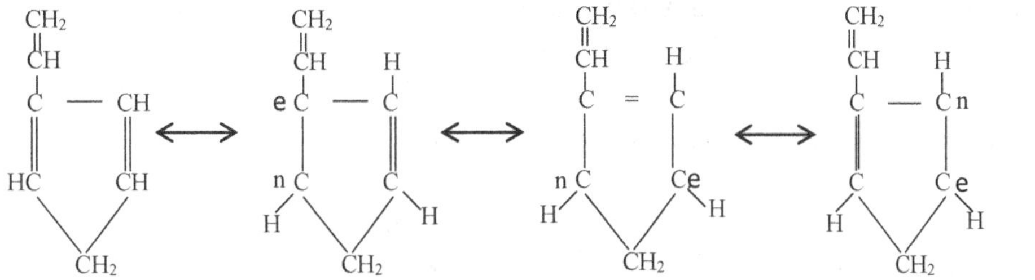

Discrete closed- loop resonance stabilization   NOT A MON-FORM

IMPOSSIBLE RESONANCE STABILIZATION

that which is impossible, since the vinyl group regardless what it is carrying will always be the least nucleophilic center.

(Laws of Creations for 2-Vinyl Cyclopentadiene)

Rule 853: This rule of Chemistry for **2-Vinyl cyclopentadiene,** states that, when made to undergo molecular rearrangements as shown below, only the first kind can be favored with no possibility of

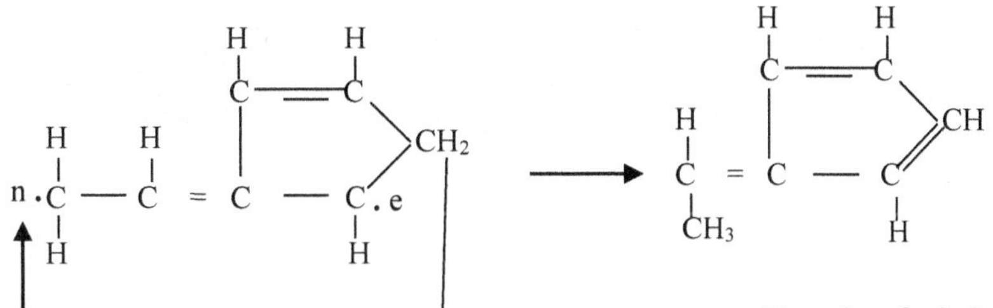

(No point of scission)

opening of the ring.

(Laws of Creations for 2-Vinyl cyclopentadiene)

Rule 854: This rule of Chemistry for **Methyl 2- vinyl cyclopentadiene,** states that, when activated, the followings are to be expected in terms of placement of radicals-

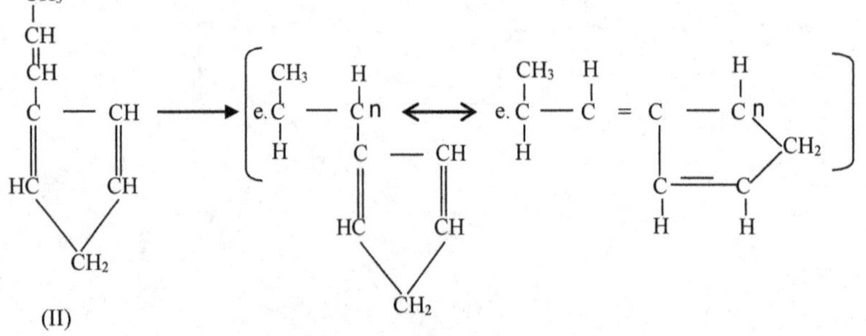

(II)

NOT FAVORED   (a)

(II)

(II)a   (II)b

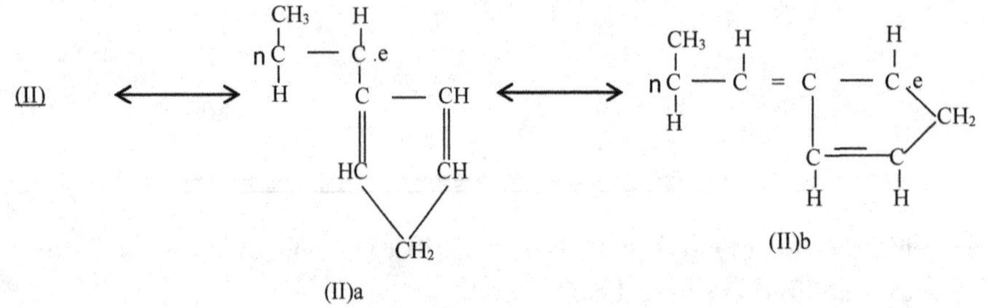

FAVORED   (b)

in view of the influence of H_2C group whose radical-pushing capacity is far greater than R.

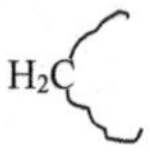

(Laws of Creations for Methyl 2- vinyl cyclopentadiene)

Rule 855: This rule of Chemistry for **1-Vinyl Cyclopentadiene of the type shown below,** states that, when activated, this favors *Opened/full Closed-loop resonance stabilization* as shown below-

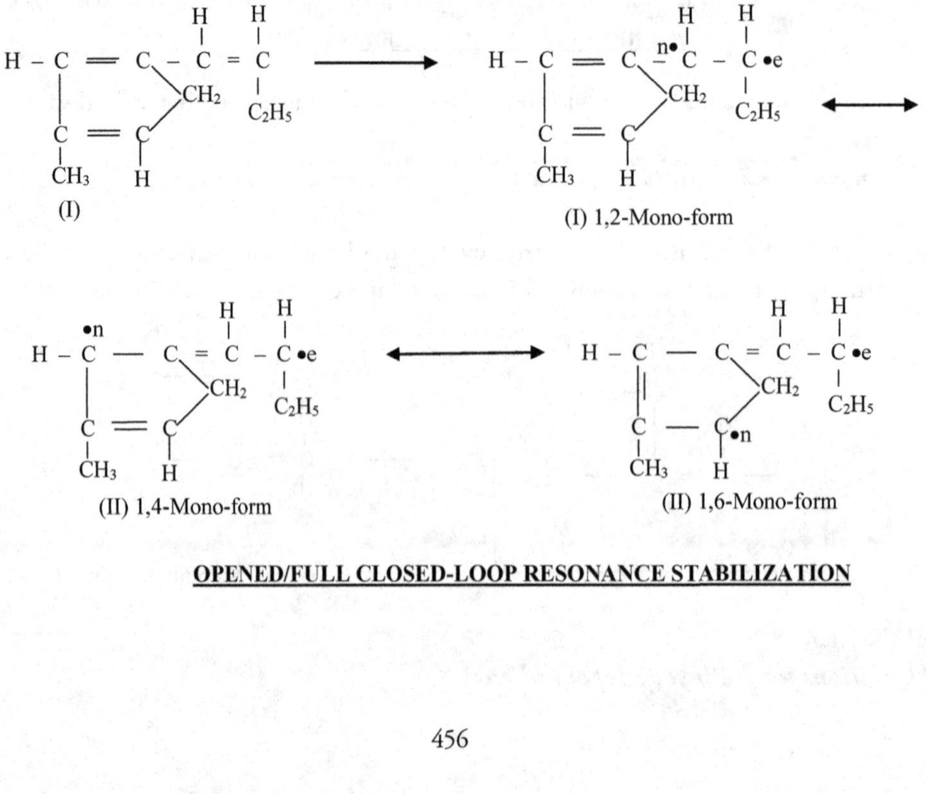

(I)   (I) 1,2-Mono-form

(II) 1,4-Mono-form   (II) 1,6-Mono-form

OPENED/FULL CLOSED-LOOP RESONANCE STABILIZATION

for which unlike 2-vinyl cyclopentadiene which has two mono-forms, here there are three mono-forms, noting here also that CH_2 group inside the ring and CH_3 group are fully resonance stabilized.
(Laws of Creations for 1-Vinyl cyclopentadiene)

Rule 856: This rule of Chemistry for **1-Vinyl cyclopentadiene of the type shown below,** states that, this can undergo, molecular rearrangement of the third kind as shown below to give a Di- vinyl acetylene (IV), noting the type of product obtained, making it look as if the externally located alkene

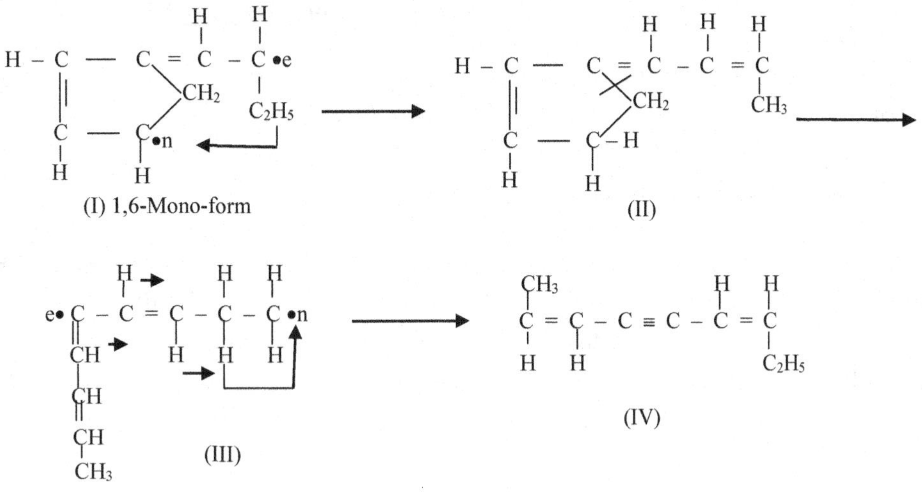

(I) 1,6-Mono-form

(II)

(III)

(IV)

group was not tempered with, which indeed is not the case since rearrangement stops when C_2H_5 is changed to C_3H_7.
(Laws of Creations for 1-Vinyl cyclopentadienes)

Rule 857: This rule of Chemistry for **Cyclopentadiene acetylenic type of resonance stabilization group,** states that, the followings are to be expected, for which the monomer is opened/half closed

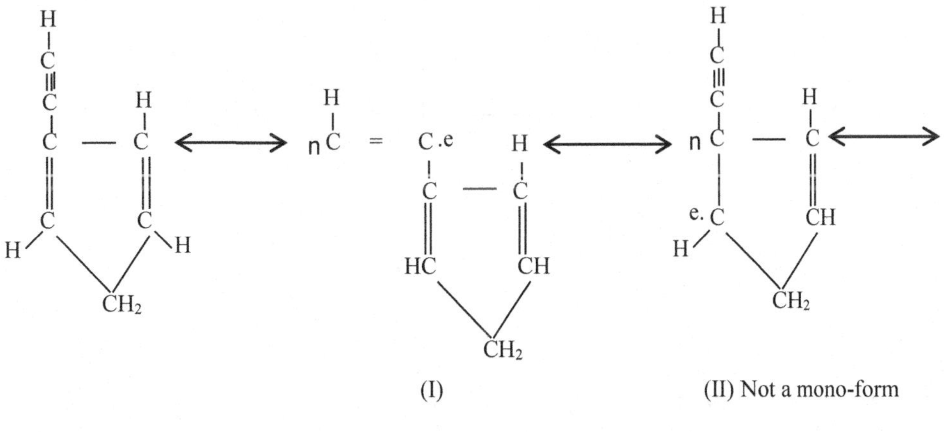

(I)

(II) Not a mono-form

457

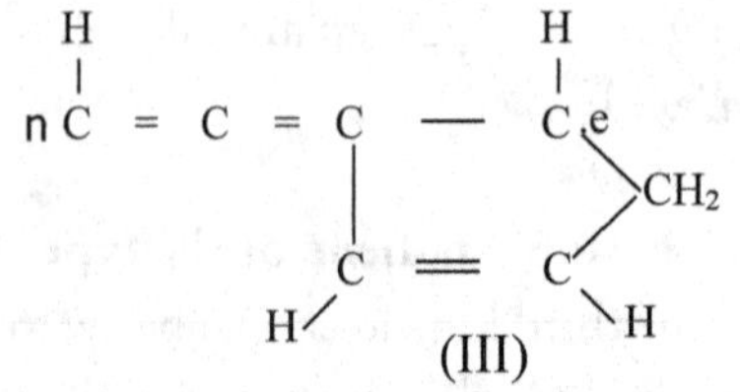

$$n \overset{H}{\underset{}{C}} = C = \overset{}{\underset{}{C}} \quad — \quad \overset{H}{\underset{}{C}} e$$

(III)

Opened/Half Closed-loop resonance stabilized

loop resonance stabilized, with no molecular rearrangement of the first type, but of the tenth type with no opening of the ring possible, noting that only (I) and (III) are the mono-forms and in fact, there may be need to replace the H atom on acetylene to keep the monomer stable.

(Laws of Creations for 2-Acetylenic cyclopentadiene)

Rule 858: This rule of Chemistry for **Cyclopentadiene carrying acetylene with CH₃ and higher groups in 2-position,** states that, while (II)a, and (II)b shown below are the opened/half closed-loop resonance stabilized mono-forms, (II)d which is discrete closed loop resonance cannot exist, since acetylene is the resonance stabilization provider, and (II)c is not a mono-form, being the most nucleophilic center.

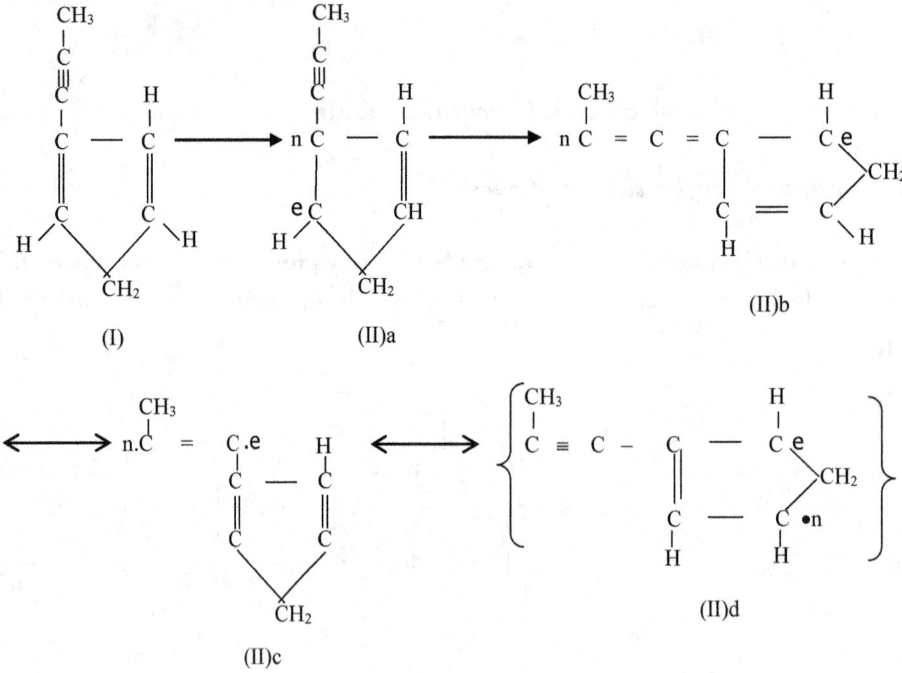

(I)   (II)a   (II)b

(II)c   (II)d

(Laws of Creations for 2-Acetylenic cyclopentadienes)

Rule 859: This rule of Chemistry for1-**Acetylenic cyclopentadiene,** states that, the following is valid for it when activated-

458

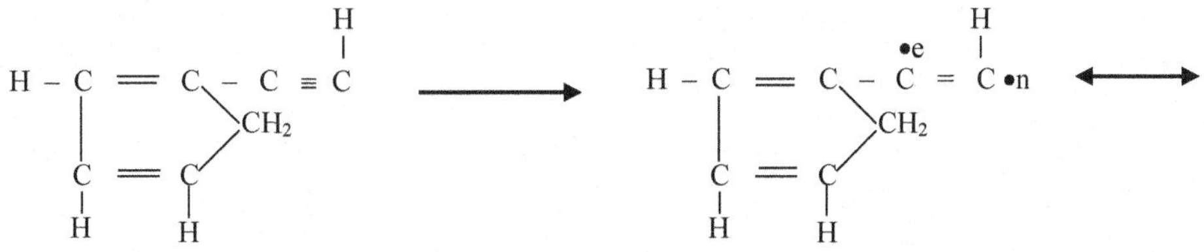

(I)   (II) 1,2-Mono-form

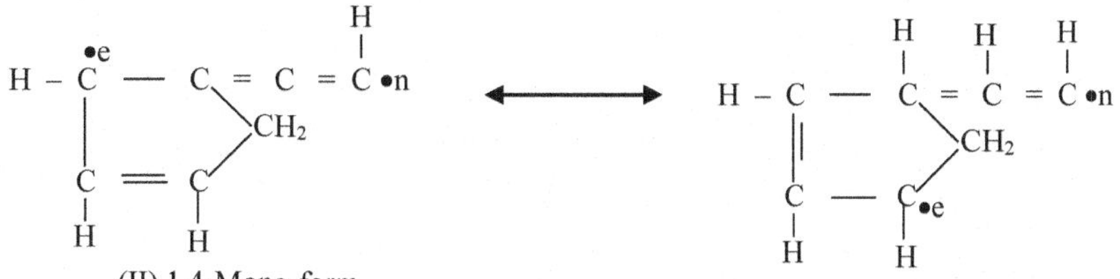

(II) 1,4-Mono-form   (II) 1,6-Mono-form

noting that this cannot undergo any rearrangement, and in fact highly unstable, always in Equilibrium state of existence, the H being held coming from the ring just like cyclopropene, and like cyclopropene, the CH_2 group is fully resonance stabilized.

(Laws of Creations for 1-Acetylenic cyclopentadiene)

Rule 860: This rule of Chemistry for **1-Acetylenic cyclopentadienes wherein acetylene is carrying a radical-pushing alkylane group,** states that, in view of the fact that CH_2 group is resonance stabilized, the following is valid-

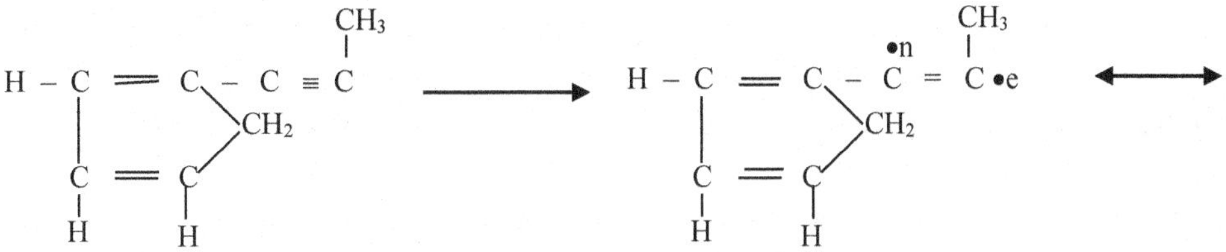

(I)   (II) 1,2-Mono-form

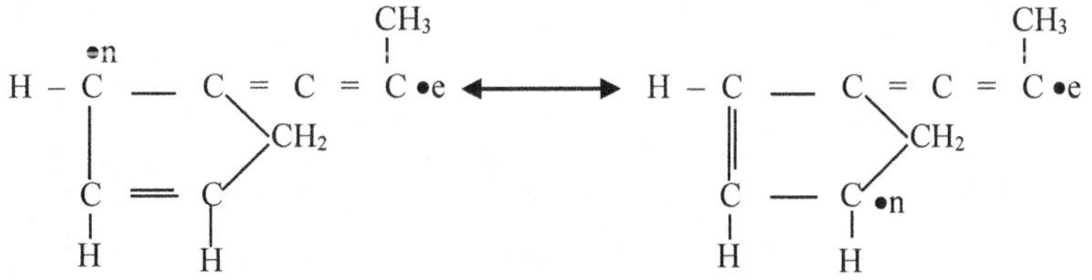

(II) 1,4-Mono-form   (II) 1,6-Mono-form

noting that this can undergo molecular rearrangement to give a more Nucleophilic monomer- $HC \equiv C - C \equiv C - CH = CH(C_2H_5)$ with limitation on the size of CH_3 being C_3H_7.

(Laws of Creations for 1-Vinyl cyclopentadienes)

Rule 861: This rule of Chemistry for **Cyclopentadiene,** states that, based on the diene character, the following is valid when used as resonance or non-resonance stabilization groups –

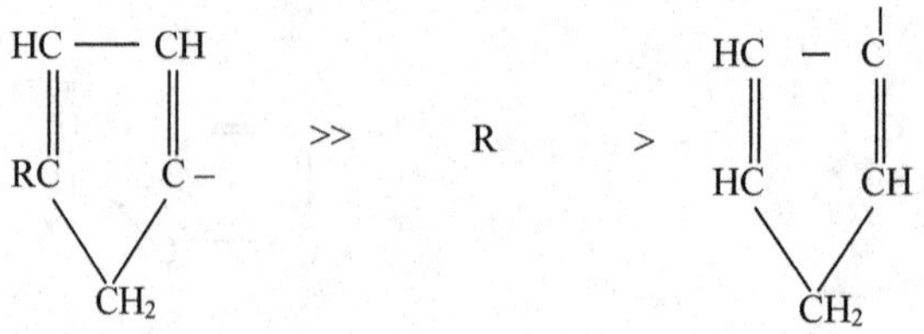

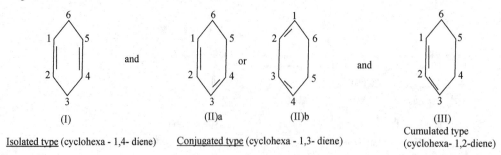

Order of radical-pushing capacity of resonance stabilization groups

(Laws of Creations for Cyclopentadiene)

Rule 862: This rule of Chemistry for **Cyclodienes, Cycloalkenes and Cycloalkanes,** states that, based on their characters, there is no doubt that the following is valid-

$$Cyclodienes \quad > \quad Cycloalkenes \quad > \quad Cycloalkanes$$

Order of nucleophilicity

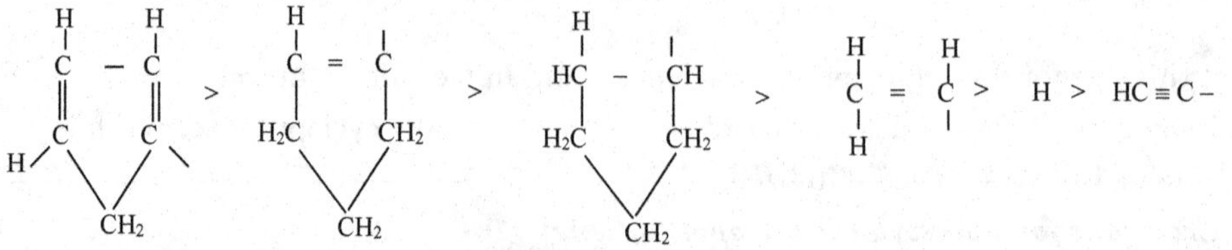

Order of radical- pushing capacity

(Laws of Creations for Cyclodienes, Cycloalkenes, and Cycloalkanes)

Rule 863: This rule of Chemistry for **Cyclopropene, Cyclopentadiene, Cycloheptatrienes, and higher look-alikes,** states that, no matter what these are carrying internally or externally, their rings can never be opened instantaneously, since they do not have any point of scission.
(Laws of Creations for Cyclopropenes, Cyclopentadienes and Cycloheptatrienes)

Rule 864: This rule of Chemistry for **Cyclohexadiene,** states that, there are three types of cyclohexadiene, depending on the location of the double bonds.

(I)   and   (II)a   or   (II)b   and   (III)

Isolated type (cyclohexa - 1,4- diene)   Conjugated type (cyclohexa - 1,3- diene)   Cumulated type (cyclohexa- 1,2-diene)

for which, while (I) has no point of scission, the (II)s, have one point of scission- the 5,6--bond, (III) which is not popularly known to exist has one point of scission- 5,6- or 4,5- bond.
(Laws of Creations for Cyclohexadiene)

Rule 865: This rule of Chemistry for **Cyclohexa-1,3-allene,** states that, this cumulative type of Cyclohexadiene is not popularly known to exist, because it readily molecularly rearranges to Cyclohexyne as shown below-

And (III)   Cyclohexyne

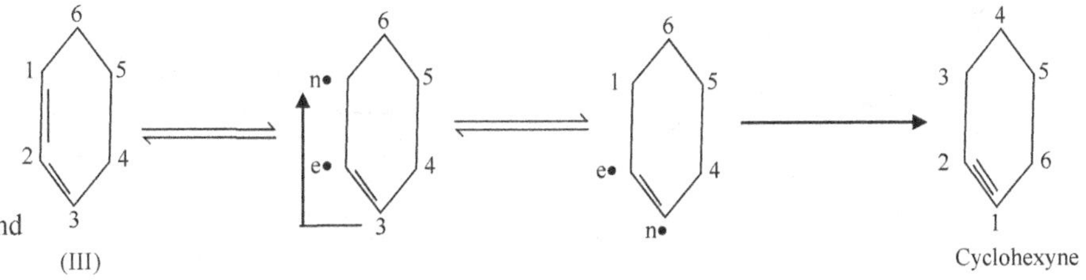

(Laws of Creations for Cyclohexa-1,3-Allene)

Rule 866: This rule of Chemistry for **Cyclohexa-1,3-Diene,** states that, when the ring is opened instantaneously based on the operating conditions, the followings is to be expected-

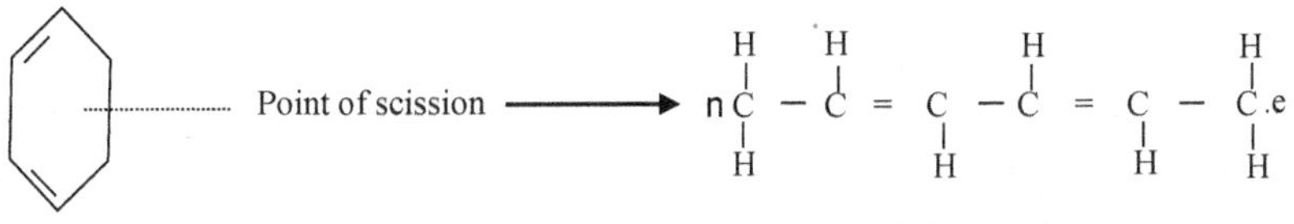

1, 3, 5- hexatriene

for which this is more favored radically than chargedly and that this is very difficult to accomplish since the radical potential difference at the only point of scission is zero, noting that radically a resonance stabilized hexatriene is obtained when deactivated.

(Laws of Creations for Cyclohexa-1,3-Diene)

Rule 867: This rule of Chemistry for **Cyclohexadiene where the two double bonds are isolatedly placed,** states that, when activated, the followings are to be expected, for which molecular rearrangement of the first kind takes place to give a conjugatedly placed diene (Cyclohexa-1,3-

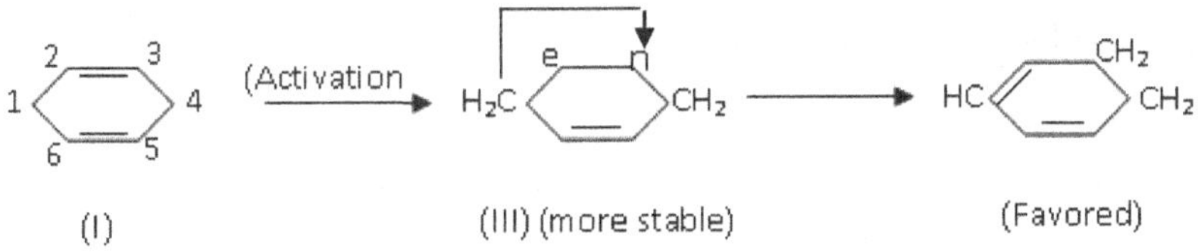

(I)   (III) (more stable)   (Favored)

Diene) which can continue further rearrangement to give same cyclohexa-1,3-diene.

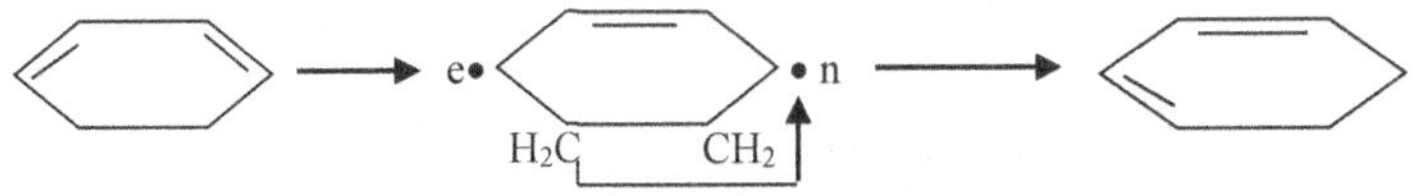

(Laws of Creations for Cyclohexa-1,4-Diene)

Rule 868: This rule of Chemistry for **Isolatedly placed cyclohexadiene with NH$_2$ type of group,** states that, when activated, the less nucleophilic center which is the first to be activated cannot be used for rearrangement as shown below-

461

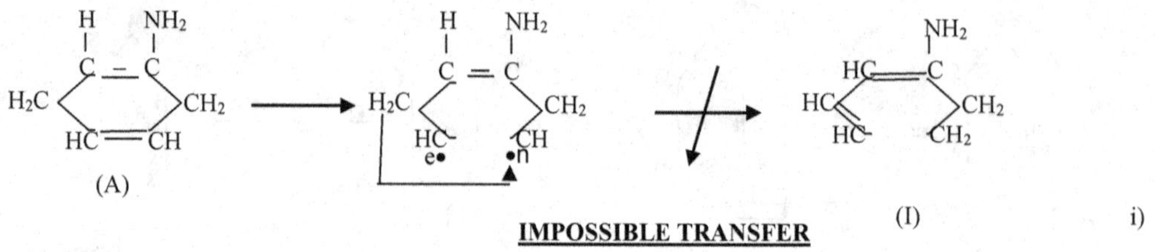

IMPOSSIBLE TRANSFER (I) i)

for which if (I) had been obtained, it rearranges as shown below to give an Electrophile which is not resonance stabilized-

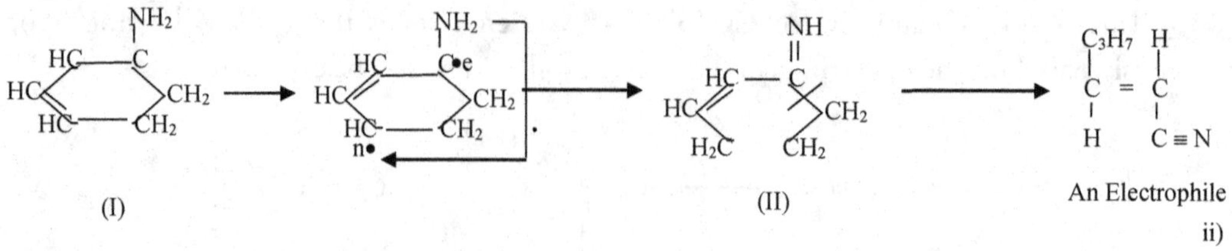

(I) (II) An Electrophile ii)

noting that unlike 1,3-cyclohexadiene, the monomer above (A) cannot rearrange.
(Laws of Creations for 1,3- and 1,4- Cyclohexadienes)

Rule 869: This rule of Chemistry for **Radical-pushing groups**, states that, the followings are valid-

$$C_3H_7 - CH = CH - \; > \; C_2H_5 - CH = CH - CH_2 - \; > \; CH_3 - CH = CH - CH_2 - CH_2 - \; > \; C_3H_7$$

ORDER OF RADICAL- PUSHING CAPACITY

(Laws of Creations for Radical-pushing groups).

Rule 870: This rule of Chemistry for **Isolated Cyclohexadienes (1,4- Cyclohexadienes) carrying R groups,** states that, when activated the possibility of opening the ring does exist as shown below-

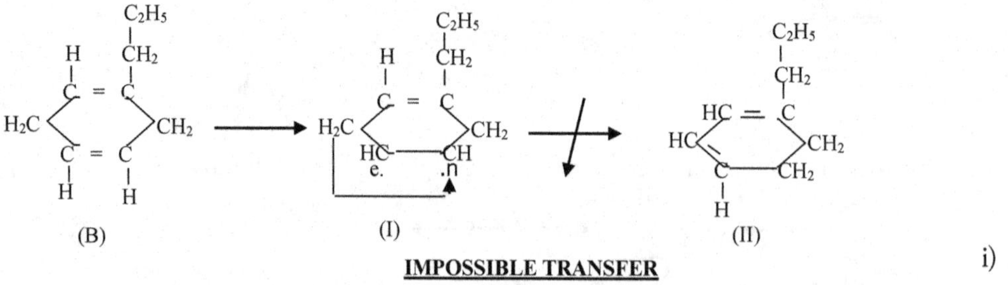

(B) (I) (II) i)
IMPOSSIBLE TRANSFER

for which if R is less than C_2H_5, but not H, the situation will remain the same; the (II) unlike (B) can however rearrange as shown below to give a resonance stabilized Nucleophile, with limitations placed at C_4H_9.

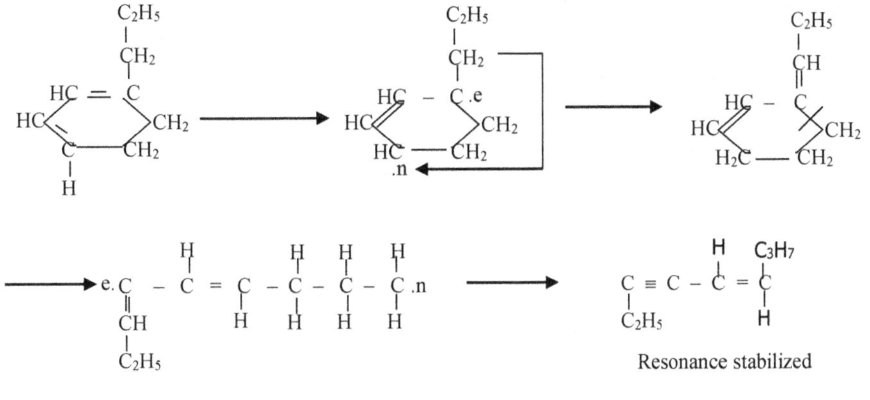

ii)

Resonance stabilized

(Laws of Creations for 1,3- and 1,4- Cyclohexadienes)

Rule 871: This rule of Chemistry for **Cyclohexa-1,3-Diene,** states that, when activated chargedly in the presence of a weak coordination initiator, the followings are obtained-

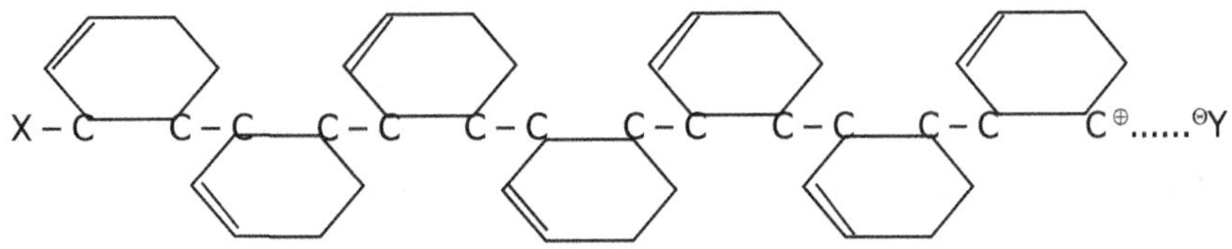

Syndiotactic-1,2-Poly (Cyclohexadiene)

for which when the ring is not opened instantaneously, chargedly only the 1,2 –mono-form can exist along the chain as shown above since resonance stabilization cannot take place chargedly, noting that the carrier of the chain X could be Rh but not an ionic metal; and in the absence or presence of vacant orbitals or reservoirs for the monomers and presence of steric limitations, the placement can only be syndio-tactic.
(Laws of Creations for Cyclohexa-1,3-diene)

Rule 872: This rule of Chemistry for **Cyclohexa-1,3-Diene,** states that, when activated, presence of 1,2- and 1,4- mono-forms along the chain can only be obtained free-radically and not chargedly as shown below when a weak initiator is used-

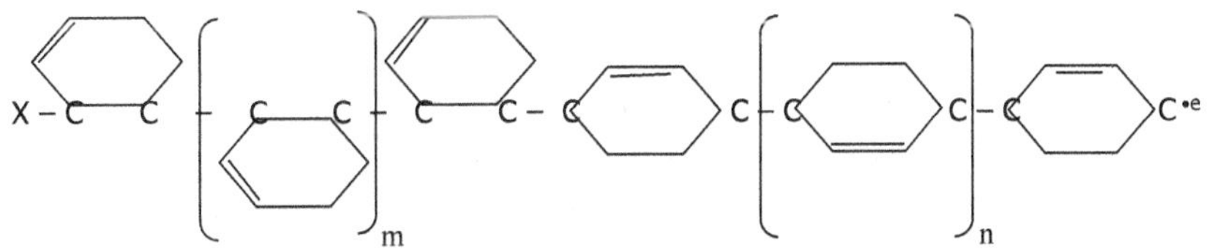

Poly(Cyclohexadiene)-1,2-/1,4- mono-forms

noting the presence of the 1,2-mono-form at the beginning of growth when the initiator is weak and as the active growing center becomes stronger and stronger in strength, 1,4- mono-forms begin to appear along the chain continuously, all monomer units sydiotactically placed; noting that the weaker the initiator, the more the presence of 1,2- mono-form and the stronger the initiator, the more the presence of 1,4-mono-form along the chain.
(Laws of Creations for Cyclohexa-1,3-Diene)

Rule 873: This rule of Chemistry for **Cyclohexa-1,3-Diene,** states that, it like cyclobutene will favor instantaneous opening of its ring in the presence of strong coordination initiators both radically and chargedly as shown below radically-

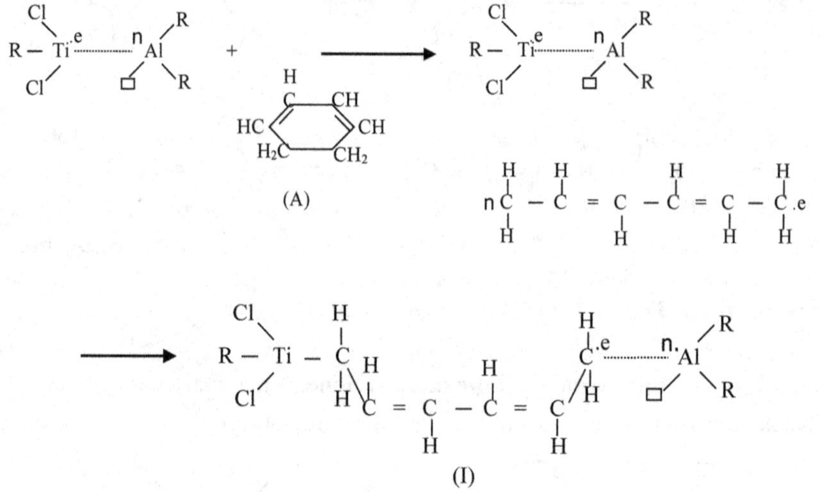

(A)

$$n\overset{H}{\underset{H}{C}} - \overset{H}{\underset{}{C}} = C - \overset{H}{\underset{}{C}} = C - \overset{H}{\underset{H}{C}}.e$$

(I)

with absence of transfer species and presence of cis-placement of the monomer units along the chain.
(Laws of Creations for Cyclohexa-1,3-Diene)

Rule 874: This rule of Chemistry for **Cyclohexa-1,3-Diene carrying R group,** states that, like cyclobutene, it will favor instantaneous opening of its ring in the presence of strong coordination initiators, which should not be sterically hindered; for which the followings are to be expected-

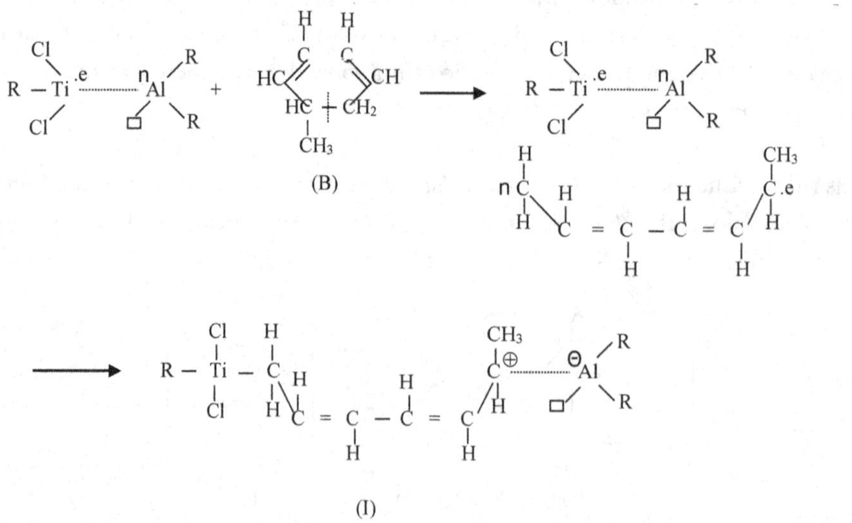

(B)

(I)

for which only one route is favored- the electro-free-radical or positively charged route and the placement is cis-isotactic with transfer species of the first kind.
(Laws of Creations for Cyclohexa-dienes)

Rule 875: This rule of Chemistry for **Cyclohexa-1,3-Diene,** states that, in the presence of weak initiators with two vacant orbitals, the followings can be obtained only electro-free-radically when resonance stabilization takes place-

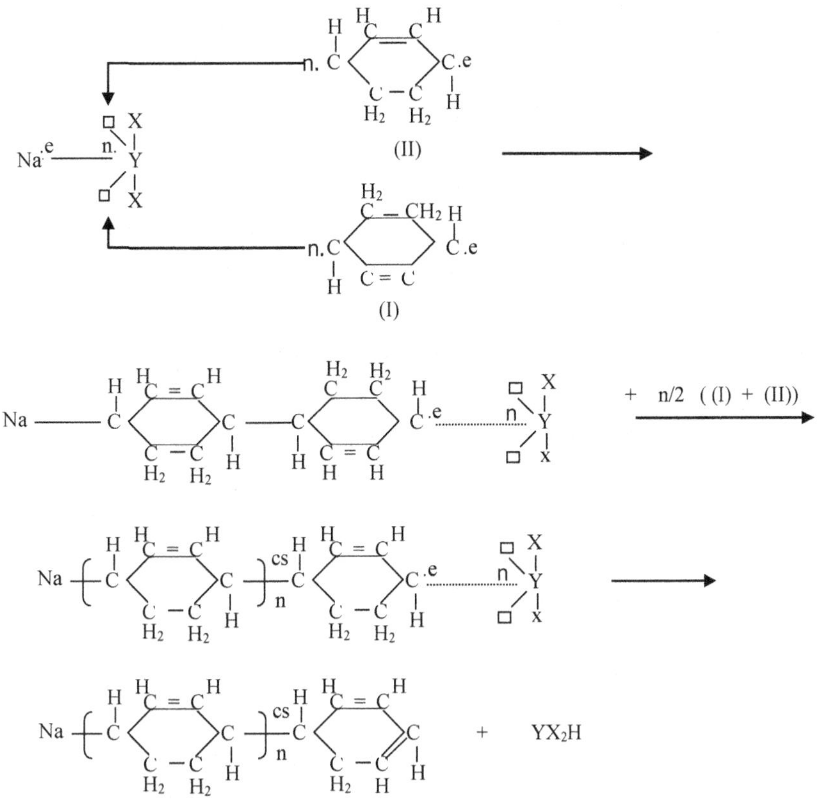

Transfer species of first kind of the tenth type

for which the polymer obtained is polycyclohexadiene with a terminal cyclic double bond when terminated naturally.

(Laws of Creations for Cyclohexa-1,3-Diene)

Rule 876: This rule of Chemistry for **Cyclohexadiene which polymerizes to give polycyclohexadiene with a dead terminal cyclic cyclohexadiene double bond,** states that, the chain can be fully dehydrogenated only radically to produce polyphenylene as shown below; for which the mechanism is via Equilibrium with many stages as also shown,

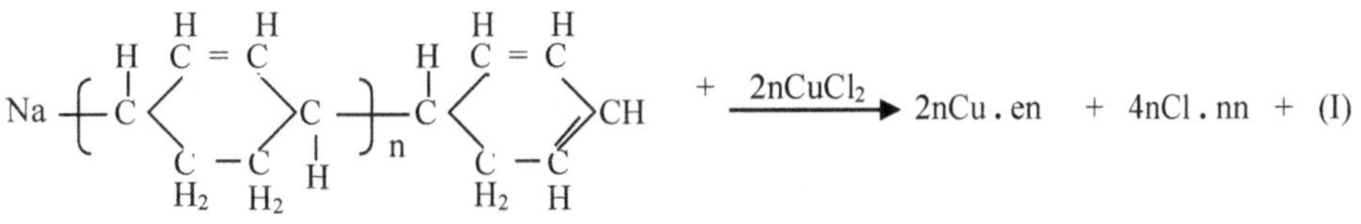

(I)

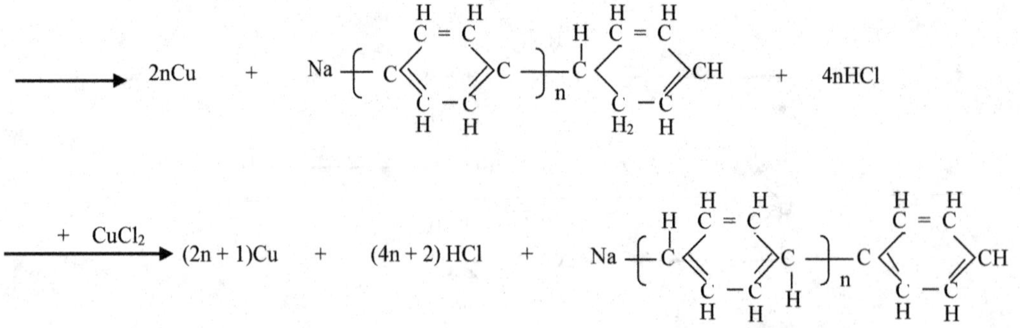

$$\xrightarrow{\hspace{2cm}} 2nCu \; + \; \text{Na}\left[\!\!{\Large(}\!\!C\!\!\underset{\underset{\text{H}}{\text{C}}-\underset{\underset{\text{H}}{\text{C}}}{}}{\overset{\overset{\text{H}}{\text{C}}=\overset{\text{H}}{\text{C}}}{}}\!\!{\Large)}\!\!C\right]_n C\underset{\underset{\text{H}_2}{\text{C}}-\underset{\underset{\text{H}}{\text{C}}}{}}{\overset{\overset{\text{H}}{\text{C}}=\overset{\text{H}}{\text{C}}}{}}\!\!CH \; + \; 4n\text{HCl}$$

$$\xrightarrow[\hspace{1.5cm}]{+\ CuCl_2} (2n+1)Cu \; + \; (4n+2)\,\text{HCl} \; + \; \text{Na}\left[\!\!{\Large(}\!\!C\!\!\cdots{\Large)}\!\!C\right]_n C\cdots CH$$

in which each monomer unit is dehydrogenated sequentially to the last but one monomer unit in four stages as shown below.

Stage 1: $CuCl_2 \; \xrightleftharpoons[\text{Existence}]{\text{Equilibrium State of}} \; ClCu \bullet en \; + \; nn\bullet Cl$

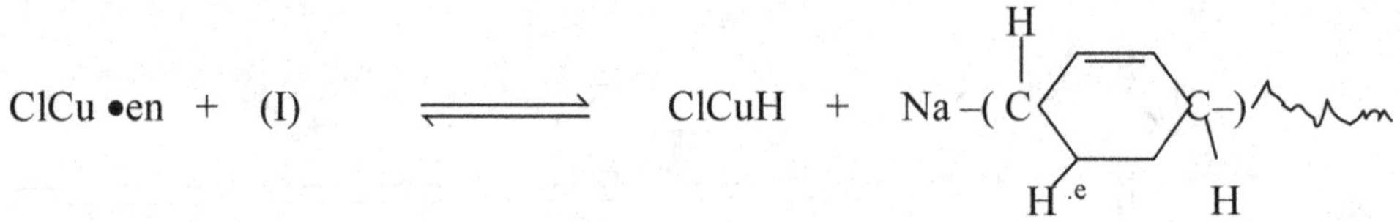

$ClCu \bullet en \; + \; (I) \; \rightleftharpoons \; ClCuH \; + \; $ Na–(C ⋯ C–)⌇⌇

(II)

$(II) \; \rightleftharpoons \; H\bullet e \; + \; $ Na –(${}^{n}\bullet$C ⋯ C–)⌇⌇

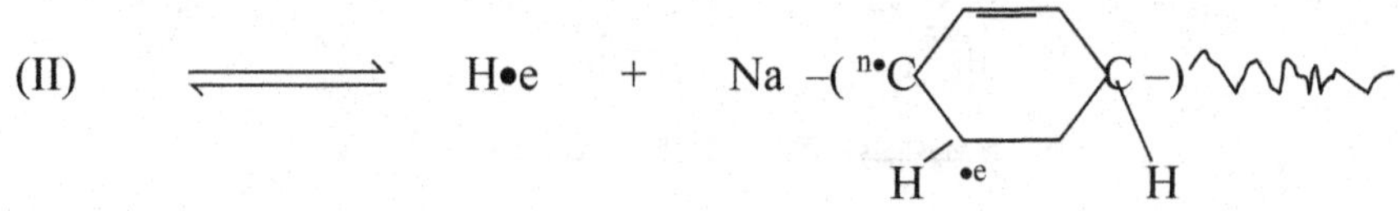

(III)

$H \bullet e \; + \; nn \bullet Cl \; \rightleftharpoons \; HCl$

$(III) \; \xrightarrow{\text{Deactivation}} \; $ Na –(C ⋯ C–)⌇⌇ + Heat

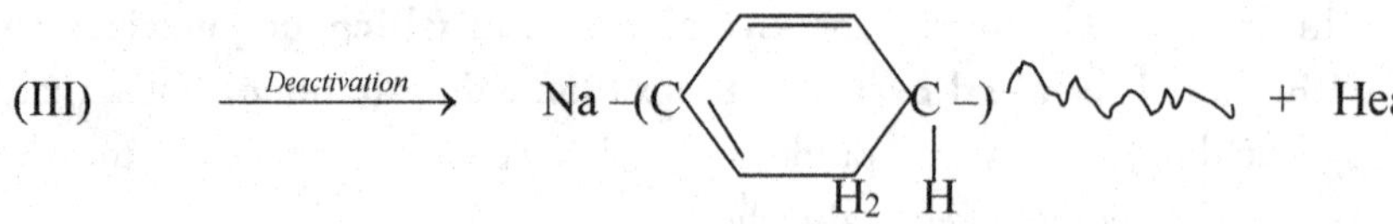

Stage 2: ClCuH $\updownarrow$ ~Equilibrium State of~ Existence Cl •nn + en• CuH

en• CuH $\updownarrow$ ^^^^^^^ en• Cu •nn + H •e

H •e + Cl •nn $\updownarrow$ ^^^^^^^ HCl

en• Cu •nn $\xrightarrow[Copper]{}$ $Cu:$ |

Stage 3: Similar to Stage 1 for the third center to give

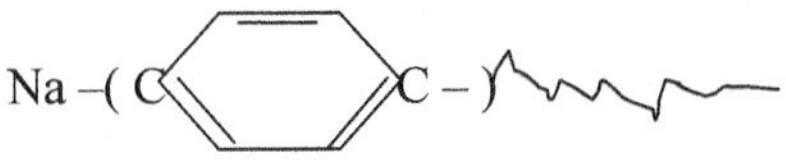

(IV)

Stage 4: Same as Stage 2.

Overall Equation: (I) + 2CuCl$_2$ $\longrightarrow$ (IV) + 4HCl + 2Cu

(Laws of Creations for Cyclohexa-1,3-Diene)

Rule 877: This rule of Chemistry for **Conjugated cyclohexadienes carrying R types of groups,** states that, when the group is internally located (i.e. 2-position), it is shielded for which when activated, it rearranges as shown below to give 1- alkylane substituted cyclohexadiene.

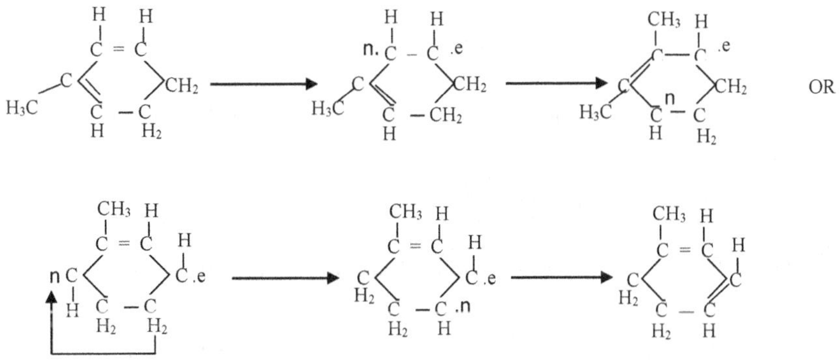

(Laws of Creations for 2-Methyl Cyclohexa-1,3-Diene)

Rule 878: This rule of Chemistry for **Conjugated cyclohexadienes carrying R types of groups,** states that, when the group is externally located the ring can readily be opened if the right size of R group is carried as shown below, for which for example beginning with CH_3, rearrangement is not

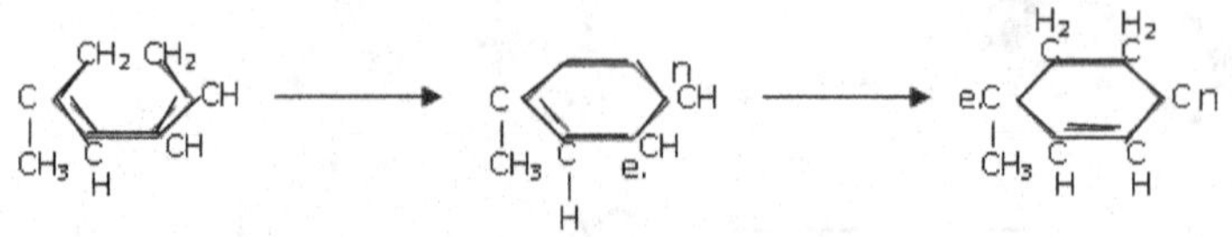

FAVORED PLACEMENT

favored, since the capacity of CH_3 is less than what is inside the ring; for when the CH_3 group is replaced with C_2H_5 group the rearrangements begin followed by opening of the ring, and stops at C_4H_9 and the product obtained in the narrow range is a vinyl acetylene. [See Rule 870]
(Laws of Creations for Cyclohexa-1,3-Diene)

Rule 879: This rule of Chemistry for **Activation of Activation centers,** states that, when a compound or monomer carries more than one activation center of the same capacity and character, i.e., all X, then they can all be activated at the same time, provided they are *isolatedly placed.*
(Laws of Creations for Activation of Activation Centers)

Rule 880: This rule of Chemistry for **Activation of Activation centers,** states that, when a compound or monomer carries more than one activation center *cumulatively or conjugatedly placed,* only one center can be activated, whether the centers are of the same nucleophilic capacity or not.
(Laws of Creations for Activation of Activation Centers)

Rule 881: This rule of Chemistry for **Activation of Activation centers,** states that, when a compound or monomer carries more than one activation center of different capacities but same character, i.e., all X, then they cannot be activated at the same time, but one at a time, and the first to be activated all the time is the least nucleophilic center, no matter how placed.
(Laws of Creations for Activation of Activation Centers)

Rule 882: This rule of Chemistry for **Activation of Activation centers,** states that, when a compound or monomer carries more than one activation center of different capacities and character, i.e., X and Y *conjugatedly placed*, both centers can be activated at the same time depending on the operating conditions or one at a time depending on the type of activating force; for if the force is a nucleo-radical, Y is first activated and if the force is an electro-radical, then X is first activated only if it is less nucleophilic than Y, for if not, then both centers can be activated at the same time.
(Laws of Creations for Activation of Activation Centers)

Rule 883: This rule of Chemistry for **Activation of Activation centers,** states that, when a compound or monomer carries more than one activation center of different capacities and character, i.e., X and Y *cumulatively placed,* both can be activated at the same time separately on two monomers, i.e., only one at a time on one monomer.
(Laws of Creations for Activation Centers)

Rule 884: This rule of Chemistry for **Activation of Activation Centers,** states that, when two or more compounds or monomers carrying one activation center which can only be X, are present together, they can all be activated at the same time based on the operating conditions or one at time beginning with the monomer carrying the least nucleophilic center depending on the operating conditions.
(Laws of Creations for Activation of Activation Centers)

Rule 885: This rule of chemistry for **Conjugated cyclohexadienes carrying radical-pushing groups of lower capacity than H such as CF$_3$,** states that, the para-placement of the radicals is now reversed when activated whether the group is internally or externally located-

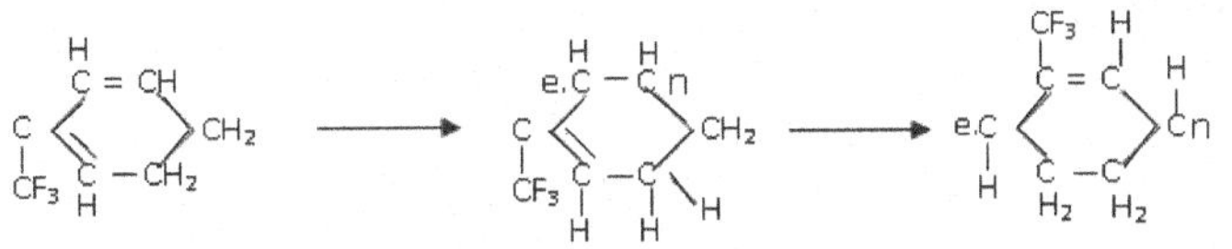

(C)

i)

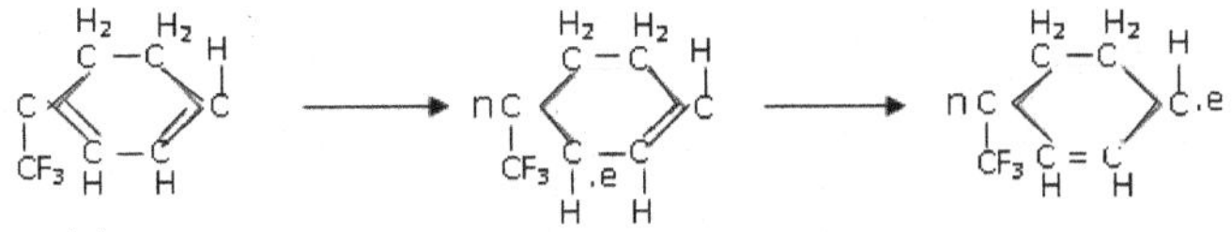

(D)

(VI) para - placement of radicals

ii)

with no possibility for rearrangement and favoring any route when externally located, even when paired free-radical coordination initiators are used to polymerize the monomer, noting that while CF$_3$ in (C) is resonance stabilized, that in (D) is not resonance stabilized; and the fact that the wrong center was activated above for the first case.
(Laws of Creations for Cyclohexa-1,3-Diene)

Rule 886: This rule of Chemistry for **Cyclohexadienes,** states that, based on the saturated part of the ring which controls the domains of the compound during rearrangements, the following is valid for the family-

$$H_2C - CH_2 \qquad \geq \qquad C_2H_5$$
CLOSED

Order of Radical-pushing capacity

and the same should apply to larger sized trienes, tetraenes and higher.
(Laws of Creations for Cyclohexadienes)

Rule 887: This rule of Chemistry for **2-Vinyl Cyclohexa-1,3-Diene,** states that, when activated, the followings are to be expected-

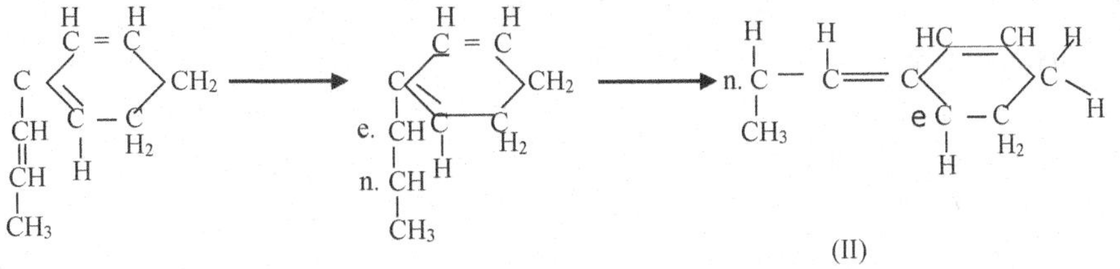

(I)

Opened/half closed loop resonance stabilization

(II)

for which only two mono-forms exist for its opened/half-closed loop resonance stabilization system, with rearrangement taking place when the alkylane group is C_4H_9 to give a more nucleophilic Di-vinyl acetylene $[(C_3H_7)CH = CH - C \equiv C - CH = CH(C_3H_7)]$, whose only route of polymerization is electro-free-radical route where fully resonance stabilized mono-form is the monomer unit in the product as strong as steel.

(Laws of Creations for 2-Vinyl Cyclohexa-1,3-Diene)

Rule 888: This rule of Chemistry for **1-Vinyl Cyclohexa-1,3-Diene (that in which the capacity of the alkylane group on the vinyl center is greater than that in the ring),** states that, when activated, the followings are to be expected-

(I)

Opened/closed loop reonsance stabilization

(II) (1, 6 - para - placement)

(VI)a Resonance stabilized

FAVORED

for which, three mono-forms exist for its opened/full closed-loop resonance stabilization system, with favored rearrangement to give Di-vinyl acetylenes, whose only route of polymerization is electro-free-radical route, and when fully resonance stabilized mono-form is the monomer unit in the product, the product is as strong as steel.

(Laws of Creations for 1-Vinyl Cyclohexa-1,3-Diene)

Rule 889: This rule of Chemistry for **Acetylenic Cyclohexadienes,** states that, the same as for Vinyl cyclohexadienes almost apply, with the following differences - *the states of activations, the order of nucleophilicity of the centers based on the capacity and types of groups carried, and a resulting difference in the types of products obtained when the ring based on the groups carried is made to undergo rearrangement followed by ring opening-* to give more nucleophilic products with two triple bonds and one double bond.

(Laws of Creations for Acetylenic Cyclohexadienes)

Rule 890: This rule of Chemistry for **Vinyl and Acetylenic Cyclohexa1,3-Dienes,** states that, as far as *the instantaneous opening of the ring is concerned,* only those carrying the vinyl ($H_2C = CH-$) type when placed in the 1- or 2- position will favor its ring being opened instantaneously for as long as part of the ring remains the source of resonance stabilization.

(Laws of Creations for Cyclohexa-1,3-Dienes)

Rule 891: This rule of Chemistry for **Cycloheptadiene,** states that, unlike cyclo-hexadiene, there is no para-placement of free radicals, due to lack of symmetry, and like cyclohexadiene, there are three types -

(a)   One in which the double bonds are isolatedly placed (1,5- or 1,4-).

(b)   One in which the two double bonds are conjugatedly placed (1,3-).

(c)   One in which the two double bonds are cumulatively placed (1,3-).

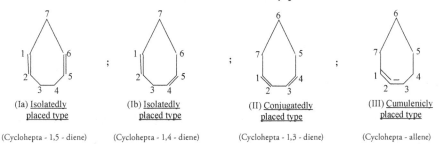

(Ia) <u>Isolatedly</u> <u>placed type</u>	(Ib) <u>Isolatedly</u> <u>placed type</u>	(II) <u>Conjugatedly</u> <u>placed type</u>	(III) <u>Cumulenicly</u> <u>placed type</u>
(Cyclohepta - 1,5 - diene)	(Cyclohepta - 1,4 - diene)	(Cyclohepta - 1,3 - diene)	(Cyclohepta - allene)

noting that while (Ia) has only one point of scission- the 3,4-bond, (II) has two identical points of scission- 5,6- or 6,7- bonds and (III) which is unstable and readily rearranges to cycloheptayne, has two points of scission- 4,5- or 6,7- bonds.
(Laws of Creations for Cyclohepta-Diene)

Rule 892: This rule of Chemistry for **Cyclohepta-1,3-Diene,** states that, when the ring is opened instantaneously under mild operating conditions, the followings are obtained only free-radically-

for which either (III) is obtained and used as a monomer unit or (IV) or (V) is obtained after one type of resonance stabilization takes place in the absence of an initiator.
(Laws of Creations for Cyclohepta-1,3-Diene)

Rule 893: This rule of Chemistry for **Cyclohepta-1,5 or 1,4--Diene,** states that, when the ring is opened instantaneously under mild operating conditions, the followings are obtained only free-radically-

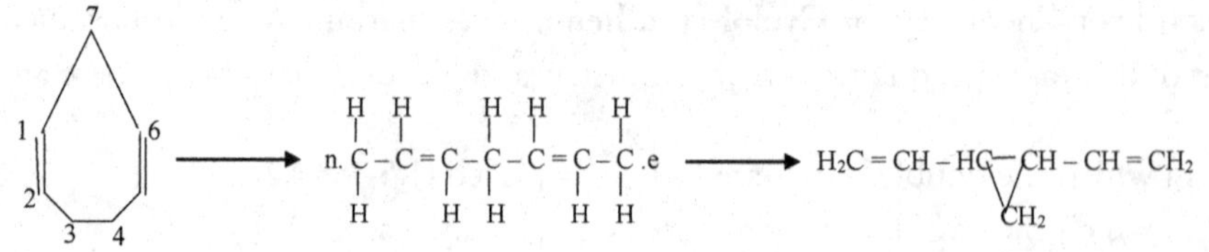

(I) <u>Isolatedly placed type</u>
(Cyclohepta - 1,5 - diene)

(II)

(III)

for which either (II) is obtained and used as a monomer unit or (III) is obtained after both types of resonance stabilization take place in the absence of an initiator.
(Laws of Creations for Cyclohepta-1,5- or 1,4-Diene)

Rule 894: This rule of Chemistry for **Cyclohexadienes and Cycloheptadienes,** states that, since Equilibrium states of existence of compounds are very important, shown below are theirs-

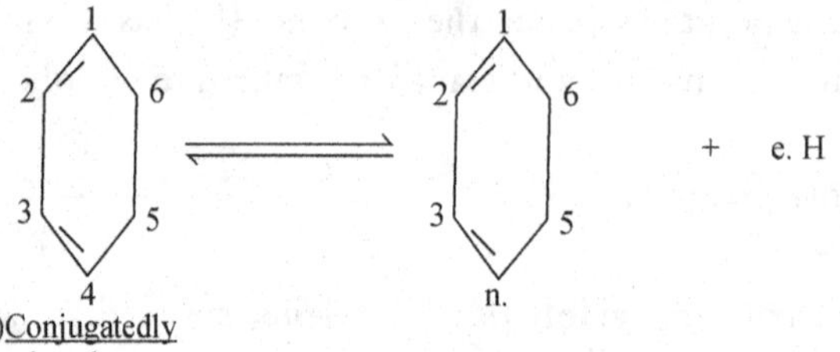

(I)<u>Conjugatedly placed type</u>
(Cyclohexa- 1,3 - diene)

+ e. H

(a)

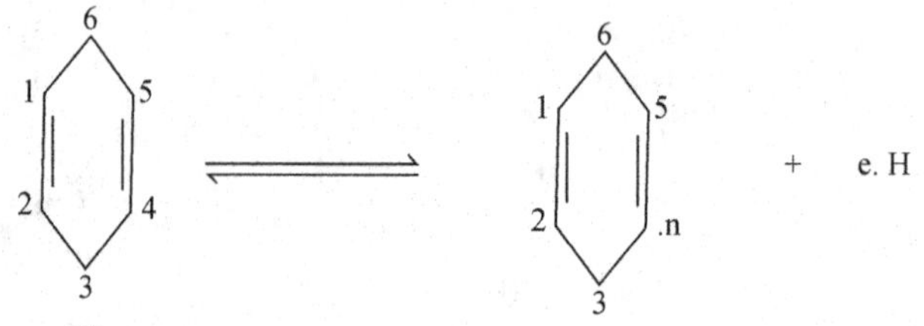

(II)
<u>Isolated type</u> (cyclohexa - 1,4- diene)

+ e. H

(b)

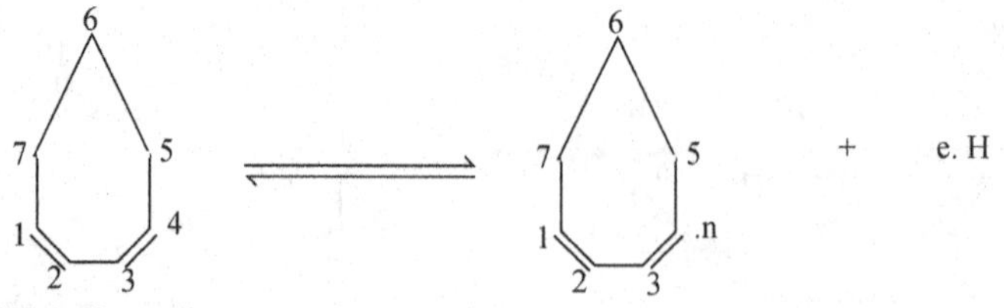

(III)<u>Conjugatedly placed type</u>
(Cyclohepta - 1,3 - diene)

+ e. H

(c)

(Laws of Creations for Equilibrium states of existence)

Rule 895: This rule of Chemistry for **1-methyl Cyclohexa-1,3-Diene,** states that, when thermally decomposed at temperatures above 600°C in a quartz flow system, the followings are to be expected-

472

Stage 1:

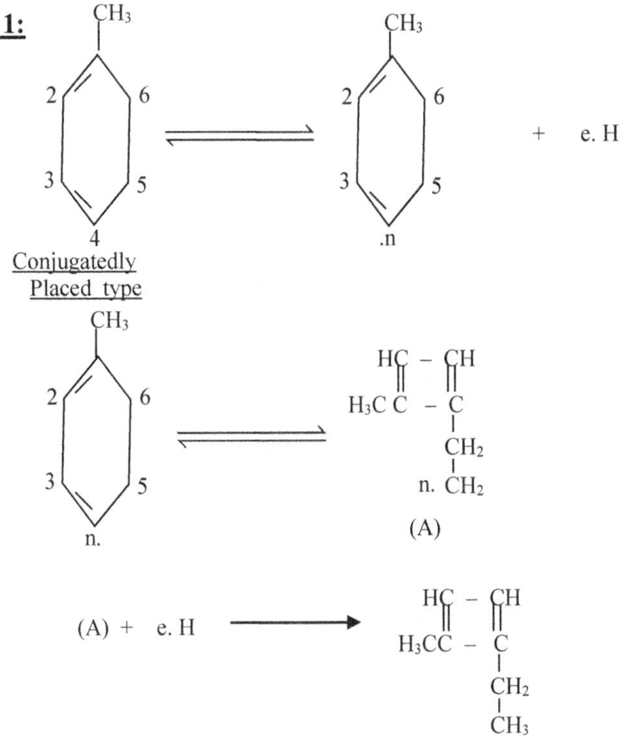

Conjugatedly
Placed type

+ e. H

(A)

(A) + e. H →

(a)

Stage 2:

+ H .e

+ e. H →

(A)

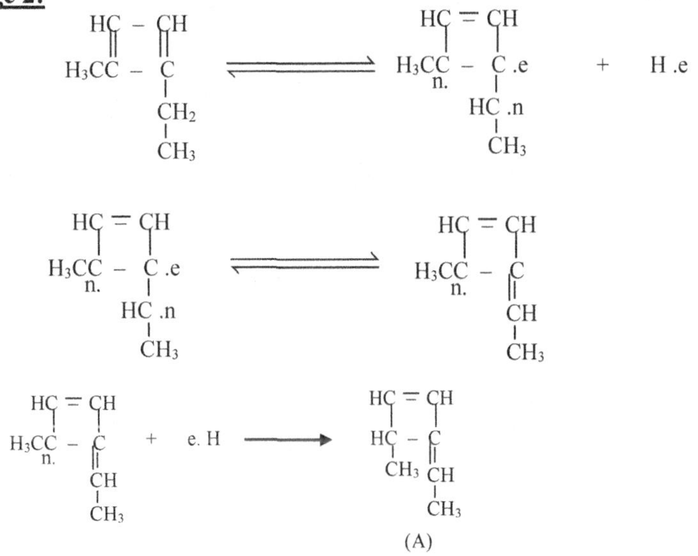

(b)

Stage 3a:

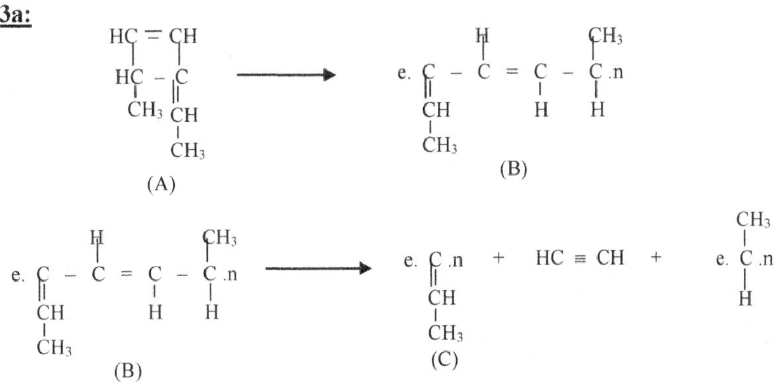

(A) → (B)

(B) → e. C.n + HC ≡ CH + e. C.n (C)

$$\underset{H}{\overset{CH_3}{e.\ C.n}} \quad + \quad \underset{CH_3}{\overset{\overset{C.n}{\parallel}}{CH}} \quad \longrightarrow \quad \underset{CH_3}{\overset{H}{C}} = \overset{e.}{C} - \underset{H}{\overset{CH_3}{C.n}}$$

(C)

$$\underset{CH_3}{\overset{H}{C}} = \overset{e.}{C} - \underset{H}{\overset{CH_3}{C.n}} \quad \longrightarrow \quad \underset{CH_3}{\overset{H}{C}} = C = \underset{H}{\overset{CH_3}{C}}$$

(D).

(c)

Stage 3b:

$$\begin{array}{c} HC = CH \\ HC - C \\ CH_3\ CH \\ CH_3 \end{array} \longrightarrow \underset{\substack{CH \\ CH_3}}{\overset{H}{e.\ C}} - C = C - \underset{H}{\overset{CH_3}{C.n}}$$

(A) (B)

$$\underset{\substack{CH \\ CH_3}}{\overset{H}{e.\ C}} - \overset{}{C} = \overset{}{C} - \underset{H}{\overset{CH_3}{C.n}} \quad \longrightarrow \quad \underset{H}{\overset{CH_3}{C}} \equiv C - \underset{H}{\overset{C_2H_5}{C}} = \underset{H}{\overset{}{C}}.$$

(B) (E)

(d)

Overall equation: 2 Methyl cyclohexadiene $\longrightarrow$ **(D)** + **(E)** + $HC \equiv CH$

(e)

Stage 4:

$$\underset{H}{\overset{CH_3}{C}} \equiv C - \underset{H}{\overset{C_2H_5}{C}} = \overset{}{C}. \quad \rightleftharpoons \quad \underset{H}{\overset{CH_3}{C}} \equiv C - \overset{\bullet n}{\underset{H}{C}} - \overset{CH_2}{\underset{}{C.e}}$$

(E) (F)

$$\longrightarrow \quad \underset{H}{\overset{CH_3}{C}} \equiv C - \underset{H}{\overset{H}{C}} - \underset{H}{\overset{}{C}} = \underset{H}{\overset{CH_3}{C}}$$

(G)

(f)

Stage 5:

$$\underset{H}{\overset{CH_3}{C}} \equiv C - \underset{H}{\overset{H}{C}} - \underset{H}{\overset{}{C}} = \underset{H}{\overset{CH_3}{C}} \quad \rightleftharpoons \quad \underset{H}{\overset{CH_3}{C}} \equiv C - \underset{H}{\overset{H}{C}} - \underset{H}{\overset{}{C}} = \overset{H_2C\bullet n}{\underset{H}{C}} + H \bullet e$$

(G) (H)

$$(H) \quad \rightleftharpoons \quad \underset{H}{\overset{CH_3}{C}} \equiv C - \underset{H}{\overset{H}{C}} - \overset{n\bullet}{\underset{H}{C}} - \underset{H}{\overset{H}{C}} = \overset{}{C}$$

(H)

$$(H) \ + \ e.H \longrightarrow$$

$$\underset{H_3C-C \equiv C-\overset{\displaystyle H}{\underset{\displaystyle H}{C}}-\overset{\displaystyle H}{\underset{\displaystyle H}{C}}-C = \overset{\displaystyle H}{\underset{\displaystyle H}{C}}}{}$$

(I)

(g)

Stage 6:

$$H_3C-C \equiv C-\overset{H}{\underset{H}{C}}-\overset{H}{\underset{H}{C}}-C = \overset{H}{C} \rightleftharpoons H_3C-C \equiv C-\overset{H}{\underset{H}{C}}-\overset{H}{\underset{H}{C}}-C = \overset{H}{C} \bullet n \ + \ e \bullet H$$

(I)   (I)

$$H_3C-C \equiv C-\overset{H}{\underset{H}{C}}-\overset{H}{\underset{H}{C}}-C = \overset{H}{C} \bullet n \rightleftharpoons HC \equiv CH \ + \ e \bullet \overset{H}{\underset{H}{C}}-\overset{H}{\underset{H}{C}} \bullet n \ + \ H_3CC \equiv C \bullet n$$

(I)

$$H_3CC \equiv C \bullet n \ + \ e \bullet H \rightleftharpoons H_3CC \equiv CH$$

$$e \bullet \overset{H}{\underset{H}{C}} - \overset{H}{\underset{H}{C}} \bullet n \longrightarrow H_2C = CH_2$$

(h)

Stage 7: **(D)** molecularly rearranges to $H_3CC \equiv CC_2H_5$ via **Equilibrium mechanism**

(i)

Overall overall equation: 2 Methyl Cyclohexadiene $\longrightarrow 2HC \equiv CH \ + \ H_2C = CH_2$

$$+ \ HC \equiv CCH_3 \ + \ H_5C_2C \equiv CCH_3$$

(j)

Stage 8:

$$HC \equiv CCH_3 \rightleftharpoons n. \overset{\displaystyle H}{\underset{\displaystyle CH_3}{C}} = C .e$$

(I)

$$(i) \ + \ HC \equiv CH \rightleftharpoons n. \overset{H}{\underset{CH_3}{C}} = C - C = \overset{H}{\underset{H}{C}} .e$$

(J)

$$n. \overset{H}{\underset{CH_3}{C}} = C - C = \overset{H}{\underset{H}{C}} .e \ + \ HC \equiv CH \rightleftharpoons n. \overset{H}{\underset{CH_3}{C}} = C - \overset{H}{\underset{H}{C}} = C - C = \overset{H}{\underset{H}{C}} .e$$

(J)   (K)

$$n. \overset{H}{\underset{CH_3}{C}} = C - \overset{H}{\underset{H}{C}} = C - C = \overset{H}{\underset{H}{C}} .e \longrightarrow$$

(K)   TOLUENE

(k)

Overall overall equation: 2 Methyl Cyclohexadiene $\longrightarrow$ Toluene + $H_3CC \equiv CC_2H_5$

$$+ \quad H_2C = CH_2 \tag{l}$$

Stage 9: $H_2C = CH_2$ decomposes further to give $HC \equiv CH + H_2$ (m)

Overall overall equation: 2 Methyl Cyclohexadiene $\longrightarrow$ $HC \equiv CH$ + $H_5C_2C \equiv CCH_3$ +

$$\text{Toluene} + H_2 \tag{n}$$

noting that the order of nucleophilicity of some acetylenes when they are present is as follows.

$$H_3CC \equiv CC_2H_5 \quad > \quad HC \equiv CC_2H_5 \quad > \quad HC \equiv CCH_3 \quad > \quad HC \equiv CH$$

Order of Nucleophilicity

(Laws of Creations for 1-methyl cyclohexa-1,3-Diene)

Rule 896: This rule of Chemistry for **5-Methyl-, 1-Methyl- and 2-Methyl-Cyclohexadienes,** states that, when rings are opened instantaneously, the followings are obtained-

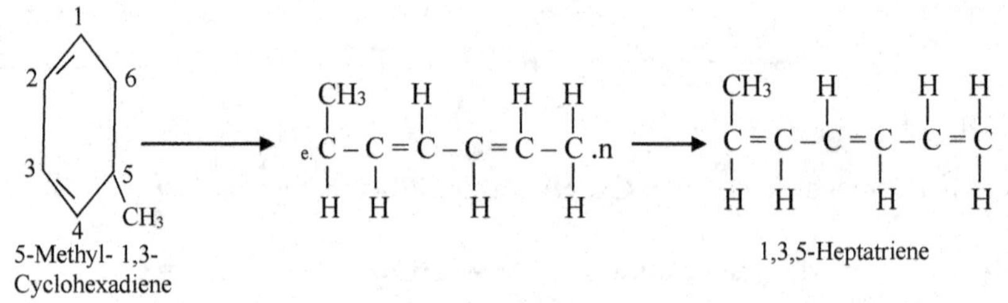

(A)

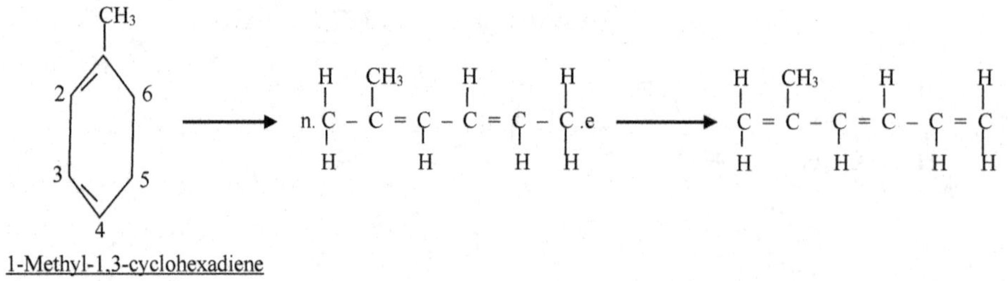

(B)

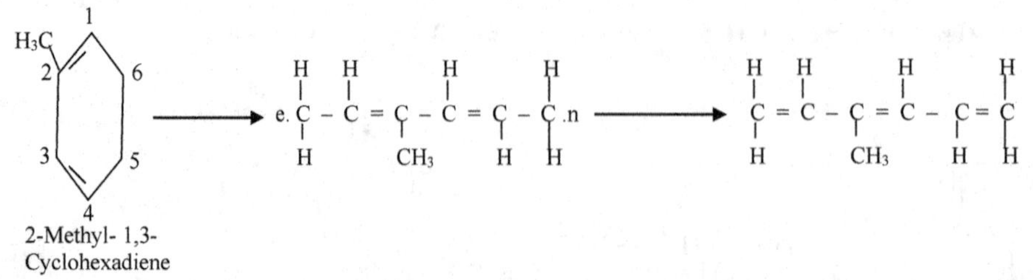

(C)

for which only the activated state of (A) will favor further decomposition to give smaller products in the absence of deactivation.

(Laws of Creations for Methyl cyclohexadienes)

Rule 897: This rule of Chemistry for **Tri-enes obtained from Cyclohexadienes,** states that, the followings are their Equilibrium states of existence-

476

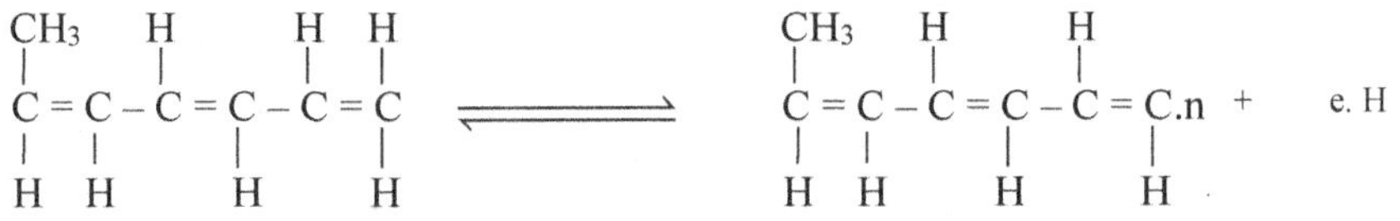

$$CH_3 \quad H \qquad H \quad H$$
$$C=C-C=C-C=C \quad \Longrightarrow \quad C=C-C=C-C=C.n \quad + \qquad e.\,H$$

1,3,5-Heptatriene

(A)

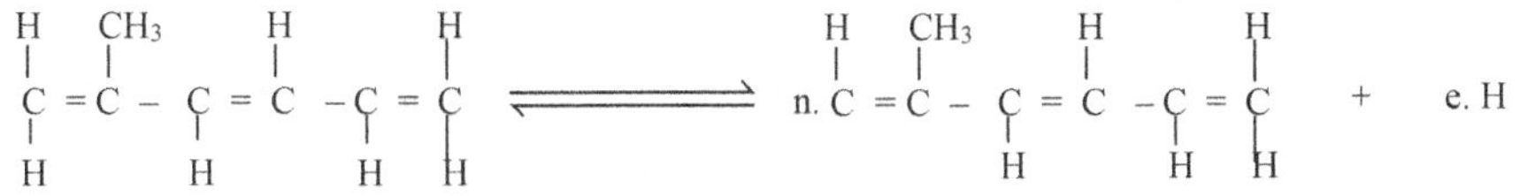

2-Methyl-hexatriene From 1-methyl-cyclohexadiene

(B)

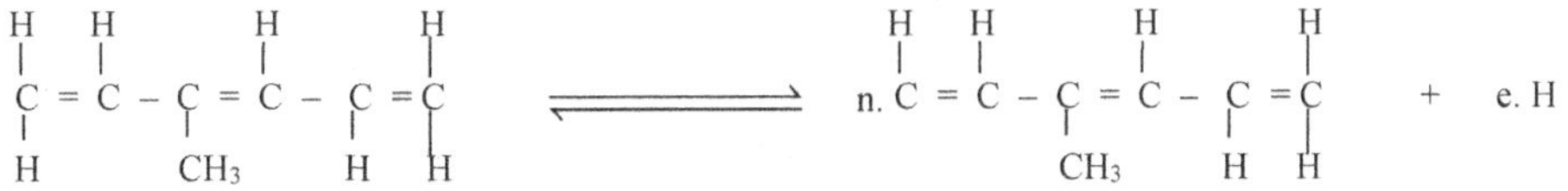

3-Methyl-hexatriene From 2-methyl-cyclohexadiene

(C)

for which only (A) and (B) can be further decomposed to give smaller products at higher operating conditions via Equilibrium mechanism.
(Laws of Creations for Trienes)

Rule 898: This rule of Chemistry for **1-Methyl-Cyclohexa-1,4-diene,** states that, since its ring cannot be opened instantaneously, the only way it can be opened is when it is in Equilibrium state of existence as shown below-

Stage 1:

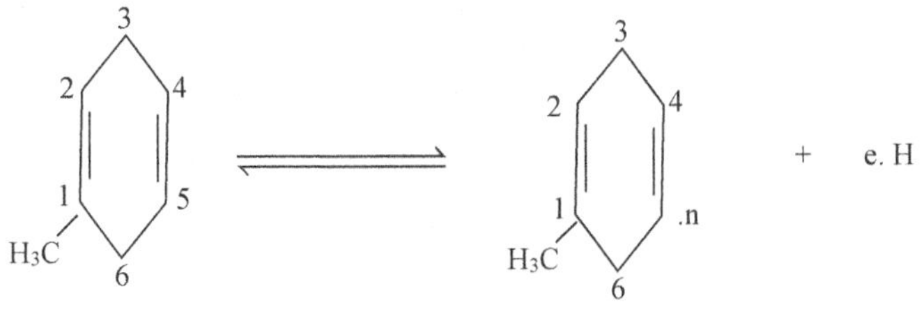

(I)

1-Methyl-Cyclohexa - 1,4- diene

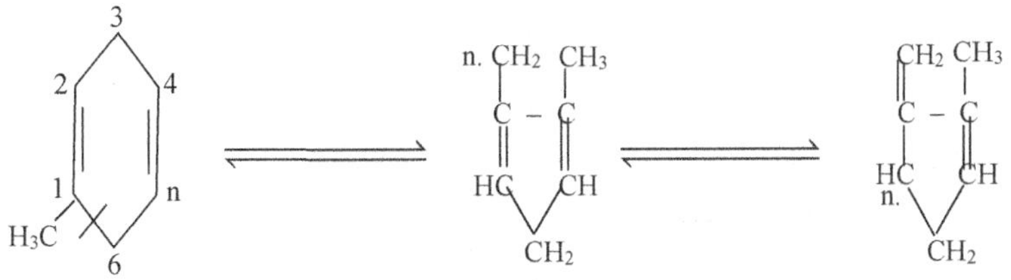

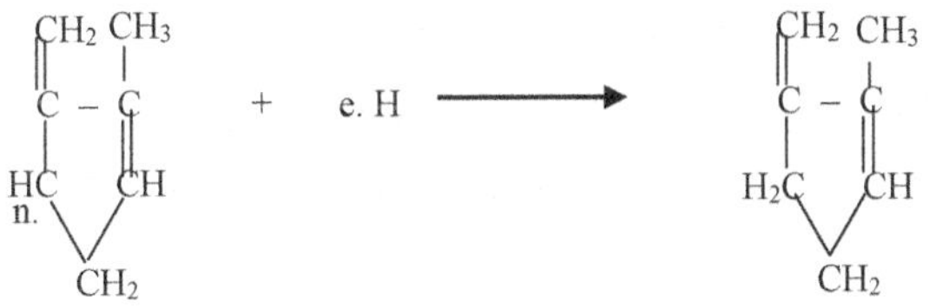

(A)

(a)

477

Stage 2:

$$2 \quad \overset{\overset{\displaystyle CH_2}{\underset{\displaystyle \|}{}} \overset{\displaystyle CH_3}{\underset{\displaystyle \|}{}}}{\underset{\displaystyle H_2C \quad CH}{C - C}} \overset{\displaystyle }{\underset{\displaystyle CH_2}{\diagup}} \quad \longrightarrow \quad 2 \text{ e.} \overset{CH_3}{\underset{CH_2}{C}} - \overset{}{C} = \overset{H}{\underset{H}{C}} - \overset{H}{\underset{H}{C}} - \overset{H}{\underset{H}{C}}\text{.n}$$

$$2 \quad \text{e.} \overset{CH_3}{\underset{CH_2}{C}} - C = \overset{H}{\underset{H}{C}} - \overset{H}{\underset{H}{C}} - \overset{H}{\underset{H}{C}}\text{.n} \quad \longrightarrow \quad 2HC \equiv CCH_3 \quad + \quad 2\,H_2C = CH_2 \quad + \quad 2\text{ e.} \overset{C}{\underset{CH_2}{\|}}\text{.n}$$

$$2\text{ e.} \overset{C}{\underset{CH_2}{\|}}\text{.n} \quad \longrightarrow \quad H_2C = C = C = CH_2$$

(b)

Stages 3 & 4: The Cumulene is broken down as already shown in Stages 8 and 9 of Rule 843.

(c)

Overall equation: $3(\text{1-Methyl-cyclohexadiene}) \longrightarrow \text{(A)} + 2HC \equiv CCH_3 +$

$$2C + 3H_2C = CH_2$$

(d)

Stage 5: Ethenes break down to give $\quad 3HC \equiv CH + 3H_2$ (e)

Stage 6:

$$\overset{\overset{\displaystyle CH_2}{\underset{\displaystyle \|}{}} \overset{\displaystyle CH_3}{\underset{\displaystyle \|}{}}}{\underset{\displaystyle H_2C \quad CH}{C - C}} \overset{}{\underset{\displaystyle CH_2}{\diagup}} \quad \longrightarrow \quad \text{e.} \overset{CH_3}{\underset{CH_2}{C}} - C = \overset{H}{\underset{H}{C}} - \overset{H}{\underset{H}{C}} - \overset{H}{\underset{H}{C}}\text{.n}$$

$$\text{e.} \overset{CH_3}{\underset{CH_2}{C}} - C = \overset{H}{\underset{H}{C}} - \overset{H}{\underset{H}{C}} - \overset{H}{\underset{H}{C}}\text{.n} \quad \longrightarrow \quad \overset{H}{\underset{}{C}} \equiv C - \overset{H}{\underset{}{C}} = \overset{C_2H_5}{\underset{CH_3}{C}}$$

(f)

Stage 7:

$$\overset{H}{\underset{}{C}} \equiv C - \overset{H}{\underset{}{C}} = \overset{C_2H_5}{\underset{CH_3}{C}} \quad \rightleftharpoons \quad \text{n.} \; C \equiv C - \overset{H}{\underset{CH_3}{C}} = \overset{C_2H_5}{\underset{}{C}} \quad + \quad \text{e. H}$$

$$\text{n.} \; C \equiv C - \overset{H}{\underset{CH_3}{C}} = \overset{C_2H_5}{\underset{}{C}} \quad \rightleftharpoons \quad \text{n.} \; \overset{H}{\underset{CH_3}{C}} = \overset{C_2H_5}{\underset{}{C}} \quad + \quad \text{n.} \; C \equiv C \text{.e}$$

$$\text{n.} \; \overset{H}{\underset{CH_3}{C}} = \overset{C_2H_5}{\underset{}{C}} \quad \rightleftharpoons \quad HC \equiv CCH_3 \quad + \quad \text{n.} \; CH_2 - CH_3 + \text{Heat}$$

$$n.\ CH_2 - CH_3 \quad \rightleftharpoons \quad CH_2 = CH_2 \quad + \quad n.\ H \ + \ Heat$$

$$H\ .n \quad + \quad e.\ H \quad \rightleftharpoons \quad H_2$$

$$n.\ C \equiv C\ .e \quad \longrightarrow \quad 2C\ [Carbon\ black] \tag{g}$$

Stage 8: Ethene breaks down to $\ HC \equiv CH \ + \ H_2 \ + \ $ Heat $\tag{h}$

Overall equation: (A) $\longrightarrow HC \equiv CH \ + \ 2H_2 \ + \ 2C \ + \ HC \equiv CCH_3 \ + \ $ Heats $\tag{i}$

Stage 9: Toluene was obtained from the methyl acetylene and acetylene. $\tag{j}$

Overall overall equation: 3(1-Methyl-1,4-Cyclohexadiene) $\longrightarrow$ 2Toluene $+$ 4C

$$+ \quad Heats \quad + \quad HC \equiv CCH_3 \quad + \quad 5H_2 \tag{k}$$

for which if the Equilibrium state of existence had not been as shown in the first stage, these products will not be obtained..
(Laws of Creations for 1-Methyl-1,4-Cyclohexa-1,4-Diene)

Rule 899: This rule of Chemistry for **Iso-Toluenes,** states that, these are non-aromatic ringed isomers of Toluene, for which four of them are shown below-.

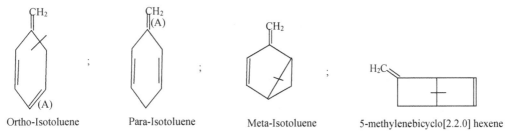

| Ortho-Isotoluene | Para-Isotoluene | Meta-Isotoluene | 5-methylenebicyclo[2.2.0] hexene |

wherein the weakest nucleophilic center in the ortho- and para-, have been identified with (A), noting that while the ortho- first rearranges to the para-, the para- rearranges straight to toluene, their ability for rearrangement depending on the operating conditions and only the first, third, and the last have only one point of scission.
(Laws of Creations for Iso-Toluenes)

Rule 900: This rule of Chemistry for **Cyclohexa-1,3-Diene,** states that, when <u>at particular operating conditions</u>, a fraction is kept in Equilibrium state of existence, leaving the remaining fraction in Stable state of existence, the followings are obtained-

Stage 1:

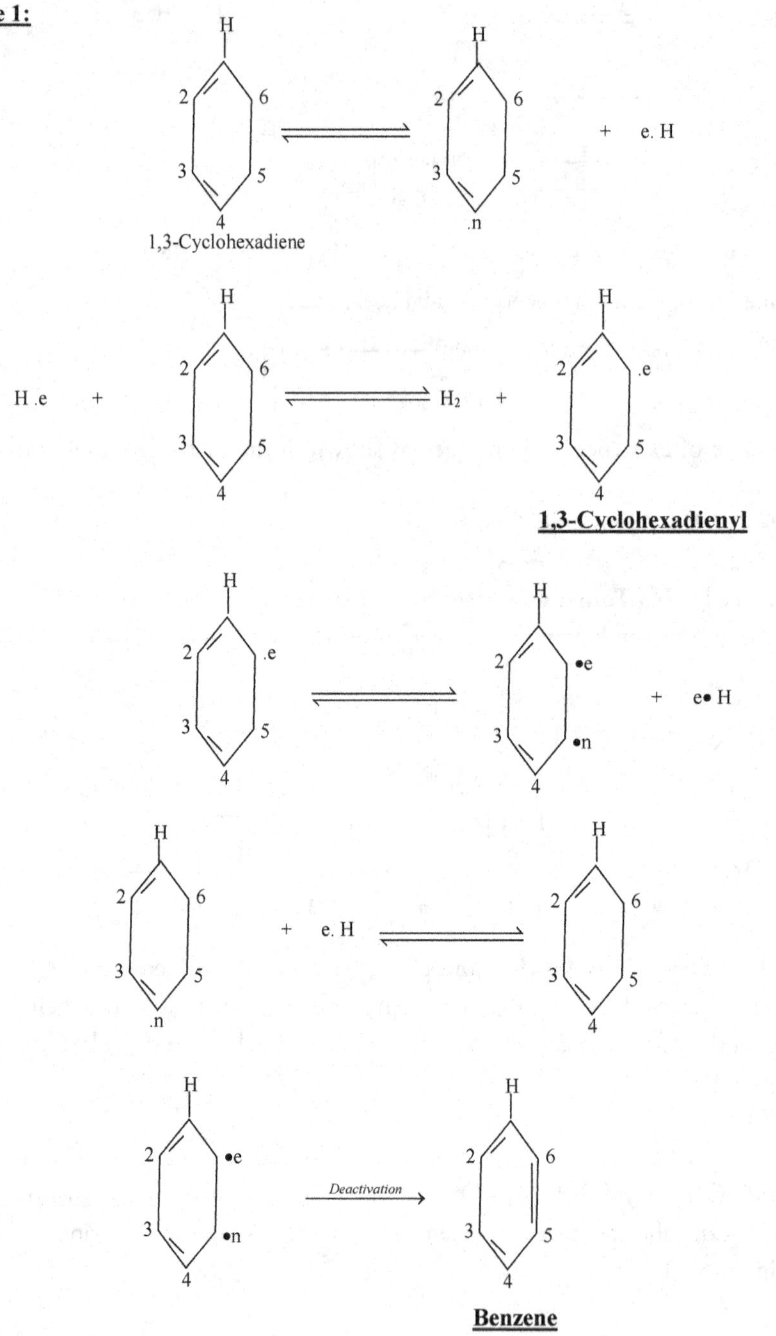

1,3-Cyclohexadiene

+ e. H

1,3-Cyclohexadienyl

Deactivation

Benzene (a)

Overall equation: 2(1,3-Cyclohexadiene) ⟶ 1,3-Cyclohexadiene + H_2 +
Benzene (b)

wherein *the monomer has dehydrogenated itself by using itself to do the job* in a single stage.
(Laws of Creations for Cyclohexadiene)

Rule 901: This rule of Chemistry for **Cyclohepta-1,3-Diene,** states that, when <u>at particular operating conditions,</u> a fraction is kept in Equilibrium state of existence, leaving the remaining fraction in Stable state of existence, the followings are obtained-

Stage 1:

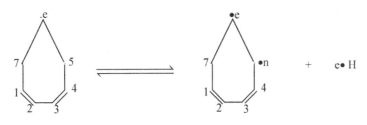

(Cyclohepta - 1,3 - diene)

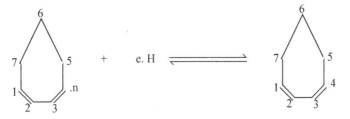

1,3-Cycloheptadienyl

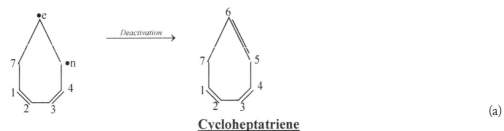

Cycloheptatriene

(a)

Overall equation: 2(1,3-Cycloheptadiene) ⟶ 1,3-Cycloheptadiene + H_2 + Cycloheptriene

(b)

wherein the monomer has dehydrogenated itself by using itself to do the job in a single stage.
(Laws of Creations for Cyclohepta-1,3-Diene)

Rule 902: This rule of Chemistry for **Cyclooctadiene,** states that, there are three types of this family of monomer as shown below, for which there are two types of isolatedly placed double bonds-

481

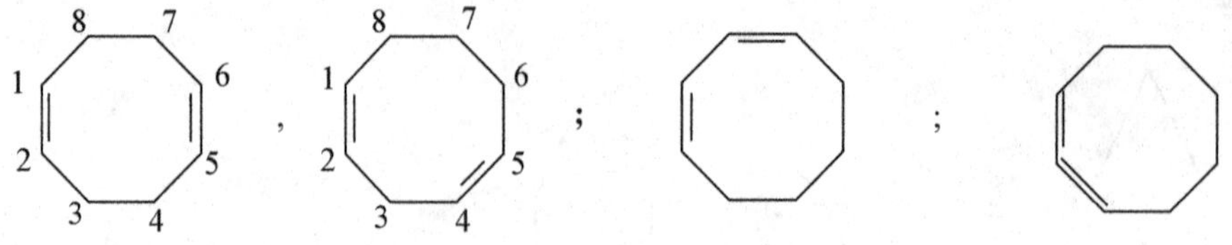

(I)a	(I)b	(II)	(III)
Isolatedly placed double bonds		Conjugatedly placed double bonds	Cumulatively placed Double bonds

symmetrically placed (I)a and non-symmetrically placed (I)b, which cannot molecularly rearrange to (I)a when activated and unlike cyclohexadiene of type (II), para-placement of the radicals in (II) cannot take place, and (III) which if cannot be isolated, rearranges to Cyclooctayne.
(Laws of Creations for Cyclooctadiene)

Rule 903: This rule of Chemistry for **1,3-Cyclooctatriene,** states that, when this was prepared from 1,5-cyclooctadiene, the reaction was said to need the reagent H_2 and catalyst Polymer bound Ni^2 catalyst at temperature of 100^0C for its operating conditions; and the followings were obtained-

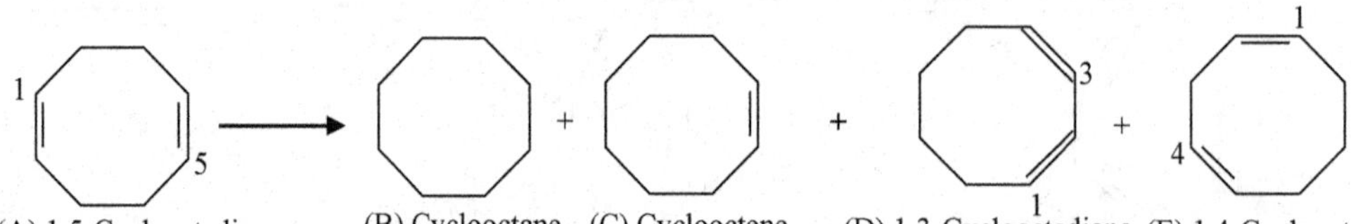

(A) 1,5-Cyclooctadiene   (B) Cyclooctane (C) Cyclooctene   (D) 1,3-Cyclooctadiene (E) 1,4-Cyclooctadiene

for which the mechanisms for the reaction can be explained as follows- with the presence of the Ni complex, hydrogenation of the 1,5-cycloocyadiene took place to produce (B) in two stages and (C) in one stage using three H_2 molecules, noting that if there had been excess H_2 in the system, (D) and (E) will never be obtained and this was then followed by the formation of (E) via molecular rearrangement as follows-

Stage 1:

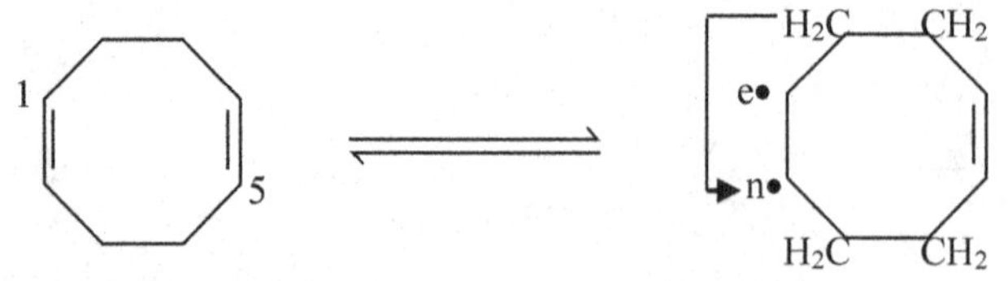

(A) 1,5-Cyclooctadiene

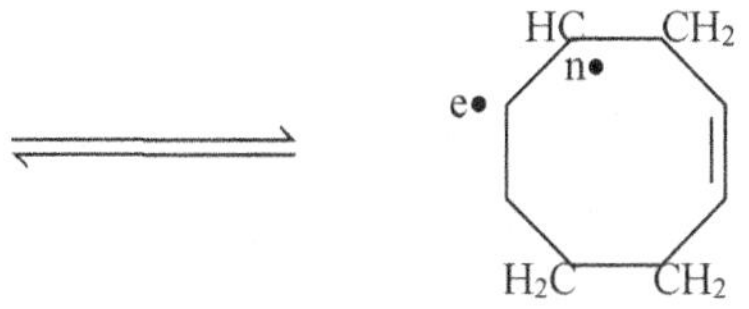

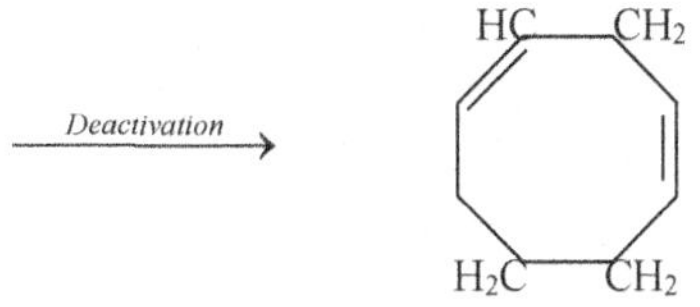

$$\rightleftharpoons$$

(structure with HC—CH$_2$, e•, n•, H$_2$C—CH$_2$)

$$\xrightarrow{\text{Deactivation}}$$

(E)

Overall equation: 1,5-Cyclooctadiene $\longrightarrow$ 1,4-Cyclooctadiene

and from a fraction of (E), (D) was formed as follows -

Stage 2a: **ELECTRORADICALIZATION**

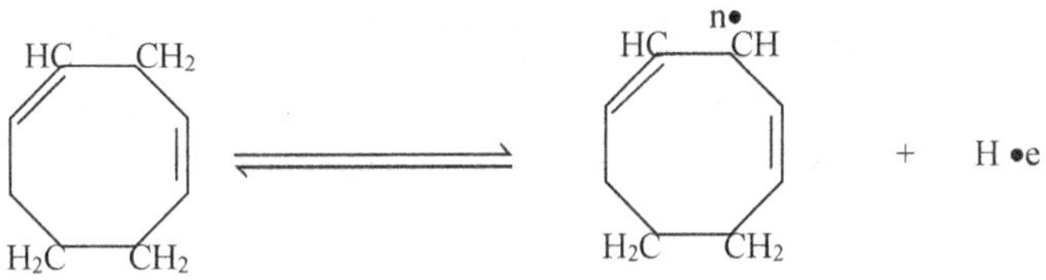

(E)

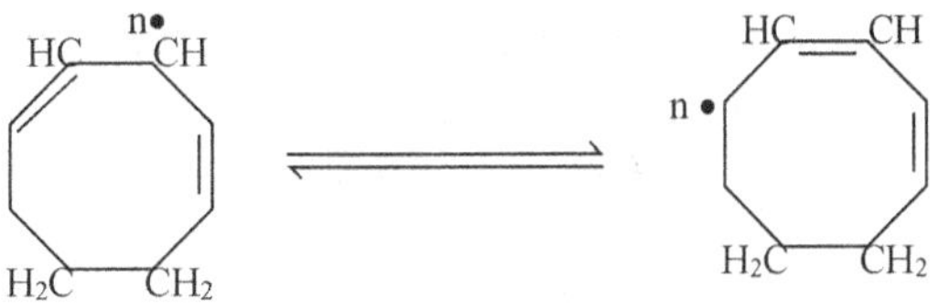

$$\rightleftharpoons \qquad + \quad H \bullet e$$

$$\rightleftharpoons$$

(F)

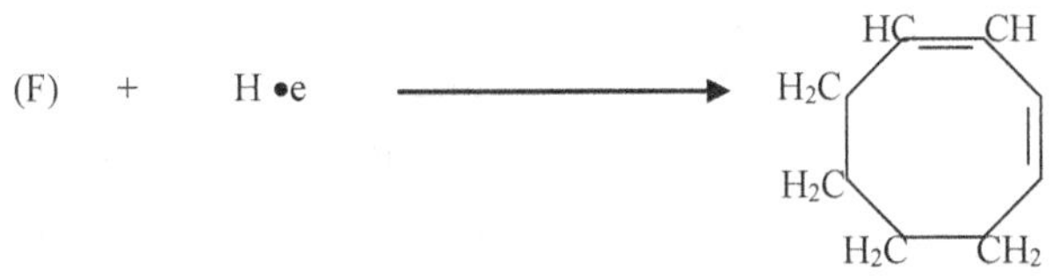

$$(F) \quad + \quad H \bullet e \quad \longrightarrow$$

(D)

Overall equation: 1,4-Cyclooctadiene $\longrightarrow$ 1,3-Cyclooctadiene

Stage 2b:

$$H_2 \quad \rightleftharpoons \quad H \cdot e \quad + \quad H \cdot n$$

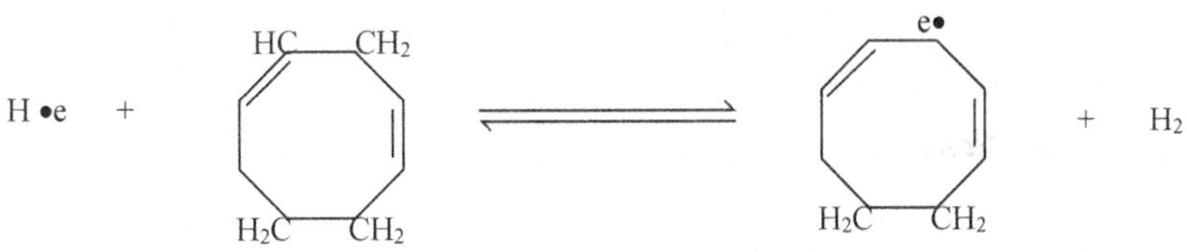

$$H \bullet e \quad + \quad \rightleftharpoons \qquad + \quad H_2$$

(E)

(a)
(b)

(c)

(d)

483

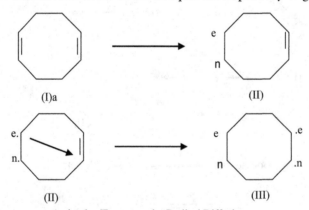

(d)

Overall overall equation: 6 (1,5-Cyclooctadiene)   +   3H₂ ⟶ Cyclooctane   +

Cyclooctene   +   1,4-Cyclooctadiene   +   3(1,3-Cyclooctadiene)   (e)

noting that 1,4-cyclooctadiene cannot rearrange to 1,3-cyclooctadiene or 1,5-cyclooctadiene via activation, but can with use of H₂ be made to produce 1,3-cyclootadiene as shown in Stage 2b, the favored stage; and most importantly note that the wrong Equilibrium state of existence for 1,4- cyclooctadiene has been used in Stage 2a.

(Laws of Creations for 1,4-Cyclooctadiene)

Rule 904: This rule of Chemistry for **1,5- Cyclooctadiene,** states that, when activated the following is obtained and because of the distance between the isolatedly placed double bonds, the electro-free-radical end of one center diffuses to the next activation center of equal nucleophilicity to give (III)

Activation/Trans-annular Radical Diffusion

only when a weak initiator is used for activation, for which when a stronger initiator is used, both centers can be activated at the same time since they are of the same nucleophilicity.

(Laws of Creations for Cycloocta-1,5-Diene)

Rule 905: This rule of Chemistry for **1,5-Cyclooctadiene,** states that, when used as a monomer, the followings are obtained in the presence of two different strong initiators, for which

(A) – NaX/YX₃ combination

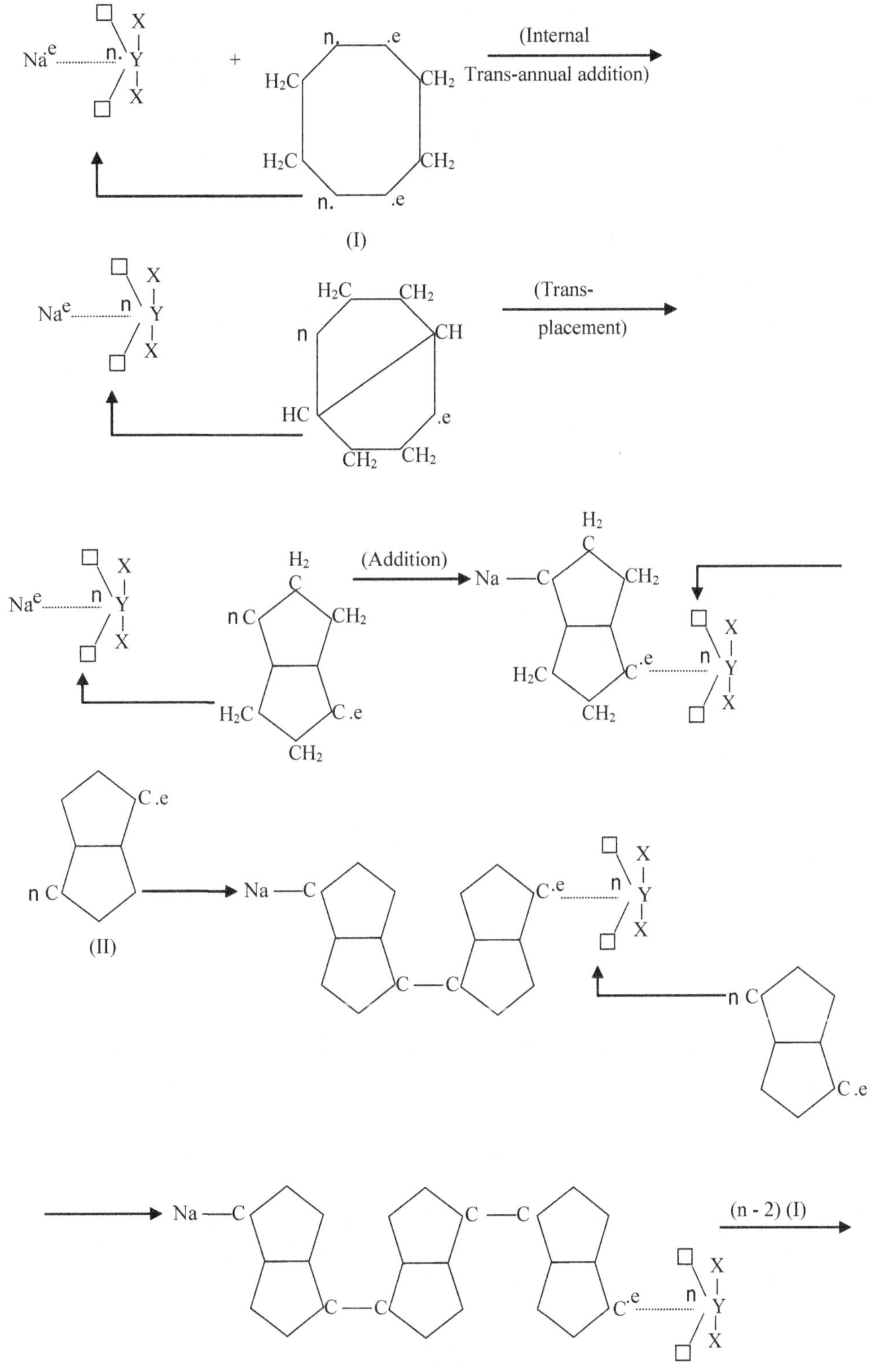

Na^e ⋯⋯ n. $\begin{array}{c} \square \\ X \\ | \\ Y \\ | \\ \square \quad X \end{array}$ + (ring I with H$_2$C, CH$_2$, H$_2$C, CH$_2$, n., .e) → (Internal Trans-annual addition)

(I)

Na^e ⋯⋯ n $\begin{array}{c} \square \\ X \\ | \\ Y \\ | \\ \square \quad X \end{array}$ (ring with H$_2$C, CH$_2$, CH, HC, CH$_2$, CH$_2$, n, .e) → (Trans-placement)

Na^e ⋯⋯ n $\begin{array}{c} \square \\ X \\ | \\ Y \\ | \\ \square \quad X \end{array}$ (bicyclic structure: C H$_2$, n C, CH$_2$, H$_2$C, C .e, CH$_2$) →(Addition)→ Na—C (bicyclic with H$_2$C, CH$_2$, C.e) ⋯ n $\begin{array}{c} \square \\ X \\ | \\ Y \\ | \\ \square \quad X \end{array}$

(II) bicyclic: C.e, n C → Na—C (two fused bicyclic units) C.e ⋯ n $\begin{array}{c} \square \\ X \\ | \\ Y \\ | \\ \square \quad X \end{array}$ ← n C (bicyclic with C.e)

→ Na—C (three fused bicyclic units) C.e ⋯ n $\begin{array}{c} \square \\ X \\ | \\ Y \\ | \\ \square \quad X \end{array}$ → (n - 2) (I)

485

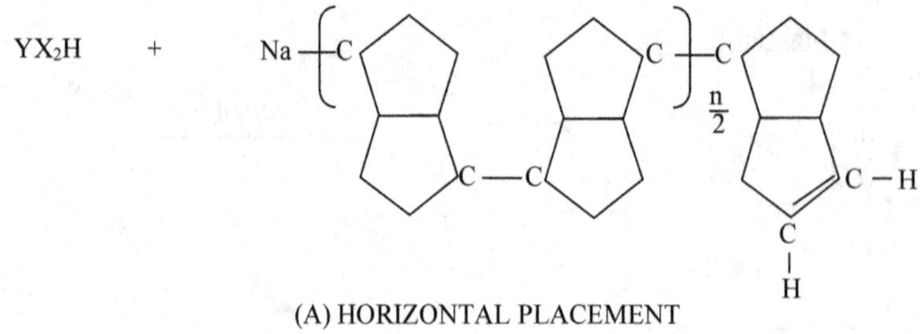

YX_2H +

(A) HORIZONTAL PLACEMENT

(B) AlR₃/TiCl₄

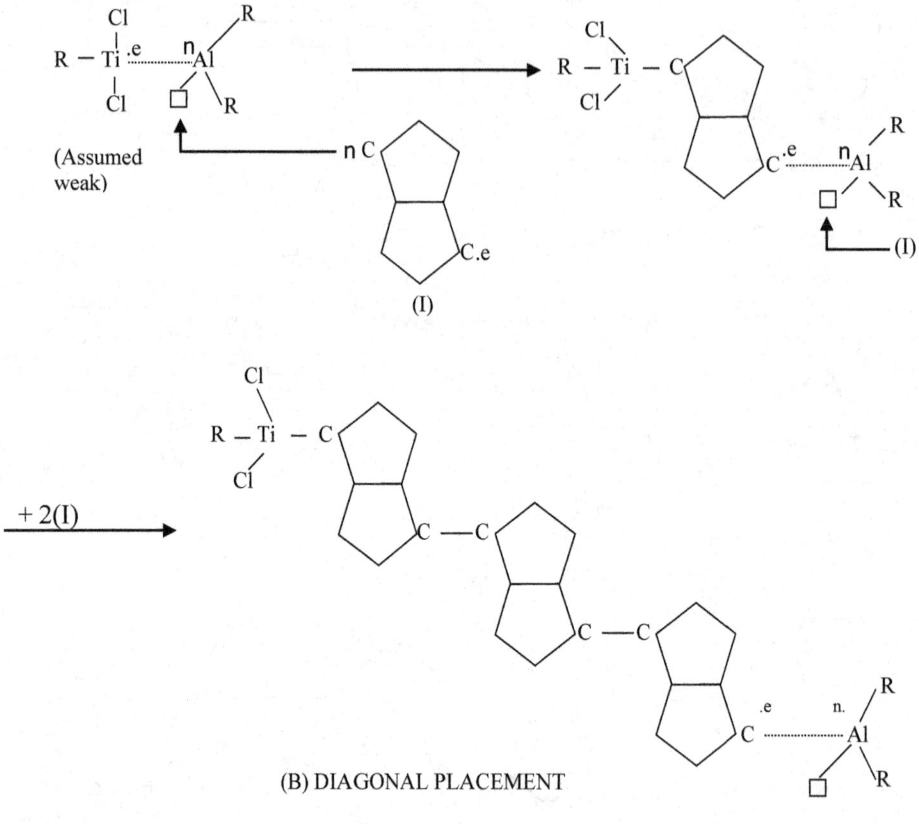

(B) DIAGONAL PLACEMENT

one can observe the different types of polymeric products obtained when one or two vacant orbitals are present, noting that the five membered fused ring does not have large strain energy in them to demand instantaneous opening of the ring.

(Laws of Creations for Cycloocta-1,5-Diene)

Rule 906: This rule of Chemistry for **Conjugated Cyclohexadiene, the six-membered counter-part of the cyclooctadiene,** states that, it may favor trans-annular addition as shown below; noting that Minimum required stain energy and points of scission already exist in the condensed four mem-

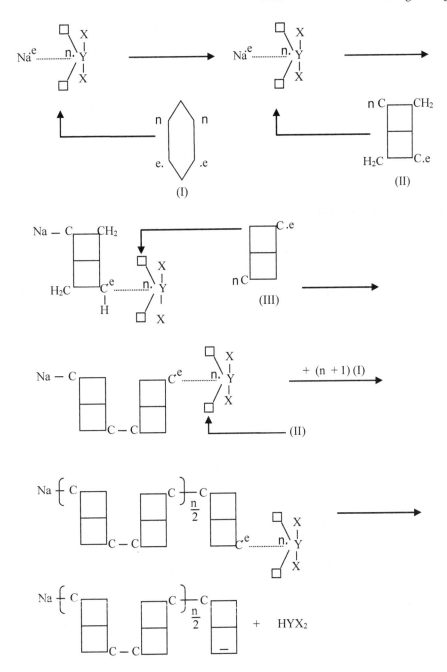

bered rings, and this is further worsened by the terminal double bond obtained after killing the chain.
(Laws of Creations for Cyclohexa-1,4-Diene)

Rule 907: This rule of chemistry for **Cyclo-Dienes,** states that, just as 1,4-Cyclohexadiene, 1,5-Cyclooctadiene will favor trans-annular addition, so also ten-membered, twelve-membered rings and above (even-membered rings) will favor this phenomenon as shown below for a ten-membered ring, for which the initiator has to be strong in strength for trans-annular addition to take place

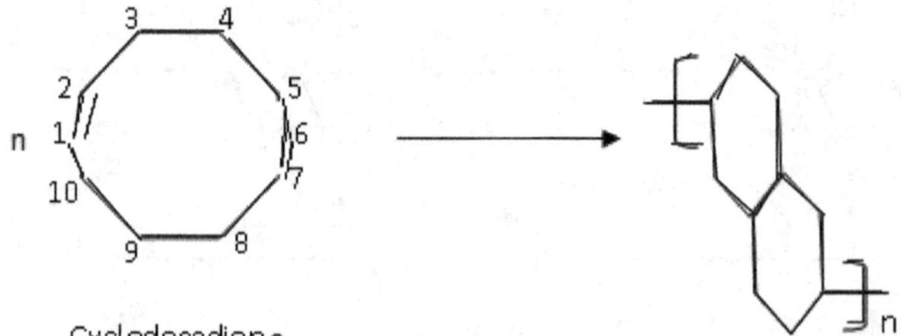

Cyclodecadiene

(Laws of Creations for Trans-annular Addition)

Rule 908: This rule of Chemistry for **Cyclo-Dienes,** states that, for the seventh and eight membered rings to favor molecular rearrangement of the first kind, one of the active centers must carry radical-pushing groups externally located with capacity greater than C_3H_7 and C_4H_9 respectively; for which after the molecular rearrangement, the rings can now be readily opened,

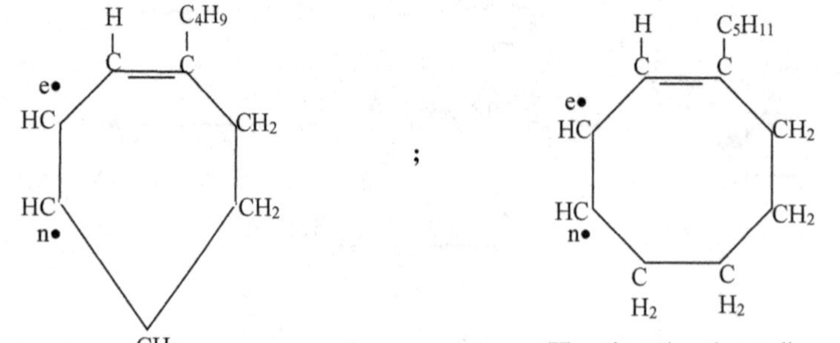

(I) activated cycloheptadiene

(II) activated cyclooctadiene

followed by molecular rearrangement of the third kind as shown below-

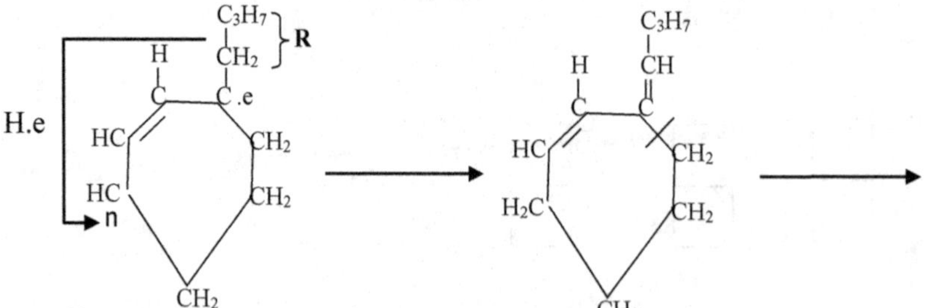

Resoance stabilized

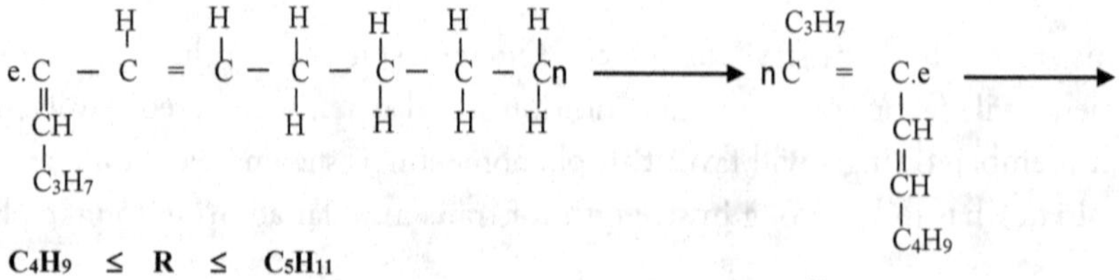

$$C_4H_9 \;\leq\; R \;\leq\; C_5H_{11}$$

(I)

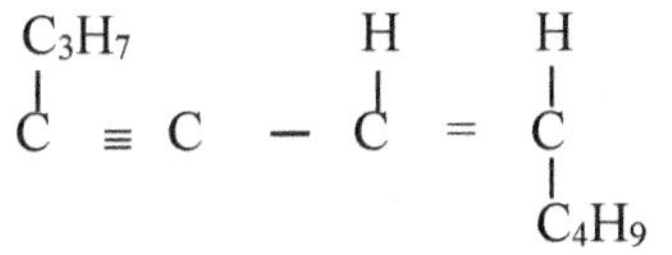

$$C_3H_7 - C \equiv C - C(H) = C(H) - C_4H_9$$

(II)

(Resonance stabilized)

(Laws of Creations for Cyclo-1,3-Dienes)

Rule 909: This rule of Chemistry for **Cycloocta-1,3-Diene,** states that, when conditions for instantaneous opening of the ring exist, the followings are obtained-

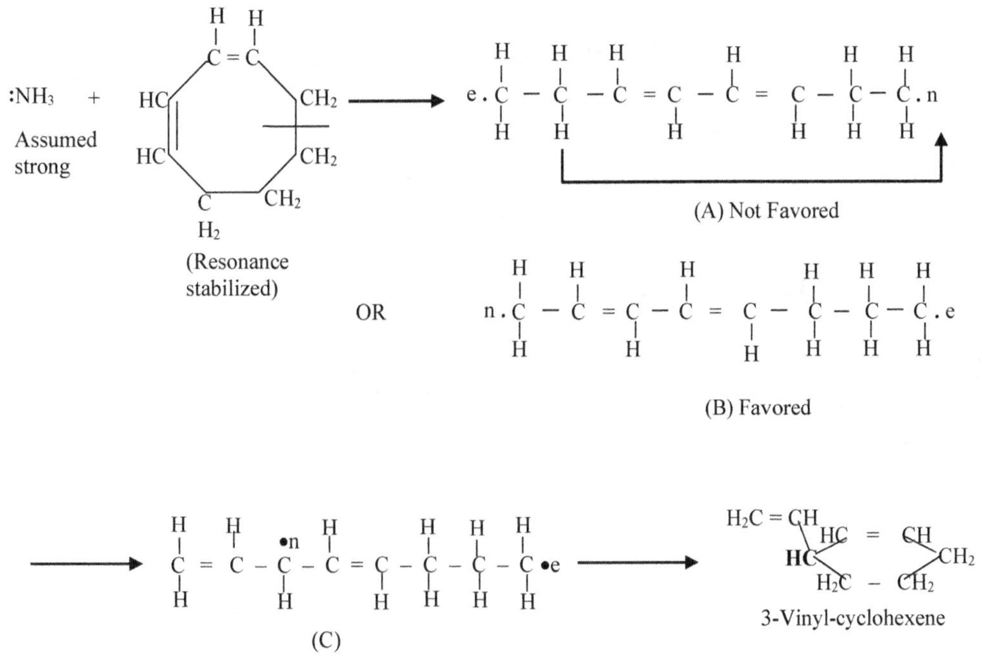

$$:NH_3 \;+\; \text{(Resonance stabilized ring)} \longrightarrow$$

Assumed strong

$$e \cdot C(H) - C(H) - C(H) = C - C(H) = C - C(H) - C(H) \cdot n$$

(A) Not Favored

OR

$$n \cdot C(H) - C(H) = C - C(H) = C - C(H) - C(H) - C(H) \cdot e$$

(B) Favored

$$\longrightarrow C(H) = C(H) - C \cdot n \quad H \quad C(H) - C = C - C(H) - C(H) - C(H) \cdot e \longrightarrow$$

(C)

3-Vinyl-cyclohexene

OR Butadienyl cyclobutane

noting that (A) is not favored, since the radical potential difference between the two carbon centers where scission took place is zero; the favored one being (B) which cannot undergo molecular rearrangement of the third kind to give a triene or deactivate and which after resonance stabilization, 3-Vinyl-cyclohexene or Butadienyl cyclobutane may be obtained and in absence of resonance stabilization, can be used as monomer units nucleo- and electro-free-radically.

(Laws of Creations for Cycloocta-1,3-Diene)

Rule 910: This rule of Chemistry for **Cycloocta-1,3-Dienes,** states that, the following is to be expected for this type of ring when a radical-pushing group is placed on the 1-position.

489

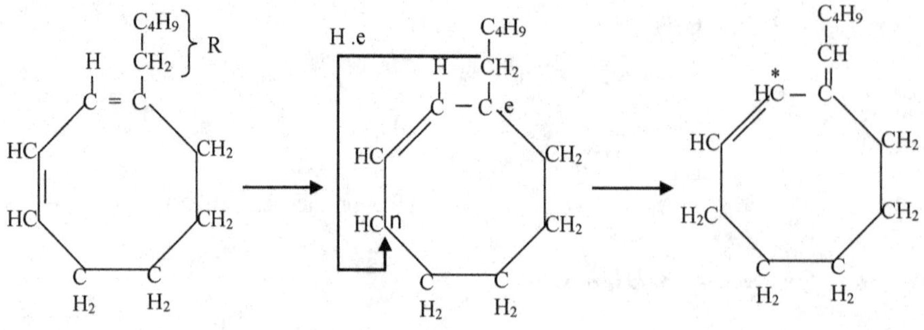

$$C_5H_{11} \ \leq \ R \ \leq \ C_6H_{13}$$

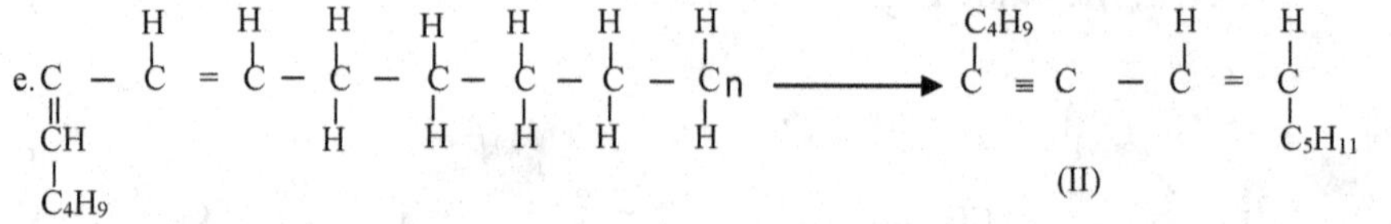

(II)

(Resonance stabilized)

(Laws of Creations for Cycloocta-1,3-Dienes) [See Rule 908]

Rule 911: This rule of Chemistry for **Cycloocta-1,5-Diene,** states that, when the ring is opened instantaneously at fairly **high operating conditions**, the followings are to be expected-

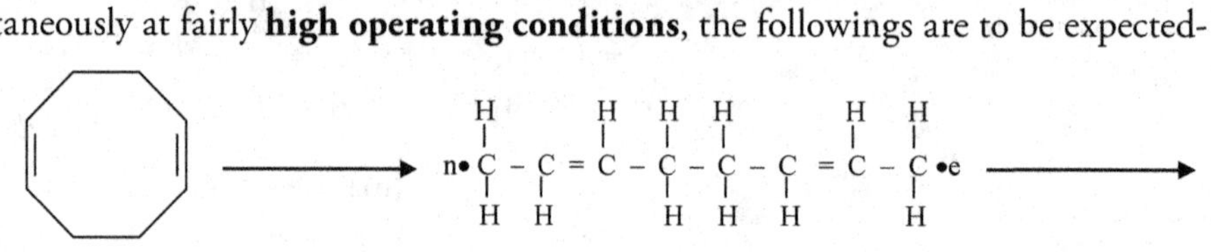

(I)a

(A)

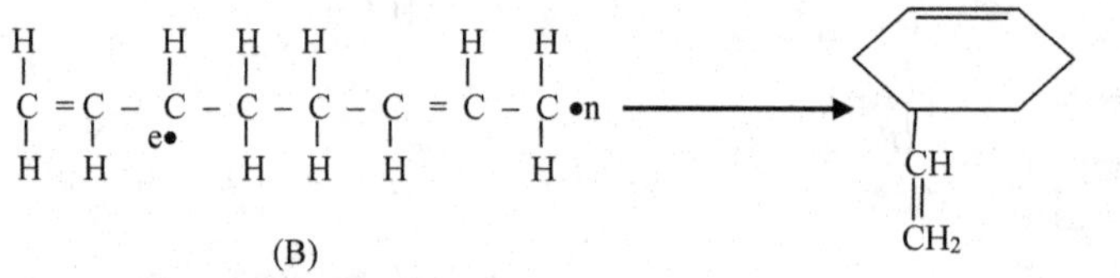

(B)

4-Vinylcyclohexene

OR 1,2 –Di-vinylcyclobutane
OR
2 Butadiene

for which 4-Vinyl cyclohexene is obtained when allowed to deactivate, the same as was obtained in the production of the cyclooctadiene from butadiene via 1,4- to 1,2 addition, or preferably 1,2–di-vinylcyclobutane.

(Laws of Creations for Cycloocta-1,5-Diene)

Rule 912: This rule of Chemistry for **Cycloocta-1,4-diene,** states that, the followings are to be expected when the ring is opened instantaneously.

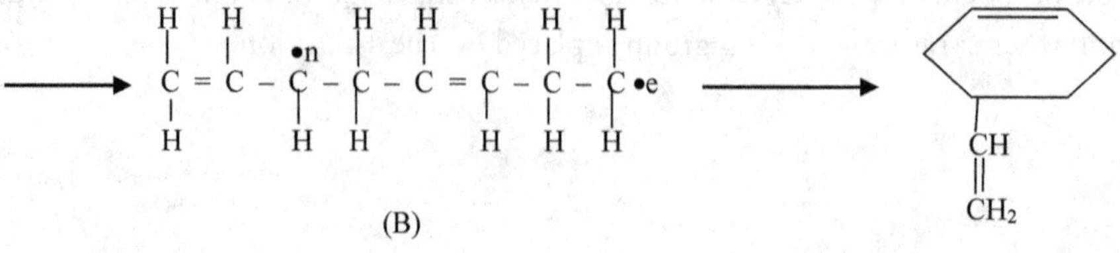

(B)

4-Vinylcyclohexene

490

noting that in the monomer unit there are two isolatedly placed double bonds, that which cannot be deactivated to give three double bonds, and in the absence of a strong initiator, it may be deactivated to form a 4-Vinylcyclohexene alone.
(Laws of creations for Cycloocta-1,4-Diene)

Rule 913: This rule of Chemistry for **Cycloocta-1,5-Diene,** states that, when made to react with SCl_2 or similar reagent, the reaction is a two-stage Equilibrium Addition mechanism system to give 2,6-dichloro-9-thiabicyclo[3,3,1] nonane

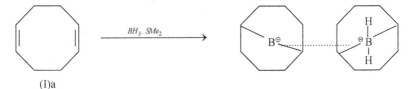

that is that in which $Cl - S\bullet en$ does the activation of one center, adds to it and in the second stage, the same S center does the second activation of the remaining center and add to it.
(Laws of Creations for Cycloocta-1,5-Diene)

Rule 914: This rule of Chemistry for **Cycloocta-1,5-Diene,** states that, when made to react with borane, the reaction is a three stage Equilibrium Addition mechanism system to give 9-borabicyclo[3,3,1] nonane (9-BBN), a very important reagent used for hydroboronations-

TRANS-PLACEMENT OF BH₃

noting the structure of the 9-borobicyclo nonane, wherein there is an electrostatic bond between the boron centers.
(Laws of Creations for Cycloocta-1,5-Diene)

Rule 915: This rule of Chemistry for **Cyclodienes,** states that, like cycloalkanes, cycloalkenes, the order of nucleophilicity of the members of cyclodienes follows the same pattern as the others.

Cyclooctadiene > Cycloheptadiene > Cyclohexadiene > Cyclopentadiene > Cyclobutadiene
Order of Nucleophilicity

noting that this is the reverse of the order of strain energy in their rings,
(Laws of Creations for Cyclodienes)

Rule 916: This rule of Chemistry for **Dienes, Trienes, and Tetraene,** states that, since between resonance stabilized and non-resonance stabilized members of cyclohexadiene members and above, the non-resonance stabilized members are of lesser nucleophilic than their resonance stabilized counterparts, hence the following order is clearly valid.

> Tetraenes > Trienes > Dienes > Ethenes
> Order of Nucleophilicity

for which one should know the order to expect for triynes, diynes and ethynes (acetylenes).
(Laws of Creation for Poly-enes)

Rule 917: This rule of Chemistry for **Cyclo-Tri-enes,** states that, these are monomers/compounds with three double bonds in their rings, for which due to the influence of strong presence of Maximum strain energy (MaxRSE), existence of 3-, 4-,and 5-membered rings are impossible as shown below-

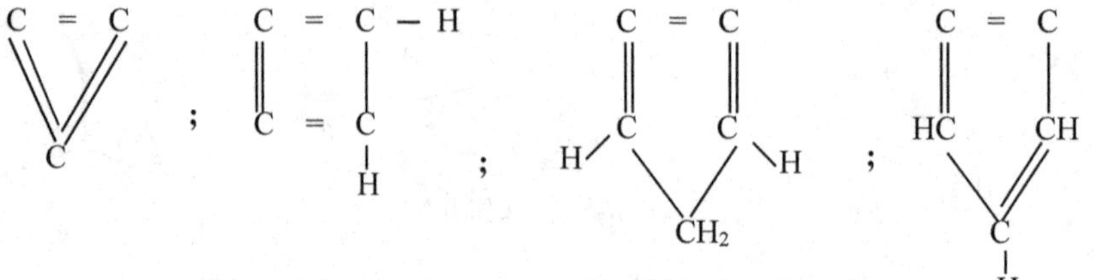

Very highly strained – CANNOT EXIST

(Laws of Creations for Cyclo-trienes)

Rule 918: This rule of Chemistry for **Benzene,** states that, when activated, the resonance stabilization phenomenon here is a two-step process as shown below-

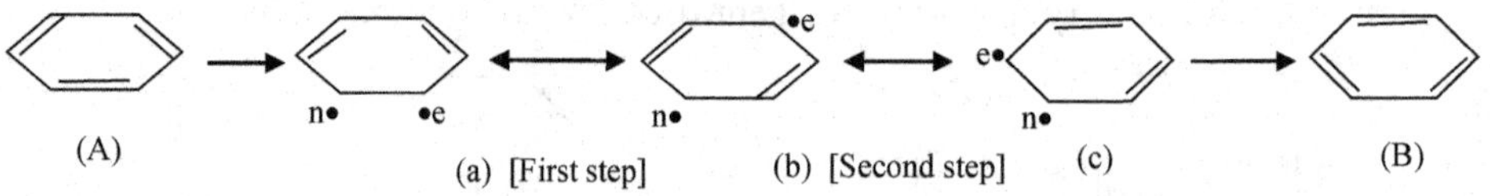

(A) (a) [First step] (b) [Second step] (c) (B)

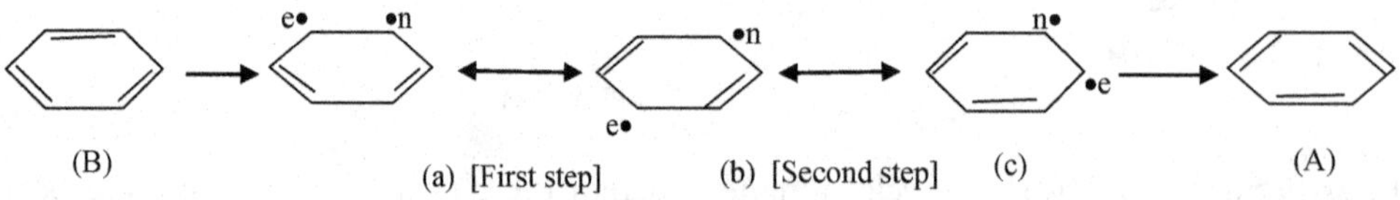

(B) (a) [First step] (b) [Second step] (c) (A)

for which the isolated cases shown above, the (b)s, are only transient in character; noting that ***Trans-placement of the radicals is impossible because of the presence of an adjacently located double bond in A CLOSED-LOOP system,*** and the type of resonance stabilization is of the closed-loop type and continuous in character, since deactivation takes place after every two-steps movement from (A) to (B) and begins again in the next stage from (B) to (A) the original ring – a two by two process.
(Laws of Creations for Benzene)

Rule 919: This rule of Chemistry for **Benzene,** states that, in view of the Closed-loop character

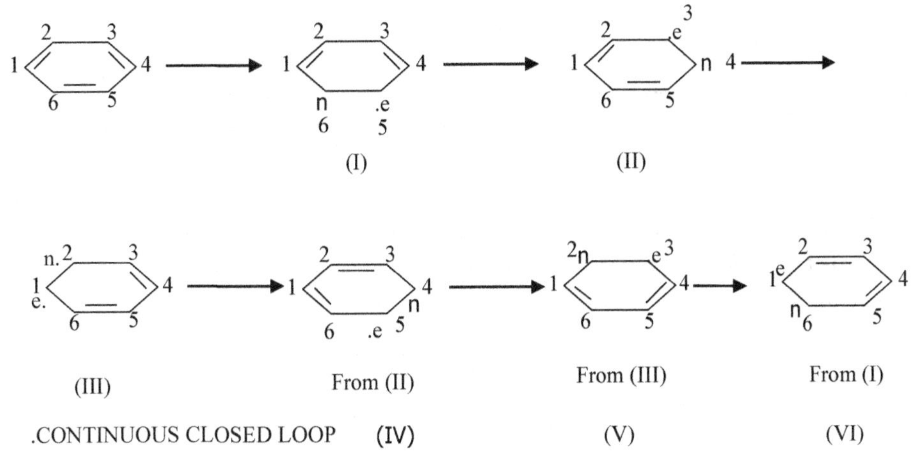

(I)

(II)

(III)

.CONTINUOUS CLOSED LOOP   **(IV)**

From (II)

(V)

From (III)

(VI)

From (I)

and the electro-free-radical movement, there is only ONE mono-form for Benzene.
(Laws of Creations for Benzene)

Rule 920: This rule of Chemistry for **Benzene with radical-pushing and pulling group (s),** states that, when activated, the followings are obtained for them-

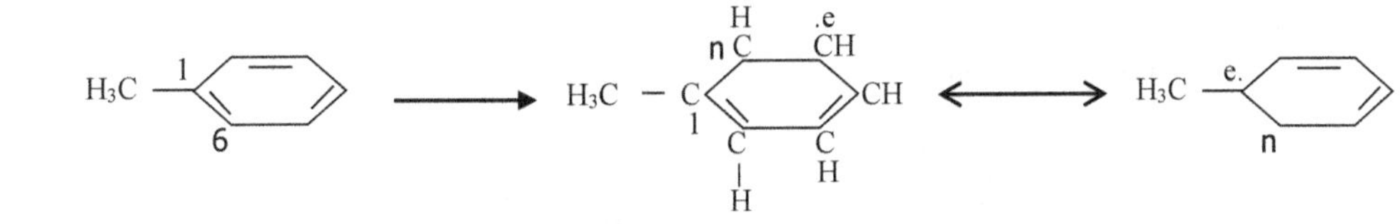

(a)

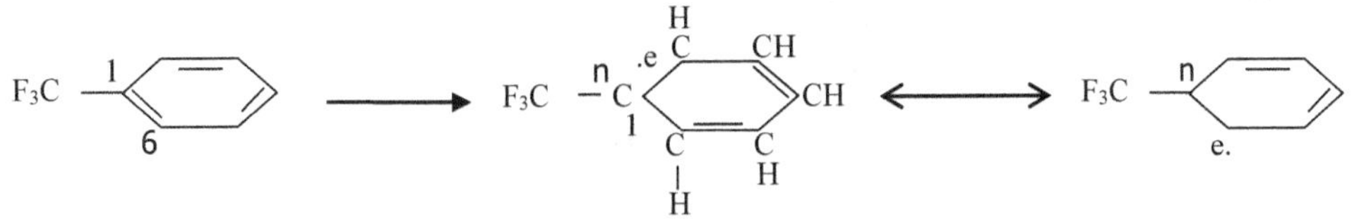

Nucleophile

(b)

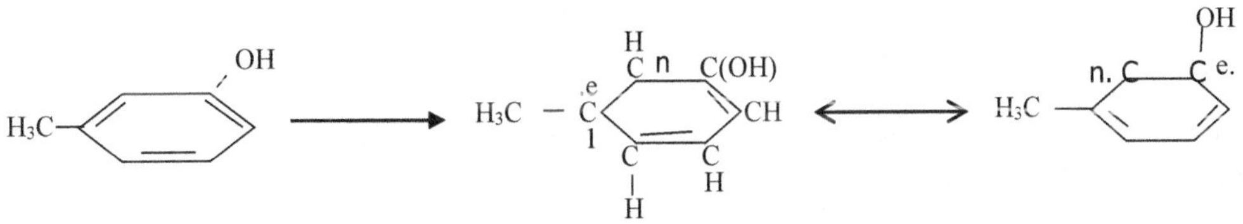

(c)

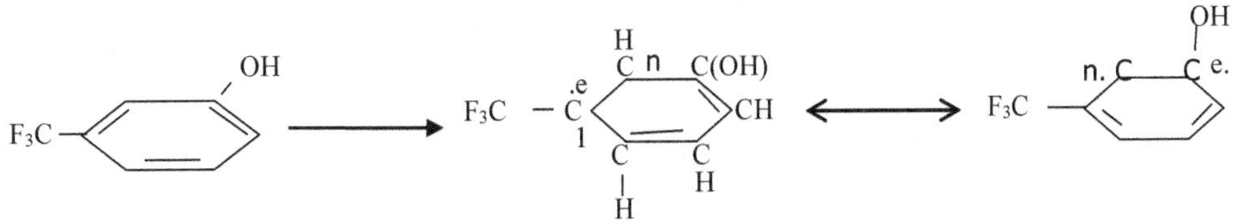

(d)

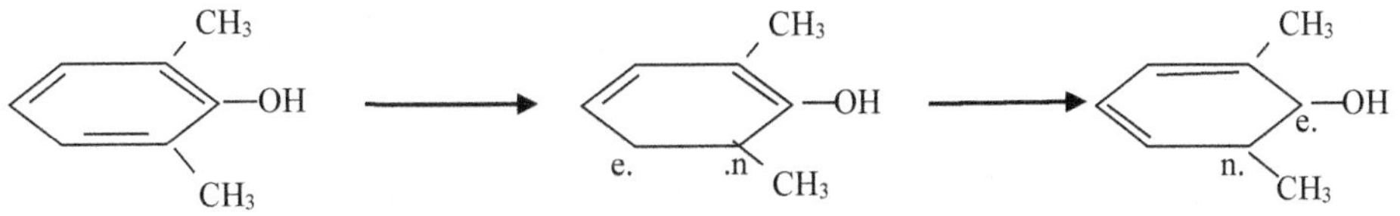

(e)

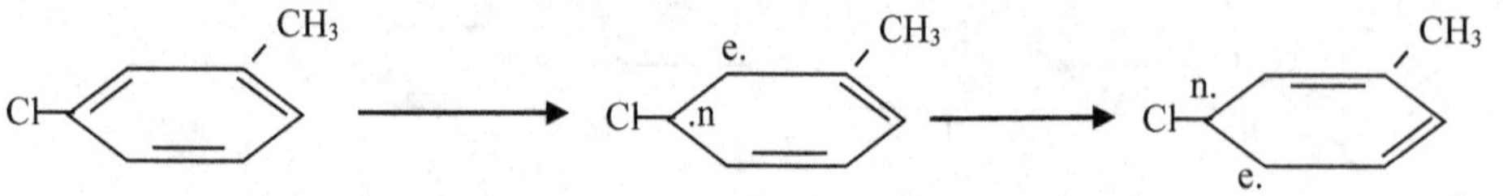

(f)

taking note of the fact that activation begins with the least nucleophilic center, followed by movement of the visible electro-free-radical in two steps, back to another center.
(Laws of Creation for Benzene compounds)

Rule 921: This rule of Chemistry for **Benzenes,** states that, when activated, the groups carried by it are resonance stabilized, that is, they cannot be tempered with as shown below-

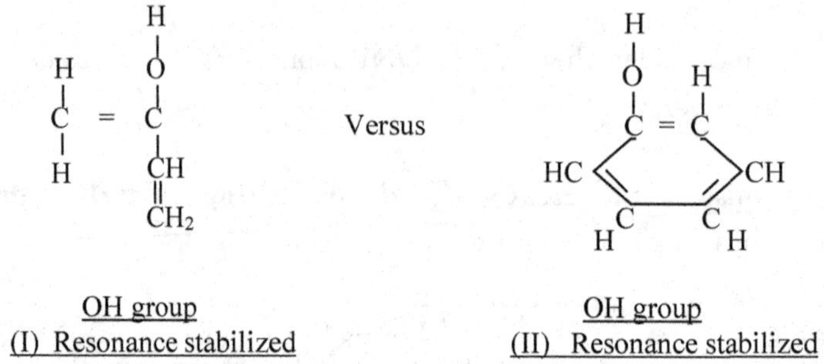

OH group
(I) Resonance stabilized

OH group
(II) Resonance stabilized

(a)

Female   Male

Not Favored

Female   Male

Not favored

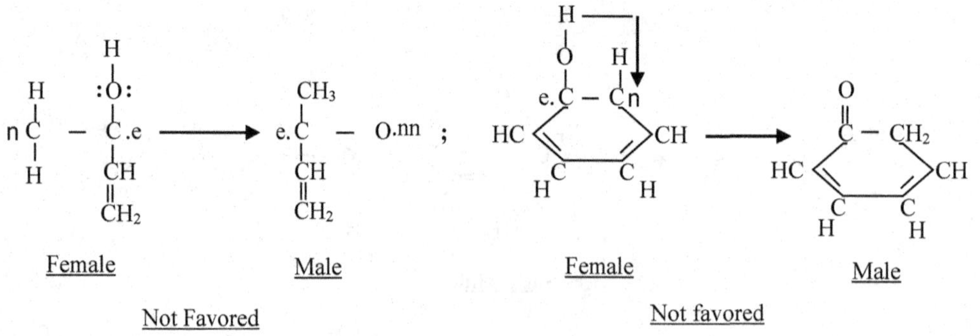

(b)

for which just as the OH group in (I) is resonance stabilized, so also is the OH group in (II), despite the continuous closed loop and (discrete) opened-loop nature of both systems.
(Laws of Creations for Benzene compounds)

Rule 922: This rule of Chemistry for **Benzenes,** states that, when not activated or placed in Energized state of existence, the groups carried by it are not resonance stabilized, for which they can be made where possible to exist in Equilibrium state of existence as shown below-

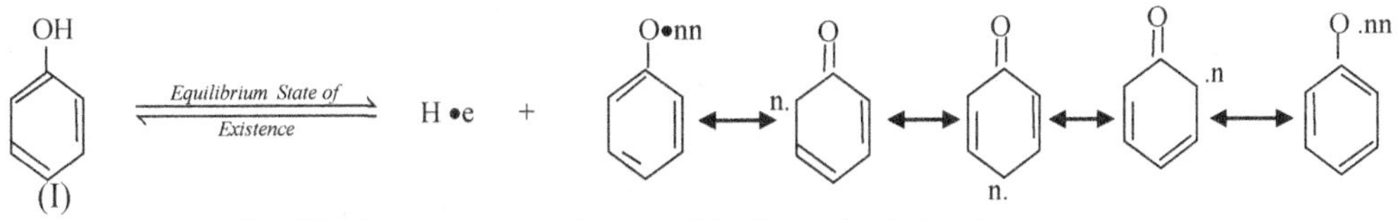

Equilibrium state of existence [De-Energized state] (a)

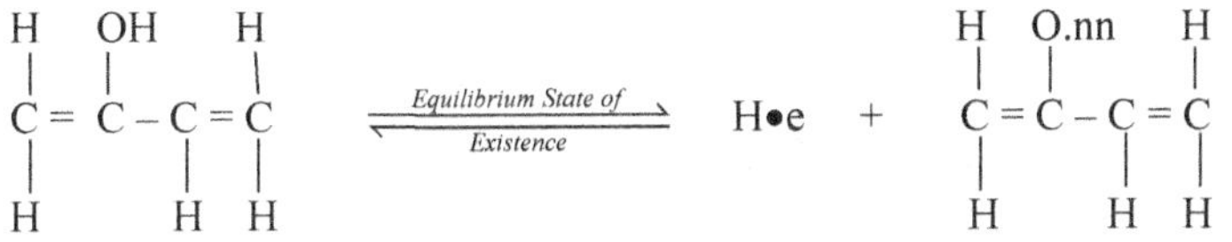

(II)   **Equilibrium state of existence** (b)

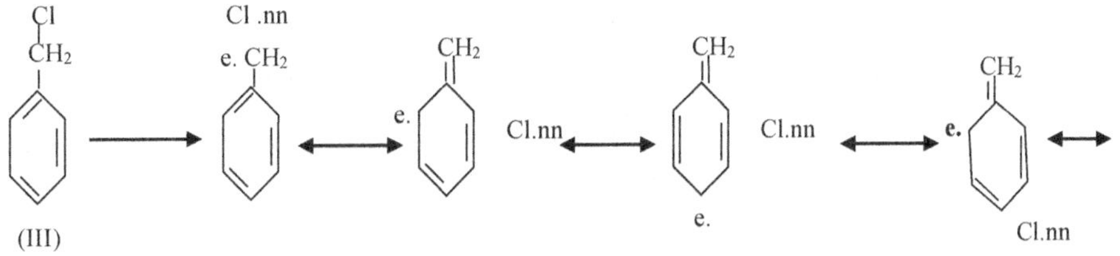

(III)   **Resonance Stabilization/Equilibrium State of Existence** (c)

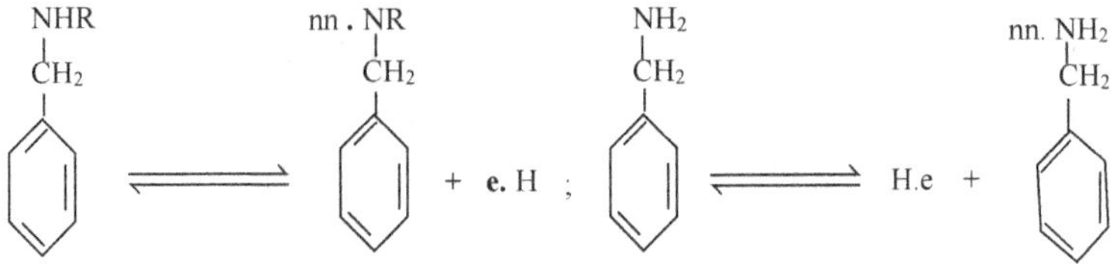

(IV)   **Cannot be resonance stabilized**
Equilibrium State of Existence (d)

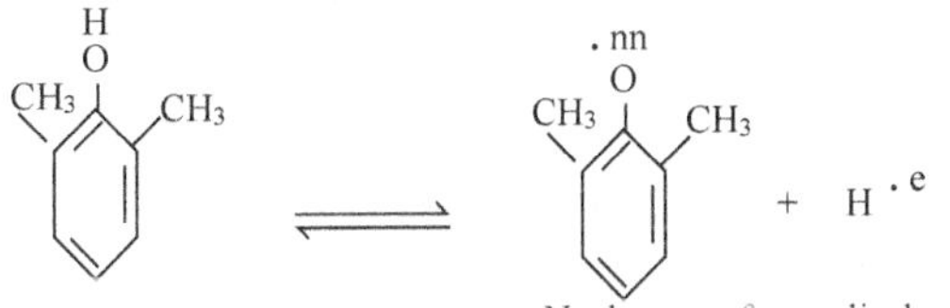

(V)   Nucleo-non-free-radical
(Can be resonance stabilized)
Equilibrium State of Existence (e)

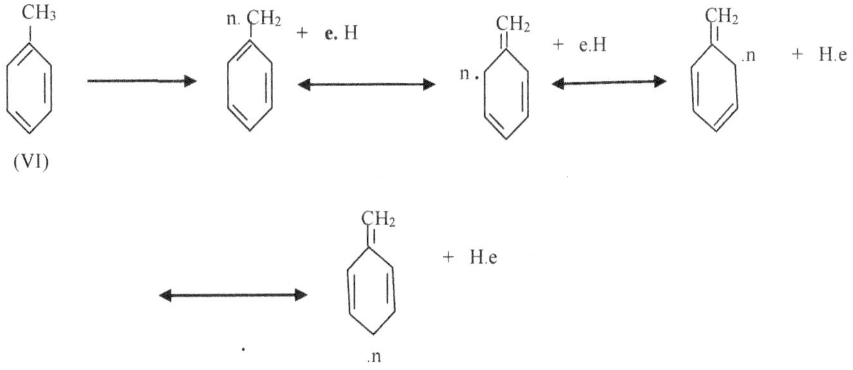

(VI)

Electroradicalization/Equilibrium State of Existence (f)

495

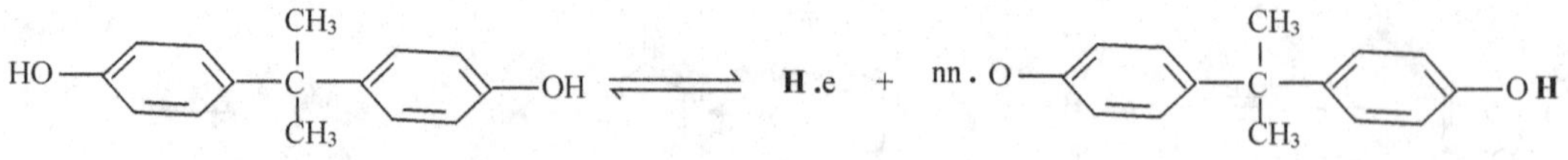

(VII) Nucleo-non-free-radical
(Can be resonance stabilized)

Equilibrium State of Existence (g)

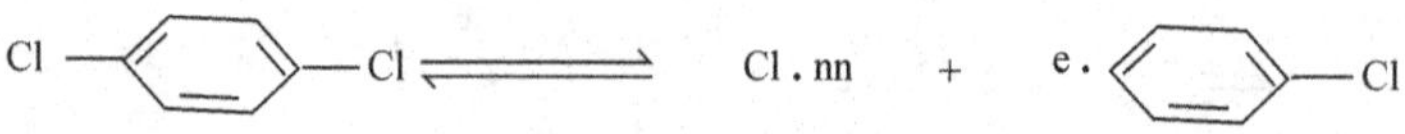

(VIII) <u>Non-resonance stabilized</u>

Equilibrium State of Existence (h)

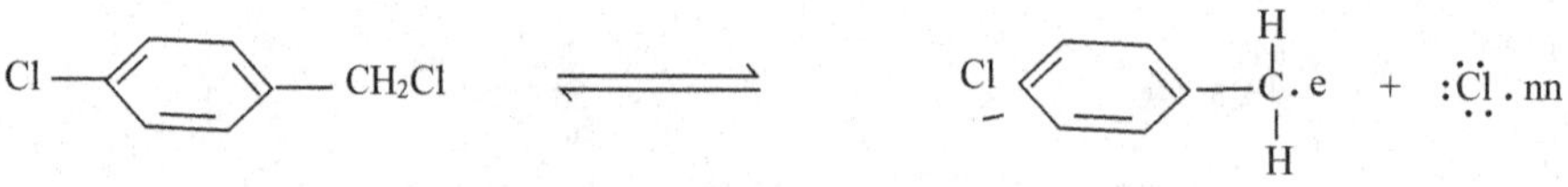

(IX) <u>Resonance stabilized monomer</u>

Equilibrium State of Existence (i)

noting cases where the electro-free-radicals, nucleo- and nucleo-non-free-radicals were said to be resonance stabilized, like in styrene and type of movement wherein the original state is obtained, otherwise no movement such as in (a) will take place.

(Laws of Creations for Benzene compounds)

Rule 923: This rule of Chemistry for **Benzene compounds,** states that, when benzene carries one or more groups, there are two Equilibrium states of existence – the ***De-energized and the Energized Equilibrium states of existence***, as shown below-

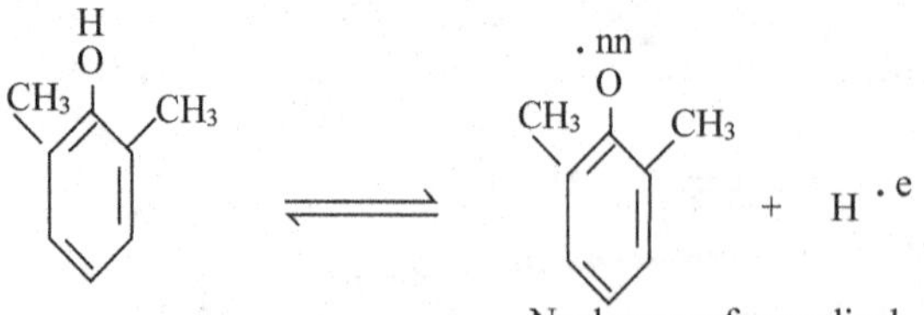

(I) Nucleo-non-free-radical
(Can be resonance stabilized)
De-Energised Equilibrium State of Existence (1a)

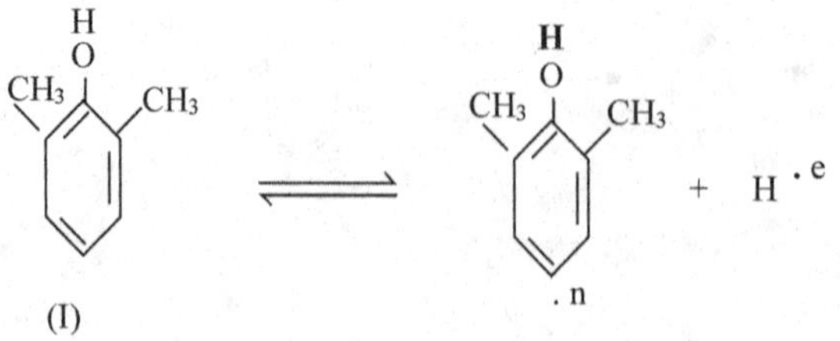

(I)

Equilibrium State of Existence in the Energized state (1b)

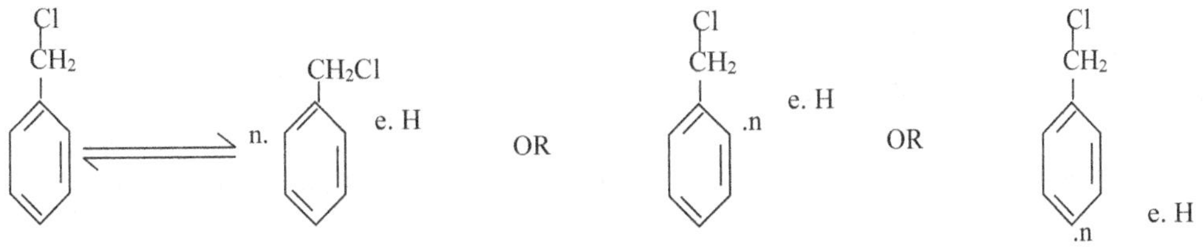

Equilibrium State of Existence in the Energized State (2a)

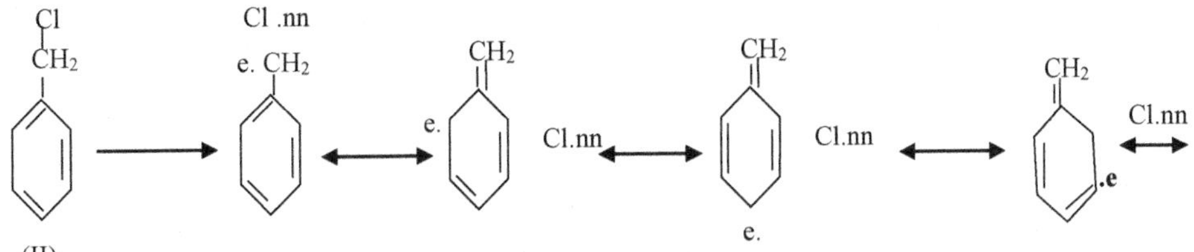

(II)

De-energized Equilibrium State of Existence (2b)

(Laws of Creations for Benzene compounds)

Rule 924: This rule of Chemistry for **Benzene,** states that, this has only one Equilibrium state of existence in the Energized state of existence, while in the De-energized states of existence, it is stable as shown below-

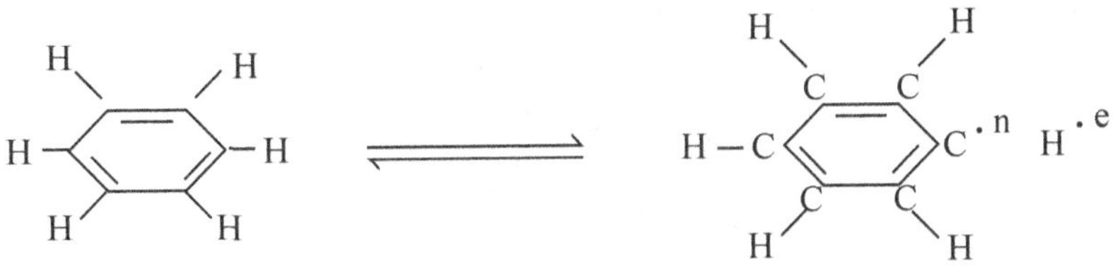

Energized Equilibrium state of existence

for which when the hydrogen held is replaced by another group, the group takes over control of the ring as to which H should next be held in Equilibrium state of existence.

(Laws of Creations for Benzene)

Rule 925: This rule of Chemistry for **Some specific Benzene compounds,** states that, in the absence of activation of an activation center in the ring, some benzenes bi-functional in character can be used as Step monomers, in which the substituted groups act as functional groups as shown below-

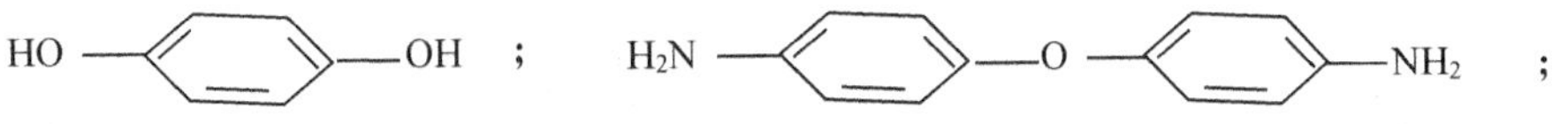

Non - resonance stabilized monomers when De-energized

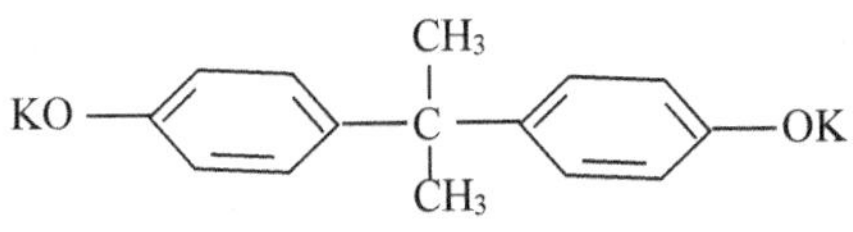

497

when they are in De-energized state of existence; existing in Equilibrium state of existence from the terminals.
(Laws of Creations for Benzene compounds).

Rule 926: This rule of Chemistry for **p-Dichlorobenzene,** states that, this like benzene, has only one Equilibrium state of existence-

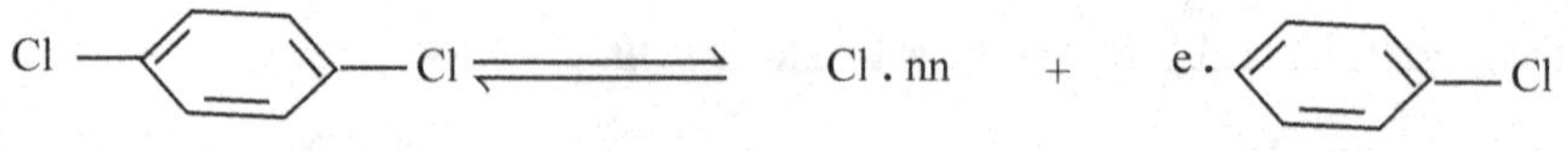

Energised and De-energised Equilibrium State of Existence

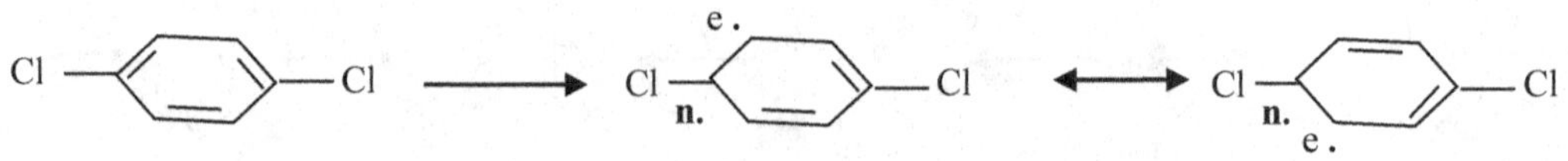

Resonance stabilization in Activated state

for which the meta-positions are not usually readily available for use, that is, the H atom is not held in Equilibrium state of existence unlike with other radical-pulling groups, which are free.
(Laws of Creations for p-Dichlorobenzene)

Rule 927: This rule of Chemistry for **Benzene and Acetylene,** states that, like in acetylene, after replacement of one hydrogen atom in benzene with a radical-pushing group, a second hydrogen atom is still slightly loosely bonded to a carbon center, but unlike acetylene, this depends more on the capacity of the group already carried by the ring, and not that of the activating species and unlike in acetylene, benzene is resonance stabilized and also unlike in acetylene, when they are activated, while in benzene the H or groups carried are shielded, this is not the case with acetylene.
(Laws of Creations for Benzene/Acetylene)

Rule 928: This rule of Chemistry for **Benzene,** states that, the type of group in terms of pulling and pushing abilities and capacities initially carried by benzene, determine how, ortho-, meta - and para-placements are obtained.
(Laws of Creations for Benzene)

Rule 929: This rule of Chemistry for **Benzene,** states that, when the group carried is a radical-pushing group, then the position where another hydrogen atom is replaced is the ortho- or para-position as shown below beginning first with the two ortho- positions-

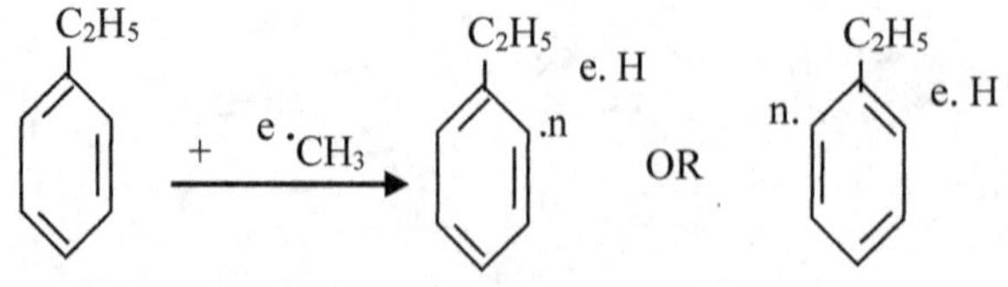

$(C_2H_5 > CH_3)$

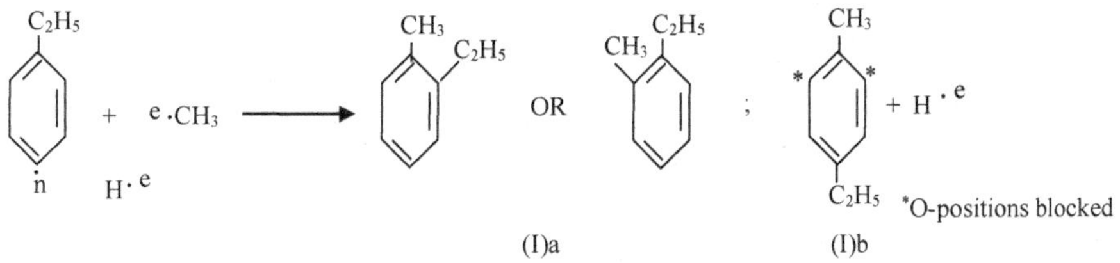

(a)

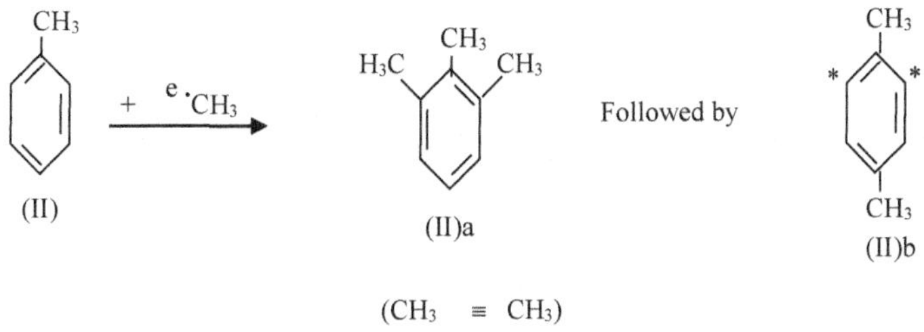

(b)

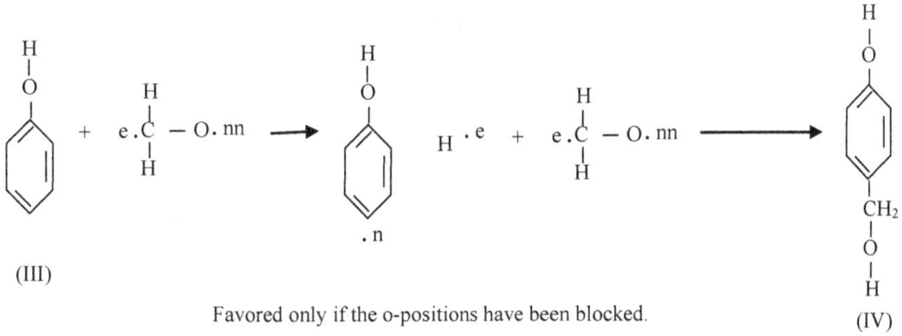

Favored only if the o-positions have been blocked.

(c)

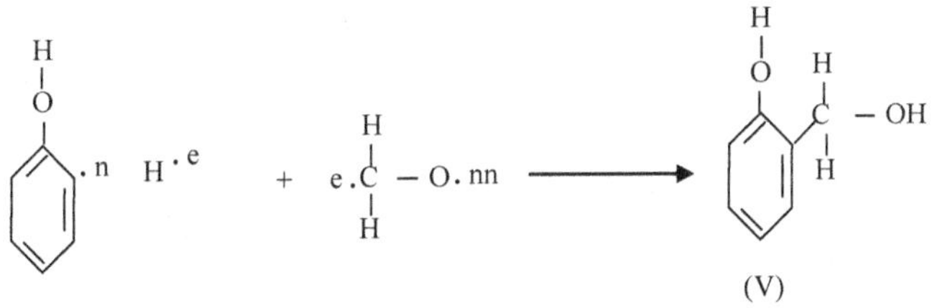

(d)

noting that the above reactions involving displacement reactions are favored, only when the group coming in is of equal or less-radical pushing capacity than the group already present on the ring, for example CH_2OH group from the formaldehyde is of weaker radical-pushing capacity than that of OH group, and for all of them, the ortho- positions must be be blocked before the para- positions can be used, if they have not been filled.

(Laws of Creations for Benzenes)

Rule 930: This rule of Chemistry for **Benzene,** states that, unlike in acetylene, after the second replacement, there is a third one - the two ortho-positions if the p-position has been filled by blocking and deblocking or one para-position if the two o-positions have been filled and so on as shown below, noting that what controls the ring in terms of the next position is the group with the largest radical-pushing capacity, and the incoming group must be of lower or equal capacity than the controlling group-

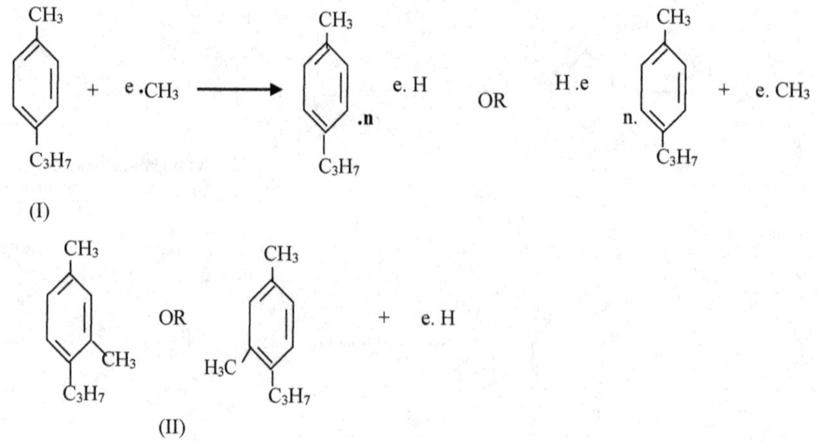

(I)

(a)

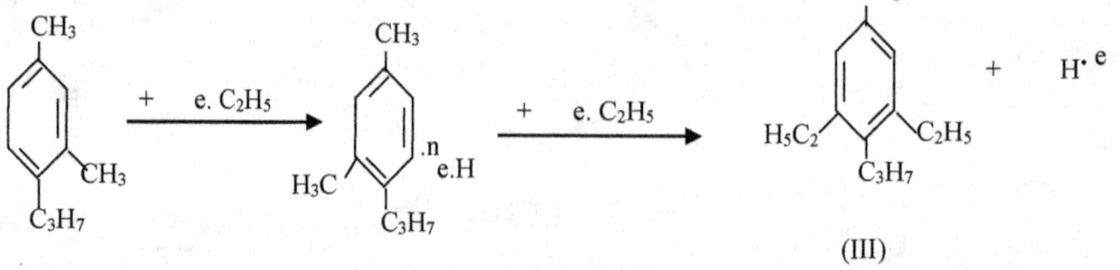

(II)

Favored placements

(III)

(b)

(II)

(IV)

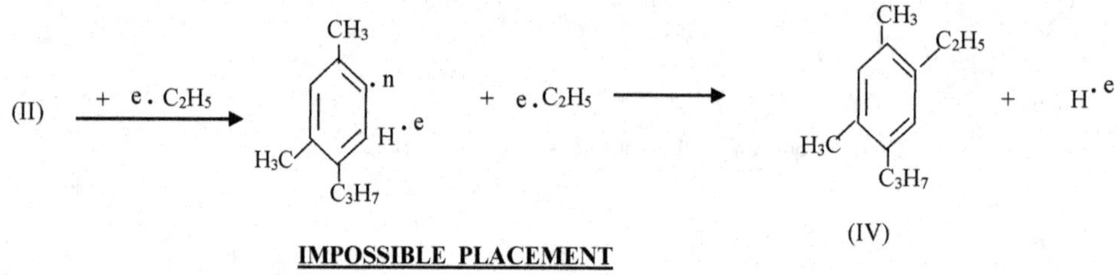

IMPOSSIBLE PLACEMENT

(c)

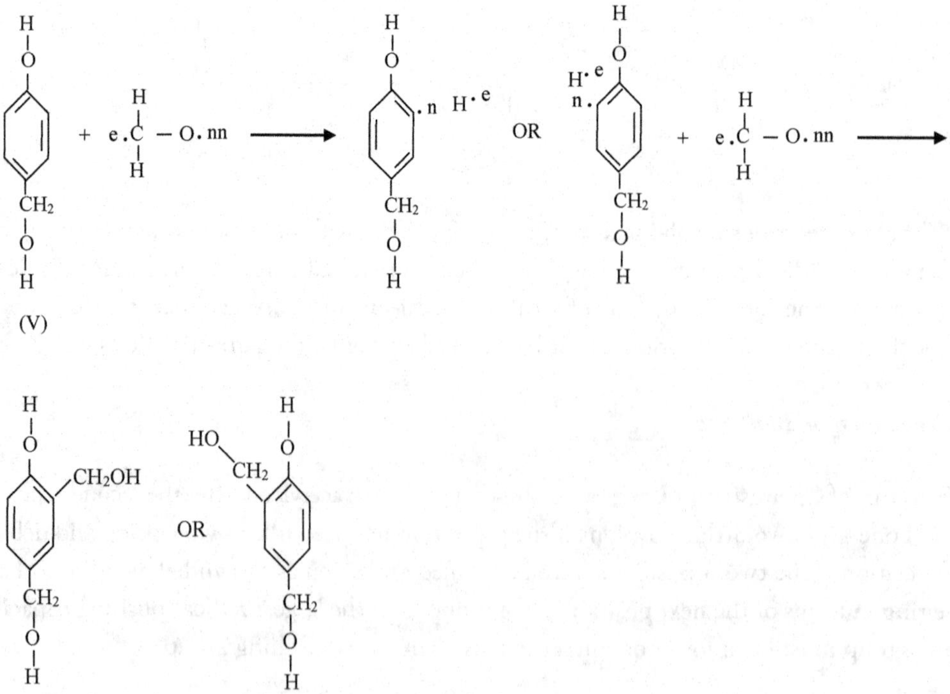

(V)

(VI) (Resonance stabilized)

(d)

noting that (V) was obtained after blocking the o-positions, then followed by deblocking.
(Laws of Creations for Benzene)

Rule 931: This rule of Chemistry for **Benzene,** states that, when there are more than one group in the ring, it is the largest group in capacity that decides the new para- and ortho- positions, for which the followings are valid free-radically.

$$-OH \;>\; -C(OH)_3 \;>\; -CH(OH)_2 \;>\; -C(OH)H_2 \;>\; -CH_3$$

Order of radical-pushing capacity

$$-CH_3 \;>\; -CH_2F \;>\; CHF_2 \;>\; -H \;>\; -CF_3$$

Order of radical-pushing capacity

(Laws of Creations for Radical-pushing groups)

Rule 932: This rule of Chemistry for **Benzene,** states that, when two same radical-pushing groups are symmetrically placed on the ring, all four positions in the ring become available for replacement when energized, in view of its symmetric character, and examples of such cases include o-Xylene, Resorcinol and so on, as shown below using resorcinol-

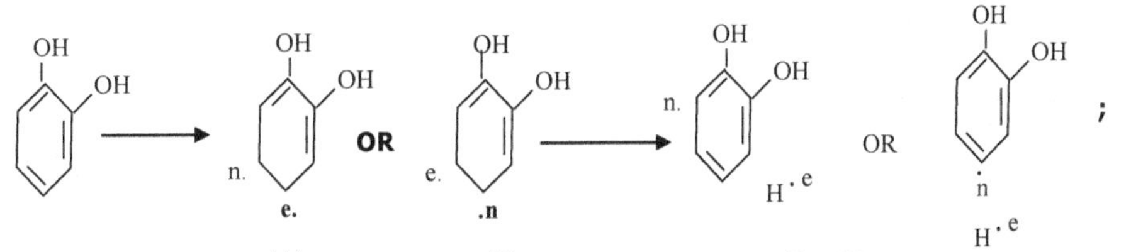

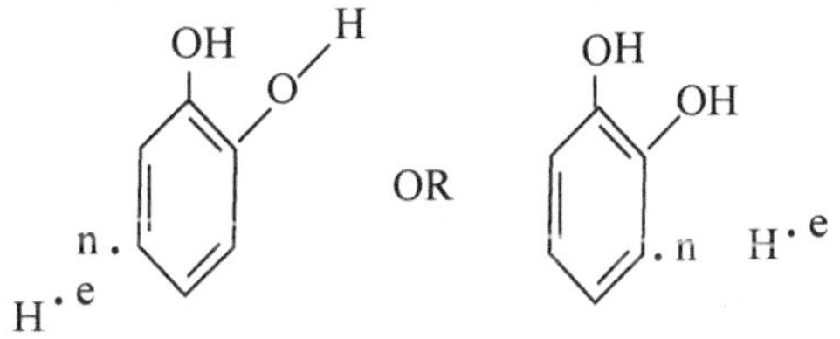

For (B)

noting that due to the symmetry, the two active centers in (A) and (B) above can carry any free-radical, and hence, the four positions which include meta-, ortho- and para- positions are available for use
(Laws of Creations for Benzenes)

Rule 933: This rule of Chemistry for **Benzene,** states that, when groups such as NO_2, SO_2R etc. [which were identified as strong radical-pulling groups (Chargely) and weak radical-pushing groups (Radically) with respect to H and F] are present on the ring alone, because their capacity is less than that of hydrogen atom, the ortho- and para-positions are no longer available for use; instead it is the meta- positions that are now made available for use as shown below-

501

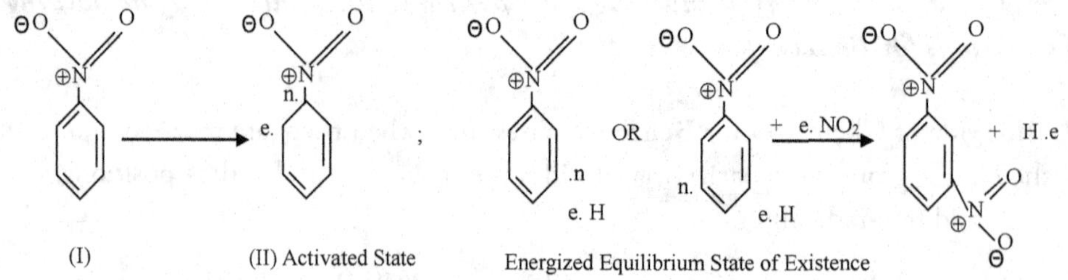

(I) (II) Activated State Energized Equilibrium State of Existence

noting very importantly that just as a radical-pushing group when present on the ring, makes the ortho- and para- positions available, so also a radical-pulling group or radical-pushing group of lower capacity than H when present on the ring, makes only the meta-position available.
(Laws of Creations for Benzenes)

Rule 934: This rule of Chemistry for **Benzene,** states that, when two nitro- groups are present on the ring, meta-(Natural position), ortho- and para- placed, and attempts are to be made to replace one of them by abstraction, it is only those placed in the position not natural to them that can be replaced as shown below-

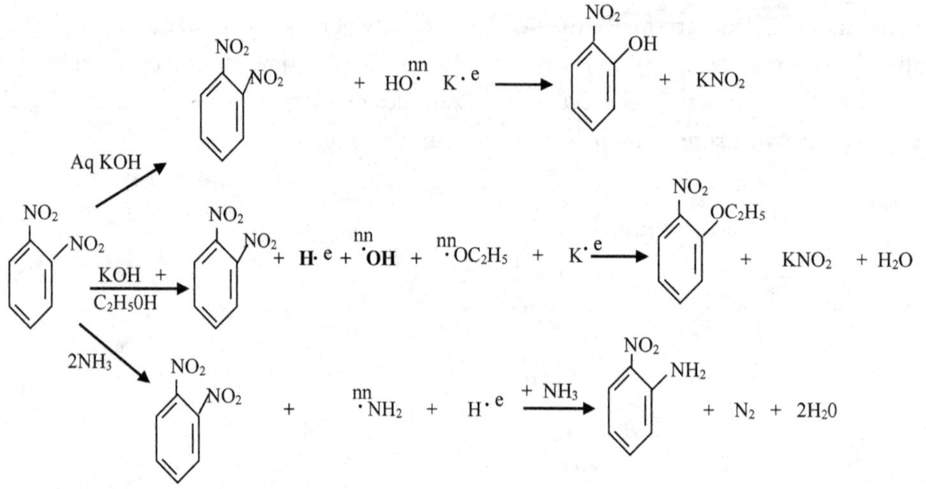

(a)

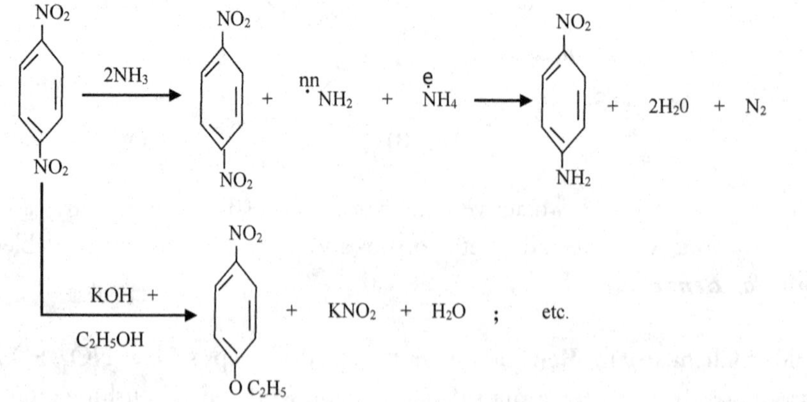

(b)

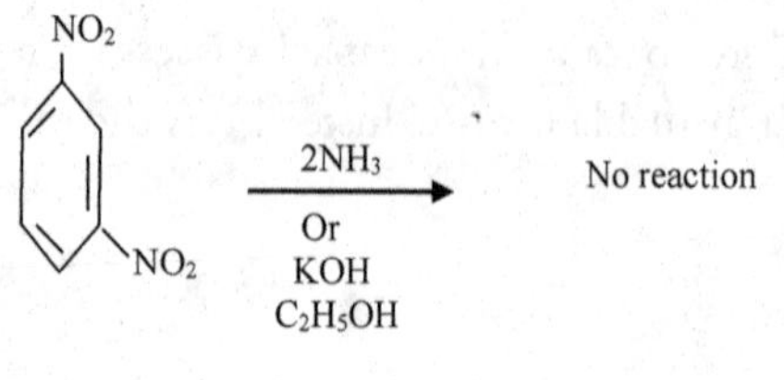

(c)

502

for which it is very important to note that all the reactions are radical in characters, different from what has been known to be the case for many years and all of them are Equilibrium mechanism systems of one, two or more stages.
(Laws of Creations for Benzenes)

Rule 935: This rule of Chemistry for **Benzene,** states that, when two nitro- groups are placed on the ring in their natural positions, attempts to introduce a third nitro group is impossible, because the ring with two nitro- groups can no longer exist in Equilibrium state of existence, that is, the last H in the meta- position can no longer be held; otherwise special catalysts may be required to keep it in Equilibrium state of existence.
(Laws of Creations for Benzenes)

Rule 936: This rule of Chemistry for **Benzene,** states that, when two nitro-groups are placed on the ring, attempt to place a third nitro- group can be done as follows-

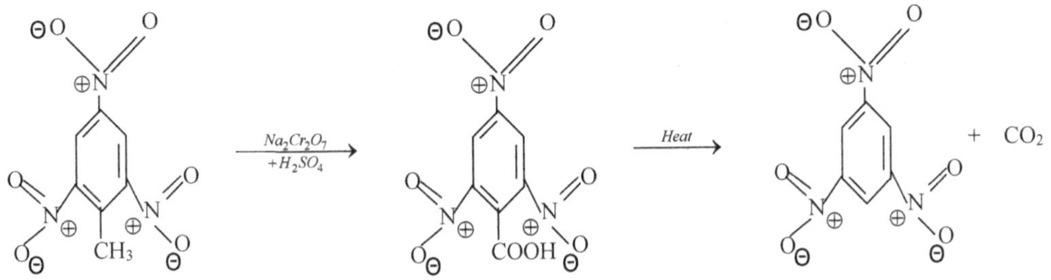

(A)   (B)   (C)

for which in (A) above, through the CH_3 group initially placed on the ring, three NO_2 groups were first easily placed on the ring to give the 2,4,6-Trinitrotoluene (TNT) an explosive (A); and with the use of oxidizing agents, $Na_2Cr_2O_7$, the CH_3 group was changed to COOH to give (B) which with heat and the presence of COOH, the benzene ring was put into the De-energized state to release CO_2 to give (C)-1,3,5-Trinitrobenezene a more powerful explosive than TNT.
(Laws of Creations for Benzenes)

Rule 937: This rule of Chemistry for **Benzene,** states that, when nitrobenzene is heated with potassium hydroxide in the presence of air, to give o-nitrophenol, the mechanism of the reaction is as follows-
Stage 1:

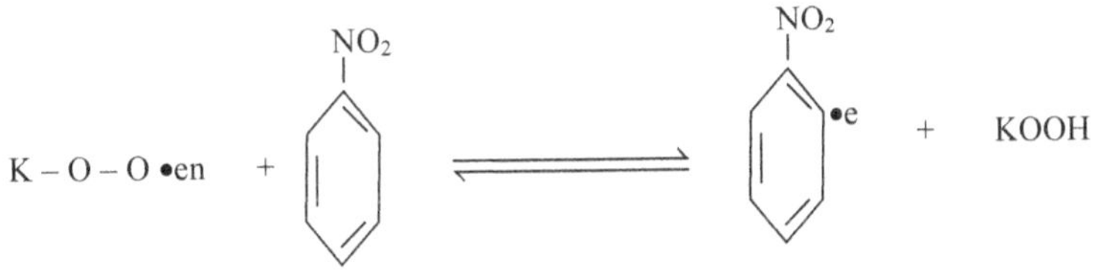

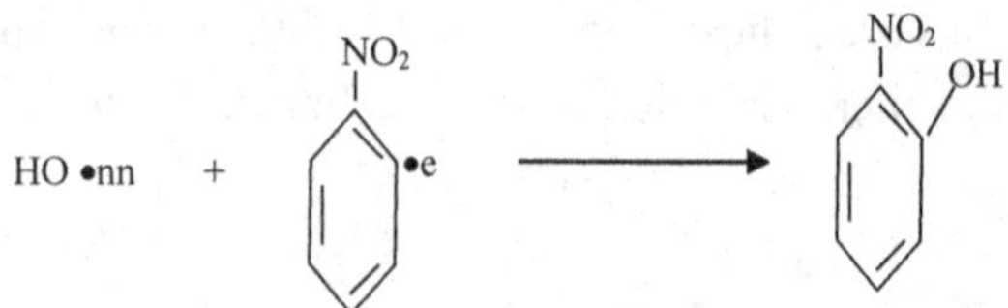

$$\text{HO} \bullet nn \quad + \quad \text{(NO}_2\text{ ring with }\bullet e\text{)} \quad \longrightarrow \quad \text{(NO}_2\text{, OH ring)} \quad (B)$$

(a)

Overall equation: (I) + KOH + O$_2$ ⟶ (B) + KOOH

(b)

Stage 2:

$$2\text{KOOH} \rightleftharpoons 2\text{KO}\bullet nn \quad + \quad 2en \bullet \text{OH}$$

$$2\text{H}-\text{O}\bullet en \rightleftharpoons 2en \bullet \text{O} \bullet nn \quad + \quad 2e\bullet \text{H} \quad + \quad \text{Heat}$$

$$2\text{KO}\bullet nn + 2e\bullet \text{H} \longrightarrow 2\text{KOH}$$

(c)

Overall equation: 2(I) + 2KOH + 2O$_2$ ⟶ 2(B) + 2KOH + $\underline{\bm{O_2}}$

$$\qquad\qquad\qquad\qquad\text{(AIR)}\qquad\qquad\qquad\qquad\qquad\text{(Oxidizing Oxygen)}$$

(d)

noting that this is a two stage Equilibrium mechanism system in which the H removed is that located on the ortho- position, because the meta- position is the natural position for NO$_2$ group, and more.
(Laws of Creations for Benzenes)

Rule 938: This rule of Chemistry for **Benzenes,** states that, when a single nitro-group is in the ring, the nitro- group can be replaced by hydrogenation of the group to give aniline as shown below-

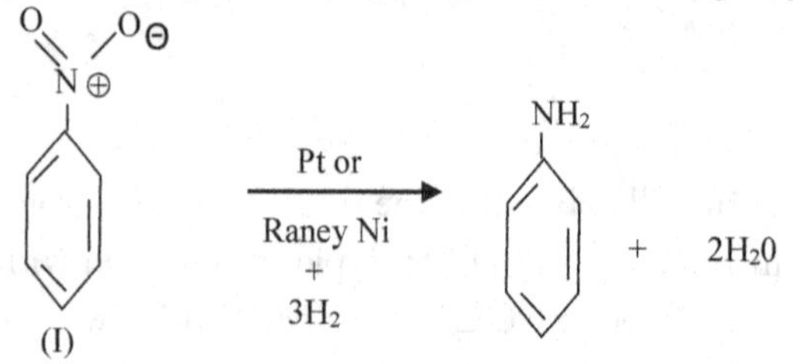

$$\xrightarrow[\substack{\text{Raney Ni}\\+\\3\text{H}_2}]{\text{Pt or}}$$

(I)   +   2H$_2$0

(a)

Stage 1:

$$\text{H}_2 \rightleftharpoons \text{H}\bullet e \quad + \quad n\bullet \text{H}$$

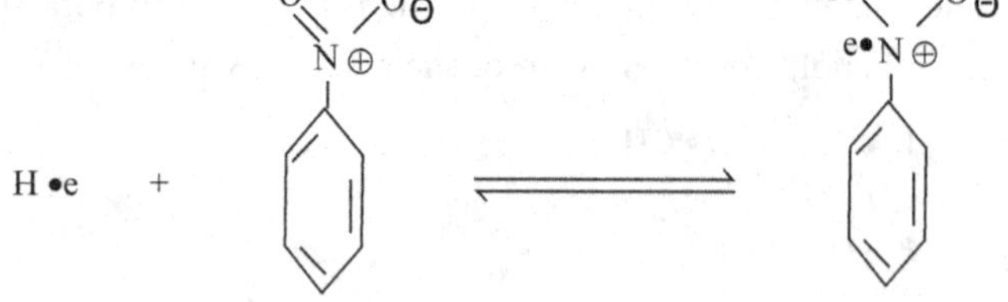

$$\text{H}\bullet e \quad + \quad \text{(I)} \rightleftharpoons$$

504

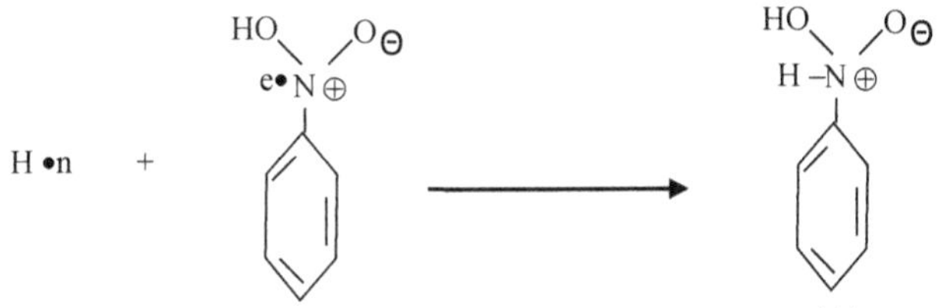

$H \bullet n \quad + \quad$ [structure with $HO\!-\!N^{\oplus}\!-\!O^{\ominus}$, $e\bullet N^{\oplus}$ on benzene ring] $\quad\longrightarrow\quad$ [structure with $HO\!-\!N^{\oplus}\!-\!O^{\ominus}$, $H\!-\!N^{\oplus}$ on benzene ring] (A)

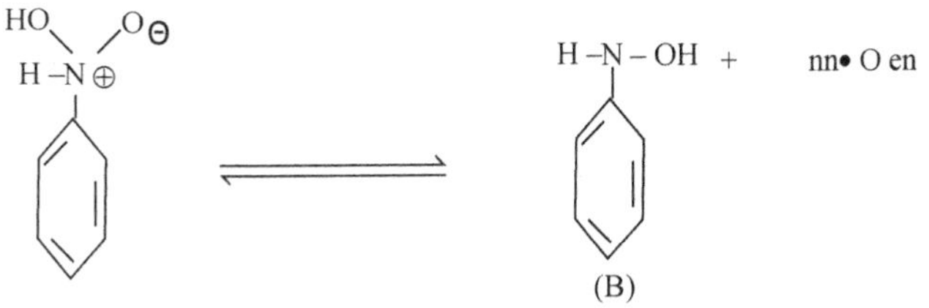

(b)

Overall equation: (I) $+$ H_2 $\longrightarrow$ (A)

(c)

Stage 2:

[structure $HO\!-\!N^{\oplus}\!-\!O^{\ominus}$, $H\!-\!N^{\oplus}$ on benzene ring] $\quad\rightleftharpoons\quad$ $H\!-\!N\!-\!OH$ [on benzene ring] (B) $\quad+\quad$ $nn\bullet O\ en$

$nn\bullet O \bullet en \quad + \quad H_2 \quad\rightleftharpoons\quad HO\bullet nn \quad + \quad e\bullet H$
$\longrightarrow \quad H_2O$

(d)

(e)

Overall equation: (I) $+$ $2H_2$ $\longrightarrow$ (B) $+$ H_2O

Stage 3:

$H_2 \quad\rightleftharpoons\quad H\bullet e \quad + \quad H \bullet n$

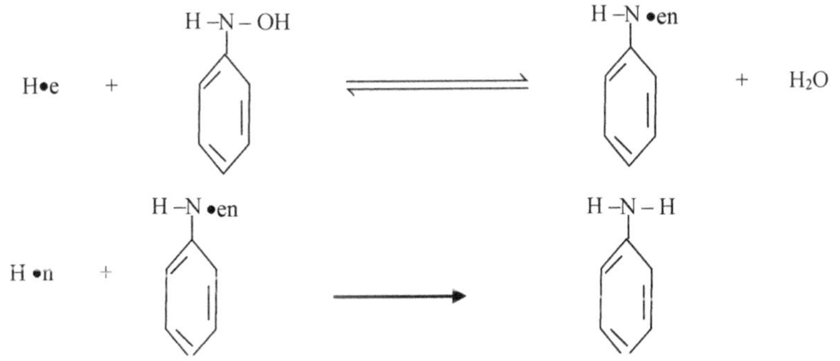

$H\bullet e \quad + \quad$ [$H\!-\!N\!-\!OH$ on benzene ring] $\quad\rightleftharpoons\quad$ [$H\!-\!N\bullet en$ on benzene ring] $\quad+\quad H_2O$

$H \bullet n \quad + \quad$ [$H\!-\!N\bullet en$ on benzene ring] $\quad\longrightarrow\quad$ [$H\!-\!N\!-\!H$ on benzene ring]

(f)
(g)

noting that it is a three stage Equilibrium mechanism system wherein the benzene ring was kept in a De-energized state of existence throughout the three stages, and since the N = O bond is least nucleophilic center, activation started with it.
(Laws of Creations for Benzenes)

Rule 939: This rule of Chemistry for **Nitro-benzene compounds,** states that, while the ortho- and para-placed nitro-benzenes cannot be reduced by hydrogenation, because these positions are not natural when a nitro- group is on the ring, the meta- placed nitro- benzene can be reduced, because this position is natural to the nitro-benzene as shown below.

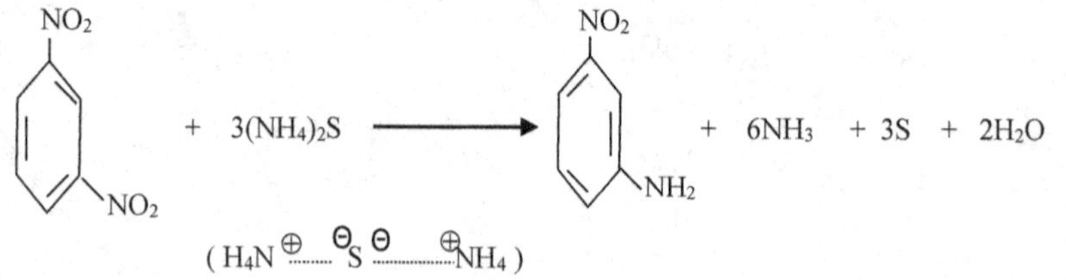

$$+ \; 3(NH_4)_2S \longrightarrow \qquad + \; 6NH_3 \; + \; 3S \; + \; 2H_2O$$

$$(H_4N \overset{\oplus}{\text{.......}} \overset{\ominus}{S} \overset{\ominus}{\text{.......}} \overset{\oplus}{N}H_4)$$

(a)

Stage 1: $\quad H_4N^{\oplus}...^{\ominus}S^{\ominus}....^{\oplus}NH_4 \; \underset{\text{Existence}}{\overset{\text{Equilibrium State of}}{\rightleftharpoons}} \; H \bullet e \; + \; nn.NH_3^{\oplus}......^{\ominus} S^{\ominus}....^{\oplus}NH_4$

(A)

$$(A) \quad \rightleftharpoons \quad NH_3 \; + \quad nn.\overset{\bullet\bullet\;\ominus}{\underset{\bullet\bullet}{S}} .....^{\oplus}NH_4$$

(B)

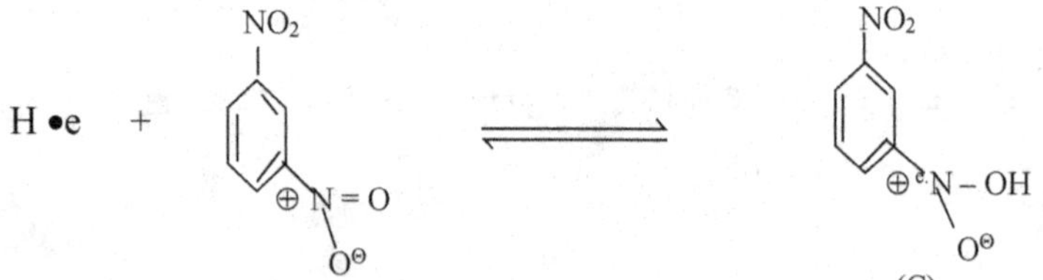

$$H \bullet e \; + \qquad \rightleftharpoons \qquad$$

(C)

$$(B) \; + \; (C) \longrightarrow$$

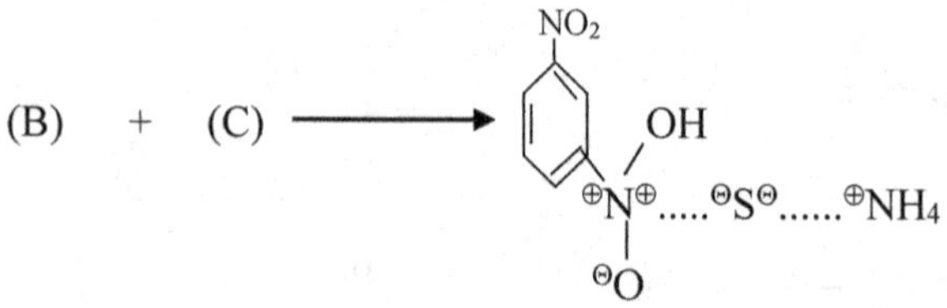

(D)

(b)

Stage 2: $\quad H_4N^{\oplus}...^{\ominus}S^{\ominus}....^{\oplus}NH_4 \; \underset{\text{Existence}}{\overset{\text{Equilibrium State of}}{\rightleftharpoons}} \; H \bullet e \; + \; nn.NH_3^{\oplus}......^{\ominus} S^{\ominus}....^{\oplus}NH_4$

(A)

$$(A) \quad \rightleftharpoons \quad NH_3 \; + \quad nn.\overset{\bullet\bullet\;\ominus}{\underset{\bullet\bullet}{S}} .....^{\oplus}NH_4$$

(B)

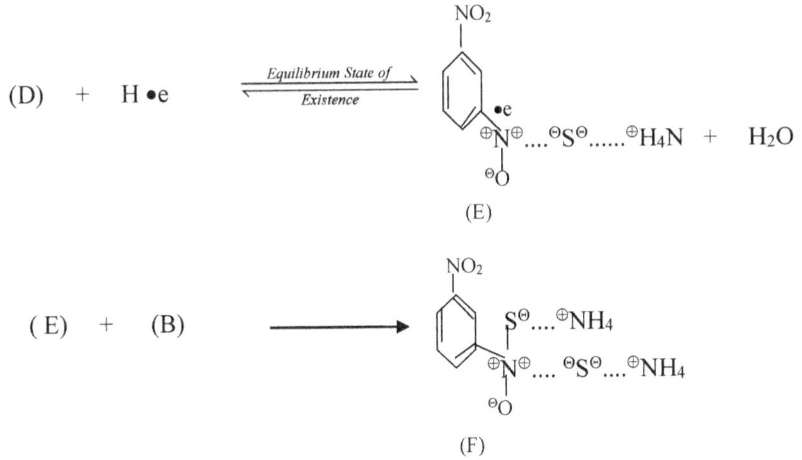

(D) + H •e $\xrightleftharpoons[\text{Existence}]{\text{Equilibrium State of}}$

NO₂ structure with •e, ⊕N⊕ ....⊖S⊖......⊕H₄N + H₂O

(E)

(E) + (B) ⟶

NO₂ structure: S⊖....⊕NH₄ / ⊕N⊕.... ⊖S⊖....⊕NH₄ / ⊖O

(F)

(c)

Stage 3: H₄N⊕...⊖S⊖....⊕NH₄ $\xrightleftharpoons[\text{Existence}]{\text{Equilibrium State of}}$ H •e + nn.NH₃⊕......⊖ S⊖....⊕NH₄

(A)

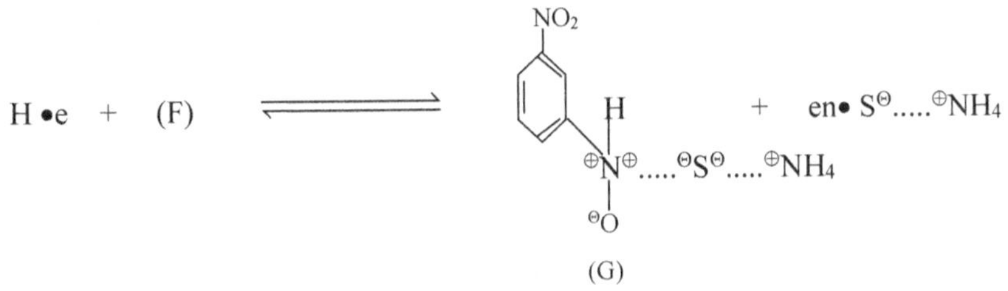

H •e + (F) ⇌

NO₂ structure: H / ⊕N⊕.....⊖S⊖.....⊕NH₄ / ⊖O   + en• S⊖.....⊕NH₄

(G)

(A) ⇌ NH₃ + nn.S̈ ⊖ .....⊕NH₄

(B)

en• S⊖.......⊕NH₄ ⇌ S + NH₃ + H •e

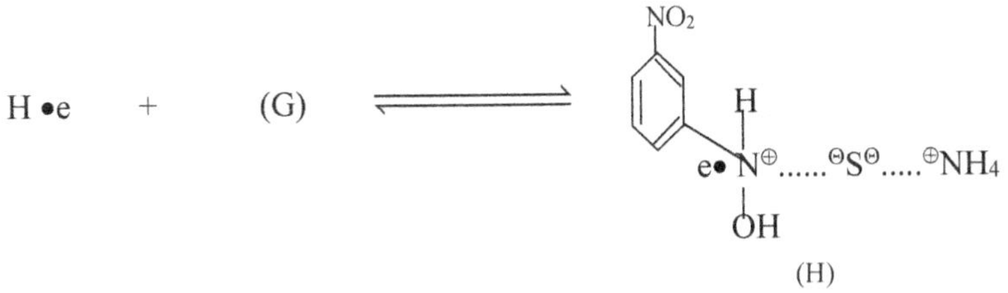

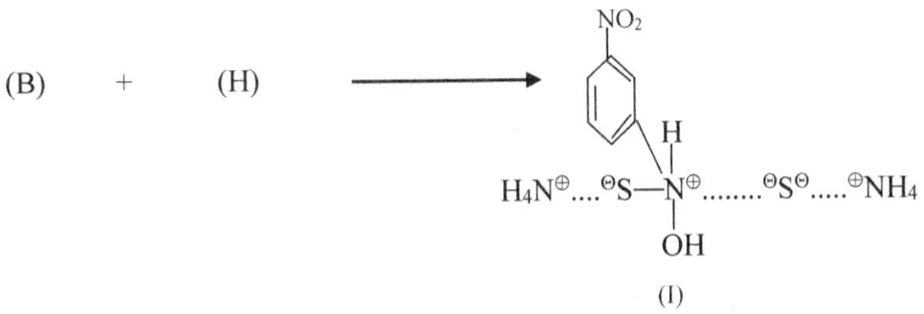

H •e + (G) ⇌

NO₂ structure: H / e•N⊕......⊖S⊖.....⊕NH₄ / OH

(H)

(B) + (H) ⟶

NO₂ structure: H / H₄N⊕....⊖S—N⊕........⊖S⊖.....⊕NH₄ / OH

(I)

(d)

Stage 4:

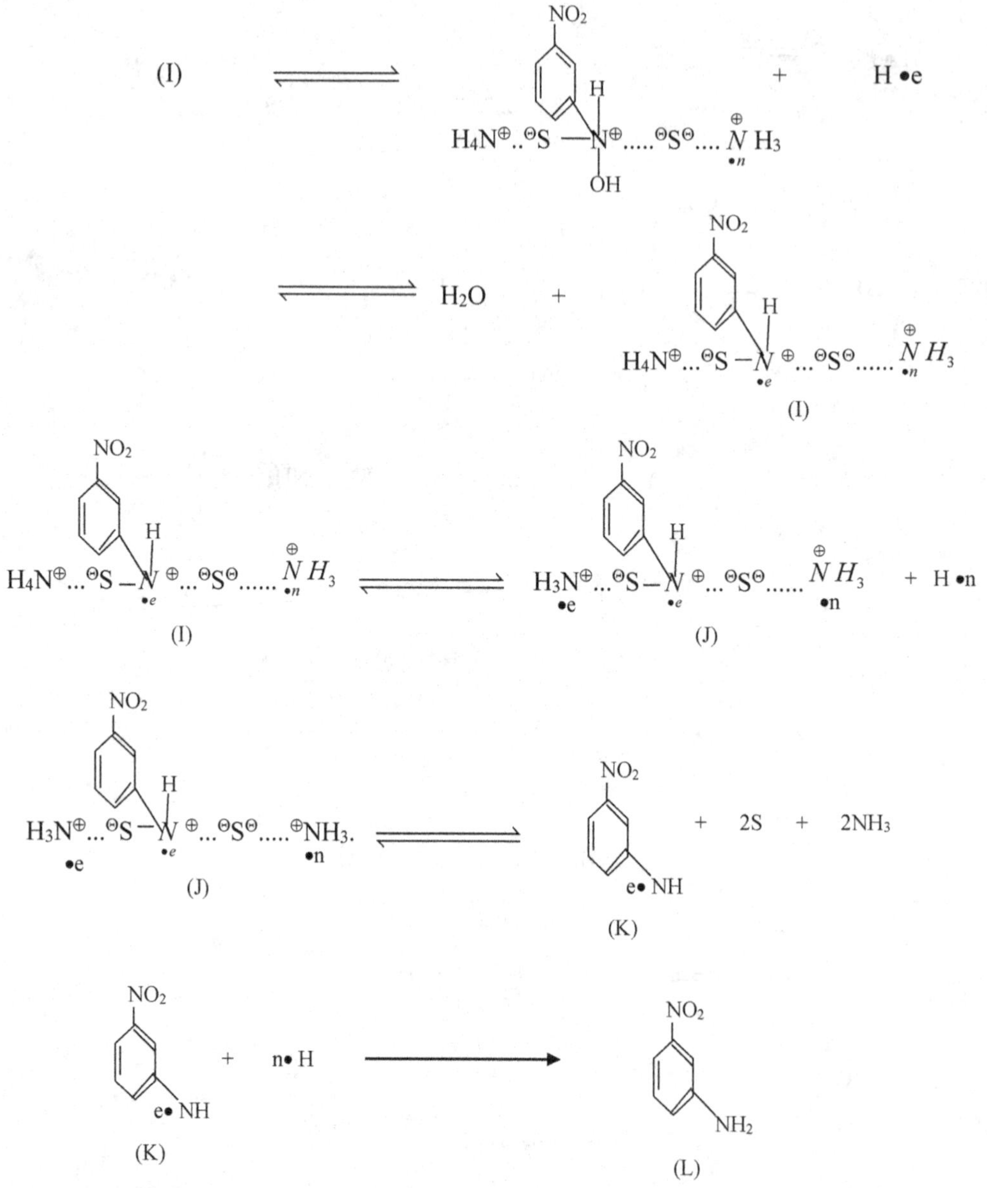

Overall Equation: Same as Equation (a)   (e)
(Laws of Creations for Nitro-benzenes)

Rule 940: This rule of Chemistry for **Reduction of nitro groups (NO_2) to amine group (NH_2),** states that, when the nitro- group meta-placed to another nitro- group has been reduced, attempts to reduce the second nitro- group is impossible, because once the NH_2 is put in place on the ring, it takes control over the actions of the ring, being far more radical-pushing than the nitro group, in order words, if the ring is kept in the Energized state of existence, the positions available for use will then be the ortho- and para- positions with respect to NH_2.
(Laws of creations for Nitro-benzenes)

Rule 941: This rule of Chemistry for **Benzene carrying radical-pushing groups such as OR,** states that, molecular rearrangement can only take place after partial hydrogenation of the ring using Na in liquid ammonia/Ethanol, followed by hydrolysis to give an alcohol which rearranges either via Electroradicalization or molecular rearrangements to give α,β- unsaturated ketone an Electrophile, which when higher operating conditions are used, opens the ring instantaneously to allow molecular rearrangement of the third kind to take place to give a 1,4-Ketene also an Electrophile.

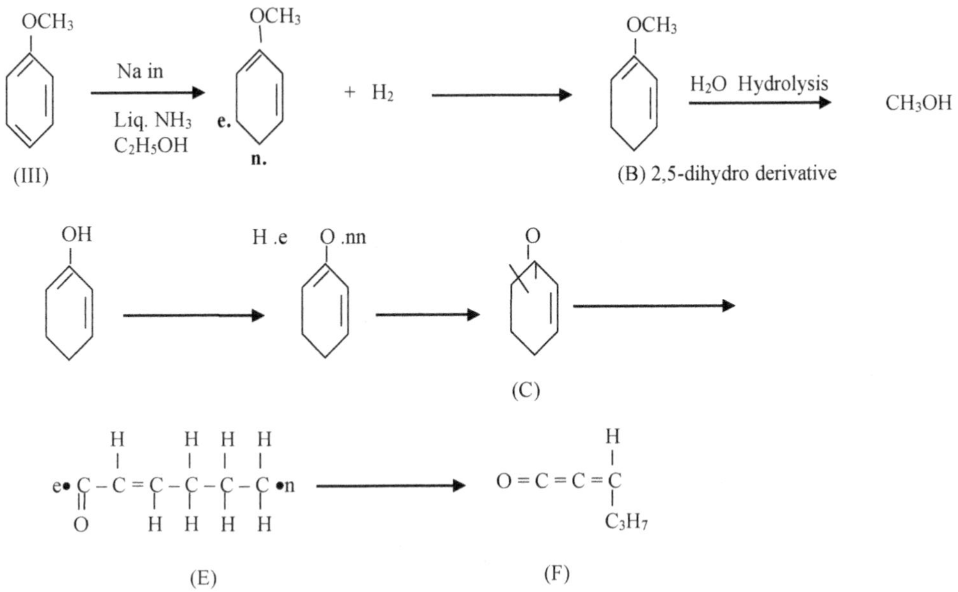

(Laws of Creations for Benzenes)

Rule 942: This rule of Chemistry for **Ring expansion of Benzene,** states that, when diazomethane is used, only a seven-membered cycloheptatriene can be obtained as shown below-

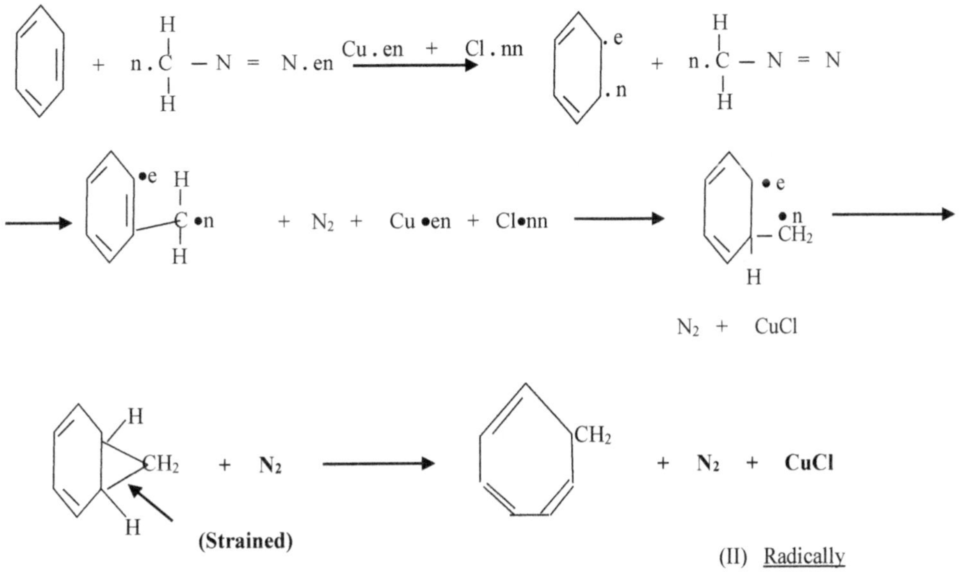

for which the ring was not opened, but expanded via activation, noting that this cannot place chargedly. *(Laws of Creations for Benzene)* [See Rule 840]

Rule 943: This rule of Chemistry for **Benzene,** states that, when this compound is used as monomer for polymerization under vibratory milling conditions, the followings are obtained-

Stage 1:

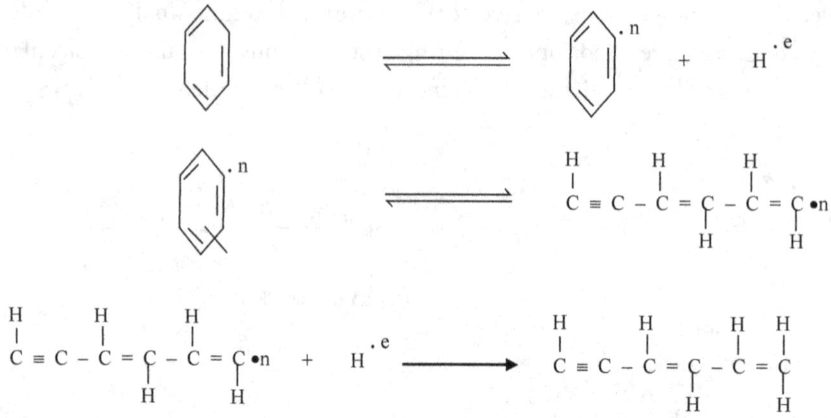

$$C \equiv C - \underset{\underset{H}{|}}{C} = \underset{\underset{H}{|}}{C} - \underset{\underset{H}{|}}{C} = \underset{H}{C} \bullet n \quad + \quad H^{\bullet e} \longrightarrow \quad C \equiv C - \underset{\underset{H}{|}}{C} = \underset{\underset{H}{|}}{C} - \underset{\underset{H}{|}}{C} = \underset{\underset{H}{|}}{C}$$

(A) Butadienyl acetylene   (a)

Overall equation: 20 Benzene $\longrightarrow$ 20 Butadienyl acetylene   (b)

Stage 2:

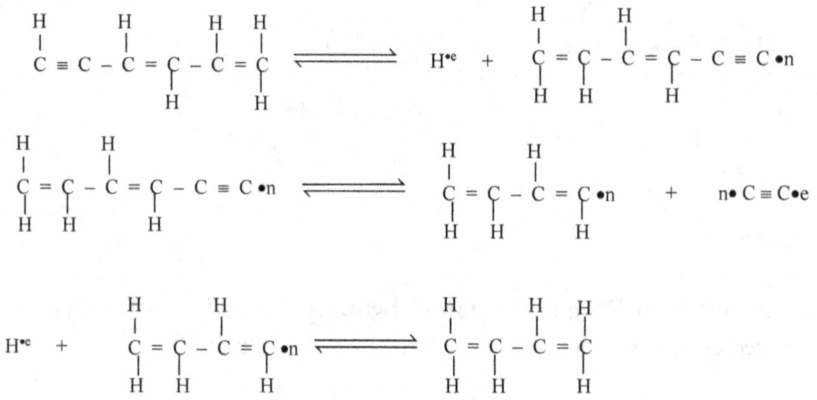

(B) Butadiene

$$n\bullet C \equiv C\bullet e \quad \longrightarrow \quad 2C \text{ (Carbon black)}$$ (c)

Overall equation: 10 Butadienyl acetylene $\longrightarrow$ 20C + 10 Butadiene   (d)

Stage 3:

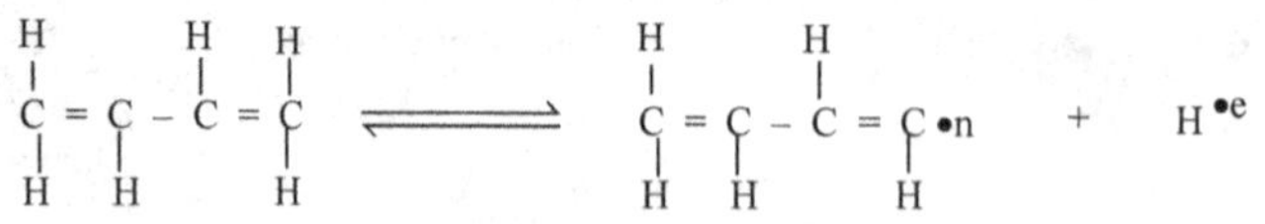

(C)

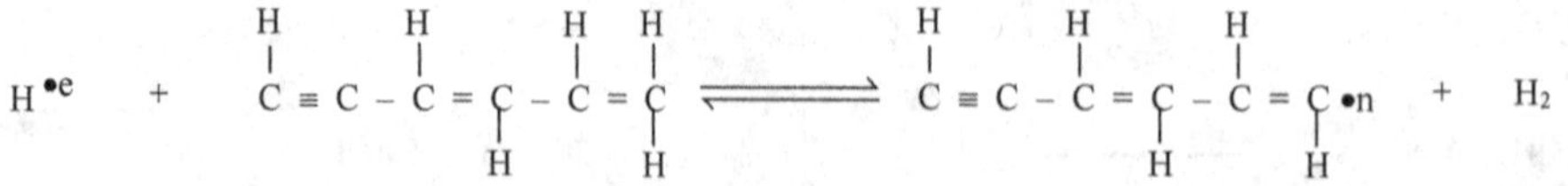

(D)

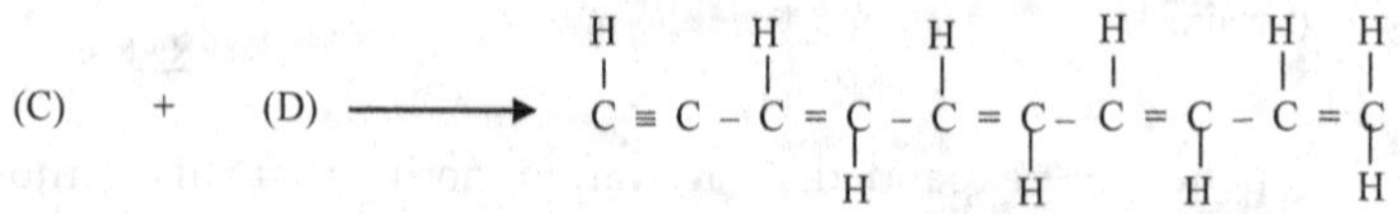

(C) + (D) $\longrightarrow$

(E) Octatetraenyl acetylene   (e)

510

Overall equation: 10 Butadiene + 10 Butadiene acetylene ——→ 10 Octatetraenyl acetylene

+   10 H$_2$ (f)

Stage 4:

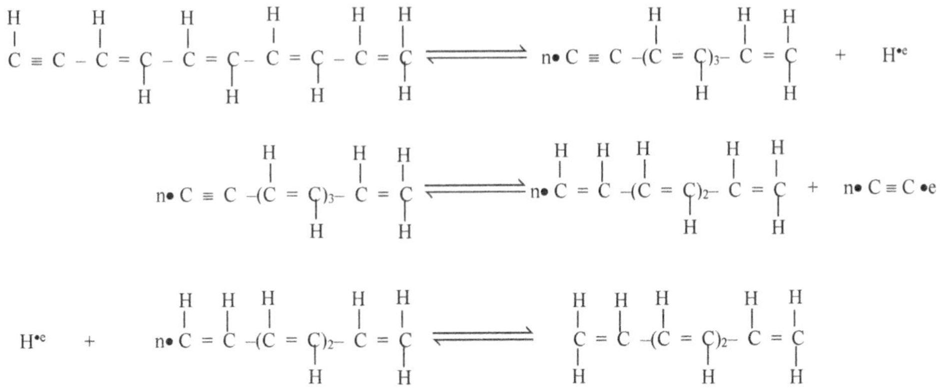

$$H-C \equiv C - C = C - C = C - C = C - C = C \rightleftharpoons n\bullet C \equiv C -(C = C)_3 - C = C \quad + \quad H^{\bullet e}$$

$$n\bullet C \equiv C -(C = C)_3 - C = C \rightleftharpoons n\bullet C = C -(C = C)_2 - C = C \quad + \quad n\bullet C \equiv C \bullet e$$

$$H^{\bullet e} \quad + \quad n\bullet C = C -(C = C)_2 - C = C \rightleftharpoons C = C -(C = C)_2 - C = C$$

(F)

$$n\bullet C \equiv C \bullet e \quad \longrightarrow \quad 2C \text{ (Carbon black)}$$ (g)

Overall equation: 5 Octatetraenyl acetylene ——→10 C   +   5 Octatetraene   (h)

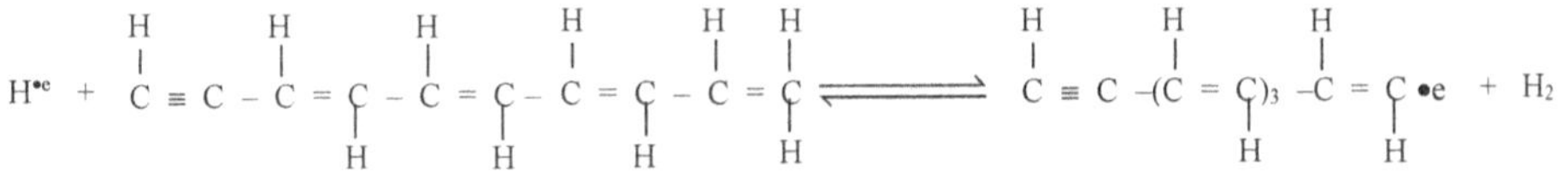

Stage 5:

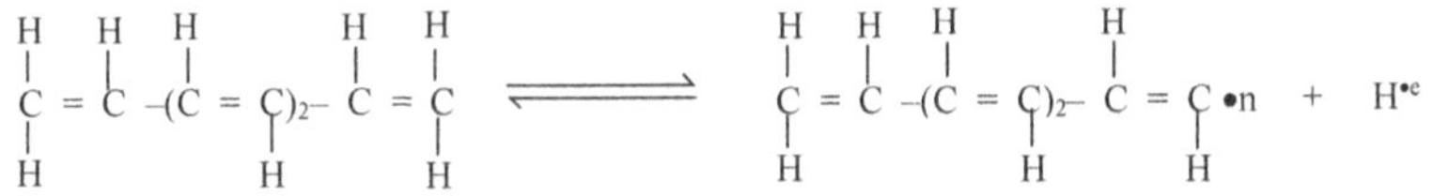

$$C = C -(C = C)_2 - C = C \rightleftharpoons C = C -(C = C)_2 - C = C \bullet n \quad + \quad H^{\bullet e}$$

(G)

$$H^{\bullet e} + C \equiv C - C = C - C = C - C = C - C = C \rightleftharpoons C \equiv C -(C = C)_3 -C = C \bullet e \quad + \quad H_2$$

(H)

$$(G) \quad + \quad (H) \quad \longrightarrow \quad C \equiv C -(C = C)_7 - C = C$$

(I)   (i)

Overall equation: 5 Octatetraene + 5 Octatetraenyl acetylene ——→5H$_2$ + 5Polyacetylene   (j)

Overall overall equation: 20 Benzene ——→30 C   +   15H$_2$ +   5Polyacetylene   (k)

for which, one can see the number of stages involved in producing the Living polyacetylene from 20 molecules of benzene, noting that *this is a Step polymerization system wherein all the stages take place via Equilibrium mechanism (Mechano-Chemical system).*
(Laws of Creations for Benzene as a monomer)

Rule 944: This rule of Chemistry for **Benzene,** states that, when used as a monomer for polymerization in the presence of ionic metals such as K, the followings are to be expected-

Stage 1:

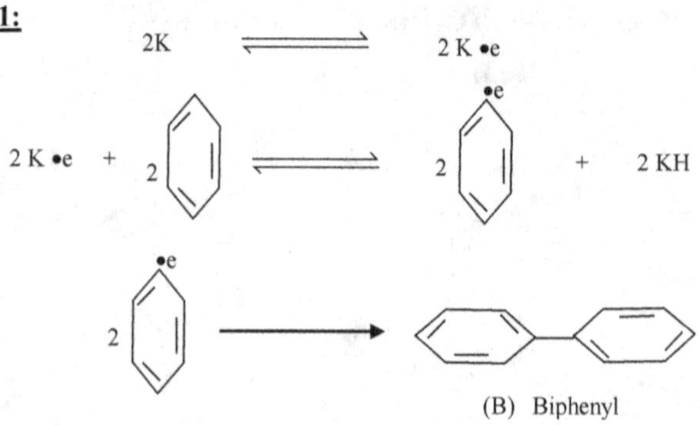

(B) Biphenyl

(a)

Overall equation: 2 K + 2 Benzene ⟶ 2KH + Biphenyl

(b)

for which after the first stage, with no K in the system, the followings take place-

Stage 2:

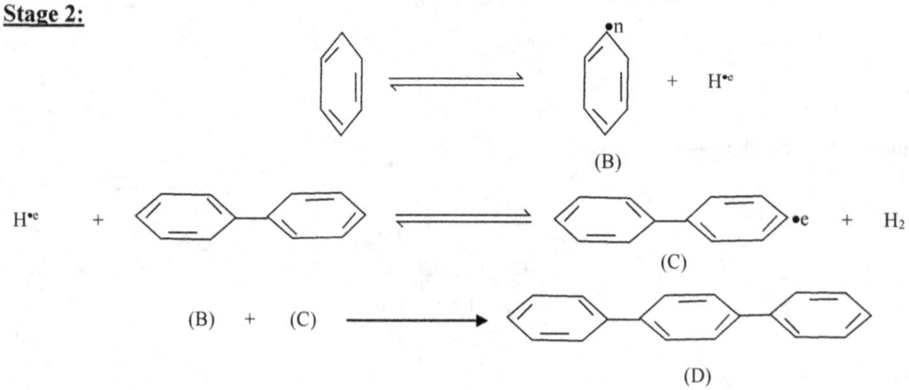

(B)

(C)

(B) + (C) ⟶

(D)

(c)

Overall equation: Benzene + Biphenyl ⟶ Terphenyl + H_2

(d)

Stage 3:

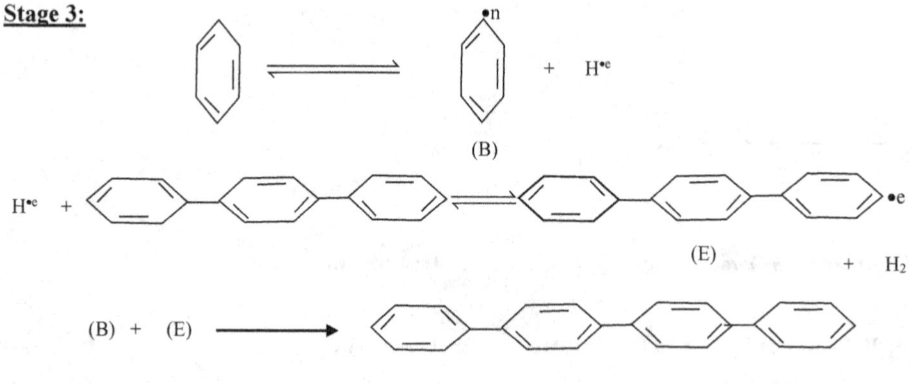

(B)

(E)   + H_2

(B) + (E) ⟶

(e)

Overall equation: Benzene + Terphenyl ⟶ Quaterphenyl + H_2

(f)

Overall overall equation: 4 Benzene + K ⟶ KH + $2H_2$ + Quaterphenyl

(g)

and with presence of more benzene in the polymerization system, a long chain of p-polyphenyl is obtained, noting that the H abstracted is not that ortho-placed, but that para-placed *and that* this is *a STEP polymerization system.*

(Laws of Creations for Benzene as a monomer)

512

Rule 945: This rule of Chemistry for **Benzene,** states that, when used as a monomer for polymer-ization in the presence of *$FeCl_3$,* the followings are to be expected when the molar ratio of the $FeCl_3$ to Benzene is 2 to 1-

Stage 1:

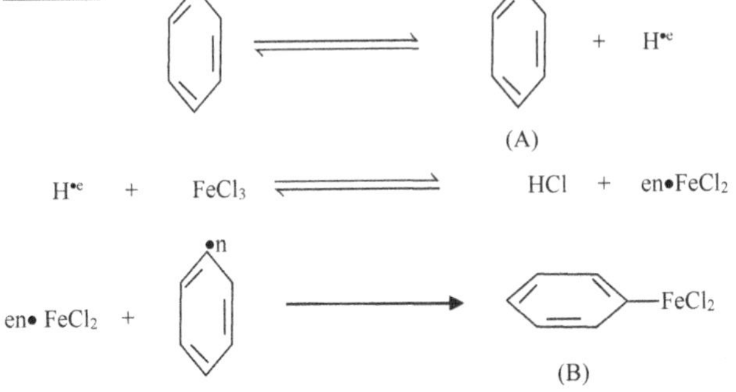

(A)

$H^{\bullet e}$ + $FeCl_3$ ⇌ HCl + en•$FeCl_2$

en• $FeCl_2$ + ⟶ —$FeCl_2$

(B)

(a)
(b)

Overall equation: Benzene + $FeCl_3$ ⟶ HCl + (B)

Stage 2:

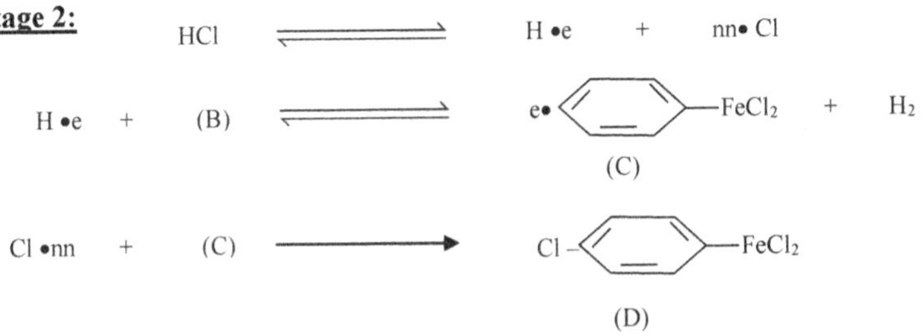

HCl ⇌ $H \bullet e$ + nn• Cl

$H \bullet e$ + (B) ⇌ e• —$FeCl_2$ + H_2

(C)

$Cl \bullet nn$ + (C) ⟶ Cl— —$FeCl_2$

(D)

(c)

Overall equation: HCl + (B) ⟶ H_2 + (D)

(d)

Overall overall equation: 100 Benzene + 200 $FeCl_3$ ⟶ **100H_2 + 100(D) + 100 $FeCl_3$**

(e)

and with no benzene any more in the system, the polymerization route proceeds differently as shown below-

Stage 3: (D) ⇌ Cl— •n + en• $FeCl_2$

(E)

$Cl_2Fe \bullet en$ + (D) ⇌ e• —$FeCl_2$ + $FeCl_3$

(F)

(E) + (F) ⟶ Cl— — —$FeCl_2$

(F)

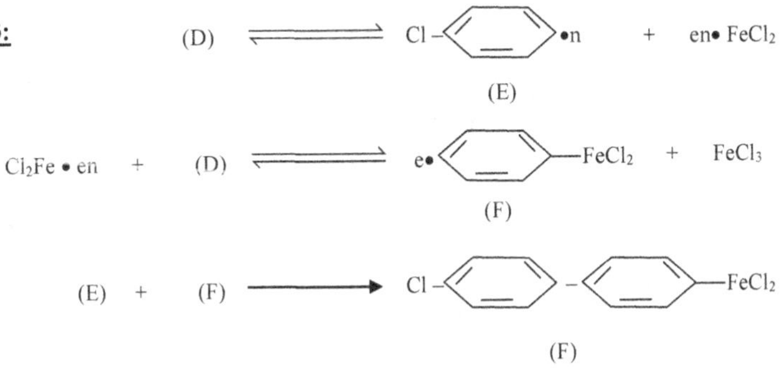

(f)

Stage 4: (D) ⇌ Cl— •n + en• $FeCl_2$

(E)

$Cl_2Fe \bullet en$ + (F) ⇌ e• — —$FeCl_2$ + $FeCl_3$

(G)

(E) + (G) ⟶ Cl— — — —$FeCl_2$

(H)

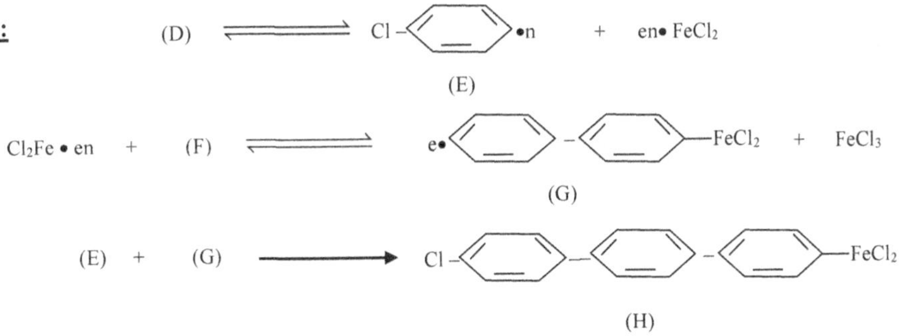

(g)

addition continuing in a stepwise fashion until all (D)s are consumed (STEP POLYMERIZATION), all taking 102 stages to give the following polymeric chain and overall equation.

$$Cl-\underset{}{\bigcirc}\left(\bigcirc\right)_{98}\bigcirc-FeCl_2$$

(I)

Overall overall equation: 100 Benzene + 200 FeCl$_3$ $\longrightarrow$ 100 H$_2$ + (I) + 199 FeCl$_3$

(h)

and since FeCl$_3$ is a hydrogenation catalyst, H$_2$ formed can readily be kept in Equilibrium state of existence, for which the reactions continue as follows-

Stage 103:

$$2H_2 \rightleftharpoons 2\,H\bullet e + 2\,n\bullet H$$
$$2H\bullet e + 2FeCl_3 \rightleftharpoons 2HCl + 2\,en\bullet FeCl_2$$
$$2en\bullet FeCl_2 \rightleftharpoons 2\,FeCl_2$$
$$2n\bullet H \longrightarrow H_2$$

(i)

Overall equation: 100 H$_2$ + 100 FeCl$_3$ $\longrightarrow$ 100 HCl + 100 FeCl$_2$ + 50H$_2$ (j)

Stage 104: This is the same as Stage 103 above to give the overall equation below.

Overall equation: 50 H$_2$ + 50 FeCl$_3$ $\longrightarrow$ 50 HCl + 50 FeCl$_2$ + 25 H$_2$ (k)

Stage 105: This is also the same as Stage 103 above to give the overall equation below.

Overall equation: 24 H$_2$ + 24 FeCl$_3$ $\longrightarrow$ 24 HCl + 24 FeCl$_2$ + 12H$_2$ (l)

Stage 106: This is also the same as Stage 103 above to give the overall equation below.

Overall equation: 12H$_2$ + 12 FeCl$_3$ $\longrightarrow$ 12 HCl + 12 FeCl$_2$ + 6 H$_2$ (m)

Stage 107: This again is the same as Stage 103 above with the following overall equation.

Overall equation: 6H$_2$ + 6 FeCl$_3$ $\longrightarrow$ 6 HCl + 6 FeCl$_2$ + 3H$_2$ (n)

Recalling that we left one H$_2$ molecule from Stage 105, we now have 4H$_2$ molecules left.

Stage 108: This is the same as the last stage.

Overall equation: 4H$_2$ + 4FeCl$_3$ $\longrightarrow$ 4HCl + 4FeCl$_2$ + 2H$_2$ (o)

Stage 109: This is the same as Stage 103.

Overall equation: 2H$_2$ + 2FeCl$_3$ $\longrightarrow$ 2HCl + 2FeCl$_2$ + H$_2$ (p)

Overall overall equation: 100 H$_2$ + 199 FeCl$_3$ $\longrightarrow$ 198 HCl + 198 FeCl$_2$ + H$_2$ + FeCl$_3$

(q)

Overall overall equation: 100 C$_6$H$_6$ + 200 FeCl$_3$ $\longrightarrow$ (I) + 198HCl + 198 FeCl$_2$ + H$_2$ + FeCl$_3$

(r)

Stage 110:

$$H_2 \;\rightleftharpoons\; H \bullet e \;+\; n \bullet H$$

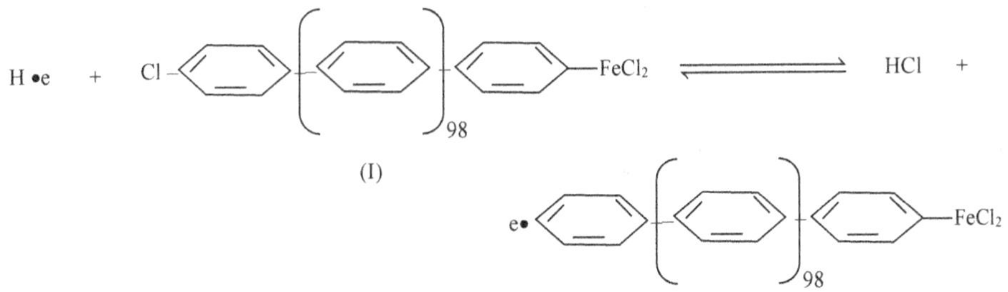

$$H \bullet e \;+\; Cl-\!\!\left[\text{ring}\right]\!\!\left(\text{ring}\right)_{98}\!\!\left[\text{ring}\right]-FeCl_2 \;\rightleftharpoons\; HCl \;+$$

$$e\bullet \left[\text{ring}\right]\!\!\left(\text{ring}\right)_{98}\!\!\left[\text{ring}\right]-FeCl_2$$

(J)

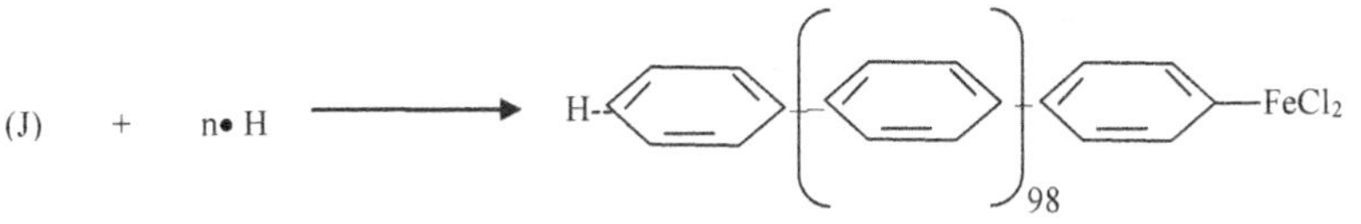

$$(J) \;+\; n \bullet H \;\longrightarrow\; H-\!\!\left[\text{ring}\right]\!\!\left(\text{ring}\right)_{98}\!\!\left[\text{ring}\right]-FeCl_2$$

(K)

(s)

Stage 111:

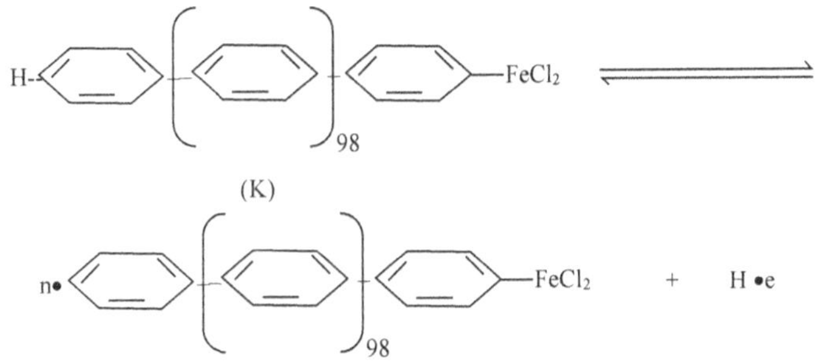

$$H-\!\!\left[\text{ring}\right]\!\!\left(\text{ring}\right)_{98}\!\!\left[\text{ring}\right]-FeCl_2 \;\rightleftharpoons\;$$

(K)

$$n\bullet \left[\text{ring}\right]\!\!\left(\text{ring}\right)_{98}\!\!\left[\text{ring}\right]-FeCl_2 \;+\; H \bullet e$$

(L)

$$H \bullet e \;+\; FeCl_3 \;\rightleftharpoons\; HCl \;+\; en\bullet FeCl_2$$

$$Cl_2Fe \bullet en \;+\; (L) \;\longrightarrow\; Cl_2Fe-\!\!\left[\text{ring}\right]\!\!\left(\text{ring}\right)_{98}\!\!\left[\text{ring}\right]-FeCl_2$$

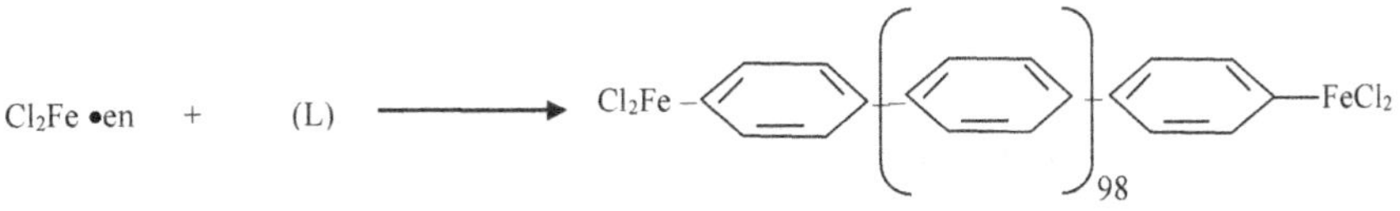

(M)

(t)

Overall equation: $H_2 \;+\; (I) \;+\; FeCl_3 \;\longrightarrow\; 2HCl \;+\; (M)$

Final overall equation: $100\,C_6H_6 \;+\; 200\,FeCl_3 \;\longrightarrow\; 200HCl \;+\; (M) \;+\; 198FeCl_2$

(u)

(Laws of Creations for Benzene as a monomer)

Rule 946: This rule of Chemistry for **Benzene,** states that, when used as a monomer for polymerization using *MoCl₅* at equi-molar ratio, polymerization proceeds in the same manner as with $FeCl_3$ until Stage 102, and from Stage 103 the use of $FeCl_3$ begins to differ from the use of $MoCl_5$ as shown below.

515

FeCl$_3$ until Stage 102, and from Stage 103 the use of FeCl$_3$ begins to differ from the use of MoCl$_5$ as shown below.

Overall equation: 100 Benzene + 100MoCl$_5$ ⟶ 100 H$_2$ + (I) + 99MoCl$_5$

(a)

where (I) is as shown below

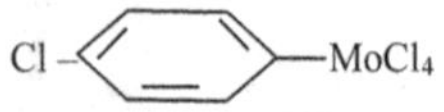

(D) – For MoCl$_5$

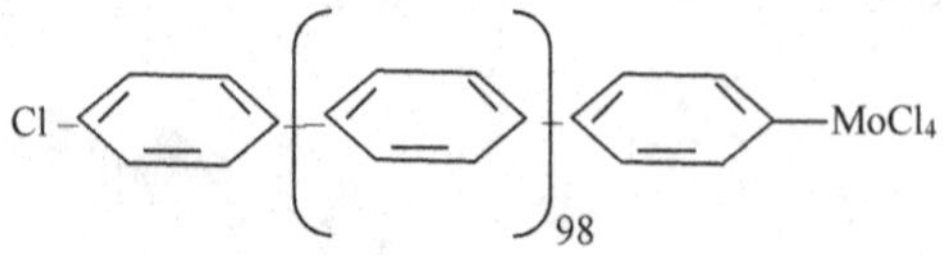

(I)

Stage 103:

$2H_2 \rightleftharpoons 2\,H \bullet e + 2\,n\bullet H$

$2H\bullet e + 2MoCl_5 \rightleftharpoons 2HCl + 2\,en\bullet MoCl_4$

$2en\bullet MoCl_4 \rightleftharpoons MoCl_3 + MoCl_5$

$2n\bullet H \longrightarrow H_2$

(b)

Overall equation: 100 Benzene + 100 MoCl$_5$ ⟶ 51H$_2$ + 50 MoCl$_5$ + 49MoCl$_3$ + . (I) + 98HCl

(c)

Stage 104: Same as above.

Overall equation: 100 Benzene + 100 MoCl$_5$ ⟶ 26 H$_2$ + 25 MoCl$_5$ + 74 MoCl$_3$ + (I) + 148 HCl

(d)

Stage 105: Same as above.

Overall equation: 100 Benzene + 100 MoCl$_5$ ⟶ 14 H$_2$ + 13 MoCl$_5$ + 86 MoCl$_3$ + (I) + 172 HCl

(e)

Stage 106: Same as above.

Overall equation: 100 Benzene + 100 MoCl$_5$ ⟶ 8H$_2$ + 7MoCl$_5$ + 92 MoCl$_3$ + (I) + 184 HCl

(f)

Stage 107: Same as above.

Overall equation: 100 Benzene + 100 MoCl$_5$ ⟶ 5H$_2$ + 4MoCl$_5$ + 95 MoCl$_3$ + .(I) + 190 HCl

(g)

Stage 108: Same as above.

Overall equation: 100 Benzene + 100 MoCl$_5$ ⟶ 3 H$_2$ + 2MoCl$_5$ + 97 MoCl$_3$ + . (I) + 194 HCl

(h)

Stage 109: Same as above.

Overall equation: 100 Benzene + 100 MoCl$_5$ ⟶ 2H$_2$ + MoCl$_5$ + 98 MoCl$_3$ + (I) + 196HCl

(i)

and like the case of $FeCl_3$, the followings are obtained

Stage 110:

$$H_2 \rightleftharpoons H\bullet e + n\bullet H$$

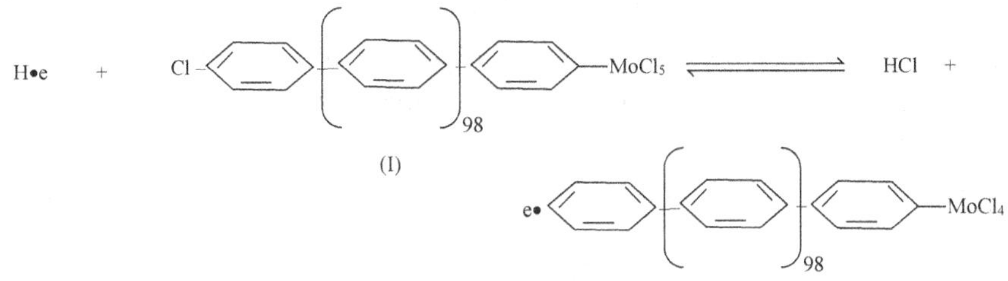

(I)

(J)

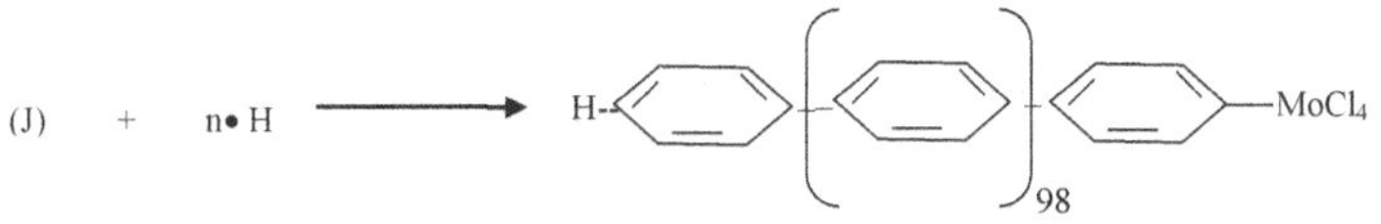

$$(J) + n\bullet H \longrightarrow \quad (K)$$

(K)

(j)

Stage 111:

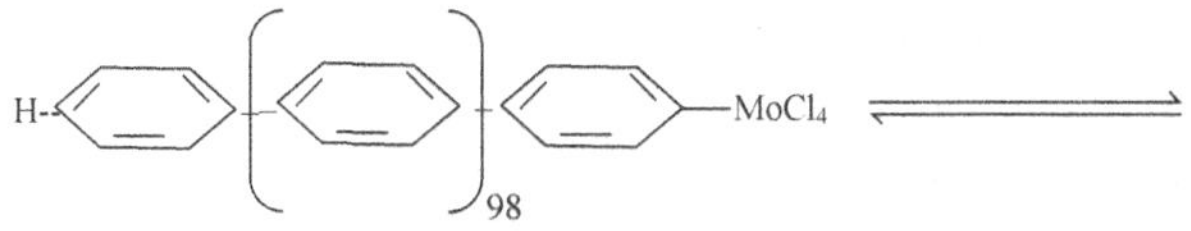

(K)

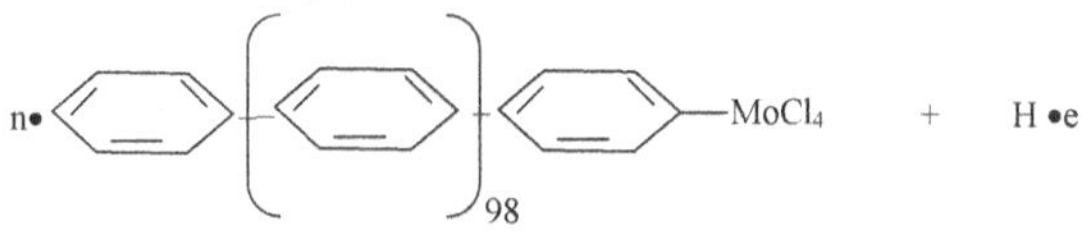

(L)

$$H\bullet e + MoCl_5 \rightleftharpoons HCl + en\bullet MoCl_4$$

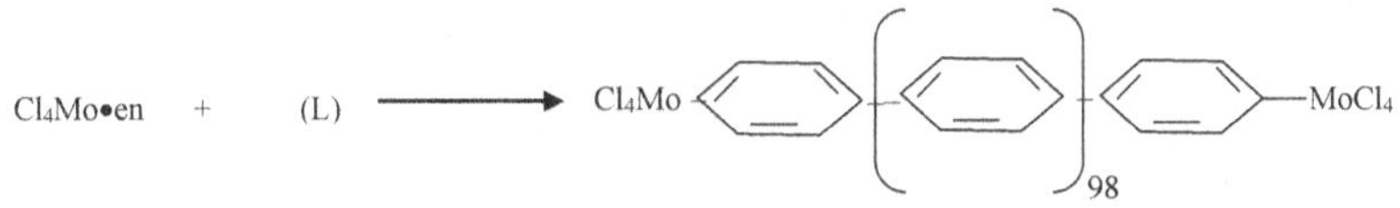

$$Cl_4Mo\bullet en + (L) \longrightarrow \quad (M)$$

(M)

(k)

Overall equation: $H_2 + MoCl_5 + (I) \longrightarrow 2HCl + (M)$ (l)

Overall overall equation: $100\ C_6H_6 + 100\ MoCl_5 \longrightarrow H_2 + 198\ HCl + 98\ MoCl_3$

$+ (M)$ (m)

(Laws of Creations for Benzene as a monomer)

Rule 947: This rule of Chemistry for **Benzene,** states that, when used as a monomer for polymerization using *AlCl$_3$,* the followings are to be expected- just as with the use of $FeCl_3$ and $MoCl_5$, after the production of the same type of (D) for Fe in the first two Stages, Step polymerization commences in Stage 3 in the same fashion until all (D)s are consumed in 102 stages to give the following polymeric chain and overall equation.

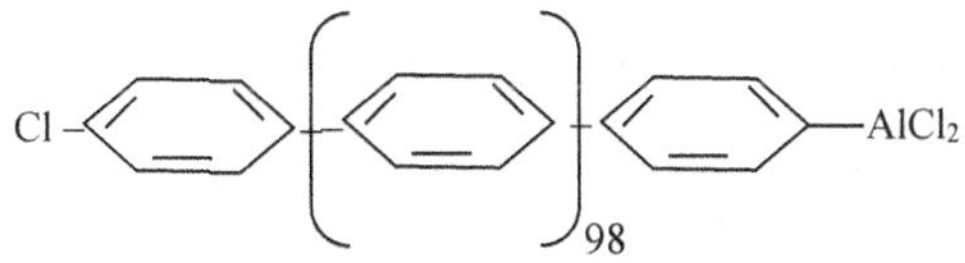

(I)

Overall overall equation: $100\ Benzene + 100\ AlCl_3 \longrightarrow 100\ H_2 + (I) + 99\ AlCl_3$

517

noting that the chain cannot react with H_2 or $AlCl_3$ or both and be productive and in the absence of a passive hydrogen catalyst, nothing can be done at the operating conditions.
(Laws of Creations for Benzene as a monomer)

Rule 948: This rule of Chemistry for **Benzene,** states that, when used as a monomer for polymerization using coordination initiators, no polymerization is favored, until the ring is made to carry one radical-pushing group of greater capacity than H as shown below-

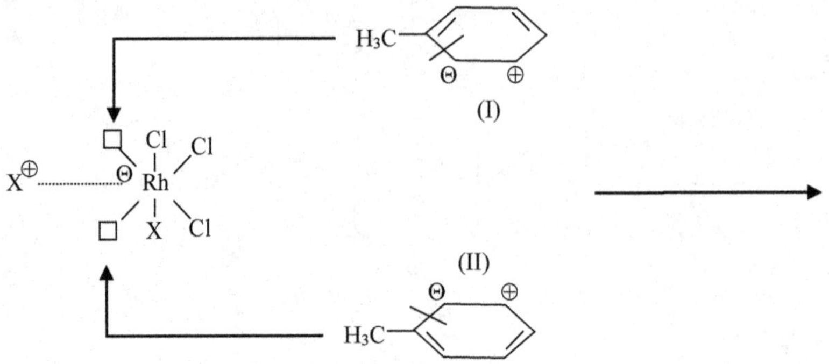

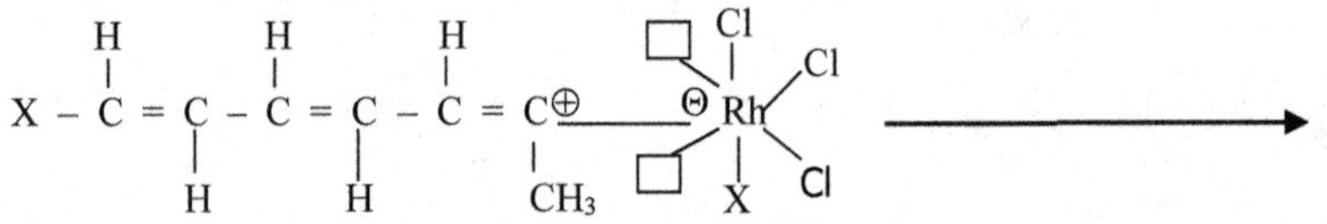

Syndiotactic poly(Methyl-tri-acetylene)

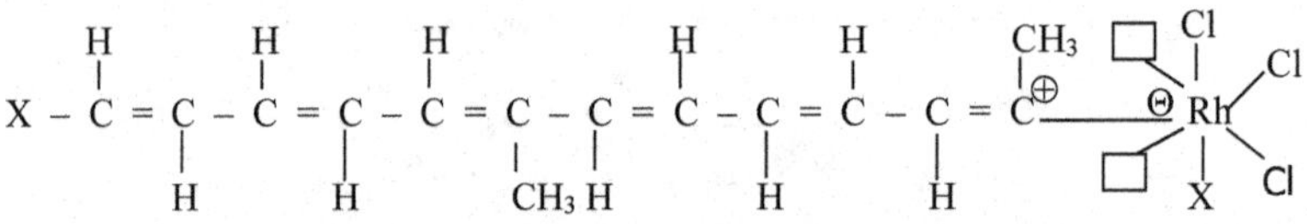

noting that, it is the center carrying the radical-pushing group that made it possible to open the ring instantaneously in this manner and the larger the radical-pushing capacity of the group placed on the ring, the easier it is for the ring to be opened and this is Free-radical or Positively charged polymerization by Combination mechanism; also noting that the reaction is also favored by Cyclooctatetraene, ***noting that though benzene cannot readily be used chargedly, unlike acetylene, it can be activated chargedly;*** for if otherwise, the charged initiator must be radical in character.
(Laws of creations for Benzene as a monomer)

Rule 949: This rule of Chemistry for **Scission between two conjugatedly placed double bonds in a ring,** states that, scission can only take place, either when the monomer or compound is in Equilibrium state of existence or in Activated state of existence either in the presence of an initiator and a radical-pushing group on the ring or in its absence with no radical-pushing group.
(Laws of Creations for Point of Scission in a ring)

Rule 950: This rule of Chemistry for **Cycloheptatriene,** states that, like benzene, there is only one type of cycloheptatriene and unlike benzene, it is discretely closed-loop resonance stabilized, with no point of scission in the ring, and when activated, the followings are obtained-

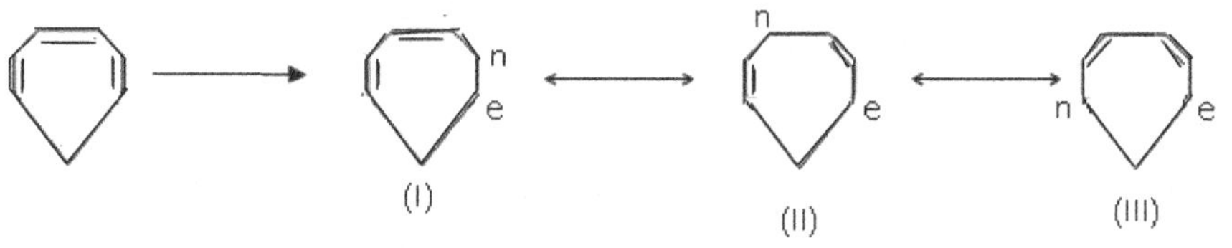

(I)   (II)   (III)

in which (III) is the stable mono-form, which however cannot readily be used for Addition polymerization, due to steric limitations.
(Laws of Creations for Cycloheptatriene)

Rule 951: This rule of Chemistry for **Cycloheptatriene,** states that, when a radical-pushing group is internally placed anywhere on the ring, the followings are to be expected.

(I)

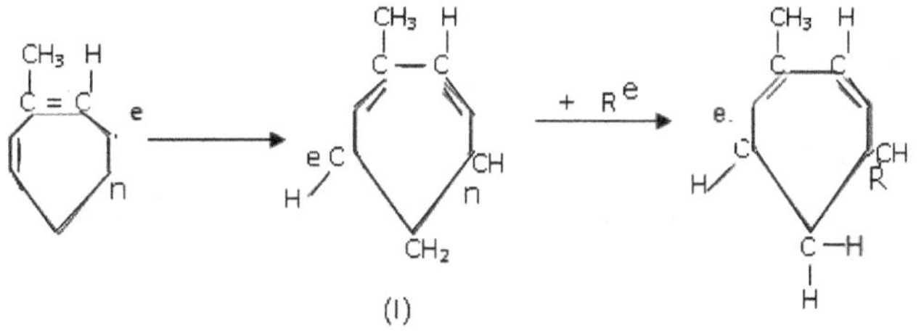

for which the CH_3 group is resonance stabilized and cannot be tempered with and as a strong Nucleophile, there is only transfer species inside the ring- the CH_2 group whose capacity is greater than any R carried externally located to the double bonds in the ring.
(Laws of Creations for Cycloheptatrienes)

Rule 952: This rule of Chemistry for **Cycloheptatriene,** states that, when a radical-pushing group is externally placed on the ring, the followings are to be expected-

center of activation

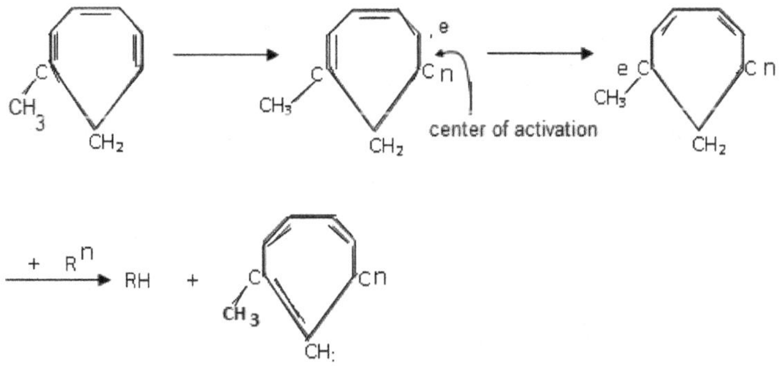

for which though the group is not resonance stabilized, it cannot still provide transfer species.
(Laws of Creations for Cycloheptatrienes)

Rule 953: This rule of Chemistry for **Cycloheptatriene,** states that, when made to carry radical-pushing group of lesser capacity than H, (e.g. CF_3). the following are to be expected-

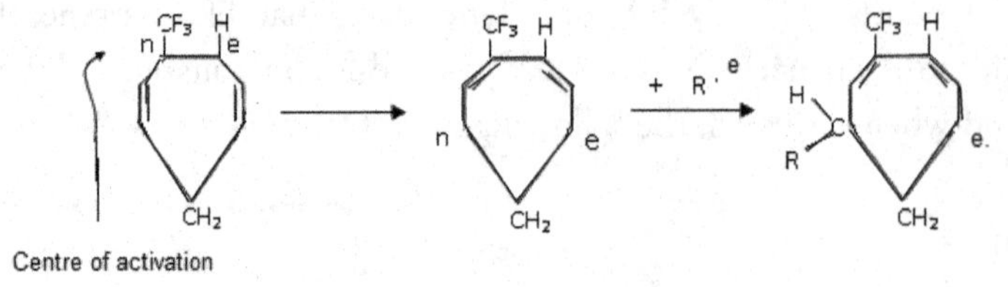

(a)

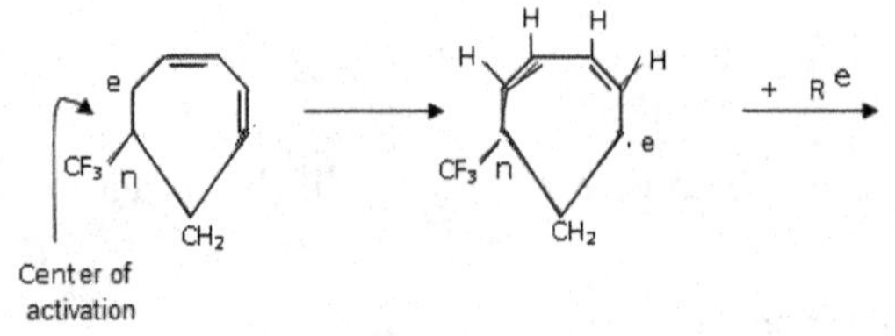

(b)

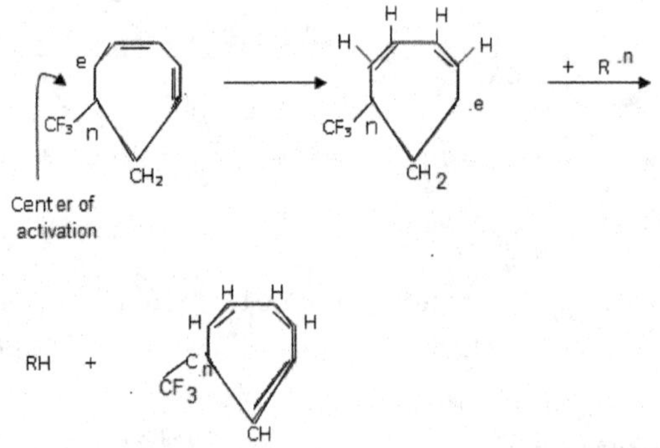

(c)

noting that, this monomer is neither a Nucleophile nor an Electrophile.
(Laws of Creations for Cycloheptatrienes)

Rule 954: This rule of Chemistry for **Trienes, Dienes and Enes,** states that, by their characters and type of resonance stabilized products obtained from them, there is no doubt that the following is valid-

Cyclotrienes   >   Cyclodienes   >   Cycloalkenes   >   etc.

Order of Nucleophilicity

and within the trienes themselves, the traditional order should also apply here, while the order of strain energy present in the rings is the reverse.

Cyclooctatriene   >   Cycloheptatriene   >   Cyclohexatriene   >   etc.

Order of nucleophilicity

Rule 955: This rule of Chemistry for **Cycloheptatriene,** states that, the possibility of cyclohep-tatriene cracking in the same way cyclopentadiene did to give Naphthalene, may be possible here to give what is shown below.

OR/AND

(I)   (II)

(Laws of Creations for Cycloheptatriene)

Rule 956: This rule of Chemistry for **Cyclooctatriene,** states that, there are two types of cyclooctatriene -

(i)   One in which the three double bonds are conjugatedly placed, (I)

(ii)   One in which two are conjugatedly placed and one is isolatedly placed, (II).

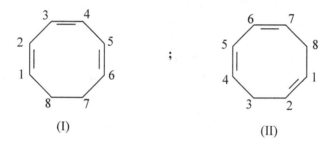

for which (I) has one point scission and (II) has none.

(Laws of Creations for Cyclooctatriene)

Rule 957: This rule of Chemistry for **Cyclooctatriene,** states that, with easy provision of minimum required strain energy for the ring, the following should be obtained when opened instantaneously for (I) using strong initiators or initiating force.

$$n\ \overset{\overset{H}{|}}{\underset{|}{C}} - \overset{\overset{H}{|}}{C} = C - \overset{\overset{H}{|}}{\underset{|}{C}} = C - \overset{\overset{H}{|}}{C} = C - \overset{\overset{H}{|}}{\underset{|}{C}}.e$$

(I)   1,3,5,7 - Octatetraene

noting that while cyclobutene favors the existence of 1,3 - butadiene and cyclohexadiene favors the existence of 1,3,5- hexatriene, cyclooctatriene favors the existence of 1,3,5,7- octatetraene or 1,3,5- substituted cyclohexadienes, all of them only radically and not chargedly.

(Laws of Creations for Cyclooctatriene)

Rule 958: This rule of Chemistry for **Conjugatedly placed Cyclooctatriene,** states that, when made to carry alkylane group externally located, the followings are to be expected-

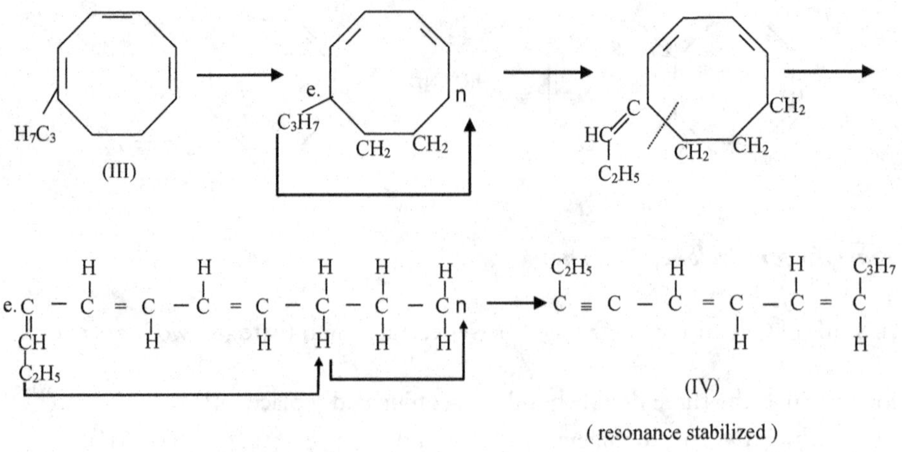

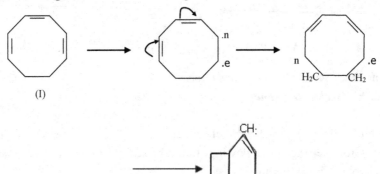

noting the size of the group chosen, with limitations at C_4H_9.
(Laws of Creations for Cyclooctatriene)

Rule 959: This rule of Chemistry for **Cyclooctatriene,** states that, when activated, this will favor trans-annular addition to give four-membered ring fused to a six-membered cyclohexadiene.

(Laws of Creations for Cyclooctatriene)

Rule 960: This rule of Chemistry for **Cycloocta-1,3,6-Triene,** states that, when activated, the isolatedly placed center is the first to be activated being the least nucleophilic, with no possibility of trans-annular addition to form a stable compound.
(Laws of Creations for Cycloocta-1,3,6-Triene)

Rule 961: This rule of Chemistry for **Cycloocta-1,3,6-Triene,** states that, when activated with a weak initiator, it can molecular rearrange to Cycloocta-1,3,5-Triene as shown below-

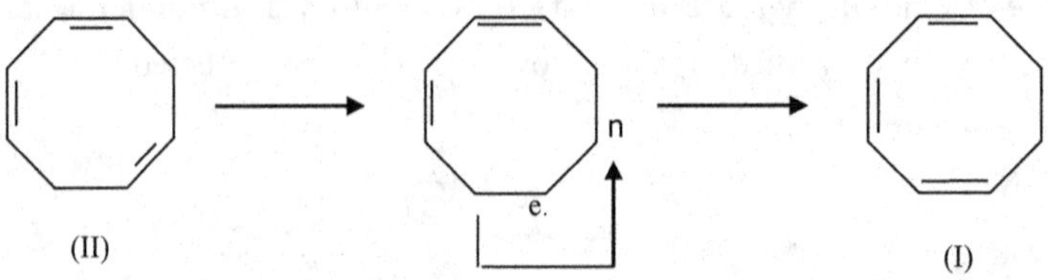

(Laws of Creations for Cycloocta-1,3,6-Triene)

Rule 962: This rule of Chemistry for **Cycloocta-1,3,6-Triene,** states that, when made to carry an alkylane group internally located in the resonance stabilized region, then will trans-annular addition take place as shown below if and only if the externally located double is the least nucleophilic center-

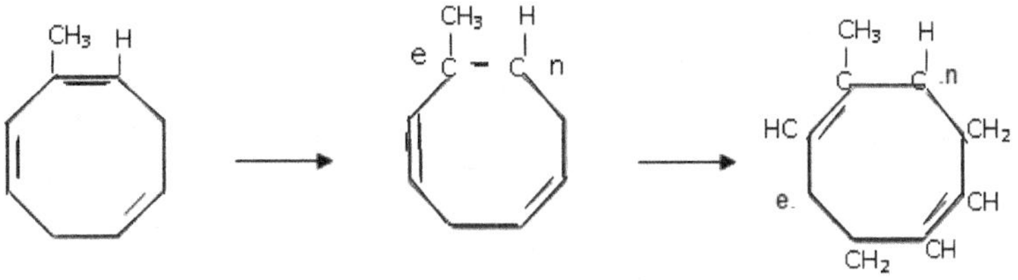

(I)   (I)a

NOT FAVORED

(I)

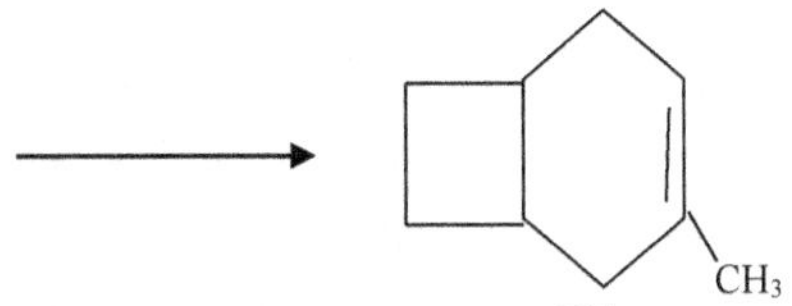

(I)b

for which in the first case, the wrong center has been activated, and in the second case, (I)b is obtained when the right center is activated.

(Laws of Creations for Cycloocta-1,3,6-Triene)

Rule 963: This rule of Chemistry for **Cycloocta-1,3,6-Triene,** states that, when an alkylane group is placed on the isolatedly placed double bond, the followings take place

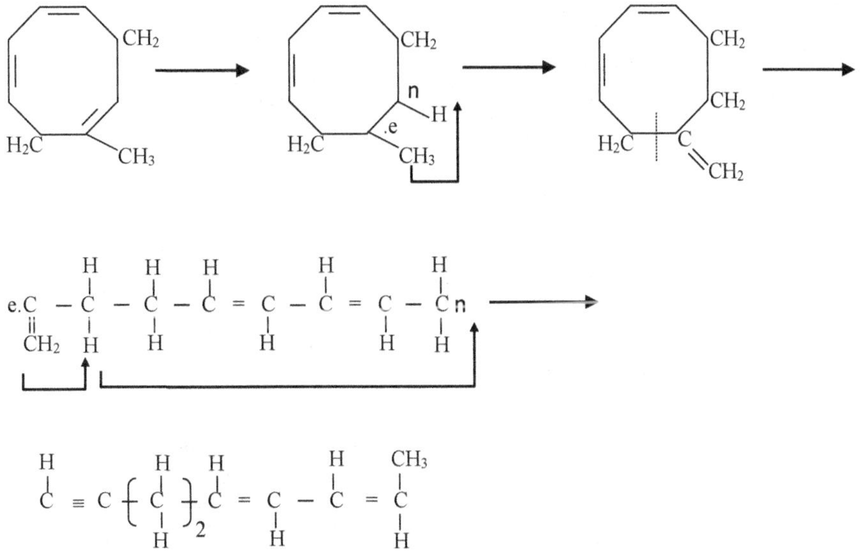

(III)b Resonance stabilized

FAVORED

for which if the center activated above is the least nucleophilic center, then the ring opening above is favored; taking important note of the point of scission in the ring.

(Laws of Creations for Cycloocta-1,3.6-Triene)

523

Rule 964: This rule of Chemistry for **Cyclooctatetraene**, states that, when compared with benzene by *artificial activation* of all the activation centers, which according to one of the rules is impossible in the presence of resonance stabilization, the followings are obtained-

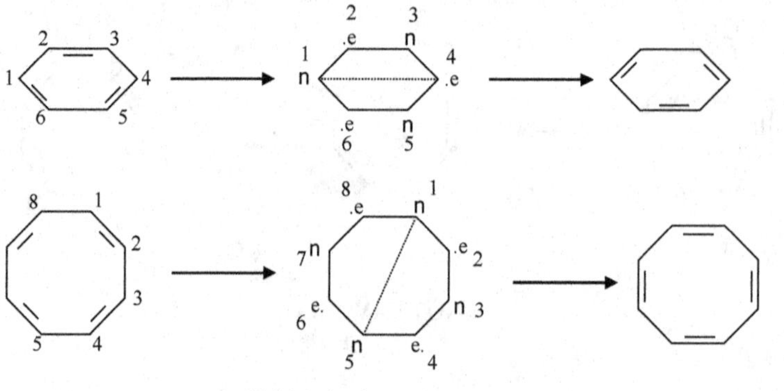

(a)

Artificial Activations

(b)

thus, while in benzene the linearly externally located active centers (para-placement) carry opposite radicals, in cyclooctatetraene the linearly externally located active centers (1,5 - centers) cannot carry opposite radicals, hence inside a fully unsaturated ring, movement of electro-free-radicals cannot go beyond two step; for which cyclo-octatetraene can be seen as weakly resonance stabilized.
(Laws of Creations for Resonance stabilization phenomena inside a ring)

Rule 965: This rule of Chemistry for **Cyclooctatetraene,** states that, when activated the followings readily take place-

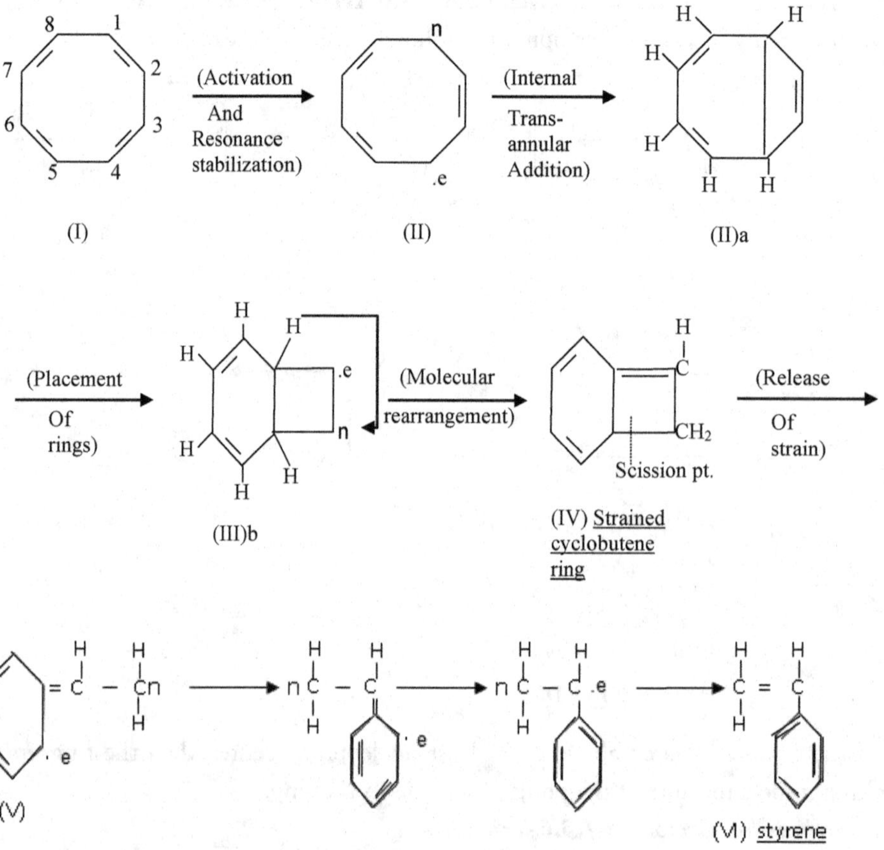

(I) (Activation And Resonance stabilization) (II) (Internal Trans-annular Addition) (II)a

(Placement Of rings) (III)b (Molecular rearrangement) (IV) Strained cyclobutene ring (Release Of strain)

Scission pt.

(V) (VI) styrene

for if (I) had favored full or continuous close-loop resonance stabilization, the existence of (II) would have been impossible; in order words one can confidently say that ***complete continuous closed-loop resonance stabilization is impossible regardless the strength of initiators,*** and therefore, one can clearly see a clear distinction between benzene (Continuous) and Cyclootatetraene (Discrete)- the "continuous" being the origin of aromaticity.
(Laws of Creations for Cyclooctatetraene)

Rule 966: This rule of Chemistry for **Cyclodecapentaene,** states that, when activated, the followings are obtained-

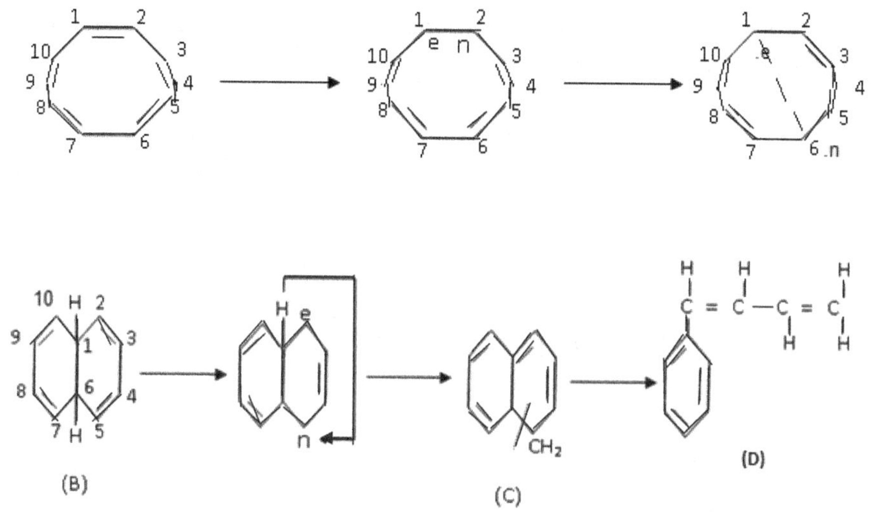

for which the favored existence of (D) depends on the strength of the activating initiator and the temperature, noting that not only is cyclooctatetraene unstable; so also is cyclodecapentaene and higher members, due to the limit placed by six-membered rings.
(Laws of Creations for Cyclodecapentaene)

Rule 967: This rule of Chemistry for **Cyclodecapentaene, Cyclooctatetrane, Cyclohexatriene,** states that, the followings are valid-

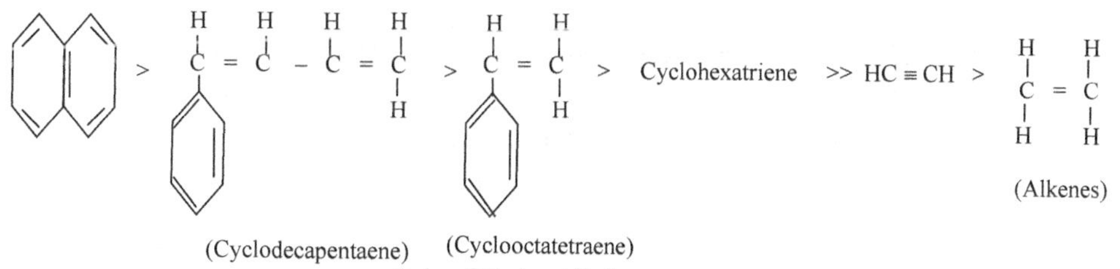

all of which can provide resonance stabilization when used as a group.
(Laws of Creations for Cyclodecapentaene, Cyclooctatetraene, Cyclohexatriene)

Rule 968: This rule of Chemistry for **Resonance stabilization groups,** states that, the following is valid for them-

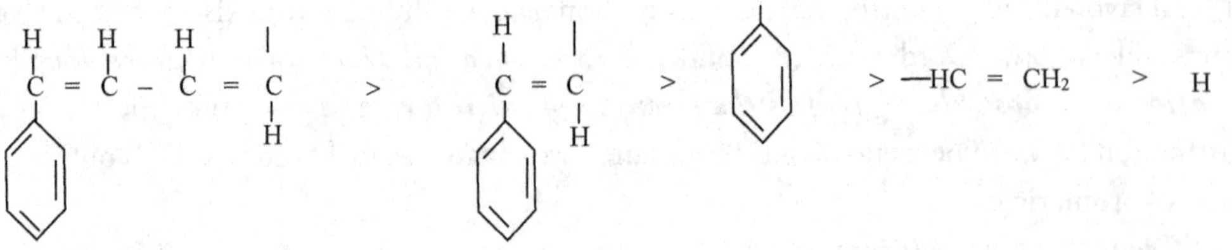

(Cyclodecapentaene)   (Cyclooctatetraene)

Order of Radical-pushng capacity of Resonance stabilization groups

(Laws of Creations for Resonance stabilization groups)

Rule 969: This rule of Chemistry for **Non-continuous closed loop resonance stabilization group,** states that, the followings are valid for them-.

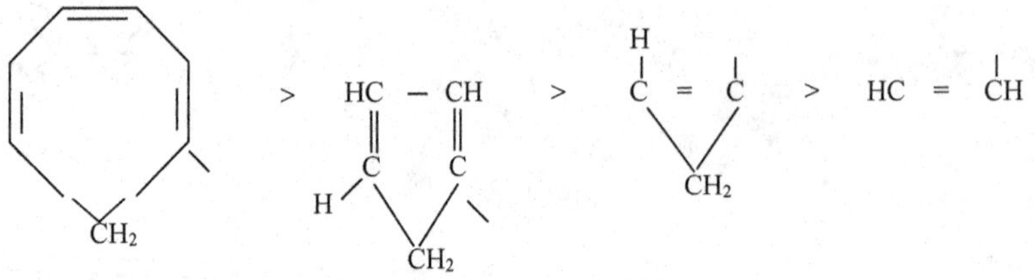

Order of Radical-pushing capacity of substituent groups

(a)

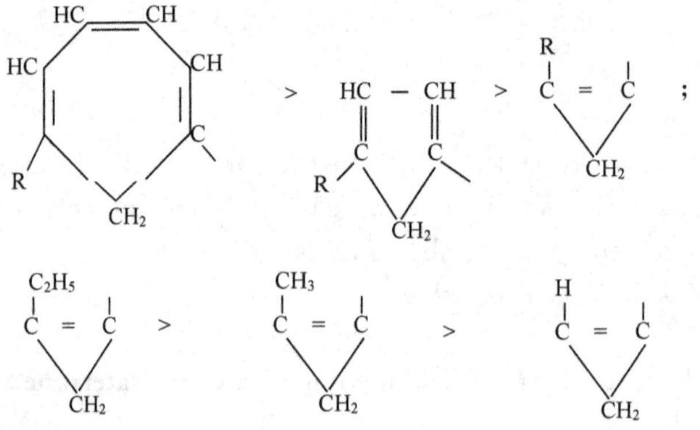

Order of radical-pushing capacity of substituent groups

(b)

(Laws of Creations for Discrete closed-loop resonance stabilization groups)

Rule 970: This rule of Chemistry for **Cyclooctatetraene,** states that, when made to carry radical Pushing group of lower capacity than H, the followings are obtained-

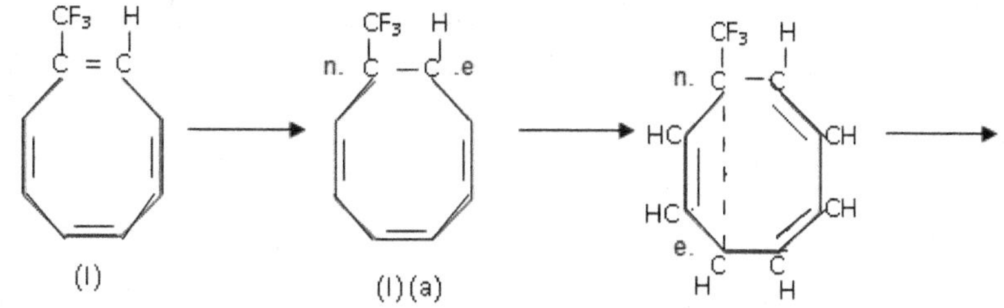

(I)   (I)(a)

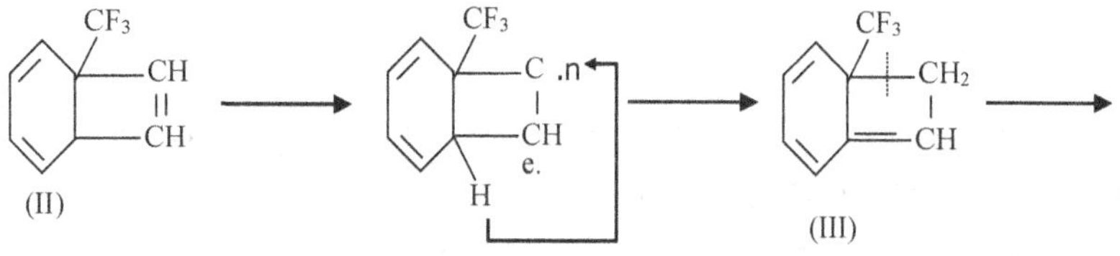

(II)   (III)

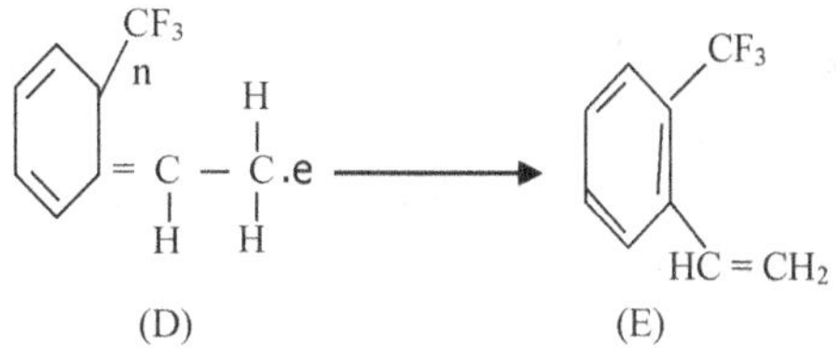

(D)   (E)

noting the type of styrene above.
(Laws of Creations for Cyclooctatetraenes)

Rule 971: This rule of Chemistry for **Cyclooctatetraene,** states that, when made to carry one radical-pushing group, the followings are obtained.

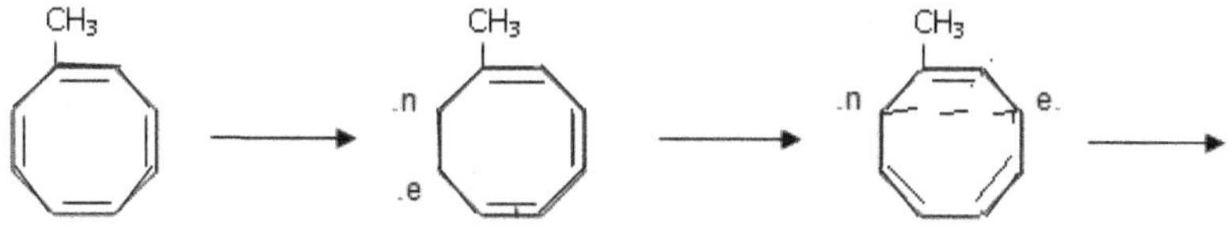

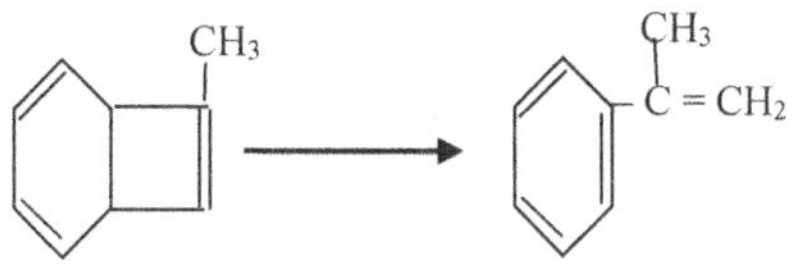

the product being 1-Methyl styrene.
(Laws of Creations for Cyclooctatetraene)

Rule 972: This rule of Chemistry for **Fully unsaturated rings,** states that, the following is the order of their abilities to favor Equilibrium state of existence-

[Cyclobuta-di-ene] > Cyclohexa-tri-ene > Cycloocta-tetra-ene > Cyclodeca-penta-ene > Cycloduodeca-hexa-ene >

ORDER OF ABILITY TO EXIST IN EQUILIBRIUM STATE OF EXISTENCE

for which it is only benzene that favors continuous closed-loop resonance stabilization in this family and <u>this is one of the origins if not the only one of its</u> <u>***Aromaticity***</u>.
(Laws of Creations for Fully Unsaturated rings)

Rule 973: This rule of Chemistry for **Cyclooctatetraene,** states that, though cyclooctatetraene is known not to be aromatic, its so-called "dianion" $C_8H_8^{-2}$, called cyclooctatetraenide is aromatic, and this can be explained as follows-

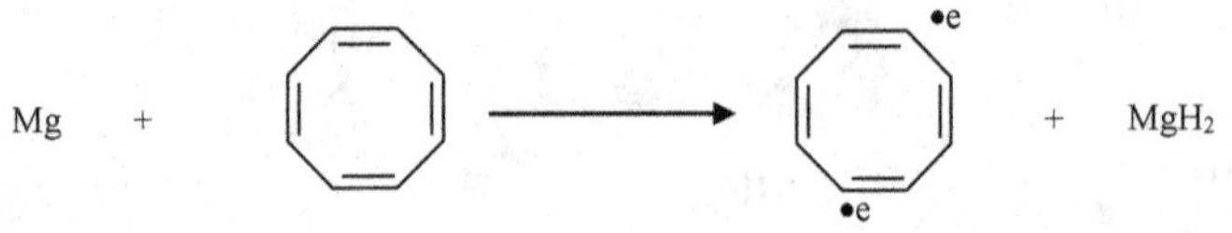

Cyclooctatetra-dienide

noting that, with two electro-free-radicals placed on the ring, resonance stabilization in the closed system becomes continuous and no longer discrete, since the two radicals do the movement at the same time to go round once and change their positions.; hence, it shows the aromatic character, noting also that the ring cannot carry negative charges or anions on the C centers, but only nucleo-free- or electro-free-radicals or positive charges.

(Laws of Creations for Cyclooctatetraene/Aromaticity)

Rule 974: This rule of Chemistry for **Aromatic Cyclobutadiene-iron tricarbonyl,** states that, shown below is the real structure of cyclobutadiene-iron tricarbonyl, which has been identified to be "aromatic" in character-

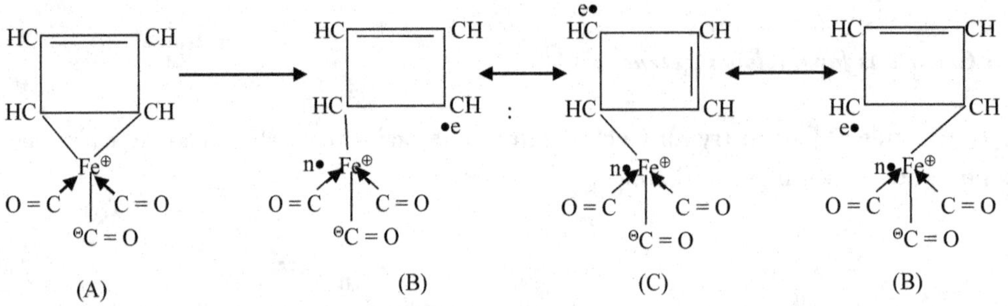

Cyclobutadiene-iron tricarbonyl

though there are three holes on the iron center, (A) is resonance stabilized when activated to give (B) in which the electro-free-radical on the C center can only move discretely to form (C); hence as it seems, (A) is not the real structure of the so-called aromatic cyclobutadiene-iron tricarbonyl, but that shown below which is indeed aromatic-

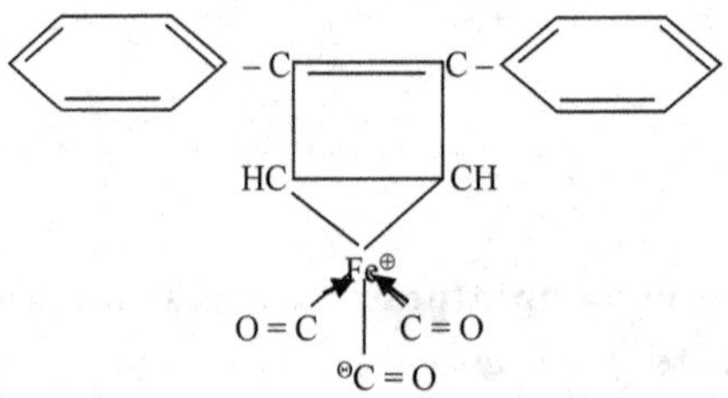

(D) Cyclo-Di-phenyl-butene-iron tricarbonyl

(Laws of Creation for Aromatic Cyclobutadiene –iron tricarbonyl)

Rule 975: This rule of Chemistry for **Iron carbonyls,** states that, shown below are the structures of two of them-

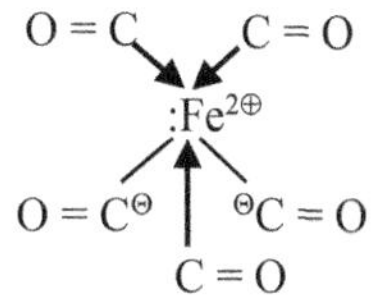

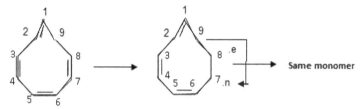

Iron pentacarbonyl [Fe(CO)₅] 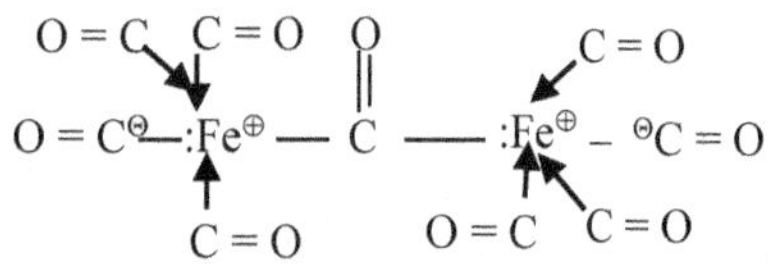   Di-iron nonacarbonyl [Fe₂(CO)₉]

wherein most of the eight radicals and the vacant orbitals on the last shell of Fe centers are fully used datively, covalently and polarly, while the COs have been used as radical-donating molecules for dative bonding, covalently and polarly.
(Laws of Creations for Iron carbonyls)

Rule 976: This rule of Chemistry for **Cyclononatetraene,** states that, this is a resonance stabilized compound/monomer, with no point of scission, and like Cyclopropene, Cyclopentatadiene, Cyclo-

heptatriene, the CH_2 group in the ring is still of greater capacity than any R group and it molecularly rearranges back to itself as shown above.
(Laws of Creations for Cyclononatetraene)

Rule 977: This rule of Chemistry for **Cyclononatetraene,** states that, it unlike Cyclooctatetraene, is stable when activated, since it cannot produce styrene derivatives as shown below, in the presence of a weak coordination initiator.

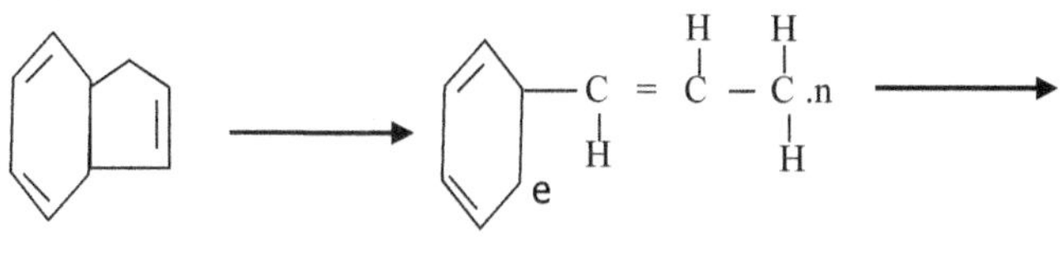

(Laws of Creations for Cyclononatetraene)

Rule 978: This rule of Chemistry for **Cyclotetraenes,** states that, for them, the following is the order of their nucleophilicity-

➢   Cyclodecatetraene   >   Cyclononatetraene   >   Cyclootatetraene

Order of Nucleophilicity

(Laws of Creations for Cyclotetraenes)

Rule 979: This rule of Chemistry for **Norbornene,** states that, when used as a monomer, the polymer shown below is the product, and how it was obtained is as shown below-

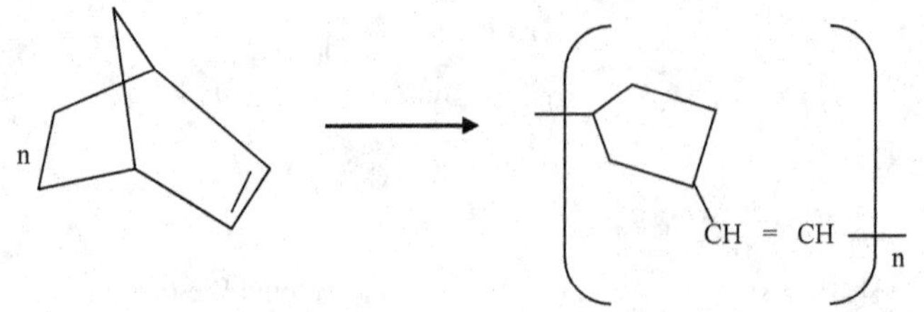

(a)

for which during Initiation, the norbornene first rearranges as shown below-

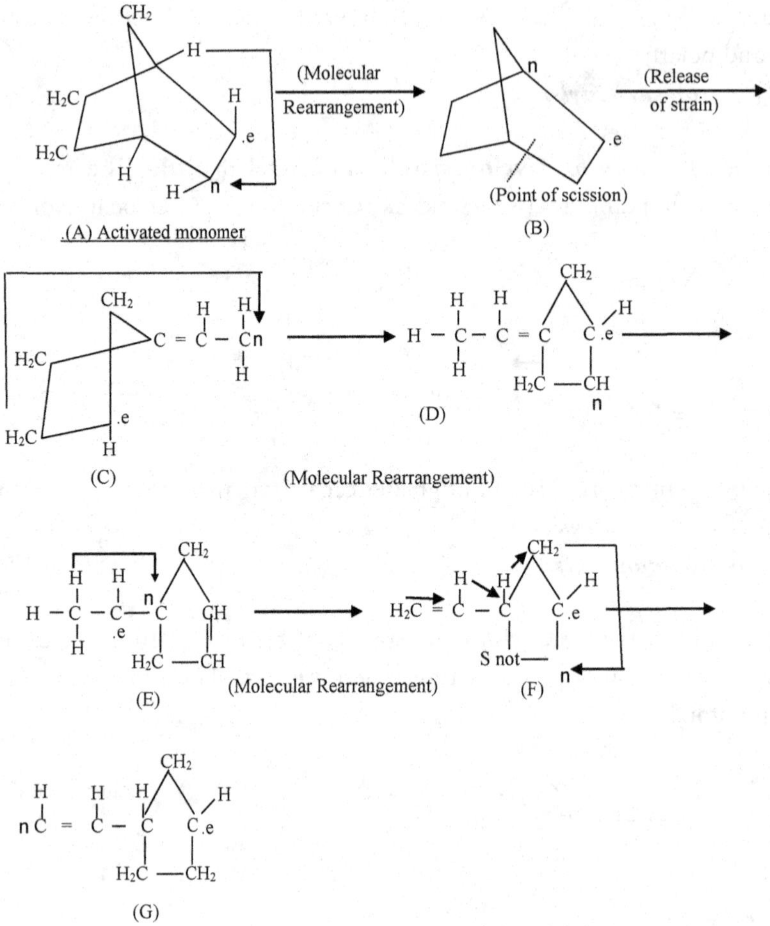
(Molecular Rearrangement)
(Release of strain)
(Point of scission)
(A) Activated monomer
(B)
(C)
(D)
(Molecular Rearrangement)
(E)
(Molecular Rearrangement)
(F)
(G)

(b)

and with this monomer unit whose presence is not questionable radically and chargedly, addition takes place and this continues for every monomer unit added-

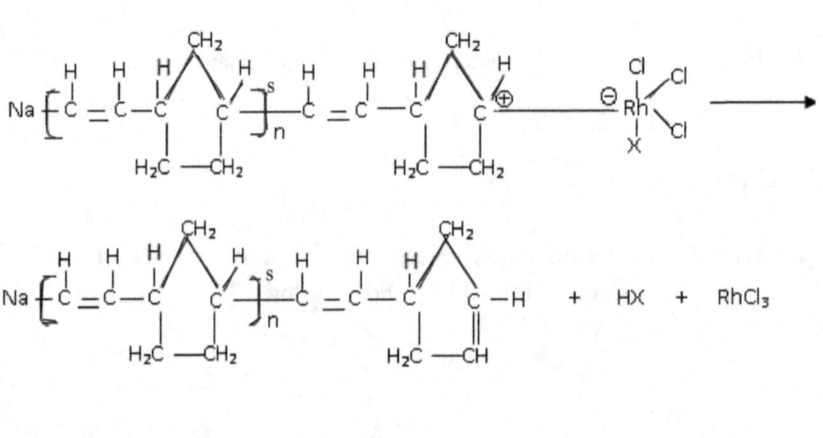
+ HX + RhCl3

(c)

for which one can see that the polymerization time will be long.
(Laws of Creations for Norbornene as a monomer)

Rule 980: This rule of Chemistry for α-**Pinene,** states that, when used as a monomer at low polymerization temperature in the presence of aluminum chloride as initiator, the followings are obtained

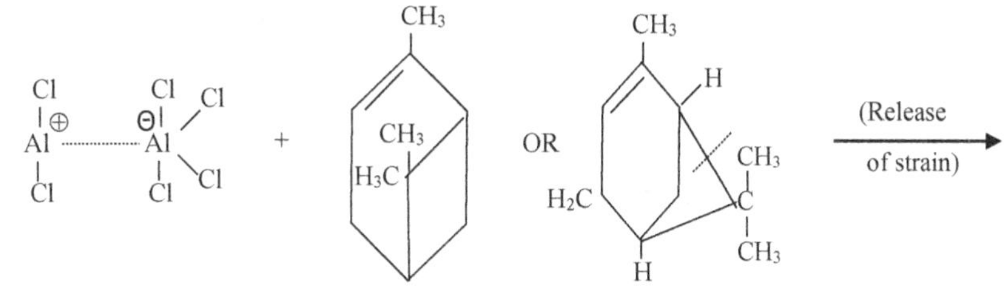

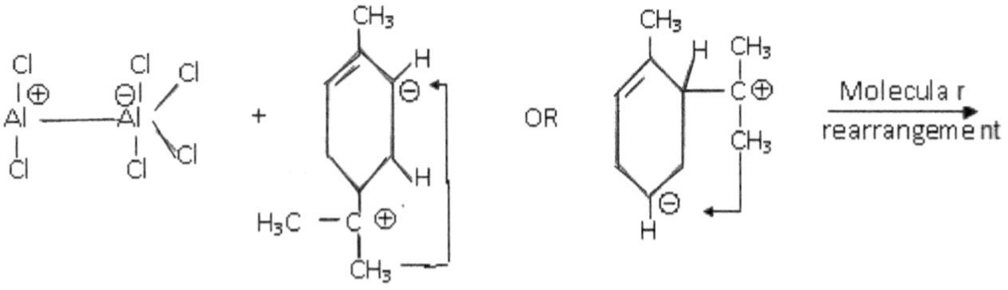

(I) (Not favored)   (II) Not favored

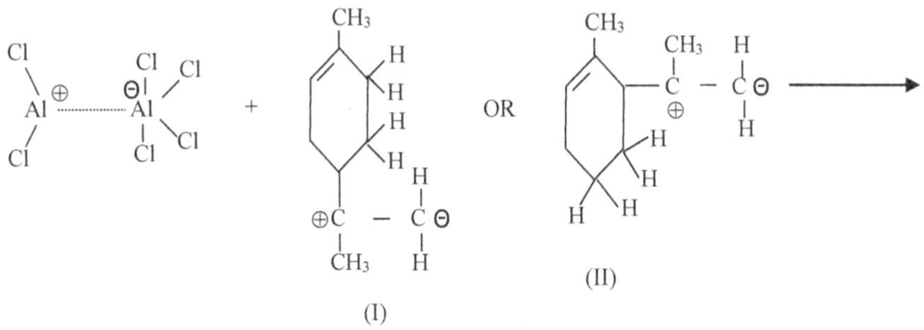

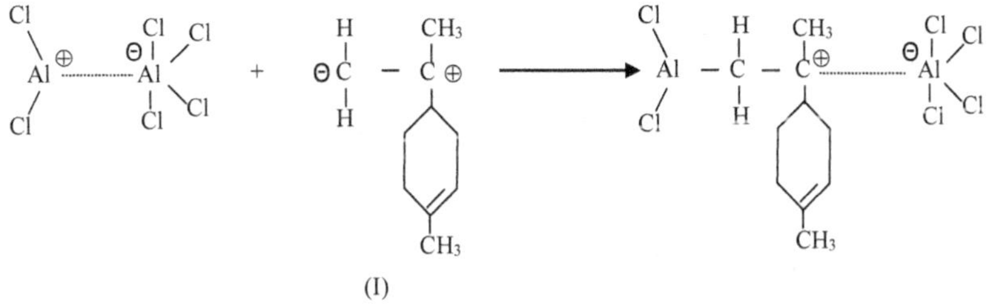

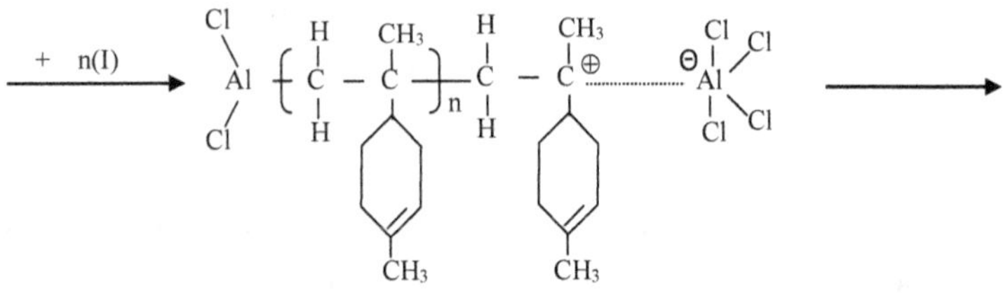

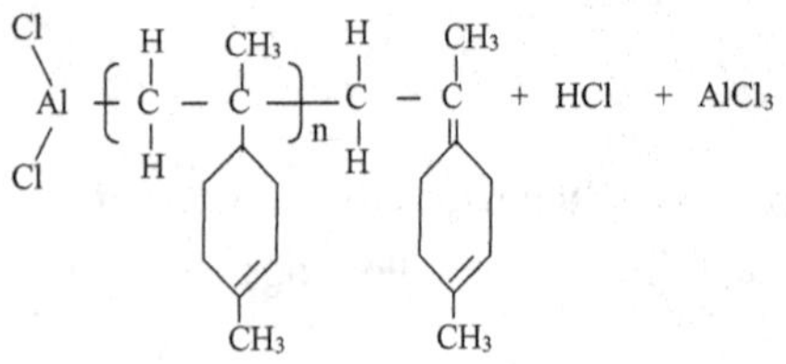

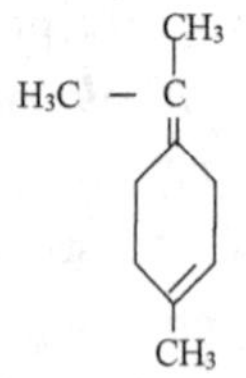

(III) Monomer from α-Pinene

noting that it is (III) that is favored and not (I), for which the polymers produced are oligomers and not polyterpene.
(Laws of Creations for α-Terpenes as a Monomer)

Rule 981: This rule of Chemistry for β-**Pinene,** states that, when used as a monomer in the presence of $AlCl_3$ as initiator, the following are obtained in the absence of rearrangement-

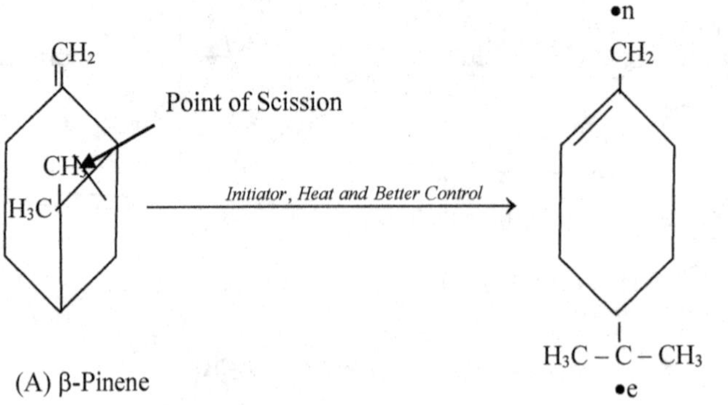

(A) β-Pinene

Point of Scission

Initiator, Heat and Better Control

(B) Monomer Unit

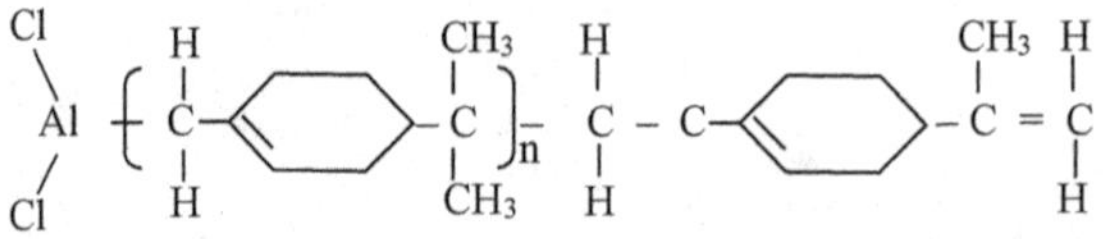

Dead polymers from β-Pinene

(Laws of Creations for β-Pinene as a monomer)

Rule 982: This rule of Chemistry for α- **and** β-**Pinenes,** states that, while β- pinene cannot molecularly rearrange to α- Pinene, so also α-pinene when mildly activated without opening of the ring, cannot rearrange to give β-pinene, like the case of methyl cyclopropene and methylene cyclopropane.
(Laws of Creations for α- and β-Pinenes)

Two hundred and thirty nine additional rules have been proposed to complete carbon-carbon-ringed monomers or compounds- cycloalkenes, cyclodienes, cyclotrienes, cyclotetraenes and so on. From these foundation rules, other developments which were not completely covered herein, can readily be obtained. One has tried to cover a very broad ground for this purpose, so that, most of what have never been known before could readily be explained without questions or doubts or confusions. Nevertheless, it has never been an easy task. To the students of Metaphysics, one can begin to appreciate the fact that rules in chemical systems are natural laws. The significance of benzene ring in placing the limit to so many phenomena is worthy of note. It is metaphysical in nature. What make Benzene to be aromatic has partly been explained, partly in the sense that the origin of the aroma has not yet been shown (Physics of SMELL)

For so many years, the phenomena of resonance stabilization were thought to be clearly understood. But from all the considerations so far, it can be observed that very little or nothing was known about them. Even the mechanisms by which they operate were unknown. The existence of charges, blurred the understanding of the phenomena. How can one remove the positive charge on $H^{\text{Å}}$ which is a hole, to leave what behind?

For the first time, it is being realized that all cases where double or triple bonds are conjugatedly placed exhibit resonance stabilization phenomena.. For the first time, it is being realized that some 1,4- dienes or 1,5- dienes can be resonance stabilized. For the first time, it is being realized that resonance stabilization is not only limited to ringed compounds, but also to linear compounds. It is only continuous in one and that one is benzene and those that carry it. For the first time in general, we have begun to fully see the Real and Imaginary sides of Chemistry. Hence there are so many forbidden reactions in all present day chemistry textbooks and literatures and all related disciplines, particularly in the medical related disciplines, where the foundations being largely laid are dedicated to.

References

1. G. Odian, "Principles of Polymer System", McGraw-Hill, Book Company, (1970), Pgs. 580 - 581.

2. Chr. P. Pinazzi, J. Brossas, Die MaKromoleculare Chemie, Vol 122, Issue 1, pages 105-122, 14ᵗʰ March 1969.

3. M. M. Frey, J. Metcalfe, M. Pope, "Thermal isomerization of cyclobutenes, Part 17- 1-Ethyle-2-Vinylcyclobutene and 1,2,3-trimethylcyclobutene", Transactions of the Faraday Society, (Jan. 1971), 67, pg750-753.

4. Albert J. Fry, Richard H. Moore, J. Org. Chem, 1968, 33 (!), pp425-426.

5. Chao-Wan Chang, Ying-Chih Lin, Gene-Hsiang Lee, Yu Wang, "Reactions of ruthenium acetylide complexes with benzylidenemalonitrile", J. Chem. Soc.,Dalton Trans., 1999, 4223-4230.

6. W. Tsang, "Decyclization of Cyclohexene, 4-Methylcyclohexene and 4-Vinylcyclohexene in a Single-Pulse Shock Tube", J. Chem.Phys., 42, 1805 (1965).

7. C. R. Noller, "Textbook of Organic Chemistry", W. B. Saunders Company, (1966), Pgs. 634.

8. C. R. Noller, "Textbook of Organic Chemistry", W. B. Saunders Company, (1966), Pgs. 643.

9. Makoto Ouchi, Masami Kamigaito, Mitsuo Sawamoto, "Synthesis of end-functionalized polymers and copolymers of cyclopentadiene with vinyl ethers by cationic polymerization", Journal of Polymer Science, Part A: Polymer Chemistry (2001) Volume 39, Issue 3, Pages: 398-407.

10. Makoto Ouchi, Masmi Kamigaito, Mitsuo Sawamoto, "Cationic Polymerization of Cyclopentadiene with $SnCl_4$; Control of Molecular Weight and Narrow Molecular Weight Distribution", Macromolecules, 2001, 34 (10), pp. 3176-3181.

11. Aso, C., Kunitake, T., Ishimoto, Y., (1968) "Studies of Polymers from cyclic dienes. V. Cationic polymerization of cyclopentadiene, influence of Polymerization conditions on the Polymer structure" . Journal of Polymer Science Part A-1: Polymer Chemistry, Volume 6, Issue 5, pages 1175-1194.

12. A. F. Donate, J. W. Bozzelli, K. S. Dennis, "Anionic polymerization of Cis- and Trans-1,3-Pentadiene from a mixture of saturated and unsaturated hydrocarbons", Patent No US 4482771, (1984).

13. "Anionic Polymerization and Termination Process", Patent No US3458491, (1969)

14. Itaru Natori, Shohel Inoue, 'Anionic Polymerization of 1,3-Cyclohexadiene with Alkyllithium/Amine Systems. Characteristics of n-Butyllithium/N,N,N^1,N^1-Tetramethyl-ethylenediamine System for Living Anionic Polymerization", Macromolecules, 1998, 31 (15), pp. 4687-4694.

15. Zilkha, A, and Ottolenghi, A. (1962), "Anionic polymerization of Acrylonitrile by cyclopentadienyl sodium", Journal of Polymer Science, 56: 539-554.

16. Karin Roy, Christof Horn, Peter Frank, Vladislav G. Slutsky, Thomas Just, Symposium (International) on Combustion, Volume 27, Issue 1, 1998, Pages 329-336.

17. S. Nakra, R. J. Green, S. L. Anderson, "Thermal decomposition of JP-10 studied by micro flowtube pyrolysis-mass spectrometry", Combustion and Flame (2006), Volume: 144, Issue: 4, Pages 662-674.

18. Burcat, A. and Dvinyaninov, M. (1997), "Detailed Kinetics of cyclopentadiene decomposition studied in a shock tube", International Journal of Chemical kinetics, Volume 29, Issue 7, pages 505-514.

19. Tokmakov, I. V., Moskaleva, L. V. and Lin, M. C. (2004) "Quantumchemical/vRRKM Study on the thermal decomposition of Cyclopentadiene" International Journal of Chemical kinetics, Volume 36, Issue 3, pages 139 – 151.

20. George, B. Bacskay and John, C, Mackie, "The pyrolysis of cyclopentadiene: quantum chemical and kinetic modeling studies of the acetylene plus propyne/allene decomposition channels" Phys. Chem. Chem. Phys., 2001, 3, 2467-2473.

21. Do Hyong Kim, Jeong-Kwon Kim, Seong-Ho Jang, James A. Mulholland, and Jae-Yong Ryu, "Thermal Formation of Polycyclic Aromatic Hydrocarbons from Cyclopentadiene (CPD)", Environ. Eng. Res. Vol. 12, No. 5, pp. 211-217, (2007).

22. McGivern, W. S., Mannion, J. A., Tsang, W., "Ring expansion reactions in the thermal decomposition of tert-butyl-1,3-cyclopentadiene", J. Phys. Chem A. 2006 Nov 30, 110 (47): 12822-31.

23. Faina Dubnikokova and Assa Lifshitz, "Ring Expansion in Methycyclopentadiene Radicals. Quantum Chemical and Kinetics Calculations". J. Phys. Chem. A, 2002, 106 (35), pp. 8173-8183.

24. A. T. Bottini, F. P. Corson, R. Fitzgerald, K. A. Frost II, "Reactions of 1-halocyclohexenes and methyl substituted 1-halocyclohexenes with potassium t-butoxide", Tetrahedron, Vol. 26, Issue 19, (1972), Pages 4883-4904.

25. Yousufari, A. H. K., Endo, K. and Otsu, T., "Monomer-isomerization polymerization, XIII. Monomer-isomerization polymerization of 1,4-cyclohexadiene with Ziegler-Natta catalyst", Journal of Polymer Science: Polymer Chemistry Edition, Volume 13, Issue 7, pages 1601-1605, (1975).

26. Natori, I. and Sato, H., "Oxidation of poly(1,3-cyclohexadiene): Influence of the Polymer chain structure", Journal of Polymer Science Part A: Polymer Chemistry, Vol. 44, Issue 2, pages 837-845, (2006).

27. S. M. Marvel,Soc. Plastics Engrs., Stability of Plastics, Washington, D.C. June 1964.

28. P. Kovocic and A. Kyriakis, J. Am. Chem. Soc., 83 : 1967 (1961).

29. Hofmann, J., Zimmermann, G. and Kopinke, F-D., "The thermal Aromatization of Methyl-1,3-cyclohexadienes-an important argument against commonly accepted sigma- tropic 1,7-H shift reaction", Journal fur Praktishe Chemie/Chemiker-Zeitung, Volume 336, Issue 3, pages 201-206, (1994).

30. C. D. MacPherson, D. Q. Hu, M. Doan, K. T. Leung, "Catalytic dehydrogenation and thermal chemistry of Cyclohexene and 1-methyl-1,4-cyclohexadiene on Si (111)7 x 7", Surface Science, Volume 310, Issues 1-3, (1994), pages 231-242.

31. J. J. Gajewski, A. M. Gortva, "Bimolecular reactions of 3-methylene-1,4-cyclohexadiene (p-1sotoluene), 5-methylene-1,3-cyclohexadiene (o-isotoluene), 1-methylene-1,4-dihydro- naphthalene (benzo-p-isotoluene), and 9-methylene-9,10-dihydroanthracene (dibenzo-p-isotoluene)", J. Org. Chem., (1989), 54 (2), pp. 373-378.

32. Horning, E. C., Chem. Rev. 1943, 33, 89.

33. Bartmess, J. E., J. Am. Chem. Soc., 1982, 104, 335.

34. Plieninger, H., Marler-Borst, W., Angew.Chem., Int. Ed. Eng., 1964, 3, 62.

35. Ray, M., Huber, U. A., Dreiding, A. S., Tetrahedron Lett., 1988, 3683.

36. Kopecky, K. R., Lau, M. P., J.Org. Chem. 1978, 43, 525.

37. http://www.lookchem.com/1,3-Cyclooctadiene/

38. Yuri, V. Korshak, Mikchail, A. Teinkopatchev, Boris, A. Dolgoplosk, Elene, G. Avdelkina, Dimitry, F. Kutepov, "Intra- and intermolecular metathesis reactions in the formation and degradation of unsaturated of unsaturated polymers", Journal of Molecular catalysis, Vol.15, Issues 1-2, (1982), pages 207-218

39. Wolfgang Holtrup, Roland Streck, Wolfgang Zaar, Dieter Zerpner, "Preparation and polymerization of 1,4-cyclooctadiene, Journal of Molecular catalysis", Vol. 36, Issues 1- 2, (1986), pages 127-133.

40. Marc A. Hillmyer, Robert H. Grubbs, "Chain transfer in the Ring-Opening Metathesis Polymerization of Cyclooctadiene Using Discrete Metal Alkylidenes" Macromolecules (1995), 28, (25), pp. 8662-8667.

41. Lehnert, et. al., Process for the ring opening polymerization of cyclooctadiene (1,5)- Patent No. 3935178; http://www. patentgenius.com/patent/3935178.html.

42. David Diaz, Converso, Antonella, Sharpless, K. Barry, Finn, M. G., (2006) 2,6-Dichloro-9-thiabicyclo[3,3,1] nonane : Multigram Display of Azide and Cyanide components on a Versatile Scaffold". Molecules, 11 (4): 212-218.

43. John A. Soderquist and Alvin Negron, (1998), "9-Borabicyclo[3,3,1] nonane Dimer"Org . Synth, http://www.orgsyn.org/orgsyn/orgsyn/prepContent.asp?prep=cv9p0o95; Coll. Vol. 9:95.

44. Thomas Schiffer, Georg Oenbrink, "Cyclododecatriene, Cyclooctadiene, and 4-Vinyl cyclohexene" Ullmann's Encyclopedia of Industrial Chemistry, 2005, Wiley-VCH, Weinheim.

45. C. R. Noller,"Textbook of Organic Chemistry", W. B. Saunders Company, (1966), Pg. 454.

46. C, R. Noller,"Textbook of Organic Chemistry",W.B. Saunders Company,(1966), Pg.648.

47. Oprea, C. V. and Popa M. (1984)," Opening of aromatic rings in Mechanochemical polymerization of Benzene and Pyridine", Die Angewandte Makromoleculare Chemie, Vol.127, Issue 1, pages 49-58.

48. Boluk, M. Y. and Akovalli, (1981), "Some studies on plasma polymerization of Benzene and Cyclohexane in the gas phase", Polymer Engineering & Science, Vol. 21, Issue 11, pages 664-667.

49. Kyozi Kaeriyama, Mass-aki Sato, Kazuo Someno and Susumu Tanaka, J. Chem.Soc.,Chem. Commun., (1984), 1199-1200.

50. Peter Kovacic, Alexander Kyriakis, "Polymerization of Benzene to p-Polyphenyl by Aluminum Chloride-Cupric Chloride", Journal of American Chemical Society, (1963) Vol. 85, Issue 4: Publisher: American Chemical Society, Pages 454-458.

51. Peter Kovacic, Fred. W. Koch, "Polymerization of Benzene to p-Polyphenyl by Ferric Chloride", J. Org. Chem. (1963), 28(7), pp. 1864-1867.

52. Kovacic, P. and Wu, C. (1960), "Reaction of ferric chloride with Benzene", Journal of Polymer Science, Vol. 47, Issue 149, pages 45-54.

53. Kovacic, P., Koch, F. W. and Stephan, C. E., "Water catalysis in the polymerization of Benzene by Ferric chloride", Journal of Polymer Science Part A: General Papers, Vol.2, Issue 3, pages 1193-1203.

54. Peter Kovacic, Richard M. Lange, "Polymerization of Benzene to polyphenyl by Molybdenum Pentachloride", J. Org. Chem., (1963), 28(4), pp. 968-972.

55. 55. Susumu Matsuzaki, Masaharu Taniguchi, Mizuka Sano, "Polymerization of Benzene occluded in Graphite-alkali metal intercalation compound", Synthetic Metals, Volume 16, Issue 3, (1986), pages 343-348.

56. Fumio Teraoka & Toshisada Takabashi, "Morphology of poly-p-phenylene formed during the polymerization of Benzene by molybdenum pentachloride or Ferric chloride", Journal of Macromolecular Science, Part B, Vol.18, Issue 1, (1980), pgs. 73-82.

57. http://www.ilpi.com./organomet/romp.html.

58. J. Thomas, J. Katz, "The Cyclooctatetraenyl dianion", J. Am.Chem. Soc., (1960); 82(14); 3784-3785.

59. Floyd L. Klavetta, Robert H. Grubbs, "Polyacetylene through metathesis polymerization of Cyclooctatetraene", Chemistry: Miscellaneous Papers, (1987), Vol. 28, Issue 2, pages 425-426.

60. R. H. Grubbs, C. B. Gorman, E. J. Ginsburg, Joseph W. Perry and Seth R. Marder, "New Polymeric Materials with Cubic Optical Nonlinearities Derived from Ring-Opening Metathesis Polymerization" Materials for Nonlinear Optica, Chapter 45, pp 672-682.

61. S. N. E. Omorodion, "Size Exclusion Chromatography", PhD thesis, McMaster University, Hamilton, Canada, (1980).

62. C. R. Noller,"Textbook of Organic Chemistry", W. B. Saunders Company, (1966), Pgs. 638 - 658.

63. Lewis Watts and Rowland Pettit, "Chemistry of Cyclobutadiene-iron Tricarbonyl",Advances in Chemistry, Vol. 62. (1967).

64. G. F. Emerson, L. Watts, R.Pettit, "Cyclobutadiene- and Benzocyclobutadiene-Iron Tricarbonyl Complexes", J. Am. Chem. Soc.; (1965), 87(1), 131-133.

65. G. F. Emerson, L. Watts, R.Pettit, "Cis-Dichlorocyclobutene", Organic Syntheses, Coll. Vol. 6, p.36 (1970).

66. R. Pettit and J. Henery, "Iron, tricarbonyl (η 4-1,3-cyclobutadiene", Organic Syntheses, Coll. Vol. 6,p 310 (1988).

67. J. D.Fitzpatrick, L. Watts, G. F. Emerson, R. Pettit, "Cyclobutadiene iron tricarbonyl, A New Aromatic System", J. Am. Chem. Soc.; 1965; 87(14); 3254-3255.

68. Floyd I. Klavetter, Robert H. Grubbs, "Polyacetylene through metathesis polymerization of Cyclooctatetraene", Chemistry: Miscellaneous Papers, (1987), Vol. 28, Issue:2, Pg.425-426.

69. Eric J. Moorhead and Anna G. Wenzel, "Two Undergraduate Experiments in Organic Chemistry: The preparation of Polyacetylene via Ring-Opening Metathesis Polymerization", Journal of Chemical Education, Vol. 86, No 8, (2009), 973

70. C. B. Gorman, E. J. Ginsburg, M. J. Sailor, J. S. Moore, T. H. Jozefiak, N. S. Lewis, R. H. Grubbs, S. R. Marder, J. W. Perry, "Substituted polyacetylene through the ring opening metathesis polymerization (ROMP) of substituted cyclooctatetraene", Synthetic Metals, (1991), Vol. 41, Issue 3, pgs. 1033-1038.

71. Xia Meng, Guang-Rong Tang and Guo-Xin Jin, "Vinyl and ring-opening metathesis polymerization of norbornene with novel half-sandwich iridium (III) complexes bearing hydroxyindanimine ligands", Chem. Commun. Issue 27, (2008), 3178-3180.

72. V. Dragutan, A. T. Balaban, M. Dimonic, "Olefin Metathesis and Ring-Opening Polymerization of Cycloolefins, Wiley-Interscience, Chichester, 1985.

73. Hue Liang, Jingyu Liu, Xizofang Li, Yuesheng Li, "Synthesis,structure and norbornene polymerization behavior of neutral palladium complexes", Polyhedron, Vol. 23, Issue 9, (2004), pp. 1619-1627.

74. Yee-On Yeung, Anthony C. H. Ng, Dennis K. P Ng, "Ring-opening polymerization of norbornene by h-cycloheptatrienyl molybdenum and tungsten complexes", Inorganica Chimica Acta, Volume 288, Issue 2, (1999), Pages 226-228.

75. Li, X. -F. and Li, Y. –S, "Vinylic polymerization of norbornene by neutral nickel(II)-based catalysts", Journal of Polymer Science Part A: Polymer Chemistry, Volume 40, Issue 15, pages 2680-2685, (2002).

76. Charles R. Chambers Jr., 'COMA Polymers for 193nm applications", The University of Texas at Austin., (2001).

77. 'Radical polymerization of norbornene", J Macromol. SCI.-Chem., A11 (5), pp.1053-1070, (1997).

78. Ring opening metathesis polymerization-Wikipedia, the free encyclopedia.

79. R.B. Seymour, "Introduction to Polymer Chemistry", International Student Edition, McGraw-Hill Book Company, (1971,) Pgs. 169 - 170.

80. Higashimura, T., Lu, J., Kamigaito, M. and Deng, Y. –X., (1992), "Cationic polymerization of a-pinene with binary catalyst $AlCl_3/SbCl_3$", Die Makromoleculare Chemie, Vol. 193, Issue 9, pages 2311-2321.

81. Higashimura, T., Lu, J., Kamigaito, M., Sawamoto, M. and Deng, Y. –X. (1993), "Cationic polymerization of a-pinene with aluminum-based binary catalysts, 3. Effects of added base", Die Makromoleculare Chemie, Volume 194, Issue 12, pages 3455-3465.

82. Raquel, P. F. Guine, Jose, A. A. M. Castro, "Polymerization of β-pinene with Ethylaluminum Dichloride ($C_2H_5AlCl_3$)", Journal of Applied Polymer Science, Vol. 82, pgs. 2558-2565, (2001).

83. 83. F. Cataldo, J. R. Brucato, Y. Keheyan, "Gamma-radiation induced polymerization of a chiral monomer: a new way to produce chiral amplification", Orig.Life Evol. Biosph, (2006), 36(5 8), 477-85.

84. Marvel, C. S., Hanley, J. R., and Longone, D. T., (1959), "Polymerization of β-pinene with Ziegler type catalysts", Journal of Polymer Science, Volume 40, Issue 137, pages 551-555.

Problems

11.1   Distinguish between the different types of molecular rearrangement phenomena so far identified.

11.2   Identify the following terms using examples.

    (a)   Para-placement of charges or free-radicals.
    (b)   Trans-placement of activated monomers.
    (c)   Trans-annular addition of charges and trans-placement of charges.
    (d)   Ring expansion phenomena.
    (e)   (Ring reduction phenomena.

11.3   Identify the following terms using examples.

    (a)   Linear or opened-loop free-radical resonance stabilization phenomena (discrete).
    (b)   Discrete closed-loop free-radical resonance stabilization phenomena.
    (c)   Continuous closed-loop free-radical resonance stabilization phenomena.
    (d)   Opened/Half discrete closed-loop free-radical resonance stabilization phenomena.
    (e)   Opened/Full discrete closed-loop free-radical resonance stabilization phenomena.
    (f)   Opened/Continuous closed-loop free-radical resonance stabilization phenomena.

11.4   Discuss extensively on the laws of conservation of transfer of transfer species with respect to resonance stabilization phenomena.

11.5   Explain the mechanism of free-radical resonance stabilization phenomena when double bonds are conjugatedly placed in a closed-loop compound.

11.6   When radical-pushing and pulling groups are internally and externally located in opened and closed- loop systems, explain what happens when they are activated and attacked by nucleo-free- and electro-free-radicals.

11.7   Discuss on the unique features of benzene ring.

11.8   (a)   Can a compound be resonance stabilized when not activated?
    (b)   Why is it that all groups carried by Benzene are resonance stabilized in its Energized state of existence?
    (c)   What is the origin of aromaticity in Benzene ring?

11.9. (a) Under what conditions can some 1,4- and 1,5 - dienes be made resonance stabilized monomers?

(b) Under what conditions are benzene rings suited for use as step monomers?

11.10 (a) Shown below are two vinylcyclohexadienes.

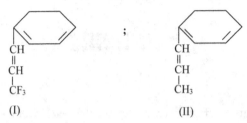

CH
‖
CH
|
CF₃

(I)

CH
‖
CH
|
CH₃

(II)

(i) Show that (II) is more nucleophilic than (I).

(ii) What type of resonance stabilization is exhibited by the monomers above?

(b) What happens when a radical-pushing group is now internally and externally located inside the ring?

11.11. (a) Show how cyclopentadiene dimerizes slowly on standing to dicyclopentadiene.

(b) Shown below are two isolatedly placed double bonds in cyclooctadienes.

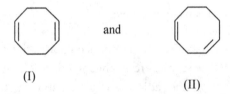

and

(I)   (II)

Show how (II) will rearrange to (I) and how they can be used as Addition monomers.

11.12. (a) Why does the corresponding six-membered ring of (I) of Q 11.11(b) not favor being used as a monomer when not rearranged?

(b) Show whether a twelve-membered cycloduodecahexaene will undergo ring reduction phenomena (⊞)? If it cannot be reduced, explain why?

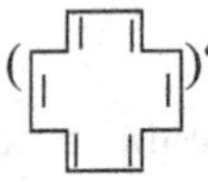

(c) Why does an eighteen-membered cyclooctadecanonaene (annulene)

favor any stable existence, despite the size of the ring (✳)? Explain using the laws stated so far.

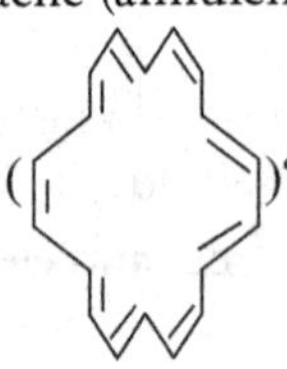

11.13. (a) When nitrobenzene is heated with potassium hydroxide in the presence of air, a hydroxynitrobenzene (O - nitrophenol) is obtained.

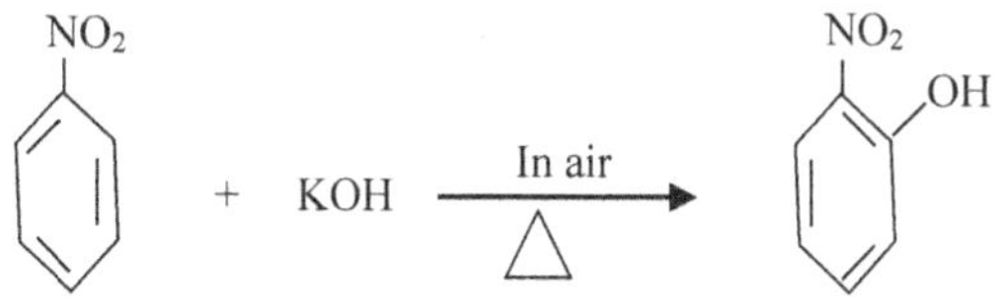

(I)

Explain the mechanism of the reaction tentatively.

(b) What type of resonance stabilization is favored by (I) above?

(c) Shown below are three types of dinitrobenzenes

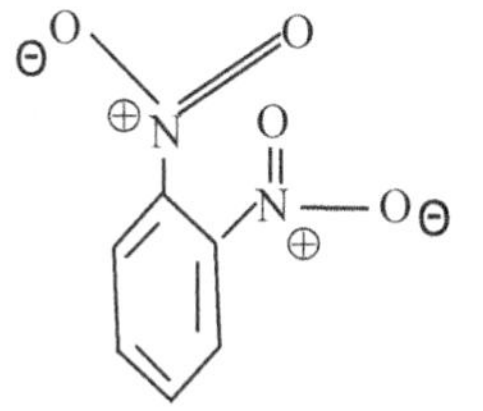

(II) 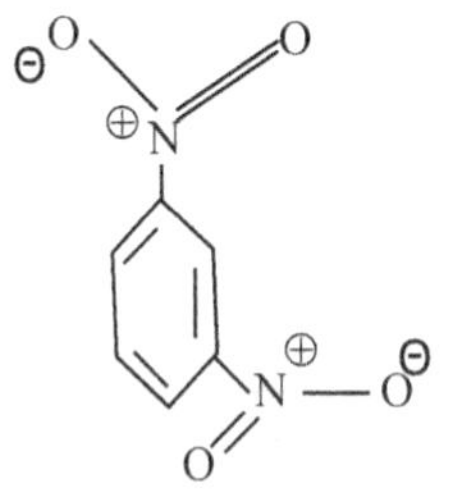 ; (III) 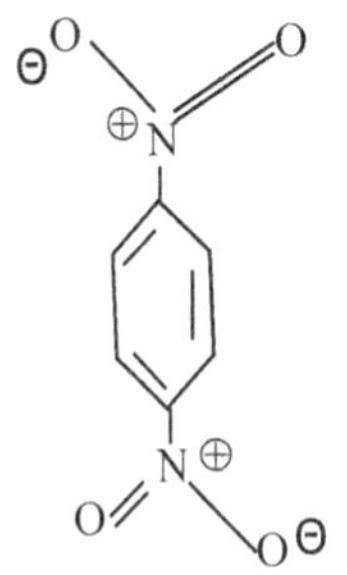 ; (IV)

Distinguish between the three types.

11.14. (a) Distinguish between the two monomers shown below in terms of resonance stabilization phenomena favored by them.

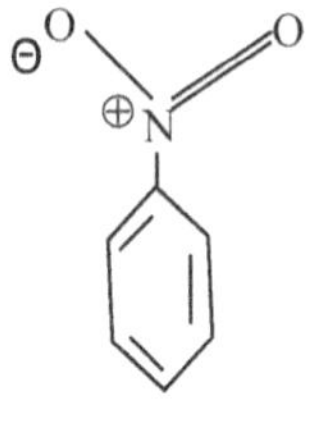

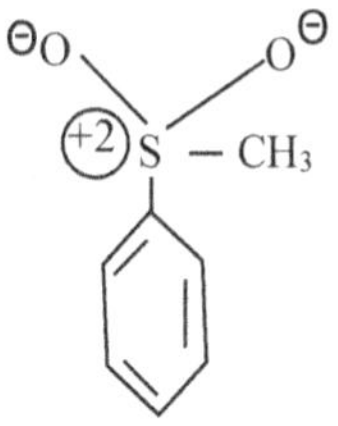

(b) Explain the mechanism of the following reactions.

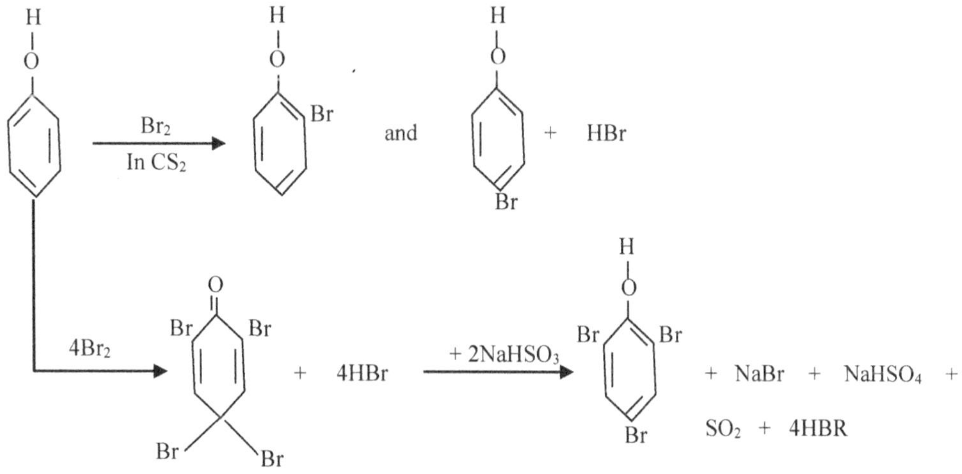

based on current level of development.

1.15   Explain the mechanisms of the following reactions based on current level of developments.

(a)

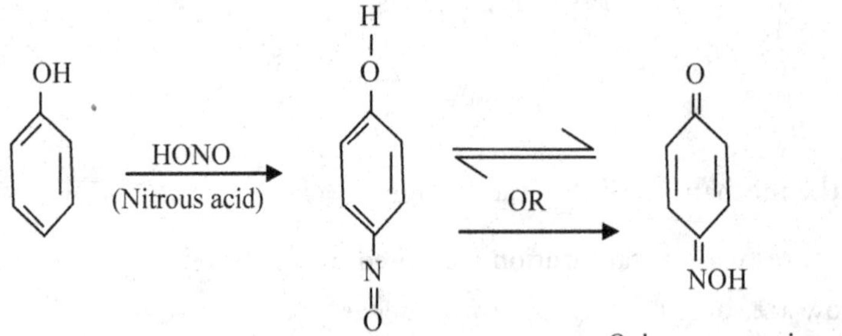

P -Nitrosophenol   Quinone monoxime

(b)

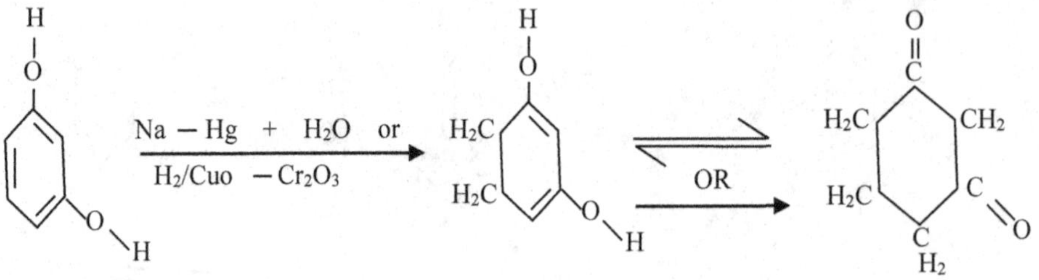

Dihydroresorcinol   1,3 - Cyclohexanediene

(c)

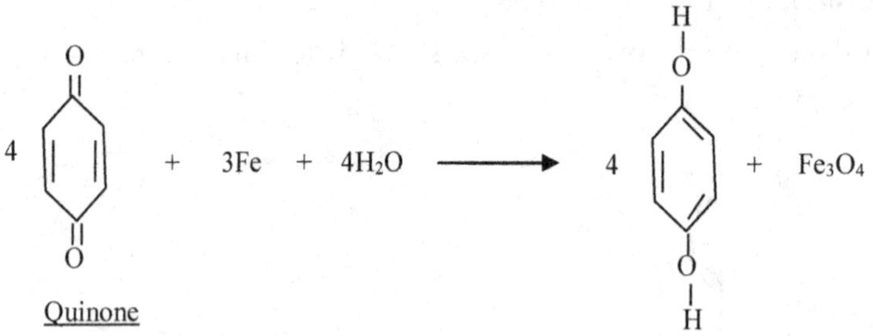

Quinone   Hydroquinone

Index

V conts

W

X

Y

Z

Printed in the United States
By Bookmasters